Annual Reports in

MEDICINAL CHEMISTRY

VOLUME **46**

Annual Reports in

MEDICINAL CHEMISTRY

VOLUME **46**

Sponsored by the Division of Medicinal Chemistry of the American Chemical Society

Editor-in-Chief

JOHN E. MACOR
Neuroscience Discovery Chemistry
Bristol-Myers Squibb R&D
Wallingford, CT, United States

Section Editors

ROBICHAUD • STAMFORD • WEINSTEIN •
McALPINE • PRIMEAU • LOWE • DESAI

Amsterdam • Boston • Heidelberg • London • New York • Oxford
Paris • San Diego • San Francisco • Singapore • Sydney • Tokyo
Academic Press is an imprint of Elsevier

Academic Press is an imprint of Elsevier
Linacre House, Jordan Hill, Oxford OX2 8DP, UK
84 Theobald's Road, London WC1X 8RR, UK
Radarweg 29, PO Box 211, 1000 AE Amsterdam, The Netherlands
225 Wyman Street, Waltham, MA 02451, USA
525 B Street, Suite 1900, San Diego, CA 92101-4495, USA

First edition 2011

ISBN: 978-0-12-386009-5
ISSN: 0065-7743

For information on all Academic Press publications
visit our website at www.elsevierdirect.com

Printed and bound in USA

11 12 13 14 15 10 9 8 7 6 5 4 3 2 1

CONTENTS

PART II: Cardiovascular and Metabolic Diseases

Section Editor: Andy Stamford, Merck Research Laboratories, Kenilworth, New Jersey

PART IV: Oncology

Section Editor: Shelli R. McAlpine, School of Chemistry, University of New South Wales, Sydney, Australia

PART V: Infectious Diseases

Section Editor: John L. Primeau, AstraZeneca, Waltham, Massachusetts

PART VI: Topics in Biology

Section Editor: John Lowe, JL3Pharma LLC, Stonington, Connecticut

PART VIII: Trends and Perspectives

CONTRIBUTORS

Jeffrey M. McKenna	135	Britt-Marie Swahn	53
Nicholas A. Meanwell	263	David C. Swinney	301
Scot Mente	33	Roger D. Tung	403
James Morelli	419	Eugeni Vaisberg	369
Philippe G. Nantermet	19	Travis T. Wager	33
D. Obrecht	245	Gerhard Wagner	189
James Offord	33	Michael A. Walters	385
Jeffrey F. Ohren	33	Yanping Xu	69
Michael Pollastri	385	Melvin J. Yu	227
Jay P. Powers	171	Wanjun Zheng	227
Boris M. Seletsky	227	Yajun Zheng	89
Mark S. Sundrud	155		

PREFACE

When I accepted the role of editor-in-chief for *Annual Reports in Medicinal Chemistry*, I worked with the Executive Committee of the Division of Medicinal Chemistry (the actual owners of the book) to redefine the role of editor-in-chief. Through these efforts, we have made the position a defined responsibility for the Executive Committee and a position answerable to the Executive Committee. We changed the role of editor-in-chief to a position elected by the Executive Committee with a 5-year term and with an *ad hoc* appointment on the Executive Committee of the Medicinal Chemistry Division of the ACS. My term expires with the delivery of this volume, the 46th in the series. I want to thank Dave Rotella for recruiting me into this position, and the other members of the Executive Committee for supporting me throughout my 5-year term. It went by all too quickly. The new editor-in-chief (for Volumes 47–51) elected by the Executive Committee of our Division will be Manoj Desai, a friend and former colleague of mine and the very capable section editor of the "Topics in Drug Discovery" section for many years. I have no doubt that he will do an outstanding job continuing the excellence that has defined *Annual Reports in Medicinal Chemistry*. I wish him the very best in his new role. Thank you again to all who supported me in my role with "The Book", especially you, the readers.

Annual Reports in Medicinal Chemistry has reached Volume 46. Its longevity continues to be a testament to the vitality of the field of medicinal chemistry, and I hope that *Annual Reports in Medicinal Chemistry* continues to be *the* review resource for medicinal chemists. Volume 46 upholds the traditions of *Annual Reports in Medicinal Chemistry* with 26 chapters covering the themes of central nervous system disease, cardiovascular and metabolic diseases, inflammation/pulmonary/GI, oncology, infectious disease, topics in biology, topics in drug design and discovery, and finally our review of new drugs introduced in 2010 in the "To Market, To Market" chapter.

With the field of medicinal chemistry suffering continued contraction in 2010 and 2011, it was a challenge to maintain the high energy of *Annual Reports in Medicinal Chemistry*, albeit one certainly worth the effort. A number of section editors and chapter authors have been affected by the consolidation of the pharmaceutical industry and the field of medicinal chemistry in general, led by the business decisions of Pfizer, Merck, and other companies. The contraction of the pharmaceutical industry in the past 10–15 years has been dramatic. However, small molecules are still a main driver in extending and improving human health, and I don't doubt that a day in the near future will come when the industry realizes that it has cut back too much and that experienced medicinal chemists are worth their weight in gold.

Volume 46 of *Annual Reports in Medicinal Chemistry* is the result of the efforts of a stellar team comprised entirely of volunteers. I'd first like to thank all of the chapter authors in Volume 46 for their dedication and talent. They enthusiastically provided a consistent quality to the book. Helping bring all of this together were the section editors: Albert Robichaud, Andrew Stamford, David Weinstein, Shelli McAlpine, John Primeau, Manoj Desai, and John Lowe. I would like to thank them for their time and constant dedication. I'd like to welcome and thank a new team for our "To Market, To Market" chapter which covers the new molecular entities (NMEs) introduced in the past year (2010). Joanne Bronson and her coauthors Murali Dhar, Rick Ewing, and Nils Lonberg have taken on their new role in *Annual Reports in Medicinal Chemistry* with gusto and have closed the book with an excellent review of the drugs introduced in 2010. It has been my pleasure to be able to work with such a dedicated and talented group of section editors for my term. I cannot thank them enough for the time they have given to *Annual Reports in Medicinal Chemistry*. Finally, I'd like to thank my administrative assistant Cathy Hathaway for doing whatever has been needed to get the job done for the book.

Helping the section editors and me were a team of reviewers/proofreaders who have done another spectacular job behind the scenes as well. I would like to acknowledge these reviewers/proofreaders by listing their names below as a demonstration of our appreciation for their time and effort.

AstraZeneca—Greg Bisacchi and Art Patten

Bristol-Myers Squibb—Stephen Adams, Kenneth Boy, Peter Cheng, James J.-W. Duan, Andrew Degnan, Derek Denhart, Carolyn Dzierba, Bruce Ellsworth, Matthew Hill, Siew Peng Ho, John Hynes, John Kadow, George Karageorge, James Kempson, Dalton King, Lawrence Marcin, Richard Olson, James Sheppeck, Lawrence Snyder, Steven Spergel, Drew Thompson, William Washburn, Ryan Westphal, Mark Wittman, and Christopher Zusi

Eli Lilly and Company—Paul Ornstein

Gilead Sciences—Randall Halcomb, James Taylor, and Will Watkins

Lundbeck—Allen T. Hopper, John M. Peterson, and Andrew D. White

Monclair State University—David Rotella
Mnemosyne Pharmaceuticals—Bert Chenard
Pfizer—George Chang, Robert Dow, Brian O'Neill, and Ralph Robinson

In summary, I have tremendously enjoyed delivering Volumes 42–46 of *Annual Reports in Medicinal Chemistry* to you all, and I hope you have gotten much out of the series. I know that *Annual Reports in Medicinal Chemistry* will continue to be a key reference for your medicinal chemistry pursuits. Thank you to all who have contributed to this endeavor and its success.

John E. Macor, Ph.D.
Editor-in-Chief, *Annual Reports in Medicinal Chemistry* (Volumes 42–46)
Executive Director, Neuroscience Discovery Chemistry
Bristol-Myers Squibb, R&D
Wallingford, CT, USA
August 27th, 2011

PART I:

Central Nervous System Diseases

Editor: Albert J. Robichaud
Chemical & Pharmacokinetic Sciences
Lundbeck Research USA
Paramus
New Jersey

CHAPTER 1

Progress in the Medicinal Chemistry of Group III Metabotropic Glutamate Receptors

Kevin G. Liu and **Darío Doller**

Contents

1. INTRODUCTION

G-protein-coupled receptors (GPCRs), among the largest and most diverse protein families in mammalian genomes and accounting for 30% of all modern medicinal drugs, constitute a superfamily of proteins whose

Lundbeck Research USA, Chemical & Pharmacokinetic Sciences, 215 College Road, Paramus, NJ 07652, USA

Annual Reports in Medicinal Chemistry, Volume 46
ISSN: 0065-7743, DOI: 10.1016/B978-0-12-386009-5.00026-6

primary function is to transmit extracellular stimuli into intracellular signals. Sequence comparison of different GPCRs revealed the existence of at least six classes (A through F) [1]. Metabotropic glutamate (mGlu) receptors belong to the class C GPCR family and are activated by L-glutamate, a major excitatory neurotransmitter of the mammalian brain. To date, eight subtypes of mGlu receptors have been identified and divided into three groups based on sequence homology, signal transduction, and pharmacology. Group I includes mGlu1 and mGlu5; group II includes mGlu2 and mGlu3; and group III includes mGlu4, mGlu6, mGlu7, and mGlu8 receptors [2]. The widespread expression of mGlu receptors and their role in synaptic signaling throughout the central nervous system (CNS) have made them attractive therapeutic targets [3,4]. While significant progress has been made in identifying specific ligands for and in elucidating the function of group I and group II receptors, the evolution of group III mGlu receptors has been slow, in large part due to the lack of selective tool compounds. However, superior ligands for receptors in this group are emerging, for the mGlu4 receptor in particular. These chemical tools are beginning to unveil the important role of mGlu group III receptors, and their modulation has been shown to provide therapeutic potential for a number of indications such as Parkinson's disease (mGlu4) [5] and cognition impairment (mGlu7) [6]. This review focuses on the recent progress on identification of group III mGlu receptor ligands.

2. GROUP III ORTHOSTERIC LIGANDS

2.1. Group III agonists

The endogenous ligand L-glutamate (**1**) is a nonselective agonist, activating all mGlu receptors, as well as ligand-gated ionotropic (iGlu) receptors, and glutamate transporters. Depending on the specifics of the assay, the *in vitro* potency of L-glutamate is in the single or double-digit micromolar range at all mGlu receptors except mGlu7, at which its potency is in the millimolar range [7–9]. Other known mGlu group III-selective agonists (*e.g.*, **2–8**) also display much lower affinity at the mGlu7 receptor than at other mGlu receptors. This has caused speculation on the presence of a surrogate endogenous ligand for mGlu7, as well as suggested that this receptor may only be activated under specific conditions where elevation in the extracellular glutamate concentration is higher than during normal synaptic transmission.

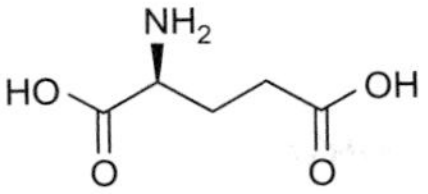

1 L-Glu **2** L-AP_4 **3** L-SOP

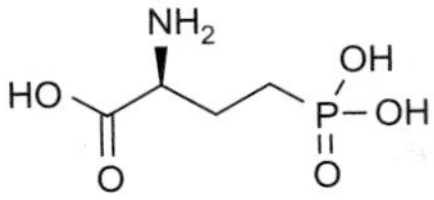

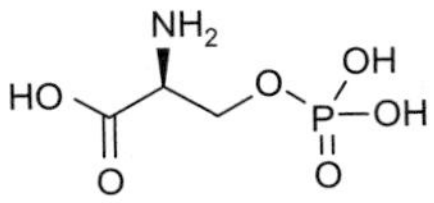

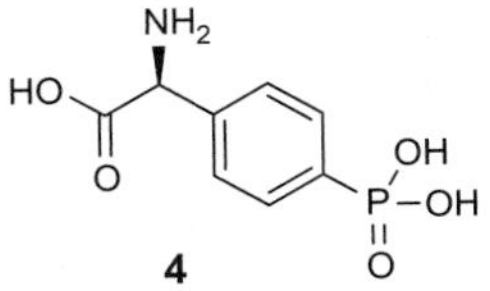

4 (*S*)-PPG **5** ACPT-1

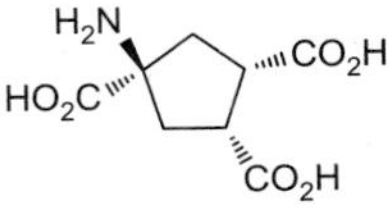

To date, all of the known group III-selective agonists identified are amino acid derivatives bearing additional acidic groups to mimic L-glutamate and lack subtype selectivity. L-AP_4 (**2**), a γ-phosphonic acid, has been known as the most potent group III agonist for many years and is highly selective against group I and group II mGlu and iGlu receptors (EC_{50} =0.08, 2.08, 440, and 0.128 μM at mGlu4, mGlu6, mGlu7, and mGlu8, respectively) [10]. L-SOP, **3**, (*R,S*)-PPG, **4**, and ACPT-1, **5**, are some other early group III agonists [7–9,11]. Efforts by Acher and coworkers led to several new and potent group III agonists (**6–9**) in recent years. L-thio-AP_4, **6** was shown to be slightly more potent at all group III receptors (EC_{50} = 0.039, 0.73, 197, and 0.054 μM at mGlu4, mGlu6, mGlu7, and mGlu8, respectively) than L-AP_4 [10]. The enhanced potency of **6** was hypothesized to be derived from its stronger acidity of the thiophosphonate moiety (pK_a = 5.56) than that of the corresponding phosphonate of L-AP_4 (pK_a = 6.88), as determined by ^{31}P NMR.

In an effort to identify subtype-selective group III agonists, the strategy of conformationally constraining L-AP_4 was further explored [12]. Of the four possible stereoisomers of cyclopropane derivative **7**, the (1*S*,2*R*)-isomer was reported to be the most potent. Additional constrained analogs such as the cyclobutyl and cyclopentyl derivatives were significantly less potent [12,13]. Nonetheless, **7** displays very similar pharmacology to that of L-AP_4 and, therefore, is not subtype selective.

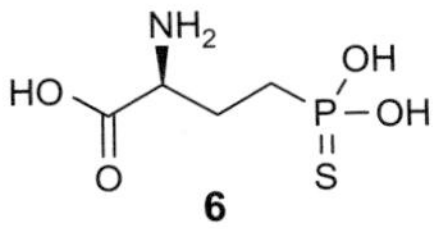

6 L-thio-AP_4 **7** (1*S*,2*R*)-APCPr

In another attempt to identify subtype-selective agonists, longer-chain analogs of L-AP_4 were designed to test the hypothesis that these may interact with less-conserved, remote regions in the glutamate binding pocket [14,15]. Based on this concept, a virtual high-throughput screen (vHTS) was carried out and agonist **8** was identified [16]. Structure–activity relationship (SAR) studies led to the identification of several compounds, such as **9**, with improved potency. Interestingly, **9** and some of its closely related analogs activate mGlu7 receptor more potently than known agonists. This is of particular interest, but for other reasons, as the mGlu7 receptor is difficult to activate [7–9]. Nevertheless, subtype-selective agonists were not identified from this approach. Another class of compounds with an aromatic moiety (*e.g.*, **10**) was also reported by the same group, which displays a similar subtype selectivity profile [15,17].

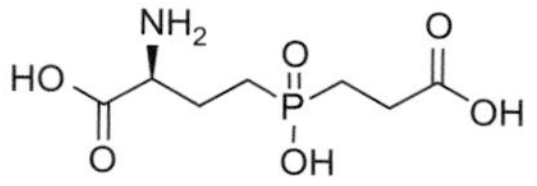

8 (*S*)-PCEP **9** LSP1-3154 **10** LSP1-2111

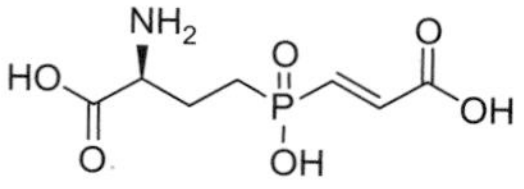

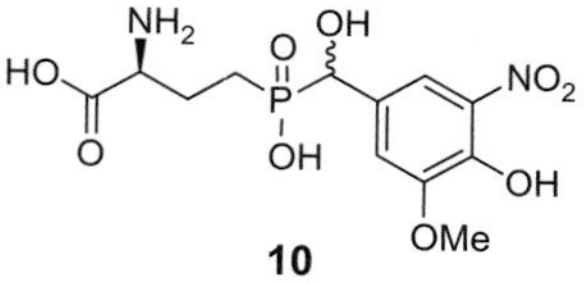

2.2. Group III antagonists

Following the report of the compound MCPG (**11**) as the first group III antagonist more than a decade ago [18], a number of antagonists have subsequently been identified. These include **12** [19], **13** [20], **14** [21], **15** [22,23], **16** [24], and **17** [25,26]. All these orthosteric antagonists are α-methyl or α-alkyl analogs of the group III agonists (*e.g.*, **15** *vs.* **2** and **16** *vs.* **3**). Although these compounds have been used as chemical tools to probe group III mGlu receptor function, their utility is limited by their low group III receptor affinity and paucity of selectivity over group I and group II receptors.

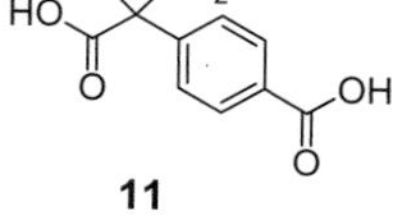

11 MCPG **12** MPPG **13** CPPG **14** UBP1112

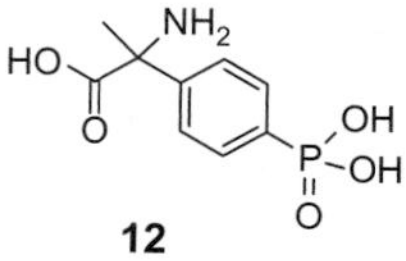

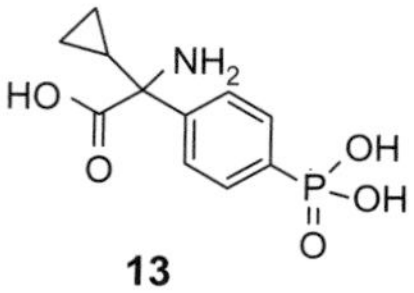

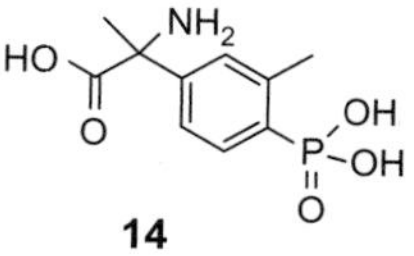

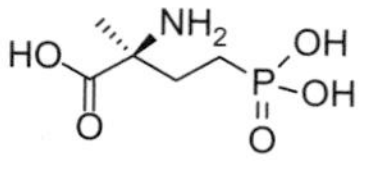

15 MAP4 **16** MSOP **17** LY341495

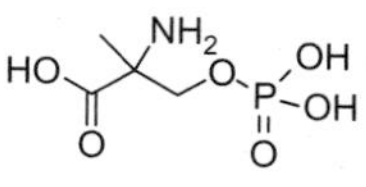

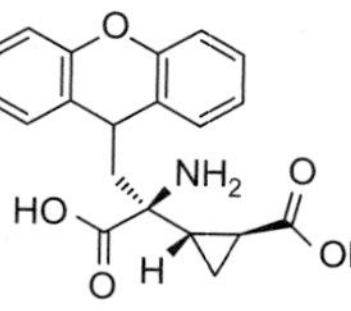

3. mGlu4 RECEPTOR ALLOSTERIC LIGANDS

3.1. mGlu4 receptor positive allosteric modulators

Due to the difficulty in identifying selective orthosteric ligands, as well as the opportunity to explore ligand/receptor interactions in regions of the receptor that may afford certain advantages from the target selectivity and CNS drug design perspectives [27], efforts were directed toward finding positive allosteric modulators (PAMs). Unlike the orthosteric ligands binding at the glutamate binding site, PAMs bind to a distinct region (the allosteric site) which in most cases is believed to be in the seven transmembrane region. This allosteric interaction enhances the activity of the receptor in the presence of an orthosteric agonist such as glutamate. Of the variety of research performed in the past few years, good progress has been recently made with the mGlu4 receptor in particular. Notably, the first subtype-selective mGlu4 PAM PHCCC (**18**) was independently discovered by two research groups [28,29]. Initially, racemic (±)-PHCCC was reported to be an mGlu1 receptor antagonist [30] and although the mGlu4-active enantiomer (−)-PHCCC still partially antagonizes the mGlu1b receptor ($IC_{50} = 3.4$ μM), it is inactive against other mGlu subtypes at 10 μM and is selective against a panel of 28 relevant CNS receptors [29]. (−)-PHCCC was shown to potentiate the responses of both human and rat mGlu4 receptors to L-glutamate or L-AP_4 with EC_{50} values in the range of 2.0–4.1 μM [28]. The study with mGlu4/1b chimeras clearly demonstrated that (−)-PHCCC binds to the seven transmembrane region (the allosteric binding motif) [29]. Being the first subtype-selective mGlu4 ligand, (−)-PHCCC has been used as a tool compound to probe mGlu4 functions in numerous *in vivo* studies [28,31–37]. Due to the suboptimal physicochemical properties and difficulties achieving central exposure levels upon peripheral administration of (−)-PHCCC, several of these studies were carried out through brain local injections, thus limiting its utility as a potential drug candidate.

Aimed at improving the potency and physicochemical properties of previous ligands, and to address the selectivity against mGlu1b of (−)-PHCCC, researchers at Vanderbilt University carried out SAR studies around the PHCCC scaffold [38,39]. In their efforts, the 2-pyridyl group was identified to be superior to the original phenyl ring, affording VU0359516 (**19**) with >fourfold improved potency ($EC_{50} = 380$ nM) over (−)-PHCCC and without mGlu1 activity. Replacement of the 2-pyridyl group with the classical pyridine mimetic 2-thiazole led to a dramatic reduction in potency (>5 μM) underscoring the ubiquitously narrow SAR reported for several structurally diverse series of mGlu4 PAMs (*vide infra*). The PAM activity of **19** was further characterized by a 22-fold leftward shift of the glutamate dose–response curve at 30 μM.

Unlike (−)-PHCCC, **19** did exhibit some mGlu4 agonist activity at high concentrations (>30 μM) showing this ligand to be a mixed allosteric modulator and orthosteric agonist in a calcium mobilization assay using Chinese hamster ovary (CHO) cells expressing human mGluR4 and a chimeric G-protein Gqi5.

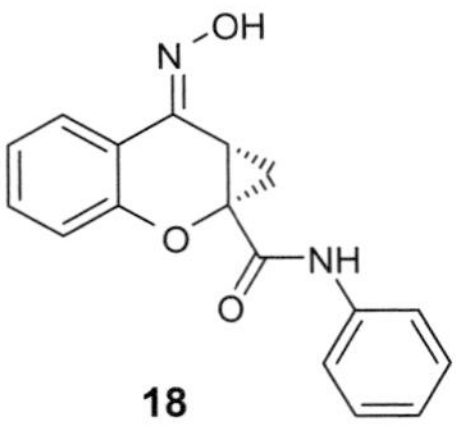

18
(−)-PHCCC

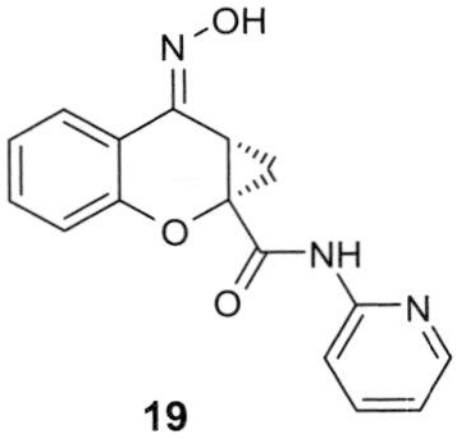

19
VU0359516

The same group of scientists at Vanderbilt University later described the identification of a class of cyclohexane-dicarboxylic monoamides as represented by **20** (hEC_{50} = 798 nM) [39]. Further investigation showed the *trans*-isomers of **20** to be inactive. The (+)-(1*R*,2*S*)- and (−)-(1*S*,2*R*)-*cis*-enantiomers were originally reported [39] to be of equal potency and efficacy, but it was recently discovered that the activity primarily resides on the (**1***R*,**2***S*) isomer [40]. Unlike (−)-PHCCC, which is a pure PAM, **20** also displayed some intrinsic agonist activity in a calcium mobilization assay, which could not be blocked by the known orthosteric antagonist LY341495 (**17**) indicating that **20** may not interact with the endogenous glutamate binding site. Compound **20** did not show activity at 10 μM in a panel of 67 GPCR, ion channel, and transporter selectivity assays. However, **20** was shown not to penetrate the brain upon peripheral administration, and its use as an *in vivo* CNS pharmacology tool is limited to intracerebral (icv) dosing [39]. Continuing efforts were focused on the SAR studies in all three ring systems of **20** and essentially all modifications resulted in inactive compounds or compounds with significantly reduced potency [41]. A notable exception discovered independently by researchers at Lundbeck is the primary amide Lu AF21934 (**21**), which has comparable potency to **20** and is brain penetrant (brain/plasma 0.8) [42]. Again, this work underscores the generally narrow SAR of mGlu4 PAMs and potential challenges in the lead optimization of these compounds.

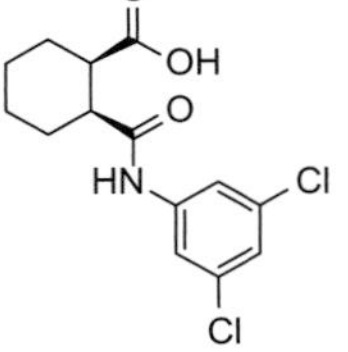

20
(±)-VU0155041

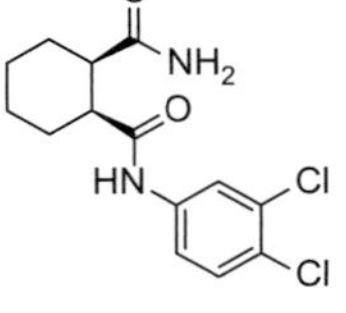

21
(+)-Lu AF21934

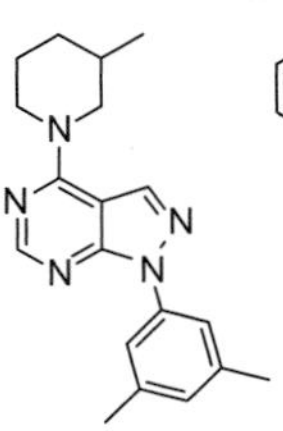

22
VU0080421

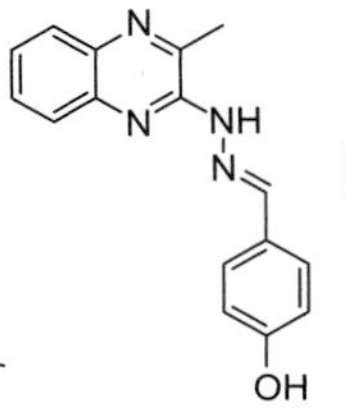

23
VU0001171

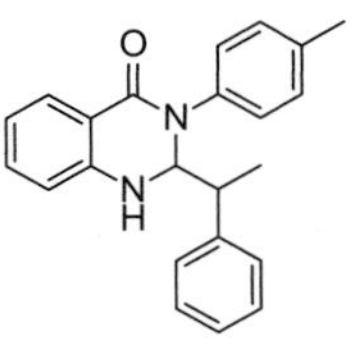

24
VU0092145

Among additional mGlu4 PAMs subsequently reported by the Vanderbilt group are **22–24** [43,44]. All of these new ligands were directly identified from their extensive high-throughput screen (HTS) campaign. The adenine derivative **22** potentiated glutamate with $EC_{50} = 5\ \mu M$ and caused a leftward shift of the glutamate response curve of 12- to 27-fold. However, **22** is also a full antagonist of mGlu1 receptor ($IC_{50} = 2.6\ \mu M$) and has low stability in a liver microsomes preparation [43]. A general and efficient microwave-assisted synthesis was therefore developed to explore the SAR and to optimize this chemical series [45]. Disappointingly, all 126 analogs synthesized were uniformly inactive except a very small subset showing reduced potency compared to the original hit **22**. HTS hit **23** was found to be a relatively potent mGlu4 PAM ($EC_{50} = 650$ nM, Glu E_{max} 141%) and to induce a 36-fold shift of glutamate potency. This compound was also selective against other mGlu receptors in both activation (agonist/PAM) and inhibition (antagonist) assays. However, the unattractive and potentially labile hydrazone moiety prevented it from further advancement. All attempts to replace the hydrazone with other chemical moieties were reportedly unsuccessful. The identification of compound **24** represents another new chemotype which has a potency ($EC_{50} = 3.0\ \mu M$) similar to that of **18** [44]. There are also some structural similarities between these two compounds, and two small libraries were synthesized in this series for SAR development and attempted optimization. Again the results were not fruitful as only two additional active compounds were identified, both with significantly weaker activity compared to the original hit **24**. Not to overstate it, but the now commonly occurring narrow SAR of mGlu4 PAMs (*vide infra*) represents a significant hurdle for medicinal chemists.

Adding to the list of new mGlu4 PAMs, the Vanderbilt group disclosed a class of heterobiarylamides as represented by ligands **25–27** [46]. The furan derivative **25** was one of the initial hits in its class with weak mGlu4 PAM activity ($EC_{50} > 5\ \mu M$), and extensive optimization was then carried out for both the furyl and aniline regions. Compound **26** was among the most potent ($EC_{50} = 240$ nM) compounds identified with good brain penetration (brain/plasma 4.1) and reasonable drug free fraction (plasma free fraction 2.2%). However, **26** was characterized by high *in vivo* clearance in rat (894 mL/min/kg), consistent with its poor *in vitro* microsomal stability. The same group also disclosed a related diamide series as represented by **27** [47]. 2-Pyridyl again was shown to be the optimal group for mGlu4 potency. Compound **27** and some of its analogs are reportedly potent ($EC_{50} = 100$ nM), brain penetrant, and highly protein bound (plasma protein binding > 99.5%).

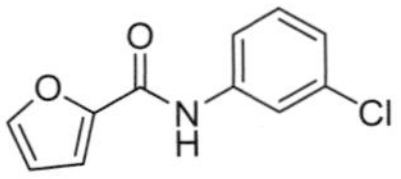

25 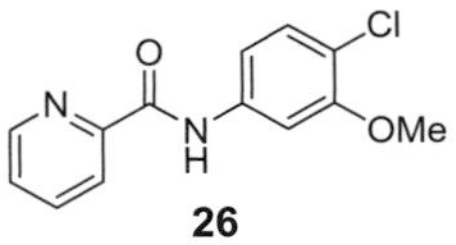 **26** 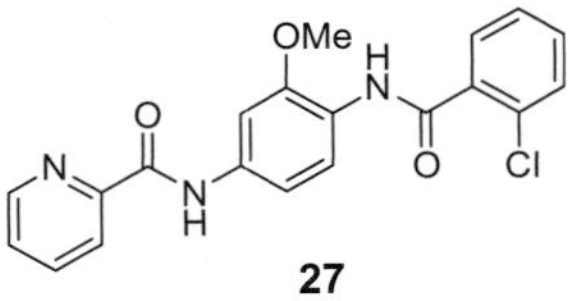 **27**
VU0415374

More recently, the Vanderbilt University group reported a class of biarylsulfonamides as potent mGlu4 PAMs [45]. Again, the lead compound (**28**, $EC_{50} = 6\ \mu M$) was directly identified from their HTS campaign. Based on the SAR learned from the heterobiarylamide series (**25** and **26**), the 2-furyl was replaced by 2-pyridyl which resulted in 30-fold increase in potency. Further optimization led to identification of **29** ($EC_{50} = 20$ nM), which represents one of the most potent mGlu4 PAMs reported to date. Although being very potent, this class of compounds exhibited high protein binding and poor microsomal stability in both human and rat species, thus limiting its utility. Finally, another sulfonamide series represented by **30** was reported [48], and although this class of compounds is not highly potent (*e.g.*, **30** $EC_{50} = 3.5\ \mu M$), it is characterized by relatively low protein binding (plasma free fraction > 3%), presumably related to improved aqueous solubility resulting from the nonplanar core and the presence of a basic nitrogen.

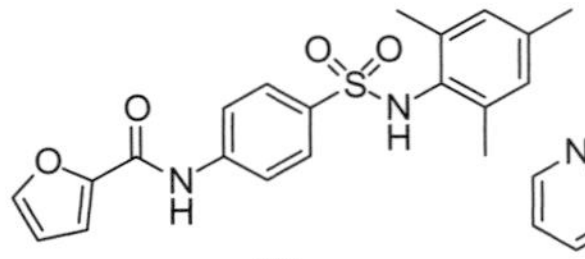

28 **29** **30**
VU0130734 VU0364439

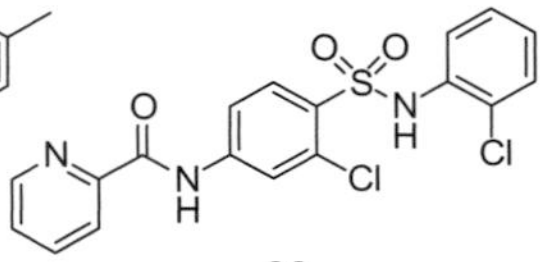

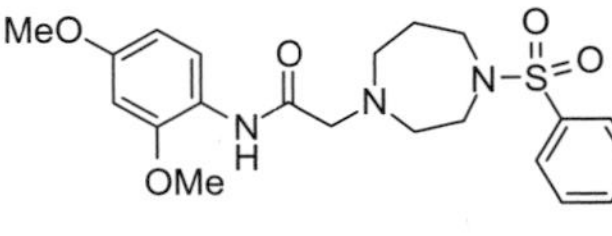

Recently, scientists at Merck reported the exact same biarylsulfonamide derivative class [49]. They also disclosed a series of phthalimides as mGlu4 PAMs [50]. Ultimately, two potent compounds were tritiated to provide radioligands **31** and **32** (34 and 160 nM, respectively) [51]. In a competition binding analysis, **31** and **32** reportedly displaced each other, indicating their allosteric binding sites are related. However, VU0155041 (**20**) enhances rather than displaces binding of **31** suggesting the existence of more than one allosteric binding site for various ligands.

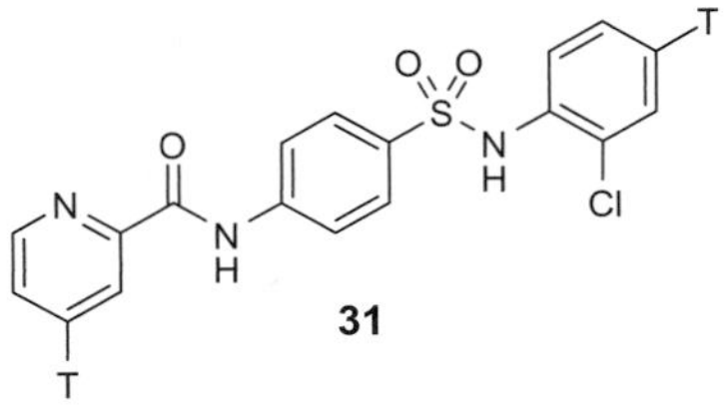

31 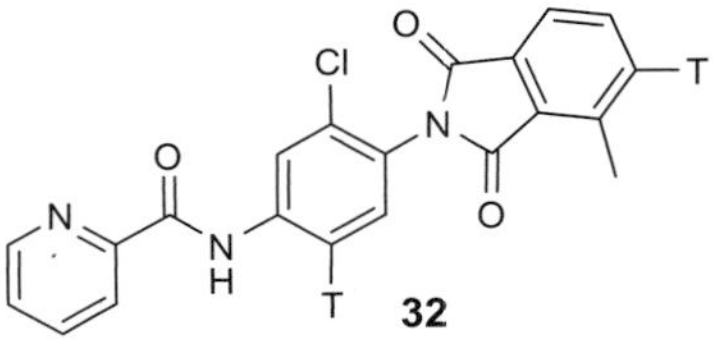 **32**

Compound **33**, structurally similar to mGlu5 NAMs SIB-1893 and MPEP (which also show weak mGlu4 PAM activity [52]), was recently reported by East *et al.* [53]. This compound was identified as an intermediate during resynthesis of a HTS hit and displayed an EC_{50} of 1 μM against both human and rat mGlu4 receptors. It is reportedly selective against mGlu5 (>10-fold) and has less than 50% activity at 10 μM for a panel of 68 targets. Though **33** is characterized by rapid *in vivo* clearance (75 mL/min/kg in rat), its high brain penetration (brain/plasma 2.9) and good oral bioavailability together with good CNS drug-like physicochemical properties have qualified it as a tool compound for acute *in vivo* proof of concept studies *via* peripheral administration.

Most recently, scientists at Lundbeck reported a class of tricyclic thiazolopyrazoles (*e.g.*, **35**) as potent, selective, and orally bioavailable mGlu4 PAMs. These compounds were designed to improve upon previous derivatives and to expand the chemical diversity among ligands for mGlu4 [54]. The tricyclic derivatives have similar potency compared to their acyclic analogs (*e.g.*, $EC_{50} = 13$ nM for **34** and 9 nM for **35**), and compound **35** is a highly subtype-selective mGlu4 PAM (>1000-fold against mGlu1, 2, 3, 5, 6, 7 receptors) with a brain/plasma ratio of *ca.* 1.2 [55].

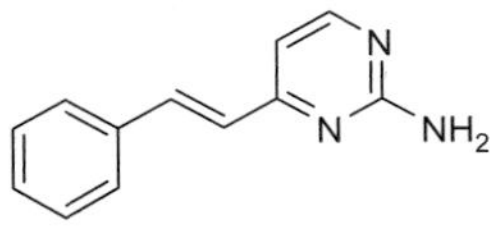

33 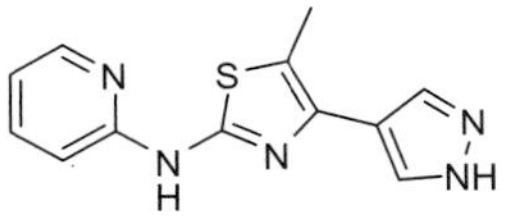 **34** 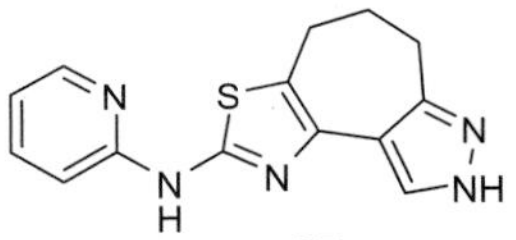 **35**

3.2. mGlu4 receptor negative allosteric modulators

While several groups, both in industry and in academia, have successfully identified numerous mGlu4 PAMs from HTS efforts, identification of mGlu4 NAMs has been shown to be challenging and none have been reported to date. Lacking any progress in this area, its significance remains another hard to interpret characteristic of allosteric mGlu4 modulators.

4. mGlu6 RECEPTOR LIGANDS

Unlike every other mGlu receptor, which is broadly expressed, the mGlu6 receptor is exclusively located in the retina [2]. The only selective mGlu6 ligand reported to date is **36** [56], an isoxazolyl bioisostere of 2-aminoadipic acid. Though compound **36** is a weakly potent orthosteric agonist ($EC_{50} = 82$ μM), it is selective against iGlu and other mGlu receptors. It was subsequently reported that all of the mGlu6 activity resides in the (*S*)-enantiomer [57].

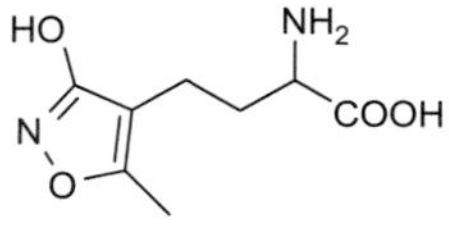

36
Homo-AMPA

5. mGlu7 RECEPTOR LIGANDS

AMN082, **37**, is the only mGlu7 agonist reported to date [58]. This compound was shown to be a full agonist in both GTPγS binding ($EC_{50} = 260$ nM) and cAMP ($EC_{50} = 64$ nM) assays and to act *via* an allosteric site, as supported by results of chimeric receptor studies and competitive binding with known group III orthosteric agonists and antagonists. However, it has been shown that the mGlu7 activity of **37** is highly context dependent and could be demonstrated in some pathways and cell backgrounds but not in others, thus complicating any determination [58–61]. In addition, though **37** was initially reported to be selective over 30 targets [58], it has been found to interact with many other targets (26 of 71) in a broader screen [61]. Finally, a metabolite was identified *in vivo* in rodents with monoaminergic activity [62]. Since many *in vivo* studies have been done with **37** to probe mGlu7 receptor function prior to that elucidation, the results should be carefully interpreted.

More recently, a class of isoxazolopyridone derivatives represented by **38** and **39** were reported as potent and subtype-selective mGlu7 antagonists [63,64]. The lead compound MDIP (**38**) was identified from HTS and was subsequently optimized to generate **39** with comparable potency but with improved physicochemical and ADME properties. Compound **39** was shown to antagonize L-AP_4-induced responses in both Ca^{2+} mobilization ($IC_{50} = 26$ nM, rat mGlu7) and cAMP assays (EC_{50} values of 220 and 610 nM, in rat and human mGlu7 receptors, respectively) [63]. Compound **39** is reportedly selective against other mGlu receptors and a panel of 168 targets. In addition, **39** displayed a favorable rat pharmacokinetic profile characterized by good oral bioavailability (65%), low clearance (1.0 mL/min/kg), and good brain penetration (brain/plasma 1.0), making it a valuable tool compound to study mGlu7 receptor function. However, like **37**, the mGlu7 activity of **39** was also found to be context dependent [60]. For example, **39** did not block L-AP_4-induced inhibition of cAMP accumulation in HEK cells. Furthermore, **39** was unable to block agonist-mediated responses at the Schaffer collateraL-CA1 synapse, a location at which neurotransmission has been shown to be modulated by mGlu7 receptor activity [60]. The context dependence of both agonist **37** and antagonist **39** further reveals the complexity of mGlu7 pharmacology.

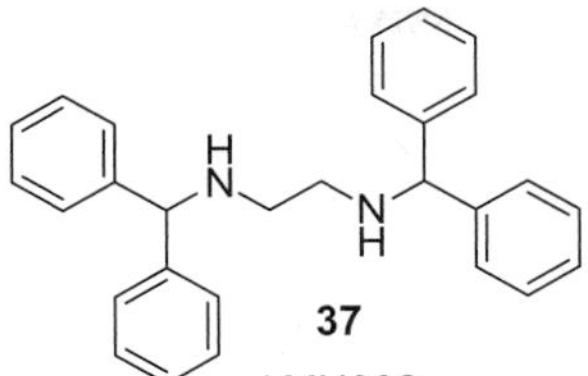

37
AMN082

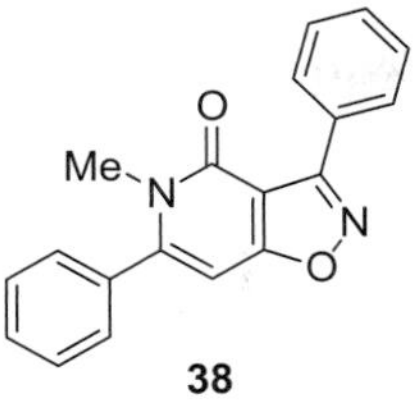

38
MDIP

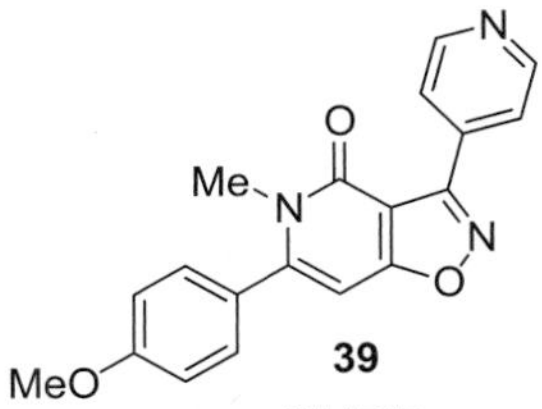

39
MMPIP

6. mGlu8 RECEPTOR LIGANDS

(*S*)-3,4-DCPG (**40**), originally synthesized to search for potential selective ionotropic glutamate receptor (iGlu) antagonists [65], was found to be a subtype-selective mGlu8 agonist ($EC_{50} = 31$ nM, >100-fold selectivity over other mGlu receptors) [66] and has been widely used as a tool compound to selectively activate this receptor. Though effects on several regions of the brain were observed through systemic administrations (*e.g.*, ip dosing) [67], the majority of the *in vivo* studies with **40** were carried out *via* central administration due to the poor brain penetration of this polar amino acid with additional hydrophilic carboxylic acid groups [68–72]. To date there are no additional reports of mGlu8 ligands.

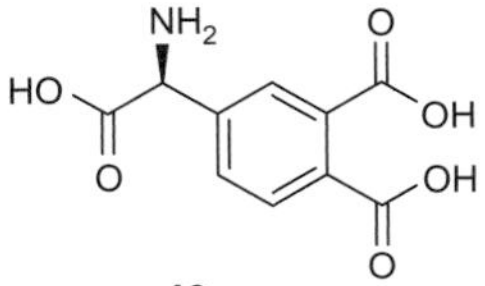

40
(*S*)-3,4-DCPG

7. CONCLUSIONS

While progress has been made over the past several years in identifying subtype-selective and systemically available ligands for some of the group III mGluRs, this field is still in its infancy. Most mGlu4 PAMs reported to date have undesired physicochemical and pharmacokinetic properties, limiting their utility as tool compounds and preventing them from drug candidate consideration. It is obvious from the various studies that the notoriously narrow SAR of mGlu4 PAMs, which leaves little room for lead optimization, is the greatest challenge. Concurrently, scientific evidence continues to support the biological hypothesis that potentiation of the mGlu4 receptor may provide potential treatment for motor symptoms in Parkinson's disease as well as mood and neuroinflammation disorders. In addition, evidence is beginning to support the biological rationale for the use of mGlu7 modulation in cognition and mood disorders. With the limitations found for AMN082 and MMPIP, additional tool compounds need to be discovered to fully unravel the potential utility of this target. Likewise, the same is true for the mGlu8 receptor. Future progress and clinical validation will ultimately afford a clear perspective on the value of these receptors as potential drug targets.

REFERENCES

[1] T. K. Attwood and J. B. C. Findlay, *Protein Eng.*, 1994, **7**, 195.
[2] F. Nicoletti, J. Bockaert, L. Collingridge Graham, P. J. Conn, F. Ferraguti, D. D. Schoepp, J.T. Wroblewski and J. P. Pin, *Neuropharmacology*, 2011, **60**, 1017.
[3] C. M. Niswender and P. J. Conn, *Annu. Rev. Pharmacol.*, 2010, **50**, 295.
[4] M. J. O'Neill, M. J. Fell, K. A. Svensson, J. M. Witkin and S. N. Mitchell, *Drugs Future*, 2010, **35**, 307.
[5] C. R. Hopkins, C. W. Lindsley and C. M. Niswender, *Future Med. Chem.*, 2009, **1**, 501.
[6] R. M. O'Connor, B. C. Finger, P. J. Flor and J. F. Cryan, *Eur. J. Pharmacol.*, 2010, **639**, 123.
[7] D. D. Schoepp, D. E. Jane and J. A. Monn, *Neuropharmacology*, 1999, **38**, 1431.
[8] H. Lavreysen and F. M. Dautzenberg, *Curr. Med. Chem.*, 2008, **15**, 671.
[9] M. Recasens, J. Guiramand, R. Aimar, A. Abdulkarim and G. Barbanel, *Curr. Drug Targets*, 2007, **8**, 651.
[10] C. Selvam, C. Goudet, N. Oueslati, J.-P. Pin and F. C. Acher, *J. Med. Chem.*, 2007, **50**, 4656.
[11] F. C. Acher, F. J. Tellier, R. Azerad, I. N. Brabet, L. Fagni and J.-P. R. Pin, *J. Med. Chem.*, 1997, **40**, 3119.
[12] P. Sibille, S. Lopez, I. Brabet, O. Valenti, N. Oueslati, F. Gaven, C. Goudet, H.-O. Bertrand, J. Neyton, M. J. Marino, M. Amalric, J.-P. Pin and F. C. Acher, *J. Med. Chem.*, 2007, **50**, 3585.
[13] R. L. Johnson and K. S. S. P. Rao, *Bioorg. Med. Chem. Lett.*, 2005, **15**, 57.
[14] C. Selvam, N. Oueslati, I. A. Lemasson, I. Brabet, D. Rigault, T. Courtiol, S. Cesarini, N. Triballeau, H.-O. Bertrand, C. Goudet, J.-P. Pin and F. C. Acher, *J. Med. Chem.*, 2010, **53**, 279.
[15] F. Acher, C. Selvam, N. Triballeau, J.-P. Pin and H.-O. Bertrand, Preparation of hypophosphorous acid derivatives, particularly substituted hypophosphorous acid-containing glycine derivatives, as agonists and antagonists of metabotropic glutamate receptors, especially mGluR4, *Patent WO 2007052169*, 2007.
[16] N. Triballeau, F. Acher, I. Brabet, J.-P. Pin and H.-O. Bertrand, *J. Med. Chem.*, 2005, **48**(7), 2534.
[17] C. Beurrier, S. Lopez, D. Revy, C. Selvam, C. Goudet, M. Lherondel, P. Gubellini, L. Kerkerian-LeGoff, F. Acher, J.-P. Pin and M. Amalric, *FASEB J.*, 2009, **23**, 3619.
[18] M. Kemp, P. Roberts, P. Pook, D. Jane, A. Jones, P. Jones, D. Sunter, P. Udvarhelyi and J. Watkins, *Eur. J. Pharmacol., Mol. Pharmacol. Sect.*, 1994, **266**, 187.
[19] D. E. Jane, K. Pittaway, D. C. Sunter, N. K. Thomas and J. C. Watkins, *Neuropharmacology*, 1995, **34**, 851.
[20] D. E. Jane, N. K. Thomas, H. W. Tse and J. C. Watkins, *Neuropharmacology*, 1996, **35**, 1029.

[21] J. C. Miller, P. A. Howson, S. J. Conway, R. V. Williams, B. P. Clark and D. E. Jane, *Br. J. Pharmacol.*, 2003, **139**, 1523.
[22] D. E. Jane, P. L. S. J. Jones, P. C. K. Pook, H. W. Tse and J. C. Watkins, *Br. J. Pharmacol.*, 1994, **112**, 809.
[23] M. C. Kemp, D. E. Jane, H.-W. Tse and P. J. Roberts, *Eur. J. Pharmacol.*, 1996, **309**, 79.
[24] N. K. Thomas, D. E. Jane, H. W. Tse and J. C. Watkins, *Neuropharmacology*, 1996, **35**, 637.
[25] A. E. Kingston, P. L. Ornstein, R. A. Wright, B. G. Johnson, N. G. Mayne, J. P. Burnett, R. Belagaje, S. Wu and D. D. Schoepp, *Neuropharmacology*, 1998, **37**, 1.
[26] S. M. Fitzjohn, Z. A. Bortolotto, M. J. Palmer, A. J. Doherty, P. L. Ornstein, D. D. Schoepp, A. E. Kingston, D. Lodge and G. L. Collingridge, *Neuropharmacology*, 1998, **37**, 1445.
[27] M. Rocheville and S. L. Garland, *Drug Discovery Today Technol.*, 2010, **7**, E87.
[28] M. J. Marino, D. L. Williams Jr., J. A. O'Brien, O. Valenti, T. P. McDonald, M. K. Clements, R. Wang, A. G. DiLella, J. F. Hess, G. G. Kinney and P. J. Conn, *Proc. Natl. Acad. Sci. U.S.A.*, 2003, **100**, 13668.
[29] M. Maj, V. Bruno, Z. Dragic, R. Yamamoto, G. Battaglia, W. Inderbitzin, N. Stoehr, T. Stein, F. Gasparini, I. Vranesic, R. Kuhn, F. Nicoletti and P. J. Flor, *Neuropharmacology*, 2003, **45**, 895.
[30] H. Annoura, A. Fukunaga, M. Uesugi, T. Tatsuoka and Y. Horikawa, *Bioorg. Med. Chem. Lett.*, 1996, **6**, 763.
[31] A. M. Canudas, V. Di Giorgi-Gerevini, L. Iacovelli, G. Nano, M. D'Onofrio, A. Arcella, F. Giangaspero, C. Busceti, L. Ricci-Vitiani, G. Battaglia, F. Nicoletti and D. Melchiorri, *J. Neurosci.*, 2004, **24**, 10343.
[32] G. Battaglia, C. L. Busceti, G. Molinaro, F. Biagioni, A. Traficante, F. Nicoletti and V. Bruno, *J. Neurosci.*, 2006, **26**, 7222.
[33] E. S. Choe, E. H. Shin and J. Q. Wang, *Neurosci. Lett.*, 2006, **394**, 246.
[34] K. Stachowicz, K. Klak, A. Klodzinska, E. Chojnacka-Wojcik and A. Pilc, *Eur. J. Pharmacol.*, 2004, **498**, 153.
[35] K. Stachowicz, E. Chojnacka-Wojcik, K. Klak and A. Pilc, *Pharmacol. Rep.*, 2006, **58**, 820.
[36] K. Klak, A. Palucha, P. Branski, M. Sowa and A. Pilc, *Amino Acids*, 2007, **32**, 169.
[37] L. Iacovelli, A. Arcella, G. Battaglia, S. Pazzaglia, E. Aronica, P. Spinsanti, A. Caruso, E. De Smaele, A. Saran, A. Gulino, M. D'Onofrio, F. Giangaspero and F. Nicoletti, *J. Neurosci.*, 2006, **26**, 8388.
[38] R. Williams, Y. Zhou, C. M. Niswender, Q. Luo, P. J. Conn, C. W. Lindsley and C. R. Hopkins, *ACS Chem. Neurosci.*, 2010, **1**, 411.
[39] C. M. Niswender, K. A. Johnson, C. D. Weaver, C. K. Jones, Z. Xiang, Q. Luo, A.L. Rodriguez, J. E. Marlo, T. de Paulis, A. D. Thompson, E. L. Days, T. Nalywajko, C.A. Austin, M. B. Williams, J. E. Ayala, R. Williams, C. W. Lindsley and P. J. Conn, *Mol. Pharmacol.*, 2008, **74**, 1345.
[40] C. Christov, P. Gonzalez-Bulnes, F. Malhaire, T. Karabencheva, C. Goudet, J.-P. Pin, A. Llebaria and J. Giraldo, *ChemMedChem*, 2011, **6**, 131.
[41] R. Williams, K. A. Johnson, P. R. Gentry, C. M. Niswender, C. D. Weaver, P. J. Conn, C.W. Lindsley and C. R. Hopkins, *Bioorg. Med. Chem. Lett.*, 2009, **19**, 4967.
[42] D. Doller, M. A. Uberti, S.-P. Hong, M. T. Nerio, R. M. Brodbeck, N. Breysse, M. A. Merciadez, T. Vialeti, M. DenBlekyer, S. Fallon, M. A. Bacolod, M. E. Nattini, M. Cajina, K. G. Liu, S. W. Topiol, M. Sabio, D. Andersson, T. N. Sager and K. Fog, Lu AF21934, a brain penetrant mGlu4 receptor positive allosteric modulator tool compound. 655.28/M10, 40th Annual Society for Neuroscience, San Diego, CA, November, 2010.
[43] C. M. Niswender, E. P. Lebois, Q. Luo, K. Kim, H. Muchalski, H. Yin, P. J. Conn and C.W. Lindsley, *Bioorg. Med. Chem. Lett.*, 2008, **18**, 5626.

[44] R. Williams, C. M. Niswender, Q. Luo, U. Le, P. J. Conn and C. W. Lindsley, *Bioorg. Med. Chem. Lett.*, 2009, **19**, 962.
[45] D. W. Engers, P. R. Gentry, R. Williams, J. D. Bolinger, C. D. Weaver, U. N. Menon, P.J. Conn, C. W. Lindsley, C. M. Niswender and C. R. Hopkins, *Bioorg. Med. Chem. Lett.*, 2010, **20**, 5175.
[46] D. W. Engers, C. M. Niswender, C. D. Weaver, S. Jadhav, U. N. Menon, R. Zamorano, P.J. Conn, C. W. Lindsley and C. R. Hopkins, *J. Med. Chem.*, 2009, **52**, 4115.
[47] D. W. Engers, J. R. Field, U. Le, Y. Zhou, J. D. Bolinger, R. Zamorano, A. L. Blobaum, C.K. Jones, S. Jadhav, C. D. Weaver, P. J. Conn, C. W. Lindsley, C. M. Niswender and C.R. Hopkins, *J. Med. Chem.*, 2011, **54**, 1106.
[48] Y.-Y. Cheung, R. Zamorano, A. L. Blobaum, C. D. Weaver, P. J. Conn, C. W. Lindsley, C.M. Niswender and C. R. Hopkins, *ACS Comb. Sci.*, 2011, **13**, 159.
[49] J. A. McCauley, J. W. Butcher, J. W. Hess, N. J. Liverton and J. J. Romano, Sulfonamide derivatives as mGluR4 receptor ligands, *Patent WO 2010033350*, 2010.
[50] J. A. McCauley, J. W. Hess, N. J. Liverton, C. J. McIntyre, J. J. Romano and M. T. Rudd, Preparation of phthalimidephenylpyridinecarboxamide derivatives for use as metabotropic glutamate R4 modulators, *Patent WO 2010033349*, 2010.
[51] J. Hess, M. J. Clements, T. P. McDonald, N. Liverton, J. A. McCauley, J. W. Butcher, K. Nguyen, D. Charvin, B. Campo, C. Bolea, E. Le Poul and I. J. Reynolds, Identification and characterization of radioligands that bind to an allosteric modulator site on the mGlur4 receptor. 643.22/F23, 40th Annual Society for Neuroscience, San Diego, CA, November, 2010.
[52] J. M. Mathiesen, N. Svendsen, H. Brauner-Osborne, C. Thomsen and M. T. Ramirez, *Br. J. Pharmacol.*, 2003, **138**, 1026.
[53] S. P. East, S. Bamford, M. G. A. Dietz, C. Eickmeier, A. Flegg, B. Ferger, M. J. Gemkow, R. Heilker, B. Hengerer, A. Kotey, P. Loke, G. Schaenzle, H.-D. Schubert, J. Scott, M. Whittaker, M. Williams, P. Zawadzki and K. Gerlach, *Bioorg. Med. Chem. Lett.*, 2010, **20**, 4901.
[54] C. Bolea and S. Celanire, Preparation of novel heteroaromatic derivatives and their use as positive allosteric modulators of metabotropic glutamate receptors, *Patent WO 2009010455*, 2009.
[55] S.-P. Hong, K. G. Liu, G. Ma, M. Sabio, J. Peterson, M. A. Uberti, M. D. Bacolod, Z. Z. Zou, A. J. Robichaud and D. Doller, *J. Med. Chem.*, 2011, **54**, 5070.
[56] H. Braeuner-Osborne, F. A. Slok, N. Skjaerbaek, B. Ebert, N. Sekiyama, S. Nakanishi and P. Krogsgaard-Larsen, *J. Med. Chem.*, 1996, **39**, 3188.
[57] H. Ahmadian, B. Nielsen, H. Braeuner-Osborne, T. N. Johansen, T. B. Stensbol, F. A. Slok, N. Sekiyama, S. Nakanishi, P. Krogsgaard-Larsen and U. Madsen, *J. Med. Chem.*, 1997, **40**, 3700.
[58] K. Mitsukawa, R. Yamamoto, S. Ofner, J. Nozulak, O. Pescott, S. Lukic, N. Stoehr, C. Mombereau, R. Kuhn, K. H. McAllister, H. van der Putten, J. F. Cryan and P. J. Flor, *Proc. Natl. Acad. Sci. USA*, 2005, **102**, 18712.
[59] J. E. Ayala, C. M. Niswender, Q. Luo, J. L. Banko and P. J. Conn, *Neuropharmacology*, 2008, **54**, 804.
[60] C. M. Niswender, K. A. Johnson, N. R. Miller, J. E. Ayala, Q. Luo, R. Williams, S. Saleh, D. Orton, C. D. Weaver and P. J. Conn, *Mol. Pharmacol.*, 2010, **77**, 459.
[61] R. M. Brodbeck, F. Scotty, X. Pu, M. Uberti, L. K. Issac, T. Schulenburg, B. Chen and M. Didriksen, 159.23/T12. Characterization of AMN082: A rich pharmacology which does not appear to include a measurable stimulation of the mGlu7 receptor, 40th Annual Neuroscience Meeting, San Diego, California, November 13–17, 2010.
[62] S. Sukoff Rizzo, S. K. Leonard, A. Gilbert, P. Dollings, D. L. Smith, M.-Y. Zhang, L. Di, B. J. Platt, S. Neal, C. Bender, J. Zhang, T. Lock, D. Kowal, A. Kramer, A. Randall,

S. Sears, C. Huselton, K. Vishwanathan, J. Butera, R. H. Ring, S. Rosenzweig-Lipson, Z. A. Hughes and J. Dunlop, 643.28/F29. The mGluR7 allosteric agonist AMN082 is a monoaminergic agent in disguise!, 40th Annual Neuroscience Meeting, San Diego, California, November 13–17, 2010.
[63] G. Suzuki, N. Tsukamoto, H. Fushiki, A. Kawagishi, M. Nakamura, H. Kurihara, M. Mitsuya, M. Ohkubo and H. Ohta, *J. Pharmacol. Exp. Ther.*, 2007, **323**, 147.
[64] M. Nakamura, H. Kurihara, G. Suzuki, M. Mitsuya, M. Ohkubo and H. Ohta, *Bioorg. Med. Chem. Lett.*, 2010, **20**, 726.
[65] N. K. Thomas, P. Clayton and D. E. Jane, *Eur. J. Pharmacol.*, 1997, **338**, 111.
[66] N. K. Thomas, R. A. Wright, P. A. Howson, A. E. Kingston, D. D. Schoepp and D. E. Jane, *Neuropharmacology*, 2001, **40**, 311.
[67] A.-M. Linden, M. Bergeron, M. Baez and D. D. Schoepp, *Neuropharmacology*, 2003, **45**, 473.
[68] J. Folbergrova, R. Druga, R. Haugvicova, P. Mares and J. Otahal, *Neuropharmacology*, 2008, **54**, 665.
[69] I. Marabese, F. Rossi, E. Palazzo, V. de Novellis, K. Starowicz, L. Cristino, D. Vita, L. Gatta, F. Guida, V. Di Marzo, F. Rossi and S. Maione, *J. Neurophysiol.*, 2007, **98**, 43.
[70] F. L. Jiang, Y. C. Tang, S. C. Chia, T. M. Jay and F. R. Tang, *Epilepsia*, 2007, **48**, 783.
[71] K. Stachowicz, K. Klak, A. Pilc and E. Chojnacka-Wojcik, *Pharmacol. Rep.*, 2005, **57**, 856.
[72] I. Marabese, V. de Novellis, E. Palazzo, M. A. Scafuro, D. Vita, F. Rossi and S. Maione, *Neuropharmacology*, 2007, **52**, 253.

CHAPTER 2

Recent Advances Toward Pain Therapeutics

Philippe G. Nantermet and **Darrell A. Henze**

Contents

1. INTRODUCTION

Chronic pain includes a variety of indications, all of which share a common feature that sufferers experience long-term periods of pain interfering with normal life. The treatment of all forms of chronic pain remains a huge unmet medical need. It is estimated that there are over 160 million people suffering from chronic pain conditions in the G7 countries, with

Merck Research Laboratories, West Point, PA 19486, USA

Annual Reports in Medicinal Chemistry, Volume 46
ISSN: 0065-7743, DOI: 10.1016/B978-0-12-386009-5.00025-4

those numbers increasing as the population ages. Unfortunately, none of the currently available treatments appear to be able to help even half of patients achieve at least a 50% reversal of their pain without significant adverse events. Thus, although pain sufferers annually spend over $20 billion, many seek new therapies that have both better efficacy and tolerability than existing treatments.

The dominant types of chronic pain include osteoarthritis (OA), chronic lower back pain (CLBP), fibromyalgia, and neuropathic pain. These various forms are believed to have complex and overlapping mechanisms of action. For example, inflammatory cytokines in degenerating OA joints activate nociceptive nerve endings in the periphery. However, the central pain pathways receiving input from the OA joint are also likely to be sensitized in ways similar to that which occurs during various types of neuropathic pain. Conversely, neuropathic pain may involve long-term increases in the spontaneous activity of peripheral nerves that drive central pain pathways sensitized by the chronic excitatory drive [1]. In either case, the signals that are transmitted ultimately to the brain are persistent and strongly activate areas of the brain involved in pain perception. Thus the potential list of mechanisms to target for novel pain therapies is very diverse [2]. Targets in the periphery include those involved in sensory transduction and central nervous system (CNS) transmission. In the CNS, potential targets are involved in signal processing and setting the relative excitability in the pain processing pathways.

Two of the most promising novel targets actively pursued for chronic pain are the voltage-gated sodium (Nav) channels, in particular Nav 1.3, 1.7, and 1.8, and nerve growth factor (NGF/tropomyosin-related kinase receptor A (TrkA). Both of these targets are predominantly localized in the periphery. They are believed to play key roles in the sensitization and excitability of the nociceptors that takes place during chronic pain. The achievement of good target selectivity with small molecules is also a common challenge of both pathways. This review will provide a summary of recent drug development for these two pathways, highlighting recent developments and progress made toward finding truly selective inhibitors.

A third target, fatty acid amide hydrolase (FAAH), has been the focus of much work in the past several years. FAAH is an enzyme that catabolizes a family of molecules called fatty acid amides (FAAs). A subset of FAAs is known to have agonist activity at cannabinoid receptors. Therefore, inhibition of FAAH is predicted to raise the levels of the endocannabinoids and provide analgesic relief. In contrast to sodium channels and NGF/TrkA, FAAH is believed to play a role in both the periphery and the CNS. This review will cover some recently disclosed novel noncovalent inhibitors as well as some recent clinical data that raise questions about the validity of the target.

2. SODIUM CHANNEL BLOCKERS

The use of sodium channel blockers for the treatment of pain has been extensively reviewed in the past 5 years [3–13]. This chapter will very briefly summarize background and key findings from this period and then focus on recent developments.

2.1. Nav subtypes as pain targets

Even before their molecular target was known, nonselective sodium channel blockers were used as analgesics. Voltage-gated sodium (Nav) channels are formed by a pore forming alpha subunit and an auxiliary beta subunit. Cloning studies have resulted in the identification of nine alpha subtypes of the sodium channel (Nav1.1–9) sharing $>50\%$ sequence identity in the extracellular loops and membrane domains. Of these subtypes, Nav1.3, 1.7, 1.8, and 1.9 are expressed in peripheral nerves and have been considered as potential targets to develop novel analgesics. Further interest by the pharmaceutical industry to identify selective Nav1.7 inhibitors has arisen from the recent discovery that Nav1.7 mutations associated with loss-of-function are linked to congenital indifference to pain [14], while gain-of-function mutations [15] result in primary erythermalgia. However, in order to achieve a robust therapeutic window when treating chronic pain, blockade of the following subtypes must be avoided: Nav1.2 (CNS side effects), Nav1.4 (paralysis), Nav1.5 (cardiac side effects), and Nav1.6 (neuromuscular side effects) [6]. Although structurally diverse, the known Nav inhibitors are able to bind all Nav subtypes. Furthermore, mutagenesis studies have demonstrated that a diverse set of known Nav blockers appears to bind to a common binding site which is unfortunately highly conserved across various Nav subtypes [3]. The high subtype homology displayed within the known anesthetic binding site suggests that other chemotypes, binding to an alternate site, would have to be discovered to achieve significant binding selectivity.

2.2. Pharmacological versus functional selectivity *via* state-dependent inhibition

Despite being able to block all Nav subtypes *in vitro*, a variety of Nav blockers such as lidocaine and mexilitine can be used systemically in the clinic with appropriate monitoring. The presence of a therapeutic margin is believed to arise from an interaction between state-dependent binding of these inhibitors, variations in the rate- and voltage-dependence of the transitions between channel states, and tissue-specific expression of Nav subtypes [3,5]. State-dependent blockers are most potent while cells are strongly depolarized when Nav channels are predominantly in the inactivated state. These state-dependent blockers have also been referred to as "use dependent" since the Nav blockade preferentially takes place on cells firing action potentials at nominally high rates. Thus, during high activity states such as those evoked in pain pathways relevant to chronic pain (>10 Hz), Nav channels accumulate in the inactivated state and allow state-dependent inhibitors to bind. In comparison, the cardiac activity rate is much lower (1–2 Hz), cells rest at more negative resting potentials, resulting in Nav1.5 channels not accumulating in the inactivated state and experiencing very little inhibition. The overall state dependency of each of the nonselective blockers is directly related to their *in vivo* efficacy and safety. State-dependent inhibitors that can provide functional selectivity with minimal CNS and cardiac effects are still actively pursued.

In contrast to state-dependent selectivity, achieving pharmacological selectivity across Nav subtypes has remained an elusive goal. Pharmacological selectivity can be defined as the relative potency of a compound to inhibit various Nav subtypes under a common relative activation state of the channels. A typical approach is to use electrophysiological protocols specific to each Nav subtype to cause ~50% of channels to be in the inactivated state and then determine the relative IC_{50}s [16–18]. To date, there are very few compounds that have been demonstrated to have any true pharmacological selectivity for subtypes in pain over those critical for cardiac function (see below).

2.3. Subtype-selective inhibitors

The quest for pharmacologically selective Nav blockers has been greatly hampered by the limited throughput techniques necessary to control for channel activation states. However, efforts to find leads with increased subtype selectivity have recently been catalyzed by improvements in electrophysiological screening technologies. IonWorks and PatchXpress are current examples of automated patch clamp systems. In the recent past, a large number of inhibitors with improved potency have been disclosed, but selectivity is usually lacking or not documented appropriately [4,7,8]. Unfortunately, definitions of selectivity have not been standardized and have mostly been based solely on state-dependent properties. In the absence of a standardized approach, IC_{50}s for state-dependent selectivity should be accompanied by a suitable description of the assay to allow more informed comparisons across structural series. Following are the examples of inhibitors with demonstrated subtype selectivity against one of the Nav subtypes targeted for pain.

A-803467 (**1**) has been described as a 100- to 1000-fold pharmacologically selective Nav1.8 state-dependent blocker displaying analgesia in a variety of pain models. However, clinical development has been hampered by poor bioavailability [16,17]. Mutation studies suggest a binding site that partially overlaps with nonselective Nav blockers [19]. Further studies led to inhibitors with improved oral bioavailability [20,21]. A-887826 (**2**) is a representative example with ~30-fold pharmacological selectivity against the Nav1.5 channel [22].

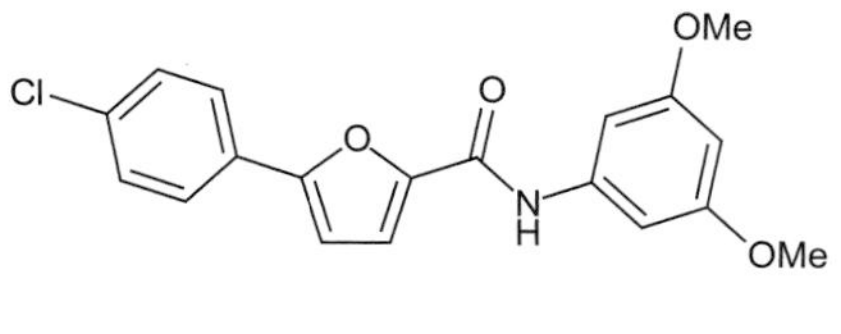

1 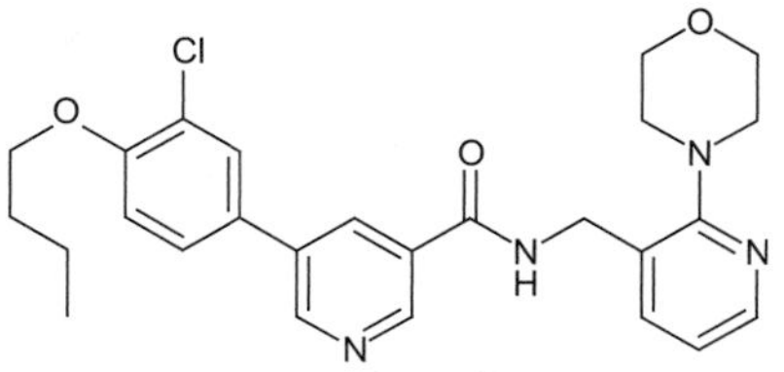 **2**

The search for selective Nav1.7 blockers has experienced a recent surge in activity. Benzazepinone **3** has been reported to be a 90/680 nM state-dependent inhibitor of the Nav1.7/Nav1.8 channels while maintaining ~10-fold use-dependent selectivity against Nav1.5 [23]. Although compounds from this series are efficacious in animal models, their moderate bioavailability limits their potential for clinical development. Binding experiments suggested that these inhibitors bind to or very close to the original local anesthetic binding site.

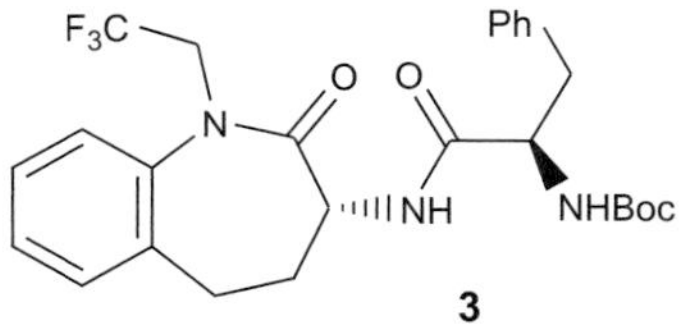

3

In 2010, a patent application disclosed a series of potent Nav1.7 blockers. The most preferred examples were claimed to display up to 100-fold pharmacological selectivity against the Nav1.5 channel and unspecified pharmacological selectivity against the Nav1.3 channel [18]. Estimated IC_{50} values for Nav1.7 using a PatchXpress platform were reported in the single nanomolar to picomolar range for a number of examples. Two examples, among those that were prepared in multigram quantities, are represented below (**4** and **5**). Interestingly, shortly after the publication of this patent, the originating corporations, Icagen and Pfizer, announced their engagement in human clinical studies with potent and selective Nav1.7 blockers [24]. The original microdosing pharmacokinetics study (July 2010) included four compounds and concluded with the selection of one, PF-05089771, that is now the subject of further clinical studies (December 2010). Consultation of the NIH clinical trials website will direct the reader to two clinical trials that could correspond to the Icagen press releases mentioned above [25].

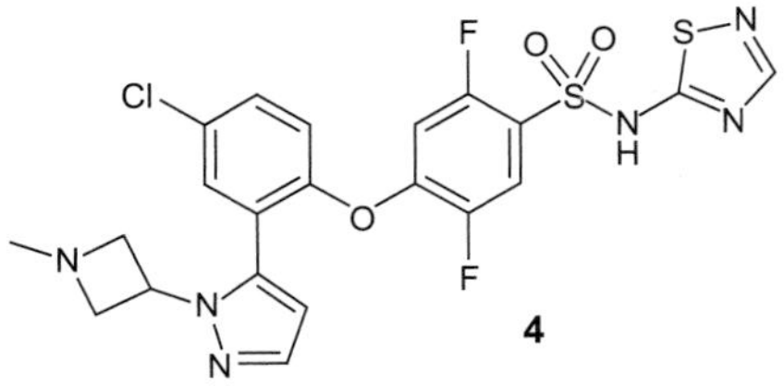

Another recent patent application has disclosed a series of aryl substituted carboxamides as blockers of T-type calcium channels or sodium channels [26]. Preferred compounds are claimed to bind potently to Nav1.7 and Nav1.3 and to be selective against the Nav1.5 channel. A representative example is illustrated below (**6**). A more recent application from the same group has claimed picolinamides **7** as potent blockers of Nav1.7 and Nav1.3 with selectivity against the Nav1.5 channel [27].

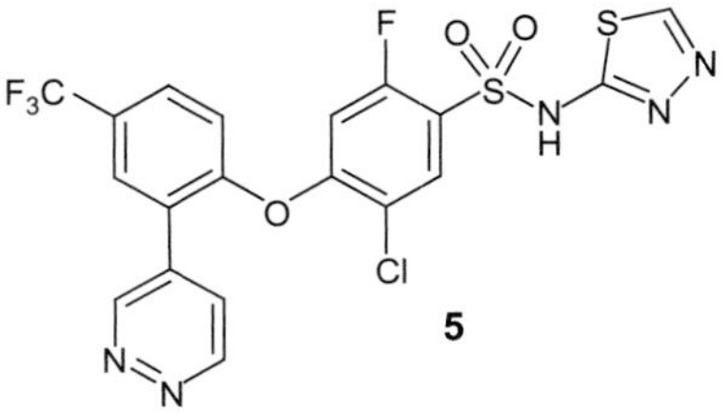

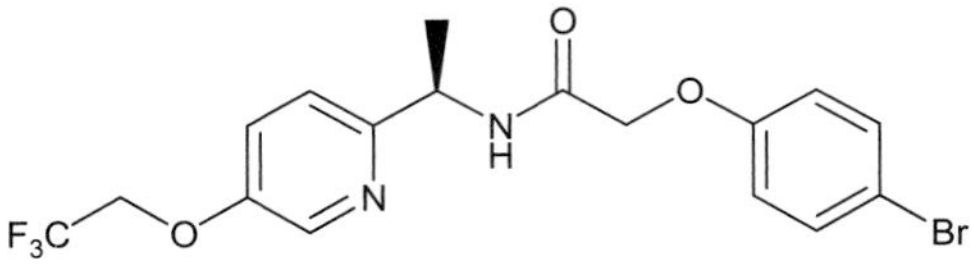

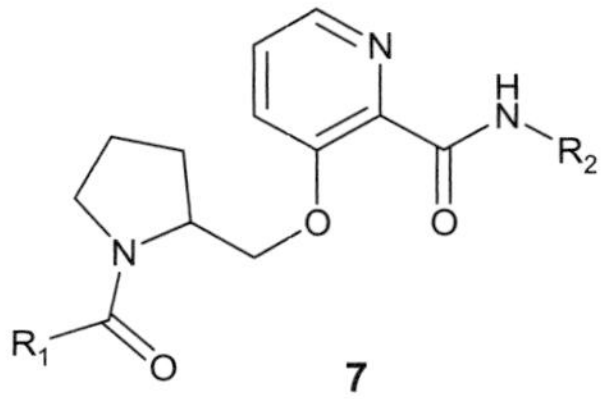

A series of benzamides related to **4** and **5** have been claimed as pharmacologically selective Nav1.3 blockers over Nav1.5 [28]. For example, **8** is a 15-nM inhibitor of Nav1.3 with >1000-fold selectivity over Nav1.5.

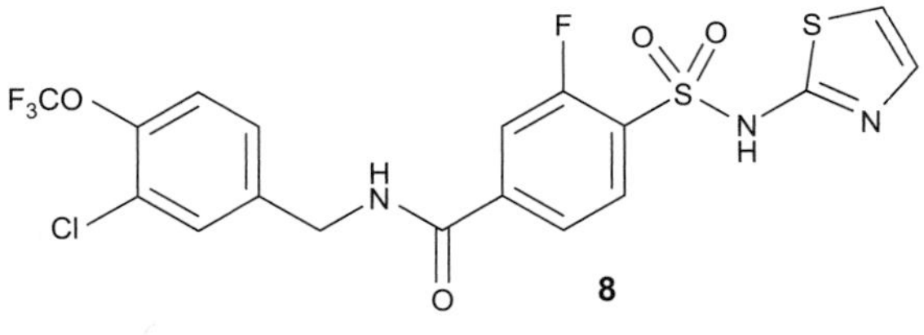

3. NGF/TrkA PATHWAY INTERFERENCE

Strategies targeting blockade of NGF activation of the TrkA receptor for the treatment of inflammatory and neuropathic pain have been recently reviewed [29]. This chapter will very briefly summarize various approaches and then focus on the most recent developments.

3.1. NGF sequestration

The NGF/TrkA pathway is one of the few new targets with clinical proof of concept against chronic pain [30]. The recent reports of positive PhII and PhIII trials against OA, chronic low back pain, and interstitial cystitis using neutralizing anti-NGF antibodies (Tanezumab [30], REGN475 [31]) have all demonstrated good efficacy and tolerability and have driven an increased focus on developing novel therapeutics in this pathway. However, it is important to note that the FDA has imposed a clinical hold on all active anti-NGF programs in the USA to explore concerns related to potential acceleration and exacerbation of joint degeneration. The release of the actual clinical findings driving these concerns will be necessary to understand the impact of NGF sequestration with antibodies on safety and efficacy. It is important to note that all the companies with anti-NGF mAbs in late stage clinical programs are purported to be working with the FDA to resolve the concerns and proceed with development.

Other preclinical efforts at NGF sequestration include using a variety of engineered peptide fragments including those derived from the extracellular NGF-binding domain of the TrkA receptor to bind and sequester NGF. These efforts have been recently reviewed [29]. These peptide approaches have largely been discontinued in favor of the anti-NGF antibody approaches due to concerns about neoepitope-induced immunogenicity which is a risk of the engineered peptides of these classes.

3.2. Inhibition of NGF/TrkA interaction

MNAC-13, also known later as BXL-1H5 or GBR-900, is a specific TrkA monoclonal antibody that has shown improvements in various pain models [32]. It has recently been in-licensed by Glenmark Pharmaceuticals to be advanced toward clinical development.

Small peptide mimetics of NGF are capable of functionally blocking the NGF–TrkA interaction. One such peptide, IPTRK3, has recently been demonstrated to suppress both thermal and mechanical hyperalgesia induced by CFA upon local injection [33]. Other peptides, NL1L4 and L1L4, have been shown to reduce neuropathic behavior and restore neuronal function in a rat model of pain [34].

Triazine-diketopiperazine-based peptidomimetics have recently been studied as antagonists of the TrkA [35] and TrkC [36] receptors.

Selectively disrupting the interaction between NGF and TrkA using a small molecule as opposed to an antibody or peptide remains an attractive goal from a drug discovery standpoint. ALE-0540 (**9**) was one of the first nonpeptidic molecules to specifically inhibit the binding of NGF to TrkA and had some limited efficacy in models of neuropathic and inflammatory pain [37]. In 2009, Painceptor Pharma claimed derivatives of

ALE-0540 to be efficient in animal models of chronic pain, with the compound PPC-1807 being the closest to clinical development [38]. PD90780 (**10**) represents another class of antagonists of the NGF–TrkA interaction [39,40]. More recent contributions by the same group of scientists are illustrated by thioxothiazolidine Y1036 (**11**) which prevents neurotrophin-mediated differentiation of dorsal-root ganglion sensory neurons [41].

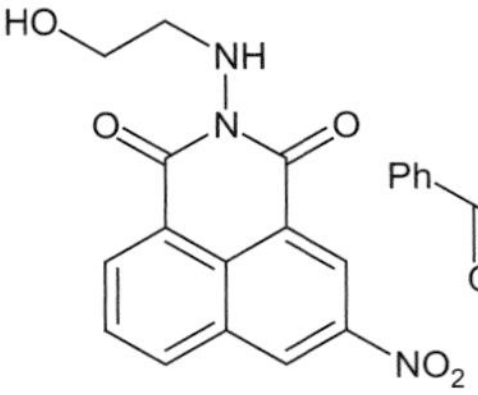

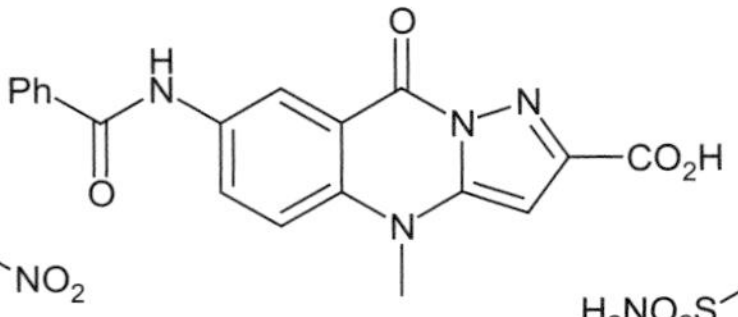

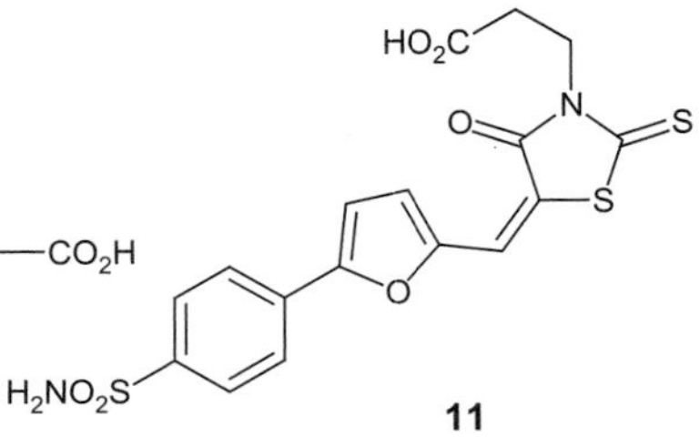

3.3. TrkA kinase inhibitors

Numerous small-molecule inhibitors of Trk receptor kinases have been described or claimed, mostly for the treatment of cancer. They have recently been reviewed and very few of these are Trk selective [42]. Lestaurtinib (CEP-701, **12**), an indolocarbazole staurosporine derivative inhibitor of multiple kinases including TrkA, and CEP-751, a pan-Trk inhibitor, have entered clinical development for various cancers. Another example, AZ-23 (**13**) has been characterized as a 2/8 nM inhibitor of TrkA/B with anticancer potential but still retains significant activity against a number of other kinases such as Flt3 and FGFR1 [43,44].

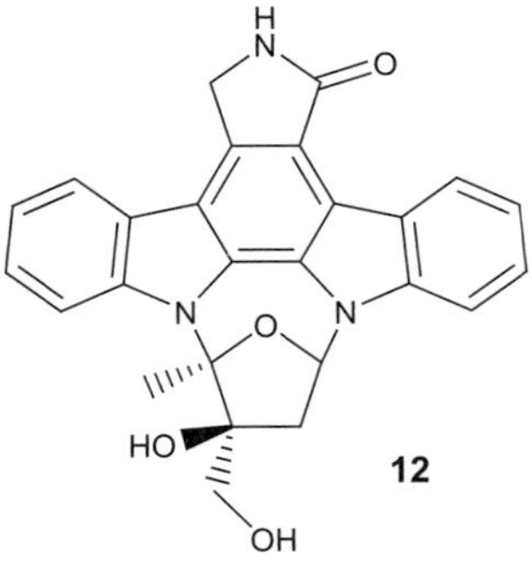

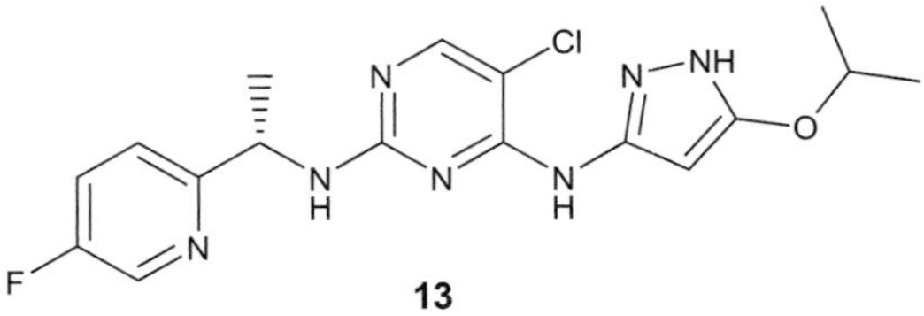

A recent patent application disclosed a series of isoform-selective TrkA kinase inhibitors with activities in the 10–100 nM range against TrkA and >10 μM against other kinases including TrkB and C [45].

One example (compound B, structure not identified) was claimed to reduce mechanical allodynia in the chronic-constriction injury rat neuropathic pain model in a dose-dependent and more significantly efficacious manner than gabapentin. The examples illustrated in the application are quite structurally diverse. Compound **14**, corresponding to example 201 or compound A, is claimed to be an 85-nM inhibitor of TrkA.

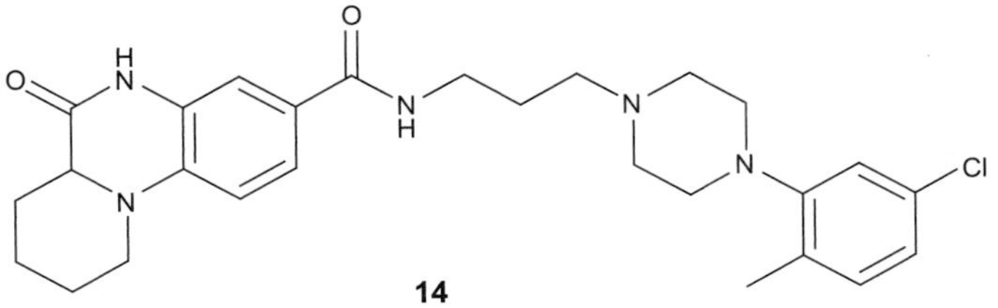

14

In the past 3 years, scientists from Array Biopharma have delivered several presentations related to small-molecule Trk inhibitors that are efficacious in various preclinical models of pain [46]. While they have indicated that they are pursuing TrkA selective inhibitors [47], thus far they have only disclosed data related to pan-Trk inhibitors. AR00457470 (formerly designated AR-872, structure not disclosed) has been characterized as a "first in class" pan-Trk inhibitor, with activities in the 10 nM range against TrkA,B,C. It exhibits $>$100-fold selectivity against a very diverse panel of kinases, and $>$1000-fold selectivity versus a diverse panel of receptors and ion channels. AR00457470 is claimed to have drug-like physical properties (MW $<$ 450, cLogP $<$ 3, PSA $<$ 100, $>$1 mg/mL solubility at pH 1.2–7.4) and to present low potential for drug–drug interactions. This compound achieves good oral exposures in rats and is restricted to the periphery (high PgP efflux). AR00457470 has been described to be efficacious in various pain models. It inhibits thermal hyperalgesia under acute or prophylactic dosing paradigms in the rat CFA or CIA models, inhibits CFA-induced mechanical allodynia, and is superior to Naproxen, Valdecoxib, or Rofecoxib in a CFA-induced model of monoarthritis, as measured by guarding index and difference in weight bearing. AR00457470 has also been demonstrated to have no effect on normal thermal responses or normal nociceptive behaviors in preclinical species. AR00457470 reduces nonmalignant skeletal pain in the adult mouse without interfering with sensory and sympathetic nerve fibers or early fracture healing [48]. Recent patent applications from the same group of scientists claim Trk kinase inhibitors to be useful for treating pain [49–51]. Representative structures of such amides Trk kinase inhibitors are illustrated below (**15–17**).

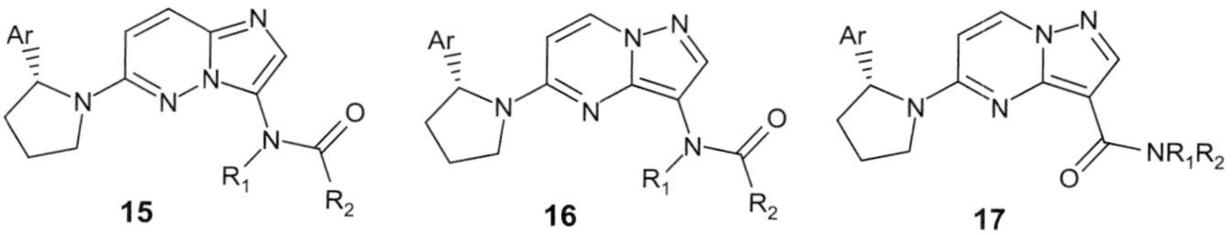

15 **16** **17**

4. FAAH INHIBITORS

The role of endocannabinoids in pain relief and the potential value of FAAH inhibitors have been reviewed [52–54]. This chapter will focus on recent developments including both new chemical series and recent clinical data.

4.1. Covalent FAAH inhibitors

The majority of compounds pursued as FAAH inhibitors have been of the irreversible covalent type. An irreversibly inhibited enzyme would not show any subsequent competition for binding by accumulated endogenous substrates. This has been proposed to be a necessary feature of FAAH inhibitors to enable and maintain the essentially complete inhibition of the enzyme that appears to be required for analgesic efficacy in preclinical pain models. The most advanced covalent compound, PF-4457845 (**18**), is reported to have a $k_{\text{inact}}/K_{\text{i}}$ of 40,200 M^{-1} s^{-1} against human FAAH [55]. This compound is effective in preclinical models of pain as long as near maximum FAAH inhibition is achieved. PF-4457845 was recently taken into a PhII clinical study for OA. Although the compound achieved good peripheral target engagement, it failed to show any analgesic efficacy versus a placebo group [56]. A number of other irreversible covalent compounds of unknown structure are also in clinical development including IPI-940, IW-6118, and V-158866.

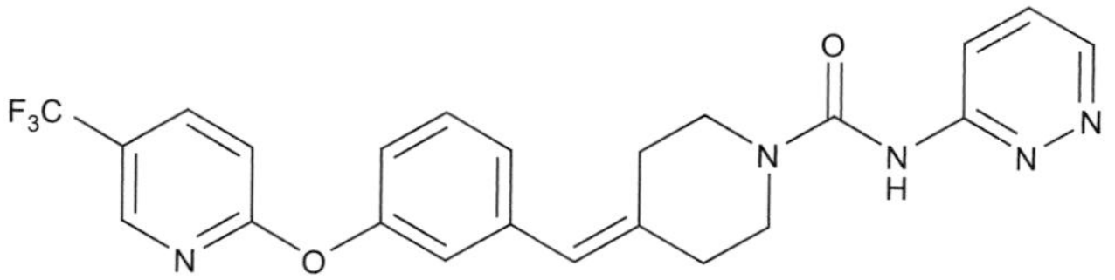

18

4.2. Noncovalent FAAH inhibitors

Recently, a few novel scaffolds have been disclosed as the root for potent, reversible noncovalent inhibitors of the FAAH enzyme. Aminopyrimidine **19** is reported to have an IC_{50} against recombinant human FAAH of ~14 nM with reasonable CNS penetration [57]. Sulfonamide **20** has been described as a 19-nM reversible and selective FAAH inhibitor that might act as a transition-state analogue [58]. A series of patent applications have claimed FAAH inhibitors based on oxazole, imidazole, and pyrazole templates as illustrated by generic structures **21–23** [59–62]. In no case has the *in vivo* analgesic efficacy been reported for these compounds. However, assuming these compounds are indeed active *in vivo*, they would provide an alternative approach to irreversible covalent inhibitors that might eliminate potential safety concerns over the creation of long-lived covalent adducts between compounds and the FAAH enzyme.

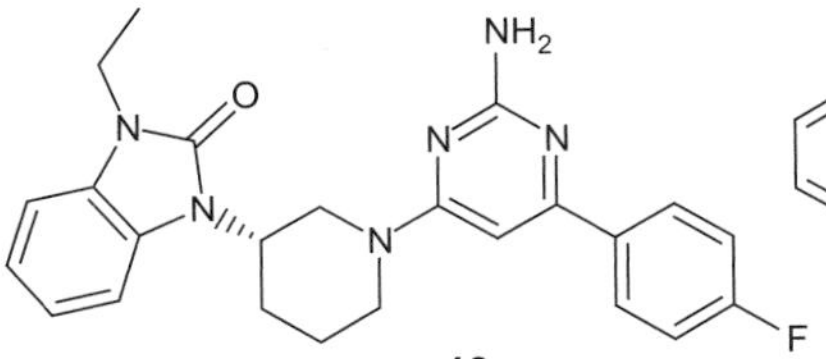

19 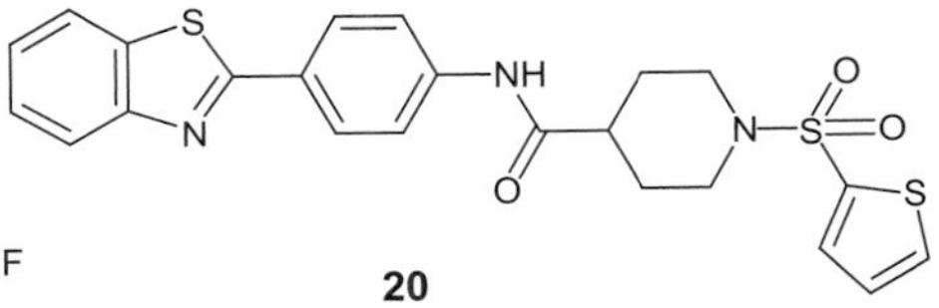 **20**

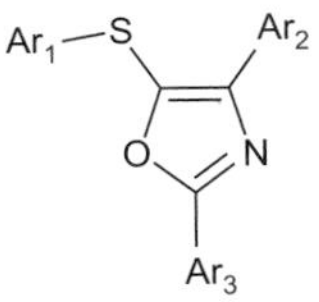

21 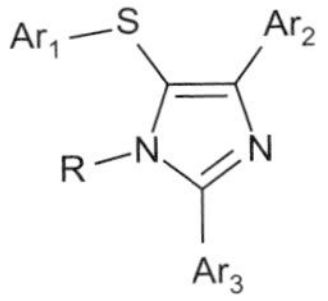 **22** 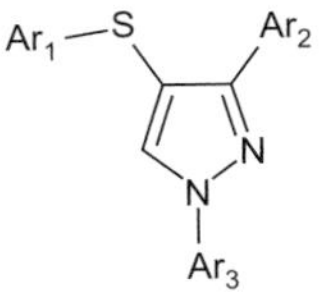 **23**

5. CONCLUSIONS

The treatment of chronic pain is potentially on the cusp of entering a new era with a number of novel mechanistic approaches in clinical development. Many of these novel mechanisms present unique medicinal chemistry challenges in developing safe and effective medicines. This review has touched on three of the leading targets being actively pursued across the pharmaceutical industry. For the Nav inhibitor approach, the recent identification of compounds with true pharmacological selectivity for Nav1.7 versus Nav 1.5 has provided evidence that alternatives to state-dependent selective inhibition are possible. The next step will be to see if these frontrunner Nav compounds can demonstrate the necessary efficacy and safety in the clinic against chronic pain. For the NGF/TrkA pathway, thus far selective inhibition has only been achieved with antibodies. It remains to be seen whether small-molecule TrkA kinase or NGF-binding inhibitors with adequate selectivity and potency can be discovered and developed. Finally, in contrast to Nav and TrkA, FAAH is one of the few targets that appears to impact pain processing both in the periphery and in the CNS. Despite this, the FAAH inhibitor approach has failed its first attempt to establish clinical proof of concept to treat OA pain. However, sufficient questions remain unanswered in this trial, such as whether there was adequate CNS target engagement or whether OA pain is the most relevant clinical pain population. The remaining active programs will be watched closely to see how they inform on the clinical relevance of this target.

Taken as a group, these three targets demonstrate both the opportunity and challenges faced in developing novel analgesics and the next few years will be very interesting to watch how these programs evolve. Finally, it will be interesting to see what additional novel targets will be brought forward and the medicinal chemistry challenges they will present.

REFERENCES

[1] C. J. Woolf, *Pain*, 2011, **152**, S2.
[2] F. Marchand, M. Perretti and S. B. McMahon, *Nat. Rev. Neurosci.*, 2005, **6**, 521.
[3] J. J. Clare, *Expert Opin. Invest. Drugs*, 2010, **19**, 45.
[4] V. Zuliani, M. Rivara, M. Fantini and G. Costantino, *Expert Opin. Ther. Patents*, 2010, **20**, 755.
[5] B. T. Priest, *Curr. Opin. Drug Discov. Dev.*, 2009, **12**, 682.
[6] B. Bear, J. Asgian, A. Termin and N. Zimmermann, *Curr. Opin. Drug Discov. Dev.*, 2009, **12**, 543.
[7] M. A. Matulenko, M. J. C. Scanio and M. E. Kort, *Curr. Top. Med. Chem.*, 2009, **9**, 362.
[8] V. Zuliani, M. K. Patel, M. Fantini and M. Rivara, *Curr. Top. Med. Chem.*, 2009, **9**, 396.
[9] S. England, *Expert Opin. Invest. Drugs*, 2008, **17**, 1849.
[10] N. J. Hargus and M. K. Patel, *Expert Opin. Invest. Drugs*, 2007, **16**, 635.
[11] B. Marron, *Annu. Rep. Med. Chem.*, 2006, **41**, 59.
[12] M. I. Kemp, *Prog. Med. Chem.*, 2010, **49**, 81.
[13] S. England and D. Rawson, *Future Med. Chem.*, 2010, **2**, 775.
[14] J. J. Cox, F. Reimann, A. K. Nicholas, G. Thornton, E. Roberts, K. Springell, G. Karbani, H. Jafri, J. Mannan, Y. Raashid, L. Al-Gazali, H. Hamamy, E. M. Valente, S. Gorman, R. Williams, D. P. McHale, J. N. Wood, F. M. Gribble and C. G. Woods, *Nature*, 2006, **444**, 894.
[15] Y. Yang, Y. Wang, S. Li, Z. Xu, H. Li, L. Ma, J. Fan, D. Bu, B. Liu, Z. Fan and G. Wu, *J. Med. Genet.*, 2004, **41**, 171.
[16] M. F. Jarvis, P. Honore, C. C. Shieh, M. Chapman, S. Joshi, X. F. Zhang, M. Kort, W. Carroll, B. Marron, R. Atkinson, J. Thomas, D. Liu, M. Krambis, Y. Liu, S. McGaraughty, K. Chu, R. Roeloffs, C. Zhong, J. P. Mikusa, G. Hernandez, D. Gauvin, C. Wade, C. Zhu, M. Pai, M. Scanio, L. Shi, I. Drizin, R. Gregg,

M. Matulenko, A. Hakeem, M. Gross, M. Johnson, K. Marsh, P. K. Wagoner, J.P. Sullivan, C. R. Flatynek and D. S. Krafte, *Proc. Natl. Acad. Sci. USA*, 2007, **104**, 8520.

[17] M. E. Kort, I. Drizin, R. J. Gregg, M. J. C. Scanio, L. Shi, M. F. Gross, R. N. Atkinson, M.S. Johnson, G. J. Pacofsky, J. B. Thomas, W. A. Carroll, M. J. Krambis, D. Liu, C. C. Shieh, X. Zhang, G. Hernandez, J. P. Mikusa, C. Zhong, S. Joshi, P. Honore, R. Roeloffs, K.C. Marsh, B. P. Murray, J. Liu, S. Werness, C. R. Faltynek, D. S. Krafte, M. F. Jarvis, M.L. Chapman and B. E. Marron, *J. Med. Chem.*, 2008, **51**, 407.

[18] S. Beaudoin, M. C. Laufersweiler, C. J. Markworth, D. S. Millan, D. J. Rawson, S. M. Reister, K. Sasaki, R. I. Storer, P. A. Stupple, N. A. Swain, C. W. West and S. Zhou, *Patent Application WO 2010/079443-A1*, 2010.

[19] L. E. Browne, F. E. Blaney, S. P. Yusaf, J. J. Clare and D. Wray, *J. Biol. Chem.*, 2009, **284**, 10523.

[20] M. J. C. Scanio, L. Shi, I. Drizin, R. J. Gregg, R. N. Atkinson, J. B. Thomas, M. S. Johnson, M. L. Chapman, D. Liu, M. J. Krambis, Y. Liu, C. C. Shieh, X. Zhang, G. H. Simler, S. Joshi, P. Honore, K. C. Marsh, A. Knox, S. Werness, B. Antonio, D. S. Krafte, M. F. Jarvis, C.R. Faltynek, B. E. Marron and M. E. Kort, *Bioorg. Med. Chem.*, 2010, **18**, 7816.

[21] M. E. Kort, R. N. Atkinson, J. B. Thomas, I. Drizin, M. S. Johnson, M. A. Secrest, R. J. Gregg, M. J. C. Scanio, L. Shi, A. H. Hakeem, M. A. Matulenko, M. L. Chapman, M. J. Krambis, D. Liu, C. C. Shieh, X. Zhang, G. Simler, J. P. Mikusa, C. Zhong, S. Joshi, P. Honore, R. Roeloffs, S. Werness, B. Antonio, K. C. Marsh, C. R. Faltynek, D. S. Krafte, M. F. Jarvis and B. E. Marron, *Bioorg. Med. Chem. Lett.*, 2010, **20**, 6812.

[22] X. F. Zhang, C. C. Shieh, M. L. Chapman, M. A. Matulenko, A. H. Hakeem, R. N. Atkinson, M. E. Kort, B. E. Marron, S. Joshi, P. Honore, C. R. Faltynek, D. S. Krafte and M. F. Jarvis, *Neuropharmacology*, 2010, **59**, 201.

[23] B. S. Williams, J. P. Felix, B. T. Priest, R. M. Brochu, K. Dai, S. B. Hoyt, C. London, Y. S. Tang, J. L. Duffy, W. H. Parsons, G. J. Kaczorowski and M. L. Garcia, *Biochemistry*, 2007, **46**, 14693.

[24] For press releases, see http://ir.icagen.com/phoenix.zhtml?c=178443&p=irol-news&nyo=1.

[25] See http://clinicaltrials.gov/ct2/show/NCT01165736?term=PF-05089771&rank=2 and http://clinicaltrials.gov/ct2/show/NCT01259882?term=PF-05089771&rank=1.

[26] T. Inoue, S. Watanabe, T. Yamagishi, Y. Arano, M. Morita and K. Shimada, *Patent Application WO 2010/137351-A1*, 2010.

[27] H. Noguchi, T. Inoue, M. Morita and Y. Arano, *Patent Application WO 2011/016234-A1*, 2011.

[28] C. J. Markworth, B. E. Marron and N. A. Swain, *Patent Application WO 2010/035166-A1*, 2010.

[29] J. J. Watson, S. J. Allen and D. Dawbarn, *Biodrugs*, 2008, **22**, 349.

[30] A. Cattaneo, *Curr. Opin. Mol. Ther.*, 2010, **12**, 94.

[31] L. Macdonald, R. Torres, M. R. Morra, J. H. Martin, J. C. Reinhardt and P. Tiseo, *Patent Application US 2011/0014208 A1*, 2011.

[32] G. Ugolini, S. Marinelli, S. Covaceuszach, A. Cattaneo and F. Pavone, *Proc. Natl. Acad. Sci. USA*, 2007, **104**, 2985.

[33] K. Ueda, M. Hirose, E. Murata, M. Takatori, M. Ueda, H. Ikeda and K. Shigemi, *J. Pharmacol. Sci.*, 2010, **112**, 438.

[34] A. M. Colangelo, M. R. Bianco, L. Vitagliano, C. Cavaliere, G. Cirillo, L. De Gioia, D. Diana, D. Colombo, C. Redaelli, L. Zaccaro, G. Morelli, M. Papa, P. Sarmientos, L. Alberghina and M. Martegani, *J. Neurosci.*, 2008, **28**, 2698.

[35] F. Brahimi, J. Liu, A. Malakhov, S. Chowdhury, E. O. Purisima, L. Ivanisevic, A. Caron, K. Burgess and H. U. Saragovi, *Biochim. Biophys. Acta*, 2010, **1800**, 1018.

[36] J. Liu, F. Brahimi, H. U. Saragovi and K. Burgess, *J. Med. Chem.*, 2010, **53**, 5044.

[37] J. B. Owolabi, G. Rizkalla, A. Tehim, G. M. Ross, R. J. Riopelle, R. Kamboj, M. Ossipov, D. Bian, S. Wegert, F. Porreca and D. K. H. Lee, *J. Pharmacol. Exp. Ther.*, 1999, **289**, 1271.
[38] R. Vohra, G. Dube, Z. Gan, N. Gill and X. Cui, *Patent Application US 2009/0082368-A1*, 2009.
[39] A. Colquhoun, G. M. Lawrance, I. L. Shamovsky, R. J. Riopelle and G. M. Ross, *J. Pharmacol. Exp. Ther.*, 2004, **310**, 505.
[40] J. C. Jaen, E. Laborde, R. A. Bucsh, B. W. Caprathe, R. J. Sorenson, J. Fergus, K. Spiegel, J. Marks, M. R. Dickerson and R. E. Davis, *J. Med. Chem.*, 1995, **38**, 4439.
[41] J. K. Eibl, S. A. Chapelsky and G. M. Ross, *J. Pharmacol. Exp. Ther.*, 2010, **332**, 446.
[42] T. Wang, D. Yu and M. L. Lamb, *Expert Opin. Ther. Patents*, 2009, **19**, 305.
[43] T. Wang, M. L. Lamb, D. A. Scott, H. Wang, M. H. Block, P. D. Lyne, J. W. Lee, A. M. Davies, H. Zhang, Y. Zhu, F. Gu, Y. Han, B. Wang, P. J. Mohr, R. J. Kaus, J. A. Josey, E. Hoffmann, K. Thress, T. MacIntyre, H. Wang, C. A. Omer and D. Yu, *J. Med. Chem.*, 2008, **51**, 4672.
[44] K. Thress, T. MacIntyre, H. Wang, D. Whitson, Z. Y. Liu, E. Hoffmann, T. Wang, J.L. Brown, K. Webster, C. Omer, P. E. Zage, L. Zeng and P. A. Zweidler-McKay, *Mol. Cancer Ther.*, 2009, **8**, 1818.
[45] J. J. Wu and L. Wang, *Patent Application WO 2010/077680-A2*, 2010.
[46] S. W. Andrews, Small Molecule Trk Inhibitors in Pre-Clinical Models of Pain, Ninth Annual World Pharmaceutical Congress, Philadelphia, PA, USA, June 15, 2010.
[47] K. Foster, *IDrugs*, 2009, **12**, 478.
[48] J. R. Ghilardi, K. T. Freeman, J. M. Jimenez-Andrade, W. G. Mantyh, A. P. Bloom, K. S. Bouhana, D. Trollinger, J. Winkler, P. Lee, S. W. Andrews, M. A. Kuskowsi and P.W. Mantyh, *Bone*, 2011, **48**, 389.
[49] S. W. Andrews, J. Haas, Y. Jiang and G. Zhang, *Patent Application WO 2010/033941-A1*, 2010.
[50] J. Haas, S. W. Andrews, Y. Jiang and G. Zhang, *Patent Application WO 2010/048314-A1*, 2010.
[51] S. Allen, S. W. Andrews, K. R. Condroski, J. Haas, L. Huang, Y. Jiang, T. Kercher and J. Seo, *Patent Application WO 2011/006074-A1*, 2011.
[52] K. Ahn, D. S. Johnson and B. F. Cravatt, *Expert Opin. Drug Discov.*, 2009, **4**, 763.
[53] S. Petrosino and V. Di Marzo, *Curr. Opin. Invest. Drugs*, 2010, **11**, 51.
[54] M. Seierstad and J. G. Breitenbucher, *J. Med. Chem.*, 2008, **51**, 7327.
[55] D. S. Johnson, C. Stiff, S. E. Lazerwith, S. R. Kesten, L. K. Fay, M. Morris, D. Beidler, M.B. Liimatta, S. E. Smith, D. T. Dudley, N. Sadagopan, S. N. Bhattachar, S. J. Kesten, T.K. Nomanbhoy, B. F. Cravatt and K. Ahn, *ACS Med. Chem. Lett.*, 2010, **2**, 91.
[56] S. R. Langman, T. S. Smart, G. L. Li, M. Boucher, L. Taylot, T. J. Young and J. P. Huggins, Poster Presentation at 13th World Congress on Pain, Montreal, Quebec, Montreal, 2010.
[57] D. J. Gustin, Z. Ma, X. Min, Y. Li, C. Hedberg, C. Guimaraes, A. C. Porter, M. Lindstrom, D. Lester-Zeiner, G. Xu, T. J. Carlson, S. Xiao, C. Meleza, R. Connors, Z. Wang and F. Kayser, *Bioorg. Med. Chem. Lett.*, 2011, **21**, 2492.
[58] X. Wang, K. Sarris, K. Kage, D. Zhang, S. P. Brown, T. Kolosa, C. Surowy, O. F. El Kouhen, S. W. Muchmore, J. D. Brioni and A. O. Stewart, *J. Med. Chem.*, 2009, **52**, 170.
[59] H. Chobanian, L. S. Lin, P. Liu, M. D. Chioda, R. J. De Vita, R. P. Nargund and Y. Guo, *Patent Application US 2011/0021531-A1*, 2011.
[60] H. Chobanian, L. S. Lin, P. Liu, M. D. Chioda, R. J. De Vita, R. P. Nargund and Y. Guo, *Patent Application WO 2010/017079-A1*, 2010.
[61] L. S. Lin, M. D. Chioda, P. Liu and R. P. Nargund, *Patent Application WO 2009/152025-A1*, 2009.
[62] L. S. Lin, L. L. Chang, H. Chobanian and R. P. Nargund, *Patent Application WO 2009/151991-A1*, 2009.

CHAPTER 3

Central Modulation of Circadian Rhythm *via* CK1 Inhibition for Psychiatric Indications

Paul Galatsis*, Travis T. Wager*, James Offord, George J. DeMarco†, Jeffrey F. Ohren*, Ivan Efremov*** and **Scot Mente***

Contents

1. INTRODUCTION

The field of circadian biology involves the study of circadian rhythms (CRs) which are periodic behavioral and physiologic processes that cycle approximately every 24 h. CRs are primarily regulated by an organism's

* Worldwide Medicinal Chemistry, Pfizer Worldwide Research and Development, Eastern Point Road, Groton, CT 06340, USA
** Neuroscience Research Unit, Pfizer Worldwide Research and Development, Eastern Point Road, Groton, CT 06340, USA
† Worldwide Comparative Medicine, Pfizer Worldwide Research and Development, Eastern Point Road, Groton, CT 06340, USA

Annual Reports in Medicinal Chemistry, Volume 46
ISSN: 0065-7743, DOI: 10.1016/B978-0-12-386009-5.00024-2

environmental photoperiod (day/night), thus CRs likely function to synchronize critical biologic processes with light/dark cycles. It has been well established that CRs in higher organisms are generated by a master clock located in the suprachiasmatic nucleus (SCN) of the brain. The master clock synchronizes central and peripheral clocks in the body regulating numerous and diverse functions including sleep–wake cycles, hormone release, and body temperature. There is a growing body of evidence linking disrupted CRs to the pathophysiology of neuropsychiatric disease including insomnia, depression, seasonal affective disorder, and bipolar disorder.

2. DESCRIPTION OF CIRCADIAN CLOCK

All organisms have an endogenous timing system which synchronizes biologic functions to exogenous daily cycles [1,2]. The circadian timing system is synchronized by the action of zeitgebers (''time-givers'' or synchronizers), the most powerful of which is light. In most animals, light is transduced into a circadian neural signal by specialized nonvisual cells in the retina, the melanopsin expressing retinal ganglion cells [3]. Signals from the retinal ganglion cells travel down the retinohypothalamic tract, a monosynaptic projection from the retina, which innervates a small paired structure within the hypothalamus called the SCN [4]. The SCN is comprised of about 20,000 cells and is generally accepted as the site of the master biological clock [4]. The SCN also receives input from the intergeniculate leaflet and serotonergic pathways in the brain that mediate the action of nonphotic zeitgebers on circadian cycles [1].

The time period it takes for one circadian cycle to occur is called tau (τ) and results from a core loop of translational, transcriptional, and post-translational events in the SCN. In the absence of light and other effective zeitgebers, the molecular events that govern clock periodicity run at an endogenous rate, a condition called ''free-running.'' The τ for animals under free-running conditions, endogenous τ, may be less than, equal to, or greater than 24 h. In mouse, endogenous τ is strain dependent and ranges from 23.3 to 23.9 h [5]. In humans, endogenous τ appears to be more variable (24–25 h) which may reflect difficulties in establishing true free-running conditions for people [6]. Since light cycles are the primary regulator of CRs, most animals are synchronized to a 24-h photoperiod by light resetting the timing of molecular events in the SCN on a daily basis.

Mammalian genetics has played a large part in defining the components of the circadian timing system. In humans, familial advanced sleep phase syndrome (FASPS) is a disease in which patients show a daily advance in the time when they go to sleep. Genetic association studies showed that these patients carry a mutation in the protein Period2 (Per2), which removes a phosphorylation site [7]. Per2 is a member of the PAS (Per-ARNT-Sim) family of genes and is a homologue of a protein that had been identified in *Drosophila* as a component of the circadian clock. Per2 expression has been shown to be expressed in a pattern that oscillates with the circadian time. The link between Per2 phosphorylation and FASPS provided a link between Per protein phosphorylation and CR in mammals.

The first mammalian mutation linked to CR was identified in golden hamsters (*Mesocricetus auratus*). The mutation was an autosomal allele first described in the late 1980s [8] and was identified using positional cloning [9]. These animals displayed a shortened period length (20 h in homozygous animals) and were thus designated as τ mutants. Cloning revealed that the τ mutation was in casein kinase (CK) 1ε and decreased the maximal phosphorylation rates of CK1ε.

There are several genes and their respective proteins that are involved with CR and a brief description of them follows. Cryptochrome (Cry) is a flavoprotein that acts as a blue light photoreceptor, directly modulating photo-input into the circadian clock. It has been shown that Cry associates with Per proteins [7] and is homologous to the protein Timeless (tim), first identified from *Drosophila* [1]. A mouse mutation generated through chemical mutagenesis and then screened for mutations, which alter CR, was mapped and the locus cloned using positional cloning [10,11]. This protein was called Clock (Circadian Locomotor Output Cycles Kaput) and shown to be required for normal functioning of the circadian timing system. The Clock mutation in mice is a deletion of exon 19 and encodes a shorter protein [11,12]. Clock interacts with a protein called Bmal1 (Brain and Muscle Aryl hydrocarbon receptor nuclear translocator ARNT-Like 1), which contains a basic helix–loop–helix protein domain. Bmal1 interacts with the Clock protein to form a heterodimer which activates transcription from genes containing an E-box in their promoters [7].

Periodicity in the SCN is generated by cyclic transcriptional/translational positive and negative feedback-loops of which Per, Cry, Clock, Bmal1, and their respective proteins are key components. The process by which light synchronizes (entrains) SCN periodicity to light/dark cycles is summarized as follows (Figure 1). Light signals, transmitted by the retinohypothalamic tract, initiate the transcription of Per by glutamate mediated activation of calcium channels and pituitary adenylate cyclase-activating polypeptide (PACAP) activation of the vasoactive intestinal peptide receptor 1 (VIPR-1) in SCN neurons [13]. VIPR-1 activation, and Ca influx through L channels, initiates phosphorylation of CREB, which subsequently binds to a CRE in the Per gene inducing its transcription [14]. As a result, cytoplasmic Per (protein) levels accumulate during the day and heterodimerize in the cytoplasm with the protein Cry. The Per-Cry complex translocates into the nucleus and inhibits the expression of

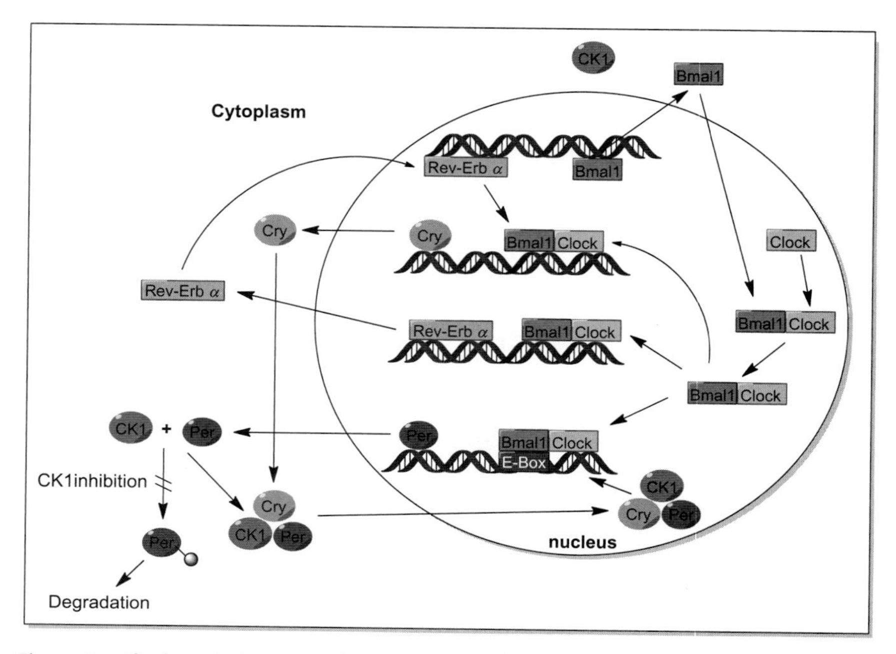

Figure 1 Clock cycle in mammalian SCN. (See Color Plate 3.1 in Color Plate Section.)

Clock and Bmal1. The Clock and Bmal1 proteins heterodimerize to form a complex, which activates the transcription of genes containing an E-box structure, including the genes for Per and Cry. As Per levels increase throughout the day in the cytoplasm, the protein is phosphorylated by CK1ε or CK1δ and targeted for degradation by the ubiquitin ligase system [1,15]. Blockade of CK1 prevents Per degradation and increases Per levels in the cytoplasm and the nucleus. It is this series of transcriptional, translational, and posttranslational activities that sets the timing of the endogenous clock.

A secondary loop exists and functions to stabilize the core loop [15]. The stabilizing loop is driven by expression of the nuclear receptors Rev-erb α (also known as NR1D1—Nuclear Receptor subfamily 1, group D, member 1) and RORβ (RAR-related orphan receptor β, also known as NR1F2) [15,16]. Rev-erb α functions as a negative regulator of Bmal1. Bmal1 interacts with Clock to drive expression of a set of genes with a common E-box promoter. One of these proteins is RORβ, which positively affects transcription of Bmal1, completing the stabilizing loop [15,16]. Stabilization of the core loop is thought to preserve synchronized SCN function over time periods longer than a day [15,16]. By inhibiting CK1, phosphorylation of Per could be blocked and this should alter the timing of the circadian clock, thus changing the rhythm of biological processes.

Other enzymes outside of the circadian clock also have effects on CR. Lithium, a drug used to treat bipolar disorder, has been shown to lengthen τ in a variety of systems [17]. While the molecular mechanism whereby lithium exerts its therapeutic effect is unclear, it is known to be an inhibitor of the enzyme glycogen synthase kinase 3beta (GSK-3β) [17]. Lithium at therapeutic concentrations has been shown to inhibit GSK-3β and this subsequently inhibits the phosphorylation of Rev-erb α [18].

3. CK1 INHIBITORS

3.1. Structure and function of CK1 enzymes

There are seven mammalian isoforms of CK1 (α, α2, γ1, γ2, γ3, δ, and ε) that belong to the superfamily of serine/threonine protein kinases [19]. CK1δ and CK1ε are closely related with respect to their sequence; they are predicted to be nearly identical within the kinase active site and share greater than 90% sequence identity within the kinase domain. In contrast to many other protein kinases, the CK1 isoforms do not require phosphorylation of the kinase activation segment by an upstream kinase to maintain the kinase in catalytically active form. CK1δ and CK1ε activity in the cell is regulated by a complex set of molecular events including phosphorylation on their respective carboxy-terminal regulatory domains [20,21]. Subcellular localization [22] and protein–protein interactions are also involved in the regulation of CK1δ and CK1ε activity and their coordination with the circadian cycle. Finally, an additional level of regulation for CK1δ and CK1ε activity is their strong preference for a primed or acidic substrate [22].

Chemical proteomics studies have shown that several other protein kinases are involved in setting and maintaining daily clock cycle. Interestingly, gene knockdown experiments show that suppression of CK1δ and CK1ε protein expression only induced a delay in sleep onset of a few hours while treating mice with a multitargeted kinase inhibitor like longdaysin **33** shifted the clock by more than 12 h [23]. Other protein kinases that may be involved in modulating CRs are CK1α, GSK3β, and p38 kinase.

3.2. Mechanism of action of CK1 inhibitors

The first structural determination of the kinase domain of human CK1δ occurred in 1996 [24]; however since then, there have been few other reported structural studies of the CK1 isoforms [25,26]. Therefore, the unliganded structure of CK1δ was used to enable ligand docking studies and structure-guided design of CK1 inhibitors [24]. CK1δ, like most protein kinases, retains the highly conserved features of a smaller

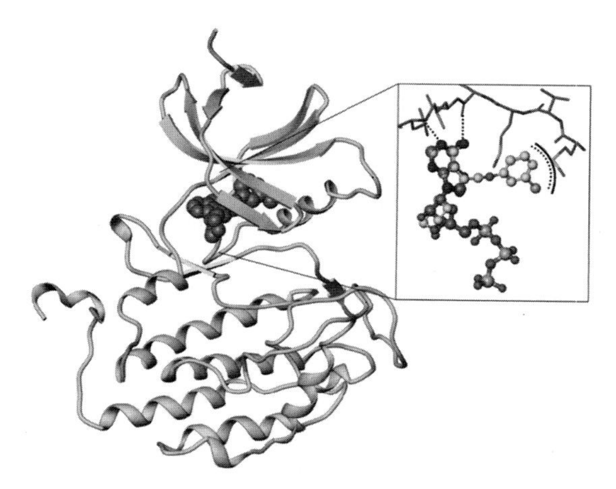

Figure 2 Model of PF-4800567 (**35**) bound in the ATP-binding site of CK1δ. (See Color Plate 3.2 in Color Plate Section.)

primarily β-strand amino-terminus lobe and a larger primarily α-helical carboxyl-terminal lobe, linked by a hinge region containing the kinase active site [27]. Figure 2 shows human CK1δ with PF-4800567 (**35**) modeled into the conserved kinase ATP-binding site demonstrating the ATP-competitive nature of the compound. This binding mode is fully supported by biochemistry experiments and displacement assays. The modeling studies suggest that the CK1 inhibitor binds primarily through H-bonding interactions with the kinase hinge region, with multiple hydrophobic interactions in the CK1δ active site as well. An overlay of **35** with ATP from the crystal structure of CK1i [26] demonstrates that the interactions of the ligand with the kinase are likely to be highly similar to those utilized by the adenine moiety of ATP with the exception of the inhibitor *m*-methoxy phenyl group which binds in a hydrophobic pocket adjacent to the CK1δ gatekeeper residue Met82 (Figure 2).

3.3. CK1 inhibitor chemical scaffolds

The disclosed CK1 chemical matter generally resides in chemical space favorable for optimization of small molecule drug candidates. Figure 3 demonstrates the distribution of CNS multiparameter optimization (CNS MPO desirability) score for patented CK1 inhibitors [28]. A significant fraction of the exemplified compounds possess a profile typical of CNS-penetrant drugs. While kinase targets are generally considered difficult for CNS indications, in part due to the significant number of polar functional groups and hydrogen bond donors required for efficient binding in the ATP site, this analysis of brain availability, in general, bodes well for development of neurotherapeutic CK1 inhibitors. Figure 4 provides a

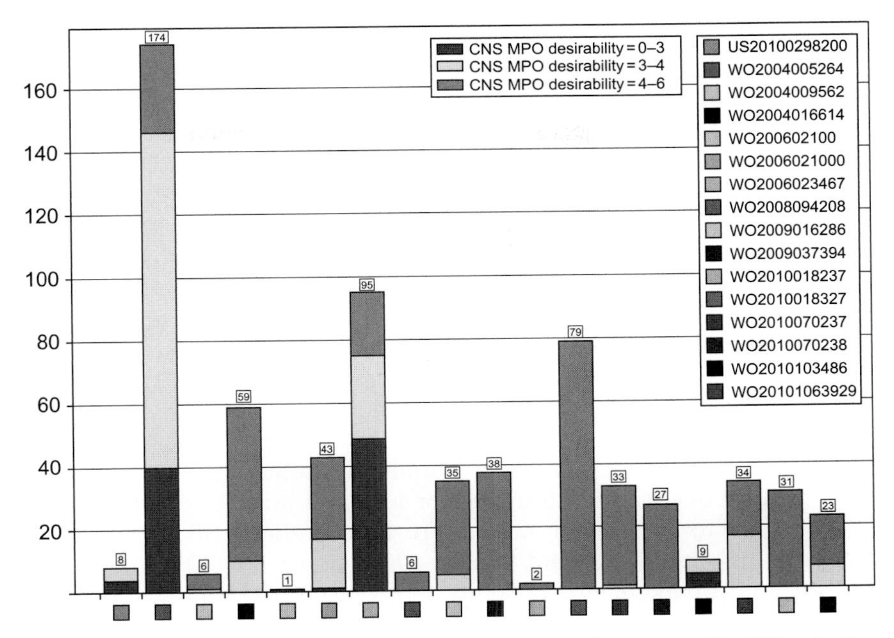

Figure 3 CNS MPO desirability analysis of 703 compounds representing CK1 patents (2004–2010). (See Color Plate 3.3 in Color Plate Section.)

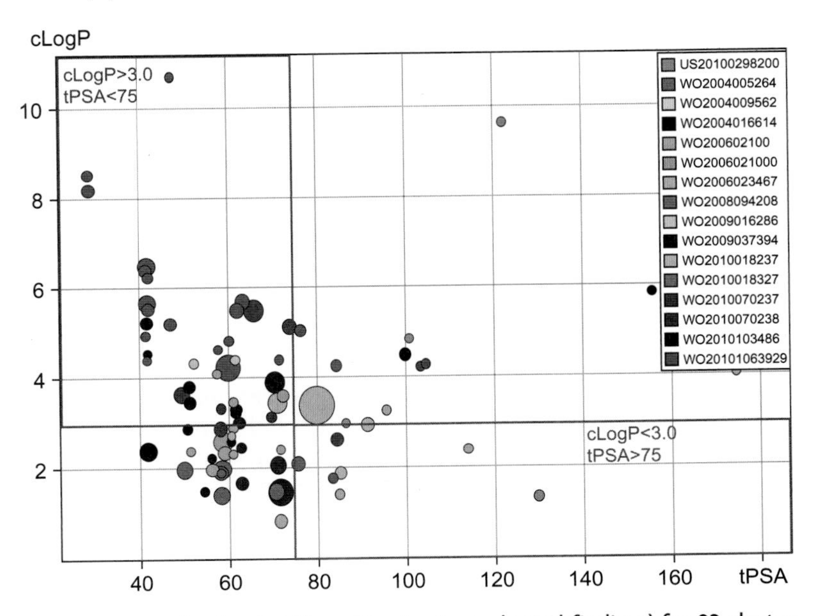

Figure 4 Toxicity plot (probability of *in vivo* toxicological findings) for 82 cluster centroids representing CK1 patent chemical space (2004–2010) sized by number of compounds in a cluster. (See Color Plate 3.4 in Color Plate Section.)

clustered view of claimed CK1 inhibitors in regard with the probability of general *in vivo* organ toxicological findings, not addressing issues related to kinase selectivity. The CK1 chemical space does not appear to be overly biased toward high lipophilicity/low polar surface area which is a profile associated with higher incidence of preclinical *in vivo* toxicity [29]. Of the 82 clusters analyzed 11% fell into the quadrant of lowest safety risk (clogP < 3 and tPSA > 75). Ligand efficiency (LE) [30] and lipophilic efficiency (LLE) [31] trends of CK1 inhibitors disclosed in 2004–2010 are presented in Figure 5. Such an analysis highlights chemotypes with higher potential for medicinal chemistry optimization efforts [32]. From this perspective, CK1 can be viewed as a target which is very amenable to small molecule drug discovery. Correlation between highly LE and LLE inhibitors and CNS MPO desirability score is also noteworthy.

Compound CK1-7 **1** appears to be the first reported CK1 inhibitor as identified by screening a series of isoquinolinesulfonamides [33]. Selectivity for CK1 over CK2 determined for **1** was found to be 8.2-fold with a $K_i = 8500$ nM for CK1. An inhibitor of IL-1 biosynthesis, SB-202190 **2**, was disclosed [34] and later found to have CK1 activity [35]. This compound displayed improved *in vitro* potency, relative to **1**, with an $IC_{50} = 600$ nM (enzyme assay). A final, early entry into the field of CK1 inhibitors with a unique chemical scaffold was IC261 **3** [36]. It was found to be equipotent at CK1δ and CK1ε with IC_{50}s $= 1000$ nM. The ATP-site binding interaction was confirmed by X-ray crystallography [25].

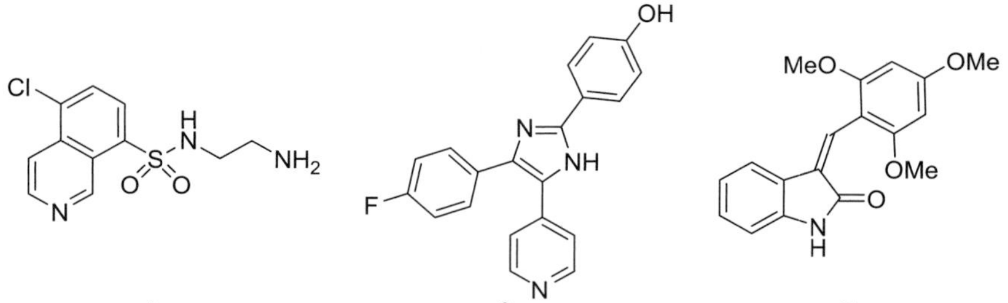

The triarylimidazole scaffold was identified by broad kinase screening activities of inhibitors from other disease indications. Thus, the ALK5 inhibitor [37] SB-431542 (ALK5 $IC_{50} = 94$ nM) led to the identification of D4476 **4** as a relatively selective CK1 inhibitor ($IC_{50} = 300$ nM), in addition to the ALK5 activity [38]. The selectivity of these CK1 inhibitors was profiled against a panel of 70–80 kinases [39]. CK1δ inhibition by **4** was 20- to 30-fold more potent than PKD1 or p38a MAPK, whereas **1** and **3** were 5- to 10-fold less potent inhibitors of CK1 and also inhibited several other kinases, including PIM1 and PIM3, while **1** also hit ERK8, MNK1, AMPK, and SGK1. Anthraquinones were later identified as CK1

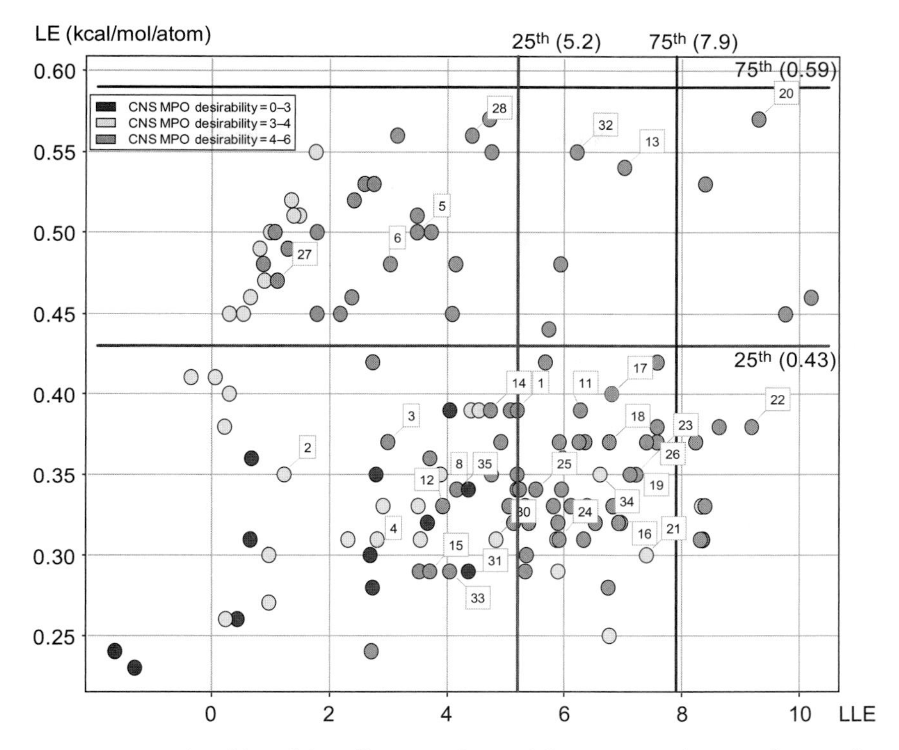

Figure 5 Ligand and lipophilic efficiency of CK1 inhibitors, using the LE and LLE 25th and 75th percentiles defined by marketed drug analysis [31]. Markers flagged by compound numbers in the body of the text. (See Color Plate 3.5 in Color Plate Section.)

inhibitors from a structure-based virtual screen method [40]. This approach showed that **5** (CK1δ IC_{50} = 330 nM, CK1γ_1 IC_{50} = 34,000 nM, and CK1α IC_{50} = 4000 nM) and **6** (CK1δ IC_{50} = 660 nM, CK1γ_1 IC_{50} = 26,200 nM, and CK1α IC_{50} = 4000 nM) were potent at CK1δ.

4 **5** **6**

CK1 inhibitors containing the imidazole core were also identified from profiling p38α kinase inhibitors derived from an isoxazole [41]. Scaffold hopping and subsequent elaboration of the isoxazole hit **7** (% inhib at 10 μM p38α = 80; CK1δ = 94) gave imidazole **8** (p38α IC_{50} = 19 nM; CK1δ-KD IC_{50} = 4 nM; CK1δ-GST IC_{50} = 5 nM; CK1ε IC_{50} = 73 nM).

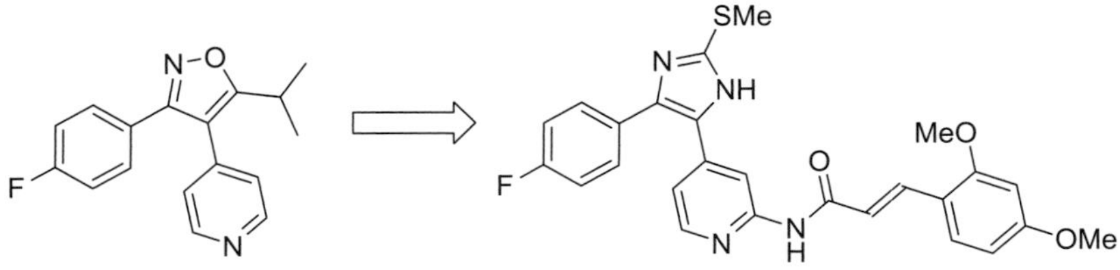

7 **8**

Related disclosures revealed substituted imidazoles as effective inhibitors of CK1α, CK1δ, and CK1ε, but the targeted indication was for hepatitis C viral (HCV) infections [42]. Compound **9** was reported to be safe by the authors, in a cell viability assay up to 10 μM and found to have an IC_{50} = 9 μM in a phenotypic whole cell assay of viral activity.

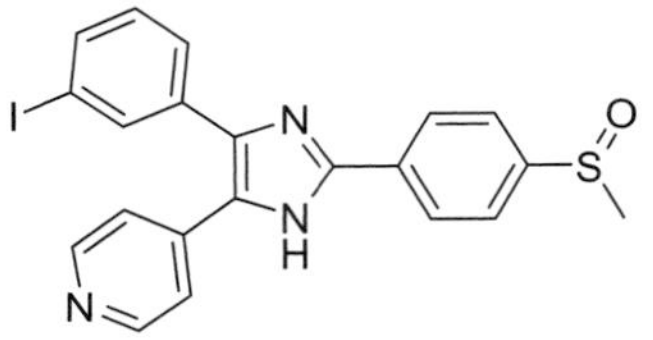

9 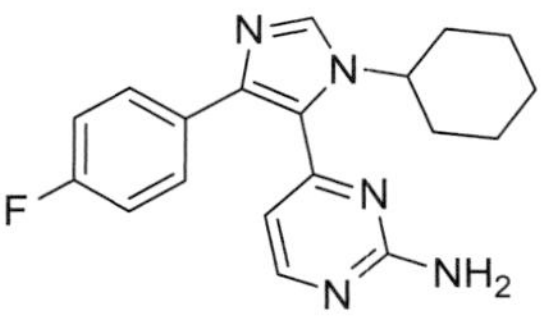 **10**

The design of imidazoles with aligned physicochemical properties to maximize brain penetration led to the disclosure of PF-670462 **10** [43]. This compound was very potent against the enzyme (CK1ε IC_{50} = 7.7 nM), exhibited potent cell-based activity (EC_{50} = 290 nM) in a Per nuclear translocation assay, and produced phase shifts in cynomolgus monkeys (see Section 4).

Triazine derivatives, initially focused on inhibition of CDK, GSK, and VEGF, were shown to also possess CK1 inhibition [44]. Derivatives **11** (CDK1 IC_{50} = 39 nM; GSK IC_{50} = 5 nM; VEGF IC_{50} = 2560 nM; CK1 IC_{50} = 115 nM) and **12** (CDK1 IC_{50} = 16 nM; GSK IC_{50} = 17 nM; VEGF IC_{50} = 2570 nM; CK1 IC_{50} = 1410 nM) illustrated how the CK1 SAR could be modulated to effect potency.

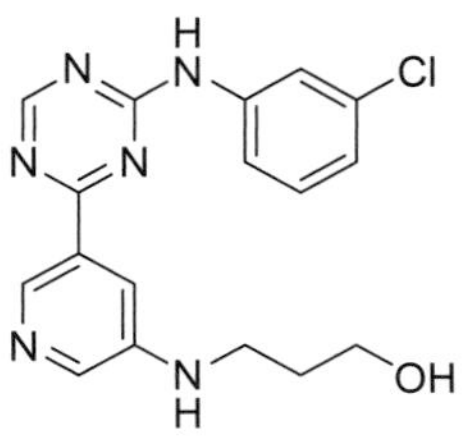

11 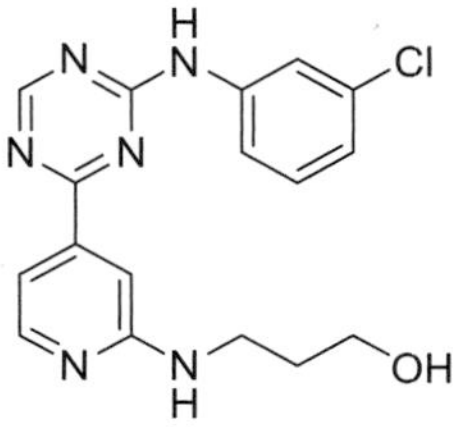 **12**

With the validation of CK1 as an authentic target to pursue, discovery programs were established to specifically develop inhibitors affecting the human circadian system. Substituted indoles were assessed against CK1δ/CK1ε and in an *in vivo* rat model of CR by measuring the mean body temperature circadian shifts. Assessment of the daily displacement of core body temperature in animals exposed to a constant dark cycle (0/24 h light/dark), which removes light as a zeitgeber, provided a measure of the compounds effectiveness in resetting τ ($EC_{\Delta\tau+1h}$). Two examples [45] of these structures are illustrated with **13** (CK1ε $K_i = 7$ nM; $EC_{\Delta t+1h} = 440$ nM) and **14** (CK1ε $K_i = 340$ nM; $EC_{\Delta t+1h} = 3340$ nM).

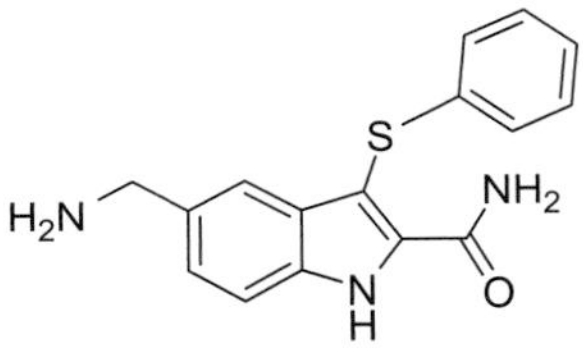

13 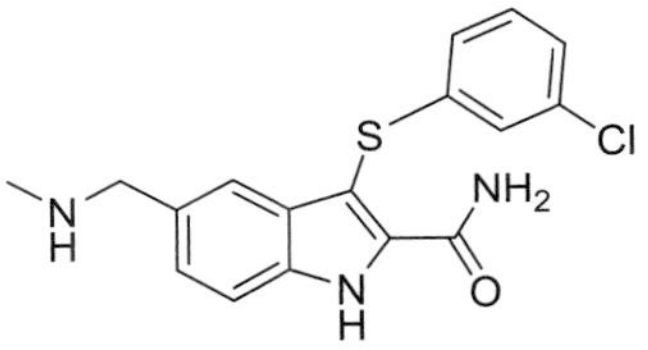 **14**

Transforming the known imidazole core into an aminopyridazine scaffold afforded dual p38α/CK1δ inhibitors [46]. Compound **15** (p38α $IC_{50} = 800$ nM; CK1δ $IC_{50} = 300$ nM) is a representative from this chemical series.

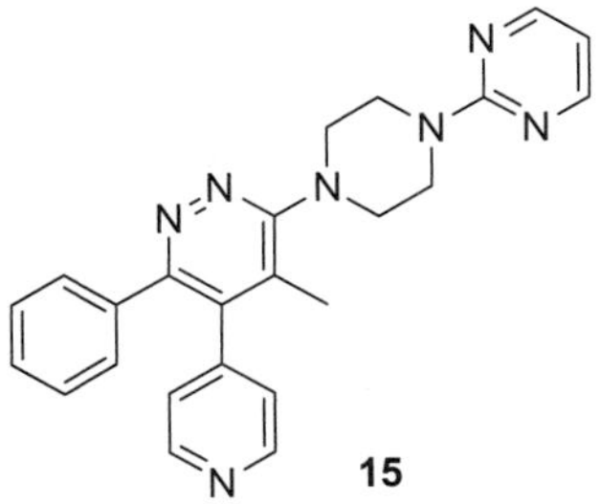

15

Multiple variations on the imidazopyridazine scaffold have been reported. The initial disclosure revealed the pyridyl derivatives as the hinge binding moiety [47]. Example **16** (CK1ε $IC_{50} = 69$ nM; CK1δ $IC_{50} = 52$ nM) illustrates that this template is equipotent at the two CK1 isoforms of interest, while **17** (CK1δ $IC_{50} = 5$ nM; $EC_{\Delta t+1h} = 28$ nM) and **18** (CK1ε $IC_{50} = 8$ nM; $EC_{\Delta t+1h} = 6$ nM) possess high enzyme potency which translated into excellent *in vivo* activity.

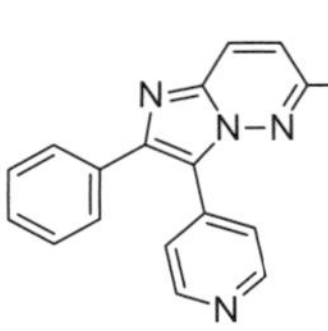

16 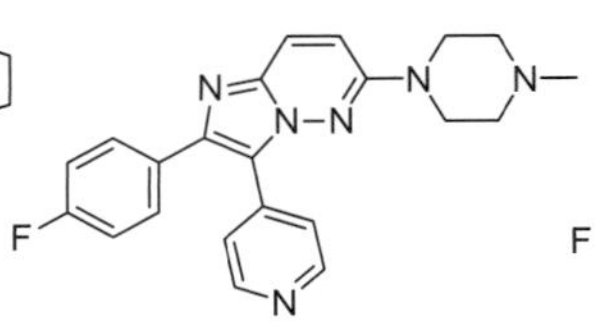 **17** 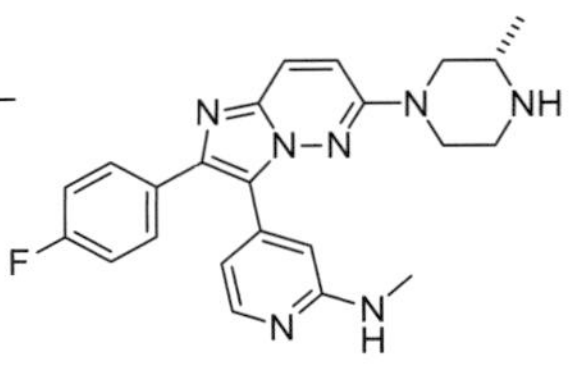 **18**

The utility of this template was probed with several modifications. The modulation of the hinge interaction and physicochemical properties were examined by transforming the pyridine ring to a pyridazine [48]. Compound **19** (CK1ε $IC_{50} = 24$ nM; CK1δ $IC_{50} = 48$ nM; $EC_{\Delta t+1h} = 5$ nM) showed that CK1 potency could be maintained without eroding *in vivo* function. Additionally, there are examples where *in vivo* potency is maintained through the evaluation of potentially subtype-selective compounds [49]. Subtype selectivity, at least at the enzyme level, of approximately 60-fold could be achieved by **20** (CK1ε $IC_{50} = 57$ nM; CK1δ $IC_{50} < 1$ nM; $EC_{\Delta t+1h} = 20$ nM) with excellent effects on CR. Exposure data were not disclosed, thus a complete assessment of subtype selectivity on CR was not possible.

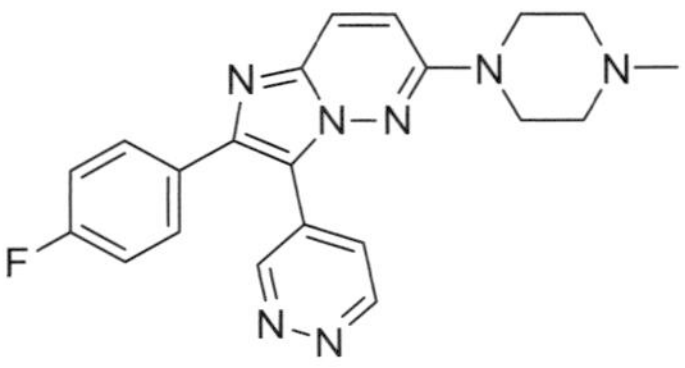

19 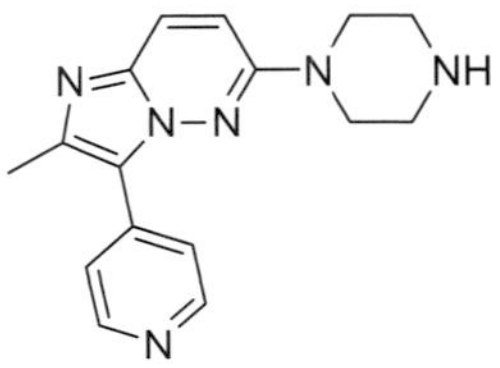 **20**

Additional hinge binding interactions, as exemplified by an indole moiety **21** (CK1ε $IC_{50} = 7$ nM; CK1δ $IC_{50} = 19$ nM; $EC_{\Delta t+1h} = 4$ nM), retained potency and functional activity [50]. Compound **22** (CK1ε $IC_{50} = 15$ nM; $EC_{\Delta t+1h} = 17$ nM) showed that increased polarity in the hydrophobic pocket, by making the fluorophenyl to furan substitution, could be tolerated [51].

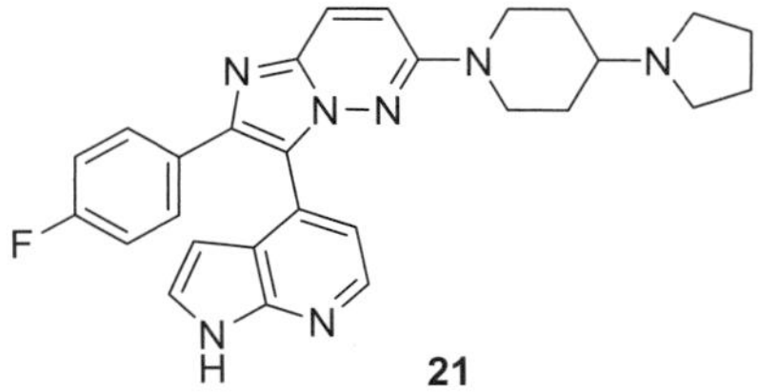

21 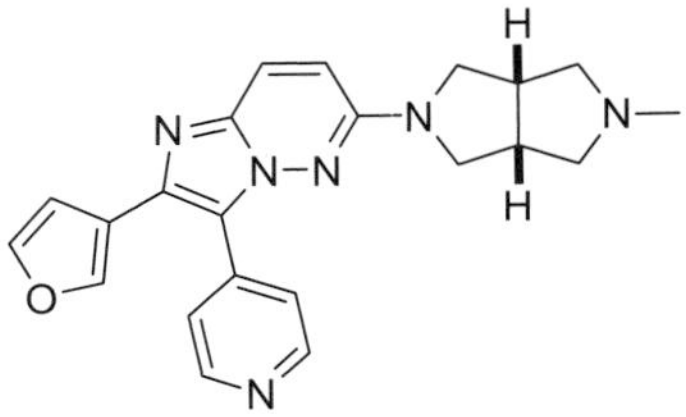 **22**

However, positioning pyridyl-substituents in the hydrophobic pocket, as with **23** (CK1ε IC_{50} = 160 nM; CK1δ IC_{50} = 107 nM) and **24** (CK1ε IC_{50} = 290 nM; CK1δ IC_{50} = 287 nM), clearly demonstrated too much polarity would not be tolerated [52].

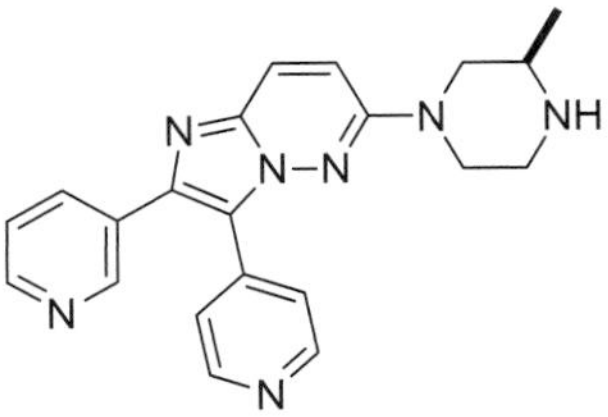

23 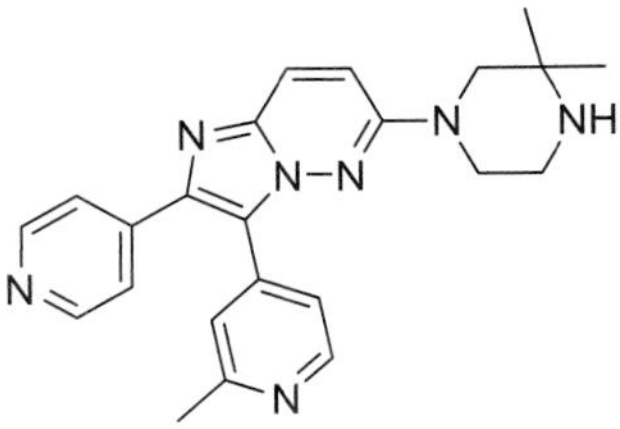 **24**

From a synthetic standpoint, these imidazopyridazines can be prepared using a four-step synthetic route. Each of the steps, organometallic deprotonation and substitution, bromination, cyclization, and S_NAr reaction, has been fully automated and can be run using flow microreactors. Using this technology, a collection of 20 diverse analogs was prepared [53].

Further modification of this scaffold core to imidazothiazoles, while maintaining similar substituents, showed that the SAR was partially transferable to this new template [54]. Compound **25** (CK1ε IC_{50} = 40 nM; CK1δ IC_{50} = 300 nM) showed a modest subtype selectivity (7.5-fold) favoring CK1ε, while **26** (CK1ε IC_{50} = 155 nM; CK1δ IC_{50} = 4 nM) displayed a distinct preference (39-fold) for CK1δ.

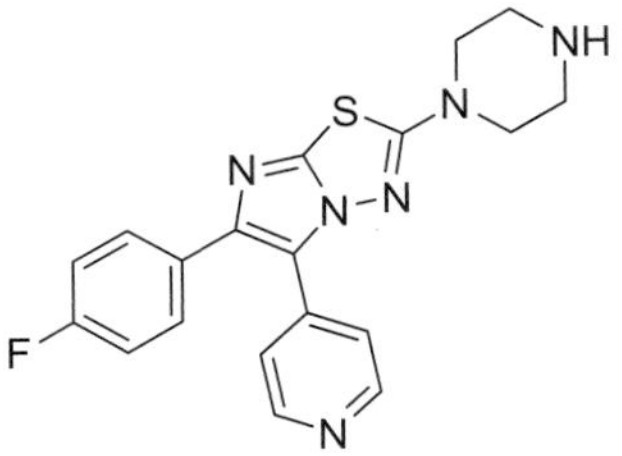

25 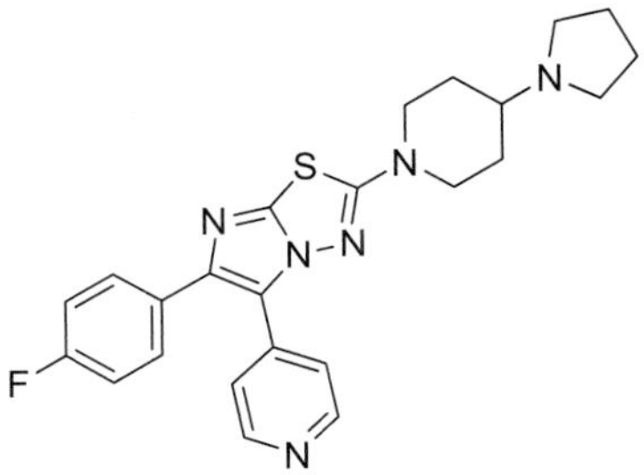 **26**

The structurally related pyrrolothiophenes and pyrrolothiazoles provide access to a novel template with CK1 activity [55]. Compound **27** (CK1ε IC_{50} = 811 nM; $EC_{\Delta t+1h}$ = 874 nM) exhibited weak enzyme potency and *in vivo* activity, while **28** (CK1ε IC_{50} = 10 nM; $EC_{\Delta t+1h}$ = 662 nM) had good enzyme potency, it also displayed weak *in vivo* activity.

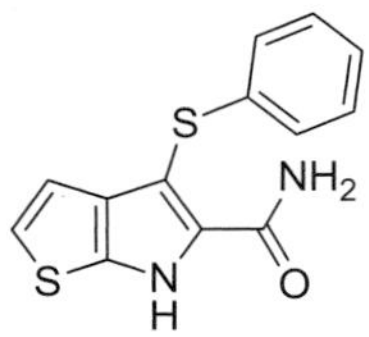

27 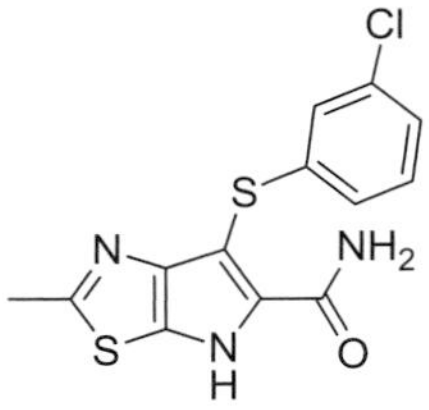 **28** 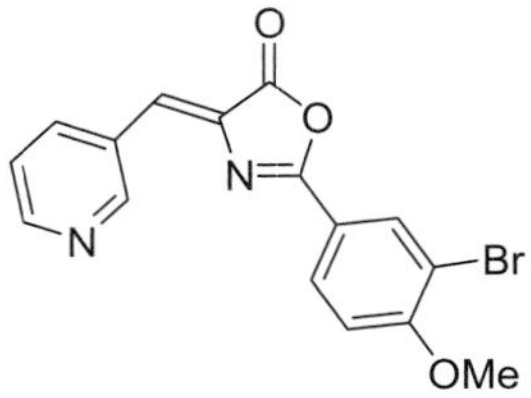 **29**

Recently disclosed [56], derivatives of 5-oxazolones, such as **29**, exhibited 94% and 99% inhibition for CK1δ and CK1ε, respectively, in an assay where the kinase concentration was 300 ng/mL.

Pyrazolotriazines, first identified with CDK activity, have now been shown to also possess activity at CK1 [57]. Compound **30** (CDK1/cyclin B $IC_{50} = 70$ nM; CDK2/cyclin A $IC_{50} = 62$ nM; CDK5/p25 $IC_{50} = 53$ nM; CDK9/cyclin T $IC_{50} = 68$ nM; CK1 $IC_{50} = 92$ nM) and **31** (CDK1/cyclin B $IC_{50} = 42$ nM; CDK2/cyclin A $IC_{50} = 42$ nM; CDK5/p25 $IC_{50} = 44$ nM; CDK9/cyclin T $IC_{50} = 53$ nM; CK1 $IC_{50} = 120$ nM) displayed a mixed profile of kinase activities including CK1 that appeared to be equipotent at all the targets. The ability to separate these activities to generate selective inhibitors will be challenging and would be required to achieve a safe compound for a CNS indication.

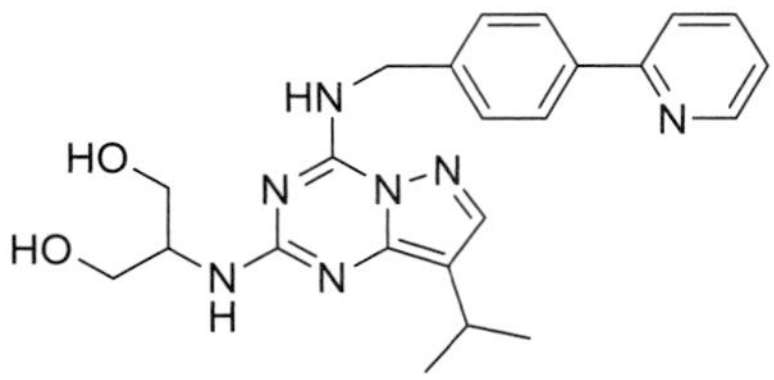

30 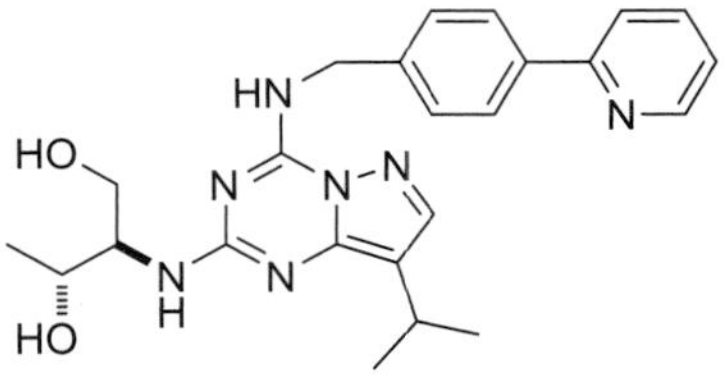 **31**

A novel pyrroloazepinone scaffold was identified when the natural product hymenialdisine, **32** (CK1 $IC_{50} = 35$ nM; CDK1/cyclin B $IC_{50} = 22$ nM; CDK5/p25 $IC_{50} = 28$ nM; GSK3 $IC_{50} = 10$ nM), was isolated from a marine sponge [58]. While it was very potent at CK1, it also showed equal potency at three additional kinases (CDK1, CDK5, and GSK3).

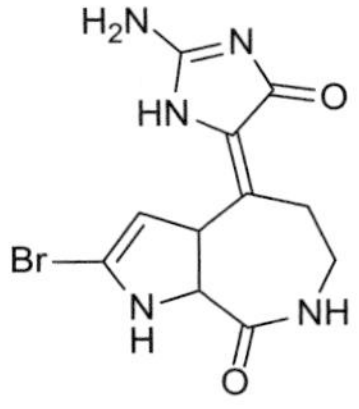

32

Imidazopyrimidines were identified by conducting a circadian screen of 120,000 compounds on human cells [59]. The compound longdaysin, **33**, (CK1δ IC_{50} = 8800 nM; EC_{50} = 9700 nM; CK1α IC_{50} = 5600 nM; EC_{50} = 9200 nM), was identified as lengthening the period of a variety of cells and was found to interact with CK1δ and CK1α by affinity chromatography.

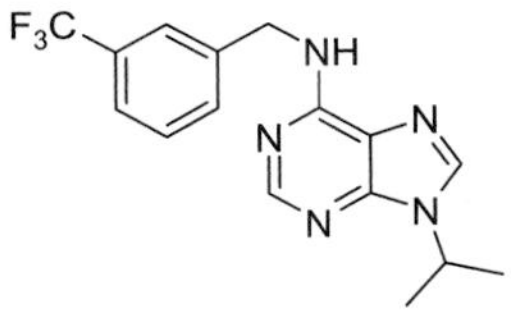

33 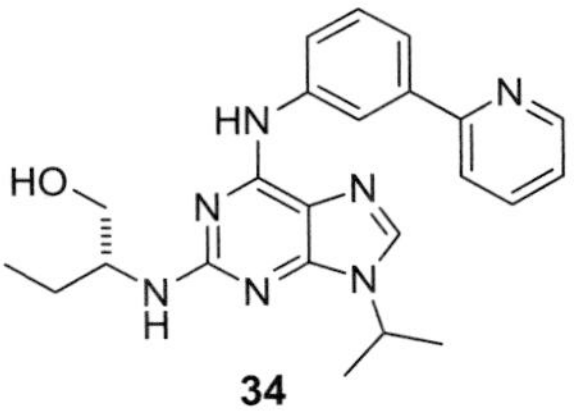 34 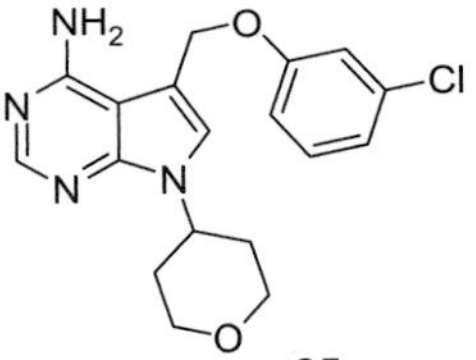 35

A structurally related compound **34** (CK1 IC_{50} = 14 nM; CDK5 IC_{50} = 80 nM) was found to display CDK5 activity in addition to CK1 [60].

CK *in vitro* subtype selectivity has been reported with pyrazolopyrimidine templates exemplified by PF-4800567, **35**, [61]. This CK1ε over CK1δ preferring compound was measured against the enzyme (CK1ε IC_{50} = 32 nM, CK1δ IC50 = 711 nM) and in a whole cell assay (CK1ε IC_{50} = 130 nM, CK1δ IC50 = 2650 nM). This pharmacological tool compound was used to elucidate the isoform *in vitro* and *in vivo* function with respect to circadian function. The CK1ε preferring compound **35** minimally altered circadian period when compared to the non-subtype-selective CK1ε/CK1δ compound **10** (see Section 4).

4. PRECLINICAL ANIMAL MODELS

The pharmacologic inhibition of CK1ε/CK1δ by PF-670462 **10** has consistently shown modulation of circadian activity in rat, mouse, and nonhuman primate models. Under 12:12 light:dark conditions, **10** (10 and 30 mg/kg) administered at zeitgeber time (ZT) 11 generated robust dose-dependent phase delays in rat [62]. In the same study, Sprouse *et al.* also demonstrated that administration of **10** at ZT 6 produced a smaller phase shift which was not significantly different from controls at the 10 mg/kg dose [62]. These results are consistent with the phase response curve reported for **10** [43] and suggest that the pharmacodynamics of CK1 inhibition follows cyclic changes in period protein (PER) levels [63]. These data further suggest that if used in patients, the clinical efficacy of a CK1ε/CK1δ inhibitor may be influenced by the timing of drug administration.

The effects of **10** on circadian activity in socially housed *Macaca fasicularis;* a diurnal nonhuman primate species with a circadian period similar to humans has been investigated. Administration of **10** (10 and 32 mg/kg) at ZT 10:30 under 12:12 light:dark conditions elicited dose-dependent phase delays in *M. fasicularis* [64]. These data are consistent with studies in singly house monkeys [65] and suggest that CK1 inhibition would induce phase delays in other diurnal species, including humans.

The aforementioned models demonstrate the effects of CK1 inhibition in animals with normal CRs. Although these observations are important, central to the development of CK1 inhibitors as therapeutic agents for psychiatric indications is the ability to modulate circadian activity in arrhythmic or disrupted circadian paradigms. Meng *et al.* have established that **10** can restore circadian rhythmicity in two different mouse models of circadian asynchrony [66]. The vasoactive intestinal peptide receptor 2 knockout mouse (*Vipr2*$^{-/-}$) exhibits an arrhythmic or weakly rhythmic circadian phenotype due to disrupted signal transduction in the SCN [67,68]. Robust entrainment of *Vipr2*$^{-/-}$ mice can be induced and sustained by daily **10** treatment [65]. Meng *et al.* also demonstrated that treatment with **10** prevented constant light-induced circadian asynchrony in wild-type mice [66].

In summary, inhibition of CK1ε/CK1δ produces phase delays in both nocturnal (rodents) and diurnal (nonhuman primate) species and restores rhythmicity under conditions that disrupt SCN function. These lines of evidence support the contention that a CK1ε/CK1δ inhibitor would be of therapeutic benefit in neuropsychiatric conditions that are associated with disrupted circadian function.

5. CONCLUSIONS

The depth of understanding of CR has greatly increased over the past decade. Confirmation of CK1 as a drug-able target has been validated by identification of chemical matter consistent with drug-like attributes, including good LE, LLE, and brain availability. Moreover, these compounds have provided the *in vivo* translation of the science and have shown CR entrainment and phase shifts in nonhuman primates, in addition to rodents. In particular, the dissection of CK1 subtype selectivity was demonstrated with **35** (CK1ε selective), with **10** (CK1ε/CK1εδ dual inhibitor) exhibiting *in vivo* efficacy at modulating of CR. Selective CK1 tool compounds have enabled a greater understanding of the role this kinase plays in modulating CR. Although no CK1 compound has been reported in the clinic; clinical evaluation of the CR hypothesis using selective CK1 inhibitors is now within the realm of possibility.

REFERENCES

[1] P. L. Lowrey and J. S. Takahashi, *Ann. Rev. Genomics Hum. Genet.*, 2004, **5**, 407.
[2] J. S. Takahashi, H. K. Hong, C. H. Ko and E. L. McDearmon, *Nat. Rev. Genet.*, 2008, **9**, 764.
[3] J. Hannibal, *Cell Tissue Res.*, 2002, **309**, 73.
[4] E. S. Maywood, J. S. O'Neill, J. E. Chesham and M. H. Hastings, *Endocrinology*, 2007, **148**, 5624.
[5] W. J. Schwartz and P. Zimmerman, *J. Neurosci.*, 1990, **10**, 3685.
[6] M. R. Smith, H. J. Burgess, L. F. Fogg and C. I. Eastman, *PLoS ONE*, 2009, **4**, e6014.
[7] Y. Xu, K. L. Toh, C. R. Jones, J.-Y. Shin, Y.-H. Fu and L. J. Ptacek, *Cell*, 2007, **128**, 59.
[8] M. R. Ralph and M. Menaker, *Science*, 1988, **241**, 1225.
[9] P. L. Lowrey, K. Shimomura, M. P. Antoch, S. Yamazaki, P. D. Zemenides, M. R. Ralph, M. Menaker and J. S. Takahashi, *Science*, 2000, **288**, 483.
[10] M. H. Vitaterna, D. P. King, A. M. Chang, J. M. Kornhauser, P. L. Lowrey, J. D. McDonald, W. F. Dove, L. H. Pinto, F. W. Turek and J. S. Takahashi, *Science*, 1994, **264**, 719.
[11] D. P. King, Y. Zhao, A. M. Sangoram, L. D. Wilsbacher, M. Tanaka, M. P. Antoch, T. D. L. Steeves, M. H. Vitaterna, J. M. Kornhauser, P. L. Lowrey, F. W. Turek and J. S. Takahashi, *Cell*, 1997, **89**, 641.
[12] M. P. Antoch, E. J. Song, A. M. Chang, M. H. Vitaterna, Y. Zhao, L. D. Wilsbacher, A. M. Sangoram, D. P. King, L. H. Pinto and J. S. Takahashi, *Cell*, 1997, **89**, 655.
[13] S. M. Reppert and D. R. Weaver, *Nature*, 2002, **418**, 935.
[14] D. Gau, T. Lemberger, C. von Gall, O. Kretz, N. Le Minh, P. Gass, W. Schmid, U. Schibler, H. W. Korf and G. Schutz, *Neuron*, 2002, **34**, 245.
[15] A. C. Liu, W. G. Lewis and S. A. Kay, *Nat. Chem. Biol.*, 2007, **3**, 630.
[16] P. Emery and S. M. Reppert, *Neuron*, 2004, **43**, 443.
[17] H. Klemfuss, *Pharm. Ther.*, 1992, **56**, 53.
[18] X. Yin, L. J. Wang, P. S. Klein and M. A. Lazar, *Science*, 2006, **311**, 1002.
[19] U. Knippschild, A. Gocht, S. Wolff, N. Huber, J. Lohler and M. Stoter, *Cell. Signal.*, 2005, **17**, 675.
[20] P. R. Graves and P. J. Roach, *J. Bio. Chem.*, 1995, **270**, 21689.
[21] S. A. Brown, D. Kunz, A. Dumas and P. O. Westermark, *Proc. Natl. Acad. Sci. USA*, 2008, **105**, 1602.

[22] H. Flotow, P. R. Graves, A. Wang, C. J. Fiol, R. W. Roeske and P. J. Roach, *J. Biol. Chem.*, 1990, **265**, 14264.
[23] T. Hirota, J. W. Lee, W. G. Lewis, E. E. Zhang, G. Breton, X. Liu, M. Garcia, E. C. Peters, J.-P. Etchegaray, D. Traver, P. G. Schultz and S. A. Kay, *PLoS Biol.*, 2010, **8**, e1000559.
[24] K. L. Longenecker, P. J. Roach and T. D. Hurley, *J. Mol. Biol.*, 1996, **257**, 618.
[25] N. Mashhoon, A. J. DeMaggio, V. Tereshko, S. C. Bergmeier, M. Egli, M. F. Hoekstra and J. Kuret, *J. Biol. Chem.*, 2000, **275**, 20052.
[26] R.-M. Xu, G. Carmel, J. Kuret and X. Cheng, *Proc. Natl. Acad. Sci. USA*, 1996, **93**, 6308.
[27] G. Manning, D. B. Whyte, R. Martinez, T. Hunter and S. Sudarsanam, *Science*, 1912, **2002**, 298.
[28] T. T. Wager, X. Hou, P. R. Verhoest and A. Villalobos, *ACS Chem. Neurosci.*, 2010, **1**, 435.
[29] J. D. Hughes, J. Blagg, D. A. Price, S. Bailey, G. A. Decrscenzo, R. V. Devraj, E. Ellsworth, Y. M. Fobian, M. E. Gibbs, R. W. Gilles, N. Greene, E. Huang, T. Krieger-Burke, J. Loesel, T. Wager and Y. Zhang, *Bioorg. Med. Chem. Lett.*, 2008, **18**, 4872.
[30] A. L. Hopkins, C. R. Groom and A. Alex, *Drug Disc. Today*, 2004, **9**, 430.
[31] T. Ryckmans, M. P. Edwards, V. A. Horne, A. M. Correia, D. R. Owen, L. R. Thompson, I. Tran, M. F. Tutt and T. Young, *Bioorg. Med. Chem. Lett.*, 2009, **19**, 4406.
[32] T. T. Wager, R. Y. Chandrasekaran, X. Hou, M. D. Troutman, P. R. Verhoest, A. Villalobos and Y. Will, *ACS Chem. Neurosci.*, 2010, **1**, 420.
[33] T. Chijiwa, M. Hagiwara and H. Hidaka, *J. Biol. Chem.*, 1989, **264**, 4924.
[34] T. F. Gallagher, S. M. Fler-Thompson, R. S. Garigipati, M. E. Sorenson, J. M. Smietana, D. Lee, P. E. Bender, J. C. Lee, J. T. Laydon, D. E. Griswold, M. C. Chabot-Fletcher, J. J. Breton and J. L. Adams, *Bioorg. Med. Chem. Lett.*, 1995, **5**, 1171.
[35] N. P. Shanware, L. M. Williams, M. J. Bowler and R. S. Tibbetts, *BMB Rep.*, 2009, **42**, 142.
[36] (a) L. Behrend, D. M. Milne, M. Stoter, W. Deppert, L. E. Campbell, D. W. Meek and U. Knippschild, *Oncogene*, 2000, **19**, 5303; (b) N. Mashhoon, A. J. DeMaggio, V. Tereshko, S. C. Bergmeier, M. Egli, M. F. Hoekstra and J. Kuret, *J. Biol. Chem.* 2000, **275**, 20052; (c) N. Hottecke, M. Kiebeck, K. Baumann, R. Schubenel, E. Winkler, H. Steiner and B. Schmidt, *Bioorg. Med. Chem. Lett*, 2010, **20**, 2958.
[37] J. F. Callahan, J. L. Burgess, J. A. Fornwald, L. M. Gaster, J. D. Harling, F. P. Harrington, J. Heer, C. Kwon, R. Lehr, A. Mathur, B. A. Olsen, J. Weinstock and N. L. Laping, *J. Med. Chem.*, 2002, **45**, 999.
[38] G. Rena, J. Bain, M. Elliot and P. Cohen, *EMBO Rep.*, 2004, **5**, 60.
[39] J. Bain, L. Plater, M. Elliot, N. Shpiro, J. Hastie, H. McLauchlan, I. Klevernic, J. S. C. Arthur, D. R. Alessi and P. Cohen, *Biochem. J.*, 2007, **408**, 297.
[40] G. Cozza, A. Gianoncelli, M. Montopoli, L. Caparrotta, A. Venerando, F. Meggio, L. A. Pinna, G. Zagotto and S. Moro, *Bioorg. Med. Chem. Lett.*, 2008, **18**, 5672.
[41] C. Pfeifer, M. Abadleh, J. Bischof, D. Hauser, V. Schattel, H. Hirner, U. Knippschild and S. Laufer, *J. Med. Chem.*, 2009, **52**, 7618.
[42] K. Salassidis, A. Kurtenbach, H. Daub and S. Obert, *Patent Application WO 2004/005264-A2*, 2004.
[43] L. Badura, T. Swanson, W. Andamowicz, J. Adams, J. Cianfrogna, K. Fisher, J. Holland, R. Kleiman, F. Nelson, L. Reynolds, K. St. Germain, E. Schaffer, B. Tate and J. Sprouse, *J. Pharmacol. Exp. Therap.*, 2007, **322**, 730.
[44] G.-H. Kuo, A. DeAgnelis, A. Wang, Y. Zhang, S. L. Emanuel and S. A. Middleton, *Patent Application WO 2004/009562-A1*, 2004.
[45] W. A. Metz Jr. and F.-X. Ding, *Patent Application WO 2006/023467-A1*, 2006.
[46] M. D. Watterson and L. J. VanEldik, *Patent Application WO 2008/094208-A2*, 2008.
[47] A. Almario Garcia, M. Barrague, P. Burnier, C. Enguehard, Z. Gao, P. George, A. Gueiffier, A.-T. Li and F. Puech, *Patent Application WO 2009/016286-A2*, 2009.
[48] P. Burnier, Y. Chiang, S. Cote-des-Combers, A.-T. Li and F. Puech, *Patent Application WO 2009/037394-A2*, 2009.

[49] A. Almario Garcia, P. Burnier, S. Cote-des-Combers, J.-F. Gilbert, C. Pacaud, F. Puech, Y. Chiang, L. Davis, Z. Gao and Q. Zhao, *Patent Application WO 2010/018327-A1*, 2010.
[50] C. Pacaud and F. Puech, *Patent Application WO 2010/063929-A1*, 2010.
[51] Y. Chiang, C. Enguehard, P. George, A. Gueiffier, F. Puech, M. Sevrin and Q. Zhao, *Patent Application WO 2010/070237-A1*, 2010.
[52] M. Barraue, Y. Chiang, P. George, W. A. Metz and M. Sevrin, *Patent Application WO 2010/070238-A1*, 2010.
[53] F. Venturoni, N. Nikbin, S. V. Ley and I. R. Baxendale, *Org. Biomol. Chem.*, 2010, **8**, 1798.
[54] R. Oudot, C. Pacaud and F. Puech, *Patent Application WO 2010/130934-A2*, 2010.
[55] D. M. Fink, Y. Chiang and N. D. Collar, *Patent Application WO 2006/021000-A2*, 2006.
[56] M. Okamoto and K. Takayama, *Patent Application WO 2010/092660-A1*, 2010.
[57] L. Meijer, H. Galons, B. Joseph, F. Popwycz and N. Oumata, *Patent Application WO 2010/ 103486-A1*, 2010.
[58] L. Meijer, A.-M. W. H. Thunnissen, A. W. White, M. Garnier, M. Nikolic, L.-H. Tsai, J. Walter, K. E. Cleverley, P. C. Salinas, Y.-Z. Wu, J. Biernat, E.-M. Mandelkow, S.-H. Kim and G. R. Petit, *Chem. Biol.*, 2000, **7**, 51.
[59] T. Hirota, J. W. Lee, W. G. Lewis, E. E. Zhang, G. Breton, X. Liu, M. Garcia, E. C. Peters, J.-P. Etchegaray, D. Traver, P. G. Schultz and S. A. Kay, *PLoS Biol.*, 2010, **8**, e1000559.
[60] N. Oumata, K. Bettayeb, Y. Ferandin, L. Demange, A. Lopez-Giral, M.-L. Goddard, Y. Myrianthopoulos, E. Mikros, M. Flajolet, P. Greengard, L. Meijer and H. Galons, *J. Med. Chem.*, 2008, **51**, 5229.
[61] K. M. Walton, K. Fisher, D. Rubitski, M. Marconi, Q.-J. Meng, M. Sladek, J. Adams, M. Bass, R. Chandrasekaran, T. Butler, M. Griffor, F. Rajamohan, M. Serpa, Y. Chen, M. Claffey, M. Hastings, A. Loudon, E. Maywood, J. Ohren, A. Doran and T. T. Wager, *J. Pharmacol. Exp. Ther.*, 2009, **330**, 430.
[62] J. Sprouse, L. Reynolds, R. Kleiman, B. Tate, T. A. Swanson and G. E. Pickard, *Psychopharmacology (Berl.)*, 2010, **210**, 569.
[63] S. Amir, E. W. Lamont, B. Robinson and J. Stewart, *J. Neurosci.*, 2004, **24**, 781.
[64] G. J. DeMarco and T. A. Swanson, *Soc. NeuroSci. Abs.*, 2010 35.
[65] J. Sprouse, L. Reynolds, T. A. Swanson and M. Engwall, *Psychopharmacology (Berl.)*, 2009, **204**, 735.
[66] Q. J. Meng, E. S. Maywood, D. A. Bechtold, W. Q. Lu, J. Li, J. E. Gibbs, S. M. Dupre, J. E. Chesham, F. Rajamohan, J. Knafels, B. Sneed, L. E. Zawadzke, J. F. Ohren, K. M. Walton, T. T. Wager, M. H. Hastings and A. S. Loudon, *Proc. Natl. Acad. Sci. USA*, 2010, **107**, 15240.
[67] A. J. Harmar, H. M. Marston, S. Shen, C. Spratt, K. M. West, W. J. Sheward, C. F. Morrison, J. R. Dorin, H. D. Piggins, J. C. Reubi, J. S. Kelly, E. S. Maywood and M. H. Hastings, *Cell*, 2002, **109**, 497.
[68] S. J. Aton, C. S. Colwell, A. J. Harmar, J. Waschek and E. D. Herzog, *Nat. Neurosci.*, 2005, **8**, 476.

CHAPTER 4

Recent Progress in the Discovery of Kv7 Modulators

Ismet Dorange and **Britt-Marie Swahn**

Contents

1. INTRODUCTION

1.1. Function

The Kv7 ion channels belong to the voltage-gated ion channel superfamily [1–3]. There are to date five known members, Kv7.1–Kv7.5. These potassium-selective channel proteins are found in cardiac tissue (Kv7.1), the central and peripheral nervous system (Kv7.2, Kv7.3, and Kv7.5) and in the inner ear and some auditory nuclei (restricted to Kv7.4) [4–6]. Kv7 channels can homo- or heteromultimerize to form various tetramers

AstraZeneca R&D, SE-151 85 Södertälje, Sweden

Annual Reports in Medicinal Chemistry, Volume 46
ISSN: 0065-7743, DOI: 10.1016/B978-0-12-386009-5.00023-0

having different pharmacological properties, and can also associate with other auxiliary proteins, further diversifying the biological response. Kv7.2, Kv7.3, and Kv7.5 channels are activated at a lower-threshold membrane potential than required to initiate the firing of neurons (subthreshold activation) and underlie the M current, a non-inactivating current that hyperpolarizes the neuronal membranes. In turn, this has the effect to reduce or prevent the firing of an action potential. Thus, Kv7 ion channels play a crucial role in regulating the excitability of membranes, and in fact, Kv7.2 or Kv7.3 inherited loss-of-function mutations in humans are associated with benign familial neonatal convulsion syndrome [7], whereas Kv7.1 mutations are associated with cardiac arrhythmias and deafness, [8–10] and Kv7.4 mutations with hearing loss [11]. It is only quite recently that these channels, and in particular Kv7.2/Kv7.3, have attracted attention as potential therapeutic targets. Judging by the amount of recent publications, the current main interest consists of finding modulators that have the potential to treat epilepsy and pain.

1.2. Structural biology

All Kv7 channels share a similar structure that is formed by four subunits (homo- or heteromeric tetramers). Each of the tetramers consists of six segments (S1–S6) spanning across the membrane. The S1–S4 fragments act as the voltage-sensitive domain (VSD), wherein the fourth helix that bears many alternating positively charged arginine residues has the prime role, while the S5–S6 fragments form the potassium permeable pore [12–16].

2. Kv7.1 CHANNELS

Kv7.1 channels (KCNQ1) are mainly involved in the repolarization phase and the duration of the cardiac action potential [17]. Together with KCNE1 (MinK), an auxiliary subunit, Kv7.1 associates to form the slow delayed rectifier Iks cardiac potassium current [18]. Human mutations of Kv7.1 are associated with potentially severe conditions, such as short (gain of function) and long QT syndrome (loss of function), and atrial fibrillation (gain of function). As a therapeutic target, the blockade of Kv7.1 channels has attracted some attention, and Azimilide (**1**) has reached phase 3 in clinical development as a dual inhibitor of Iks (Kv7.1/KCNE1) and the rapid delayed rectifier Ikr (Kv11.1) for the treatment of arrhythmias. As far as openers of Kv7.1 are concerned, no recent report is found in the literature, and in fact, the only existing reports date from the late 1990s.

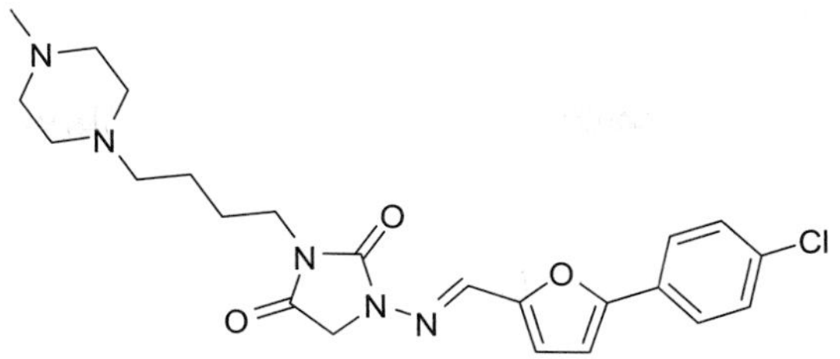

3. Kv7.2–Kv7.5 CHANNELS

3.1. Function

Subunits Kv7.2 and Kv7.3 coassemble and form a tetramer that underlies the M current [19]. It is noteworthy that other subunits (Kv7.4 and Kv7.5) are also associated, albeit to a lesser extent, with M current characteristics [20,21]. The M current which is activated at a lower-threshold membrane potential than would normally activate neuronal cells, hyperpolarizes the cell membrane, and consequently reduces the firing of action potential. In other words, modulation of these channels may control neuronal excitability. Recognizing that neuronal hyperexcitability is the cause of several clinical disorders such as epilepsy and pain, modulation of these channels represents an appealing approach for the treatment of such conditions.

3.2. Binding sites

Even though no crystal structure has been published to date for any of the Kv7 ion channels, mutagenesis, crystallographic, and molecular docking studies have identified two distinct binding sites [13]. The first is located in the pore of the channel, namely in the S5–S6 region, wherein a conserved Tryptophan (Trp 236 in Kv7.2 and Trp 265 in Kv7.3) from Kv7.2 to Kv7.5 was shown to be crucial to maintain sensitivity to retigabine (**2**, *vide infra*), a non-subunit (Kv7.2–Kv7.5)-selective Kv7 activator [22]. Another conserved amino acid (G 301) in the S6-gating hinge region proved to be essential for the efficacy of retigabine [23].

The other binding site is located in the VSD formed by the S1–S4 helices. ICA-27243 (**4**, *vide infra*), a subtype-selective Kv7 activator was shown to bind in this binding pocket [24,25]. More precisely, the compound was shown to bind to the pocket that is formed by at least two helices, S2 and S3. The low degree of sequence homology in this region is consistent with ICA-27243 being a Kv7 subunit-selective compound.

3.3. Compounds in clinical development: Flupirtine, retigabine, and ICA105665

In the recent years, there has been an increased interest in developing positive modulators (activators) of Kv7.2–Kv7.5 channels. The first identified channel opener was retigabine [1,2]. Retigabine is a non-subtype-selective activator of Kv7.2–Kv7.5 ion channels (except Kv7.1) that was shown to be efficacious in a broad range of preclinical seizure models [26]. Recently, this year, retigabine was preapproved as an adjunctive therapy in adult epilepsy patients with partial-onset seizures (POS) [27]. Retigabine was also efficacious in various animal models of pain [28,29]. However, a phase 2a clinical trial aimed at treating pain associated with post-herpetic neuralgia failed to meet the primary end point of pain intensity reduction. The compound is still listed in the company website as being in a phase 2 clinical trial for the treatment of pain (https://www.valeant.com).

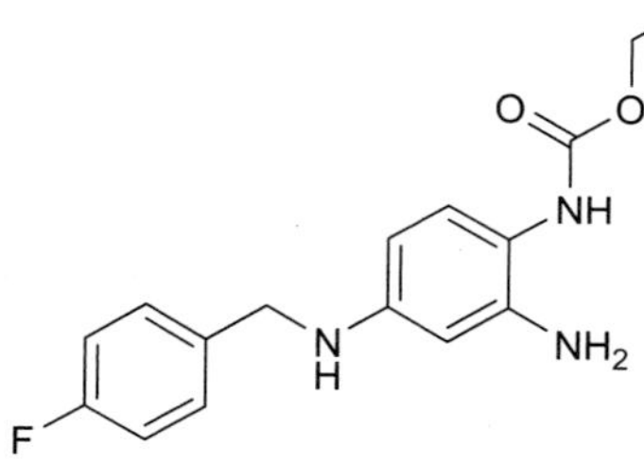

A close structural analog to retigabine that also activates Kv7.2–Kv7.5 channels, flupirtine (**3**), has been on the market for more than two decades in certain countries (Brazil, Estonia, Germany, Italy, Latvia, Lithuania, Portugal, Slovakia, and Russia) for the treatment of various types of pain [30]. Flupirtine is currently in a clinical trial (phase 2) for the treatment of fibromyalgia, the end point being the reduction of musculoskeletal pain and overall symptoms of fibromyalgia (reduction of severity of mood, fatigue, cognitive symptoms, and sleep disturbance). Flupirtine is also being investigated in a clinical trial (phase 2a) for the treatment of neuropathic pain as an adjunct to current opioid drug therapy (https://www.delevarepharma.com).

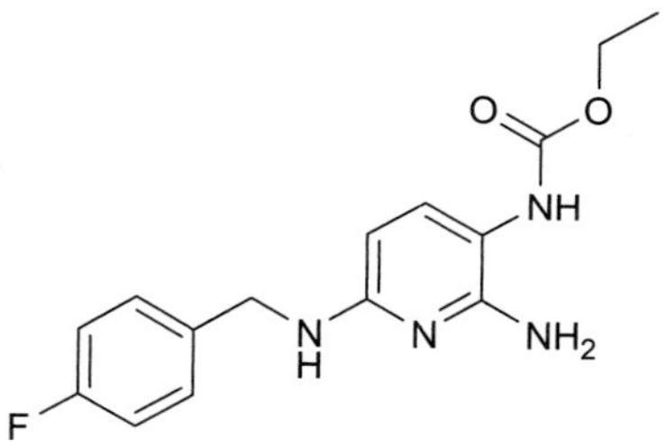

Although both these compounds modulate Kv7 channels [1,2], they lack Kv7 subtype selectivity (except for Kv7.1). Moreover, flupirtine is also an indirect NMDA receptor antagonist [30], and flupirtine and retigabine have shown GABA A agonistic effects, albeit in 10- to 100-fold higher concentrations than required to activate Kv7 channels [31–34] (plasma concentrations for flupirtine required for analgesic activity correlate well with *in vitro* activation concentrations of 2.5–6.5 μM). In order to avoid potential side effects, current efforts consist of developing selective Kv7.2–Kv7.3 channel openers. ICA-105665, a potentially more selective compound, is currently in a clinical trial (phase 2) for the treatment of epilepsy. While the clinical trial was suspended in September 2010 following occurrence of serious adverse effects in the high dose group (600 mg), the FDA has since lifted the hold (February 2011). This compound is also being evaluated in a clinical trial for the treatment of pain. Unfortunately, in a phase 1b pain study, the ability of ICA-105665 to decrease the sensation of pain in response to the intradermal injection of capsaicin (the red chili pepper pungent irritant component) or to a UV-simulated sunburn was not observed at the dose tested of 200 mg. As of today, the structure of ICA-105665 has not been revealed.

3.4. Compounds in drug discovery

The relative success of the compounds in the clinic and the discovery of ICA-27243, a subtype-selective Kv7.x channel modulator [35], currently used as a tool compound, have boosted the interest of pharmaceutical companies to develop small molecule activators of Kv7 channels. Several preclinical drug discovery programs have emerged and these efforts can be classified into two main themes. The first consists of using the retigabine/flupirtine scaffold as the starting point for the design of new compounds, and the second uses a more rational drug design approach (starting point identified by compound screening) such as in the discovery of **4**.

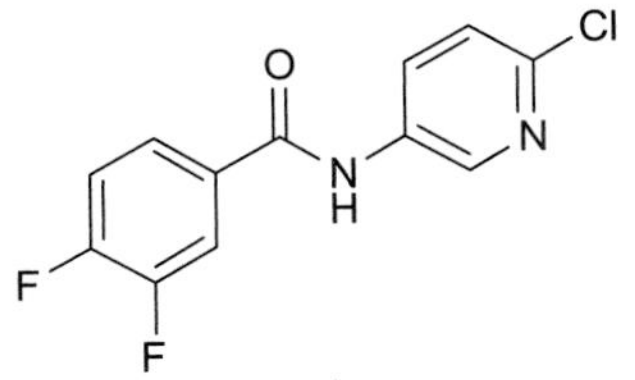

O Cl N N H F F

4

3.4.1. Efforts using retigabine/flupirtine as template

The development of Kv7 openers based upon the retigabine scaffold (**2**) is described elsewhere and only a brief summary will be given here, followed by a complementary update [2]. Generally, the 4-amino-1-carbonylaminophenyl moiety of retigabine is preserved (highlighted in structures **5–8**), whereas replacement of the 2-amino group with small substituents such as methyl, halogen or cyano **6**, or/and isosteric substitution of the benzylamine phenyl in retigabine such as in **5** or **6** proved to be beneficial for improving *in vitro* activity. The amino group in the 4-position has also been explored, and its incorporation into a ring such as in the indoline **7** also improves activity [36,37]. These modifications were all reported to exhibit an $EC_{50} < 2\ \mu M$ in a Kv7.2 rubidium flux assay (Rb^+ assay), and many analogs had an $EC_{50} < 200$ nM. Restriction of conformational changes by replacing the flexible benzylamine moiety with the more rigid 1,2,3,4-tetrahydroisoquinoline moiety such as in **8** [38] or with the corresponding naphthyridine analogs [39] brought a substantial increase in the *in vitro* activity, with EC_{50}s < 50 nM in a Kv7.2–Kv7.3 Rb^+ assay.

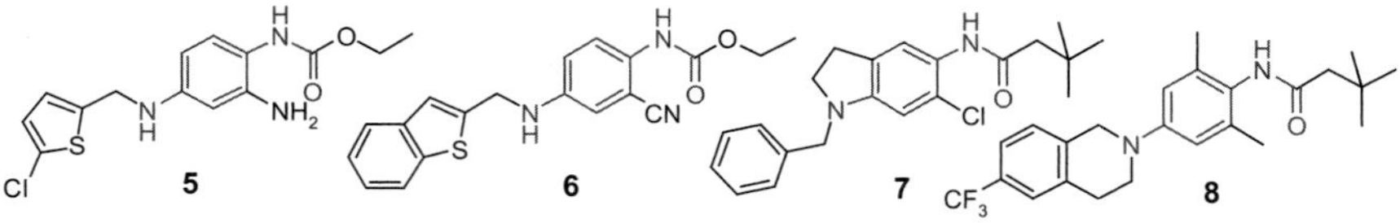

Substitution of the 4-amino moiety by a 4-ether group as in **9** yielded potent compounds with reported EC_{50}s < 10 nM in the Kv7.2–Kv7.3 Rb^+ assay [40]. It is noteworthy that the use of a second small substituent α to the amide bond such as in **8** and **9** led to compounds with a 10- to 100-fold improvement in potency. The use of flupirtine as starting point for the development of new Kv7 openers has also been fruitful. Keeping the 1,4-diamino pattern intact and replacing the benzylamine with morpholine led to pyridine **10** and pyrimidine **11**, with the best derivatives displaying EC_{50}s < 200 nM in the Kv7.2–Kv7.3 Rb^+ assay [41,42].

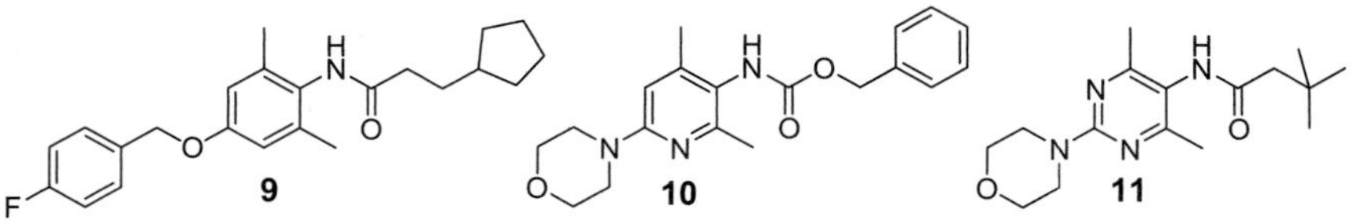

More recently, the 2-amino moiety was investigated and introduction of a morpholine such as in **12** led to a 35-nM compound in a thallium-sensitive assay using retigabine as a reference [43]. Furthermore, exchange of the benzylamine with an aliphatic amine such as in **13** (R = tetrahydropyranylmethyl) increases *in vitro* potency, with **13** reported to increase the amplitude of the current by 708% at 0.03 μM in a patch clamp assay using HEK-293 cells expressing Kv7.2–Kv7.3 [44]. Interestingly, incorporation of heteroaromatic five-membered ring carboxamides in place of the flupirtine ethyl carbamate, combined with the modifications previously discussed (*vide infra*), led to **14**, a 20-nM compound in the Tl^+ assay [45,46].

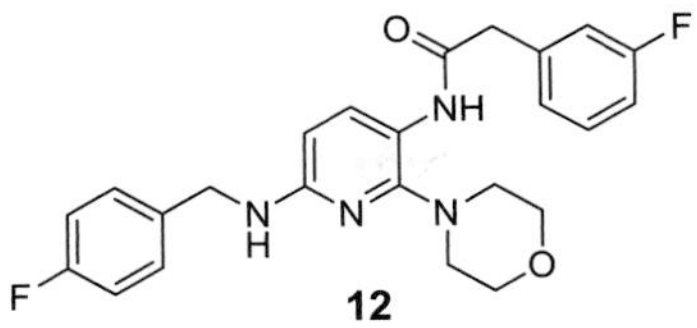

12 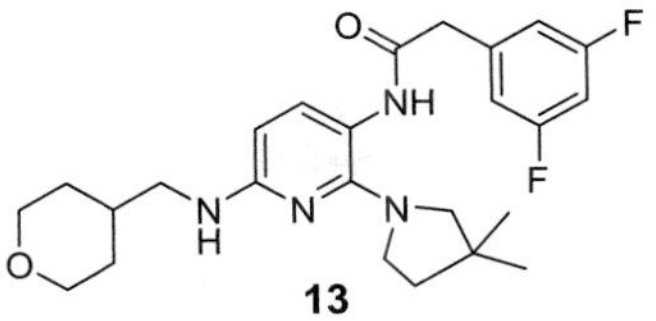 13

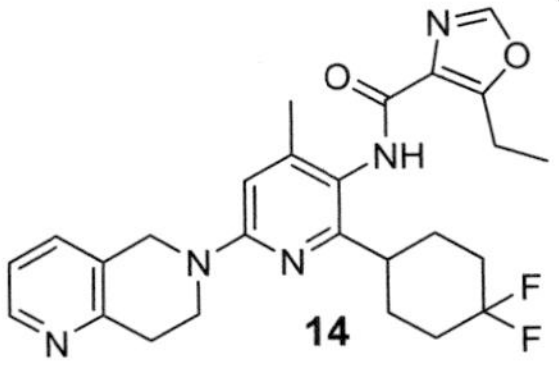

14 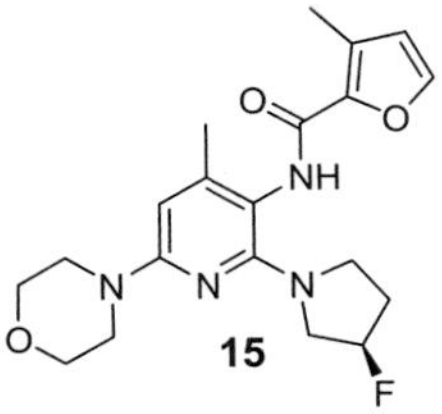 15 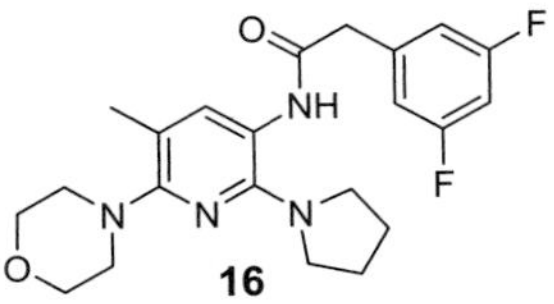 16

Moving the 4-Me (see **15**) to the 5-position of the pyridine as in **16** led to the most potent compound published to date with an $EC_{50} = 0.32$ nM in a Tl^{+}-sensitive assay using retigabine as a reference [47]. Of interest, the homomorpholine analogs were also reported active with EC_{50}'s ranging from 0.4 to 270 nM [48–50].

3.4.2. Other Kv7 activators

A very important compound for the discovery of subtype-selective Kv7 openers is ICA-27243 (*vide supra*). This compound is part of a series of compounds consisting of pyridin-3-yl benzamides that were discovered by rational drug design [51,52]. Compound **4** was shown to enhance activation of heteromeric Kv7.2–Kv7.3 but had little effect on homomeric Kv7.1, Kv7.3, and Kv7.4 channels as well as on heteromeric Kv7.3–Kv7.5 channels [24]. Derivative **4** has been extensively studied and was shown to alleviate pain as well as epileptic seizures in many animal models [35,53]. SAR analysis demonstrated that the left-hand difluorophenyl could be replaced by different heterocycles such as the methylpyrazole in **17**. The 2-Cl atom on the pyridine was the most suited substituent for activity, whereas other substitutions on the ring were detrimental for potency. Amide bond bioisosteres such as in indazole **18** or benzisoxazoles were tolerated transformations [54].

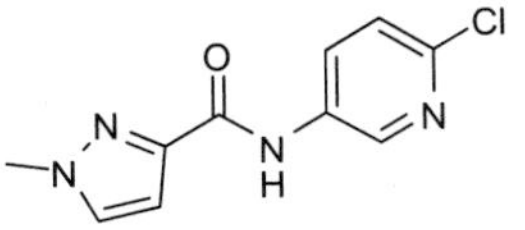

17 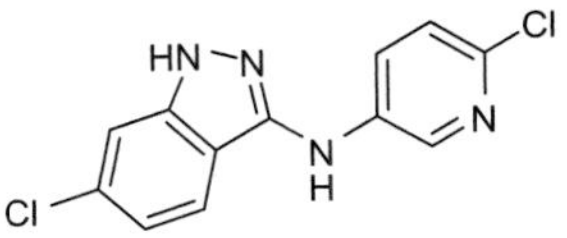 18

Quinazolinone derivatives were another class of compounds structurally different from the retigabine scaffold. The first series of this class to be disclosed consisted of 2-thio quinazolinones [55], and the most potent example **19** (*vide infra*) displays an activity of 70 nM against the M current (NG-108 FLIPR assay). The potentially labile thioether group was successfully replaced by a variety of other groups such as alkyl, cycloalkyl, and aryls as in **20–23** [56]. The phenyl quinazolinone core could be modified to the more polar pyrazolopyrimidinone core as in **21** with a reported $EC_{50} < 50$ nM [57]. The linker between the amide and the phenyl group has been extensively explored [58,59] and led to ethyl derivative **22**, which enhances the current amplitude by 187% at 3 μM. In a more recent example, different central scaffolds were explored, however, the most potent compounds reported kept the quinazolinone in place, with the key modification being replacement of the benzylic moiety with a bicyclic aliphatic group [60]. This compound (**23**) has an EC_{50} of 17 nM against Kv7.2–Kv7.3 in a Tl^+ assay.

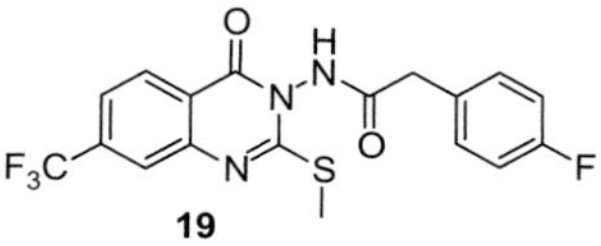

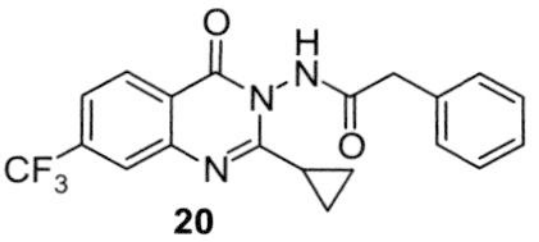

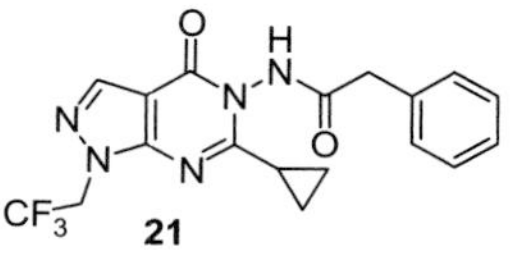

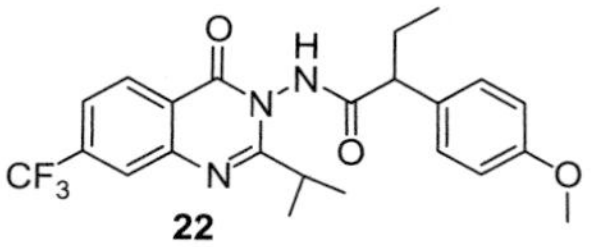

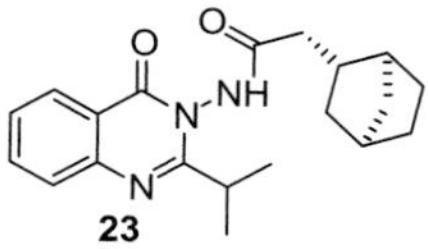

A different series of compounds with yet another modified central core, but retaining the acyl hydrazine moiety, was identified from an High Throughput Screening (HTS) effort [61]. The first identified hit (**24**, R1 = H, R2 = Me, R3 = H) had an EC_{50} of 27 nM against Kv7.2–Kv7.3 channels in a Rb^+ isotopic efflux assay and was 100-fold-selective against the Kv7.1–KCNE1 channel. This agonist exhibited anticonvulsant properties in epileptic animal models when administrated intraperitoneally, but due to poor bioavailability, ceased to show effects when given orally. Thus, the first chemistry efforts were aimed at evaluating the impact of small substituents on both activity and bioavailability. The different positions of the aryl substituents were evaluated and the only variation that showed both good *in vitro* EC_{50} (25 nM in the Rb^+ efflux assay) and *in vivo* efficacy (epileptic animal model) when given orally contained an OCF_3 at the R3 position (see **24**, R3 = OCF_3). It is noteworthy that methylation at the R1 position or replacement of the sulfur by an oxygen atom led to substantial loss of activity ($EC_{50} > 10$ and 2.7 μM, respectively). The medicinal chemistry effort then concentrated on the replacement of the adamantane moiety to give compounds of structure **25**. In particular, compounds **26**, **27**, and **28** proved to have good *in vitro* activity (*vide infra*) and *in vivo* efficacy in anticonvulsant assays (mouse maximal electroshock seizure (MES) and rat subcutaneous pentylenetetrazol seizure models) with ED_{50} values <10 mg/kg. These analogs were also efficacious in animal pain models. In fact, the effect of these compounds was comparable and similar to the effect of retigabine in the rat formalin and spinal nerve ligation (SNL) assays. Moreover, **28** demonstrated a complete reduction of tactile allodynia in the SNL model when orally administrated at 10 mg/kg, whereas retigabine had no effect at the same dose.

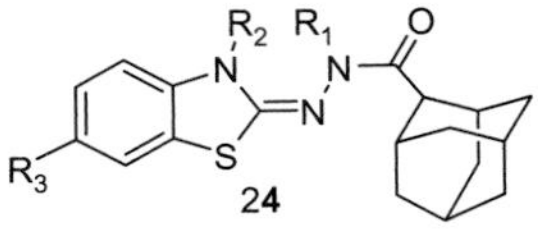

24 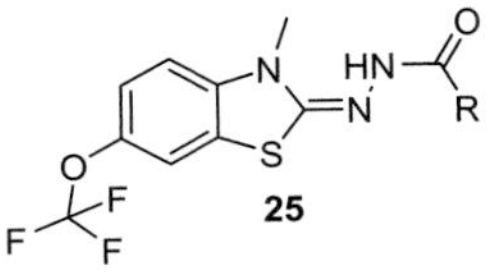 **25**

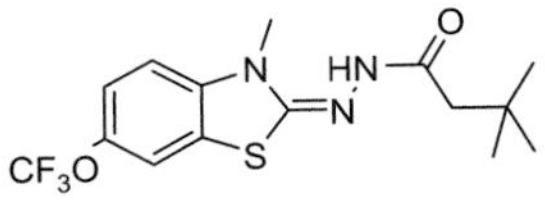

26 EC_{50} = 49 nM 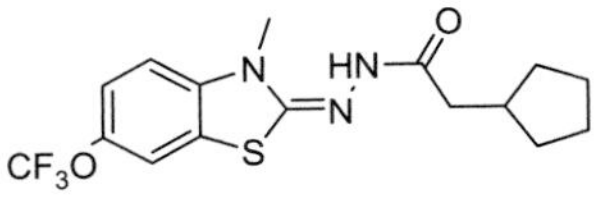 **27** EC_{50} = 44 nM 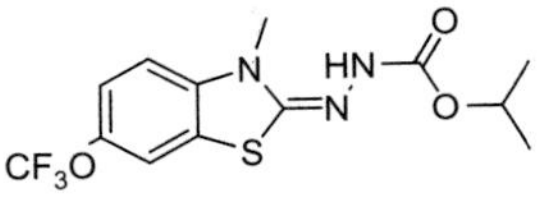 **28** EC_{50} = 95 nM

The imidazolopyridine scaffold (**29**) was disclosed as a subtype-selective Kv7 agonist [62]. An extensive SAR exercise was conducted, and while small electronegative substituents at the R2 position proved to be beneficial for activation of the Kv7.2–Kv7.3 channels, substituents at the R1, R3 and R4 positions were detrimental (see Figure 1). Variation at positions R5 and R6 has been widely investigated, and the best groups are highlighted in Figure 1.

Compounds **30** and **31** were reported to have EC_{50}s < 500 nM in a fluorometric (imaging plate reader) assay and showed protective effects in the rat MES model. Compounds **30** and **31** exhibited potency in a nerve injury model with reported ED_{50}s of 10 and 1.9 mg/kg, respectively; **31**

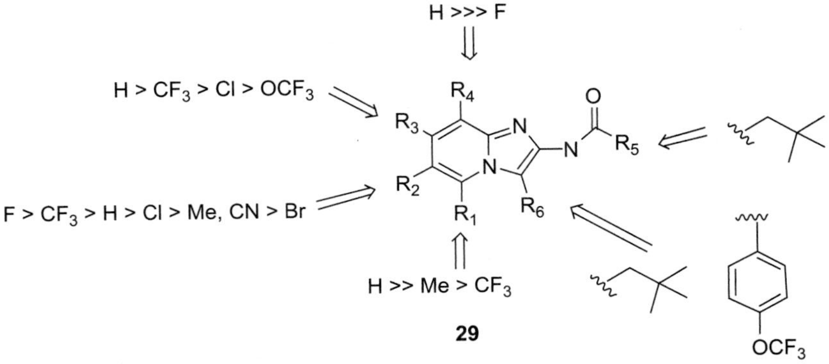

H >>> F

H > CF_3 > Cl > OCF_3

F > CF_3 > H > Cl > Me, CN > Br

H >> Me > CF_3

29

Figure 1 SAR of the imidazolopyridine series.

demonstrated potency in the carrageenan inflammatory pain model with an ED_{50} of 0.5–1 mg/kg. Further examples were disclosed [63] and revealed that the metabolic soft spot on the neopentyl moiety could be addressed by fluorination (see example **32**).

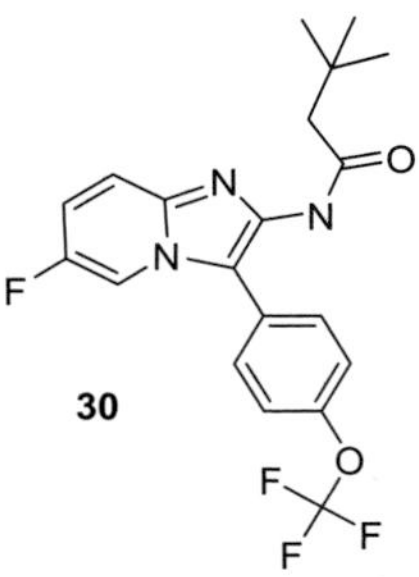

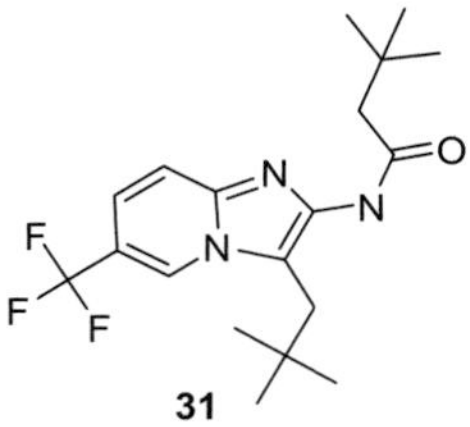

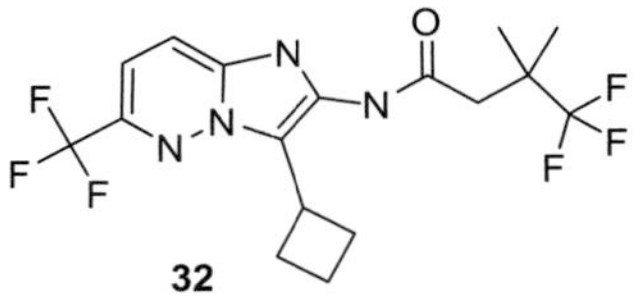

Other scaffolds were recently reported as subtype-selective agonists of Kv7.2–Kv7.3 (compounds **33–36**). Pyridine derivative **33** was shown to be active in an *in vitro* fluorometric assay with an EC_{50} of 224 nM, as well as *in vivo*, in a low-intensity tail flick rat model with an $ED_{50} = 2.7$ mg/kg [64]. The quinoline derivative **34** proved to have the exact same *in vivo* potency [65–69]. The fused bicyclic dihydropyrrolopyrazine analog **35** was also shown to exhibit activity *in vitro* with an EC_{50} of 2.4 nM in a fluorometric assay and *in vivo* to alleviate pain in the rat formalin model [68]. Recently, a pyrazolopyrimidinone core was disclosed and the most potent compound (**36**) displayed an EC_{50} of 60 nM in a Rb^+ assay. However, no *in vivo* experiments have been reported for these analogs [70].

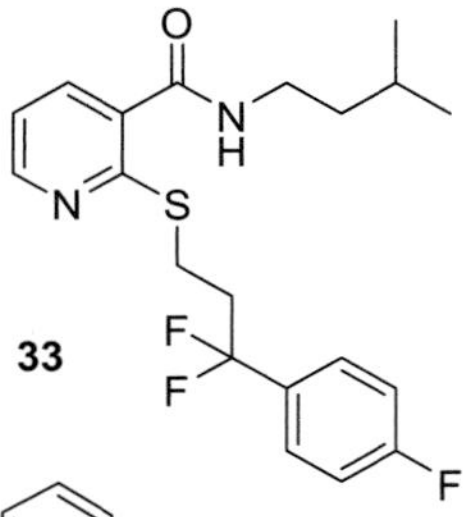

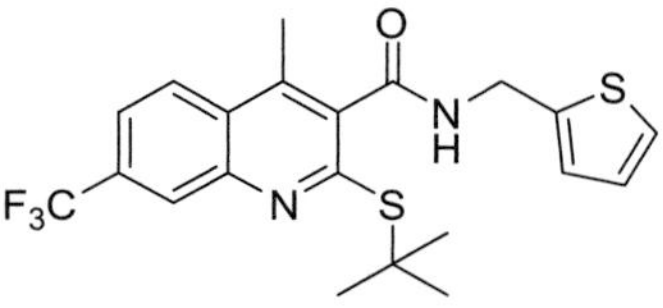

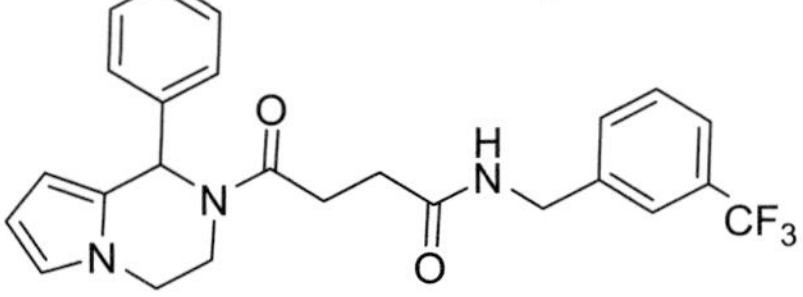

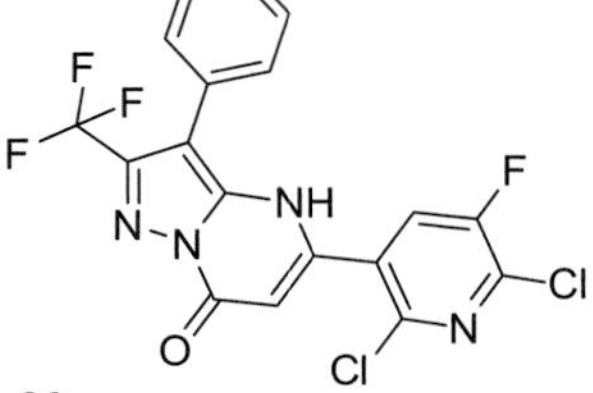

4. CONCLUSION

Kv7 channels play a central role in regulating several critical cell functions, such as regulation of the heart beat or modulation of the neuronal activity. It is thus understandable that as therapeutic targets, this class of ion channels has attracted attention. The identification of Kv7 human mutations that lead to heart disorders (Kv7.1) and to neonatal familial epilepsy (K7.2 and Kv7.3) has raised hope in finding new treatments in cardiology and epilepsy. The advancement of retigabine and ICA-105665 in the clinic is proving the validity of the target. It is somewhat more recent that Kv7 channels have shown potential for the treatment of pain. Flupirtine, a non-opoid analgesic on the market for more than 20 years has paved the way for others to follow. Moreover, the discovery of subtype-selective activators (*e.g.*, ICA-27243) has constituted another milestone in the search for analgesics having fewer side effects. However, despite strong support from animal studies, there are still no data supporting translation to the clinic. It seems more than likely that a better understanding of the Kv7 pharmacology and the translational science will be required to address some of these issues.

REFERENCES

[1] N. A. Castle, *Expert Opin. Ther. Pat.*, 2010, **20**, 1471.
[2] A. D. Wickenden and G. McNaughton-Smith, *Curr. Pharm. Des.*, 2009, **15**, 1773.
[3] G. Munro and W. Dalby-Brown, *J. Med. Chem.*, 2007, **50**, 2576.
[4] T. J. Jentsch, *Nat. Rev. Neurosci.*, 2000, **1**, 21.
[5] J. Robbins, *Pharmacol. Ther.*, 2001, **90**, 1.
[6] T. Kharkovets, J.-P. Hardelin, S. Safieddine, M. Schweizer, A. El-Amraoui, C. Petit and T.J. Jentsch, *Proc. Natl. Acad. Sci. USA*, 2000, **97**, 4333.
[7] N. A. Singh, P. Westenskow, C. Charlier, C. Pappas, J. Leslie, J. Dillon, V. Elving Anderson, M. C. Sanguinetti and M. F. Leppert, *Brain*, 2003, **126**, 2726.
[8] N. Neyroud, F. Tesson, I. Denjoy, M. Leibovici, C. Donger, J. Barhanin, S. Faure, F. Gary, P. Coumel, C. Petit, K. Schwartz and P. Guicheney, *Nat. Genet.*, 1997, **15**, 186.
[9] C. Donger, I. Denjoy, M. Berthet, N. Neyroud, C. Cruaud, M. Bennaceur, G. Chivoret, K. Schwartz, P. Coumel and P. Guicheney, *Circulation*, 1997, **96**, 2778.
[10] Q. Wang, M. E. Curran, I. Splawski, T. C. Burn, J. M. Millholland, T. J. VanRaay, J. Hen, K. W. Timothy and G. M. Vincent, *Nat. Genet.*, 1996, **12**, 17.
[11] C. Kubisch, B. C. Schroeder, T. Friedrich, B. Lutjohann, A. El-Amraoui, S. Marlin, C. Petit and T. J. Jentsch, *Cell*, 1999, **96**, 437.
[12] F. Miceli, M. V. Soldovieri, M. Martire and M. Taglialatela, *Curr. Opin. Pharmacol.*, 2008, **8**, 65.
[13] A. Peretz, L. Pell, Y. Gofman, Y. Haitin, L. Shamgar, E. Patrich, P. Kornilov, O. Gourgy-Hacohen, N. Ben-Tal and B. Attali, *Proc. Natl. Acad. Sci. USA*, 2010, **107**, 15637.
[14] W. Lange, J. Geissendorfer, A. Schenzer, J. Grotzinger, G. Seebohm, T. Friedrich and M. Schwake, *Mol. Pharmacol.*, 2009, **75**, 272.
[15] W. D. Van Horn, Carlos G. Vanoye and Charles R. Sanders, *Curr. Opin. Struct. Biol.*, 2011, **21**, 283.

[16] F. Miceli, M. V. Soldovieri, F. A. Lannotti, V. Barrese, P. Ambrosino, M. Martire, M.R. Cilio and M. Taglialatela, *Front. Pharmacol.*, 2011, **2**, 1.
[17] D. Wu, Hua Pan, Kelli Delaloye and Jianmin Cui, *Biophys. J.*, 2010, **99**, 3599.
[18] T. Jespersen, M. Grunnet and S. P. Olesen, *Physiology*, 2005, **20**, 408.
[19] D. A. Brown and G. M. Passmore, *Br. J. Pharmacol.*, 2009, **156**, 1185.
[20] B. C. Schroeder, M. Hechenberger, F. Weinreich, C. Kubisch and T. J. Jentsch, *J. Biol. Chem.*, 2000, **275**, 24089.
[21] C. Lerche, C. R. Scherer, G. Seebohm, C. Derst, A. D. Wei, A. E. Busch and K. Steinmeyer, *J. Biol. Chem.*, 2000, **275**, 22395.
[22] B. H. Bentzen, N. Schmitt, K. Calloe, W. D. Brown, M. Grunnet and S. P. Olesen, *Neuropharmacology*, 2006, **51**, 1068.
[23] T. V. Wuttke, G. Seebohm, S. Bail, S. Maljevic and H. Lerche, *Mol. Pharmacol.*, 2005, **67**, 1009.
[24] K. Padilla, A. D. Wickenden, A. C. Gerlach and K. McCormack, *Neurosci. Lett.*, 2009, **465**, 138.
[25] S. M. Blom, N. Schmitt and H. S. Jensen, *Pharmacology*, 2010, **86**, 174.
[26] V. Barrese, F. Miceli, M. V. Soldovieri, P. Ambrosino, F. A. Iannotti, M. R. Cilio and M. Taglialatela, *Clin. Pharmacol. Adv. Appl.*, 2010, **2**, 225.
[27] W. Nasreddine, A. Beydoun, S. Atweh and B. Abou-Khalil, *Expert Opin. Emerg. Drugs*, 2010, **15**, 415.
[28] G. Blackburn-Munro, W. Dalby-Brown, N. R. Mirza, J. D. Mikkelsen and R. E. Blackburn-Munro, *CNS Drug Rev.*, 2005, **11**, 1.
[29] W. Xu, Y. Wu, Y. Bi, L. Tan, Y. Gan and K. W. Wang, *Mol. Pain*, 2010, **6**, 49.
[30] J. Devulder, *CNS Drugs*, 2010, **24**, 867.
[31] M. Martire, P. Castaldo, M. D'Amico, P. Preziosi, L. Annunziato and M. Taglialatela, *J. Neurosci.*, 2004, **24**, 592.
[32] M. J. Main, J. E. Cryan, J. R. B. Dupere, B. Cox, J. J. Clare and S. A. Burbidge, *Mol. Pharmacol.*, 2000, **58**, 253.
[33] A. D. Wickenden, W. Yu, A. Zou, T. Jegla and P. K. Wagoner, *Mol. Pharmacol.*, 2000, **58**, 591.
[34] C. Rundfeldt and R. Netzer, *Neurosci. Lett.*, 2000, **282**, 73.
[35] A. D. Wickenden, J. L. Krajewski, B. London, P. K. Wagoner, W. A. Wilson, S. Clark, R. Roeloffs, G. McNaughton-Smith and G. C. Rigdon, *Mol. Pharmacol.*, 2008, **73**, 977.
[36] N. Khanzhin, M. Rottlaender and W. P. Watson, *Patent Application US 2006/0264496-A*, 2006.
[37] J.-M. Vernier, H. Chen and J. Song, *Patent Application WO 2009/023667-A1*, 2009.
[38] J.-M. Vernier, M. A. De La Rosa, H. Chen, J. Z. Wu, G. L. Larson and I. W. Cheney, *Patent Application US 2008/0139610-A1*, 2008.
[39] J.-M. Vernier, *Patent Application WO 2009/018466-A1*, 2009.
[40] J. Z. Wu, J. Vernier, H. Chen and J. Song, *Patent Application WO2010/008894-A1*, 2010.
[41] N. Khanzhin, D. R. Greve and M. Rottlaender, *Patent Application US 2007/0066612-A1*, 2007.
[42] C. W. Tornroe, N. Khanzhin, M. Rottlaender, W. P. Watson and D. R. Greve, *Patent Application WO 2006/092143-A1*, 2006.
[43] W. D. Brown, C. Jessen and D. Stroebaek, *Patent Application WO 2010/060955-A1*, 2010.
[44] C. Jessen, W. D. Brown and D. Stroebeck, *Patent Application WO 2010/026104*, 2010.
[45] W. D. Brown, C. Jessen, C. Mattsson, R. Sott and D. Stroebaek, *Patent Application WO 2010/122064-A1*, 2010.
[46] C. Jessen, W. D. Brown and D. Stroebaek, *Patent Application WO 2010/097379-A1*, 2010. See also W. Dalby-Brown, C. Jessen and D. Stroebaek, *Patent Application WO 2011/0268902*, 2011.
[47] W. D. Brown, C. Jessen and D. Stroebaek, *Patent Application WO 2010/094645-A1*, 2010.

[48] W. D. Brown, C. Jessen and D. Stroebaek, *Patent Application WO 2010/094644-A1*, 2010.
[49] W. D. Brown, C. Jessen and D. Stroebaek, *Patent Application WO 2011/026890-A1*, 2010.
[50] W. D. Brown, C. Jessen and D. Stroebaek, *Patent Application WO 2011/026891-A1*, 2010.
[51] G. A. McNaughton-Smith, M. F. Gross and A. D. Wickenden, *Patent Application WO 2001/010380-A2*, 2001.
[52] G. A. McNaughton-Smith, M. F. Gross, G. C. Rigdon and A. D. Wickenden, *Patent Application WO 2001/010381-A2*, 2001.
[53] R. Roeloffs, A. D. Wickenden, C. Crean, S. Werness, G. McNaughton-Smith and J. Stables, *J. Pharmacol. Exp. Ther.*, 2008, **326**, 818.
[54] G. A. McNaughton-Smith and G. S. Amato, *Patent Application US 2002/0193597-A1*, 2002.
[55] G. A. McNaughton-Smith, G. Andrew, J. B. Thomas and G. S. Amato, *Patent Application WO 2004/058704-A2*, 2004.
[56] G. A. McNaughton-Smith, G. Andrew, G. S. Amato and J. B. Thomas, *Patent Application WO 2005/025293-A2*, 2005.
[57] G. A. McNaughton-Smith, G. S. Amato and J. B. Thomas, *Patent Application* US 2008/ *0058319-A*, 2004.
[58] W. D. Brown, C. Jessen, J. Demnitz, T. Dyhring and D. Stroebaek, *Patent Application WO 2007/104717-A1*, 2007.
[59] W. D. Brown, L. Teuber, T. Dyhring, D. Stroebaek and C. Jessen, *Patent Application WO 2007/057447-A1*, 2007.
[60] M. J. Scanio, W. H. Bunnelle, W. A. Carroll, S. Peddi, A. Perez-Medrano and L. Shi, *Patent Application WO 2010/138828-A2*, 2010.
[61] P. C. Fritch, G. McNaughton-Smith, G. S. Amato, J. F. Burns, C. W. Eargle, R. Roeloffs, W. Harrison, L. Jones and A. D. Wickenden, *J. Med. Chem.*, 2010, **53**, 887.
[62] W. S. Mahoney, W. L. Thompson, D. J. McClure and M. S. Stay, *Patent Application WO 2009/002654-A1*, 2009.
[63] T. E. Christos, G. S. Amato, R. N. Atkinson, M. G. Barolli, L. A. Wolf-Gouveia and M. J. Suto, *Patent Application US 2010/0240663-A1*, 2010.
[64] S. Kuehnert, B. Merla, G. Bahrenberg and W. Schroeder, *Patent Application* WO 2010/ *102809-A1*, 2010.
[65] S. Kuehnert, G. Bahrenberg and W. Schroeder, *Patent Application WO 2010/102779-A1*, 2010.
[66] S. Kuehnert, G. Bahrenberg, A. Kless and W. Schroeder, *Patent Application WO 2010/ 102811-A1*, 2010.
[67] B. Merla, T. Christoph, S. Oberbörsch, K. Schiene, G. Bahrenberg, R. Frank, S. Kuhnert and W. Schröder, *Patent Application WO 2008/046582-A1*, 2008.
[68] S. Kuhnert, G. Bahrenberg, A. Kless, B. Merla, K. Schiene and W. Schröder, *Patent Application WO 2010/046108-A1*, 2010.
[69] S. Kuhnert, G. Bahrenberg, B. Merla, K. Schiene and W. Schröder, *Patent Application US 2010/0152234-A1*, 2010.
[70] J. Qi, Fan Zhang, Y. Mi, Y. Fu, W. Xu, D. Zhang, Y. Wu, X. Du, Q. Jia, K. Wang and H. Zhang, *Eur. J. Med. Chem.*, 2011, **46**, 934.

PART II:
Cardiovascular and Metabolic Diseases

Editor: Andy Stamford
Merck Research Laboratories
Kenilworth
New Jersey

CHAPTER 5

Bile Acid Receptor Modulators in Metabolic Diseases

Yanping Xu

Contents

1. INTRODUCTION

Bile acids (BAs) have for a long time been viewed as detergents able to solubilize cholesterol, fatty acids, and liposoluble vitamins, thus facilitating the digestion, transportation, and gastrointestinal absorption of nutrients. BAs have also been shown to be involved in a large variety of cellular processes. Some recent discoveries have unveiled novel actions of BAs as signaling hormones endowed with a wide array of endocrine functions. In 1999, three research groups independently identified BAs as endogenous ligands for a nuclear receptor, the farnesoid X receptor (FXR) [1–3]. In early 2000, two research groups discovered that a novel class A G-protein-coupled receptor (GPCR), TGR5 (also known as GPBAR1, GPCR19, GPR131, BG37, M-BAR, Rup 43), can be activated by BAs [4,5].

Lilly Research Laboratories, Eli Lilly and Company, Indianapolis, IN 46285, USA

Annual Reports in Medicinal Chemistry, Volume 46
ISSN: 0065-7743, DOI: 10.1016/B978-0-12-386009-5.00022-9

Highly expressed in the liver, intestine, kidney, adrenal glands, and adipose tissue, FXR is a master regulator of the synthesis and pleiotropic actions of endogenous BAs [6]. Activation of FXR by BAs or synthetic FXR agonists lowers plasma triglycerides by a mechanism involving repression of hepatic sterol regulatory element binding protein-1c (SREBP-1c) expression and the modulation of glucose-dependent lipogenic genes. Furthermore, FXR controls lipid and glucose metabolism through regulation of gluconeogenesis and glycogenolysis in the liver and through regulation of peripheral insulin sensitivity in striated muscle and adipose tissue [7–9]. Similar to effects in the liver, FXR agonists modulate lipid metabolism and promote anti-inflammatory and antifibrotic effects in the kidney, suggesting a potential use of FXR agonists to treat diabetic nephropathy and other fibrotic renal diseases [10].

TGR5 is expressed in brown adipose tissue, muscle, liver, intestine, gallbladder [11], and selected areas of the central nervous system [5]. TGR5 activation in intestinal enteroendocrine L cells stimulates secretion of the incretin glucagon-like peptide-1 (GLP-1) [12]. Activation of GLP-1 receptors by derivatives of exendin-4 or enhancement of GLP-1 half-life by dipeptidyl peptidase-4 inhibitors is clinically well-established therapeutic approaches for the treatment of type 2 diabetes [13]. By augmenting GLP-1 activity, these agents improve glycemic control in diabetic patients through increase of glucose-dependent insulin secretion and reduction of glucagon production. Treatment of mice fed a high fat diet with the TGR5 agonist oleanolic acid resulted in lower serum glucose and insulin levels and enhanced glucose tolerance [14]. In addition, administration of BAs to mice increased energy expenditure in brown adipose tissue, preventing obesity and insulin resistance *via* TGR5-mediated cAMP-dependent induction of type 2 iodothyronine deiodinase (D2), which locally stimulates thyroid hormone-mediated thermogenesis [15].

The intent of this report is to present the most recent publications on FXR agonists and to provide a comprehensive literature summary of TGR5 agonists.

2. FXR AGONISTS

Azepinol[4,5-*b*]indole **1** (hEC_{50} = 600 nM, efficacy (eff) = 100%) was identified as a FXR agonist lead from a high-throughput screening effort. Structure–activity relationship (SAR) studies around the azepine ring demonstrated that dialkyl substitution at C-1 led to a 30-fold improvement in potency. In addition, incorporation of an isopropyl ester yielded another ~3-fold boost in potency. Compound **2** represented the most potent FXR agonist within the series (hEC_{50} = 4 nM, eff = 149%) [16].

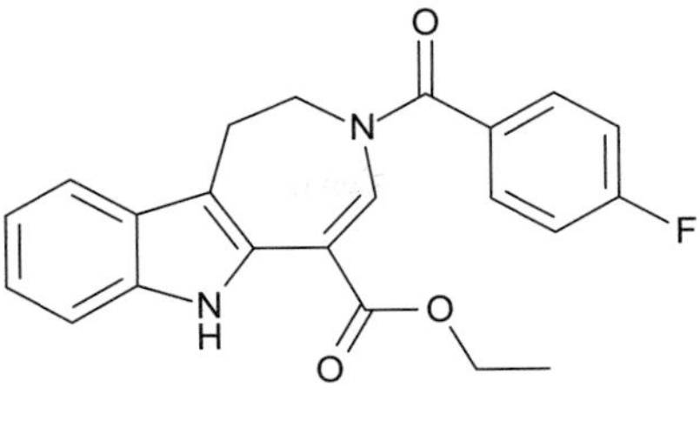

1

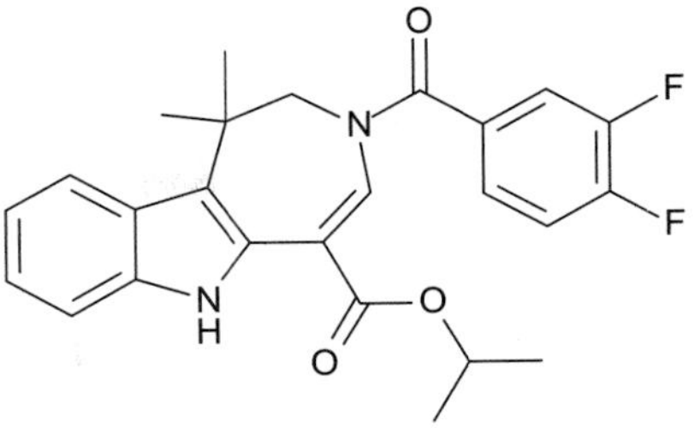

2

A rat pharmacokinetic (PK) study showed that compound **2** had good oral bioavailability ($F = 38\%$) and a long half-life ($t_{1/2} = 24$ h). Oral treatment of normal C57BL/6 mice with **2** administered at a dose of 10 mg/kg/d for 7 days yielded statistically significant reductions of triglycerides (TG, 24%) and total cholesterol (22%) levels. When administered to low-density lipoprotein receptor knockout ($LDLR^{-/-}$) mice fed a Western diet for 8 weeks, **2** lowered both TG (19% and 39% at doses of 1 and 3 mg/kg, respectively) and total cholesterol (23% and 50% at doses of 1 and 3 mg/kg, respectively). However, this molecule was poorly soluble. Guided by crystallographic data, the appended morpholine analogs **3a** and **3b** were identified and showed dramatic 400-fold improvements in equilibrium solubility measured in 0.5% methylcellulose/2% Tween-80 in water. However, bioavailability of **3a** and **3b** in rats was not improved compared to that of **2** ($F = 38\%$ and 25% for **3a** and **3b**, respectively) which is likely due to their high clearance (Cl = 52 and 64 mL/min/kg for **3a** and **3b**, respectively).

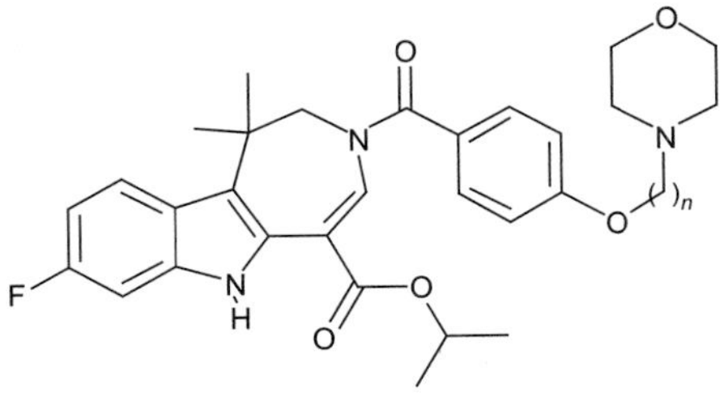

3a: $n = 3$
3b: $n = 2$

Both compounds showed potencies (**3a**: $mEC_{50} = 52$ nM, eff = 117%; **3b**: $mEC_{50} = 188$ nM, eff = 110%) at mouse FXR similar to that of compound **2** ($mEC_{50} = 152$ nM, eff = 174%). Oral administration of **3a** and **3b** to $LDLR^{-/-}$ mice caused a dose-dependent reduction of low-density lipoprotein cholesterol (LDLc). In female rhesus monkeys, **3a** given at a dose of 60 mg/kg/d po for 4 weeks resulted in a significant lowering of TG (~50%; absolute value was not reported), very low-density lipoprotein cholesterol (VLDLc ~ 50%; absolute value was not reported), and LDLc (63%) [17].

The potent FXR agonist, **4a** (GW4046, FXR transient transfection (TT) assay EC_{50} = 65 nM, eff = 100%), was unsuitable for further development due to several issues. These liabilities include poor rat PK (high clearance and low bioavailability), a potentially toxic stilbene pharmacophore, and stilbene-mediated UV light instability. SAR development at the 3- and 5-positions of the isoxazole ring revealed a preference for hydrophobic substituents [18]. In addition, some polarity was tolerated in the tether at the 3-position of the isoxazole linked to the phenyl group, for example, **4b** (FXR TT EC_{50} = 89 nM, eff = 89%). A rat PK study demonstrated that **4b** had an improved $t_{1/2}$ (2 h) and clearance (Cl = 20 mL/min/kg). Unfortunately, low oral bioavailability (F = 9%) was observed. It was hypothesized that the stilbene moiety could be the predominating detrimental structural feature irrespective of modifications elsewhere in these molecules.

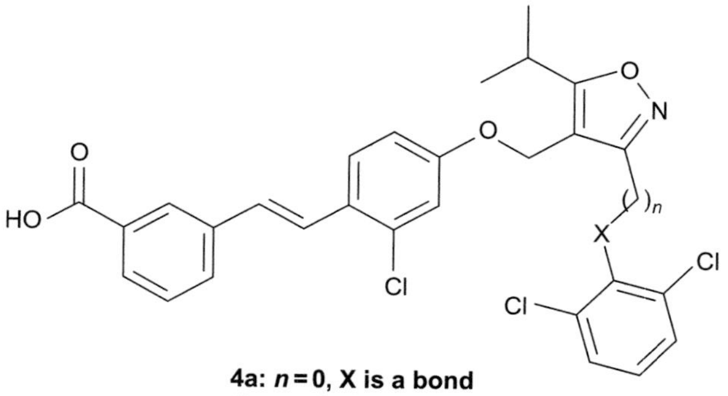

4a: *n* = 0, X is a bond
4b: *n* = 1, X = O

In an attempt to address the perceived liability of the stilbene functional group, a series of conformationally constrained analogs were explored [19]. Benzothiophene analog **5** (FXR TT EC_{50} = 32 nM, eff = 87%) was equipotent with **4a**. Indole analog **6a** showed a slight attenuation of FXR activity (FXR TT EC_{50} = 210 nM, eff = 84%). In rat PK studies, compound **5** had very high clearance (Cl = 66 mL/min/kg), a short half-life ($t_{1/2}$ = 15 min), and poor bioavailability (F = 9%).

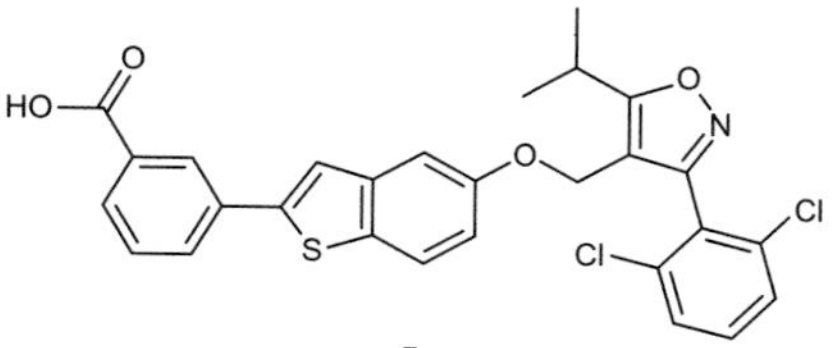

5

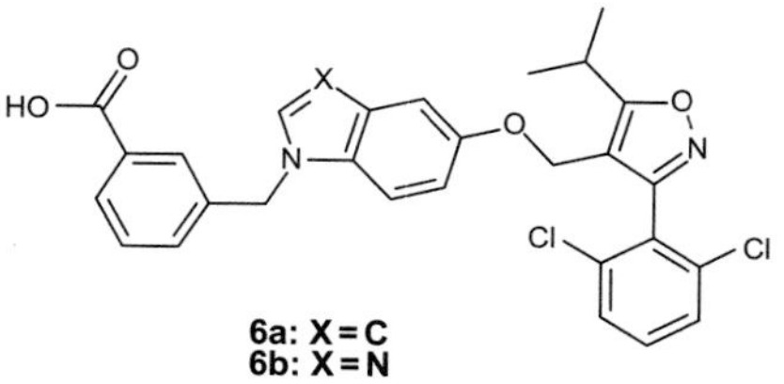

6a: X = C
6b: X = N

Indole **6a** had significantly lower clearance (Cl = 6.7 mL/min/kg); however, the half-life was modest ($t_{1/2}$ = 45 min) and the oral bioavailability was low (F = 12%). It was suggested that poor solubility of **6a** may contribute to its poor oral bioavailability. To improve solubility, a second nitrogen atom was incorporated into the indole ring to give benzimidazole **6b**. Although this compound was significantly less active at FXR (FRX TT EC_{50} = 5 μM, eff = 40%), a twofold improvement in bioavailability was achieved (F = 26%) with little alteration of clearance or half-life suggesting that low solubility of these GW 4064 analogs may limit their absorption.

GW4064 was also the starting point for a study by an independent research group. The co-crystal structure of GW4064 with FXR suggested the potential for favorable hydrogen bond interactions between the isoxazole 3-aryl group and several receptor residues such as Tyr373 and Ser336. Replacing the 2,6-dichlorophenyl with a 2,6-dichloro-4-pyridyl moiety attenuated both FXR binding and functional activity. However, combination of this pyridine moiety with an *N*-methyl indole ring, an optimized stilbene replacement, afforded **7** with good FXR binding affinity (94 nM) in a human scintillation proximity binding assay compared to 64 nM for GW4064 [20].

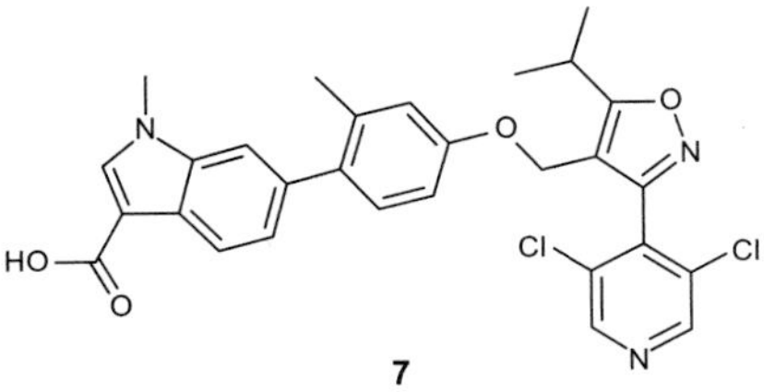

HO O N O N O Cl Cl N **7**

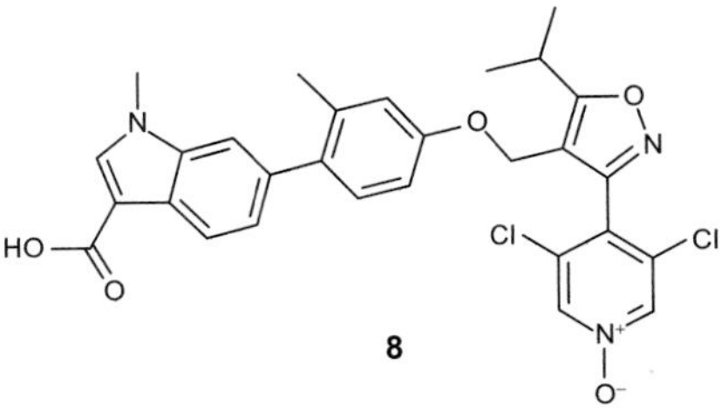

HO O N O N O Cl Cl N^+ O^- **8**

Oxidation of the pyridine to the corresponding *N*-oxide gave **8**, the most potent compound in this series, with a FXR binding affinity of 45 nM. Compared to other analogs reported in this study, **8** had the best permeability (PAMPA, 5.87 × 10^{-4} cm/s). Molecular docking suggested that the *N*-oxide oxygen most likely participates in an H-bond acceptor interaction with Tyr373 on Helix 7 and/or Ser336 on Helix 5 [20].

From a separate screening effort, benzimidazolyl acetamide **9** was discovered as a novel FXR agonist with binding affinity of 70 nM [21,22]. Attempts to improve the physical properties of this compound by replacing the cyclohexyl groups with more polar moieties proved unsuccessful. This result is consistent with the co-crystal structure of compound **9** and hFXR, where the cyclohexyl groups are oriented within highly lipophilic pockets.

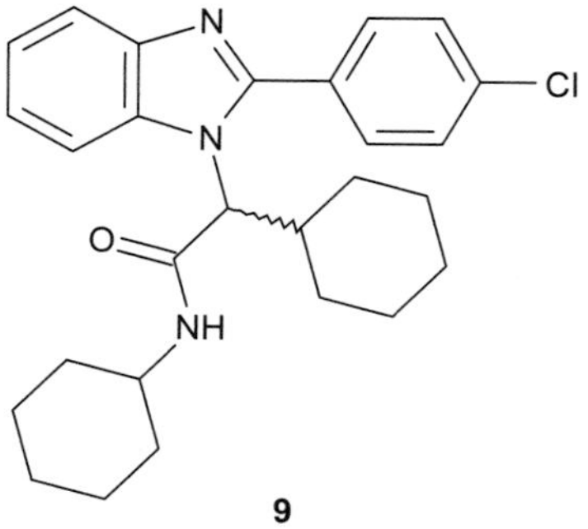

9

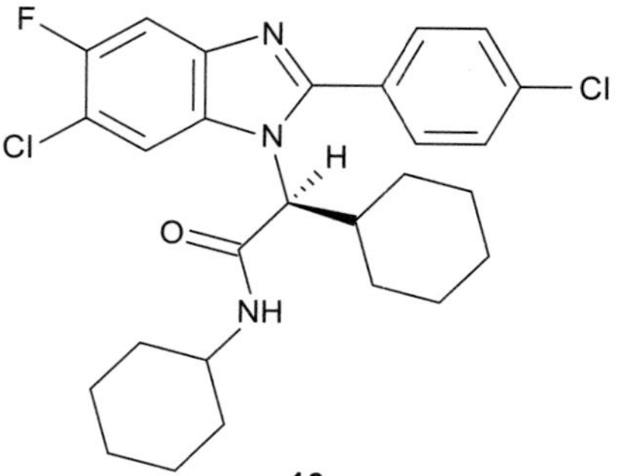

10

The lead molecule from this series, **10** (IC_{50} SPA = 13 nM), was evaluated in $LDLR^{-/-}$ mice. It significantly reduced total cholesterol (45%), LDL (48%), and TG (52%) when orally administered at a dose of 30 mg/kg/d for 5 days [21]. The poor physiochemical properties of **10**, namely high lipophilicity and low aqueous solubility, limited its potential for further development. In addition, this molecule inhibited the hERG potassium channel *in vitro* (IC_{50} = 1.6 μM). Further structural analysis revealed a more polar and yet unexplored pocket consisting of Gln267, Asn297, His298, Arg335, and three water molecules near the region where the *N*-cyclohexyl group binds [23]. Subsequent SAR efforts to replace this cyclohexyl group with a 4-carboxyphenyl ring yielded compound **11** without loss of receptor binding activity (FXR IC_{50} SPA = 50 nM) but with significantly improved solubility (88 vs. <1 μg/mL for **10**) and reduced hERG inhibition (IC_{50} > 20 μM) [23–25].

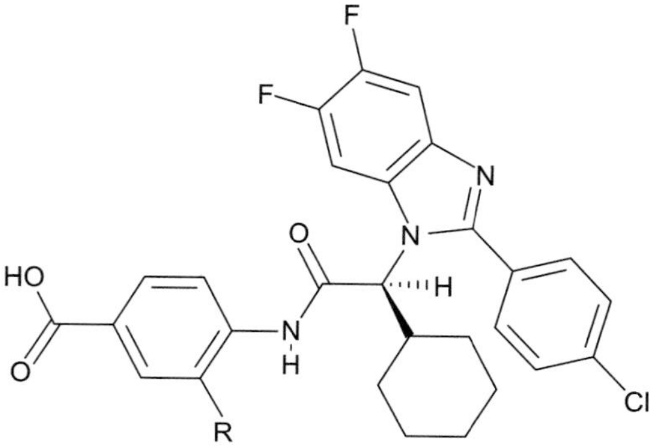

11: R=H
12: R=F

Compound **12**, a fluoro analog of **11**, showed further enhancement of FXR binding (IC_{50} SPA = 37 nM) and solubility (115 μg/mL) with no significant hERG activity (IC_{50} > 20 μM). As a result of its good murine *in vitro* potency ($IC_{50} = 290$ nM, $EC_{50} = 870$ nM, eff = 38%) and PK properties in mice (Cl = 10 mL/min/kg, $F = 33\%$), **12** was evaluated in $LDLR^{-/-}$ mice. After 5 days of treatment (10 mg/kg/d, po), statistically significant decreases in plasma total cholesterol (41%), LDLc (33%), and TG (59%) were observed [23].

There have been numerous recent patent applications disclosing FXR agonists with novel chemical scaffolds. In a 2009 U.S. patent application, hexahydropyrroloazepines exemplified by **13** were claimed as FXR agonists [26]. Using a Gal4/hFXR fusion protein expressed in the HEK293 cell line, compound **13** showed an EC_{50} of 280 nM. A related tetrahydropyrroloazepine series of FXR agonists was disclosed in a separate U.S. patent application [27]. Representative compound **14** showed an EC_{50} of 3 nM in the HEK293 cell assay.

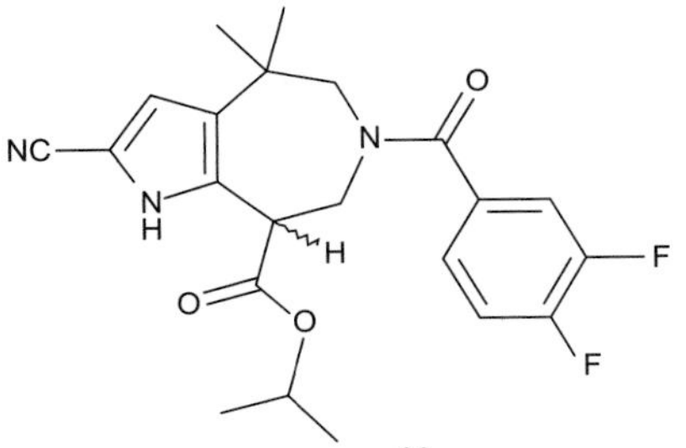

13 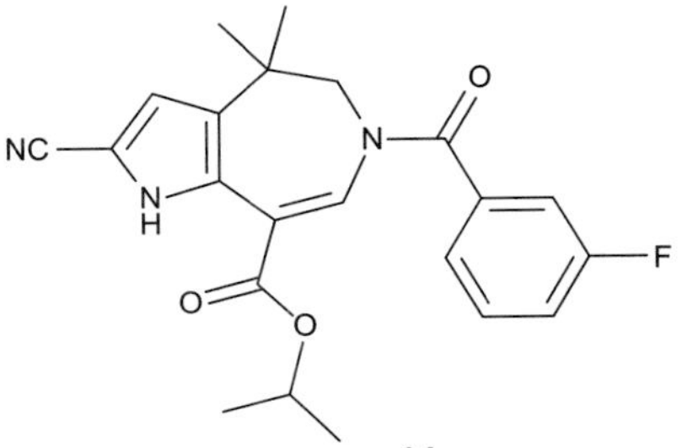 **14**

A class of benzofurane/benzothiophene/benzothiazole derivatives was described as FXR modulators. Representative compound **15** showed an EC_{50} of 1.95 μM in a hFXR transactivation assay in CV-1 cells [28,29]. Novel biaryl carboxylates were described in a 2009 patent application [30]. Compound **16** was exemplified and had an EC_{50} of <100 nM in a hFXR SRC-1 cofactor recruitment assay.

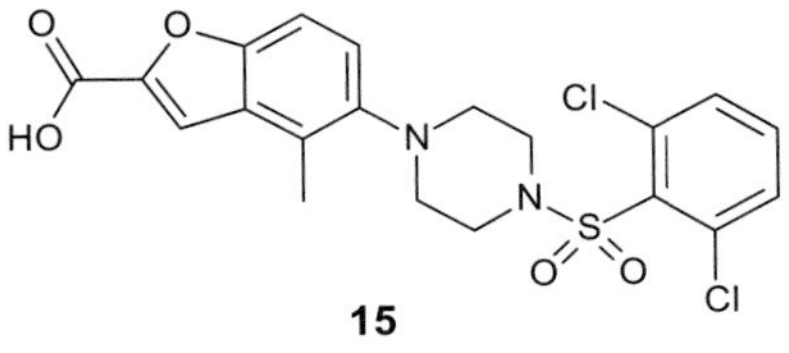

15 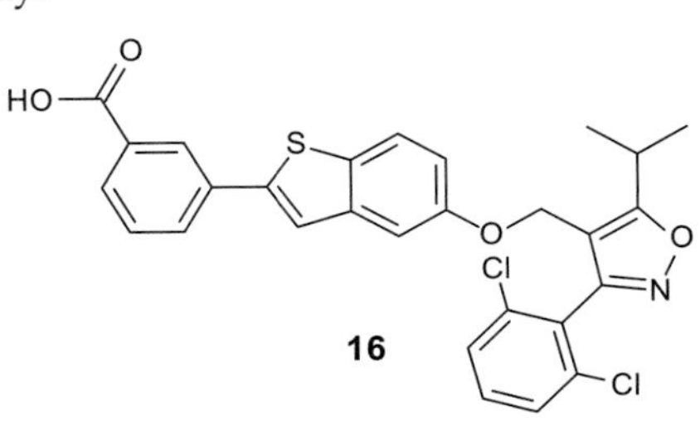 **16**

Another application described benzimidazole analogs represented by **17** which had an IC_{50} of 20 nM in a FXR binding assay [31,32].

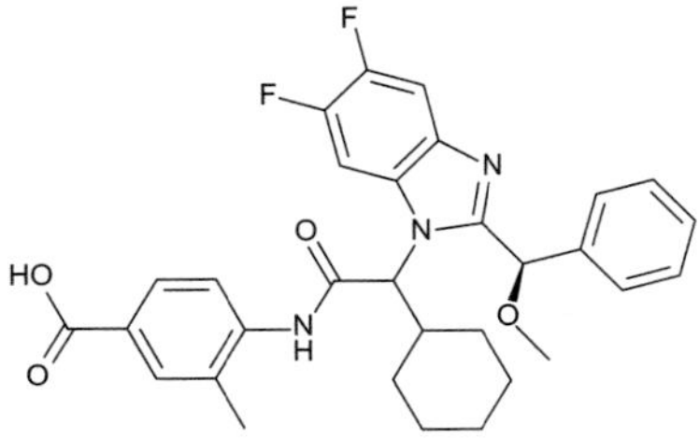

17

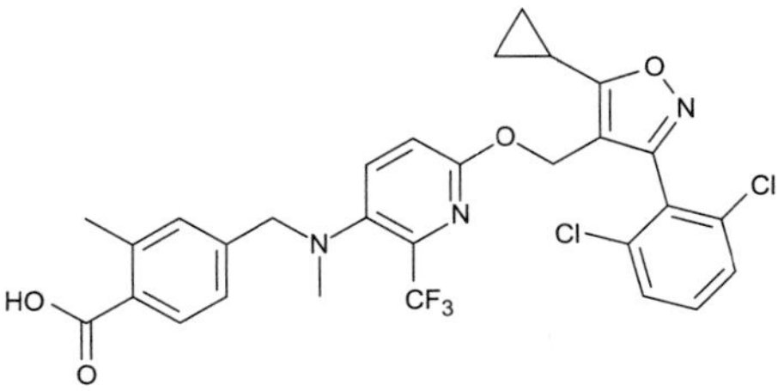

18

A series of benzoic acid analogs were reported to display FXR agonist activity. Compound **18** provided an example from this class and had an EC_{50} of 7.7 nM in a FRET functional assay [33].

3. TGR5 AGONISTS

3.1. Non-BA agonists

A high-throughput screen using a BacMam-transduced human osteosarcoma cell line (U2-OS) led to the discovery of isoxazole **19** as a TGR5 full agonist with a pEC_{50} of 5.3 ($EC_{50} = 5.0\ \mu M$) and 100% response [34]. SAR optimization on the amide phenyl ring, exemplified by compound **20** ($pEC_{50} = 7.5$, $EC_{50} = 32$ nM), suggested that *para*-substitution was preferred [35]. In melanophore cells, compound **20** was equipotent at both human ($pEC_{50} = 7.5$, $EC_{50} = 32$ nM) and canine receptors ($pEC_{50} = 7.2$, $EC_{50} = 63$ nM). In a conscious dog model, intrajejunal injection of glucose (0.125 g/kg) with co-administration of **20** at a dose of 1 mg/kg afforded a significant improvement in hepatic portal vein GLP-1 secretion and reduction in portal vein glucose levels compared to vehicle [34].

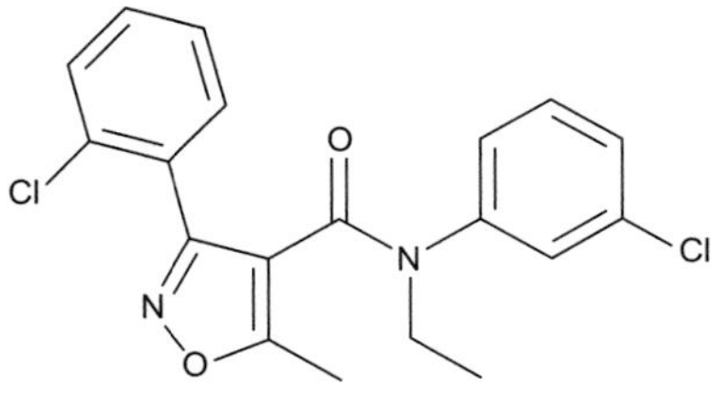

19

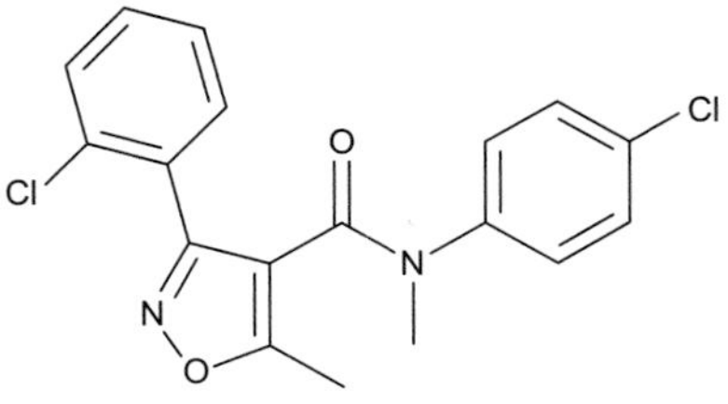

20

Compound **20** showed high *in vivo* clearance (Cl = 85 mL/min/kg) in rats and high intrinsic clearance ($Cl_{int} = 48$ mL/min/kg) in rat liver microsomes suggesting potential challenges to the developability of this series. In addition, **20** had measurable activity against two cytochrome P450 (CYP450) isoforms including 2C19 ($pIC_{50} = 6.5$, $EC_{50} = 0.3\ \mu M$) and

3A4 ($pIC_{50} = 5.9$, $EC_{50} = 1.3$ μM). SAR development at the 5-position of isoxazole revealed that some increased steric bulk was well tolerated, exemplified by **21**. Compound **21** showed improved *in vitro* potency ($pEC_{50} = 8.4$, $EC_{50} = 4$ nM) and reduced *in vitro* clearance in rat ($Cl_{int} = 10$ mL/min/kg). Interestingly, replacement of the isoxazole with 1,2,3-triazole, **22**, afforded further reduction of intrinsic clearance ($Cl_{int} = 6.5$ mL/min/kg) as well as an improved CYP450 profile ($pIC_{50} < 5.7$, $EC_{50} > 2$ μM vs. all isoforms), while maintaining good *in vitro* potency ($pEC_{50} = 7.9$, $EC_{50} = 13$ nM) [35]. However, no *in vivo* pharmacodynamic activity was reported for this compound.

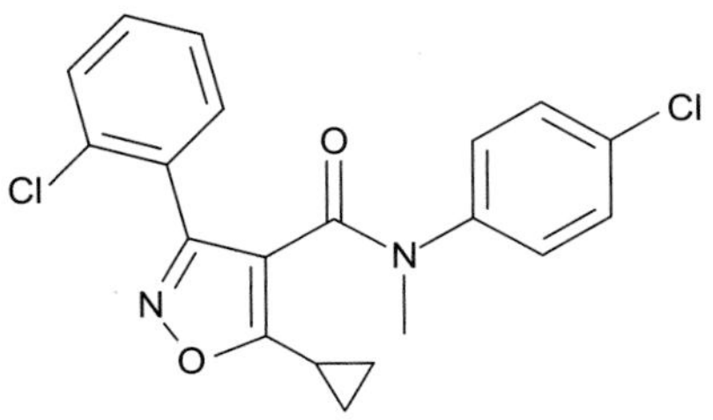

21 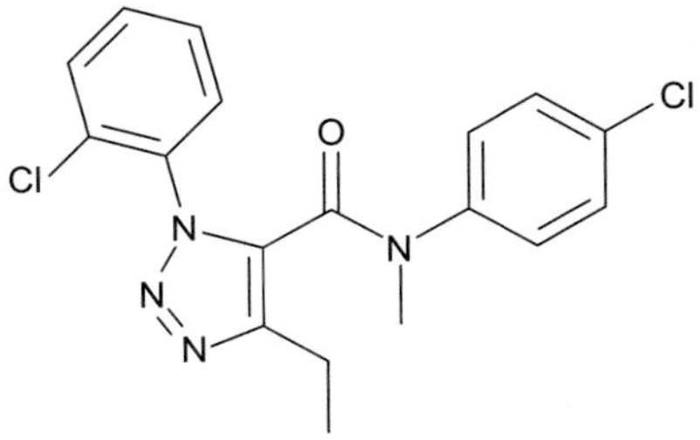 **22**

In a 2007 patent application, a series of bis-phenyl sulfonamides were disclosed as TGR5 agonists [36]. In melanophore cells transfected with human TGR5 (hTGR5), compound **23** showed agonist activity with a pEC_{50} (EC_{50}) between 6.0 (1 μM) and 6.9 (0.1 μM). Given the high TGR5 receptor expression in colon [5], it is hypothesized that maximum pharmacological effect can be achieved by local administration of drug. Thus, anesthetized CD rats were administered with **23** by intracolonic injection at a dose of 2.5 mg/kg which afforded a significant increase of plasma GLP-1 levels (measured by both active and total GLP-1). In a 16-day chronic study, conscious Goto-Kakizaki rats were dosed intracolonically with **23** (0.3 mg/kg QD). On day 16, an intravenous glucose tolerance test was performed. A significant glucose reduction was achieved in treated animals compared to vehicle control group.

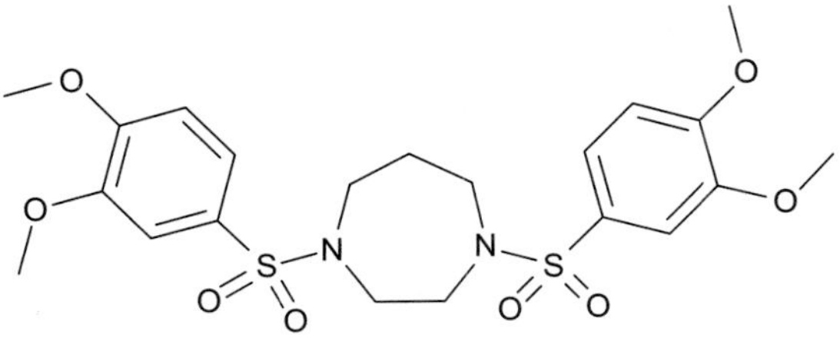

23

A class of quinoline compounds represented by **24** was discovered as TGR5 agonists [37,38]. Compound **24** was first identified as a hit from a high-throughput screen. In HEK293 cells expressing hTGR5, **24** increased cAMP production with an EC_{50} of approximately 10 μM. However, this compound was a significantly less potent agonist of mouse TGR5 ($EC_{50} > 10$ μM). Linker homologation to the phenethylamine and bromine substitution on the phenyl ring led to compound **25** affording a significant improvement in potency at both human ($hEC_{50} = 65$ nM) and mouse ($mEC_{50} = 3.2$ μM) receptors. Phenol deprotection gave analog **26**, which was considerably more active at the mouse receptor ($mEC_{50} = 0.28$ μM) but much less active at the human receptor ($hEC_{50} = 5.1$ μM). To test the hypothesis that activation of TGR5 stimulates GLP-1 release, **26** (30 mg/kg) was orally administered to diet-induced obese (DIO) mice. Following an oral glucose challenge, a statistically significant increase in plasma active GLP-1 levels was observed. Acute administration of **26** (30 mg/kg po) to DIO mice prior to an oral glucose tolerance test also resulted in a significant reduction of plasma glucose area under curve (AUC) [37]. In a separate study, treatment of C57BL/6 DIO mice with **26** for 2 weeks at doses of 3, 30, and 100 mg/kg bid reduced fasting glucose, post-prandial TG, and high-density lipoprotein (HDL) levels [38].

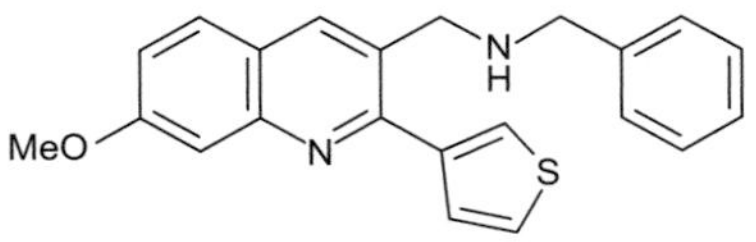

24

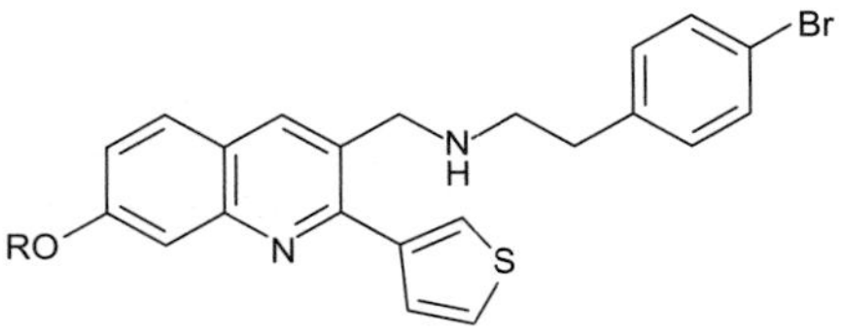

25: R = Me
26: R = H

Two separate publications described different types of arylpyridines as TGR5 agonists [39–40]. In HEK293 cells expressing hTGR5, compounds **27-29** increased cAMP production with EC_{50} values of less than 10 μM.

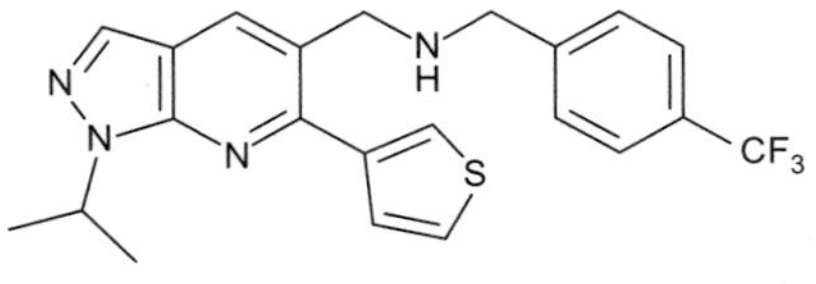

27

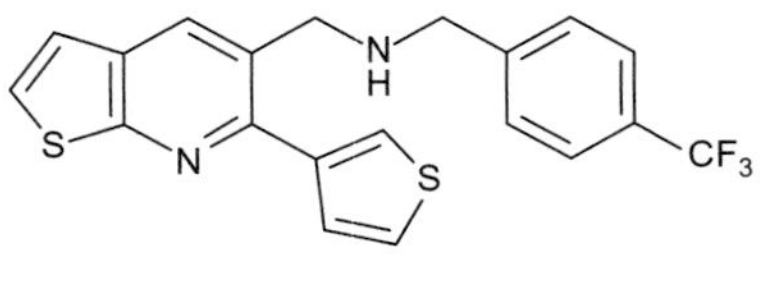

28

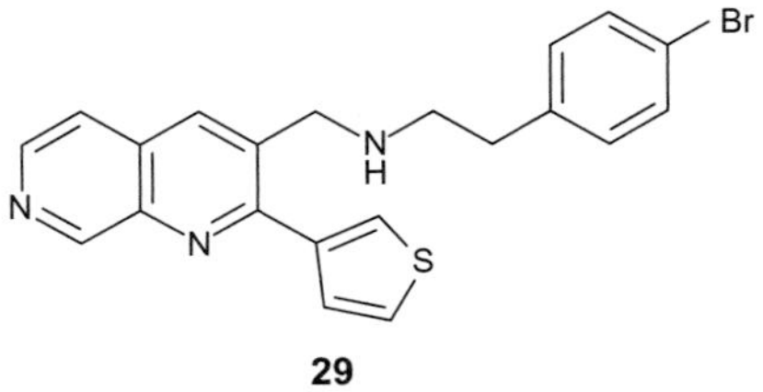

29

Recently, two patent applications were published describing different classes of pteridinone derivatives as potent TGR5 agonists [41,42]. Representative compounds **30** and **31** stimulated cAMP production with EC_{50} values of less than 10 μM in HEK293 cells expressing hTGR5.

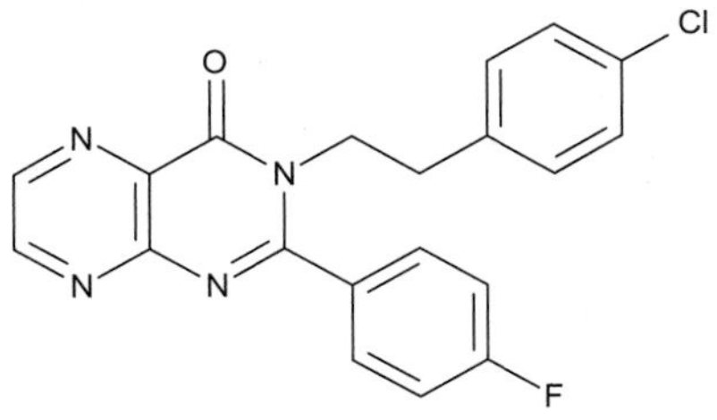

30 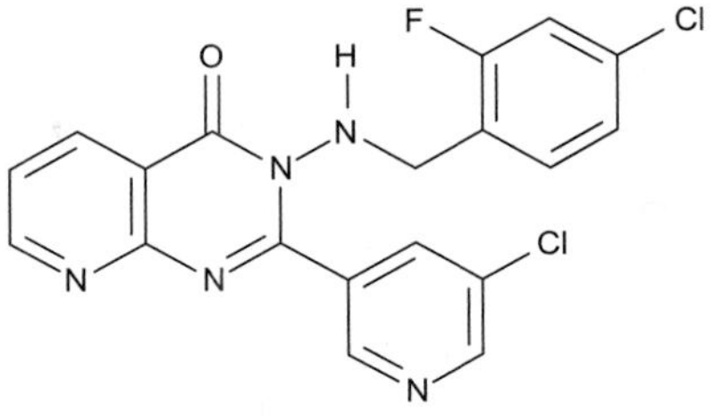 **31**

Compound **32** was reported to represent a novel class of quinazolinone TGR5 agonists [43]. In HEK293 cells expressing hTGR5, **32** stimulated cAMP production with an EC_{50} value of less than 10 μM. In an oral glucose tolerance test, mice that were dosed orally with 30 mg/kg of **32** demonstrated a 52% reduction in glucose AUC compared to a vehicle control group. This glucose reduction was accompanied by increased insulin (130%) and active GLP-1 (70%) levels compared to vehicle-treated animals [43].

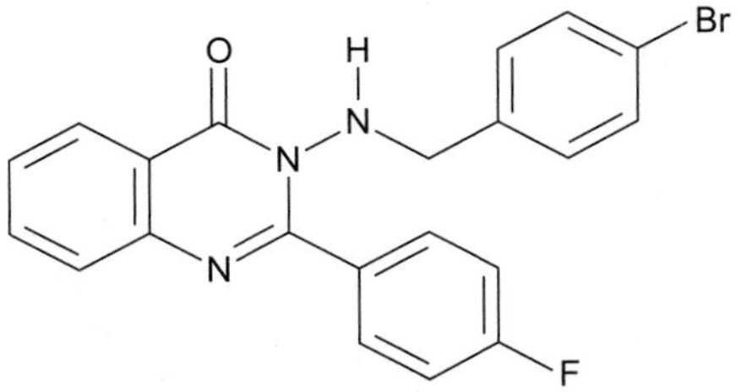

32 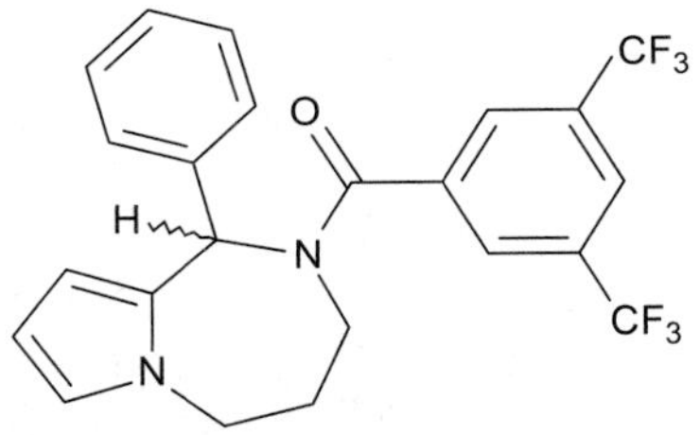 **33**

In a 2007 patent application, a family of diazepine derivatives was claimed as TGR5 agonists [44]. Representing this family, compound **33** was shown to stimulate cAMP secretion in a HEK293 cell line expressing hTGR5 with an EC_{50} value of less than 10 μM.

Related oxazepine compounds were also reported to be TGR5 agonists [45,46]. In CHO cells expressing hTGR5, compound **34** (1 μM) stimulated cAMP production by 100%. In NCI-H716 cells, an increase in cAMP production of 100% was observed in the presence of 10 μM of the *trans* isomer of compound **35**. In the same cell line, compound **35** (5 μM) was reported to increase GLP-1 secretion by 157%.

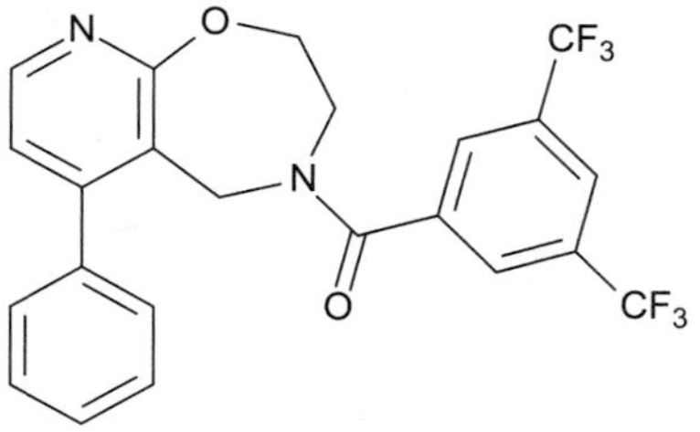

34 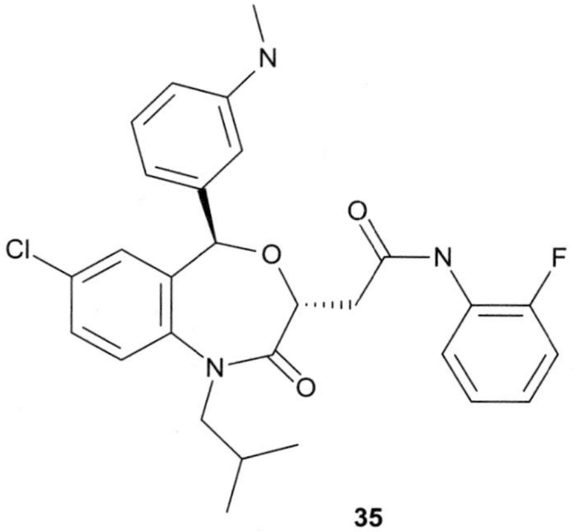 **35**

In a separate patent application [47], oxazepinone **36** (30 μM) was reported to stimulate GLP-1 secretion by 249% in NCI-H716 cells and by 360% in a rat bowel primary culture cell. To evaluate *in vivo* GLP-1 secretion and glucose-dependent insulin secretion, **36** was administered orally to male F344 rats at doses of 30 and 100 mg/kg. Animals treated with 100 mg/kg of **36** showed significant GLP-1 and insulin secretion after a glucose challenge compared to the vehicle group.

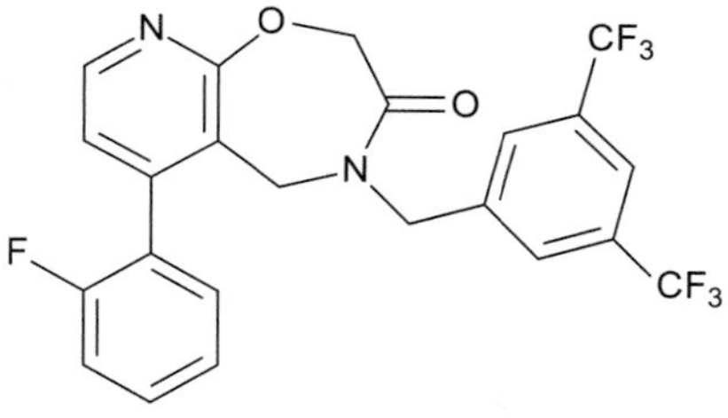

36 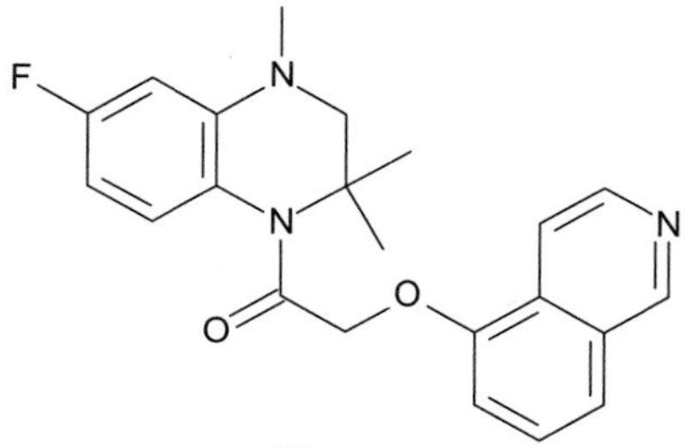 **37**

A set of heteroaryl acetamide derivatives was claimed as TGR5 agonists [48]. Among these, dihydroquinoxaline **37** (30 μM) elicited 251% GLP-1 secretion in rat bowel primary culture cells. In a rat intestine perfusion model, **37** (10 μM) gave a significant increase of portal vein GLP-1 concentration.

Recently, a series of aryl amides was discovered to have TGR5 agonist activity [49]. In CHO cells expressing hTGR5, compound **38** increased cAMP production with an EC_{50} of 7 nM.

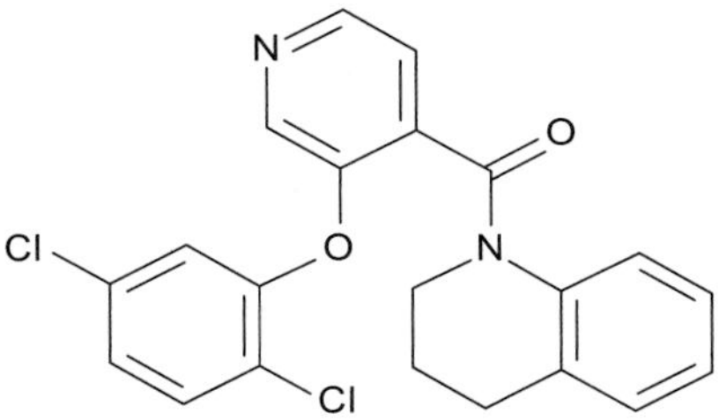

38 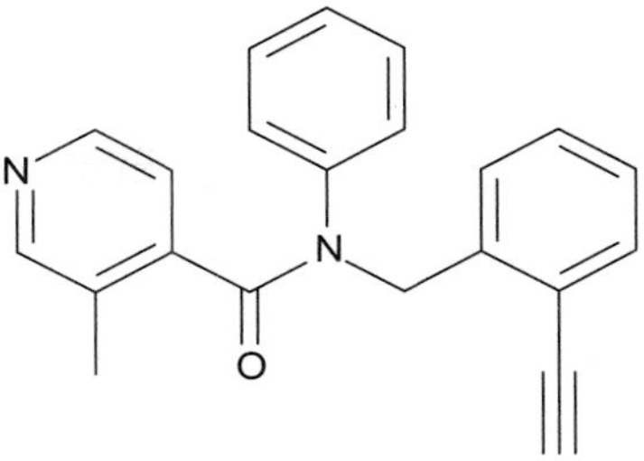 **39**

In a 2007 patent application, heterocyclic amides exemplified by compound **39** were described as TGR5 agonists [50]. No biological data were reported.

Another patent application from the same group disclosed a series of pyridazine/pyridine/pyran derivatives as TGR5 agonists [51]. Compounds **40** and **41** are shown as representative structures. However, no biological data were presented.

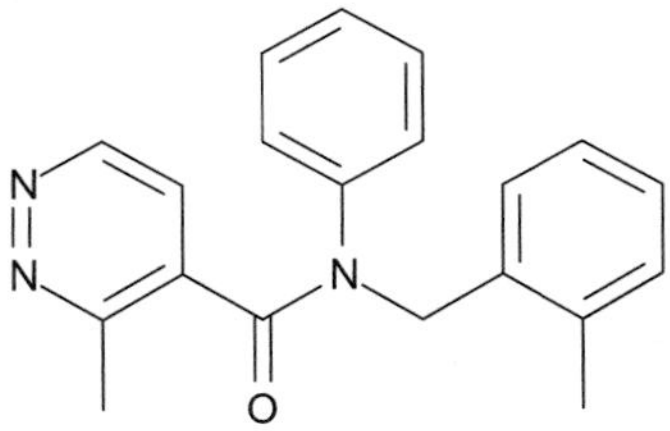

40 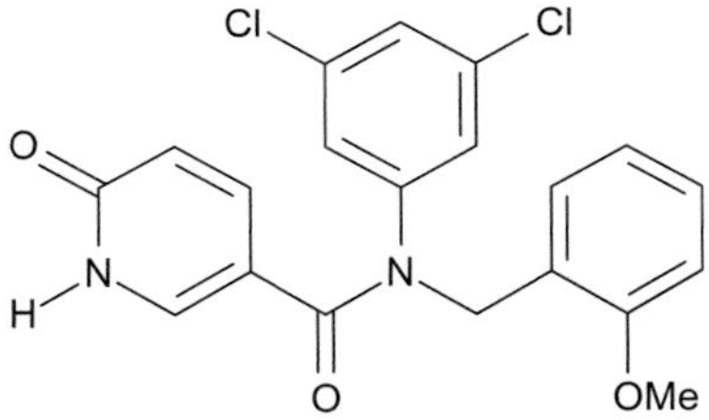 **41**

A recent patent application claimed a series of imidazole and triazole compounds exemplified by **42** and **43** [52]. In a hTGR5/CRE-luciferase assay, both compounds showed receptor activation with EC_{50} values of less than 100 nM. In mouse STC-1 cells under high glucose conditions, **42** effectively stimulated GLP-1 secretion with an EC_{50} of 17 nM. *In vivo*, a twofold increase in GLP-1 secretion was achieved when fasted C57BL/6 mice were treated with **42** at an oral dose of 30 mg/kg.

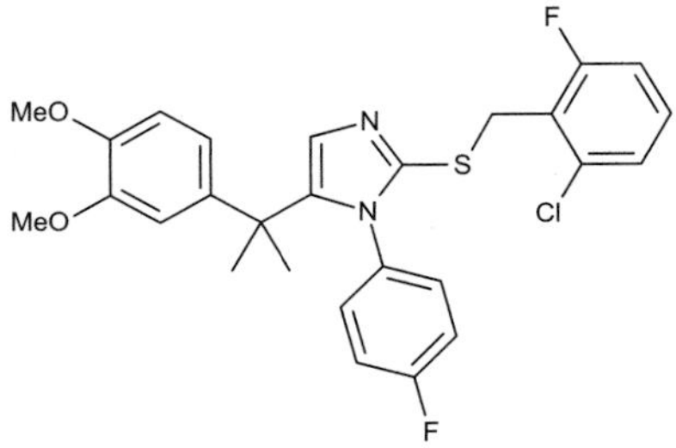

42 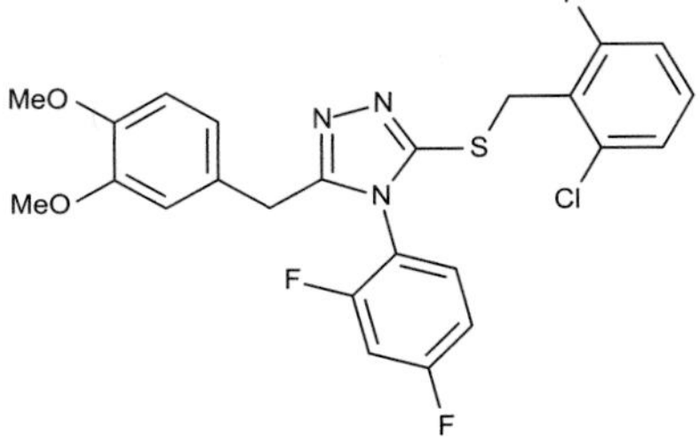 **43**

In a 2010 patent application, a class of isoquinolines was claimed to be TGR5 agonists [53]. Representative compound **44** stimulated cAMP production in HEK293 cells expressing hTGR5 with an EC_{50} of 7.37 μM. Another patent application from the same group claimed a series of isoquinolinyloxymethyl heteroaryl analogs exemplified by **45** [54]. In the same HEK293 cellular assay, compound **45** stimulated cAMP production with an EC_{50} of 229 nM.

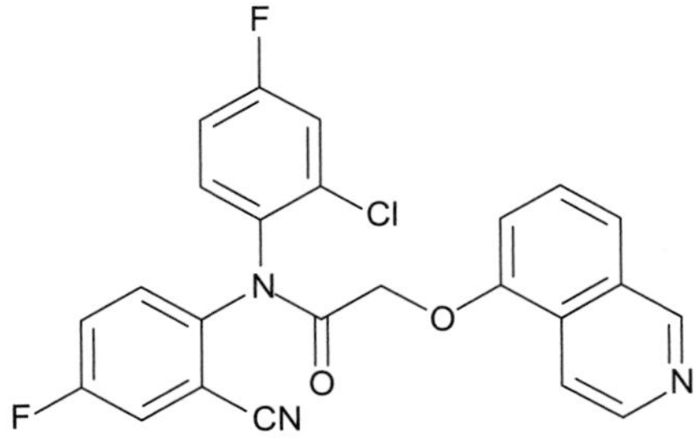

44

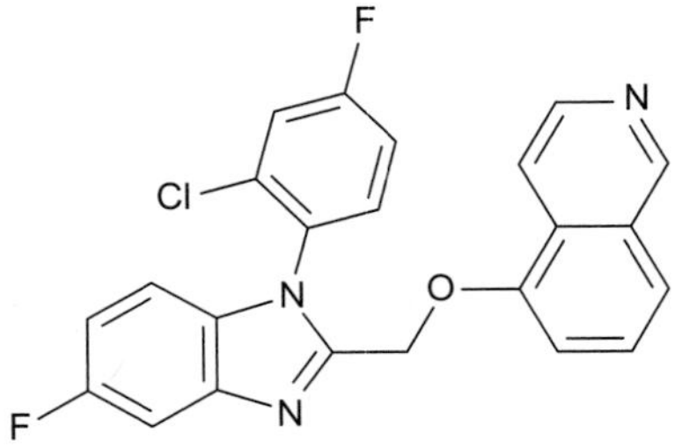

45

3.2. BA derivatives

In 2009, a semisynthetic cholic acid (CA) derivative, 6α-ethyl-23(*S*)-methyl-CA (EMCA, INT-777, **46**), was reported to be a selective TGR5 agonist [55]. Initial SAR studies unveiled that the incorporation of a methyl group at the C-23 position of CA side chain afforded the selective, albeit not very potent, TGR5 agonist **47** which had an EC_{50} of 3.58 μM (FXR $EC_{50} > 100$ μM). The *S*-configuration at C23 was critical for TGR5 potency.

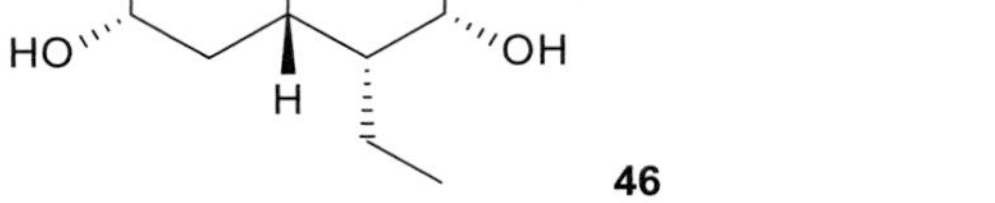

46

Improvement in TGR5 *in vitro* potency was noted by introduction of a small alkyl substituent at the C-6 position as in compound **48**. Unfortunately, this compound suffered from poor physical properties, namely low solubility and high albumin binding. Structurally, CA (**49**) differs from chenodeoxycholic acid (CDCA, **50**) at C-12 by having an additional α-hydroxyl group oriented on the polar side of the molecule.

This "minor" structural change accounts for the markedly different solubilities.

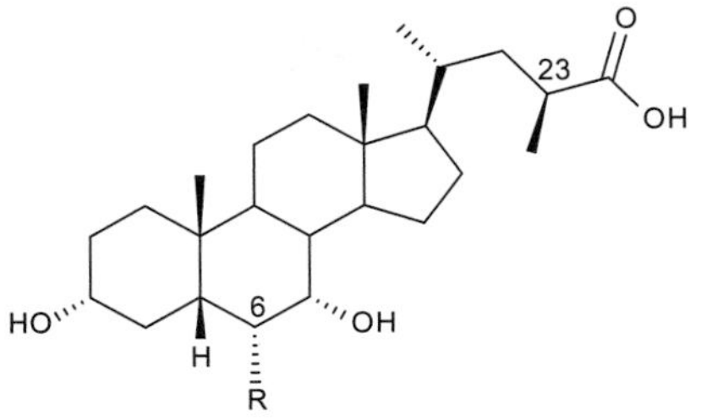

	TGR5 EC_{50} (efficacy %)	FXR EC_{50} (efficacy %)
47: R = H	3.58 μM (110%)	>100 μM (0%)
48: R = Et	0.095 μM (102%)	11.8 μM (73%)

Moreover, **49** is devoid of FXR activity ($EC_{50} > 100$ μM) while maintaining good TGR5 activity ($EC_{50} = 13.6$ μM). Introduction of the C-12 α-hydroxyl group into compound **48** afforded compound **46**, which showed potent TGR5 activity ($EC_{50} = 0.8$ μM, 166%) and excellent selectivity over FXR ($EC_{50} > 100$ μM). Compound **46** appeared to be stable to human stool broth culture, with more than 95% of compound unmodified after incubation for 12 h. It was believed that the 6α-ethyl group provided steric hindrance to the bacterial 7α-dehydroxylation process. This compound was resistant to conjugation since more than 90% of **46** was secreted into the bile in its parent form. The 23α-methyl group was thought to prevent carboxyl CoA activation and subsequent conjugation, thereby favoring the cholehepatic shunt pathway with ductular absorption and a potent choleretic effect.

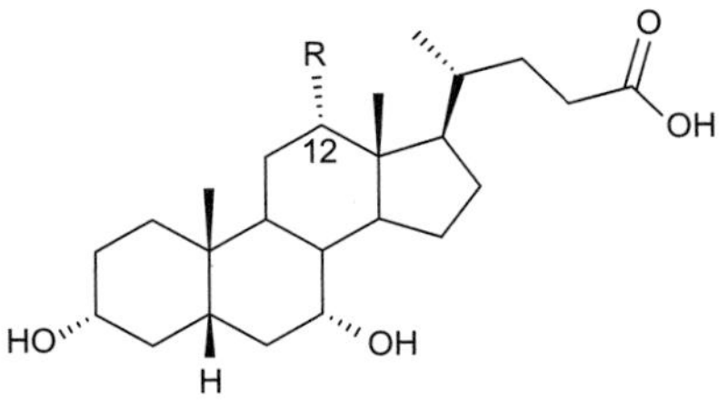

	Water solubility in 0.1 HCl	Albumin binding (%)
49: R = OH	270 μM	93
50: R = H	30 μM	54

Treatment of DIO C57BL/6 mice with compound **46** for 10 weeks at a dose of 30 mg/kg/d admixed with diet led to a significant increase in energy expenditure as determined by increases in O_2 consumption, CO_2 production, and respiratory quotient. In addition, liver function was improved as evidenced by reduction in liver steatosis. Significant reductions in plasma TG and nonesterified fatty acids were also observed. Treating these mice for 3 weeks at 30 mg/kg/d of **46** admixed with diet significantly improved glucose tolerance and insulin sensitivity [56].

4. FXR/TGR5 DUAL AGONISTS

In 2010, a FXR/TGR5 dual agonist, **51** (INT-767), was reported [57]. Using an AlphaScreen coactivator recruitment assay, the potency of **51** at FXR was 30 nM. In NCI-H716 cells, **51** stimulated intracellular cAMP secretion with an EC_{50} of 0.63 μM. Its TGR5 potency was comparable to that of the selective TGR5 agonist **46** (EC_{50} = 0.8 μM). Compound **51** also induced a dose-dependent increase of GLP-1 secretion from NCI-H716 cells.

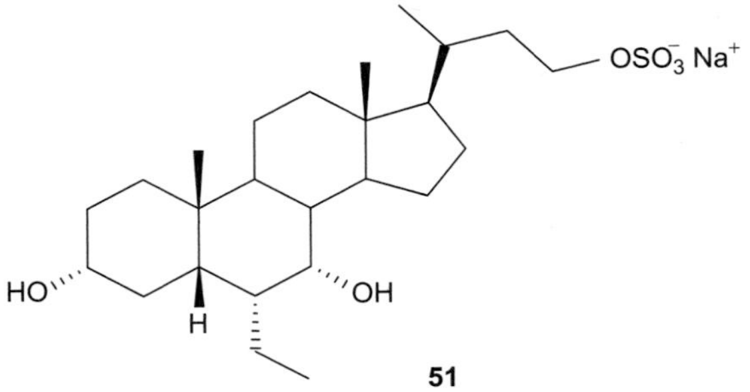

51

In DBA/2J mice, a streptozotocin-induced type 1 diabetes model, plasma cholesterol levels were significantly higher in mice fed a western diet (WD) compared with standard chow. A 3-week treatment of these mice with **51** admixed at doses of 10 or 30 mg/kg/day in the WD resulted in a significant dose-dependent decrease of plasma total cholesterol levels and a significant decrease of TG levels only at the 30 mg/kg/d dose. The marked inhibition of total cholesterol induced by compound **51** treatment was correlated with normalization of LDL cholesterol levels; HDL cholesterol levels were not affected. In db/db mice, a model of type 2 diabetes, intraperitoneal administration of **51** for 2 weeks at doses of 10 and 30 mg/kg/day significantly and dose-dependently decreased plasma total cholesterol and TG levels.

5. CLINICAL STUDIES AND OUTLOOK

To date, only limited number of BA receptor agonists have been studied in humans. In 2009, a phase II trial result was reported on a FXR agonist, INT-747, in type 2 diabetes patients with comorbid fatty acid disease [58]. From this double-blind placebo-controlled study of 64 patients, INT-747 therapy (25 and 50 mg for 6 weeks) significantly improved insulin sensitivity, induced weight loss, and reduced liver damage. In 2010, a phase II study of a TGR5 agonist, SB-756050, for treatment of type 2 diabetes was completed [59]. Further development of this compound was discontinued after the highest dose failed to meet the predetermined efficacy threshold.

With the limited clinical data available, the full therapeutic potential of molecules that modulate BA receptors has yet to be realized. Based on the prolific patent literature around these targets, it is likely that additional molecules will advance into the clinic to test their therapeutic potential over the next few years.

REFERENCES

[1] H. Wang, J. Chen, K. Hollister, L. C. Sowers and B. M. Forman, *Cell*, 1999, **3**, 543.
[2] D. J. Parks, S. G. Blanchard, R. K. Bledsoe, G. Chandra, T. G. Consler, S. A. Kliewer, J.B. Stimmel, T. M. Wilson, A. M. Zavacki, D. D. Moore and J. M. Lehmann, *Science*, 1999, **284**, 1365.
[3] M. Makishima, A. Y. Okamoto, J. J. Repa, H. Tu, R. M. Learned, A. Luk, M. V. Hull, K. D. Lusting, D. J. Mangelsdorf and B. Shan, *Science*, 1999, **284**, 1362.
[4] T. Maruyama, Y. Miyamoto, T. Nakamura, Y. Tamai, H. Okada, E. Sugiyama, T. Nakamura, H. Itadani and L. Tanaka, *Biochem. Biophys. Res. Commun.*, 2002, **298**, 714.
[5] Y. Kawamata, R. Fujii, M. Hosoya, M. Harada, H. Yoshida, M. Miwa, S. Fukusumi, Y. Habata, T. Itoh, Y. Shintani, S. Hinuma, Y. Fujisawa and M. Fujino, *J. Biol. Chem.*, 2003, **278**, 9435.
[6] P. Lefebvre, B. Cariou, F. Lien, F. Kuipers and B. Steals, *Physiol. Rev.*, 2009, **89**, 147.
[7] B. Cariou, K. van Harmelen, D. D. Sandoval, T. H. van Dijk, A. Grefhorst, M. Abdelkarim, S. Caron, G. Torpier, J.-C. Fruchart, F. J. Gonzalez, F. Kuipers and B. Staels, *J. Biol. Chem.*, 2006, **281**, 11039.
[8] K. Ma, P. K. Saha, L. Chan and D. D. Moore, *J. Clin. Invest.*, 2006, **116**, 1102.
[9] G. Rizzo, M. Disante, A. Mencarelli, B. Renga, A. Gioiello, R. Pellicciari and S. Fiorucci, *Mol. Pharmacol.*, 2006, **70**, 1162.
[10] X. Wang, T. Jiang and M. Levi, *Nat. Rev. Nephrol.*, 2010, **6**, 342.
[11] V. Keitel, K. Cupisti, C. Ullmer, T. K. Wolfram, R. Kubitz and D. Haussinger, *Hepatology*, 2009, **50**, 861.
[12] S. Katsuma, A. Hirasawa and G. Tsujimoto, *Biochem. Biophys. Res. Commun.*, 2005, **329**, 386.
[13] J. A. Lovshin and D. J. Drucker, *Nat. Rev. Endocrinol.*, 2009, **5**, 262.
[14] H. Sato, C. Genet, A. Strehle, C. Thomas, A. Lobstein, A. Wagner, C. Mioskowski, J. Auwerx and R. Saladin, *Biochem. Biophys. Res. Commun.*, 2007, **362**, 793.
[15] M. Watanabe, S. M. Houten, C. Mataki, M. A. Christoffolete, B. M. Kim, H. Sato, N. Messaddeq, J. W. Harney, O. Ezaki, T. Kodama, K. Schoonjans, A. C. Bianco and J. Auwerx, *Nature*, 2006, **439**, 484.
[16] B. Flatt, R. Martin, T.-L. Wang, P. Mahaney, B. Murphy, X.-H. Gu, P. Foster, J. Li, P. Pircher, M. Petrowski, I. Schulman, S. Westin, J. Wrobel, G. Yan, E. Bischoff, C. Daige and R. Mohan, *J. Med. Chem.*, 2009, **52**, 904.
[17] J. T. Lundquist IV, D. C. Harnish, C. Y. Kim, J. F. Mehlmann, R. J. Unwalla, K. M. Phipps, M. L. Crawley, T. Commons, D. M. Green, W. Xu, W.-T. Hum, J. E. Eta, I. Feingold, V. Patel, M. J. Evans, K. Lai, L. Borges-Marcucci, P. Mahaney and J. E. Wrobel, *J. Med. Chem.*, 2010, **53**, 1774.
[18] J. Y. Bass, R. D. Caldwell, J. A. Caravella, L. Chen, K. L. Creech, D. N. Deaton, K. P. Madauss, H. B. Marr, R. B. McFadyen, A. B. Miller, D. J. Parks, D. Todd, S. P. Williams and G. B. Wisely, *Bioorg. Med. Chem. Lett.*, 2009, **19**, 2969.
[19] A. Akwabi-Ameyaw, J. Y. Bass, R. D. Caldwell, J. A. Caravella, L. Chen, K. L. Creech, D.N. Deaton, K. P. Madauss, H. B. Marr, R. B. McFadyen, A. B. Miller, F. Navas III,

D.J. Parks, P. K. Spearing, D. Todd, S. P. Williams and G. B. Wisely, *Bioorg. Med. Chem. Lett.*, 2009, **19**, 4733.
[20] S. Feng, M. Yang, Z. Zhang, Z. Wang, D. Hong, H. Richter, G. M. Benxon, K. Bleicher, U. Grether, R. E. Martin, J.-M. Plancher, B. Kuhn, M. G. Rudolph and L. Chen, *Bioorg. Med. Chem. Lett.*, 2009, **19**, 2595.
[21] H. G. F. Richter, G. M. Benson, D. Blum, E. Chaput, S. Feng, C. Gardes, U. Grether, P. Hartman, B. Kuhn, R. E. Martin, J.-M. Plancher, M. G. Rudolph, F. Schuler, S. Taylor and K. H. Bleicher, *Bioorg. Med. Chem. Lett.*, 2011, **21**, 191.
[22] G. M. Benson, K. Bleicher, U. Grether, R. E. Martin, J.-M. Plancher, H. Richter, S. Taylor and M. Yang, *Patent Application WO2009/027264-A1*, 2009.
[23] H. G. F. Richter, G. M. Benson, K. H. Bleicher, D. Blum, E. Chaput, N. Clemann, S. Feng, C. Gardes, U. Grether, P. Hartman, B. Kuhn, R. E. Martin, J.-M. Plancher, M. G. Rudolph, F. Schuler and S. Taylor, *Bioorg. Med. Chem. Lett.*, 2011, **21**, 1134.
[24] G. M. Benson, K. Bleicher, U. Grether, R. E. Martin, J.-M. Plancher, H. Richter, S. Taylor and M. Yang, *Patent Application WO2009/062874-A2*, 2009.
[25] G. M. Benson, K. Bleicher, U. Grether, R. E. Martin, J.-M. Plancher, H. Richter, S. Taylor and M. Yang, *US Patent 0,163,552*, 2010.
[26] J. F. Mehlmann, J. T. Lundquist, P. E. Mahaney, M. L. Crawley and C. Y. Kim, *US Patent 0,131,409*, 2009.
[27] J. F. Mehlmann, J. T. Lundquist, P. E. Mahaney, M. L. Crawley and C. Y. Kim, *US Patent 0,137,554*, 2009.
[28] D. Roche, G. Mautino, I. Kober, F. Contard, S. Christmann-Franck, S. Sengupta, R. Sistla and G. Venkateshwar Rao, *Patent Application WO2009/127321-A1*, 2009.
[29] D. Roche, G. Mautino, F. Contard, S. Christmann-Franck, I. Kober, S. Sengupta, R. Sistla and G. Venkateshwar Rao, *EP Patent 2,110,374*, 2009.
[30] A. Akwabi-Ameyaw, D. N. Deaton, R. B. McFadyen and F. Navas III, *Patent Application WO2009/005998-A1*, 2009.
[31] G. M. Benson, K. Bleicher, S. Feng, U. Grether, B. Kuhn, R. E. Martin, J.-M. Plancher, H. Richter and S. Taylor, *Patent Application WO2010/043513-A1*, 2010.
[32] G. M. Benson, K. Bleicher, S. Feng, U. Grether, B. Kuhn, R. E. Martin, J.-M. Plancher, H. Richter and S. Taylor, *US Patent 0,093,818*, 2010.
[33] U. Abel and C. Kremoser, *Patent Application WO2009/149795-A2*, 2009.
[34] K. A. Evans, B. W. Budzik, S. A. Ross, D. D. Wisnoski, J. Jin, R. A. Rivero, M. Vimal, G. R. Szewczyk, C. Jayawickreme, D. L. Moncol, T. J. Rimele, S. L. Armour, S. P. Weaver, R. J. Griffin, S. M. Tadepalli, M. R. Jeune, T. W. Shearer, Z. B. Chen, L. Chen, D. L. Anderson, J. D. Becherer, M. D. L. Frailes and F. J. Colilla, *J. Med. Chem.*, 2009, **52**, 7962.
[35] B. W. Budzik, K. A. Evans, D. D. Wisnoski, J. Jin, R. A. Rivero, G. R. Szewczyk, C. Jayawickreme, D. L. Moncol and H. Yu, *Bioorg. Med. Chem. Lett.*, 2010, **20**, 1363.
[36] J. R. Szewczyk, C. P. Laudeman, K. A. Evans, Y. H. Li, S. T. Dock and Z. Chen, *Patent Application WO2007/127505-A2*, 2007.
[37] M. R. Herbert, D. L. Siegel, L. Staszewski, C. Cayanan, U. Banerjee, S. Dhamija, J. Anderson, A. Fan, L. Wang, P. Rix, A. K. Shiau, T. S. Rao, S. A. Noble, R. A. Heyman, E. Bischoff, M. Guha, A. Kabakibi and A. B. Pinkerton, *Bioorg. Med. Chem. Lett.*, 2010, **20**, 5718.
[38] A. B. Pinkerton, A. Kabakibi, M. R. Herbert and D. L. Seigel, *Patent Application WO2008/097976-A1*, 2008.
[39] M. R. Herbert, A. B. Pinkerton and D. L. Seigel, *Patent Application WO2010/016846-A1*, 2010.
[40] M. R. Herbert, A. B. Pinkerton and D. L. Seigel, *US Patent 0,054,304*, 2009.
[41] N. D. Smith, J. E. Payne and T. Z. Hoffman, *Patent Application WO2010/014739-A2*, 2010.

[42] N. D. Smith, J. E. Payne, T. Z. Hoffman, C. Bonnefous, A. B. Pinkerton and D. L. Seigel, *Patent Application WO2009/026241-A1*, 2009.
[43] A. B. Pinkerton, A. Kabakibi, T. Z. Hoffman, D. L. Seigel and S. A. Noble, *Patent Application WO2008/067219-A2*, 2008.
[44] A. B. Pinkerton, A. Kabakibi and T. C. Gahman, *Patent Application WO2008/067222-A1*, 2008.
[45] F. Itoh, S. Hinuma, N. Kanzaki, T. Miki, Y. Kawamata, S. Oi, T. Tawaraishi, Y. Ishichi and M. Hirohashi, *Patent Application WO2004/067008-A1*, 2004.
[46] F. Itoh, S. Hinuma, N. Kanzaki, T. Miki, Y. Kawamata, S. Oi, T. Tawaraishi, Y. Ishichi and M. Hirohashi, *US Patent 0,199,795*, 2006.
[47] F. Itoh, T. Tawaraishi and M. Hirohashi, *Patent Application JP2006/056881-A*, 2006.
[48] F. Itoh and H. Nicolas, *Patent Application JP2006/063064-A*, 2006.
[49] C. Bissantz, H. Dehmlow, R. E. Martin, U. O. Sander, H. Richter and C. Ullmer, *US Patent 0,105,906*, 2010.
[50] L. Arista, K. Hogenauer, N. Schmiedeberg, G. Werner and H. Jaksche, *Patent Application WO2007/110237-A2*, 2007.
[51] L. Arista, *Patent Application WO2008/125627-A1*, 2008.
[52] V. Bollu, B. C. Boren, J. E. Dalgard, B. T. Flatt, S. Hudson, R. Mohan, M. Morrissey, B. Pratt and T.-L. Wang, *Patent Application WO2010/093845-A1*, 2010.
[53] Y. Sugimoto, A. Satoh and T. Nishimura, *Patent Application WO2010/117084-A1*, 2010.
[54] K. Arakawa, Y. Sugimoto, Y. Sasaki, A. Satoh and T. Nishimura, *Patent Application WO2010/117090-A1*, 2010.
[55] R. Pellicciari, A. Gioiello, A. Macchiarulo, C. Thomas, E. Rosatelli, B. Natalini, R. Sardella, M. Pruzanski, A. Roda, E. Pastorini, K. Schoonjans and J. Auwerx, *J. Med. Chem.*, 2009, **52**, 7958.
[56] C. Thomas, A. Gioello, L. Noriega, A. Strehle, J. Oury, G. Rizzo, A. Macchiarulo, H. Yamamoto, C. Mataki, M. Pruzanski, R. Pellicciari, J. Auwerx and K. Schoonjans, *Cell Metab.*, 2009, **10**, 167.
[57] G. Rizzo, D. Passeri, F. De Franco, G. Ciaccioli, L. Donadio, G. Rizzo, S. Orlandi, B. Sadeghpour, X. X. Wang, T. Jiang, M. Levi, M. Pruzanski and L. Adorini, *Mol. Pharmacol.*, 2010, **78**, 617.
[58] Intercept Pharmaceuticals website: http://www.interceptpharma.com/ct_development.php (accessed May 11, 2011).
[59] According to Scrip's Pipeline Watch, April 2, 2010: http://www.scripintelligence.com/multimedia/archive/00090/Scrip_3490_Pipeline__90699a.pdf (accessed May 11, 2011).

CHAPTER 6

Recent Advances in Mineralocorticoid Receptor Antagonists

Katerina Leftheris, Yajun Zheng and **Deepak S. Lala**

Contents

1. INTRODUCTION

The mineralocorticoid receptor (MR) is a member of the nuclear hormone receptor (NHR) superfamily and is structurally related to the progesterone receptor (PR), androgen receptor (AR), estrogen receptor (ER),

Vitae Pharmaceuticals, 502 West Office Center Drive, Fort Washington, PA 19034, USA

Annual Reports in Medicinal Chemistry, Volume 46
ISSN: 0065-7743, DOI: 10.1016/B978-0-12-386009-5.00021-7

and glucocorticoid receptor (GR) [1]. Aldosterone is the primary physiologic steroid hormone that binds to MR and promotes sodium and water reabsorption with potassium excretion. Other endogenous steroids such as cortisol can also bind to MR with high affinity. Excessive levels of aldosterone result in deleterious conditions including hypertension and cardiac hypertrophy. There is extensive clinical validation for treating hypertension and congestive heart failure with spironolactone and eplerenone, both of which are steroid-based antagonists of MR. However, these agents have side effects such as gynecomastia (off-target), hyperkalemia (on-target) and drug–drug interactions that limit their safety and effectiveness, thus providing a need for MR antagonists with superior profiles. Recent evidence suggests that MR blockade, when given in combination with standard therapy (*e.g.*, ACE inhibitors), reduces proteinuria in patients with renal disorders such as diabetic nephropathy and chronic kidney disease. In recent years, there has been a renewed interest in identifying nonsteroid-based antagonists of MR. Significant progress has been made toward identifying agents with greater selectivity against other steroid receptors such as AR and PR and greater potency compared to known steroid-based MR antagonists [2].

2. ALDOSTERONE AND MR BIOLOGY

MR is expressed in epithelial tissues notably in the distal convoluted tubules and cortical collecting ducts of the kidney. Its expression is also detected in lung, colon, and liver. MR is also expressed in nonepithelial tissues such as heart and brain [3]. Aldosterone, through binding to MR, promotes renal sodium reabsorption and potassium secretion in the distal nephron, distal colon, and salivary and sweat glands. Under conditions of abnormally elevated aldosterone levels as in patients with congestive heart failure, aldosterone-mediated sodium and water retention leads to an elevation in blood pressure due to inappropriate intravascular volume expansion [4].

Aldosterone can also regulate blood pressure by actions in the brain and directly on the vascular wall [5]. Evidence also indicates that inappropriate levels of aldosterone, in the presence of moderate to high sodium levels, mediate significant damage in nonepithelial tissues [6]. Elevated aldosterone levels, which are normally low (<1 nmol/L) [7], are linked to endothelial dysfunction, vascular inflammation, and myocardial fibrosis [2,8]. In patients with congestive heart failure, an increase in aldosterone-mediated sodium and water retention leads to inappropriate intravascular volume expansion and clinical symptoms consistent with hypervolemia [2]. While these effects were initially thought to be due to blood pressure elevation, the realization that functional MR is expressed in blood vessels has extended the role of aldosterone beyond sodium and water balance to direct and pleiotropic effects in the vasculature [9]. The primary damaging effects of aldosterone in the vasculature appear to lie in its induction of vascular inflammation and fibrosis. These effects occur not only in the heart but also in other organs such as the kidney and brain [3,10]. Plasma aldosterone levels among patients with ST segment-elevation myocardial infarction (STEMI) are associated with early and late adverse clinical outcomes, including mortality. The association between high aldosterone levels and late mortality is independent of age, heart failure, and reperfusion status. These results underline a pivotal role for aldosterone in the setting of STEMI [11].

Aldosterone ($K_D = 1$ nM) [12] acting via MR thus plays a crucial role in the pathophysiology of hypertension and ischemic heart disease [13]. In addition to aldosterone, the endogenous glucocorticoid cortisol can also bind to MR with similar affinity [12]. *In vitro*, both aldosterone and cortisol activate MR with EC_{50} values of 1 and 3 nM, respectively [14]. In epithelial cells *in vivo*, MR is protected from cortisol activation by 11β-hydroxysteroid dehydrogenase type 2 (11β-HSD2), the enzyme that converts cortisol into inactive cortisone. Inactivation of 11β-HSD2, which occurs in the syndrome of apparent mineralocorticoid excess, allows cortisol to function as an MR agonist to increase sodium reabsorption. In some nonepithelial tissues, including the heart and specific regions of the brain, 11β-HSD2 is not coexpressed with MR. Thus in these tissues, or under conditions of glucocorticoid excess, MR is presumably occupied by the higher levels of glucocorticoids that are present raising the possibility of a role for glucocorticoids in MR activation [15,16].

3. RAAS PATHWAY, MR ANTAGONISTS VERSUS ACE, ARB THERAPY

The renin–angiotensin aldosterone system (RAAS) is the major hormonal system that regulates blood pressure by controlling salt and water homeostasis and is a major target for therapy in patients with cardiovascular disease. Renin, an aspartyl protease synthesized in the juxtaglomerular cells of the kidney, performs the first rate limiting step of the conversion of angiotensinogen to angiotensin I, which is then converted to angiotensin II (A-II) by angiotensin-converting enzymes (ACEs). A-II acts via its receptors expressed in the adrenal cortex to release aldosterone. Activation of RAAS plays a significant role in the pathophysiology of various disease states including cardiac, vascular, and renal complications due to hypertension, vascular smooth muscle and cardiac hypertrophy, and fibrosis [17].

ACE inhibitors (ACEis) and angiotensin receptor blockers (ARBs) are well-established pharmacological options for the treatment of hypertension. However, ACE and A-II receptors are upstream of aldosterone signaling, and their blockade alone is not sufficient to sustain a long-term reduction in aldosterone due to a phenomenon termed "aldosterone escape" which leads to a gradual reactivation of the aldosterone signaling cascade [9].

3.1. Steroid-based MR antagonists in the clinic—Clinical results

Spironolactone and eplerenone are the only two MR antagonists on the market at present with spironolactone being more potent on MR but less selective against other steroid receptors than eplerenone [6]. In clinical trials, the beneficial effects of spironolactone (**1**) and eplerenone (**2**) ($IC_{50} = 24$ and 990 nM, respectively), in a cell-based Gal4 response element controlled luciferase reporter assay in CHO-K1 cells [18]) have been demonstrated in the treatment of hypertension and heart failure. Eplerenone was more effective in lowering blood pressure compared to losartan in patients with low-renin hypertension [19]. Spironolactone when added to multidrug regimens that included a diuretic and an ACEi or ARB has also been shown to provide better than anticipated benefit in patients with resistant hypertension [20]. These results have suggested that MR antagonists are also useful as an add-on therapy for hypertension.

In heart failure, the Randomized Aldactone Evaluation Study (RALES) reported in 1999 demonstrated that a low dose of spironolactone, when added to therapy that included an ACEi, significantly improved survival [21]. The Eplerenone Post-Acute Myocardial Infarction Heart Failure Efficacy and Survival Study (EPHESUS) investigated the addition of eplerenone to an ACEi or ARB and beta blocker in patients with acute myocardial infarction complicated by left ventricular dysfunction and heart failure. The results from this study demonstrated a significant improvement in survival and reduced hospitalization among these patients [22]. These intervention studies support a role for aldosterone in directly contributing to the development and/or progression of cardiovascular disease.

The recent publication of the Eplerenone in Mild Patients Hospitalization and Survival Study in Heart Failure (EMPHASIS-HF) has affirmed a broader potential for clinical use of MR antagonists. In this study, eplerenone reduced both the risk of death and hospitalization among patients with chronic systolic heart failure and mild symptoms [23]. These data also suggest that investigation of MR antagonism is warranted in other cardiovascular diseases. Ongoing studies of this therapy in patients with diastolic dysfunction and acute myocardial infarction could expand the use of MR antagonists in these additional indications [24].

In addition to their use in hypertension and heart failure, recent evidence suggests MR antagonists may also be useful in treatment of diabetic nephropathy. A recent study [25] showed that addition of spironolactone to a regimen including maximal ACE inhibition provided greater renoprotection than a maximal dose of ACEi-based monotherapy in patients with diabetic nephropathy. Additionally, it was found that the benefit did not appear to be solely dependent on reduced time-integral blood pressure (BP) burden as assessed by 24-h ambulatory BP monitoring. These studies suggest that large-scale randomized trials are needed to determine whether spironolactone or other MR antagonists added onto an ACEi-based regimen will be safe and effective for reducing the incidence of end stage renal disease (ESRD) in patients with diabetic nephropathy [25].

While both compounds show significant benefit and have become a mainstay in the therapy of cardiovascular disease, both have limitations. Spironolactone is not very selective against other steroid receptors such as AR and PR. In particular, its anti-progesterone and anti-androgen properties lead to unwanted side effects such as gynecomastia, breast pain, menstrual irregularities, and impotence, thus limiting its use [2,11]. In cells (Gal4 receptor-LBDs transfected into CHO-K1), the IC_{50} values of spironolactone for MR, GR, AR, and PR are 24, 2400, 77, and 740 nM, respectively [18]). For eplerenone, the IC_{50} values are 990, 22,000, 21,000, and 31,000 nM, respectively. The sex hormone-related side effects found with spironolactone have been significantly reduced with eplerenone most likely because of its improved AR and PR profile [6]. However, it is less potent than spironolactone, is predominantly metabolized by cytochrome P450 3A4 (CYP3A4) and coadministration with drugs that inhibit CYP3A4 may require a reduction in the dose. In addition, eplerenone has been associated with gastrointestinal intolerance [5]. The major adverse effect of MR antagonism by either spironolactone or eplerenone is an induction of clinically relevant potassium levels [2].

4. STRUCTURAL FEATURES OF THE LIGAND BINDING DOMAIN OF MR

Similar to other members of the NHR family, MR has three major functional domains, namely a N-terminal activation function 1 (AF-1) domain, a DNA binding domain (DBD), and a C-terminal activation function 2 (AF-2) domain. The AF-1 domain is the least conserved among other NHRs and possesses a ligand independent transactivation function. The DBD is the most conserved among nuclear receptor members and is responsible for mediating sequence-specific DNA binding. The ligand binding domain (LBD), containing the AF-2 domain, is also relatively well conserved among other steroid receptors and is responsible for ligand binding, dimerization, and ligand-dependent activation. All three MR domains are highly conserved across species [3].

Among steroid receptors, MR has the highest homology to the GR in terms of primary amino acid sequence. The MR LBD consists of 251 amino acids and has high sequence homology with the AR, PR, and GR LBDs. Crystal structures have been reported for the wild-type and mutant MR bound to various steroid ligands including cortisone, progesterone, deoxycorticosterone, aldosterone, and spironolactone [26–29]. As expected from sequence homology [30], the three-dimensional structure of the MR LBD shares remarkable structural similarity to the crystal structures of the GR, AR, PR, and ER LBDs [31]. The LBD of MR consists of 11 α helices (H1, H3–H12) and two short β sheets arranged around a central hydrophobic pocket, with helices 3, 4, 5, 6, 7, and 11 providing the amino acids that line the binding pocket.

The ligand binding pocket in all of the reported crystal structures is fully enclosed. Even though the unliganded (apo form) crystal structure has not been reported, previous studies with other NHRs suggest that the ligand binding pocket of the apo form is partially exposed to solvent with H12 randomly distributed. Upon agonist binding, H12 adopts the position indicated in the crystal structures where it interacts with helices 3, 5, and 11. This forms a hydrophobic groove on the surface of the LBD enabling coactivator recruitment. The high stability of the agonist bound MR complexes has facilitated their purification and crystallization. However, the low stability of antagonist bound complexes has prevented characterization of the antagonist conformation of the LBD to date. Spironolactone bound crystal structures were solved using the S810L mutant where a single mutation converts the antagonist (for the wild-type MR) to an agonist (for the S810L mutant MR) [32–34].

The binding mode of spironolactone is depicted in Figure 1. The lactone carbonyl forms a hydrogen bond with the amide H–N of Asn-770, while the A-ring carbonyl oxygen forms hydrogen bonds with Gln-776, Arg-817, and a crystallographic water.

5. CURRENT MEDICINAL CHEMISTRY EFFORTS

5.1. Dihydropyridines

Recent efforts in screening compound collections for MR antagonist activity have led to the identification of several chemotypes including the 1,4-dihydropyridine (DHP) class of calcium channel blockers (CCBs) as MR antagonists. The original report demonstrated that these frequently used antihypertensive agents compete with aldosterone in binding to the MR

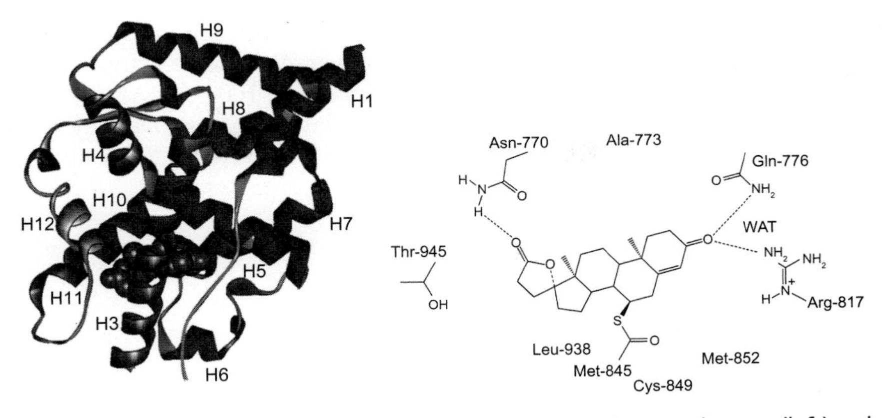

Figure 1 The 3D structure of the LBD of the MR with bound spironolactone (left) and schematic drawing of the binding mode (right). (See Color Plate 6.1 in Color Plate Section.)

LBD, block binding of coactivators such as SRC1-4a, and inhibit aldosterone-induced gene expression [35,36]. In these studies, several DHPs including mebudipine (**3**) and felodipine (**4**) were shown to possess similar IC_{50} values (126–450 nM) to eplerenone (IC_{50} = 135 nM, cell-based Gal4 response element controlled luciferase reporter assay in HUH7 cells). However, the DHP compounds as a class have significantly higher intrinsic potencies as CCBs compared to their MR antagonist potency. For example, the MR IC_{50} value of **3** is 126 nM (Gal4), and the IC_{50} for Ca^{2+} channel (CC) inhibition is 20 nM (thoracic rat cell line). Thus, blood levels of currently prescribed human doses of CCBs are probably not sufficient to greatly affect peripheral MR binding. However, it does suggest that a dual-acting compound may have synergistic benefits in BP control.

Although the first DHPs were achiral molecules, the search for vasodilators with a longer duration of action led to the discovery of chiral DHPs with nonidentical ester groups in the 3 and 5 positions (*e.g.*, **4**). While more recent DHPs are marketed as racemic mixtures, the absolute configuration required at the C4 position for CCB activity has been established [37]. In an effort to determine if a specific configuration is required for MR activity, a chiral resolution of **3** was performed and through evaluation of small molecule X-ray crystal structures, it was demonstrated that opposite enantiomers are responsible for MR and CCB activity [36]. Thus, the (+)-stereoisomer of mebudipine (**3R**) has an IC_{50} = 46 nM for MR inhibition (cell-based Gal4 response element controlled luciferase reporter assay in HUH7 cells) and an IC_{50} = 607 nM for Ca^{2+} channel (CC) inhibition (thoracic rat cell line). The (−)-enantiomer (**3S**) has an MR IC_{50} = 540 nM and a CC IC_{50} = 0.5 nM. Replacement of the methyl ester with a nitrile improved metabolic stability and along with other changes, led to the discovery of **5R**. This compound has an $IC_{50} = 10$ nM in the MR Gal4 assay and is over 220-fold selective versus the CC. NHR selectivity ranges from 30× for GR to 100× for AR/PR and 1000× for ER vs MR. Pharmacokinetic (PK) data (Sprague–Dawley (SD) rats, iv 2 mg/kg) showed moderate clearance (10 mL/min/kg) and adequate half-life (6.5 h).

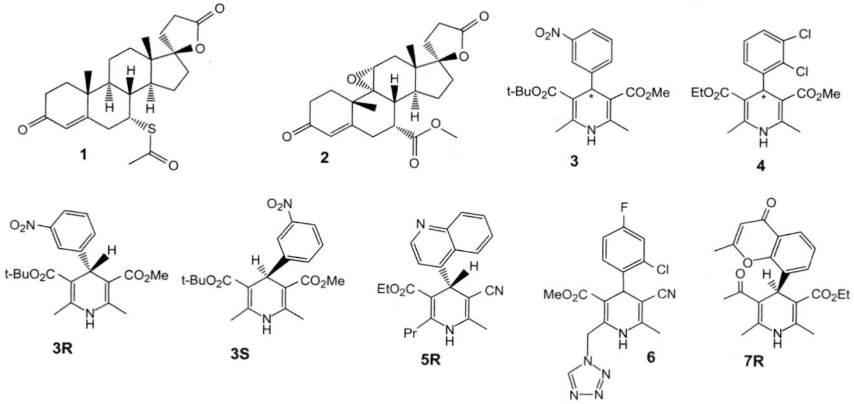

While there are no published X-ray co-crystal structures of DHPs with the MR LBD, a proposed structure based on both the wild-type and S810L MR crystal structures suggests that the N–H of the DHP ring forms an H-bond to Asn770 and the ester group fills the R-face hydrophobic pocket. The opposite (*S*) configuration cannot adapt an orientation that can accommodate these key interactions (Figure 2).

In another report by the same authors, further optimization of the cyano DHP series led to the identification of **6** (MR $IC_{50} = 64$ nM for

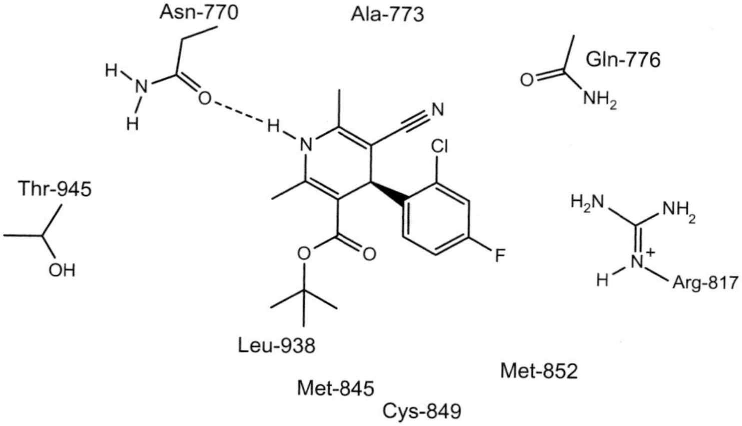

Figure 2 The proposed mode of binding for 1,4-dihydropyridines.

MR inhibition, cell-based Gal4 response element in HUH7 cells) which had improved aqueous solubility (31 μM), limited CYP inhibition liability (IC_{50} = 1.2, 1.9, >10 for CYP3A4, 2C9, 2D6, respectively) and rat liver microsomal stability (60% remaining after 30 min), although human liver microsomal stability was low (3% remaining after 30 min) [38,39]. NHR selectivity ranged from 20× for GR to 42× for AR and >156× for PR versus MR. The cell-based potency, NHR selectivity, and measured profiling parameters of **6R** were similar to that of racemic **6**. The CC inhibition data for **6** and **6R** were not reported. The PK parameters in SD rat (iv, 5 mg/kg, Cl = 14 mL/min/kg, $t_{1/2}$ = 4.8 h) were sufficiently favorable to advance compound **6** into *in vivo* studies. In these studies, **6** lowered systolic BP when administered at a dose of 60 mg/kg p.o, b.i.d. for 21 days to Dahl SS rats fed a high salt diet. A trend toward a reduction of albumin levels, although not statistically significant, was suggestive of protection against kidney damage.

In a related report by a different group, **7R** was identified as a highly potent, selective MR antagonist with *in vitro* and *in vivo* activity similar to spironolactone [18]. This compound had an IC_{50} = 28 nM in the MR Gal4 assay and was 70-fold selective versus the CC. NHR selectivity ranged from 195× versus GR, 160× versus AR, and 322× versus PR. Interestingly, it was shown that **7R** is a full, functional antagonist of both the Gal4-MR_{WT} LBD fusion protein and the Gal-4MR_{S810L} mutant LBD expressed in CHO-K1 cells. Both eplerenone and spironolactone were functional agonists when tested using the Gal-4MR_{S810L} mutant LBD. When dosed orally to conscious rats, **7R** increased the urinary Na^+/K^+ ratio in a dose-dependent manner, with a significant effect at the 1 mg/kg dose. Spironolactone showed a similar trend, but statistical significance was reached at a dose of 10 mg/kg. Finally, in the absence of a **7R** X-ray co-crystal structure, an Ala-scanning mutagenesis approach was used to determine critical residues for binding of **7R** to the MR LBD compared with spironolactone. For example, MR containing Gly instead of Ala^{773} and Ala instead of Thr^{945} gave IC_{50} (binding affinity) for BR-4628 of 960 and 30 nM, respectively. The IC_{50} (WT MR) of BR-4628 = 34 nM in this assay. For these same mutations, spironolactone IC_{50} = 84 and 444 nM, respectively (IC_{50} (WT MR) for spironolactone = 74 nM in this assay). These data collectively suggest that while key interactions (such as with Asn770) are critical for binding of both **7R** and spironolactone, **7R** does present a unique binding mode to the LBD compared with known steroids.

5.2. Pyrazoline derivatives

Pyrazolines represent another chemotype possessing MR antagonist activity identified *via* high-throughput screening efforts [40,41]. The screening hit **8** was found to have acceptable potency (MR IC_{50} = 460 nM) in an aldosterone-induced activation of a luciferase reporter driven by the MR LBD in HUH7 cells. Unwanted features included high lipophilicity (aqueous solubility $\leq 3\ \mu M$) and inhibition of the hERG channel (>30% inhibition at 10 μM in a dofetilide-Cy3B competitive binding assay). Unacceptable functional groups were replaced through optimization of each position on the pyrazoline ring ultimately leading to the discovery of **9** (MR IC_{50} 6 nM) and **10** (MR IC_{50} 9 nM). Both compounds were selective for MR versus AR (>1000×) and GR (>1000×) but were less selective versus PR (335× for **9**, 46× for **10**). The stereochemistry at C5 of the pyrazoline ring was found to be critical for MR antagonist potency. Thus, for the compounds where both isomers were tested, the (*R*)-configuration was 50- to 60-fold more potent in the MR cell assay than the corresponding (*S*)-isomer.

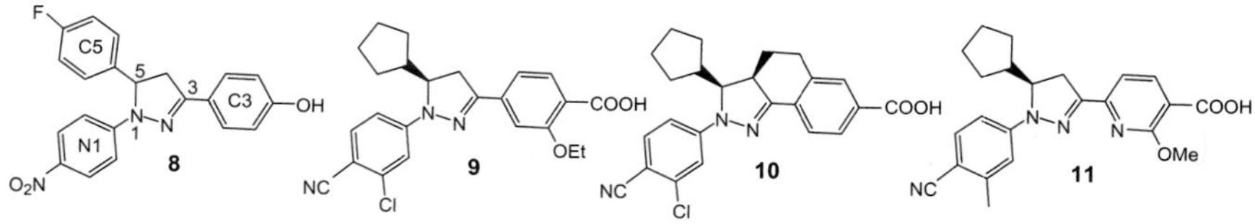

A unique feature of these compounds is the tolerance for a charged carboxylate in the LBD of the MR. Among the functional groups tested, only the carboxylate demonstrated reduced hERG channel inhibition while maintaining potent MR antagonist activity. Through investigation of an induced fit model based on an overlay of the native MR/corticosterone crystal structure with **10**, the authors postulate that the presence of the carboxylate causes a movement of Phe941, with disruption of the Leu960 and Asn770 sidechains by the cyclopentyl group. The cyanophenyl ring is believed to occupy a similar position to the A/B ring system of steroid modulators.

A recent patent application exemplifies **11** and close analogs containing a substituted pyridine ring with very limited biological data given [42].

Compound **10** (PF-3882845) administered at doses of 10, 40, and 100 mg/kg p.o. for 21 days to Dahl SS rats, a preclinical model of salt-induced hypertension and nephropathy, was shown to decrease urinary albumin, reduce blood pressure, and protect against kidney damage. Based on its PK properties and preclinical safety profile, compound **10** was chosen to advance into clinical studies for diabetic nephropathy.

5.3. Indole sulfonamides and related structures

Indoles represent a further MR antagonist chemotype identified through screening of in-house compound collections [43]. In this series, **12** was the starting point for structure–activity relationship (SAR) development leading to the identification of **13**, which displayed a preference for the

S stereoisomer for binding to the MR. A methyl sulfonamide was ultimately selected for substitution of the indole at the 7-position driven by SAR data that suggested the need for an H-bond donor. A dialkyl substituent revealed a narrowly defined hydrophobic pocket that appears to optimally accommodate a cyclopropyl(methyl) side chain. Finally, the 2,4-difluorophenyl substituent gave a threefold improvement in MR binding compared to the 4-difluorophenyl. Compound **13** potently binds the MR receptor ($K_i = 0.5$ nM), is a functional antagonist ($K_b = 19$ nM), and showed some improvement in NHR selectivity. In salt loaded, uninephrectomized, aldosterone-induced SD rats administered with **13** (10 mg/kg) for 14 days p.o., an 80% decrease in blood pressure compared to vehicle was observed. Eplerenone at the same dose gave a 30% decrease compared to vehicle.

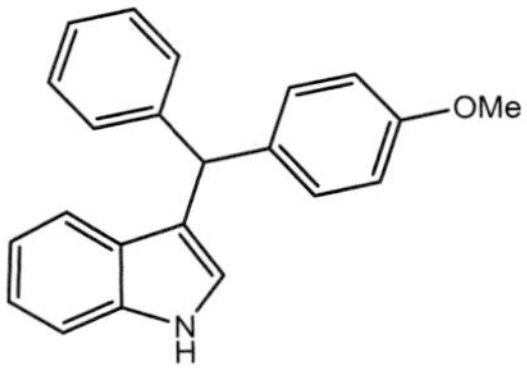

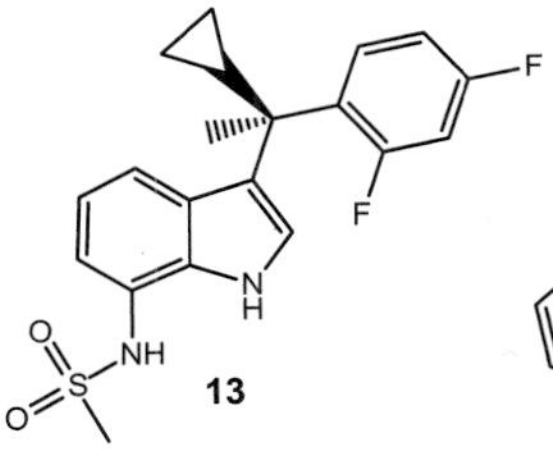

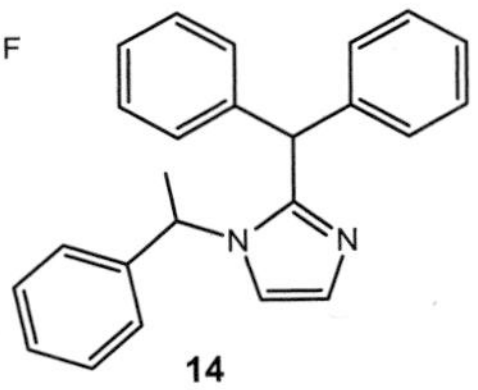

In a recent report, imidazole **14**, a modest antagonist of MR with a binding K_i of 285 nM discovered as a screening hit, afforded modest changes in Na^+/K^+ ratios when administered to SD rats at a dose of 30 mg/kg s.c. followed by administration of 0.9% saline (acute mode) [44,45].

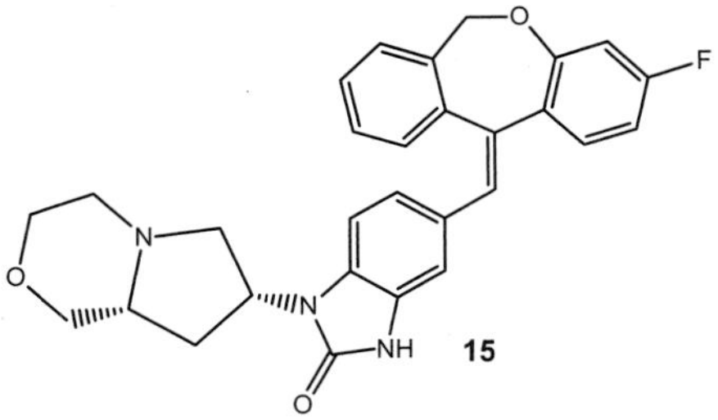

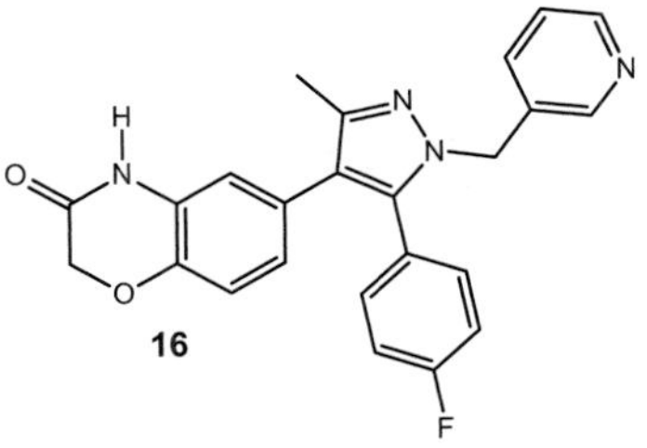

5.4. Benzimidazole derivatives

A highly selective and potent benzimidazolone MR antagonist **15** was recently reported [46]. In a MR binding assay, **15** demonstrated a K_i of 0.4 nM with >1000-fold binding selectivity versus AR, GR, and PR. In HEK293 cells singly transfected with either MR or other NHRs, **15** gave IC_{50} values of 21, 924, >10,000, and >10,000 nM for MR, PR, GR, and AR, respectively, in antagonist mode, thereby demonstrating significant NHR selectivity. In a human MR (hMR) competitive antagonist assay utilizing HEK293 cells, **15** gave a K_b of 5.1 nM, demonstrating potent hMR antagonist activity. In an *in vivo* model of adosterone-mediated renal disease using male uninephrectomized SD rats fed a high salt diet, **15** reduced urinary protein excretion by 60% when orally administered at a dose of 10 mg/kg/day for 28 days, demonstrating durable and potent *in vivo* renal protection. In the same animal model, **15** was also shown to reduce the hypertensive effects of aldosterone (administered *via* alzet pump at a rate of 0.75 μg/h, s.c.) compared to vehicle in a dose-dependent manner when given p.o. (1–30 mg/kg) for 14 days.

5.5. Other MR antagonist structures

Over the past few years, several compositions of matter patent applications covering benzofused oxazinones as NHR modulators and, specifically, MR antagonists have emerged [47–49]. In one recent narrow composition of matter application [50], limited MR binding inhibition data on selected compounds (inhibition at 10 μM) were disclosed suggesting that these compounds possess MR antagonist activity. Compound **16** is one example. Future disclosures will shed light on the value of these compounds to the field.

6. CONCLUSIONS

The past several years have seen a resurgence in targeting MR for intervention in cardiovascular disease. Known limitations of existing therapy with the steroid-based MR antagonist eplerenone and spironolactone due to off-target effects or poor efficacy have led drug discovery groups to look for highly potent, selective nonsteroid-based antagonists.

Recent findings strongly suggest that MR antagonism can play a major role not only in lowering blood pressure but also in sparing the kidney from damage due to diseases such as diabetic nephropathy and, more broadly, chronic kidney disease. Given the high rate of occurrence of type 2 diabetes in an aging, overweight population, these diseases represent a potentially huge unmet medical need. MR antagonism has also been shown to be cardioprotective, reducing mortality rates among patients recovering from acute myocardial infarction. Hyperkalemia is a potential mechanism-based side effect of MR antagonism, and it remains to be seen whether this new generation of selective, nonsteroidal MR antagonists can demonstrate an improved therapeutic window versus the steroid MR antagonists. In summary, these nonsteroidal MR antagonists will need to demonstrate a clearly superior profile to effectively displace current steroidal generics as standard of care for these indications.

REFERENCES

[1] D. J. Mangelsdorf, C. Thummel and M. Beato, *Cell*, 1995, **83**, 835.
[2] S. Viengchareun, D. L. Menuet, L. Martinerie, M. Munier, L. Pascual-Le Tallec and M. Lombès, *Nucl. Recept. Signal.*, 2007, **5**, e012.
[3] R. Rocha, C. T. Stier, I. Kifor, M. R. Ochoa-Maya, H. G. Rennke, G. H. Williams and G. K. Adler, *Endocrinology*, 2000, **141**, 3871.
[4] K. T. Weber, *N. Engl. J. Med.*, 2001, **345**, 1689.
[5] R. Rocha, A. E. Rudolph, G. E. Frierdich, D. A. Nachowiak, B. K. Kekec, E. A. Blomme, E. G. McMahon and J. A. Delyani, *Am. J. Physiol. Heart Circ. Physiol.*, 1802, **2002**, 283.
[6] E. R. Blasi, R. Rocha, A. E. Rudolph, E. A. Blomme, M. L. Polly and E. G. McMahon, *Kidney Int.*, 2003, **63**, 1791.
[7] J. W. Funder, *Prog. Cardiovasc. Dis.*, 2010, **52**, 393.
[8] C. T. Stier Jr., P. N. Chander and R. Rocha, *Cardiol. Rev.*, 2002, **10**, 97.
[9] B. A. Maron and J. A. Leopold, *Curr. Opin. Invest. Drugs*, 2008, **9**, 963.
[10] D. A. Calhoun, *Circulation*, 2006, **114**, 2572.
[11] F. Beygui, *Circulation*, 2006, **114**, 2604.
[12] J. L. Arriza, C. Weinberger, G. Cerelli, T. M. Glaser, B. L. Handelin, D. E. Housman and R. M. Evans, *Science*, 1987, **237**, 268.
[13] G. George and A. D. Struthers, *Expert Opin. Pharmacother.*, 2007, **8**, 13053.
[14] F. M. Rogerson, N. Dimopoulos, P. Sluka, S. Chu, A. J. Curtis and P. J. Fuller, *J. Biol. Chem.*, 1999, **274**, 36305.
[15] S. Viengchareun, D. L. Menuet, L. Martinerie, M. Munier, L. Pascual-Le Tallec and M. Lombès, *Nuc. Recept. Signal.*, 2007, **5**, e012.
[16] M. L. Hultman, N. V. Krasnoperova, S. Li, S. Du, C. Xia, J. D. Dietz, D. S. Lala, D. J. Welsch and X. Hu, *Mol. Endocrinol.*, 2005, **19**, 1460.
[17] K. Vijayaraghavan and K. P. Deedwania, *Cardiol. Clin.*, 2011, **29**, 137.
[18] J. Fagart, A. Hillisch, J. Huyet, L. Baerfacker, M. Fay, U. Pleiss, E. Pook, S. Schaefer, M. E. Rafestin-Oblin and P. Kolkhof, *J. Biol. Chem.*, 2010, **285**, 29932.
[19] M. H. Weinberger, *Am. Heart J.*, 2005, **150**, 426.
[20] M. K. Nishizaka, *Am. J. Hypertens.*, 2003, **6**, 925.
[21] B. Pitt, *N. Engl. J. Med.*, 1999, **341**, 709.
[22] B. Pitt, G. Bakris, L. M. Ruilope, L. DiCarlo and R. Mukherjee, *Circulation*, 2008, **118**, 1643.
[23] F. Zannad, *N. Engl. J. Med.*, 2011, **364**, 11.
[24] P. W. Armstrong, *N. Engl. J. Med.*, 2011, **364**, 1.
[25] J. Mehdi, *Am. Soc. Nephrol.*, 2009, **20**, 2641.
[26] R. K. Bledsoe, K. P. Madauss, J. A. Holt, C. J. Apolito, M. H. Lambert, K. H. Pearce, T. B. Stanley, E. L. Stewart, R. P. Trump, T. M. Willson and S. P. Williams, *J. Biol. Chem.*, 2005, **280**, 31283.
[27] J. Fagart, J. Huyet, G. M. Pinon, M. Rochel, C. Mayer and M. E. Rafestin-Oblin, *Nat. Struct. Mol. Biol.*, 2005, **12**, 554.
[28] Y. Li, K. Suino, J. Daugherty and H. E. Xu, *Mol. Cell*, 2005, **19**, 367.
[29] J. Huyet, G. M. Pinon, M. R. Fay, J. Fagart and M. E. Rafestin-Oblin, *Mol. Pharmacol.*, 2007, **72**, 563.
[30] X. Hu and J. W. Funder, *Mol. Endocrinol.*, 2006, **20**, 1471.
[31] P. Huang, V. Chandra and F. Rastinejad, *Ann. Rev. Physiol.*, 2010, **72**, 247.

[32] D. S. Geller, A. Farhi, N. Pinkerton, M. Fradley, M. Moritz, A. Spitzer, G. Meinke, F. T. Tsai, P. B. Sigler and R. P. Lifton, *Science*, 2000, **289**, 119.
[33] M. E. Rafestin-Oblin, A. Souque, B. Bocchi, G. Pinon, J. Fargt and A. Vandewalle, *Endocrinology*, 2003, **144**, 528.
[34] G. M. Pinon, J. Fagart, A. Souque, G. Auzou, A. Vandewalle and M. E. Rafestin-Oblin, *Mol. Cell. Endocrinol.*, 2004, **217**, 181.
[35] J. D. Dietz, S. Du, C. W. Bolten, M. A. Payne, C. Xia, J. R. Blinn, J. W. Funder and X. Hu, *Hypertension*, 2008, **51**, 742.
[36] G. B. Arhancet, S. S. Woodard, J. D. Dietz, D. J. Garland, G. M. Wagner, K. Iyanar, J. T. Collins, J. R. Blinn, R. E. Numann, X. Hu and H. C. Huang, *J. Med. Chem.*, 2010, **53**, 4300.
[37] G. C. Rovnyak, S. D. Kimball, B. Beyer, G. Cucinotta, J. D. DiMarco, J. Gougoutas, A. Hedberg, M. Malley and J. P. McCarthy, *J. Med. Chem.*, 1995, **38**, 119.
[38] G. B. Arhancet, S. S. Woodard, K. Iyanar, B. L. Case, R. Woerndle, J. D. Dietz, D. J. Garland, J. T. Collins, M. A. Payne, J. R. Blinn, S. I. Pomposiello, X. Hu, M. I. Heron, H. C. Huang and L. F. Lee, *J. Med. Chem.*, 2010, **53**, 5970.
[39] P. E. Brandish, M. E. Fraley, J. C. Hershey and J. T. Steen, *Patent Application WO 2009/078934*, 2009.
[40] M. J. Meyers, G. B. Arhancet, S. L. Hockerman, X. Chen, S. A. Long, M. W. Mahoney, J. R. Rico, D. J. Garland, J. R. Blinn, J. T. Collins, S. Yang, H. C. Huang, K. F. McGee, J. M. Wendling, J. D. Dietz, M. A. Payne, B. L. Homer, M. I. Heron, D. B. Reitz and X. Hu, *J. Med. Chem.*, 2010, **53**, 5979.
[41] G. B. Arhancet, A. Casimiro-Garcia, X. Chen, D. Hepworth, M. J. Meyers, W. Piotrowski and R. K. Raheja, *Patent Application WO 2010/11628*, 2010.
[42] S. Fukumoto, N. Ohyabu, T. Ohra, T. Sugimoto, T. Hasui, K. Fuji, C. Siedem and C. Gauthier, *Patent Application US 2010/0094000*, 2010.
[43] M. G. Bell, D. L. Gernert, T. A. Grese, M. D. Belvo, P. S. Borromeo, S. A. Kelley, J. H. Kennedy, S. P. Kolis, P. A. Lander, R. Richey, V. S. Sharp, G. A. Stephenson, J. D. Williams, H. Yu, K. M. Zimmerman, M. I. Steinberg and P. K. Jadhav, *J. Med. Chem.*, 2007, **50**, 6443.
[44] P. E. Brandish, H. Chen, P. Szczerba and J. C. Hershey, *J. Pharmacol. Toxicol. Methods*, 2008, **57**, 155.
[45] P. E. Brandish, J. C. Hershey, M. E. Fraley and J. T. Steen, *Patent Application WO 2008118319*, 2008.
[46] D. A. Coates, K. Gavardinas and P. K. Jadhav, *Patent Application WO 2010/104721*, 2010.
[47] P. Michellys, H. M. Petrassi, W. Richmond and W. Pei, *Patent Application WO 2006/015259*, 2006.
[48] S. Fukumoto, N. Matsunaga, T. Ohra, N. Ohyabu, T. Hasui, T. Motoyaji, S. Takashi, C. Stephen, T. Tang, D. Pisal, L. A. Demeese and C. Gauthier, *Patent Application WO 2007/077961*, 2007.
[49] J. W. Brown, A. R. Gangloff and A. A. Kiryanov, *Patent Application WO 2011/014681*, 2011.
[50] S. Fukumoto, N. Ohyabu, T. Ohra, T. Sugimoto, T. Hasui, K. Fuji, C. S. Siedem and C. Gauthier, *Patent Application US 2010/0094000*, 2010.

CHAPTER 7

SGLT2 Inhibitors for Type 2 Diabetes

Jiwen (Jim) Liu* and **TaeWeon Lee****

Contents

1. INTRODUCTION

Type 2 diabetes mellitus is a progressive disease characterized by hyperglycemia, increased peripheral insulin resistance, and declining insulin secretion. Over the past two decades, the prevalence of type 2 diabetes has increased to near epidemic proportion in both developed and developing countries [1]. As of 2011, diabetes affects 25.8 million people in the USA, or 8.3% of the population. In 2007, the estimated economic burden from diabetes in the USA was $116 billion for direct medical costs and another $58 billion for indirect costs related to disability, work loss, and premature mortality. Among all diabetic patients, greater than 90% have type 2

* Medicinal Chemistry, Amgen, Inc., 1120 Veterans Boulevard, South San Francisco, CA 94080, USA
** Metabolic Disorders, Amgen, Inc., 1120 Veterans Boulevard, South San Francisco, CA 94080, USA

Annual Reports in Medicinal Chemistry, Volume 46
ISSN: 0065-7743, DOI: 10.1016/B978-0-12-386009-5.00020-5

diabetes, while the remainder constituting type 1 diabetes are characterized by an inability to produce insulin due to destruction of pancreatic β-cells [2]. Currently available medications for type 2 diabetes are not sufficient to halt this epidemic and reduce its burden. Moreover, most current drugs are insulin dependent (improve insulin sensitivity or increase insulin levels) and lose their effectiveness to control hyperglycemia over time due to the progressive decline of β cell function. As a consequence, many patients receive multiple antidiabetic medicines and eventually require insulin therapy. The lack of sufficient control of hyperglycemia, even under therapy, contributes to the progressive nature of type 2 diabetes, which results in many burdensome complications such as diabetic retinopathy, neuropathy, nephropathy, and cardiomyopathy [3]. In addition, since a majority of diabetic patients are overweight or obese, the fraction of current therapies that are associated with weight gain exacerbates this condition.

Given the difficulty in achieving sufficient glycemic control for many diabetic patients using current therapies, there is an unmet medical need for new antidiabetic agents, especially insulin-independent therapies. Blocking glucose reabsorption in the kidney and lowering blood glucose levels through glucose excretion into the urine would provide a novel insulin-independent therapy [4]. Filtered plasma glucose is reabsorbed in the renal tubule mainly by sodium glucose cotransporter 2 (SGLT2) and reenters the systemic circulation. Recent Phase II and Phase III clinical data of SGLT2 inhibitors and genetic studies of SGLT2 mutations in humans have provided strong evidence for SGLT2 as a promising new target to treat diabetes. This potential new therapy may be used as a monotherapy or in combination with existing therapies to achieve an additive effect in controlling blood glucose levels. This review summarizes the biological rationale, the clinical trials, and preclinical research of SGLT2 inhibitors.

2. SGLT2 PHYSIOLOGY

2.1. Renal glucose reuptake by SGLT2 and SGLT1

The kidney was not previously appreciated as a diabetes target organ until the emergence of the role of SGLT2 in renal glucose recovery. Kidneys have a dynamic function in maintaining plasma glucose homeostasis through gluconeogenesis and reabsorption of glucose from the glomerular filtrate, as well as allowing overspill into the urine when glomerular glucose levels exceed renal tubule recovery capacity. In healthy humans, about 180 g of plasma glucose is filtered daily, almost all of which is reabsorbed in the kidneys. SGLT2 plays a major role in this process as shown in Figure 1. Micropuncture studies in mouse renal

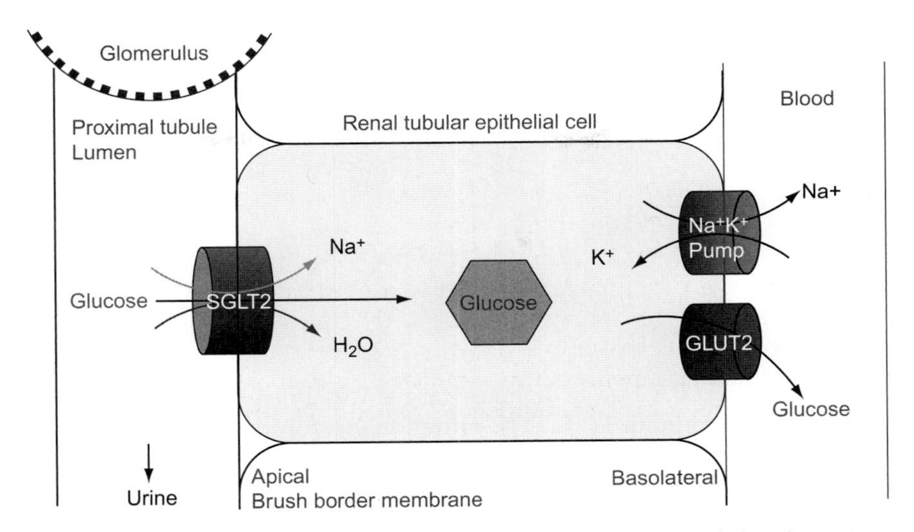

Figure 1 Tubular glucose reabsorption by SGLT2 in the kidney. (See Color Plate 7.1 in Color Plate Section.)

tubules and the phenotype of severe human SGLT2 mutations indicated that about 90% of renal glucose reabsorption is mediated by SGLT2 [5–7]. The SGLT2 localization in the S1 segment of the early proximal tubule and its high capacity for glucose transport fit well with its major role in renal glucose reabsorption. The remaining 10% of the filtered glucose is absorbed by SGLT1, a low capacity transporter which is localized in the S3 segment of the late proximal tubules [8,9]. SGLT1 and SGLT2 mediate the active transport of glucose across the apical membrane into the tubular cells via coupling with downhill cotransport of Na^+. The inward Na^+ gradient is maintained by ATP-driven Na^+/K^+ pumps. The glucose then passively diffuses out of the tubular cells and into blood stream across the basolateral membrane through facilitative glucose transporters GLUT2 and GLUT1 [9,10].

2.2. Glucosuria and regulation of plasma glucose

Phlorizin (**1**, Figure 2), a naturally occurring compound extracted from the root bark of an apple tree in 1835 and later identified as a SGLT1 and SGLT2 dual inhibitor, played a key role in elucidating the mechanism of renal glucose absorption and providing initial proof of principle for SGLT2 as a diabetes target. Treatment with phlorizin induced urinary glucose excretion (UGE) without renal abnormalities in dogs, and significantly lowered plasma glucose levels and normalized insulin sensitivity in diabetic rats [11,12].

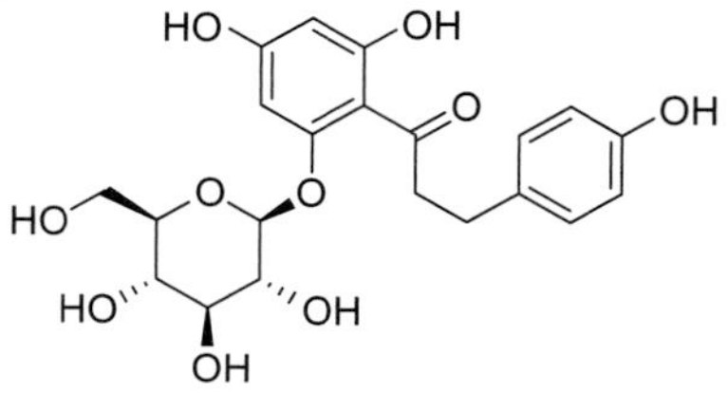

Figure 2 Phlorizin, a starting point.

Interest in SGLT2 as a diabetes target was bolstered by studies of human SGLT2 mutations. SGLT2 mutations result in familial renal glucosuria (FRG), which is asymptomatic and benign. Although patients with SGLT2 mutations excrete glucose in varied amounts into the urine (<1 to >150 g/day), they present normal blood glucose levels and no noticeable kidney dysfunction [6,7]. Cases of severe glucosuria were found in patients with homozygous or compound heterozygous SGLT2 mutations [7]. Because SGLT2 mutations do not seem to lead to any clinical consequences, pharmacological inhibition of SGLT2 to prevent glucose reabsorption could potentially be as safe. On the other hand, SGLT1 inhibition appears less attractive since human SGLT1 mutations lead to glucose/galactose malabsorption (GGM) and are associated with severe diarrhea in infants on diets containing glucose/galactose [13]. Although blocking both SGLT1 and SGLT2 could increase efficacy, the potential side effects associated with SGLT1 inhibition make the selective inhibition of SGLT2 a more appealing strategy.

3. CLINICAL TRIALS

In addition to its role in elucidating the mechanism of glucose reabsorption in kidney, phlorizin also served as a starting point for the optimization of glucoside-based SGLT2 inhibitors. Indeed, all the SGLT2 inhibitors tested in human clinical trials thus far are glucoside-based inhibitors derived from phlorizin [14]. O-glucoside SGLT2 inhibitors **2a**, **3a**, and **4a** as their respective carbonate prodrugs T-1095 (**2b**), sergliflozin (**3b**), remogliflozin etabonate (**4b**) as well as AVE2268, TS-033, and BI44847 entered development (Figure 3) but have been discontinued due to lack of sufficient stability in the gut and post-absorption [10]. All SGLT2 inhibitors currently in development are C-glucosides, which exhibit increased metabolic

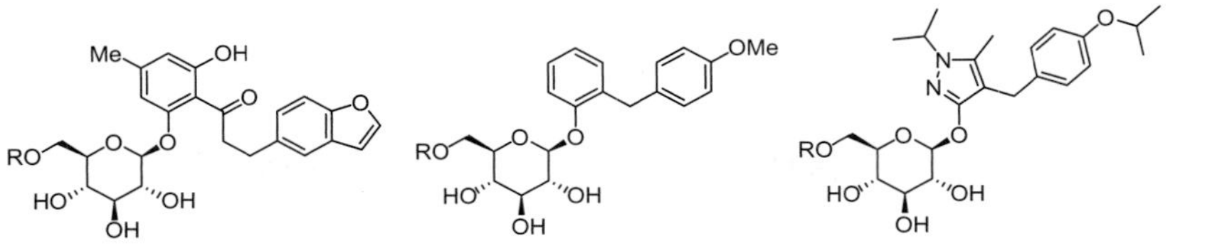

2a, R = H
2b, R = COOMe, T-1095

3a, R = H
3b, R = COOEt, Sergliflozin

4a, R = H, Remogliflozin
4b, R = COOEt, Remogliflozin etabonate

Figure 3 Disclosed structures of O-glucoside SGLT2 inhibitors previously in clinical trials.

stability, good potency, selectivity, and oral bioavailability [10,14]. The SGLT2 inhibitors currently in Phase II and Phase III clinical trials as of April 2011 are listed in Table 1. The structures of five of these inhibitors, dapagliflozin (**5**), canagliflozin (**6**), TS-071 (**7**), tofogliflozin (CSG-452, **8**), and ipragliflozin (ASP-1941, **9**), have been disclosed (Figure 4) [15,17,27–29]. Based on an analysis of the patent literature, all C-glucoside SGLT2 inhibitors in clinical trials are likely to have similar structures [14].

The clinical status, SGLT2 potency and selectivity over SGLT1, and dose ranges of inhibitors in the Phase II and Phase III clinical trials are also listed in Table 1. The *in vitro* SGLT2 potencies are similar for these compounds; all exhibit good selectivity over SGLT1 [15,17,19,21,23,26,27], except LX-4211, which is being positioned as a dual SGLT1 and SGLT2 inhibitor [16,26].

The Phase II and Phase III data of C-glucoside inhibitors was recently summarized [16]. The most advanced inhibitor in development is dapagliflozin for which data has been published for the completed pivotal Phase III clinical trials [30]. In addition to demonstrating good efficacy for lowering plasma glucose levels, dapagliflozin and other inhibitors have provided additional beneficial effects. These include insulin independence which may help β cell preservation, a low risk of hypoglycemia, and weight loss [16,31]. SGLT2 inhibitors are being tested as monotherapy and as add-on to existing antidiabetic therapies, including metformin, DPPIV inhibitors, sulfonylureas, and insulin. They induced robust hemoglobin A1c (HbA1c) reduction and moderate body weight loss as a monotherapy, and exhibited additive effects as an add-on therapy to the above-mentioned antidiabetic drugs. The SGLT2 inhibitor class also has been well tolerated, with no major safety signals; however, an increase in genitourinary infections was of significance in recent clinical trials [16]. In summary, clinical trials with multiple SGLT2 inhibitors are progressing well with no differentiating features as yet apparent in terms of efficacy and side effect profiles [16].

Table 1 SGLT2 inhibitors in Phase II and Phase III trials

Compound	SGLT2 IC_{50} (nM)	Selectivity over SGLT1	Clinical status	Dose range in the clinical studies
Dapagliflozin (BMS-512148)	1.1 [15]	1200 [15]	Phase III	2.5–10 mg q.d. (Phase III) [16]
Canagliflozin (TA-7284)	2.2 [17]	414 [17]	Phase III	100–300 mg q.d. (Phase III) [18]
BI 10773	3.1 [19]	>2500 [19]	Phase III	10–25 mg q.d. (Phase III) [20]
Ipragliflozin (ASP-1941)	7.4 [21]	255 [21]	Phase III[a]	50–300 mg q.d. (Phase II) [16,22]
Tofogliflozin (CSG-452)	2.9 [23]	2930 [23]	Phase III[a]	2.5–40 mg q.d. (Phase II) [24]
PF-04971729	n.a.	n.a.	Phase II	1–25 mg q.d. (Phase II) [25]
LX-4211	1.8 [26]	20 [26]	Phase II	150–300 mg q.d. (Phase II) [16,26]
TS-071	2.3 [27]	1765 [27]	Phase II[a]	n.a.

[a]Clinical studies in Japan.

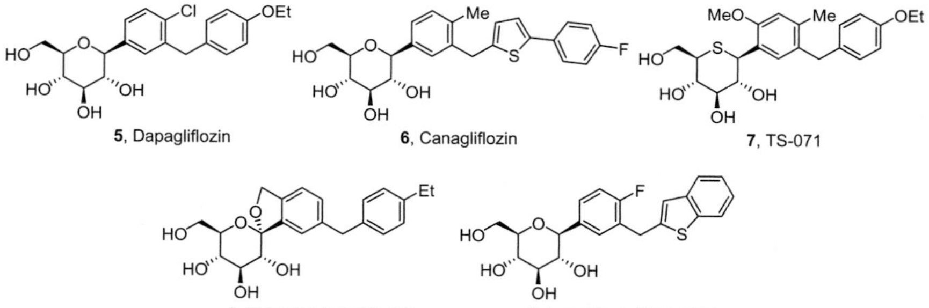

5, Dapagliflozin; 6, Canagliflozin; 7, TS-071; 8, Tofogliflozin/CSG-452; 9, Ipragliflozin/ASP-1941

Figure 4 Disclosed structures of C-glucoside SGLT2 inhibitors in clinical trials.

4. SGLT2 INHIBITORS

4.1. Glucoside-based inhibitors

Industrial research efforts on phlorizin-derived inhibitors prior to 2009 were analyzed based on patent publications and have been nicely summarized in recent reviews [10,14]. As mentioned in Section 3, O-glucoside inhibitors are generally far inferior to C-glucoside inhibitors. There were no reports of N-glucoside inhibitors moving forward in development, even though they had good potency and pharmacokinetic (PK) properties [14]. In addition to summaries of key data for representative SGLT2 inhibitors, these reviews also discussed the evolution of the C-glucoside class of inhibitors and presented a structure–activity relationship (SAR) overview.

Around the time of the publication of these reviews, peer-reviewed articles of C-glucoside inhibitors began to appear in the literature. Here, we summarize articles published after 2009. Patent applications for C-glucoside inhibitors published after 2009, which were extensions of previously reviewed patents, are not summarized here.

All C-glucoside inhibitors published thus far are structurally related to dapagliflozin. Analogous to dapagliflozin, all of the C-glucoside inhibitors have an aromatic aglycone moiety (A ring) at the C1 position of the glucoside (Figure 5). The aglycone moiety is substituted 1,3 with a glucose-like moiety and a methylene-linked second planar ring (B ring). Recently reported modifications to this structure fall into three groups: the center A ring and its substitutions, the distal B ring and its substitutions, and glucoside modifications (Figure 5).

The B ring and its *para* R^4 substituent are tolerant of many changes. The most noteworthy examples are canagliflozin and ipragliflozin, for which the B ring is a thiophene and benzothiophene, respectively [17,29]. Replacement of the ethoxy group in dapagliflozin with 2-cyclopropoxyethoxy generated EGT1442 (**10**), which was evaluated extensively in *in vivo* studies, including UGE studies in Sprague–Dawley (SD) rats and dogs, an antihyperglycemic study in db/db mice, and a study evaluating prolonged survival effect of **10** in spontaneously hypertensive stroke prone (SHRSP) rats [32]. Other B ring modifications comprising substituted pyridazine, pyrimidine, thiazole, and thiadiazole generally resulted in loss of potency compared to dapagliflozin [33–37]. Recently, disclosed variations of the A ring comprise substitution at the C6′ position (R^2, Figure 5), such as **11a** and **11b** [38], spiro connection of the A ring to the glucoside (**12** and **13**), and heterocyclic replacements [39–43]. These modifications generally afforded potent compounds (such as **11a**–**13**) with the exception of the A ring heterocyclic replacements. The glucoside moiety does not tolerate many changes. Recently published glucoside

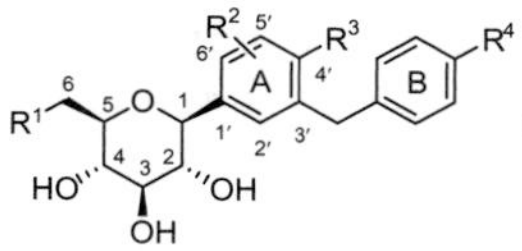

Typical modifications

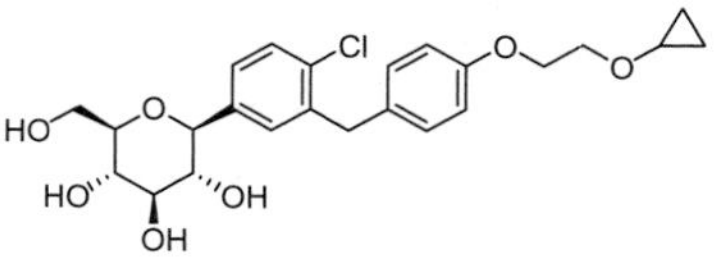

10, EGT1442, SGLT2/1 IC_{50} = 2nM/5600 nM

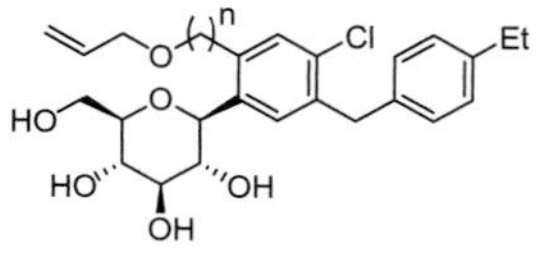

11a, n = 1, SGLT2/1 IC_{50} = 52 nM/91,000 nM
11b, n = 2, SGLT2/1 IC_{50} = 12 nM/31,000 nM

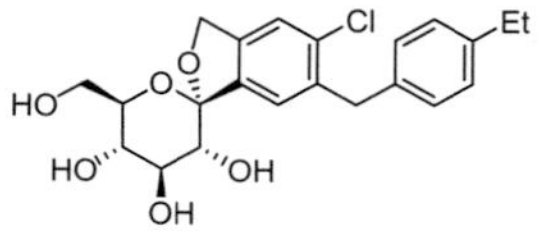

12, SGLT2/1 IC_{50} = 6.6 nM/620 nM

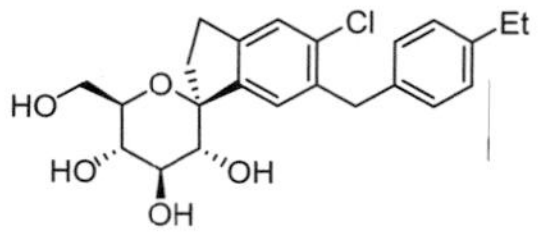

13, SGLT2/1 IC_{50} = 0.3 nM/5600 nM

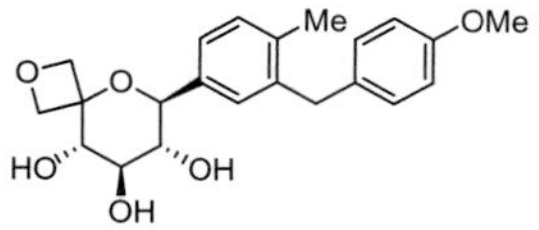

14, SGLT2/1 IC_{50} = 6.6 nM/1540 nM

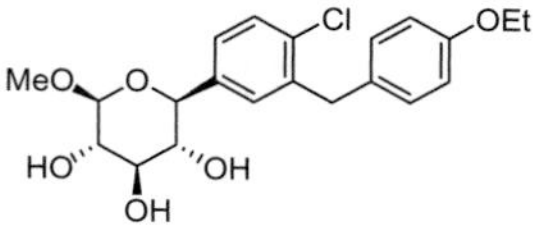

15, SGLT2/1 IC_{50} = 14 nM/1900 nM

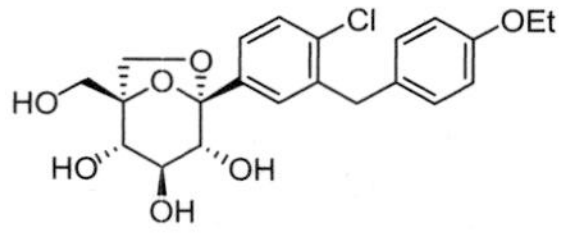

16, SGLT2/1 IC_{50} = 0.88 nM/1960 nM

Figure 5 Exemplary modifications of glucoside-based SGLT2 inhibitors.

modifications include changes at the C6 position (such as **14**) [44,45], methoxy substitution of the C5 hydroxymethyl group (L-xylose derivatives, **15**) [46], incorporation of a [1–3]-bridged ketal system (**16**), and thioglucoside replacement of the glucoside [27,47]. Thioglucoside replacement in combination with minor modifications of the aglycone moiety produced TS-071 [27].

EGT1442 and L-xylose derivatives demonstrated excellent efficacy in animals [32,46]. However, a few recently reported SGLT2 inhibitors, such as **13** and **14**, were not as efficacious at promoting UGE as dapagliflozin in SD rats, despite *in vitro* potency that was comparable or better than dapagliflozin [41,45]. The inferior *in vivo* efficacy profiles were attributed to less than optimal PK properties in rats.

4.2. Non-glucoside-based inhibitors

While hundreds of patent applications disclosing glucoside-based SGLT2 inhibitors have published, only five applications disclosing non-glucoside SGLT2 inhibitors, all from one group, have appeared [48–52]. Representative structures from these disclosures include imidazopyrimidine **17** and imidazopyridine **18** (Figure 6), and only SGLT2 IC_{50} ranges (10–1000 nM) were reported. This disparity also exists in the peer-reviewed publications: there are only two recent articles reporting one industrial

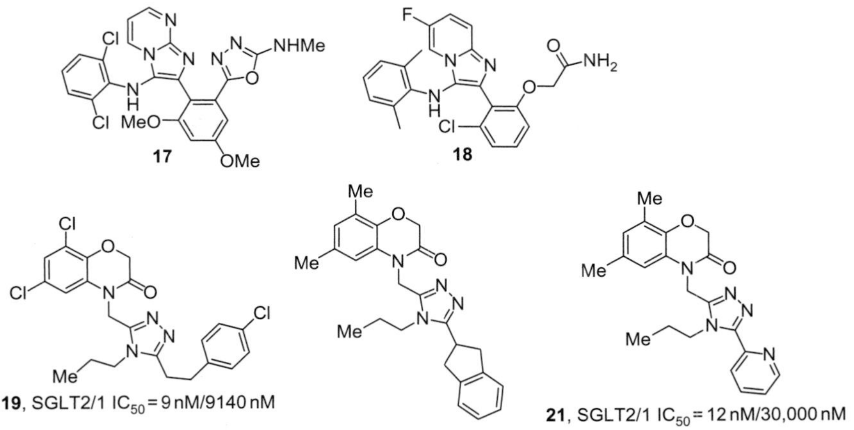

17

18

19, SGLT2/1 IC_{50} = 9 nM/9140 nM

20, SGLT2/1 IC_{50} = 5 nM/30,000 nM

21, SGLT2/1 IC_{50} = 12 nM/30,000 nM

Figure 6 Non-glucoside-based SGLT2 inhibitors.

research effort on non-glucoside-based SGLT2 inhibitors [53,54]. A class of benzooxazinone SGLT2 inhibitors was discovered through high-throughput screening (HTS). Optimization afforded compounds such as **19–21** (Figure 6), which possessed similar SGLT2 potency and better SGLT1 selectivity compared to dapagliflozin. The compounds also completely displaced [^{3}H]-dapagliflozin in a binding assay (IC_{50} 16 nM for **19**). However, these compounds were judged to have inadequate microsomal stability. No *in vivo* efficacy data were reported for these compounds [53,54].

In addition to this single HTS and optimization effort, only one other company has reported efforts to develop assays suitable for HTS against SGLT2 [55]. Given the large number of patent applications related to phlorizin-based SGLT2 inhibitors and clinical proof of concept for this mechanism, identification of non-phlorizin-based inhibitors is attractive. It seems likely that additional HTS campaigns have been conducted, but the lack of patent applications and publications suggests that tractable chemotypes structurally unrelated to phlorizin have been difficult to identify. This apparent intractability could be related to an inability to identify HTS hits, or more likely, it could be related to difficulties encountered in hit optimization. With regard to the latter, it is noteworthy that at least some SGLT2 inhibitors in clinical development possess distinct PK/pharmacodynamic (PD) characteristics.

In a Phase I single ascending dose study of dapagliflozin, a near-maximal PD response (~3 g/h UGE) was maintained in healthy volunteers for at least 24 h after a single dose of 20 mg, while the plasma concentration decreased to a range of 10–20 nM at 24 h from a C_{max} of

600–700 nM [56]. Taking into account plasma protein binding, this corresponds to an unbound dapagliflozin plasma concentration of <2 nM 24 h post-dose which is in the vicinity of its *in vitro* SGLT2 IC_{50} (1.1 nM). To the best of our knowledge, there has not been any published data suggesting that *in vitro* potency overestimates the *in vivo* drug concentration needed to inhibit SGLT2. BI 10773 has a human PK/PD profile similar to that of dapagliflozin such that, after administration of single doses of BI 10773 to healthy volunteers, rapid onset of UGE responses occurred and they were maintained long after plasma concentrations had diminished [57,58]. Plasma levels of BI 10773 peaked at about 2 h, while maximal UGE rates (5.2 g/h at the 400 mg dose) occurred at about 7 h across a range of doses (10–800 mg) and did not drop nearly as rapidly as plasma concentrations [57,58]. It is also interesting to note that following oral administration of a 1 mg/kg dose of TS-071, a close structural analog of dapagliflozin, rats exhibited kidney/plasma ratios of 35 at 4 h post-dose despite the fact that TS-071 was primarily excreted by hepatic metabolism [27].

Taken together, the available data with dapagliflozin, BI 10773, and TS-071 suggest that the glucoside-based SGLT-2 inhibitors may preferentially distribute to the site of action in the kidney and/or have a slow off-rate from SGLT2, resulting in the observed favorable PK/PD profiles. Favorable distribution to the kidneys could be a result of active renal secretion affording high local drug concentrations in the proximal tubule. However, dapagliflozin has low renal clearance that is insignificant compared to its hepatic clearance [59]. On the other hand, renal secretion delivering pharmacologically relevant concentrations of dapagliflozin to the proximal tubule could be masked by renal reabsorption [60,61]. The human metabolite profile of dapagliflozin suggests that active metabolites do not significantly contribute to the PD response especially since dapagliflozin is primarily eliminated as a pharmacologically inactive glucuronide metabolite [59].

In summary, the mechanism responsible for the favorable PK/PD characteristics of glucoside SGLT2 inhibitors in development remains unclear. However, the inherent molecular properties of the inhibitors underlying their favorable PK/PD properties may be difficult to confer to non-glucoside SGLT2 inhibitors and could explain why these have not featured prominently in the patent or primary literature to date.

5. CONCLUSION

Inhibition of renal glucose reabsorption by SGLT2 inhibitors and subsequent glucose excretion into urine is a unique mechanism of action to lower blood glucose levels. Recent clinical data demonstrate that this potential new insulin-independent antidiabetic therapy not only can reduce HbA1c levels as effectively well as existing therapeutic agents but also confers other beneficial features, such as body weight loss and low propensity for causing hypoglycemia. Overall, the available data show that SGLT2 inhibitors have demonstrated good benefit-risk profiles in human clinical trials. The U.S. Food and Drug Administration accepted a New Drug Application for dapagliflozin for review in March, 2011. It is hoped that dapagliflozin and other SGLT2 inhibitors will become important treatment options for type 2 diabetic patients.

REFERENCES

[1] M. A. Abdul-Ghani and R. A. DeFronzo, *Endocr. Pract.*, 2008, **14**, 782.
[2] Centers for Disease Control, National Diabetes Fact Sheet, United States, 2011, www.cdc.gov/diabetes/pubs/pdf/ndfs_2011.pdf (accessed April 13, 2011).
[3] R. A. DeFronzo, *Diabetes*, 2009, **58**, 773.
[4] E. M. Wright, D. D. F. Loo, B. A. Hirayama and E. Turk, *Physiology*, 2004, **19**, 370.
[5] V. Vollon, K. A. Platt, R. Cunard, J. Schroth, J. Whaley, S. C. Thomson, H. Koepsell and T. Rieg, *J. Am. Soc. Nephrol.*, 2010, **21**, 2059.
[6] S. Scholl-Burgi, R. Santer and J. H. H. Ehrich, *Nephrol. Dial. Transplant.*, 2004, **19**, 2394.
[7] R. Santer and J. Calado, *Clin. J. Am. Nephrol.*, 2010, **5**, 133.
[8] R. C. Morris and H. E. Ives, B. M. Brenner (Ed.), The Kidney, Saunders, Philadelphia, PA, 1996 p. 1764.
[9] E. M. Wright, *Am. J. Physiol. Renal Physiol.*, 2001, **290**, F10.
[10] W. N. Washburn, *J. Med. Chem.*, 2009, **52**, 1785.
[11] M. Koffler, T. Imamura, F. Santeosanio and J. H. Helderman, *Diabetologia*, 1988, **31**, 228.
[12] L. Rossetti, D. Smith, G. I. Shulman, D. Papachristou and R. A. DeFronzo, *J. Clin. Invest.*, 1987, **79**, 1510.
[13] E. M. Wright, E. Turk and M. G. Martin, *Cell Biochem. Biophys.*, 2002, **36**, 115.
[14] W. N. Washburn, *Expert Opin. Ther. Patents*, 2009, **19**, 1485.
[15] W. Meng, B. A. Ellsworth, A. A. Nirschl, P. J. McCann, M. Patel, R. N. Girotra, G. Wu, P. M. Sher, E. P. Morrison, S. A. Biller, R. Zahler, P. P. Deshpande, A. Pullockaran, D. L. Hagan, N. Morgan, J. R. Taylor, M. T. Obermeier, W. G. Humphreys, A. Khanna, L. Discenza, J. G. Robertson, A. Wang, S. Han, J. R. Wetterau, E. B. Janovitz, O. P. Flint, J. M. Whaley and W. N. Washburn, *J. Med. Chem.*, 2008, **51**, 1145.
[16] M. S. Kipnes, *Clin. Invest.*, 2011, **1**, 145–156.
[17] S. Nomura, S. Sakamaki, M. Hongu, E. Kawanishi, Y. Koga, T. Sakamoto, Y. Yamamoto, K. Ueta, H. Kimata, K. Nakayama and M. Tsuda-Tsukimoto, *J. Med. Chem.*, 2010, **53**, 6355.
[18] Clinical trials.gov, http://clinicaltrials.gov/ct2/results?term=canagliflozin; Phase III studies: NCT01106690, NCT01064414, NCT01137812, NCT01081834, NCT01106625, NCT01106677, NCT01106651, NCT01032629, NCT00968812 (accessed April 14, 2011).
[19] R. Grempler, L. Thomas, M. Eckhardt, F. Himmels-Bach, A. Sauer, M. Mark and P. Eikelmann, 69th American Diabetes Association 2009 Poster No. 521, New Orleans, Louisiana, 2009.
[20] Clinical trials.gov, http://www.clinicaltrials.gov/ct2/results?term=BI10773, Phase III studies: NCT01289990, NCT01177813, NCT01164501, NCT01210001, NCT01306214, NCT01131676, NCT01159600, NCT01167881, NCT01257334 (accessed April 14, 2011).
[21] E. Kurosaki, A. Tahara, M. Yokono, D. Yamajuku, T. Takasu, M. Imamura, T. Funatsu and Q. Li, American Diabetes Association 2010, Poster No. 570, Orlando, Florida, 2010.

[22] S. Schwartz, S. Klasen, D. Kowalski and B. Akinlade, 70th American Diabetes Association 2010, Poster No. 566, Orlando, Florida, 2010.
[23] T. Sato, M. Nishimoto, N. Taka, Y. Ohtake, K. Takano, K. Yamamoto, M. Ohmori, M. Yamaguchi, K. Takami, S. Yeu, K. Ahn, H. Matsuoka, M. Suzuki, H. Hagita, K. Ozawa, K. Yamaguchi, M. Kato and S. Ikeda, 240th American Chemical Society National Meeting 2010, Abstract MEDI-202, Boston, MA, 2010.
[24] S. Ikeda, Chugai Pharmaceutics R&D conference, December 2009, http://www.chugai-pharm.co.jp/html/meeting/pdf/091207eR&D.pdf , (accessed April 13, 2011)..
[25] Clinical trials.gov. Study NCT01096667, http://www.clinicaltrials.gov/ct2/show/NCT01096667?term=NCT01096667&rank=1 (accessed April 13, 2011).
[26] D. Powell, Endocrine Society Annual Meeting 2010 Oral presentation OR24–6, San Diego, California, http://www.lexicon-genetics.com/lexpha5/images/pdfs/LX4211_ ENDO2010_Presentation.pdf (accessed April 18, 2011).
[27] H. Kakinuma, T. Oi, Y. Hashimoto-Tsuchiya, M. Arai, Y. Kawakita, Y. Fukasawa, I. Iida, N. Hagima, H. Takeuchi, Y. Chino, J. Asami, L. Okumura-Kitajima, F. Io, D. Yamamoto, N. Miyata, T. Takahashi, S. Uchida and K. Yamamoto, *J. Med. Chem.*, 2010, **53**, 3247.
[28] Thomson Reuters Integrity, https://integrity.thomson-pharma.com/integrity/xmlxsl/pk_qcksrch.show_records?sessionID=1&history=&query=tofogliflozin&abbreviation=PRO&language=en (accessed April 13, 2011).
[29] Thomson Reuters Integrity, https://integrity.thomson-pharma.com/integrity/xmlxsl/pk_qcksrch.show_records?sessionID=1&history=&query=asp%201941&abbreviation=PRO&language=en (accessed April 13, 2011).
[30] E. Ferrannini, W. Tang, S. J. Ramos, J. F. List and A. Salsali, *Diabetes Care*, 2010, **33**, 2217.
[31] M. Pfister, J. M. Whaley, L. Zhang and J. F. List, *Clin. Pharmacol. Ther.*, 2011, **89**, 621.
[32] W. Zhang, A. Welihinda, J. Mechanic, H. Ding, L. Zhu, Y. Lu, Z. Deng, Z. Sheng, B. Lv, Y. Chen, J. Y. Robergeb, B. Seed and Y. Wang, *Pharamcol. Res.*, 2011, **63**, 284.
[33] M. J. Kim, J. Lee, S. Y. Kang, S. H. Lee, E. J. Son, M. E. Jung, S. H. Lee, K. S. Song, M. Lee, H. K. Han, J. Kim and J. Lee, *Bioorg. Med. Chem. Lett.*, 2010, **20**, 3420.
[34] J. Lee, J. Y. Kim, J. Choi, S. H. Lee, J. Kim and J. Lee, *Bioorg. Med. Chem. Lett.*, 2010, **20**, 7046.
[35] S. Y. Kang, K. S. Song, J. Lee, S. H. Lee and J. Lee, *Bioorg. Med. Chem.*, 2010, **18**, 6089.
[36] K. S. Song, S. H. Lee, M. J. Kim, H. J. Seo, J. Lee, S. H. Lee, M. E. Jung, E. J. Son, M. Lee, J. Kim and J. Lee, *ACS Med. Chem. Lett.*, 2011, **2**, 182.
[37] J. Lee, S. H. Lee, H. J. Seo, E. J. Son, S. H. Lee, M. E. Jung, M. Lee, H. K. Han, J. Kim, J. Kan and J. Lee, *Bioorg. Med. Chem.*, 2010, **18**, 2178.
[38] B. Xu, Y. Feng, B. Lv, G. Xu, L. Zhang, J. Du, K. Peng, M. Xu, J. Dong, W. Zhang, T. Zhang, L. Zhu, H. Ding, Z. Sheng, A. Welihinda, B. Seed and Y. Chen, *Bioorg. Med. Chem.*, 2010, **18**, 4422.
[39] B. Xu, B. Lv, Y. Feng, G. Xu, J. Du, A. Welihinda, Z. Sheng, B. Seed and Y. Chen, *Bioorg. Med. Chem. Lett.*, 2009, **19**, 5632.
[40] B. Lv, B. Xu, Y. Feng, K. Peng, G. Xu, J. Du, L. Zhang, W. Zhang, T. Zhang, L. Zhu, H. Ding, Z. Sheng, A. Welihinda, B. Seed and Y. Chen, *Bioorg. Med. Chem. Lett.*, 2009, **19**, 6877.
[41] B. Lv, Y. Feng, J. Dong, M. Xu, B. Xu, W. Zhang, Z. Sheng, A. Welihinda, B. Seed and Y. Chen, *ChemMedChem*, 2010, **5**, 827.
[42] C. H. Yao, J. S. Song, C. T. Chen, T. K. Yeh, M. S. Hung, C. C. Chang, Y. W. Liu, M. C. Yuan, C. J. Hsieh, C. Y. Huang, M. H. Wang, C. H. Chiu, T. C. Hsieh, S. H. Wu, W. C. Hsiao, K. F. Chu, C. H. Tsai, Y. S. Chao and J. C. Lee, *J. Med. Chem.*, 2011, **54**, 166.
[43] H. Zhou, D. P. Danger, S. T. Dock, L. Hawley, S. G. Roller, C. D. Smith and A. L. Handlon, *ACS Med. Chem. Lett.*, 2010, **1**, 19.
[44] E. J. Park, Y. Kong, J. S. Lee, S. H. Lee and J. Lee, *Bioorg. Med. Chem. Lett.*, 2011, **21**, 742.

[45] R. P. Robinson, V. Mascitti, C. M. Boustany-Kari, C. L. Carr, P. M. Foley, E. Kimoto, M. T. Leininger, A. Lowe, M. K. Klenotic, J. I. MacDonald, R. J. Maguire, V. M. Masterson, T. S. Maurer, Z. Miao, J. D. Patel, C. Preville, M. R. Reese, L. She, C. M. Steppan, B. A. Thuma and T. Zhu, *Bioorg. Med. Chem. Lett.*, 2010, **20**, 1569.
[46] N. C. Goodwin, R. Mabon, B. A. Harrison, M. K. Shadoan, Z. Y. Almstead, Y. Xie, J. Healy, L. M. Buhring, C. M. DaCosta, J. Bardenhagen, F. Mseeh, Q. Liu, A. Nouraldeen, A. G. Wilson, S. D. Kimball, D. R. Powell and D. B. Rawlins, *J. Med. Chem.*, 2009, **52**, 6201.
[47] V. Mascitti and C. Preville, *Org. Lett.*, 2010, **12**, 2940.
[48] W. Mederski, N. Beier, B. Cezanne, R. Gericke, M. Klein and C. Tsaklakidis, *Patent Application WO2007/147478-A1*, 2007.
[49] M. Klein, R. Gericke, N. Beier, B. Cezanne, C. Tsaklakidis and W. Mederski, *Patent Application WO2008/046497-A1*, 2008.
[50] W. Mederski, N. Beier, L. T. Burgdorf, R. Gericke, M. Klein and C. Tsaklakidis, *Patent Application WO2008/ 071288-A1*, 2008.
[51] L. T. Burgdorf, B. Cezanne, M. Klein, R. Gericke, C. Tsaklakidis, W. Mederski and N. Beier, *Patent Application WO2008/101586-A1*, 2008.
[52] M. Klein, W. Mederski, C. Tsaklakidis and N. Beier, *Patent Application WO2009/049731-A1*, 2009.
[53] A. Li, J. Zhang, J. Greenberg, T. Lee and J. Liu, *Bioorg. Med. Chem. Lett.*, 2011, **21**, 2472.
[54] X. Du, M. Lizarzaburu, S. Turcotte, T. Lee, J. Greenberg, B. Shan, P. Fan, Y. Ling, J. Medina and J. Houze, *Bioorg. Med. Chem. Lett.*, 2011 21, doi:10.1016/j.bmcl.2011.04.053.
[55] M. I. Lansdell, D. J. Burring, D. Hepworth, M. Strawbridge, E. Graham, T. Guyot, M. S. Betson and J. D. Hart, *Bioorg. Med. Chem. Lett.*, 2008, **18**, 4944.
[56] B. Komoroski, N. Vachharajani, D. Boulton, D. Kornhauser, M. Geraldes, L. Li and M. Pfister, *Clin. Pharmacol. Ther.*, 2009, **85**, 520.
[57] A. Port, S. Macha, L. Seman, G. Nehmiz, G. Simons, A. Koegel, D. Harder, B. Ren, M. Iovino, S. Pinnetti and K. Dugi, 70th American Diabetes Association 2010, Poster No. 569-P, Orlando, Florida, 2010.
[58] K. Dugi and M. Mark, Boehringer-Ingelheim R&D press conference, October, 17, 2008, Biberach, Germany, 2008. http://www.boehringer-ingelheim.com /content/. . . /slides_mark_dugi.pdf (accessed April 16, 2011).
[59] M. Obermeier, M. Yao, A. Khanna, B. Koplowitz, M. Zhu, W. Li, B. Komoroski, S. Kasichayanula, L. Discenza, W. Washburn, W. Meng, B. A. Ellsworth, J. M. Whaley and W. G. Humphreys, *Drug Metab Dispos.*, 2010, **38**, 405.
[60] B. Feng, J. L. LaPerle, G. Chang and M. Varma, *Expert Opin. Drug Metab. Toxicol.*, 2010, **6**, 939.
[61] M. Li, G. D. Anderson and J. Wang, *Expert Opin. Drug Metab. Toxicol.*, 2006, **2**, 505.

PART III: Inflammation/Pulmonary/Gastrointestinal Diseases

Editor: David S. Weinstein
Bristol-Myers Squibb R&D
Princeton
New Jersey

CHAPTER **8**

Recent Advances in the Discovery and Development of CRTh2 Antagonists

Laurence E. Burgess

Contents

1. INTRODUCTION

Prostaglandin D_2 (PGD_2, **1**), produced from arachidonic acid *via* cyclooxygenase and prostaglandin synthases, is a potent mediator of allergic inflammation. Primarily released from mast cells in response to immunoglobulin E (IgE)-mediated degranulation, this prostanoid undergoes rapid metabolism. Both parent PGD_2 and the resulting metabolites possess proinflammatory actions contributing to early and late phase inflammatory events. The function of these mediators appears to be largely unaffected by existing therapeutics, and it has been postulated that if the proinflammatory actions of PGD_2 could be mitigated, significant

Array BioPharma, 3200 Walnut Street, Boulder, CO 80301, USA

Annual Reports in Medicinal Chemistry, Volume 46
ISSN: 0065-7743, DOI: 10.1016/B978-0-12-386009-5.00019-9

benefit could be realized for patients suffering from allergic diseases. Indeed, the pharmacology of PGD_2 and attempts to control it *via* receptor antagonism have been reviewed [1].

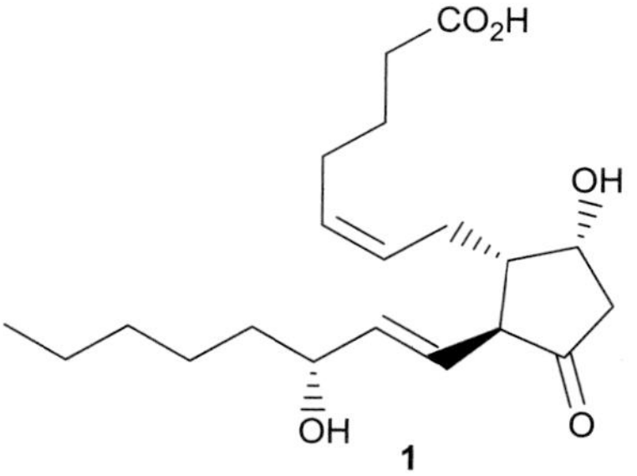

1

PGD_2 is recognized by two G-protein-coupled seven transmembrane receptors, DP (also known as DP1) and CRTh2 (*c*hemoattractant *r*eceptor-homologous molecule expressed on *Th2* cells; also known as DP2 or GPR44 or CD294). Several PGD_2 metabolites formed *in vivo* lose significant affinity for DP and retain affinity for CRTh2. Thus, of these two, CRTh2 has emerged as the relevant receptor mediating the prolonged proinflammatory activity of PGD_2 [2]. The recent disclosure that a selective DP antagonist provided no benefit in two clinical studies in asthma and allergic rhinitis further supports the focus on CRTh2 as the therapeutically relevant PGD_2 receptor in allergic disease [3]. CRTh2 is expressed on eosinophils, basophils, and Th2 effector T cells and mediates the activation and chemotaxis of these cell types. Specifically, the PGD_2/CRTh2 system regulates respiratory burst and degranulation of eosinophils [4], histamine release from basophils [5], and the production of IL-4, IL-5, and IL-13 from $CD4^+$ $CRTh2^+$ T cells without a costimulatory signal [6]. Such pathophysiology has been strongly associated with asthma, allergic rhinitis, and atopic dermatitis—three diseases that afflict millions of patients worldwide. A safe and effective CRTh2 antagonist could provide benefit to these patients.

Ramatroban (**2**), an indole propionic acid with moderate CRTh2 antagonism activity ($K_i = 137$ nM), is approved for use in allergic rhinitis in Japan [7]. Clinical studies comparing it to antihistamine treatment demonstrated that ramatroban provided a significant improvement in the control of signs and symptoms of perennial allergic rhinitis, particularly nasal obstruction [8]. While primarily a thromboxane receptor antagonist, the CRTh2 antagonism also provided by this drug has been speculated to contribute to this efficacy. Accordingly, selective CRTh2 antagonists have demonstrated efficacy in preclinical models of allergic rhinitis. In an ovalbumin (OVA)-sensitized mouse model involving intranasal challenges, both the early and late phase response as measured by changes in respiratory frequency were inhibited with a potent and selective CRTh2 antagonist, ARRY-005 (IC_{50} = 35 nM; structure not disclosed) [9]. Reduced levels of IL-4, IL-5, and IL-13 were also noted in nasal tissue compared to controls. Thus, while ramatroban has provided benefit to allergic rhinitis patients, perhaps more potent and selective CRTh2 antagonists will supply even greater benefit.

CRTh2 antagonists have also shown benefit in several rodent models of asthma; not only with OVA and cockroach antigen-based protocols but also in house dust mite induced allergic responses [10]. Unlike other allergens commonly used in preclinical models of allergy, house dust mite allergen is clinically relevant and reproduces signs and symptoms of the human disease including structural remodeling of the airway [11]. In this model, pulmonary inflammation, mucus hypersecretion and mucus cell metaplasia were inhibited by AM156 (*vide infra*) at a dose of 10 mg/kg. Clinically, several human studies have been completed or are underway to evaluate the potential for CRTh2 antagonism in allergic asthma. In a 4-week study treating mild to moderate asthmatics with daily doses of an oral CRTh2 antagonist, OC000459 (identified by others as structure **3** [12]), a significant increase in forced expiratory volume in one second (FEV_1) was noted versus placebo along with decreased serum IgE and sputum eosinophil counts [13]. These results spurred the initiation of a second trial in a greater number of patients involving 17 weeks of daily oral dosing but results have not been published [14]. Another CRTh2 antagonist, AZD1981 (structure not disclosed), is presently being evaluated in a worldwide asthma study involving over 1000 patients of varying disease severity [15]. Ultimately, patient selection criteria could significantly influence clinical study outcomes. To date, about half the trials have not selected for any degree of asthma severity among enrolled patients, while the other half have targeted mild to moderate asthmatics. Recently, levels of PGD_2 were shown to be increased in lung fluid of severe asthmatics and correlated with disease severity suggesting that these patients may benefit from a CRTh2 antagonist [16].

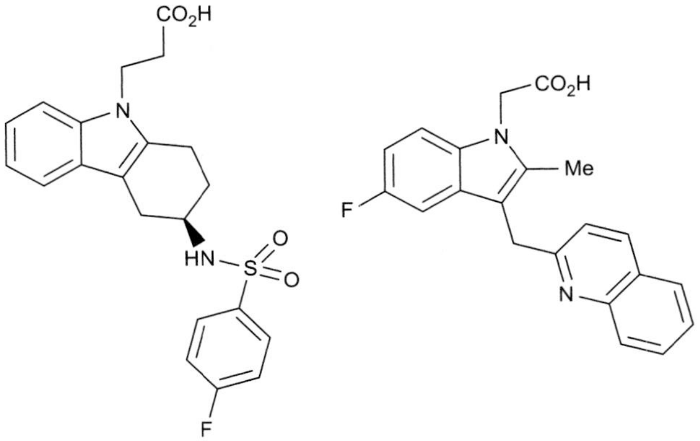

2 **3**

The cutaneous allergic response in rodents can also be controlled *via* CRTh2 modulation. In one murine model, involving epicutaneous sensitization of tape-stripped skin (mimicking scratching) in OVA-sensitized animals, skin PGD_2 levels, infiltration of inflammatory cells, and Th2 cytokine mRNA were significantly increased 24 h after the mechanical injury. Applying this same model to CRTh2−/− mice, the inflammatory infiltrate and cytokine message were significantly decreased [17]. A comparable study utilizing a selective CRTh2 antagonist in mice with functional CRTh2 produced similar results [18]. In yet another study, a selective CRTh2 antagonist controlled various skin endpoints and pruritus in the Nc/Nga mouse model of atopic dermatitis [9]. Additional data that demonstrate upregulation of CRTh2 expression in Th2 T cells and eosinophils isolated from atopic dermatitis patients compared to healthy subjects provide further support for the role of CRTh2 in skin disease; however, there have been no reports of clinical studies designed to explore CRTh2 antagonism for dermal indications [19,20].

CRTh2 antagonism may also provide benefit to patients suffering from chronic obstructive pulmonary disease (COPD). In a mouse model involving exposure to cigarette smoke, CRTh2 antagonism reduced cellular inflammation, mucus cell metaplasia, and epithelial hyperplasia in the airway [21]. While these results are encouraging, it is also important to note that murine neutrophils express CRTh2 under basal conditions while human neutrophils do not. However, based on precedence in cystic fibrosis and other human cellular studies, it might be expected that human neutrophils could upregulate CRTh2 [22]. Indeed, CRTh2 upregulation in lipopolysaccharide- and formyl peptide (fMLP)-stimulated neutrophils was recently disclosed [23]. There have also been clinical studies in COPD involving treatment with a CRTh2 antagonist, but no results have been reported [24].

The clinical studies involving CRTh2 antagonists were recently reviewed [12]. In addition to efficacy studies, several clinical trials involving CRTh2 antagonists have been designed to explore safety, specifically drug–drug interactions and CYP induction [25]. These appear to have been designed to explore compound specific properties and are probably not related to antagonism of the PGD_2/CRTh2 system.

2. CRTH2 ANTAGONISTS

CRTh2 antagonists are typically assessed for receptor affinity using standard membrane binding assays and are reported here as either an IC_{50} or a K_i. Almost all CRTh2 antagonists disclosed to date can be broadly classified as hydrophobic acids [26]. Since hydrophobic organic anions of medium size (100–600 Da) often bind nonspecifically (and strongly) to serum proteins like albumin, the concentration of exogenous protein in these binding assays is important to note when making comparisons among compounds so that protein shifts can be put into context [27]. Ideally, the use of 4% human serum albumin (HSA) in assay media would mimic a human whole blood setting [28]; however, lower concentrations of HSA or BSA (bovine serum albumin) are often employed. An additional method for evaluating the functional activity of CRTh2 antagonists is eosinophil shape change (ESC). In response to PGD_2, human eosinophils prepare for chemotaxis by activating intracellular motile machinery resulting in a shape change that is readily assessed by flow cytometry. ESC is mediated by CRTh2 and can be effectively antagonized; furthermore, when conducted in human whole blood, this assay provides meaningful context to an antagonist's potency.

2.1. Indole acetic acids

Several leads based on the structure of ramatroban have been disclosed. It was reported that contracting the carboxylic acid tether, retaining the *R* absolute stereochemistry and methylation of the sulfonamide results in a much more potent and selective CRTh2 antagonist (**4**, single enantiomer; $K_i = 1.5$ nM) [29]. Building upon those results, it was found that reversing the indole maintained potency and selectivity to provide MK-7246 (**5**, $K_i = 2.5$ nM; ESC $IC_{50} = 2.2$ nM). This compound was profiled extensively and advanced to Phase 1 clinical studies [7]. MK-7426 was characterized as a reversible, full antagonist and demonstrated moderate inhibition of CYP 2C9 *in vitro* ($IC_{50} = 9.4$ μM), low to moderate plasma clearance *in vivo* with a moderate terminal half-life and oral bioavailabilty of >57% in rat, dog, and rhesus monkey. The major metabolite of MK-7426 was acyl glucuronide (**6**). This metabolite was observed in plasma of all species dosed with MK-7426 and was most predominant in nonhuman primates (~50–150% of parent). In an attempt to reduce the potential reactivity of any resulting glucuronides, analogs bearing a substituent alpha to the carboxylic acid (**7**, **8**) were prepared but were found to lose ~100-fold affinity for CRTh2. There are very few examples of potent CRTh2 antagonists that incorporate α-substitution relative to the carboxylic acid; however, when such structural features are incorporated into carboxylic acids, lower reactivity of the corresponding glucuronide metabolites with protein or peptide nucleophiles is observed [30]. As a backup to MK-7246, aza-indole **9** was identified as having no CYP inhibition and improved off-target selectivity while retaining high potency ($K_i = 3.4$ nM; ESC $IC_{50} = 1.2$ nM) and similar pharmacokinetics upon oral dosing [31]. Compound **9** also demonstrated less nonspecific covalent binding *in vivo* as compared to MK-7426, and this was proposed to be due to the replacement of the aryl sulfonamide with a benzamide, thereby reducing potential electrophilicity. A triazole moiety (see **10**, $K_i < 5$ nM) can also replace the MK-7426 sulfonamide [32].

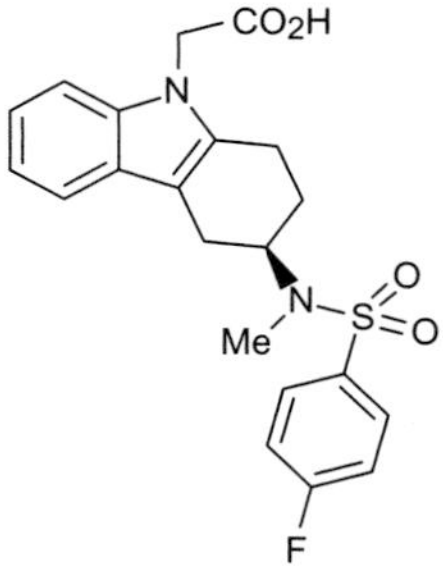

4 **9** **10**

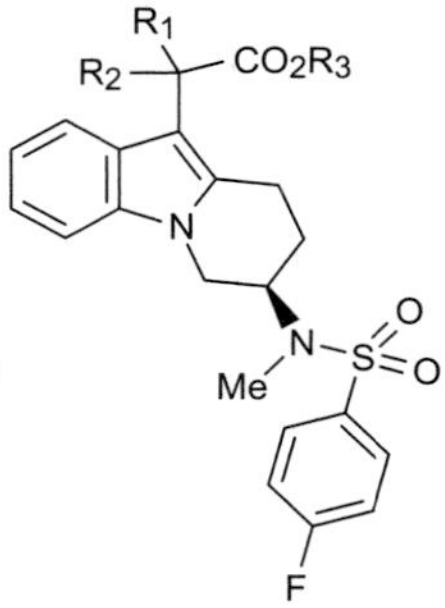

5: $R_1 = H$; $R_2 = H$; $R_3 = H$
6: $R_1 = H$; $R_2 = H$; $R_3 =$ glucuronic acid
7: $R_1 = R_2 =$ spiro-cyclopropyl; $R_3 = H$
8: $R_1 = H$; $R_2 = Me$; $R_3 = H$

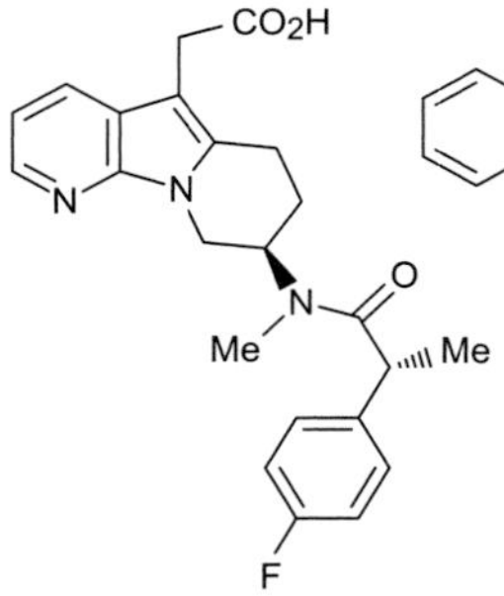

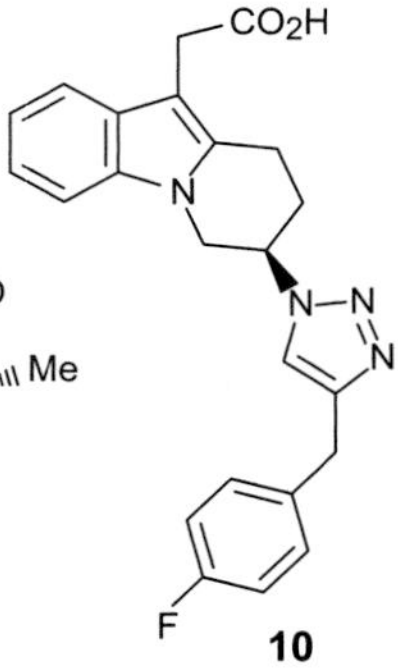

The same nitrogen shift strategy that produced MK-7426 was extended. Recently, disclosed CRTh2 antagonists **11** and **12** are also based on the tetrahydrocarbazole substructure of ramatroban, in which the nitrogen is shifted to bridgehead positions. These racemic compounds are reported to have an IC_{50} of less than 300 nM [33].

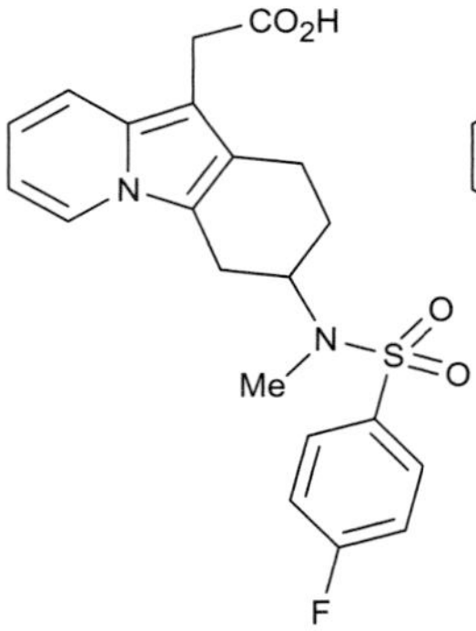

11 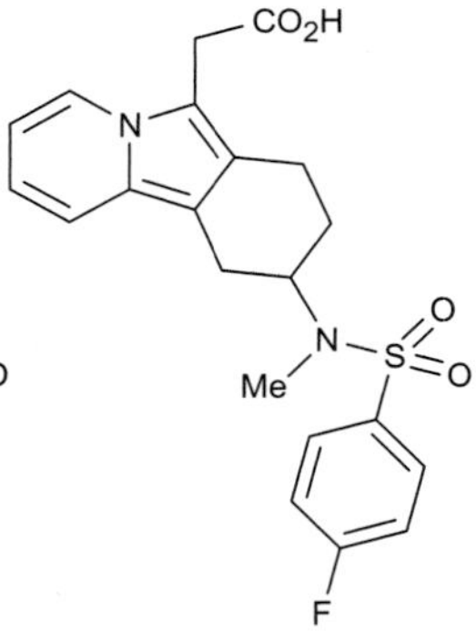 **12**

Ramatroban's tetrahydrocarbazole architecture can be disconnected in favor of less conformationally constrained substructures that still provide potent CRTh2 antagonism. Pyridyl sulfone **13** is potent ($K_i = 2$ nM; ESC $IC_{50} = 2.5$ nM) [34]. The pyridine ring can also be replaced with several other heterocycles including a thiophene (**14**; $K_i = 0.8$ nM) [35]. A related disconnection/reconnection strategy provides sultams **15** and **16** [36]. While the unsubstituted sultam, **15**, is weakly potent ($K_i = 2.8$ μM), the isoxazole substituted sultam **16** is much more so ($K_i = 12$ nM). Along with other reported SAR, this suggests the importance of a strategically placed hydrogen bond acceptor.

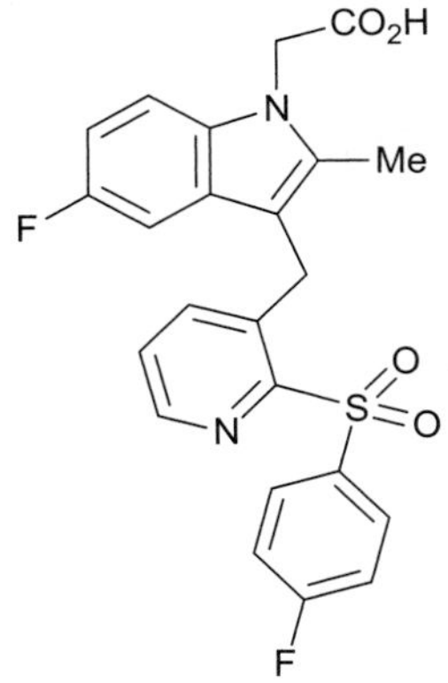

13 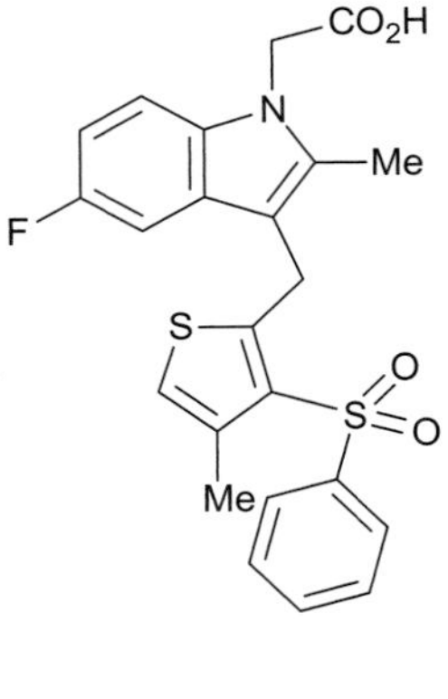 14 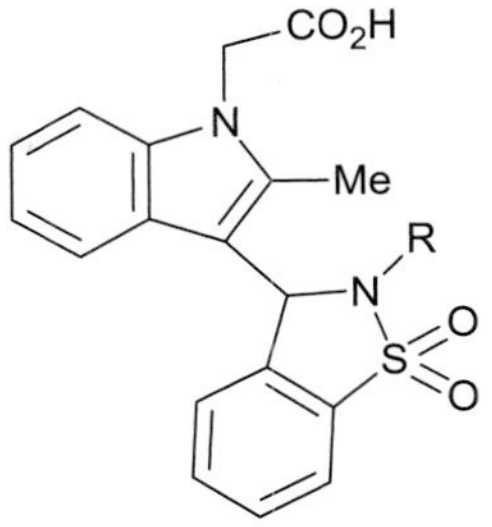 15: R = H 16: R =

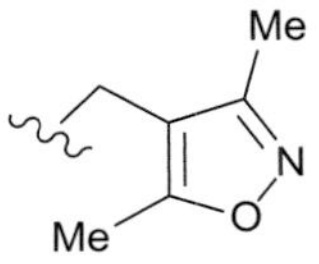

2.2. Heteroaryl acetic acids

While indole acetic acids have provided an effective template for CRTh2 antagonists, other heteroaromatic systems are also viable. Pyrrole-based antagonists such as sulfonamide **17** and sulfone **18** are potent in human whole blood assays ($IC_{50} < 5$ nM) [37]. Thiazole and pyrimidine templates (exemplified by **19** and **20**) which originated from *in silico* screening efforts have also been explored. Thiazole **19** and pyrimidine **20** demonstrated potent antagonism ($IC_{50} = 3.7$ and 1.9 nM, respectively) with no exogenous protein added [38]. Analogs of these antagonists along with molecular modeling studies have led to a proposed pharmacophore and binding alignment that involves a crucial interaction of the carboxylic acid with Lys210, a hydrogen bond to Ser266, and a hydrophobic subpocket consisting of an array of aromatic residues [39].

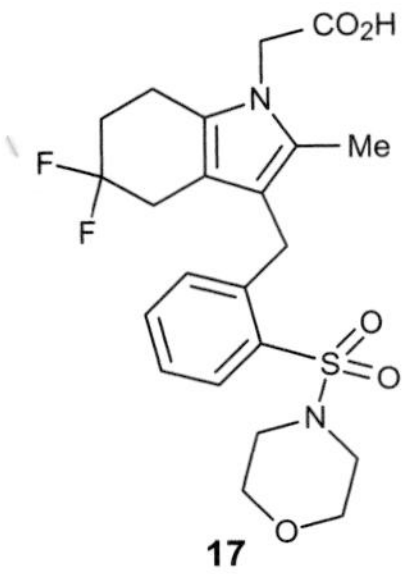

17 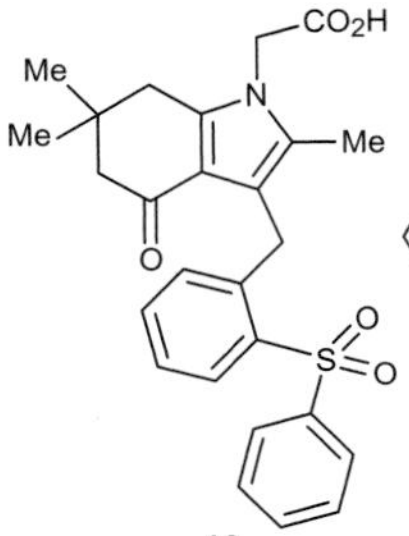 18 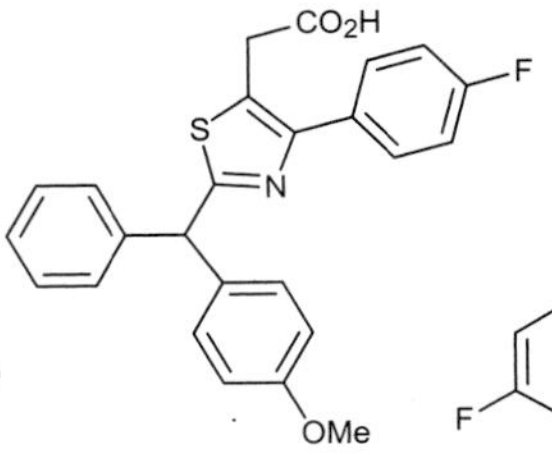 19 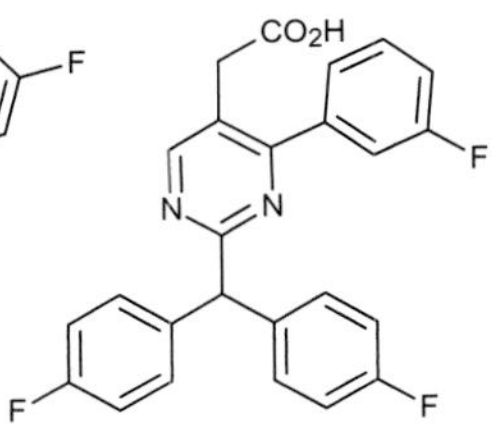 20

A series of pyrazole acetic acids have been disclosed including **21** and **22** which possess CRTh2 binding IC_{50}s of 3 nM [40]. Interestingly, several propionic acid analogs, exemplified by **23**, maintained high CRTh2 affinity (binding $IC_{50} = 3$ nM). While phenoxy acetic acids have shown high affinity for the CRTh2 receptor (*vide infra*), corresponding carbon analogs (*i.e.*, aryl propionic acids) are typically less potent.

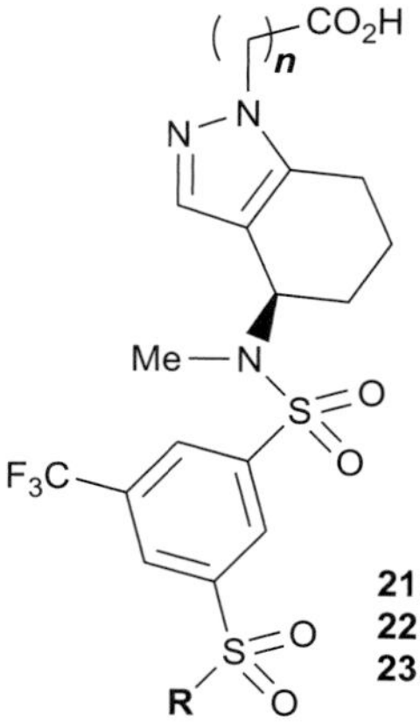

21: R = cyclopentyl; *n* = 1
22: R = (4-methyl)phenyl; *n* = 1
23: R = (4-methyl)phenyl; *n* = 2

2.3. Phenyl acetic acids

A plethora of CRTh2 antagonists based on the phenyl acetic acid template have recently been disclosed, and they fall, generally, into three structural categories: biphenyl ethers, benzophenones, and biphenyls. Of the biphenyl ether class, compound **24** ($IC_{50} = 3$ nM in buffer; $IC_{50} = 28$ nM in human plasma) was one of the first to enter clinical trials. Ultimately, compound **24** was abandoned in favor of AMG-853 (**25**) because of a reduced plasma shift. AMG-853 has been extensively profiled and also advanced to clinical studies [41]. AMG-853 is potent in both buffer ($IC_{50} = 3$ nM) and human plasma ($IC_{50} = 8$ nM) and also possesses affinity for the DP receptor ($IC_{50} = 35$ nM) [42]. In preclinical pharmacokinetic studies, this dual CRTh2/DP antagonist showed low to moderate clearance across species, excellent oral absorption, and low potential for drug–drug interactions as indicated by CYP inhibition and induction studies. In single oral dose, first-in-human clinical studies, AMG-853 pharmacokinetics were approximately dose proportional (*i.e.*, 100 mg: $C_{max} \sim 500$ ng/mL; 400 mg: $C_{max} \sim 3000$ ng/mL) and the drug was well tolerated [43]. The major metabolite was the corresponding acyl glucuronide, and this was observed at levels (based on AUC) approximating 70% of parent. As a pharmacodynamic endpoint, CRTh2 receptor internalization in response to *ex vivo* PGD_2 stimulation demonstrated that oral doses of 200 or 400 mg of AMG-853 could only suppress this phenomenon for approximately 8–10 h, suggesting that this drug may benefit from a BID dosing regimen.

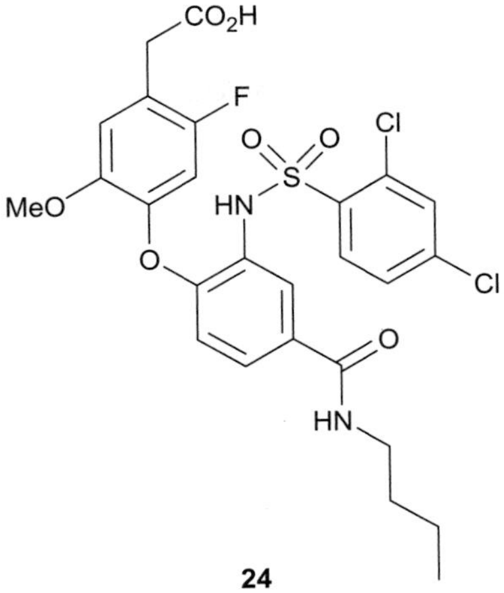

24

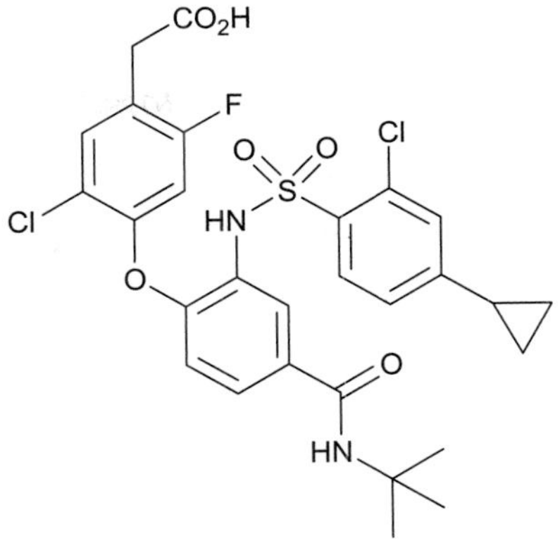

25

Another series of structurally related biphenyl ether-based phenyl acetic acid CRTh2 antagonists includes **26**, **27**, and **28** (all IC_{50}s < 300 nM) [44]. It is interesting to note that the reverse amide, **27**, and its analogous diphenyl methane, **28**, both retain CRTh2 affinity. These results suggest that the nature of the linker (benzamide *vs.* anilide or ether *vs.* alkyl) is not crucial and, comparing to AMG-853, there seems to be flexibility in the position of the linker relative to the acetic acid substituent (*i.e.*, *meta* versus *para*). The preferential use of the *tert*-butyl amide in both series of compounds also suggests a similar binding mode to AMG-853.

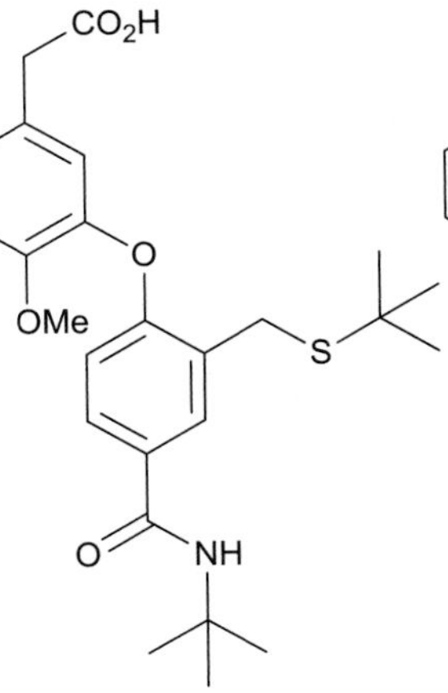

26

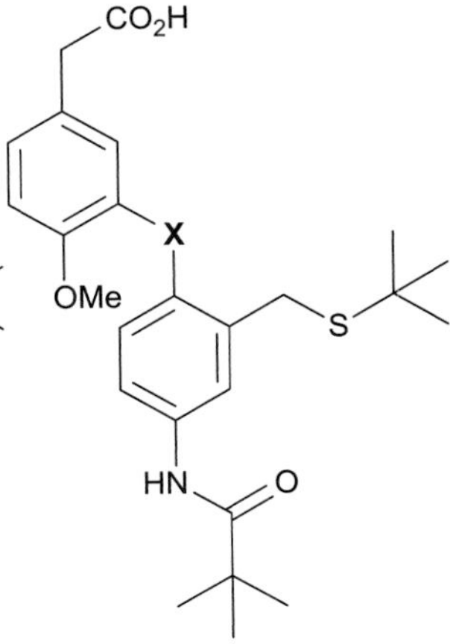

27: X = O
28: X = CH_2

A very potent series of biphenyl ether-containing phenyl acetic acid CRTh2 antagonists was recently disclosed [45]. Compounds **29–32** incorporate a 6-fluoro-3-methyl naphthalene core and utilize either an ether (**29**; IC_{50} = 2.3 nM) or a keto-linker without loss of affinity (**31**; IC_{50} = 2.1 nM) [46]. The terminal aryl sulfone can also be replaced as a piperidinesulfonamide (**32**; IC_{50} = 2.4 nM), a common functional group in many CRTh2 antagonists. In particular, compound **30**, which incorporates a bromo-substituted pyridine, demonstrates very high affinity for CRTh2 (IC_{50} = 0.02 nM).

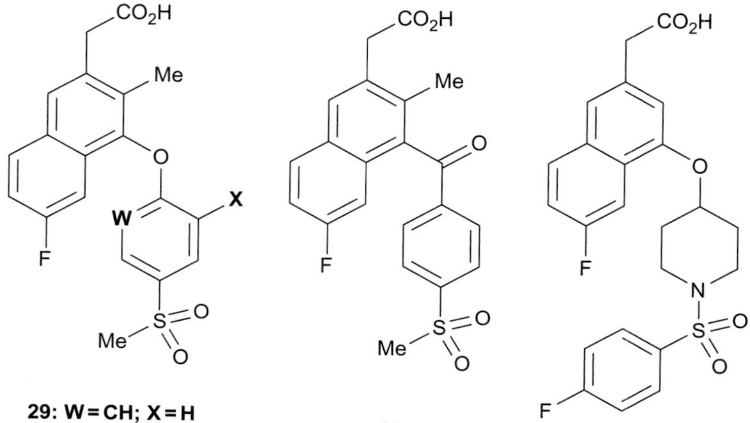

29: W = CH; X = H
30: W = N; X = Br

31

32

A series of chroman-based compounds that incorporate a biphenyl ether have also been reported to be potent CRTh2 antagonists [47]. Single enantiomers of these chiral phenyl acetic acids were prepared, and a significant difference in affinity for CRTh2 was demonstrated (**33**, eutomer; IC_{50} = 84 nM; distomer IC_{50} = 4000 nM). One of the more potent, single enantiomer antagonists in this series incorporates a pyridyl-containing biphenyl motif (**34**; IC_{50} = 22 nM).

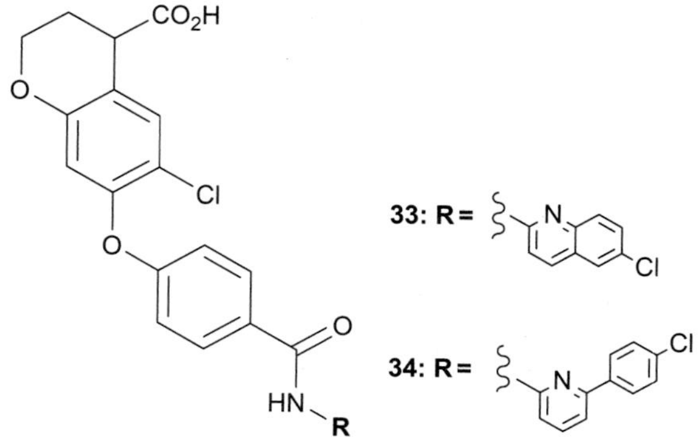

33: R =

34: R =

Additional phenyl acetic acid CRTh2 antagonists that incorporate a biaryl arrangement are represented by several active compounds related to previously disclosed AM156 (**35**; $IC_{50} = 24$ nM) [48]. CRTh2 activity is maintained when a pyridine is used in place of phenyl as evident in this series (**36**, **37**; $IC_{50} < 300$ nM). Carbamate **38** was quite potent ($IC_{50} = 8$ nM; ESC $IC_{50} = 5$ nM), demonstrated good selectivity over other prostanoid receptors and good pharmacokinetics in rat ($t_{1/2} = 8.7$ h, $F = 77\%$) and dog ($t_{1/2} = 7.4$ h, $F = 89\%$). This compound also provided similar efficacy (at a dose of 10 mg/kg) to dexamethasone (also at a dose of 10 mg/kg) in a murine model of allergic rhinitis. The carbamate was later identified as a metabolic and potential toxicological liability. In an effort to eliminate this functionality, the biphenyl was elaborated to a pyridyl-containing triarene, and the cyclopropyl amide was reintroduced to provide AM432 (**39**; $IC_{50} = 31$ nM) [49]. AM432 was considered as a clinical development compound and profiled extensively in animal models. At an orally administered dose of 10 mg/kg, AM432 provided protection from upper airway distress in a murine model of allergic rhinitis and also reduced neutrophil influx in the lower airway in a murine model of COPD involving cigarette smoke exposure. In addition to potent CRTh2 activity and selectivity over related prostanoid receptors, AM432 was also found to be devoid of agonist and antagonist activity at three isoforms of the PPAR receptor up to concentrations of 250 μM. Reports of CRTh2 antagonist selectivity screening versus nuclear hormone receptors have been rare; however, several clinical CYP induction and drug–drug interaction studies have been disclosed that may be linked to this activity [25].

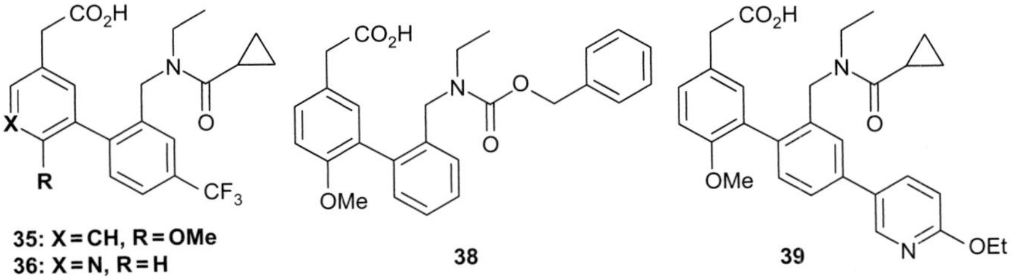

35: X = CH, R = OMe
36: X = N, R = H
37: X = N, R = OMe

38

39

2.4. Phenoxy acetic acid

A variety of phenoxy acetic acid-based CRTh2 antagonists have been recently disclosed including a series of tetrahydronaphthalene sulfonamides [50]. Unsubstituted sulfonamide **40** is potent ($IC_{50} = 6$ nM), while the methyl-substituted analog loses affinity (**41**; $IC_{50} = 74$ nM). Introduction of a sulfone (**42**) restores the CRTh2 affinity ($IC_{50} = 3$ nM).

40: $R_1 = H$, $R_2 = CF_3$
41: $R_1 = Me$, $R_2 = CF_3$
42: $R_1 = Me$, $R_2 = SO_2$-cyclopentyl

Based on the ramatroban structure, a virtual screen identified compound **43** ($IC_{50} = 16$ nM) as a CRTh2 antagonist lead [51]. Subsequent lead optimization (including side-chain modification and specific stereochemical substitutions along with physiochemical property improvements) provided amide **44** ($IC_{50} = 0.5$ nM) which demonstrated reasonable rat pharmacokinetics (Cl = 15 mL/min/kg; Vss = 1.6 L/kg; $F = 22\%$). An interesting feature of these compounds is the incorporation of a basic amine ($pK_a = 6.8$) resulting in zwitterions. This might explain the volume of distribution of **44**, which is generally higher than most CRTh2 antagonists that could be generally classified as lipophilic acids. It is worth noting that phenoxy acetic acid **44** could be substituted alpha to the carboxylic acid without much loss of affinity (**45**; $IC_{50} = 1.3$ nM); meanwhile, the carbon analog of **44**, propionic acid **46**, was significantly less potent ($IC_{50} = 79$ nM) despite having the same atom count.

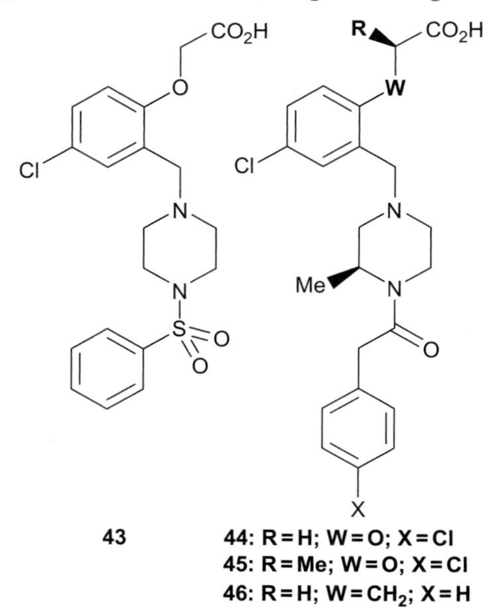

2.5. Noncarboxylic acids

There have been sporadic reports of CRTh2 antagonist structures that lack a carboxylic acid or mimic thereof. Recently, optimization of *cis*-substituted, racemic tetrahydroquinoline **47** ($IC_{50} = 43$ nM), identified through high-throughput screening, provided **48** ($IC_{50} = 26$ nM) [52]. The enantiomers of **48** were isolated and the more active single enantiomer provided potent CRTh2 affinity in the presence of 50% human plasma ($IC_{50} = 106$ nM) and good rat pharmacokinetics (Cl = 12 mL/min/kg; $F = 38\%$). Interestingly, the sulfonamide analog, **49**, lost all affinity for CRTh2 ($IC_{50} > 10{,}000$ nM) while introduction of a carboxylic acid greatly enhanced CRTh2 affinity (**50**; $IC_{50} = 5$ nM).

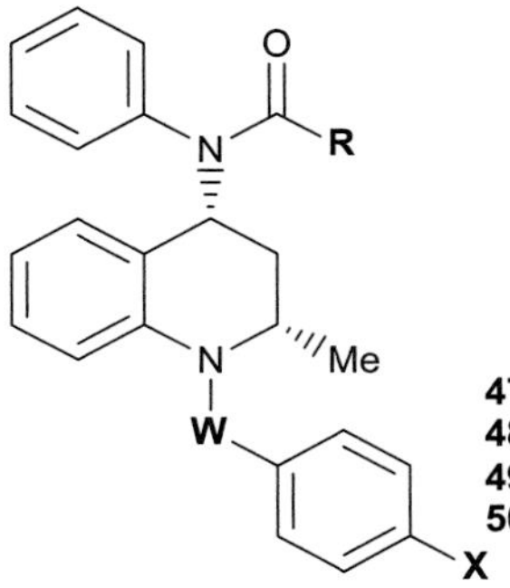

47: W = CO; X = H; R = Me
48: W = CO; X = OCF_3; R = Me
49: W = SO_2; X = H; R = Me
50: W = CO; X = H; R = $(CH_2)_2CO_2H$

3. CONCLUSIONS

While pharmacophore models have been put forward for CRTh2 antagonists, common structural features are evident from the reports summarized herein. Two lipophilic aryl ring systems oriented in an appropriate spatial arrangement relative to the carboxylic acid seem required. Linking these systems through an amide or sulfonamide is preferred and may provide a hydrogen bond interaction with the receptor. On the other hand, ether, keto, and biphenyl linkages can also be employed, with an apparent requirement for reintroduction of another amide or sulfonamide system. In general, potent affinity for CRTh2 can be engendered along with selectivity relative to other prostanoid receptors; however, additional selectivity screening is rarely reported. Differential selectivity profiles may impact the clinical progression of these antagonists. Protein binding data are also scant but will certainly play a role, particularly given the nonspecific binding typical of carboxylic acids [27]. Preclinical pharmacokinetics disclosed to date have been quite good for CRTh2 antagonists, and this seems to be translating well into human studies.

Ultimately, clinical profiles consistent with once daily dosing are expected to be advantageous over twice daily dosing regimens given the chronic nature of allergic disease and the need for patient adherence. Additional human efficacy data that could benchmark the scope of the single agent anti-inflammatory activity of CRTh2 antagonists relative to existing therapies (*i.e.*, corticosteroids) and in combination are anxiously awaited.

REFERENCES

[1] J. C. Medina and J. Liu, *Annu. Rep. Med. Chem.*, 2006, **41**, 221.
[2] R. Pettipher, T. T. Hansel and R. Armer, *Nat. Rev. Drug Discov.*, 2007, **6**, 313.
[3] G. Philip, J. van Adelsberg, T. Loeys, N. Liu, P. Wong, E. Lai, B. Dass, T. F. Reiss and J. Allergy, *Clin. Immunol.*, 2009, **124**, 942.
[4] F. G. Gervais, R. P. G. Cruz, A. Chateauneuf, S. Gale, N. Sawyer, F. Nantel, K. M. Metters, G. P. O'Neill and J. Allergy, *Clin. Immunol.*, 2001, **108**, 982.
[5] C. Yoshimura-Uchiyama, M. Iikura, M. Yamaguchi, H. Nagase, A. Ishii, K. Matsushima, K. Yamamoto, M. Shichijo, K. B. Bacon and K. Hirai, *Clin. Exp. Allergy*, 2004, **34**, 1283.
[6] L. Xue, A. Barrow and R. Pettipher, *Clin. Exp. Immunol.*, 2009, **156**, 126.
[7] M. Gallant, C. Beaulieu, C. Berthelette, J. Colucci, M. A. Crackower, C. Dalton, D. Denis, Y. Ducharme, R. W. Friesen, D. Guay, F. G. Gervais, M. Hamel, R. Houle, C. M. Krawczyk, B. Kosjek, S. Lau, Y. Leblanc, E. E. Lee, J. F. Levesque, C. Mellon, C. Molinaro, W. Mullet, G. P. O'Neill, P. O'Shea, N. Sawyer, S. Sillaots, D. Simard, D. Slipetz, R. Stocco, D. Sorensen, V. L. Truong, E. Wong, J. Wu, H. Zaghdane and Z. Wang, *Bioorg. Med. Chem. Lett.*, 2010, **21**, 288.
[8] S. Baba, *Pract. Otol. Suppl.*, 1996, **87**, 1.
[9] L. L. Carter, Y. Shiraishi, Y. Shin, L. Burgess, C. Eberhardt, A. D. Wright, N. Klopfenstein, M. McVean, A. Gomez, D. Chantry, A. Cook, K. Takeda and E. W. Gelfand, Abstract 490, AAAAI Meeting, New Orleans, LA, 2010.
[10] K. J. Stebbins, A. R. Broadhead, L. D. Correa, J. M. Scott, Y. P. Truong, B. A. Stearns, J. H. Hutchinson, P. Prasit, J. F. Evans and D. S. Lorrain, *Eur. J. Pharmacol.*, 2010, **638**, 142.
[11] J. R. Johnson, R. E. Wiley, R. Fattouh, F. K. Swirski, B. U. Gajewski, A. J. Coyle, J.-C. Gutierrez-Ramos, R. Ellis, M. D. Inman and M. Jordana, *Am. J. Respir. Crit. Care Med.*, 2004, **169**, 378.
[12] P. Norman, *Expert Opin. Invest. Drugs*, 2010, **19**, 947.
[13] N. Barnes, I. Pavord, A. Chuchalin, J. Bell, M. Hunter, T. Lewis, D. Parker, M. Payton, L. P. Collins, R. Pettipher, J. Steiner and C. M. Perkins, *Clin. Exp. Allergy*, 2011, doi: 10.1111/j.1365-2222.2011.03813.
[14] http://clinicaltrials.gov/ct2/show/NCT00890877 (accessed March 9, 2011).
[15] http://clinicaltrials.gov/ct2/show/NCT01197794 (accessed February 28, 2011).
[16] M. L. Fajt, S. Balzar, J. B. Trudeau, J. Y. Westcott and S. E. Wenzel, Abstract 1627, AAAAI Meeting, New Orleans, LA, 2010.
[17] R. He, M. K. Oyoshi, J. Y. T. Wang, M. R. Hodge, H. Jin, R. S. Geha and J. Allergy, *Clin. Immunol.*, 2010, **126**, 784.
[18] S. A. Boehme, E. P. Chen, K. Franz-Bacon, R. Sasik, L. J. Sprague, T. W. Ly, G. Hardiman and K. B. Bacon, *Int. Immunol.*, 2008, **21**, 1.
[19] M. Iwasaki, K. Nagata, S. Takano, K. Takahashi, N. Ishii and Z. Ikezawa, *J. Invest. Dermatol.*, 2002, **119**, 609.
[20] H. Yahara, T. Satoh, C. Miyagishi and H. Yokozeki, *J. Eur. Acad. Dermatol. Venereol.*, 2010, **24**, 75.

[21] K. J. Stebbins, A. R. Broadhead, C. S. Baccei, J. M. Scott, Y. P. Truong, H. Coate, N. S. Stock, A. M. Santini, P. Fagan, P. Prodanovich, G. Bain, B. A. Stearns, C. D. King, J. H. Hutchinson, P. Prasit, J. F. Evans and D. S. Lorrain, *J. Pharmacol. Exp. Ther.*, 2010, **332**, 764.
[22] K. J. Stebbins, J. F. Evans and D. S. Lorrain, *Mol. Cell. Pharmacol.*, 2010, **2**, 89.
[23] G. Bain, J. H. Hutchinson, B. A. Stearns, K. M. Schaab and J. F. Evans, *Patent Application US 2010/0173313-A1*, 2010.
[24] http://clinicaltrials.gov/ct2/show/NCT00690482 (accessed March 9, 2011).
[25] http://clinicaltrials.gov/ct2/show/NCT00859352 (accessed March 29, 2011), http://clinicaltrials.gov/ct2/show/NCT01199341 (accessed March 29, 2011), http://clinicaltrials.gov/ct2/show/NCT01056575 (accessed March 29, 2011).
[26] T. Ulven and E. Kostenis, *Expert Opin. Ther. Patents*, 2010, **20**, 1505.
[27] T. Peters, All About Albumin: Biochemistry, Genetics and Medical Applications, Academic Press, New York, NY, 1996 p. 76.
[28] B. Davies and T. Morris, *Pharm. Res.*, 1993, **10**, 1093.
[29] T. Ulven and E. Kostenis, *J. Med. Chem.*, 2005, **48**, 897.
[30] C. Skonberg, J. Olsen, K. G. Madsen, S. H. Hansen and M. P. Grillo, *Exp. Opin. Drug Metab. Toxicol.*, 2008, **4**, 425.
[31] D. Simard, Y. Leblanc, C. Berthelette, M. H. Zaghdane, C. Molinaro, Z. Wang, M. Gallant, S. Lau, T. Thao, M. Hamel, R. Stocco, N. Sawyer, S. Sillaots, F. Gervais, R. Houle and J.-F. Levesque, *Bioorg. Med. Chem. Lett.*, 2011, **21**, 841.
[32] C. Berthelette, M. Boyd, J. Coluccin, K. Villeneuve and J. Methot, *Patent Application WO 2010/099039-A1*, 2010.
[33] B. A. Stearns and R. Clark, *Patent Application WO 2010/085820-A2*, 2010.
[34] R. E. Armer, E. R. Pettipher, M. Whittaker, G. M. Wynne, J. Vile and F. Schroer, *Patent Application US 2010/0063103-A*, 2010.
[35] G. Hynd, J. G. Montana, H. Finch, R. Arienzo and S. Ahmed, *Patent Application WO 2010/142934-A1*, 2010.
[36] L. N. Tumey, M. J. Robarge, E. Gleason, J. Song, S. M. Murphy, G. Ekema, C. Doucette, D. Hanniford, M. Palmer, G. Pawlowski, J. Danzig, M. Loftus, K. Hunady, B. Sherf, R. W. Mays, A. Stricker-Krongrad, K. R. Brunden, Y. L. Bennani and J. J. Harrington, *Bioorg. Med. Chem. Lett.*, 2010, **20**, 3287.
[37] J. Jia, A. Mermerian, C. Kim, R. Lundrigan and J. Moore, *Patent Application WO 2010/039982-A1*, 2010.
[38] (a) O. Rist, M. Grimstrup, J.-M. Receveur, T. M. Frimurer, T. Ulven, E. Kostenis and T. Hogberg, *Bioorg. Med. Chem. Lett.*, 2010, **20**, 1177; (b) M. Grimstrup, J.-M. Receveur, O. Rist, T. M. Frimurer, P. A. Nielsen, J. M. Mathiesen and T. Hogberg, *Bioorg. Med. Chem. Lett.*, 2010, **20**, 1638.
[39] M. Grimstrup, O. Rist, J. M. Receveur, T. M. Frimurer, T. Ulven, J. M. Mathiesen, E. Kostenis and T. Hogberg, *Bioorg. Med. Chem. Lett.*, 2010, **20**, 1181.
[40] (a) L. Chen, F. Firooznia, P. Gillespie, Y. He, T.-A. Lin, S.-S. So and H. Y. Yun, *Patent Application WO 2010/006939-A1*, 2010; (b) L. Chen, F. Firooznia, P. Gillespie, Y. He, T.-A. Lin, S.-S. So, H. Y. Yun and Z. Zhang, *Patent Application WO 2010/006944-A1*, 2010.
[41] M. G. Johnson, J. Liu, A.-R. Li, Y. Wang, W. Shen, S. E. Lively, Y. Su, B. van Lengerich, X. Wang, S. Lai, M. Brown, S. Lawliss, Y. Sun, Q. Xu, T. Collins, J. Danao, L. Seitz, M. Grillo, J. Wait and J. Medina, Abstract MEDI-64, ACS Meeting, New Orleans, LA, 2008.
[42] J. Liu, A.-R. Li, Y. Wang, M. G. Johnson, Y. Su, W. Shen, X. Wang, S. Lively, M. Brown, S. Lai, F. Gonzalez, L. De Turiso, Q. Xu, B. van Lengerich, M. Schmitt, Z. Fu, Y. Sun, S. Lawlis, L. Seitz, J. Danao, J. Wait, Q. Ye, H. L. Tang, M. Grillo, T. L. Collins, T. J. Sullivan and J. C. Medina, *ACS Med. Chem. Lett.*, 2011, **2**, 326.

[43] C. Banfield, J. Parnes, M. Emery, L. Ni, N. Zhang and P. Hodsman, Abstract 261, AAAAI Meeting, New Orleans, LA, 2010.
[44] J. H. Hutchinson, T. J. Seiders, B. Wang, J. M. Arruda, B. A. Stearns and J. R. Roppe, *Patent Application WO 2010/003127*, 2010.
[45] L. Chen, F. Firooznia, P. Gillespie, Y. He, T.-A. Lin, E. Mertz, S.-S. So, H. Y. Yun and Z. Zhang, *Patent Application WO 2010/055004-A1*, 2010.
[46] F. Firooznia, T.-A. Lin, S.-S. So, B. Wang and H. Y. Yun, *Patent Application WO 2010/055006-A1*, 2010.
[47] A. Cook, K. W. Hunt, R. K. DeLisle, T. Romoff, C. Clark, G. Kim, C. Corrette, G. Doherty and L. E. Burgess, *Patent Application 2010/075200-A1*, 2010.
[48] J. M. Scott, C. Baccei, G. Bain, A. Broadhead, J. F. Evans, P. Fagan, J. H. Hutchinson, C. King, D. S. Lorrain, C. Lee, P. Prasit, P. Prodanovich, A. Santini and B. A. Stearns, *Bioorg. Med. Chem. Lett.*, 2011, **21**, doi:10.1016/j.bmcl.2011.01.024.
[49] N. Stock, D. Volkots, K. Stebbins, A. Broadhead, B. Stearns, J. Roppe, T. Parr, C. Baccei, G. Bain, C. Chapman, L. Correa, J. Darlington, C. King, C. Lee, D. S. Lorrain, P. Prodanovich, A. Santini, J. F. Evans, J. H. Hutchinson and P. Prasit, *Bioorg. Med. Chem. Lett.*, 2011, **21**, 1036.
[50] J.-B. Blanc, L. Chen, F. Firooznia, P. Gillespie, R. A. Goodnow, T. Lin, S. Pan, S.-S. So and H. Y. Yun, *Patent Application WO 2010/018112-A2*, 2010.
[51] T. Luker, R. Bonnert, S. W. Paine, J. Schmidt, C. Sargent, A. R. Cook, A. Cook, P. Gardiner, S. Hill, C. Weyman-Jones, A. Patel, S. Thom and P. Thorne, *J. Med. Chem.*, 2011, **54**, 1779.
[52] J. Liu, Y. Wang, Y. Sun, D. Marshall, S. Miao, F. Tonn, P. Anders, J. Tocker, H. L. Tang and J. Medina, *Bioorg. Med. Chem. Lett.*, 2009, **19**, 6840.

Developments and Advances in Gastrointestinal Prokinetic Agents

James W. Dale, Gregory J. Hollingworth and **Jeffrey M. McKenna**

Contents

Novartis Institutes for Biomedical Research, Horsham, West Sussex RH12 5AB, United Kingdom

Annual Reports in Medicinal Chemistry, Volume 46
ISSN: 0065-7743, DOI: 10.1016/B978-0-12-386009-5.00018-7

1. INTRODUCTION

The quality of life for many people suffering from functional gastrointestinal disorders (FGIDs) is significantly reduced when compared with the general population. The symptoms suffered by these people cover a wide spectrum including abdominal discomfort, bloating, nausea, regurgitation, vomiting, early satiety, fatigue, tenderness, diarrhea, and constipation. These disorders encompass several conditions: functional dyspepsia, irritable bowel syndrome with constipation (IBS-C), gastro-esophageal reflux disease (GERD), chronic constipation (CC) or chronic idiopathic constipation (CIC), and idiopathic (I-GP) or diabetic gastroparesis (D-GP). These conditions, together with post-operative ileus (POI) and the treatment of the critically ill, who often suffer delayed gastric emptying (GE) and may require naso-gastric feeding, represent the possible indications that can be targeted with prokinetics. As such, one can appreciate the need for efficacious prokinetic agents, which increase the frequency or strength of GI contractions, thereby aiding, or enhancing gut motility.

This review serves as an update to the 2006 article in these annals by Sandham and Pfannkuche [1], and as a supplement to other recent reviews in the prokinetic arena [2]. The aim of this review is to highlight the advances made in the area of gastrointestinal (GI) prokinetics in both the clinical and early discovery phases between 2006 and 2010, with an emphasis placed on the more recent developments. It does not cover in any great detail opioid receptor modulators, as this subject matter has been discussed in depth by Hipkin and Dolle in 2010 [3].

2. KEY CLINICAL DEVELOPMENTS

2.1. Serotonin 5-HT receptor modulators

2.1.1. 5-HT$_3$ receptor partial agonists

5-Hydroxytryptophan (5-HT) acts on GPCRs within the GI tract to modulate gut motility in either a pro- or antipropulsive manner, depending on the GPCR subtype and its anatomical location [4]. Pumosetrag, **1**, a 5-HT$_3$ partial agonist, has been shown to be moderately efficacious in promoting bowel motility in a small study of patients with constipation and has shown improvements in "overall response" in IBS-C patients, yet its development for these indications was reported to have been discontinued [5–7]. Nevertheless, in late 2010, it was reported that recruitment for a Phase 2 trial in GERD had begun [8]; no other 5-HT$_3$ agonists appear to be in active clinical studies for GI indications.

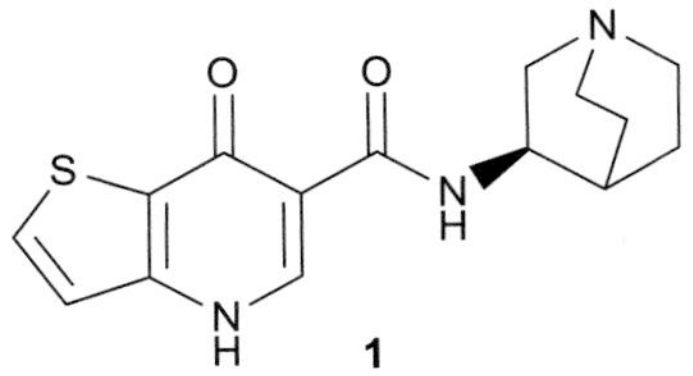

1

2.1.2. 5-HT$_4$ receptor agonists

Agonism of 5-HT$_4$ receptors is an established prokinetic mechanism, mediating smooth muscle contractions *via* action on cholinergic and ganglionic neurons. Four molecules have been marketed; however, cisapride, **2**, was withdrawn in 2000 for its well-documented cardiovascular safety profile. In the current review period, tegaserod, **3**, has also been voluntarily withdrawn due to an increased incidence of cardiovascular ischemic events in a pooled analysis of clinical trial data from circa. 11,000 patients [9]. Interestingly, an analysis carried out subsequently has provided a contrasting view of the cardiovascular risks associated with this molecule [10]. Mosapride, **4**, remains available for gastritis and GERD and has been shown in a recent study in Japan to ameliorate constipation in diabetic patients [11]. Prucalopride, **5**, the latest 5-HT$_4$ agonist to reach the market, was approved in 2010 for the treatment of CC in adults and is currently in Phase 3 development for IBS-C. Three separate 12-week Phase 3 studies consistently showed significant and improved bowel function, associated symptoms, and satisfaction in chronically constipated patients treated with prucalopride [12–14]. All trials achieved the primary endpoint of drug-treated individuals achieving three or more complete spontaneous bowel movements (SBMs) per week. Further, in an open-label follow-up study, patients from the previous double-blinded studies who continued to receive prucalopride treatment maintained their satisfaction with bowel function for up to 18 months [15]. Various other 5-HT$_4$ agonists are in late-stage clinical development. Naronapride, **6**, is an oral 5-HT$_4$ receptor agonist in Phase 2b clinical trials for GI disorders. Significant metabolism of naronapride occurs *via* esterases rather than CYP450s, so it avoids the drug–drug interactions (and high exposure risk) associated with cisapride. Modest effects of naronapride on GE have been observed in healthy volunteers [16]. Moreover, the number of SBMs were improved *versus* placebo in patients with CIC receiving an 80 mg *bid* dose of naronapride during a Phase 2b clinical study [17]. Velusetrag, **7**, is also in Phase 2 trials and was well tolerated and efficacious in patients with CIC, also achieving increased SBM frequency compared with patients receiving placebo [18]. Finally, a further 5-HT$_4$ agonist, M-0003 is reportedly entering Phase 2 for GP and GERD although its structure has not been disclosed [7].

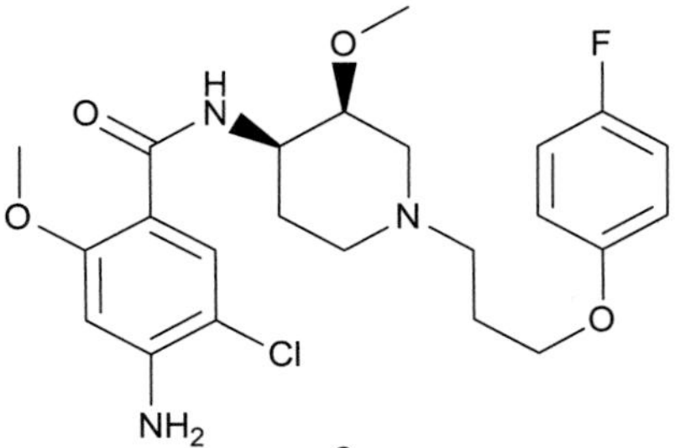

2.2. Type-2 chloride channel activators

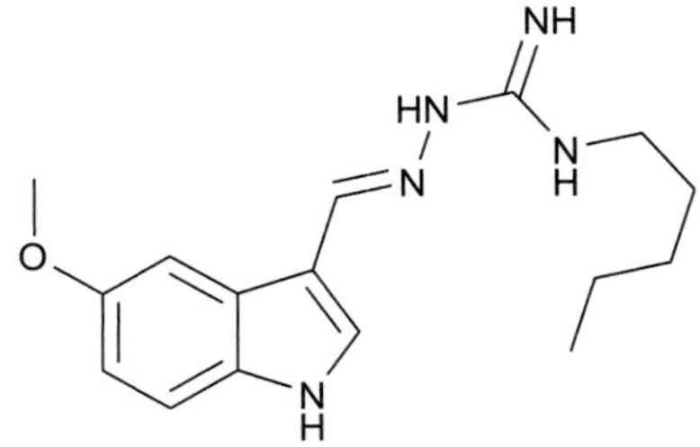

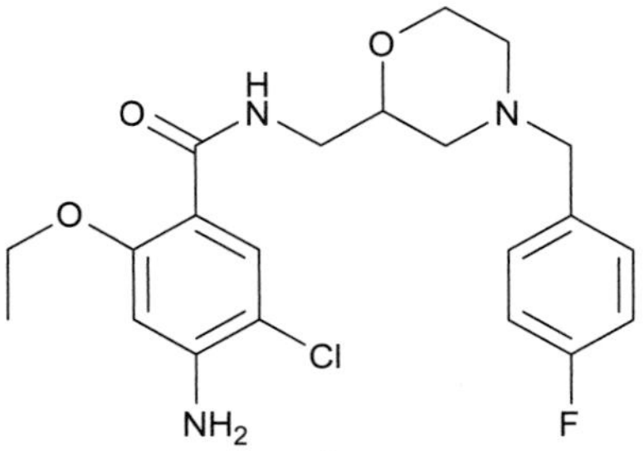

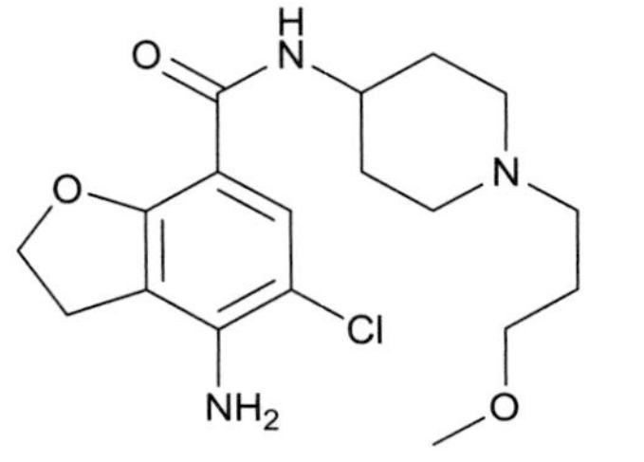

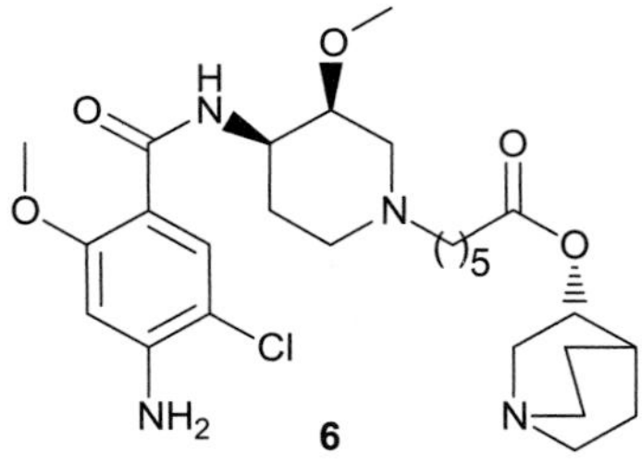

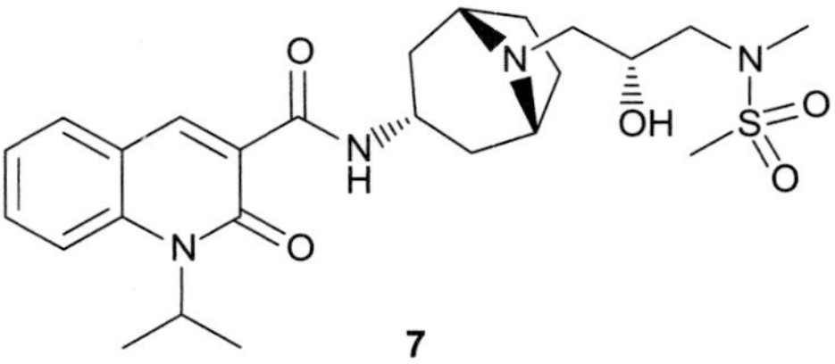

Type-2 chloride channel (ClC-2) is a chloride ion channel which is located on the apical side of epithelial cells lining the gut lumen. Activation of ClC-2 drives chloride ions into the gut lumen inducing intestinal fluid secretion, and leads to increased intestinal motility [19]. The ClC-2 activator lubiprostone, **8**, has been launched for the treatment of CC and IBS-C, and the results from several clinical trials have now been reported. Notably, in a combined analysis of two Phase 3 trials in patients with IBS-C

($n = 1171$), the number of responders based on patient-rated assessments of symptoms was significantly greater (17.9% *vs*. 10.1%; $P = 0.001$) for the lubiprostone (8 μg, *bid*) group compared with those who received placebo [20]. This data reinforced the findings of an earlier Phase 2 trial in IBS-C [21]. In trials for CC, improved SBMs were consistently observed in drug-treated groups compared with placebo. Additionally, Phase 2 studies are reported to be ongoing in opioid-induced bowel dysfunction [7,22,23].

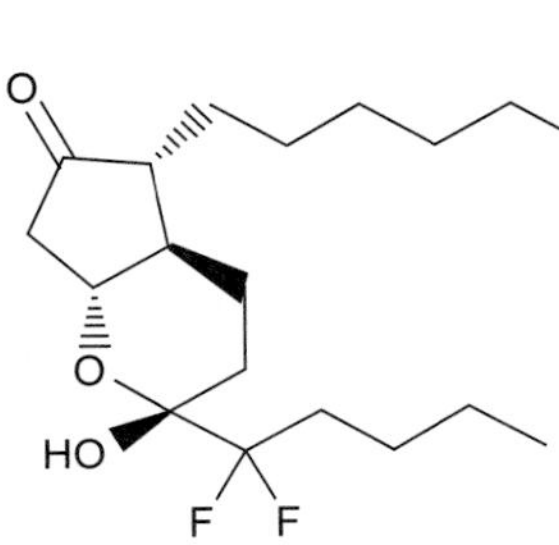

O OH O O HO F F

8

2.3. Ghrelin receptor agonists

Ghrelin, an *n*-octanoylated 28-amino acid peptide hormone, is the endogenous ligand of the growth hormone (GH) secretagogue receptor (GHS-R1a) [24,25]. Ghrelin is produced in the gastric mucosa by enteroendocrine cells and is a prokinetic which has been demonstrated to stimulate GE in healthy human volunteers when administered intravenously [26]. Moreover, several small clinical trials have shown efficacy of ghrelin in increasing GE in patients with D-GP ($n = 10$), neurogenic GP (which includes D-GP, $n = 6$), or I-GP, although no further development has been reported for ghrelin in these indications [27–29]. Nevertheless, there has been significant recent interest in the role of ghrelin receptor agonists as prokinetic agents. Ulimorelin, **9**, a potent GI prokinetic agent delivered *via* intravenous administration, is in development for the treatment of GP and POI. So far, positive clinical data in GP in diabetics has been reported in three studies. In patients with moderate to severe symptoms, a pilot study showed a statistically significant improvement in GE [30]. A larger follow-up study showed significant improvements in the patient-rated Gastroparesis Cardinal Symptom Index (GCSI) Loss of Appetite and Vomiting scores ($P = 0.034$ and 0.006, respectively) at the 80 μg/kg dose [31]. In a further study, in a subset of patients with severe nausea and vomiting, significant effects at the 80 μg/kg dose were also observed on the GCSI Vomiting score ($P < 0.008$) and the GCSI Nausea and Vomiting subscale ($P < 0.001$) *versus* placebo [32]. Additionally, the efficacy of ulimorelin has been demonstrated in POI patients after partial colectomy in a Phase 2b placebo-controlled trial, accelerating the time to first bowel movement in all treatment groups [33]. TZP-102 (structure undisclosed) is an orally delivered molecule that is also being developed. Results of a Phase 2 study stated the compound to be safe, well tolerated and effective in significantly improving disease symptoms in patients with D-GP [34]. While both ulimorelin and TZP-102 appear to be ahead in the clinic, other new molecular entities are advancing such as EX-1314, **10**, for which an investigational new drug (IND) application has been filed for use in patients with D-GP. The pentapeptide ipamorelin, **11**, completed a Phase 2 trial in 2009 to determine its safety and effectiveness in the management of POI [35]. However, at the time of writing, no results had been posted. The prokinetic effects of ghrelin agonists were the subject of a recent review [36].

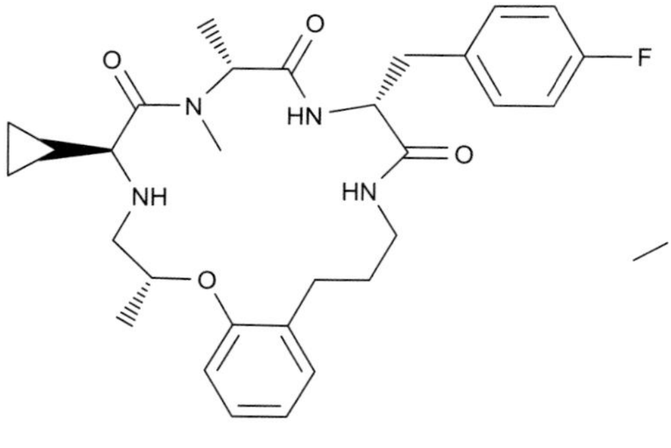

9

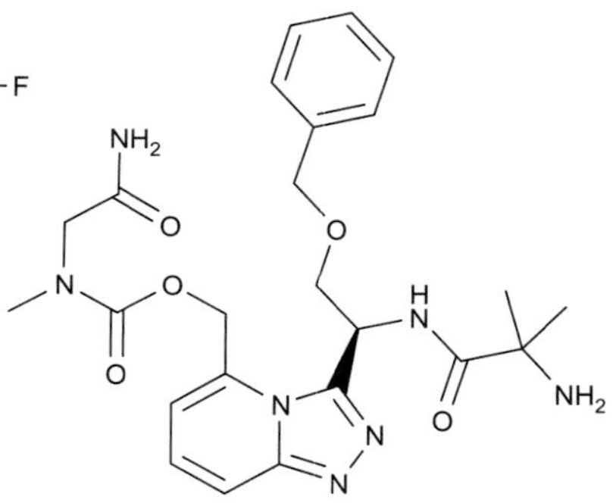

10

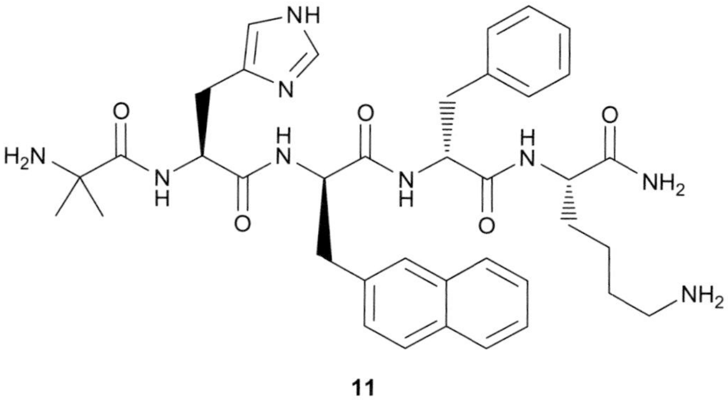

11

2.4. Motilin receptor agonists

The 22-amino acid peptide motilin, found in endocrine M-cells of the duodenum and jejunum mucosa, is released during fasting and promotes GE by stimulation of antral contractions through engagement of the motilin receptor in the gut. Agonism of the motilin receptor is thus an attractive prokinetic approach [37]. The macrolide antibiotic erythromycin, **12**, has been shown to be a potent motilin agonist making it a powerful prokinetic, and there have been numerous published clinical studies examining its use for enhancing motility. A recent example has shown it, and the related macrolide azithromycin, to be efficacious at stimulating antral activity in patients with GP [38]. Erythromycin has been widely used to treat feeding intolerance in critical care patients [39]. A study comparing erythromycin to the D2 agonist metoclopramide concluded that erythromycin is the more effective agent in the short-term treatment of feeding intolerance; successful gastric feeding had been achieved after 24 h in 87% of patients taking erythromycin *versus* 62% in the metoclopramide arm. However, tachyphylaxis can develop rapidly, and both treatments were significantly less effective after 3 days. Combination therapy was highly effective and sustained in patients who had failed monotherapy (92% achieving successful enteral feeding after 24 h) thus indicating the combination successfully prevented tachyphylaxis [40]. Concerns have been expressed that the use of erythromycin, for its prokinetic properties, may promote the emergence of macrolide-resistant bacteria. For this reason, significant efforts have been invested in the development of macrolides which retain the prokinetic activity but have reduced antibiotic activity (motilides) [41]. Clinical results to date with the motilides in patients with GP have been mixed, likely due to the induction of tachyphylaxis. For instance, a 28-day trial of mitemcinal demonstrated a significant change in post-treatment meal-retention time at 4 h for drug-treated patients compared to placebo; however, no significant improvement was seen in the gastroparetic symptom score for any treatment group compared with placebo [42]. However, in a separate 16-week study *versus* placebo, positive results were obtained for mitemcinal where a 10.6% ($P = 0.05$) increase in relieving GP symptoms was observed at a dose of 10 mg/kg [43]. No recent development has been reported for mitemcinal. Away from the macrolide arena, recent developments include the non-macrolide motilin agonist GSK-962040, **13**, which reached Phase 2 clinical trials in patients with D-GP and also those with enteral feeding intolerance in critical care (although no data has yet been reported) [44,45]. A follow-up compound, GSK-1322888, also entered Phase 1 for GP at the end of 2010 [46].

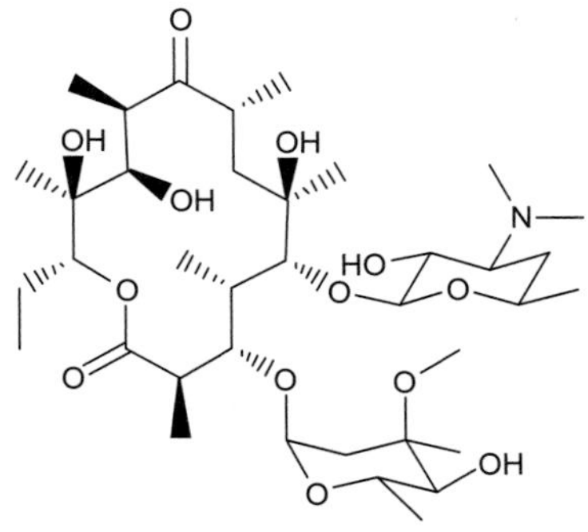

12

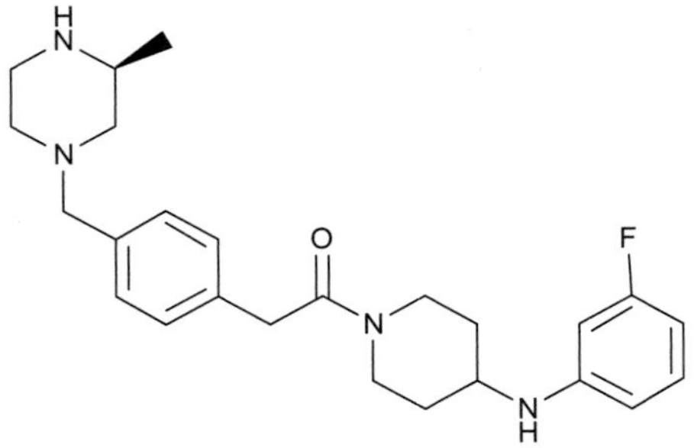

13

2.5. Cholecystokinin 1 (CCK-1) receptor antagonists

CCK is a peptide hormone which is made from prepro-CCK, a 115-aa peptide, and is found in both the periphery and the CNS [47]. In humans, CCK-8 and CCK-58 are the predominant forms [48]. CCK is released within the intestinal tract from endocrine cells and regulates gut function in response to food intake by binding to CCK receptors located on, but possibly not restricted to, vagal neurons [49]. As such, CCK antagonists represent an attractive opportunity for treating FGIDs. The development of the selective CCK-1 antagonist dexloxiglumide, **14**, had seemingly stalled after its failure to show a significant difference to placebo in two 12-week Phase 3 IBS trials. However, a further Phase 3 study was published in 2008, where in an innovative trial design, a randomized "withdrawal study" was conducted. Approximately 400 IBS-C patients, classed as "responders" to drug over an 8–12-week period, were subsequently randomized and treated with drug or placebo for a further 24 weeks. The outcome of the study was a statistically significant 16.2% difference in the maintenance of response seen for those on drug compared with those who received placebo [50]. No other significant clinical developments appear to have been reported over the review period for CCK antagonists in GI indications.

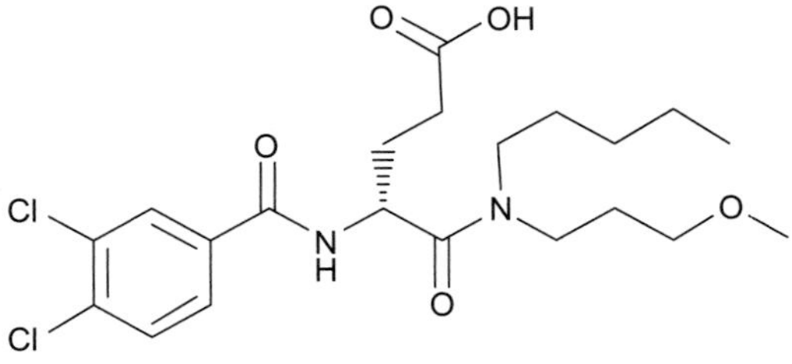

14

2.6. Guanylate cyclase 2C agonists

Uroguanylin and guanylin are peptide hormones which bind to the guanylate cyclase C (GC-C) receptor in the gut epithelium. This results in cGMP synthesis and elicits a large increase in the secretion of chloride and bicarbonate anions through activated CFTR channels, thereby increasing fluid secretion into the gut and enhancing motility [51]. These observations have stimulated interest in analogues of uroguanylin and guanylin as prokinetic agents. The orally delivered linaclotide, **15**, is a 14-mer peptidic guanylate cyclase 2C agonist in Phase 3 clinical trials for IBS-C and CC [52]. This compound is thought to act locally on the receptors at the luminal surface of the intestine and shows minimal systemic exposure. Recently published Phase 2b data demonstrated that treatment with linaclotide at doses ranging from 75 to 600 μg once daily for 12 weeks significantly improved disease symptoms, including abdominal pain and bowel symptoms, in patients with IBS-C [53]. Additionally, in a 4-week Phase 2b study in patients with CC, a significant increase in the number of SBMs (primary endpoint), as well as other symptoms, was seen for patients on linaclotide compared to placebo [54]. Moreover, recent company press releases disclose two positive Phase 3 trials for each targeted indication (IBS-C and CC), and an IND application is expected to be filed during 2011 [55]. Another guanylate cyclase 2C agonist in clinical trials is the uroguanylin analogue guanilib (plecanatide), for which a positive outcome in a Phase 2a study in CC patients was reported in October 2010 [56].

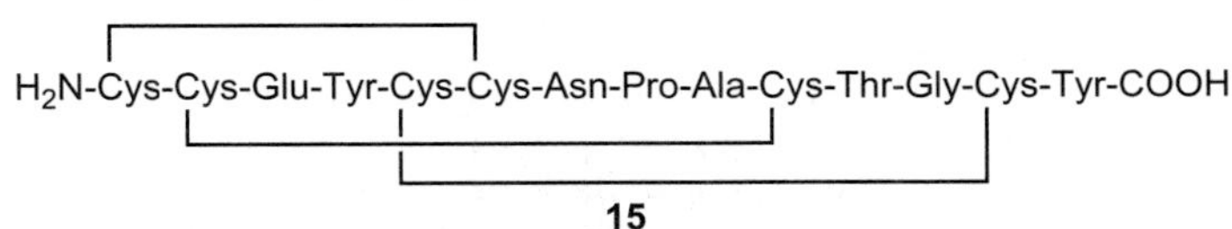

15

2.7. Ileal bile acid transporter inhibitors

The first in class ileal bile acid transporter (IBAT) inhibitor, A-3309 (structure undisclosed), has shown positive results in a double-blind placebo-controlled Phase 2b study for the oral treatment of CIC [57]. At doses of 10 and 15 mg, once daily, the primary endpoint of statistically significant increases in SBMs over baseline during treatment week 1 was achieved. The compound has minimal systemic exposure and is believed to act locally in the gut to inhibit the re-uptake of bile acids, which in turn increases motility.

2.8. μ-Opioid receptor antagonists

Several Phase 3 trials have now been reported for the peripherally acting μ-opioid antagonist alvimopan, **16**, in patients undergoing major abdominal surgery. Alvimopan counteracts the reduction in intestinal tract motility caused by endogenous opioids released in the gut as a result of the stress of surgery, as well as by administered opioids commonly given as analgesics to these patients. The poor CNS penetration of alvimopan, however, ensures that the analgesic effect of systemic opioids is not affected [58]. A pooled analysis of Phase 3 clinical studies demonstrated that alvimopan significantly accelerates GI recovery and is an effective treatment for POI following bowel resection. The compound was approved by the FDA in 2008 for this use [59].

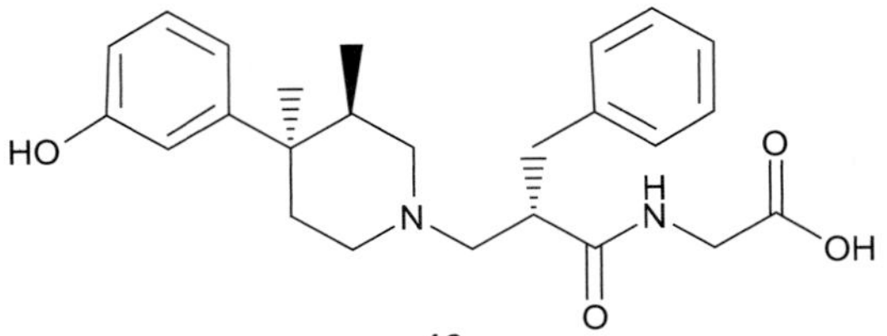

HO N H N O OH O

16

3. RECENT MEDICINAL CHEMISTRY DEVELOPMENTS

3.1. Serotonin 5-HT_4 agonists

A series of benzamide analogues have been reported in the patent literature as 5-HT_4 receptor agonists for the treatment of GP disorders. Compounds were evaluated in a rat 5-HT_4 radioligand binding assay using tissue isolated from rat striatum. Compound **17** (IC_{50}: 24 nM) has been shown to have comparable activity to **6** (IC_{50}: 23 nM) in this assay. Structure activity relationships show that the *S*-enantiomers have higher binding affinities, and optimal activity is achieved when the alkyl linker is pentyl [60].

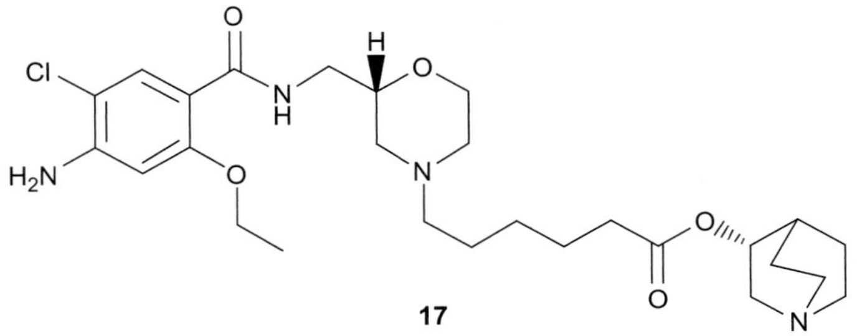

Cl O N H H O H_2N O N O O N

17

Developed using a "multivalent" optimization strategy, a family of potent 5-HT_4 receptor agonists was prepared with high selectivity over the 5-HT_3 receptor [61–63]. Initial work led to the identification of quinolinone, indazole, and benzimidazoline scaffolds while access to a secondary binding pocket was achieved *via* alkylation of the aza-bicycle. Introduction of a piperidine, **18**, or a piperazine sulfonamide, **19**, led to striking selectivity over the 5-HT_{3A} receptor (8100- and 2800-fold, respectively). A related example, **20**, was orally bioavailable in the rat (F 38%) and exhibited hERG inhibition of 38% at 3 μM. Optimization afforded, **21**, a potent (pK_i: 7.9) and selective (7400-fold) antagonist with reduced hERG activity (4% at 3 μM); its *R*-isomer, **22**, showed reduced activity (pK_i: 7.3) and selectivity (1300-fold). Compound **21** demonstrated efficacy in an isolated guinea-pig colon longitudinal muscle strip assay with pEC_{50} of 8.6 [intrinsic activity (*i.a.*): 63% of the maximal response of 5-HT] and was efficacious at 3 mg/kg (*s.c.*) in a guinea-pig model of colonic GI transit [64].

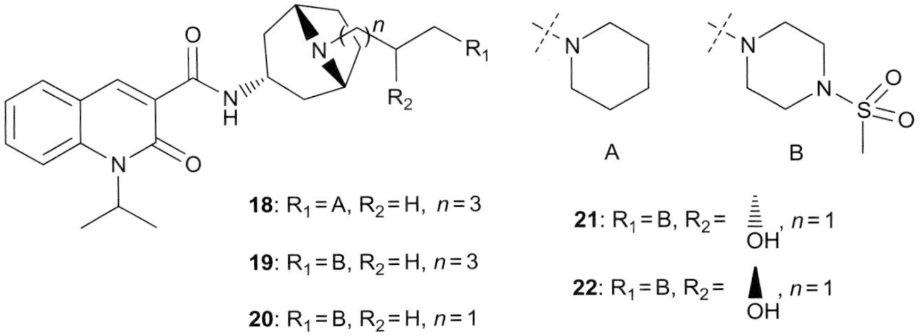

A B

18: R_1=A, R_2=H, n=3

19: R_1=B, R_2=H, n=3

20: R_1=B, R_2=H, n=1

21: R_1=B, R_2= OH, n=1

22: R_1=B, R_2= OH, n=1

3.2. Ghrelin receptor agonists

A peptidomimetic strategy has been successfully applied to the modification of the peptide frameworks of the GH-secretagogues through inclusion of amide-bond isosteres. The exploration of a 1,2,4-triazole motif within the backbone generated antagonists, partial agonists, and full agonists within a subset of closely related analogues [65]. Moreover, following the replacement of an amino-*iso*-butyryl amide (Aib) motif with an *iso*-nipecotic carboxamide, a series of full agonists was identified, including **23**. When given by *s.c.* injection, **23** was able to stimulate food intake and GH secretion in infant rats, and also potentiate hexarelin-stimulated food intake [66].

An alternative amide isostere, a 1,5-disubstituted tetrazole, was exploited upon recognition that the conformational restriction imparted by this 5-membered heterocycle could still realize potent GH-secretagogues. The retention of an Aib motif and inclusion of the tetrazole enabled the discovery of an *N1*-cyanoethyltetrazole, **24**, which had good oral bioavailability in rat and dog (*F* 56% and 75%, respectively) and was efficacious at increasing GH mean peak levels in beagle dogs at 10 mg/kg. More notably, BMS-317180, **25**, the *N1*-alkyl carbamate-based tetrazole, is a potent (EC_{50}: 1.9 nM), highly water soluble (>100 mg/mL) and orally active compound advanced into development for the treatment of cancer cachexia and wasting syndrome. The compound displays low to moderate oral bioavailability (*F* 9%, 12%, and 40%) in rat, monkey, and dog, respectively; its superior exposure in dog is attributed to a decrease in clearance. So far, development for GI indications has not been reported [67,68]. More recently, **25** has been further optimized through successful replacement of the benzyloxy-group with a *gem*-difluoropropyl benzene, giving **26**, which has an increased *in vitro* potency (EC_{50}: 0.27 nM), better efficacy, and improved exposure and oral bioavailability in rat (*F* 26%) when compared to **25** [69].

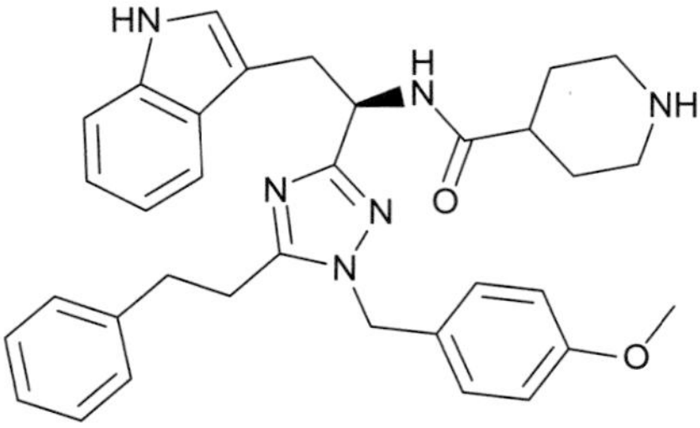

23 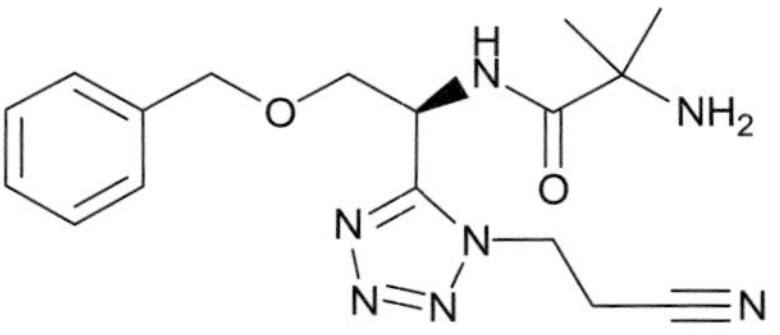 **24**

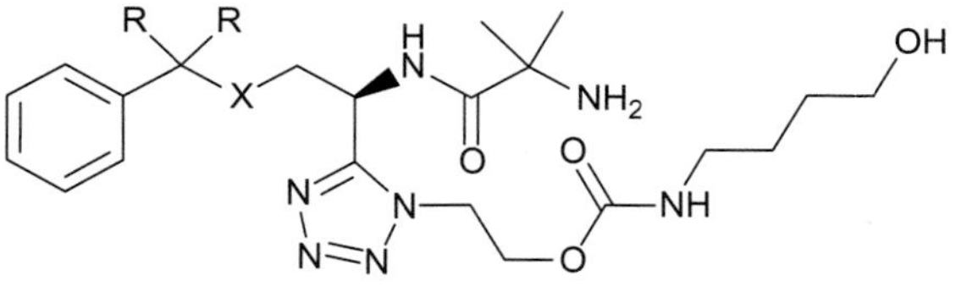

25: R = H, X = O **26**: R = F, X = CH_2

A high-throughput screening (HTS) approach was used to identify agonists of the ghrelin receptor which were devoid of the peptide-like nature of privileged fragments seen thus far [70]. The indoline hit **27**, originally a 5-HT_{1B} receptor antagonist (5-HT_{1B} pK_i: 7.9), was optimized to give SB-791016, **28**, a potent ghrelin agonist with good intrinsic activity (*i.a.*: 0.9). This compound showed significant acceleration of GE after *s.c.* dosing in rat with an ED_{50} of 0.1 mg/kg. However, after oral dosing, it had very low exposure, likely due to its high lipophilicity and low solubility. Subsequent optimization, which included ring opening of the indoline, use of an intramolecular H-bond to maintain permeability, and optimization of the sulfonamide group, led to the identification of GSK-894490A, **29**. This compound is a potent, orally bioavailable full agonist (*F* 75% in rat) which was shown to stimulate an increase in food intake in rat at 3 mg/kg following oral dosing [71].

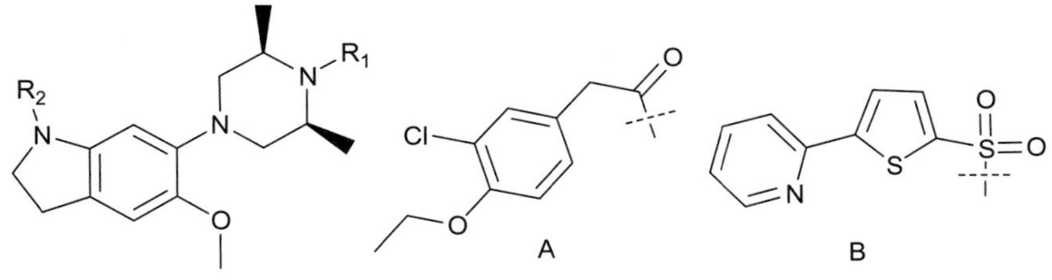

27: R_1 = Me, R_2 = A **28**: R_1 = H, R_2 = B

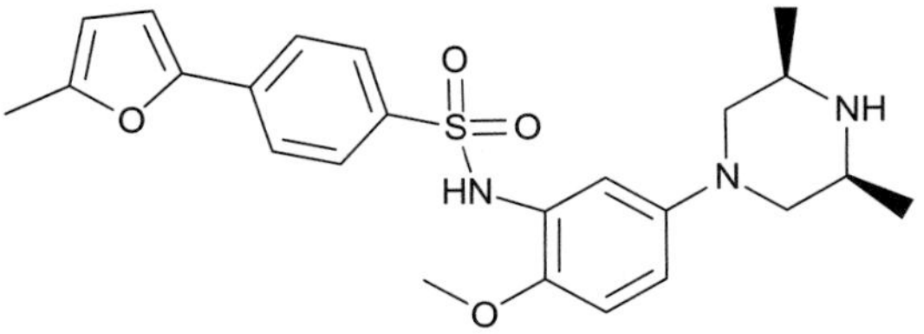

29

In a more recent publication from the same group, the spacer between the sulfonyl group and the basic piperazine has been investigated and has led to the discovery of a 3-amino-pyrrolidinyl amide derivative, **30**, that has an encouraging *i.v.* PK profile (CL 25 mL min/kg, and VD_{ss} 3.5 L/kg). Further optimization is required to achieve the necessary exposure for the *in vivo* evaluation of this class of ghrelin agonist [72].

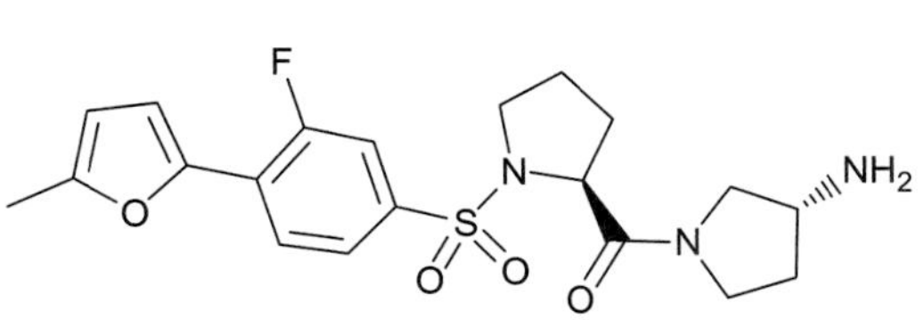

30 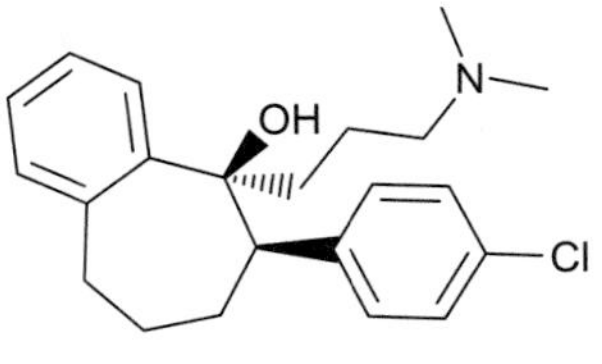 **31**

A series of potent benzocycloheptane and benzoxepine GHS-R1 agonists has been reported in the patent literature. The most efficacious of the compounds disclosed was **31**, which increased GE and stimulation of small intestine propulsion in NMRI mice at 0.63 mg/kg following *s.c.* dosing compared with vehicle-treated animals. In a further study, it was also shown that the compound had no effect on GE in GHS-1R$^{-/-}$ knock-out mice [73].

3.3. Motilin receptor agonists

A prospective 7-TM receptor pharmacophore library screened within the context of a HTS campaign yielded the hit compound **32** [74]. Subsequent modifications yielded **33**, which showed gastric prokinetic-like activity in the potentiation of neuronal-mediated contractions in isolated rabbit antrum tissue in the 0.3–10 μM range. Compound **33** also showed exposure in rat and dog after oral dosing (*F* 13%, 58%, respectively). However, **33** only shows a fivefold selectivity when compared to its activity at the ghrelin receptor (MTL-R pEC_{50}: 7.7, GHS-1R pEC_{50}: 7.0) [75]. Subsequent optimization was required to address its selectivity, CYP3A4 time-dependent inhibition (TDI), high molecular weight, and lipophilicity, and to improve its PK profile. As such, **13** (MTL-R pEC_{50}: 7.9 (*i.a.*: 0.9)) was identified with optimal oral PK properties (*F* (rat) 48%; (dog) 51%), no CYP TDI, good selectivity over the ghrelin receptor (GHS-1R pEC_{50}: <6.0), and good solubility as its HCl salt (>1 mg/mL) [44]. Compound **13** has also been demonstrated to activate the motilin receptor in rabbit and human isolated stomach and is active *in vivo* in the conscious rabbit, suggesting that it has the potential to increase GE in humans [76].

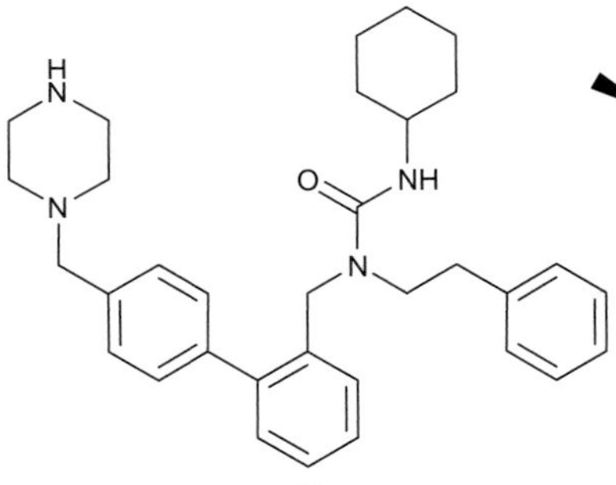

32 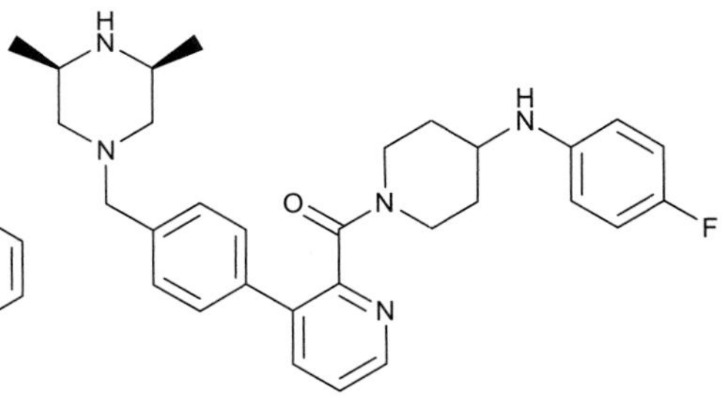 **33**

A FLIPR-based HTS, carried out by the same group, identified a series of benzazepine sulfonamides as novel motilin agonists. Initial screening activity was confirmed using tissue isolated from rabbit gastric antrum. Optimization led to the morpholine-substituted compound **34**, which displayed superagonist efficacy in the rabbit tissue strip assay (E_{max}: 476% at 1 μM); however, further optimization of this series was stopped due to persistent inhibition of CYP2D6 (<1 μM for compound **34**) [77].

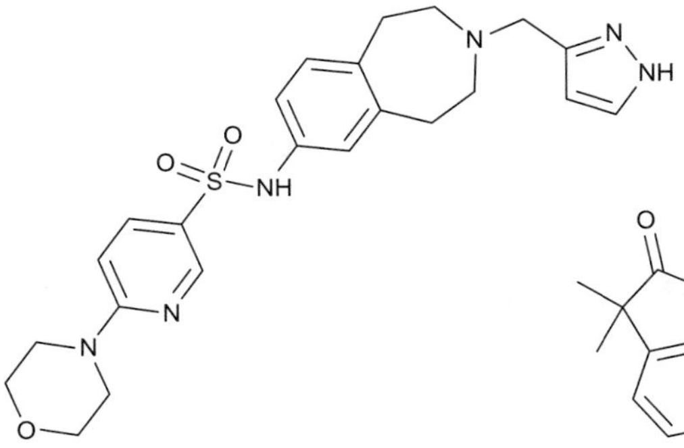

34 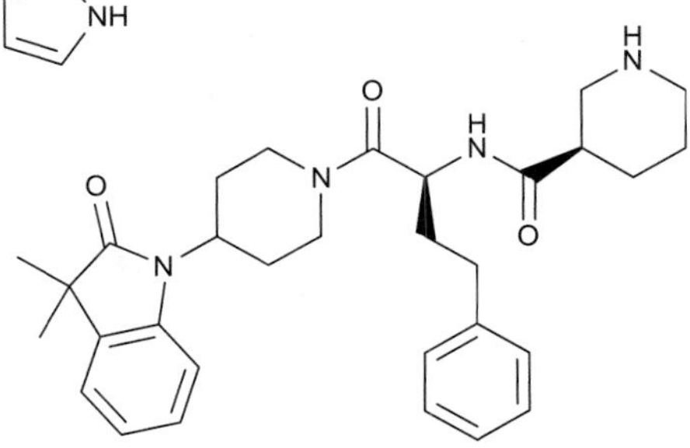 **35**

RQ-00201894 is a motilin peptidomimetic with potent agonist activity in CHO cells (EC_{50} 0.26 nM). When tested *in vivo* following oral dosing, the compound induced MMC-like contractions in the upper GI tract in fasted dogs, and dose-dependently increased GE in conscious dogs and cynomolgus monkeys up to 3 mg/kg [78]. The structure of RQ-00201894 has yet to be disclosed; however, a recent patent described a series of oxindoles, of which **35** is the most potent example [79].

3.4. Cholecystokinin dual (CCK-1/2) receptor antagonists

A series of aryl sulfonamide dual CCK-1 and CCK-2 antagonists has been reported for the potential treatment of GERD [80]. Antagonism of CCK-1 accelerates GE through improved lower esophageal sphincter function, while CCK-2 inhibition regulates gastric acid secretion. The investigators observed that selectivity between CCK-1 and CCK-2 could be modulated, and efforts were made to optimize activities at both receptors [81]. The [2,1,3]-benzothiadiazole moiety of compound **36** (CCK-1/2 pK_i: 6.8/8.0) could be replaced with a quinoxaline while maintaining potency and dual CCK activities (compound **37**, pK_i: 6.8/8.2). Although metabolism in human liver microsomes was comparable, both series suffered from high efflux in a Caco-2 permeability assay, which impacted oral bioavailability. Nevertheless, compound **36** was selected for *in vivo* studies and inhibited pentagastrin-stimulated gastric acid secretion with an ED_{50} of 1 mg/kg *p.o.* [82]. CCK-2-mediated contractility was assessed using guinea-pig corporeal muscle in the presence of 2-NAP; here compound **38** demonstrated a pK_B of 8.8 [80].

36: X = S **37**: X = CH = CH **38**

4. CONCLUSIONS

In recent years, there have been numerous clinical advances in the field of prokinetic agents, which offer great potential for patients who suffer from FGIDs. Specifically, the launches of prucalopride and lubiprostone provide new treatment options for those with CC and IBS-C. Major advances are evident in the area of GP, where the ghrelin agonists ulimorelin and TZP-102 hold much promise, and in the related motilin field, where the low molecular weight agonist GSK-962040 has also reached the clinic. In the CIC arena, the first in class IBAT inhibitor A-3309 is a highlight. The most significant setback over the review period has been the voluntary withdrawal of the 5-HT_4 agonist tegaserod. In an interesting approach, the potential of dexloxiglumide for the treatment of IBS-C was maximized through an innovative trial design. Given the complex nature of FGIDs, and specifically the low treatment responses *versus* placebo often observed in clinical trials, it is clear that the design of these trials is critical in ensuring success and ultimately enabling new molecular entities to reach patients.

REFERENCES

[1] D. A. Sandham and H. J. Pfannkuche, *Annu. Rep. Med. Chem.*, 2006, **41**, 211.

[2] D. A. Sandham, *Expert Opin. Ther. Pat.*, 2008, **18**, 501; M. D. Crowell, L. A. Harris, T. N. Lunsford and J. K. DiBaise, *Expert Opin. Emerg. Drugs*, 2009, **14**, 493.

[3] R. W. Hipkin and R. E. Dolle, *Annu. Rep. Med. Chem.*, 2010, **45**, 143.

[4] R. M. Eglen and S. S. Hegde, *Expert Opin. Invest. Drugs*, 1996, **5**, 373.

[5] T. Fujita, S. Yokota, M. Sawada, M. Majima, Y. Ohtani and Y. Kumagai, *J. Clin. Pharm. Ther.*, 2005, **30**, 611.

[6] W. G. Paterson, D. Ford, S. C. Ganguli, R. P. Reynolds, L. Pliamm, M. O'Mahony, P. Pare, S. Nurbhai, B. Feagan and S. B. Landau, *Gastroenterology*, 2008, **134**(Suppl. 1), Abst T-1397.
[7] Thomson Reuters Integrity (https://integrity.thomson-pharma.com): searched 8th March 2011.
[8] Trial NCT01161602: clinicaltrials.gov, Edusa Pharmaceuticals, 2010.
[9] www.fda.gov: Public Health Advisory: Tegaserod maleate (marketed as Zelnorm), March 30th 2007.
[10] J. Loughlin, S. Quinn, E. Rivero, J. Wong, J. Huang, J. Kralstein, D. L. Earnest and J. D. Seeger, *J. Cardiovasc. Pharmacol. Ther.*, 2010, **15**, 151.
[11] N. Ueno, A. Inui and Y. Satoh, *Diabetes Res. Clin. Pract.*, 2010, **87**, 27.
[12] M. Camilleri, R. Kerstens, A. Rykx and L. Vandeplassche, *N. Engl. J. Med.*, 2008, **358**, 2344.
[13] E. M. M. Quigley, L. Vandeplassche, R. Kirstens and J. Ausma, *Aliment. Pharmacol. Ther.*, 2009, **29**, 315.
[14] J. Tack, M. Van Outryve, G. Beyens, R. Kerstens and L. Vandeplassche, *Gut*, 2009, **58**, 357.
[15] M. Camilleri, M. J. Van Outryve, G. Beyens, R. Kerstens, P. Robinson and L. Vandeplassche, *Aliment. Pharmacol. Ther.*, 2010, **32**, 1113.
[16] M. Camilleri, M. I. Vazquez-Roque, D. Burton, T. Ford, S. McKinzie, A. R. Zinsmeister and P. Druzgala, *Neurogastroenterol. Motil.*, 2007, **19**, 30.
[17] Aryx (www.aryx.com) Press Release August 22nd 2008.
[18] M. Goldberg, Y. P. Li, J. F. Johanson, A. W. Mangel, M. Kitt, D. T. Beattie, K. Kersey and O. Daniels, *Aliment. Pharmacol. Ther.*, 2010, **32**, 1102.
[19] R. A. Borman and G. J. Sanger, *Drug Discov. Today: Ther. Strateg.*, 2007, **4**, 165.
[20] D. A. Drossman, W. D. Chey, J. F. Johanson, R. Fass, C. Scott, R. Panas and R. Ueno, *Aliment. Pharmacol. Ther.*, 2009, **29**, 329.
[21] J. F. Johanson, D. A. Drossman, R. Panas, A. Wahle and R. Ueno, *Aliment. Pharmacol. Ther.*, 2008, **27**, 685.
[22] J. F. Johanson and R. Ueno, *Aliment. Pharmacol. Ther.*, 2007, **25**, 1351.
[23] J. F. Johanson, D. Morton, J. Geenen and R. Ueno, *Am. J. Gastroenterol.*, 2008, **103**, 170.
[24] M. Kojima, H. Hosoda, Y. Date, M. Nakazato, H. Matsuo and K. Kangawa, *Nature*, 1999, **402**, 656.
[25] C. Tomasetto, S. M. Karam, S. Ribieras, R. Masson, O. Lefebvre, A. Straub, G. Alexander, M. P. Chenard and M. C. Rio, *Gastroenterology*, 2000, **119**, 395.
[26] F. Levin, T. Edholm, P. T. Schmidt, P. Grybäck, H. Jacobsson, M. Degerblad, C. Höybye, J. J. Holst, J. F. Rehfeld, P. M. Hellström and E. Näslund, *J. Clin. Endocrinol. Metab.*, 2006, **91**, 3296.
[27] C. D. R. Murray, N. M. Martin, M. Patterson, S. A. Taylor, M. A. Ghatei, M. A. Kamm, C. Johnston, S. R. Bloom and A. V. Emmanuel, *Gut*, 2005, **54**, 1693.
[28] M. Binn, C. Albert, A. Gougeon, H. Maerki, B. Coulie, M. Lemoyne, R. Rabasa Lhoret, C. Tomasetto and P. Poitras, *Peptides*, 2006, **27**, 1603.
[29] J. Tack, I. Depoortere, R. Bisschops, K. Verbeke, J. Janssens and T. Peeters, *Aliment. Pharmacol. Ther.*, 2005, **22**, 847.
[30] N. Ejskjaer, E. T. Vestergaard, P. M. Hellström, L. C. Gormsen, S. Madsbad, J. L. Madsen, T. A. Jensen, J. C. Pezzullo, J. S. Christiansen and L. Shaughnessy, *Aliment. Pharmacol. Ther.*, 2009, **29**, 1179.
[31] N. Ejskjaer, G. Dimcevski, J. Wo, P. M. Hellström, L. C. Gormsen, I. Sarosiek, E. Søfteland, T. Nowak, J. C. Pezzullo, L. Shaughnessy, G. Kosutic and R. Mccallum, *Neurogastroenterol. Motil.*, 2010, **1069**, 22.
[32] J. M. Wo, N. Ejskjaer, P. M. Hellström, R. A. Malik, J. C. Pezzullo, L. Shaughnessy, P. Charlton, G. Kosutic and R. W. McCallum, *Aliment. Pharmacol. Ther.*, 2011, **33**, 679.

[33] I. Popescu, P. R. Fleshner, J. C. Pezzullo, P. A. Charlton, G. Kosutic and A. J. Senagore, *Dis. Colon Rectum*, 2010, **53**, 126.
[34] Tranzyme Pharma (www.tranzyme.com) Press Release June 29th 2010.
[35] Trial NCT00672074: clinicaltrials.gov, Helsinn Therapeutics Inc., 2010.
[36] H. S. Sallam and J. D. Z. Chen, *Int. J. Pept.*, 2010, 1.
[37] P. Poitras and T. L. Peeters, *Curr. Opin. Endocrinol. Diabetes Obes.*, 2008, **15**, 54.
[38] B. Moshiree, R. McDonald, W. Hei and P. P. Toskes, *Dig. Dis. Sci.*, 2010, **55**, 675.
[39] K. Grant and R. Thomas, *J. Intensive Care Soc.*, 2009, **10**, 34.
[40] N. Q. Nguyen, M. J. Chapman, R. J. Fraser, L. K. Bryant and R. H. Holloway, *Crit. Care Med.*, 2007, **35**, 483.
[41] C. V. Hawkyard and R. J. Koerner, *J. Antimicrob. Chemother.*, 2007, **59**, 347.
[42] R. W. McCallum and O. Cynshi, *Aliment. Pharmacol. Ther.*, 2007, **26**, 1121.
[43] R. W. McCallum and O. Cynshi, *Aliment. Pharmacol. Ther.*, 2007, **26**, 107.
[44] S. M. Westaway, S. L. Brown, S. C. M. Fell, C. N. Johnson, D. T. MacPherson, D. J. Mitchell, J. W. Myatt, S. J. Stanway, J. T. Seal, G. Stemp, M. Thompson, K. Lawless, F. McKay, A. I. Muir, J. M. Barford, C. Cluff, S. R. Mahmood, K. L. Matthews, S. Mohamed, B. Smith, A. J. Stevens, V. J. Bolton, E. M. Jarvie and G. J. Sanger, *J. Med. Chem.*, 2009, **52**, 1180.
[45] Trials NCT00861809 and NCT01039805: clinicaltrials.gov, GlaxoSmithKline, 2009.
[46] Trial NCT01294566: clinicaltrials.gov, GlaxoSmithKline, 2010.
[47] R. J. Deschenes, L. J. Lorenz, R. S. Haun, B. A. Roos, K. J. Collier and J. E. Dixon, *Proc. Natl. Acad. Sci. USA*, 1984, **81**, 726.
[48] R. Herranz, *Med. Res. Rev.*, 2003, **23**, 559.
[49] J. N. Crawley and R. L. Corwin, *Peptides*, 1994, **15**, 731.
[50] P. J. Whorwell, F. Pace, M. D'Amato, G. Giacovelli, J. Dzieniszewski, R. Meier, T. Scholten, L. C. Rovati and C. Beglinger, *Gastroenterology*, 2008, **134**(Suppl. 1), A-157.
[51] L. R. Forte Jr., *Pharmacol. Ther.*, 2004, **104**, 137.
[52] L. A. Harris, *Nat. Rev. Gastroenterol. Hepatol.*, 2010, **7**, 365.
[53] J. M. Johnston, C. B. Kurtz, J. E. Macdougall, B. J. Lavins, M. G. Currie, D. A. Fitch, C. O'Dea, M. Baird and A. J. Lembo, *Gastroenterology*, 1877, **2010**, 139.
[54] A. J. Lembo, C. B. Kurtz, J. E. Macdougall, B. J. Lavins, M. G. Currie, D. A. Fitch, B. I. Jeglinski and J. M. Johnston, *Gastroenterology*, 2010, **138**, 886.
[55] Ironwood (www.ironwoodpharma.com) Press Releases November 1st 2010 & November 2nd 2009.
[56] Synergy (www.synergypharma.com) Press Release October 6th 2010.
[57] Albireo (www.albireopharma.com) Press Release October 12th 2010.
[58] C. P. Delaney, U. Yasothan and P. Kirkpatrick, *Nat. Rev. Drug Discovery*, 2008, **7**, 727.
[59] C. P. Delaney, B. G. Wolff, E. R. Viscusi, A. J. Senagore, J. G. Fort, W. Du, L. Techner and B. Wallin, *Ann. Surg.*, 2007, **245**, 355.
[60] X. Lei, P. C. Tang, P. Sun, and Z. Dong, *Patent Application US 2010/0210640-A*, 2010.
[61] S.-K. Choi, P. Fatheree, R. Gendron, A. A. Goldblum, L. Jiang, D. D. Long and D. Marquess, *Patent Application US 2006/0100426-A*, 2006.
[62] P. R. Fatheree, R. Gendron, A. Goldblum, D. Long, D. Marquess and D. S. Turner, *Patent Application WO 2006/069125-A*, 2006.
[63] R. Gendron, S.-K. Choi, P. R. Fatheree, A. A. Goldblum, D. D. Long, D. Marquess and S. D. Turner, *Patent Application US 2006/0276482-A*, 2006; R. Gendron, S.-K. Choi, P. R. Fatheree, A. A. Goldblum, D. D. Long, D. Marquess and S. D. Turner, *US Patent 7,317,022*, 2008.
[64] R. M. McKinnell, S. R. Armstrong, D. T. Beattie, S. K. Choi, P. R. Fatheree, R. A. L. Gendron, A. Goldblum, P. P. Humphrey, D. D. Long, D. G. Marquess, J. P. Shaw, J. A. M. Smith, S. D. Turner and R. G. Vickery, *J. Med. Chem.*, 2009, **52**, 5330.

[65] L. Demange, D. Boeglin, A. Moulin, D. Mousseaux, J. Ryan, G. Berge, D. Gagne, A. Heitz, D. Perrissoud, V. Locatelli, A. Torsello, J. C. Galleyrand, J. A. Fehrentz and J. Martinez, *J. Med. Chem.*, 1939, **2007**, 50.

[66] A. Moulin, L. Demange, G. Berge, D. Gagne, J. Ryan, D. Mousseaux, A. Heitz, D. Perrissoud, V. Locatelli, A. Torsello, J. C. Galleyrand, J. A. Fehrentz and J. Martinez, *J. Med. Chem.*, 2007, **50**, 5790.

[67] J. Li, S. Y. Chen, J. J. Li, H. Wang, A. S. Hernandez, S. Tao, C. M. Musial, F. Qu, S. Swartz, S. T. Chao, N. Flynn, B. J. Murphy, D. Slusarchyk, R. K. Seethala, M. Yang, P. Sleph, G. Grover, M. A. Smith, B. Beehler, L. Giupponi, K. E. Dickinson, H. Zhang, W. G. Humphreys, B. P. Patel, M. Schwinden, T. Stouch, P. T. W. Cheng, S. A. Biller, W. R. Ewing, D. Gordon, J. A. Robl and J. A. Tino, *J. Med. Chem.*, 2007, **50**, 5890.

[68] A. S. Hernandez, P. T. W. Cheng, C. M. Musial, S. G. Swartz, R. J. George, G. Grover, D. Slusarchyk, R. K. Seethala, M. Smith, K. Dickinson, L. Giupponi, D. A. Longhi, N. Flynn, B. J. Murphy, D. A. Gordon, S. A. Biller, J. A. Robl and J. A. Tino, *Bioorg. Med. Chem. Lett.*, 2007, **17**, 5928.

[69] J. Li, S. Y. Chen, B. J. Murphy, N. Flynn, R. Seethala, D. Slusarchyk, M. Yan, P. Sleph, H. Zhang, W. G. Humphreys, W. R. Ewing, J. A. Robl, D. Gordon and J. A. Tino, *Bioorg. Med. Chem. Lett.*, 2008, **18**, 4072.

[70] T. D. Heightman, J. S. Scott, M. Longley, V. Bordas, D. K. Dean, R. Elliott, G. Huntley, J. Witherington, L. Abberley, B. Passingham, M. Berlanga, M. De los Frailes, A. Wise, B. Powney, A. Muir, F. McKay, S. Butler, K. Winborn, C. Gardner, J. Darton, C. Campbell and G. Sanger, *Bioorg. Med. Chem. Lett.*, 2007, **17**, 6584.

[71] J. Witherington, L. Abberley, M. A. Briggs, K. Collis, D. K. Dean, A. Gaiba, N. P. King, H. Kraus, N. Shucker, J. G. A. Steadman, A. K. Takle, G. Sanger, G. Wadsworth, S. Butler, F. McKay, A. Muir, K. Winborn and T. D. Heightman, *Bioorg. Med. Chem. Lett.*, 2008, **18**, 2203.

[72] J. Witherington, L. Abberley, B. R. Bellenie, R. Boatman, K. Collis, D. K. Dean, A. Gaiba, N. P. King, J. G. A. Steadman, A. K. Takle, G. Sanger, S. Butler, F. McKay, A. Muir, K. Winborn, R. W. Ward and T. D. Heightman, *Bioorg. Med. Chem. Lett.*, 2009, **19**, 684.

[73] B. Schoentjes, A. P. Poncelet, J. G. Doyon, J. T. M. Linders and L. A. L. Verdonck, *Patent Application WO 2009/133052*, 2009.

[74] T. D. Heightman, E. Conway, D. F. Corbett, G. J. Macdonald, G. Stemp, S. M. Westaway, P. Celestini, S. Gagliardi, M. Riccaboni, S. Ronzoni, K. Vaidya, S. Butler, F. McKay, A. Muir, B. Powney, K. Winborn, A. Wise, E. M. Jarvie and G. J. Sanger, *Bioorg. Med. Chem. Lett.*, 2008, **18**, 6423.

[75] S. M. Westaway, S. L. Brown, E. Conway, T. D. Heightman, C. N. Johnson, K. Lapsey, G. J. MacDonald, D. T. MacPherson, D. J. Mitchell, J. W. Myatt, J. T. Seal, S. J. Stanway, G. Stemp, M. Thompson, P. Celestini, A. Colombo, A. Consonni, S. Gagliardi, M. Riccaboni, S. Ronzoni, M. A. Briggs, K. L. Matthews, A. J. Stevens, V. J. Bolton, I. Boyfield, E. M. Jarvie, S. C. Stratton and G. J. Sanger, *Bioorg. Med. Chem. Lett.*, 2008, **18**, 6429.

[76] G. J. Sanger, S. M. Westaway, A. A. Barnes, D. T. MacPherson, A. I. Muir, E. M. Jarvie, V. N. Bolton, S. Cellek, E. Naslund, P. M. Hellstrom, R. A. Borman, W. P. Unsworth, K. L. Matthews and K. Lee, *Neurogastroenterol. Motil.*, 2009, **21**, 657.

[77] J. M. Bailey, J. S. Scott, J. B. Basilla, V. J. Bolton, I. Boyfield, D. G. Evans, E. Fluery, T. D. Heightman, E. M. Jarvie, K. Lawless, K. L. Matthews, F. McKay, H. Mok, A. Muir, B. S. Orlek, G. J. Sanger, G. Stemp, A. J. Stevens, M. Thompson, J. Ward, K. Vaidya and S. M. Westaway, *Bioorg. Med. Chem. Lett*, 2009, **19**, 6452.

[78] N. Takahashi, N. Koba, T. Yamamoto and M. Sudo, *Gastroenerology*, 2010, **138**(Suppl. 1), S-173. https://partnering.thomson-pharma.com/partnering/partnering/reports.display?id=61260&template=Drug.

[79] M. Sudo, Y. Iwata, Y. Arano, M. Jinno, M. Ohmi and H. Noguchi, *Patent Application WO 2010/098145-A*, 2010.
[80] B. Allison, V. K. Phuong, M. C. W. Pippel, M. H. Rabinowitz and H. Venkatesan, *Patent Application WO 2006/036670-A*, 2006.
[81] M. Pippel, B. D. Allison, V. K. Phuong, L. Li, M. F. Morton, C. Prendergast, X. Wu, N. P. Shankley and M. H. Rabinowitz, *Bioorg. Med. Chem. Lett.*, 2009, **19**, 6373.
[82] M. Pippel, K. Boyce, H. Vankatesan, V. K. Phuong, W. Yan, T. D. Barrett, G. Lagaud, L. Li, M. F. Morton, C. Prendergast, X. Wu, N. P. Shankley and M. H. Rabinowitz, *Bioorg. Med. Chem. Lett.*, 2009, **19**, 6376.

CHAPTER 10

Targeting Th17 and Treg Signaling Pathways in Autoimmunity

Shomir Ghosh, Mercedes Lobera and
Mark S. Sundrud

Contents

1. INTRODUCTION

Chronic inflammatory and autoimmune disorders encompass a broad set of individual clinical diseases that share underlying pathophysiologic features. These diseases are increasingly common, represent a significant unmet clinical need, and have galvanized substantial interest and investment within the biopharmaceutical industry. Although some idiopathic inflammatory disorders are linked to innate immune dysregulation, the majority of chronic inflammatory and autoimmune diseases are driven by misguided or over-aggressive T cell responses. Accordingly, a

Tempero Pharmaceuticals (a GSK Company), 200 Technology Square, Suite 602, Cambridge, MA 02139, USA

Annual Reports in Medicinal Chemistry, Volume 46
ISSN: 0065-7743, DOI: 10.1016/B978-0-12-386009-5.00017-5

comprehensive understanding of how T cells, particularly CD4^{+} T helper (Th) cells, develop under both normal and pathologic conditions is instrumental in guiding new drug discovery.

T cells are of hematopoietic lineage; like all blood cells they develop from pluripotent progenitors resident to the marrow of long bones [1]. T cell precursors migrate from bone marrow to the thymus where they pass through a series of maturation stages to form a mature T cell that ultimately circulates through blood and lymphatic vessels [2]. In healthy individuals, the end result of T cell ontogeny is an army of mature naïve T cells, each expressing an individualized T cell antigen receptor (TCR) that will recognize unique danger signals in the form of microbe-derived protein fragments (*i.e.*, antigens), while ignoring those from host tissues or commensal microorganisms. However, even in healthy individuals, a small number of self-reactive T cells can, and do, escape the thymus [3]. In response, nature has evolved a failsafe mechanism to ensure that T cell tolerance toward host tissues is maintained, involving the parallel development of T regulatory (Treg) cells. Treg cells are distinguished from conventional naïve T cells by their constitutive expression of the transcription factor Forkhead box, winged-helix protein 3 (Foxp3); they preferentially recognize self-antigens and act to dampen the activation of local T cells through suppressive mechanisms that remain poorly elucidated [4–6]. Current paradigms suggest that the balance between conventional naïve T cell activation (and subsequent differentiation into effector subsets, see below) and Treg-mediated immune suppression controls whether immune responses are ultimately protective, ineffective, or pathogenic [7,8]. Indeed, early clinical results utilizing Treg cellular therapy supports the notion that increasing Treg numbers in autoimmune patients can support tolerance [9].

1.1. Th17 pathway

Conventional naïve T cells become activated in response to cognate antigen presented in the context of MHC class II molecules on the surface of professional antigen-presenting cells (APC), such as B cells, monocytes, macrophages, and dendritic cells. In addition to TCR signal transduction, APC also engage a number of co-receptors on the surface of T cells, which can either act to enhance or inhibit T cell activation [10]. Further, APC can produce an array of cytokines that act as tertiary signals to T cells, instructing them to differentiate into specialized effector T cell subsets (Th cells), which in turn orchestrate specific immune reactions aimed at clearing individual classes of pathogens. Originally, Th cell differentiation was thought to be bimodal, either resulting in T helper type 1 (Th1) or T helper type 2 (Th2) cell development. Whereas Th1 cells produce gamma-interferon (IFNγ) and activate phagocytic and cytolytic immunity against intracellular pathogens (*e.g.*, viruses), Th2 cells produce IL-4, IL-5, and IL-13 to induce humoral immunity against extracellular parasites [11]. However, recent advances in T cell biology have expanded the list of potential T cell subsets (see Figure 1). In particular, Th17 cells, which express IL-17A (*i.e.*, IL-17), IL-17F, and IL-22, have been implicated in mucosal immunity directed against fungal pathogens and some species of bacteria. These cells are also broadly implicated in the pathogenesis of most common autoimmune and chronic inflammatory disorders, including rheumatoid arthritis (RA), multiple sclerosis (MS), and inflammatory bowel diseases (IBD) [12,13].

The cytokines responsible for directing naïve T cell differentiation into Th17 cells include transforming growth factor β (TGF-β) and the acute phase protein IL-6 [14–16]. This combination of cytokines potently activates signal transducer activator of transcription (Stat)-3, which subsequently promotes expression of the retinoic acid-related orphan nuclear receptor RORγt (RORC in humans) [15–17]. Stat3 and RORγt subsequently function in a synergistic fashion to activate expression of IL-17. In addition, IL-23, another Stat3-activating cytokine, has been shown to act on developing Th17 cells to enforce IL-17 expression and stabilize the

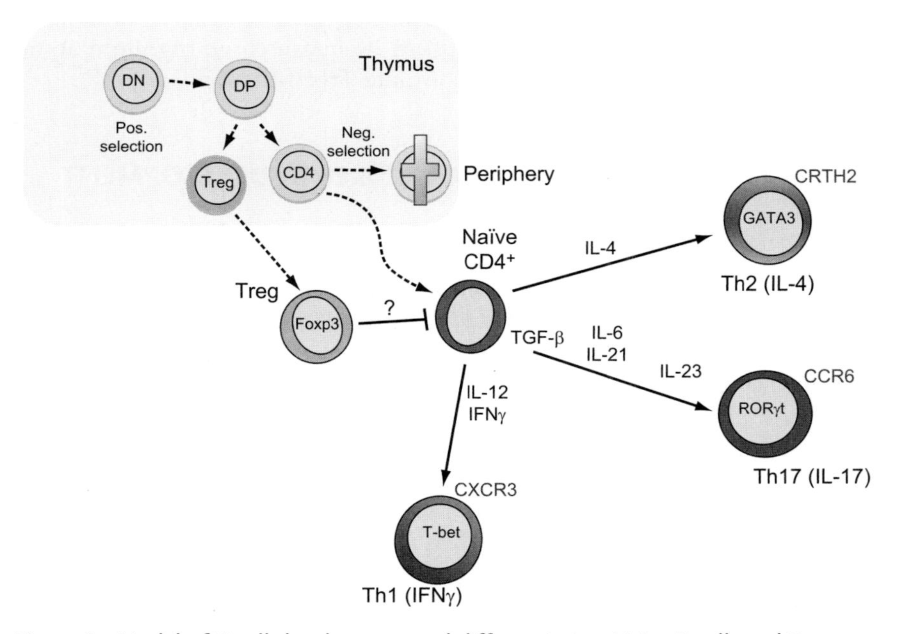

Figure 1 Model of T cell development and differentiation. Naïve T cells and Tregs develop in the thymus. Naïve T cells become activated by antigens in the periphery and can differentiate into one of three effector linages (*e.g.*, Th1, Th2, Th17). Regulatory T cells block the activation of bystander naïve T cells. (See Color Plate 10.1 in Color Plate Section.)

Th17 lineage [18]. Given the broad role of Th17 cells reported in immuno-inflammation, inhibiting Th17 cell development and/or function has profound therapeutic implications for autoimmune and chronic inflammatory indications.

At least two distinct strategies can be envisaged to target Th17-mediated inflammation: (1) blocking Th17 effector function and (2) blocking Th17 differentiation. By either strategy, targeting Th17 cells specifically represents a significant advance over current clinical modalities that act more broadly to cripple the immune system (*i.e.*, immunosuppressants, cyclosporine A, rapamycin). The drug discovery/development efforts of both these strategies will be discussed in the next section.

1.2. Treg pathway and immune homeostasis

Evidence from both autoimmune mouse models and human patients reveal that Treg cells are necessary to prevent spontaneous autoimmunity throughout life [19,20]. However, Treg cells do not necessarily distinguish among the types of T cell responses they inhibit. Recent data clearly indicates that Treg cells can also regulate immune responses to pathogens and developing tumors [21–23]. In fact, solid tumors may even actively recruit Treg cells as a means to preventing immunosurveillance [24]. Although human Treg biology is still in its infancy, we highlight some recent clinical advances that may modulate Treg function.

2. CURRENT TARGETS AND MOLECULES IN DEVELOPMENT

2.1. Th17 effector function

Several approaches have been taken to block Th17 cell cytokines. Current molecules and their development statuses are detailed below.

2.1.1. IL-17/IL-17 receptor antibodies

IL-17 and IL-17F are known to induce local cytokine and chemokine production, resulting in tissue inflammation characterized by neutrophil recruitment. IL-17 has been shown to play a key role in preclinical animal models including collagen induced arthritis (CIA) and experimental autoimmune encephalomyelitis (EAE) [25–29].

Secukinumab (AIN457), a monoclonal antibody (mAb) that neutralizes IL-17, is being evaluated for the treatment of uveitis, psoriasis, and other inflammatory conditions. In a Phase 3 study, AIN457 did not meet its primary endpoint in the treatment of non-infectious uveitis in patients with Behcets disease [30,31]. AIN457 is currently being evaluated in other uveitis studies [32]. Proof of concept has been shown in other inflammatory indications including psoriasis, RA, and ankylosing spondylitis. In psoriasis patients, AIN457 treatment showed greater benefit than placebo at all time points up to week 12, as measured by PASI50 (Psoriasis Area and Severity Index 50), reductions of histomorphological signs of acanthosis and epidermal hyperplasia, and changes in gene expression of markers of the IL-17A pathway [33]. In the RA trial, patients achieved ACR20 (American College of Rheumatology 20) response rates (50% for AIN457 and 31% for placebo) by week 4 ($P = 0.13$) which were maintained at week 16 (54% *vs.* 31%; $P = 0.08$). The 28-joint disease activity scores (DAS28) and C-reactive protein (CRP) values significantly decreased over time [33]. In a Phase 2 ankylosing spondylitis trial, AIN457 induced significantly higher ASAS20 (Assessment of SpondyloArthritis international Society 20) responses than placebo at week 6 meeting the primary endpoint [34].

Other anti-IL-17 humanized mAbs under clinical development include LY2439821 for the treatment of RA and psoriasis, RG-4934 for the treatment of RA, RO-5310074 for the treatment of psoriatic arthritis (PsA), and MEDI-571 for the treatment of RA [35–38]. AMG 827, a fully human mAb that binds to the IL-17 receptor and blocks its signaling, is currently being investigated as a treatment for a variety of inflammatory disorders including RA and psoriasis [39,40].

2.2. Th17 differentiation

Current understanding of Th17 cell differentiation and maintenance has indicated several points of therapeutic intervention. These include blocking of critical cytokines IL-6, IL-21, and IL-23 and their receptors, blocking the JAK/STAT pathway and antagonizing transcription factors such as RORγt and the aryl hydrocarbon receptor (AhR).

2.2.1. IL-6/IL-6 receptor antibodies

As discussed above, IL-6 is indispensable for the differentiation of Th17 cells from naïve precursors; it also inhibits TGF-β-induced expression of Foxp3 [16,41]. Stimulation by IL-6 in the lymph node or inflamed peripheral tissues activates JAK/STAT pathway, resulting in Stat3 activation, RORγt (and RORα) expression, and subsequent *trans*-activation of both IL-21 and the IL-23 receptor (IL-23R). Autocrine signaling through IL-21/IL-21R and through IL-23 further stabilizes Stat3 activity resulting in epigenetic modifications at the *Il17a/f* locus that allow for sustained inflammatory cytokine production [42].

Tocilizumab, a recombinant humanized mAb against IL-6 receptor (IL-6R), has been approved for the treatment of RA and Castleman disease. The antibody is currently being evaluated in Phase 3 trials for ankylosing spondylitis [43]. SAR-153191 (REGN-88), an IL-6R antibody, is in Phase 3 trials for RA and ankylosing spondylitis [44]. Olokizumab, CNTO-136, and ALD-518 are anti-IL-6 antibodies being evaluated in Phase 2 trials for RA [45–47]. CNTO-136 is also in Phase 2 trials for systemic lupus erythematosus (SLE) and lupus nephritis [48].

2.2.2. IL-23 and IL-12/23 antibodies

IL-23 is a heterodimeric cytokine produced by activated APC. It comprises IL-23p19 and IL-12p40, and signaling through the IL-23R is essential for the survival and stabilization of the Th17 phenotype. IL-23 has been implicated in several inflammatory conditions such as colitis, gastritis, arthritis, and psoriasis [49–53]. Ustekinumab, a humanized antibody targeting the p40 subunit of IL-12 and IL-23, was approved recently for the treatment of psoriasis. The antibody is in Phase 3 trials for PsA and Phase 2 trials for Crohn's disease [54,55]. SCH-90222, an anti-IL-23 antibody, is in development for psoriasis (Phase 2) [56]. An orally bioavailable small molecule inhibitor of IL-12 and IL-23 production, STA-5326, **1**, is also under development for psoriasis and Crohn's disease [57]. STA-5326 inhibits c-Rel translocation which results in inhibition of the expression of genes encoding the p40 subunit present in both IL-12 and IL-23 [58].

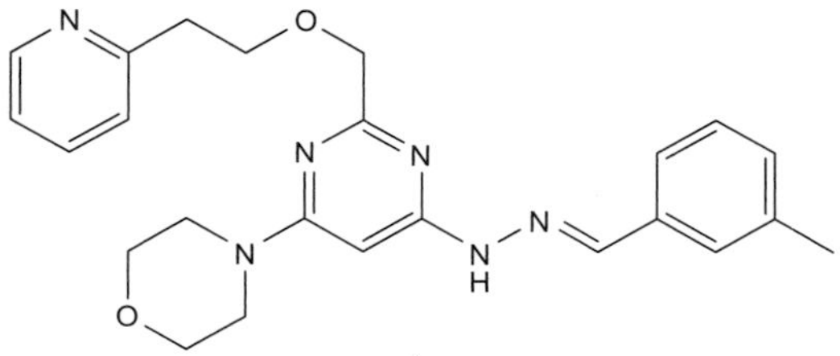

N O N N N N N H O

1

2.2.3. JAK/STAT pathway inhibitors

IL-6, IL-21, and IL-23 all regulate Th17 differentiation through their activation of Stat3. Stat3 activation downstream of these receptors is mediated by receptor associated Janus kinases (JAKs), which include JAK1, JAK2, JAK3, and TYK2. Several small molecule modulators targeting the JAK/STAT pathway have been developed that affect Th17 function and have anti-inflammatory activity; some examples are discussed below.

CP-690550, **2**, is a pan JAK inhibitor with low nanomolar potency against JAK1, JAK2, and JAK3, but with functional selectivity for JAK1/3 *versus* JAK2 in cellular assays [59,60]. CP-690550 has shown preclinical efficacy in mouse CIA and rat adjuvant-induced arthritis models [59,60]. CP-690550 is currently in Phase 3 trials for RA and Phase 2 trials for prevention of acute (renal) allograft rejection, psoriasis (oral and topical),

Crohn's disease, ulcerative colitis, and dry eye disease (topical). In the ORAL Sync Phase 3 study, in moderate-to-severe RA, CP-690550 met its primary endpoints showing statistically significant changes *versus* placebo in reducing signs and symptoms of RA, based on ACR20 response rates at 6 months and improved physical function [61,62].

INCB018424, **3**, is a selective small molecule inhibitor of JAK1 and JAK2 that potently inhibits cytokine-induced JAK signaling and function in lymphocytes and keratinocytes [63]. In an open label subtotal inunction study in 25 patients with plaque psoriasis, transcriptional changes in biopsies at baseline and following 28 days of topical INCB018424 treatment were consistent with decreased Th1 and Th17 lymphocyte activation, decreased epidermal hyperplasia and dendritic cell activation. In a subsequent Phase 2b study, the primary endpoint of total lesion score for all dose groups was decreased greater than two-fold over vehicle control at day 84 [63]. INCB028050, a selective orally bioavailable JAK1/JAK2 inhibitor, is currently under clinical evaluation for the treatment of RA [64].

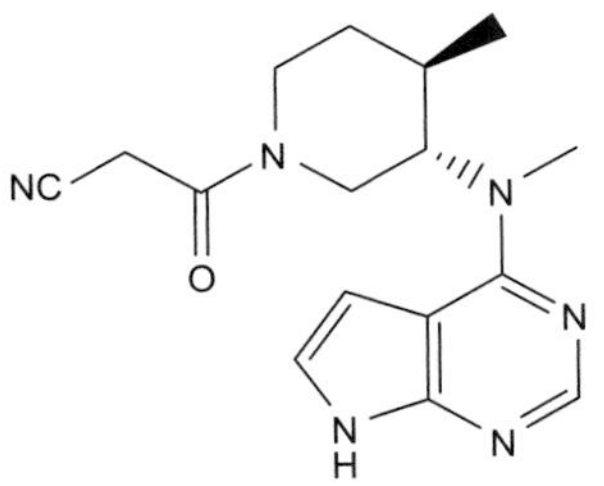

2 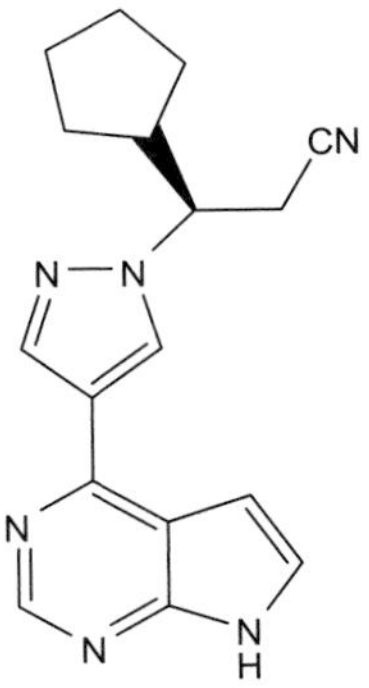 **3**

2.2.4. Inhibitors of transcription factors

Activation of STAT3 by each of the critical cytokines (IL-6, IL-21, IL-23) results in the induction of RORγt and RORα, which subsequently leads to expression of IL-17. Forced overexpression of RORγt in human naive T cells induces a Th17-like phenotype, by inducing IL-17A, IL-17F, IL-26, and CCR6 expression and downregulating IFN-γ secretion [15,65,66]. *In vivo*, RORγt-deficient mice are protected in an EAE model, show reduced susceptibility to allergen-induced airway inflammation, and are protected against crescentic glomerulonephritis [15,67,68]. In addition, it also has been shown that RORγt-deficient T cells do not induce colitis when adoptively transferred [69]. There have been very few reports in the literature that disclose RORγt inhibitors. Jetten *et al.* have shown that selective LXXLL peptides (*e.g.*, VLVEHPILGGLLSTRVDSS) bind to the ligand binding domain of RORγt and antagonize RORγt-mediated transcriptional activation [70]. It has been reported that carboxylic acid-containing compounds (*e.g.*, LE 135, **4**) structurally related to all *trans* retinoic acid (ATRA) are RORγt inhibitors which reduced IL-17 production from activated human peripheral blood mononuclear cells in a dose-dependent manner [71]. Recently, Huh *et al.* reported that digoxin, **5**, a cardiac glycoside, and two synthetic derivatives selectively inhibited RORγt activity and suppressed mouse and human Th17 differentiation. Treatment with **5** delayed onset and reduced severity of disease in a mouse EAE model [72].

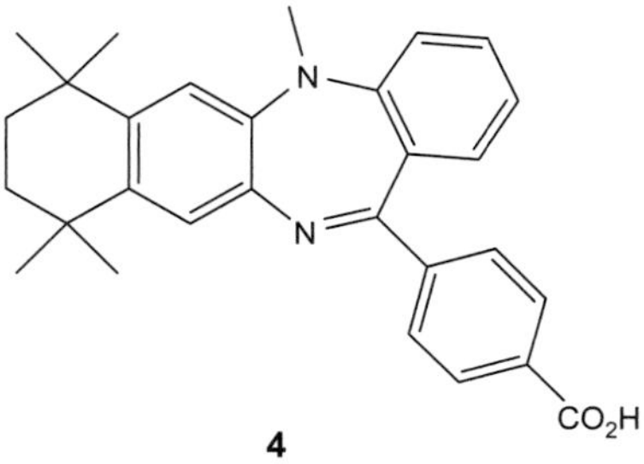

N N CO_2H

4

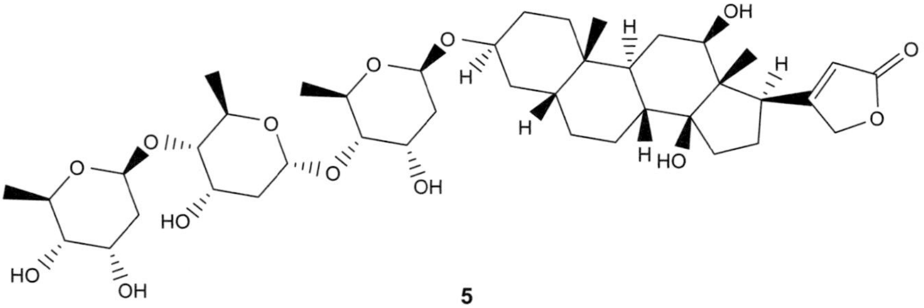

O O O O O O O OH OH HO HO OH H H H H H HO OH O O

5

AhR, a ligand-activated transcription factor, has been shown to regulate Th17-cell development and Treg differentiation in mice [73,74]. AhR expression in $CD4^+$ T cells in mice was found to be restricted to the Th17 cell subset and is essential for IL-22 production. AhR-deficient mice develop less severe disease in an EAE model. Studies indicate that AhR may also be involved in the expression of the anti-inflammatory T cell cytokine IL-10 during T cell differentiation [75,76]. Several flavonoids including apigenin, **6**, naringenin, **7**, and CH-223191, **8**, function as AhR antagonists which may be useful in the treatment of autoimmune diseases [77].

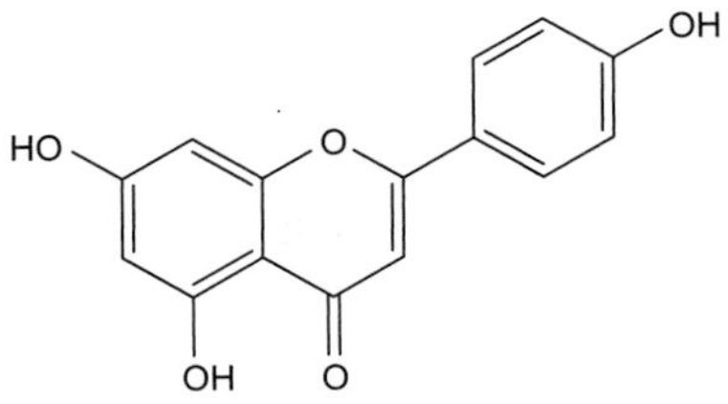

6

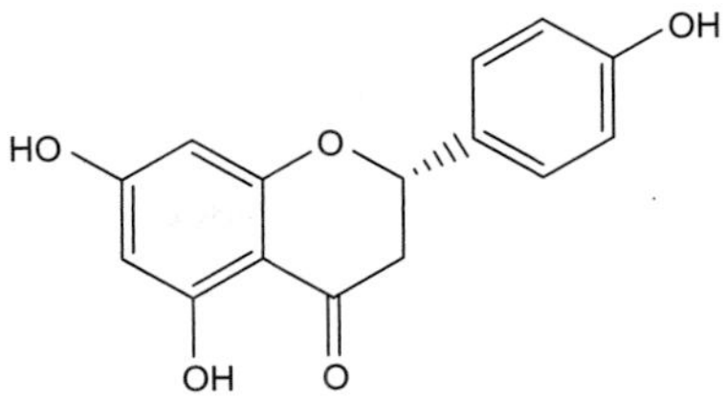

7

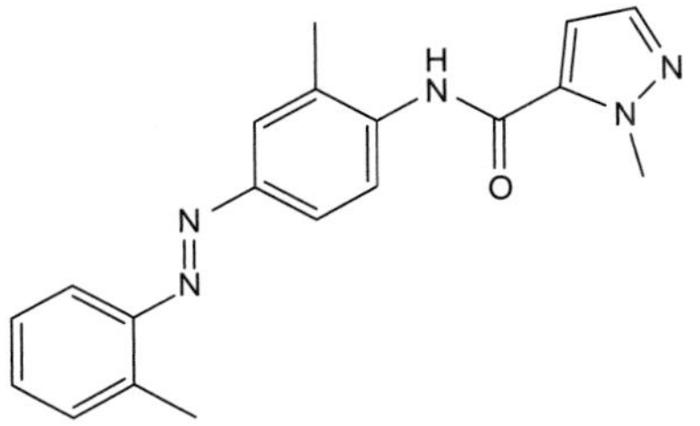

8

2.2.5. Halofuginone and the amino acid starvation response

Halofuginone, **9**, is a synthetic derivative of febrifugine, **10**, a naturally occurring alkaloid found in the root of hydrangea plants. Halofuginone has been reported to be a potent and selective inhibitor of Th17 differentiation which functions by inducing a state of nutritional stress known as the amino acid starvation response. Treatment of naïve T cells with **9** was found to block Th17 differentiation and concomitantly increase Foxp3 expression without impacting cell proliferation, or Th1 or Th2 differentiation. Administration of **9** to mice selectively reduced both Th17 differentiation and the development of Th17-driven EAE. In a second EAE model driven entirely by IFNγ-producing Th1 cells, **9** did not prevent disease onset or severity [78]. The direct cellular target of **9** remains unknown.

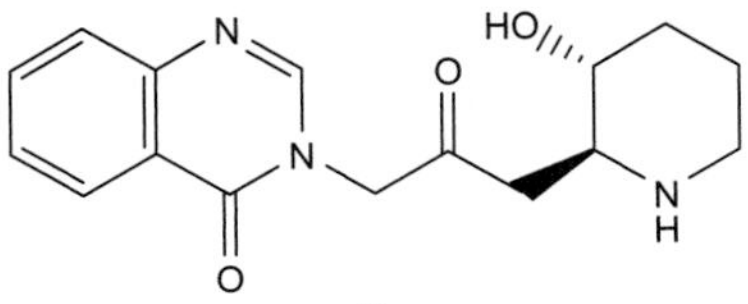

9

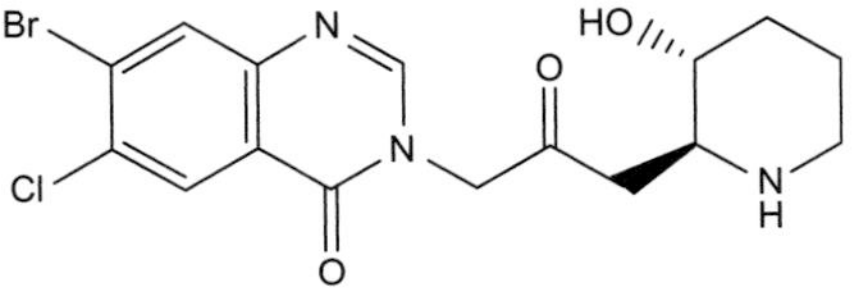

10

2.3. Treg biologics

2.3.1. Anti-CD3 antibodies

Anti-CD3 antibodies act as immunosuppressants, both by reducing the number of effector T cells and inducing the development of adaptive Tregs, although the underlying mechanism is not fully understood [79]. Muromonab-CD3 (a mouse mAb against human CD3) was approved for the prevention of renal allograft rejection, but an important side effect is CRS (cytokine release syndrome) [80,81]. Two humanized FcR nonbinding anti-CD3 antibodies (teplizumab and otelixizumab) have since been developed. In recent Phase 3 studies in patients with recent-onset Type 1 diabetes mellitus (T1DM), both teplizumab and otelixizumab failed to meet the primary end points [82,83]. Trials with otelixizumab in adolescents and adults with newly diagnosed T1DM, RA, and thyroid eye disease are ongoing [84–89]. A Phase 2 trial is currently underway to evaluate if teplizumab can help prevent or delay the onset of T1DM in relatives at high risk of developing the disease [90].

2.3.2. CTLA-4-Ig fusion protein

CTLA-4 (Cytotoxic T-Lymphocyte Antigen 4) is a cell-surface molecule that binds CD80 and CD86, resulting in an inhibitory signal that leads to suppression of T cell proliferation. Treg-specific CTLA-4-deficient mice spontaneously develop systemic lymphoproliferation, fatal T cell-mediated autoimmune disease, hyperproduction of immunoglobulin E, and enhanced tumor immunity [91].

Abatacept (ORENCIA®), a fully human fusion protein that binds to CD80/CD86 with high affinity, has been approved for RA and juvenile idiopathic arthritis [92]. Trials to evaluate abatacept in Crohn's disease, ulcerative colitis, and in non-life-threatening SLE failed to meet their primary endpoints [93–96]. Efficacy was observed in the treatment of 170 patients with PsA with 48% of patients achieving ACR20 for abatacept *versus* 19% for placebo ($P = 0.006$) [97]. Additional trials in prevention of GVHD and in the treatment of lupus nephritis are ongoing [98].

2.4. Emerging targets influencing Treg function

Epigenetic regulation and posttranslational modification of Foxp3 in Tregs have been studied by several groups. Loosdregt *et al.* reported that Foxp3 acetylation is regulated by histone acetyltransferase p300 and histone deacetylase SIRT1 [99]. *Ex vivo* treatment of $CD4^+$ T cells with SIRT1 inhibitor nicotinamide, **11**, resulted in increased Foxp3 levels and increased suppressive activity. An evolutionarily conserved CpG-rich element within the Foxp3 locus was identified by Huehn *et al.* that was selectively demethylated in natural Tregs (nTregs), but not in conventional T cells or in *in vitro* generated iTregs [100]. The methylation status of this Treg-specific demethylated region (TSDR) can be manipulated by inhibitors of DNA methyltransferase 1 (DNMT1), such as 5-azacytidine **12**. Huehn *et al.* reported that **12** promoted a more stable Foxp3 expression [101]. Hancock *et al.* have reported that Tregs isolated from HDAC9-deficient mice were more abundant and displayed increased suppressive function *in vitro* and *in vivo*; these cells also showed enhanced expression of Foxp3, CTLA-4, and GITR (glucocorticoid-induced TNFR-related protein), as well as increased acetylation of Foxp3 [102]. The HDAC9-deficient mice are also reported to be resistant to DSS-induced colitis [103]. Similar effects have been reported with Trichostatin A (**13**), a pan-HDAC inhibitor, on Treg numbers and in prevention of DSS-induced colitis [102,103].

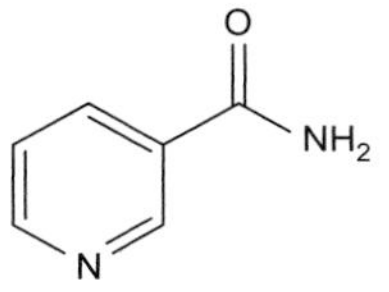

O NH2 N
11

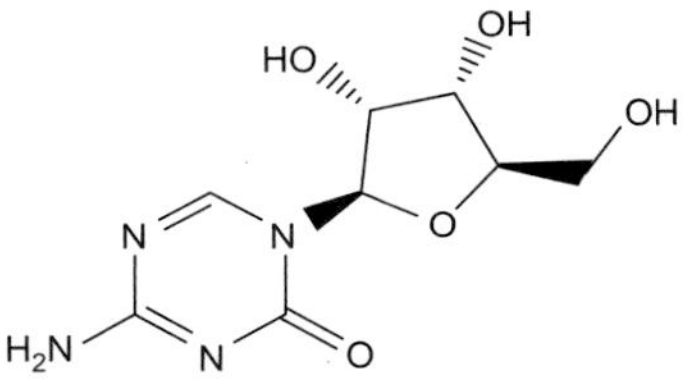

HO OH OH N N O H2N N O
12

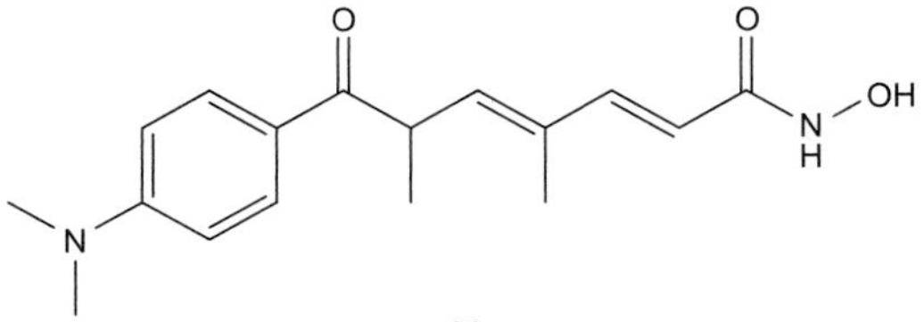

O O N OH H N
13

Two different kinases have been reported to modulate Treg function. Glycogen synthase kinase-3 (GSK-3β) regulates β-catenin, which has been shown to prolong Treg survival [104]. A GSK-3β inhibitor (SB216763, **14**) was reported to increase Treg suppressive activity and prolong Foxp3 levels [105]. *In vivo*, SB216763 treatment afforded a modest effect in prolonging islet survival in an allotransplant mouse model. Zanin-Zhorov *et al.* have reported that treatment with Protein Kinase C-theta (PKC-θ) inhibitor C20 (**15**) protected Tregs from inactivation by TNF-α, enhanced suppressive function of defective Tregs from RA patients, and enhanced the protective capabilities of Tregs in a T cell induced colitis model [106].

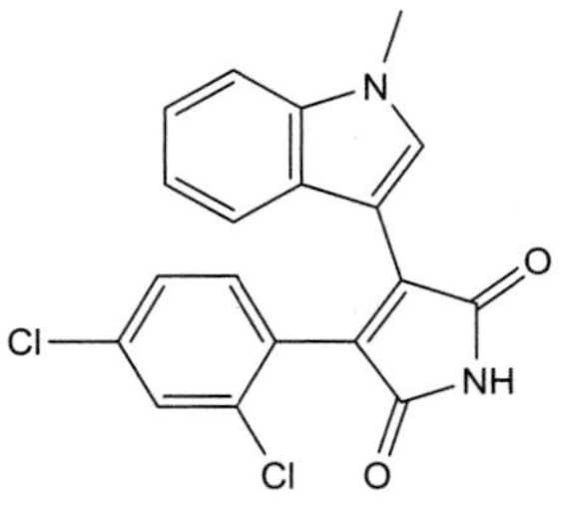

14

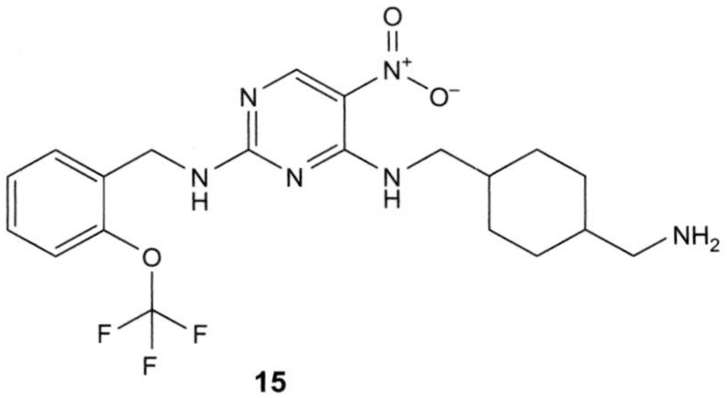

15

Recent studies have shown that fingolimod, **16**, an S1P receptor modulator, increases the functional activity of Tregs [107]. In a mouse model of colitis, treatment with fingolimod resulted in upregulation of Foxp3, IL-10, TGF-β, and CTLA-4, and it significantly suppressed the development of disease [108]. There are several reports that TLR ligands can modulate Treg function: TLR7 agonists imiquimod, **17**; gardiquimod, **18**; and flagellin (a TLR5 ligand) have been reported to enhance Treg suppressive function [109,110].

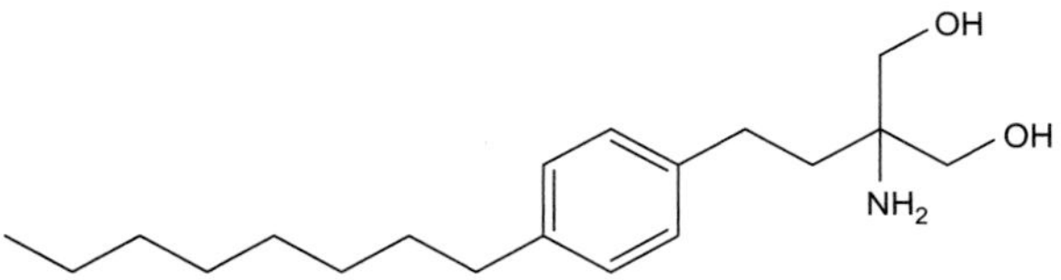

16

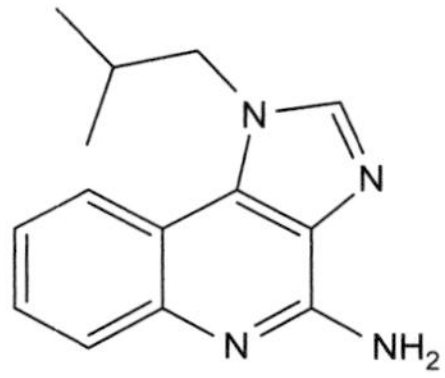

17

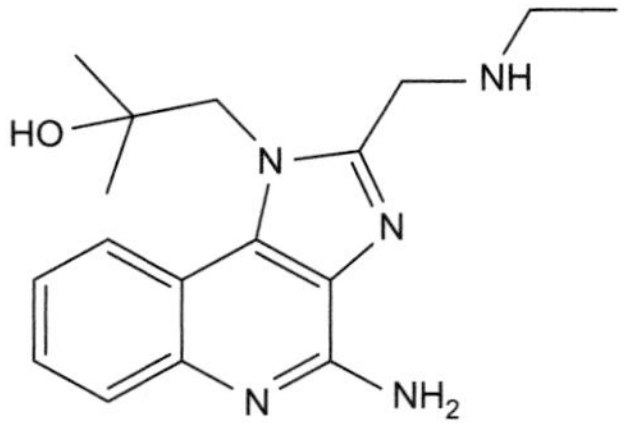

18

3. CONCLUSIONS

T cell-driven autoimmune disorders continue to present a significant unmet clinical need. Advances in our understanding of T cell activation, differentiation, and regulation have yielded novel approaches to specifically dampen pathogenic immune reactions without creating potentially dangerous states of general immune suppression. As discussed throughout, specific modulation of Th17 and Treg cells affords broad therapeutic promise for the treatment of autoimmune and chronic inflammatory conditions. Clinical results from the current biopharmaceutical therapies will provide further validation for such targeted pathways. Even though there are relatively few reports of small molecule modulators of Th17 and Treg cell function, several targets have been identified providing new opportunities for small molecule drug discovery.

REFERENCES

[1] S. H. Orkin, *Curr. Opin. Cell Biol.*, 1995, **7**, 870.
[2] A. C. Carpenter and R. Bosselut, *Nat. Immunol.*, 2010, **11**, 666.
[3] C. J. Kroger, R. R. Flores, M. Morillon, B. Wang and R. Tisch, *Arch. Immunol. Ther. Exp.*, 2010, **58**, 449.
[4] Y. Zheng and A. Y. Rudensky, *Nat. Immunol.*, 2007, **8**, 457.
[5] C. S. Hsieh, Y. Zheng, Y. Liang, J. D. Fontenot and A. Y. Rudensky, *Nat. Immunol.*, 2006, **7**, 401.
[6] S. Sakaguchi, M. Miyara, C. M. Costantino and D. A. Hafler, *Nat. Rev. Immunol.*, 2010, **10**, 490.
[7] J. H. Buckner, *Nat. Rev. Immunol.*, 2010, **10**, 849.
[8] H. Waldmann, *Nat. Rev. Nephrol.*, 2010, **6**, 569.
[9] J. L. Riley, C. H. June and B. R. Blazar, *Immunity*, 2009, **30**, 656.
[10] A. H. Sharpe, *Immunol. Rev.*, 2009, **229**, 5.
[11] I. C. Ho and L. H. Glimcher, *Cell*, 2002, **109**, S109.
[12] T. Korn, E. Bettelli, M. Oukka and V. K. Kuchroo, *Annu. Rev. Immunol.*, 2009, **27**, 485.
[13] C. T. Weaver, R. D. Hatton, P. R. Mangan and L. E. Harrington, *Annu. Rev. Immunol.*, 2007, **25**, 821.
[14] D. J. Cua and R. A. Kastelein, *Nat. Immunol.*, 2006, **7**, 557.
[15] I. I. Ivanov, B. S. McKenzie, L. Zhou, C. E. Tadokoro, A. Lepelley, J. J. Lafaille, D. J. Cua and D. R. Littman, *Cell*, 2006, **126**, 1121.
[16] E. Bettelli, Y. Carrier, W. Gao, T. Korn, T. B. Strom, M. Oukka, H. L. Weiner and V.K. Kuchroo, *Nature*, 2006, **441**, 235.
[17] X. O. Yang, A. D. Panopoulos, R. Nurieva, S. H. Chang, D. Wang, S. S. Watowich and C. Dong, *J. Biol. Chem.*, 2007, **282**, 9358.
[18] M. J. McGeachy, K. S. Bak-Jensen, Y. Chen, C. M. Tato, W. Blumenschein, T. McClanahan and D. J. Cua, *Nat. Immunol.*, 2007, **8**, 1390.
[19] J. M. Kim, J. P. Rasmussen and A. Y. Rudensky, *Nat. Immunol.*, 2007, **8**, 191.
[20] J. Kim, K. Lahl, S. Hori, C. Loddenkemper, A. Chaudhry, P. De Roos, A. Rudensky and T. Sparwasser, *J. Immunol.*, 2009, **183**, 7631.
[21] J. M. Lund, L. Hsing, T. T. Pham and A. Y. Rudensky, *Science*, 2008, **320**, 1220.
[22] Y. Furuichi, H. Tokuyama, S. Ueha, M. Kurachi, F. Moriyasu and K. Kakimi, *World J. Gastroenterol.*, 2005, **11**, 3772.
[23] K. Klages, C. T. Mayer, K. Lahl, C. Loddenkemper, M. W. Teng, S. F. Ngiow, M. J. Smyth, A. Hamann, J. Huehn and T. Sparwasser, *Cancer Res.*, 2010, **70**, 7788.
[24] A. W. Mailloux and M. R. Young, *Crit. Rev. Immunol.*, 2010, **30**, 435.
[25] S. Nakae, A. Nambu, K. Sudo and Y. Iwakura, *J. Immunol.*, 2003, **171**, 6173.

[26] E. Lubberts, M. I. Koenders, B. Oppers-Walgreen, L. van den Bersselaar, C. J. Coenen-de Roo, L. A. Joosten and W. B. van den Berg, *Arthritis Rheum.*, 2004, **50**, 650.
[27] Y. Komiyama, S. Nakae, T. Matsuki, A. Nambu, H. Ishigame, S. Kakuta, K. Sudo and Y. Iwakura, *J. Immunol.*, 2006, **177**, 566.
[28] X. O. Yang, S. H. Chang, H. Park, R. Nurieva, B. Shah, L. Acero, Y. H. Wang, K. S. Schluns, R. R. Broaddus, Z. Zhu and C. Dong, *J. Exp. Med.*, 2008, **205**, 1063.
[29] H. H. Hofstetter, S. M. Ibrahim, D. Koczan, N. Kruse, A. Weishaupt, K. V. Toyka and R. Gold, *Cell. Immunol.*, 2005, **237**, 123.
[30] http://www.novartis.com/newsroom/media-releases/en/2011/1482782.shtml.
[31] http://clinicaltrials.gov/ct2/show/NCT00995709.
[32] http://clinicaltrials.gov/ct2/show/NCT01103024.
[33] W. Hueber, D. D. Patel, T. Dryja, A. M. Wright, I. Koroleva, G. Bruin, C. Antoni, Z. Draelos, M. H. Gold, Psoriasis Study Group, P. Durez, P. P. Tak, J. J. Gomez-Reino, RA Study Group, C. S. Foster, R. Y. Kim, C. M. Samson, N. S. Falk, D. S. Chu, D. Callanan, Q. D. Nguyen, Uveitis Study Group, K. Rose, A. Haider and F. Di Padova, *Sci. Transl. Med.*, 2010, **2**, 52ra72.
[34] D. Baeten, J. Sieper, P. Emery, J. Braun, D. Van der Heijde, I. McInnes, J. M. Van Laar, R. Landewé, P. Wordsworth, J. Wollenhaupt, H. Kellner, J. Paramarta, A. P. Bertolino, A. M. Wright and W. Hueber, Presented at ACR/ARHP Annual Scientific meeting, Atlanta, November, 2010, Poster L7.
[35] (a) http://clinicaltrials.gov/ct2/show/NCT01107457; (b) http://clinicaltrials.gov/ct2/show/NCT00966875.
[36] http://www.roche.com/irp110202.pdf.
[37] http://clinicaltrials.gov/ct2/show/NCT01199809.
[38] C. Langham, C. Russell, W. Barker, M. Abbott, S. Almond, E. Kelly and S. Dawson, *Arthritis Rheum.*, 2009, **60**, Suppl. 10, Abstract 8.
[39] http://clinicaltrials.gov/ct2/show/NCT00950989.
[40] http://clinicaltrials.gov/ct2/show/NCT01101100.
[41] P. R. Mangan, L. E. Harrington, D. B. O'Quinn, W. S. Helms, D. C. Bullard, C. O. Elson, R. D. Hatton, S. M. Wahl, T. R. Schoeb and C. T. Weaver, *Nature*, 2006, **441**, 231.
[42] E. Bettelli, T. Korn and V. K. Kuchroo, *Curr. Opin. Immunol.*, 2007, **19**, 652.
[43] http://clinicaltrials.gov/ct/show/NCT01209689.
[44] http://clinicaltrials.gov/ct/show/NCT01118728.
[45] http://clinicaltrials.gov/ct/show/NCT01296711.
[46] http://clinicaltrials.gov/ct/show/NCT00718718.
[47] http://clinicaltrials.gov/ct/show/NCT00867516.
[48] http://clinicaltrials.gov/ct/show/NCT01273389.
[49] B. Oppmann, R. Lesley, B. Blom, J. C. Timans, Y. Xu, B. Hunte, F. Vega, N. Yu, J. Wang, K. Singh, F. Zonin, E. Vaisberg, T. Churakova, M. Liu, D. Gorman, J. Wagner, S. Zurawski, Y. Liu, J. S. Abrams, K. W. Moore, D. Rennick, R. de Waal-Malefyt, C. Hannum, J. F. Bazan and R. A. Kastelein, *Immunity*, 2000, **13**, 715.
[50] C. S. Lankford and D. M. Frucht, *J. Leukoc. Biol.*, 2003, **73**, 49.
[51] D. McGovern and F. Powrie, *Gut*, 2007, **56**, 1333.
[52] K. Kikly, L. Liu, S. Na and J. D. Sedgwick, *Curr. Opin. Immunol.*, 2006, **18**, 670.
[53] E. Lee, W. L. Trepicchio, J. L. Oestreicher, D. Pittman, F. Wang, F. Chamian, M. Dhodapkar and J. G. Krueger, *J. Exp. Med.*, 2004, **199**, 125.
[54] http://clinicaltrials.gov/ct2/show/NCT01009086.
[55] http://clinicaltrials.gov/ct2/show/NCT00771667.
[56] http://clinicaltrials.gov/ct2/show/study/NCT01225731.
[57] A. Billich, *IDrugs*, 2007, **10**, 53.

[58] Y. Wada, R. Lu, D. Zhou, J. Chu, T. Przewloka, S. Zhang, L. Li, Y. Wu, J. Qin, V. Balasubramanyam, J. Barsoum and M. Ono, *Blood*, 2007, **109**, 1156.
[59] M. E. Flanagan, T. A. Blumenkopf, W. H. Brissette, M. F. Brown, J. M. Casavant, C. Shang-Poa, J. L. Doty, E. A. Elliott, M. B. Fisher, M. Hines, C. Kent, E. M. Kudlacz, B. M. Lillie, K. S. Magnuson, S. P. McCurdy, M. J. Munchhof, B. D. Perry, P. S. Sawyer, T. J. Strelevitz, C. Subramanyam, J. Sun, D. A. Whipple and P. S. Changelian, *J. Med. Chem.*, 2010, **53**, 8468.
[60] L. Vijayakrishnan, R. Venkataramanan and P. Gulati, *Trends Pharmacol. Sci.*, 2011, **32**, 25.
[61] Press release, March 4, 2011, http://www.pfizer.com.
[62] http://clinicaltrials.gov/ct2/show/NCT00960440.
[63] K. Callis Duffin, M. Luchi, R. Fidelus-Gort, R. Newton, J. Fridman, T. Burn, P. Haley, P. Scherle, R. Flores, N. Punwani, R. Levy, W. Williams and A. Gottlieb, Presented at Society for Investigative Dermatology, 2010, Poster 261, May 5.
[64] J. S. Fridman, P. A. Scherle, R. Collins, T. C. Burn, Y. Li, J. Li, M. B. Covington, B. Thomas, P. Collier, M. F. Favata, X. Wen, J. Shi, R. McGee, P. J. Haley, S. Shepard, J. D. Rodgers, S. Yeleswaram, G. Hollis, R. C. Newton, B. Metcalf, S. M. Friedman and K. Vaddi, *J. Immunol.*, 2010, **184**, 5298.
[65] X. O. Yang, B. P. Pappu, R. Nurieva, A. Akimzhanov, H. S. Kang, Y. Chung, L. Ma, B. Shah, A. D. Panopoulos, K. S. Schluns, S. S. Watowich, Q. Tian, A. M. Jetten and C. Dong, *Immunity*, 2008, **28**, 29.
[66] I. I. Ivanov, L. Zhou and D. R. Littman, *Semin. Immunol.*, 2007, **19**, 409.
[67] J. F. Alcorn, C. R. Crowe and J. K. Kolls, *Annu. Rev. Physiol.*, 2010, **72**, 495.
[68] O. M. Steinmetz, S. A. Summers, P. Y. Gan, T. Semple, S. R. Holdsworth and A. R. Kitching, *J. Am. Soc. Nephrol.*, 2011, **22**, 472.
[69] M. Leppkes, C. Becker, I. I. Ivanov, S. Hirth, S. Wirtz, C. Neufert, S. Pouly, A. J. Murphy, D. M. Valenzuela, G. D. Yancopoulos, B. Becher, D. R. Littman and M. F. Neurath, *Gastroenterology*, 2009, **136**, 257.
[70] S. Kurebayashi, T. Nakajima, S. C. Kim, C. Y. Chang, D. P. McDonnell, J. P. Renaud and A.M. Jetten, *Biochem. Biophys. Res. Commun.*, 2004, **315**, 919.
[71] U. Deuschle, U. Abel, C. Kremoser, T. Schlueter, T. Hoffmann and S. Perovic-ottstadt, *WO Patent Application* 2010/049144, 2010.
[72] J. R. Huh, M. W. Leung, P. Huang, D. A. Ryan, M. R. Krout, R. R. Malapaka, J. Chow, N. Manel, M. Ciofani, S. V. Kim, A. Cuesta, F. R. Santori, J. J. Lafaille, H. E. Xu, D. Y. Gin, F. Rastinejad and D. R. Littman, *Nature*, 2011, Published online, Mar 27, 2011.
[73] F. J. Quintana, A. S. Basso, A. H. Iglesias, T. Korn, M. F. Farez, E. Bettelli, M. Caccamo, M. Oukka and H. L. Weiner, *Nature*, 2008, **453**, 65.
[74] M. Veldhoen, K. Hirota, A. M. Westendorf, J. Buer, L. Dumoutier, J. C. Renauld and B. Stockinger, *Nature*, 2008, **453**, 106.
[75] L. Apetoh, F. J. Quintana, C. Pot, N. Joller, S. Xiao, D. Kumar, E. J. Burns, D. H. Sherr, H.L. Weiner and V. K. Kuchroo, *Nat. Immunol.*, 2010, **11**, 854.
[76] R. Gandhi, D. Kumar, E. J. Burns, M. Nadeau, B. Dake, A. Laroni, D. Kozoriz, H.L. Weiner and F. J. Quintana, *Nat. Immunol.*, 2010, **11**, 846.
[77] B. Stockinger and M. Veldhoen, *WO Patent Application* 2009/115807, 2009.
[78] M. S. Sundrud, S. B. Koralov, M. Feuerer, D. P. Calado, A. E.-H. Kozhaya, A. Rhule-Smith, R. E. Lefebvre, D. Unutmaz, R. Mazitschek, H. Waldner, M. Whitman, T. Keller and A. Rao, *Science*, 2009, **324**, 1334.
[79] D. Bresson and M. von Herrath, *Sci. Transl. Med.*, 2011, **3**, 68ps4.
[80] T. T. Hansel, H. Kropshofer, T. Singer, J. A. Mitchell and A. J. T. George, *Nat. Rev. Drug. Discovery*, 2009, **9**, 325.
[81] R. S. Gaston, M. H. Deierhoi, T. Patterson, Ed. Prasthofer, B. A. Julian, W. H. Barber, D.A. Laskow, A. G. Diethelm and J. J. Curtis, *Kidney Int.*, 1991, **39**, 141.

[82] http://newsroom.lilly.com/ReleaseDetail.cfm?sh_print=yes&releaseid=521014.
[83] http://www.tolerx.com/index.php?page=prdetail&id=204.
[84] http://clinicaltrials.gov/ct2/show/NCT01222078.
[85] http://clinicaltrials.gov/ct2/show/NCT00678886.
[86] http://clinicaltrials.gov/ct2/show/NCT01123083.
[87] http://clinicaltrials.gov/ct2/show/NCT01077531.
[88] http://clinicaltrials.gov/ct2/show/NCT01101555.
[89] http://clinicaltrials.gov/ct2/show/NCT01114503.
[90] http://clinicaltrials.gov/ct2/show/NCT01030861.
[91] K. Wing, Y. Onishi, P. Prieto-Martin, T. Yamaguchi, M. Miyara, Z. Fehervari, T. Nomura and S. Sakaguchi, *Science*, 2008, **322**, 271.
[92] U. Fiocco, P. Sfriso, F. Oliviero, E. Pagnin, E. Scagliori, C. Campana, S. Dainese, L. Cozzi and L. Punzi, *Autoimmun. Rev.*, 2008, **8**, 76.
[93] BMS reportshttp://ctr.bms.com/pdf//IM101-084%20ST.pdf.
[94] http://ctr.bms.com/pdf//IM101-108%20ST.pdf.
[95] http://ctr.bms.com/OneBmsCtd/ResultDetailAction.do?prodid=23&trialid=4703.
[96] J. T. Merrill, R. Burgos-Vargas, R. Westhovens, A. Chalmers, D. D'Cruz, D. J. Wallace, S. C. Bae, L. Sigal, J. C. Becker, S. Kelly, K. Raghupathi, T. Li, Y. Peng, M. Kinaszczuk and P. Nash, *Arthritis Rheum.*, 2010, **62**, 3077.
[97] P. Mease, M. C. Genovese, G. Gladstein, A. J. Kivitz, C. Ritchlin, P. P. Tak, J. Wollenhaupt, O. Bahary, J. C. Becker, S. Kelly, L. Sigal, J. Teng and D. Gladman, *Arthritis Rheum.*, 2010, Published online 2 Dec 2010.
[98] http://clinicaltrials.gov/ct2/show/NCT00774852.
[99] J. van Loosdregt, Y. Vercoulen, T. Guichelaar, Y. Y. J. Gent, J. M. Beekman, O. van Beekum, A. B. Brenkman, D. J. Hijnen, T. Mutis, E. Kalkhoven, B. J. Prakken and P.J. Coffer, *Blood*, 2010, **115**, 965.
[100] S. Floess, J. Freyer, C. Siewert, U. Baron, S. Olek, J. Polansky, K. Schlawe, H. D. Chang, T. Bopp, E. Schmitt, S. Klein-Hessling, E. Serfling, A. Hamman and J. Huehn, *PLoS Biol.*, 2007, **5**, e38.
[101] J. K. Polansky, K. Kretschmer, J. Freyer, S. Floess, A. Garbe, U. Baron, S. Olek, A. Hamann, H. von Boehmer and J. Huehn, *Eur. J. Immunol.*, 2008, **38**, 1654.
[102] R. Tao, E. F. de Zoeten, E. Ozkaynak, C. Chen, L. Wang, P. M. Porrett, B. Li, L. A. Turka, E. N. Olson, M. I. Greene, A. D. Wells and W. W. Hancock, *Nat. Med.*, 2007, **13**, 1299.
[103] E. F. de Zoeten, L. Wang, H. Sai, W. H. Dillmann and W. W. Hancock, *Gastroenterology*, 2010, **138**, 583.
[104] Y. Ding, S. Shen, A. C. Lino, M. A. Curotto de Lafaille and J. J. Lafaille, *Nat. Med.*, 2008, **14**, 162.
[105] J. Graham, M. Fray, S. de Haseth, K. Mi Lee, M.-M. Lian, C. M. Chase, J. C. Madsen, J. Markmann, G. Benichou, R. B. Colvin, A. B. Cosimi, S. Deng, J. Kim and A. Alessandrini, *J. Biol. Chem.*, 2010, **285**, 32852.
[106] A. Zanin-Zhorov, Y. Ding, S. Kumari, M. Attur, K. L. Hippen, M. Brown, B. R. Blazar, S. B. Abramson, J. J. Lafaille and M. L. Dustin, *Science*, 2010, **328**, 372.
[107] P. J. Zhou, H. Wang, G. H. Shi, X. H. Wang, Z. J. Shen and D. Xu, *Clin. Exp. Immunol.*, 2009, **157**, 40.
[108] C. Daniel, N. Sartory, N. Zahn, G. Geisslinger, H. H. Radeke and J. M. Stein, *J. Immunol.*, 2007, **178**, 2458.
[109] N. A. Forward, S. J. Furlong, Y. Yang, T.-J. Lin and D. W. Hoskin, *J. Leukoc. Biol.*, 2010, **87**, 117.
[110] N. K. Crellin, R. V. Garcia, O. Hadisfar, S. E. Allan, T. S. Steiner and M. K. Levings, *J. Immunol.*, 2005, **175**, 8051.

CHAPTER 11

Advances in the Discovery of C5a Receptor Antagonists

Jay P. Powers, Daniel J. Dairaghi and **Juan C. Jaen**

Contents

1. INTRODUCTION

The complement system plays a central role in the generation of innate and adaptive immune responses to infectious agents, foreign antigens, virus-infected cells, and tumor cells. The complement system consists of more than 30 components which play an essential role in the responses to infection and injury. The complement cascade may be initiated *via* the classical (triggered by immune complex formation), lectin (antibody independent), or alternative pathways, all of which converge at C3 (Figure 1). Activation of the complement pathway generates biologically active fragments of complement proteins, for example, C3a, C4a, and C5a anaphylatoxins and C5b-9 membrane attack complexes (MAC), all of which

ChemoCentryx Inc., 850 Maude Avenue, Mountain View, CA 94043, USA

Annual Reports in Medicinal Chemistry, Volume 46
ISSN: 0065-7743, DOI: 10.1016/B978-0-12-386009-5.00016-3

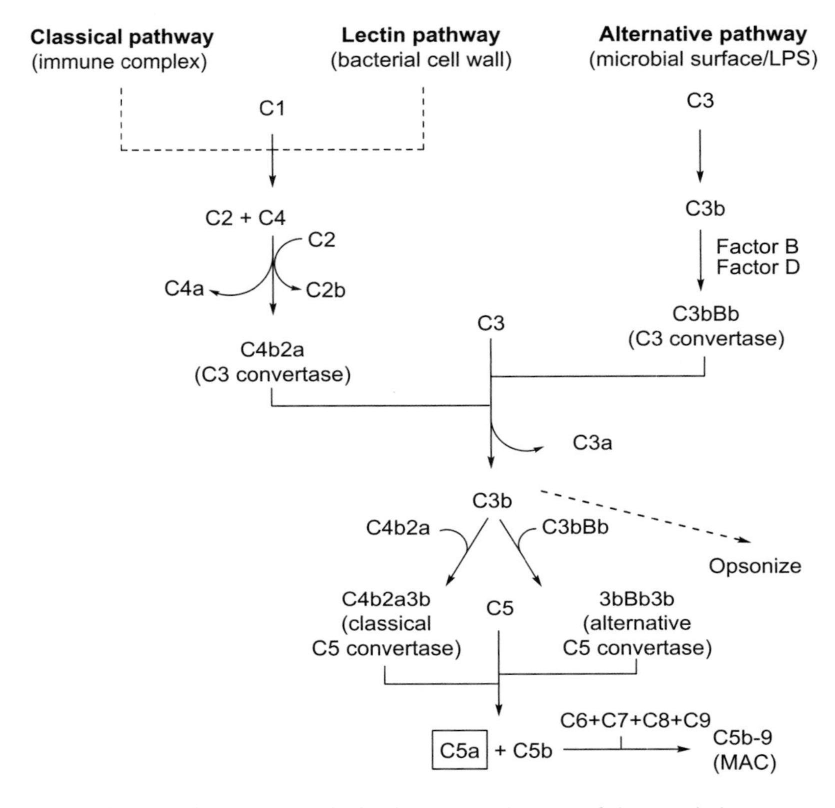

Figure 1 The complement cascade, leading to production of the anaphylatoxin C5a.

mediate inflammatory responses by inducing leukocyte chemotaxis, activating macrophages, neutrophils, platelets, mast cells, and endothelial cells and increasing vascular permeability, cytolysis, and tissue injury. Inappropriate or excessive activation of the complement system, in general, and formation of the C5a anaphylatoxin, in particular, can lead to harmful consequences due to severe inflammation and resulting tissue destruction. These consequences are clinically manifested in various human pathologies, ranging from the acute setting of septic shock and ischemia/reperfusion injury to chronic diseases such as rheumatoid arthritis (RA), systemic lupus erythematosus (SLE), transplant rejection, macular degeneration, vasculitis, and psoriasis, among others. The complement system has long been an attractive target for drug discovery [1], and interest in targeting the C5a receptor [2–4] has been widespread since initial reports on small molecule antagonists in the early 1990s. However, targeting the C5a/C5a receptor pair has been extremely challenging, with only a single entity (eculizumab, an anti-C5 antibody) advancing to approval over that same 20-year period. In this chapter, we review the recent advances in the discovery of peptide and small molecule antagonists of the C5a receptor, with a focus on results that have come to light since 2004, when the topic was last reviewed in Annual Reports [2].

1.1. The C5a receptor

C5a is one of the most potent proinflammatory mediators of the complement system, being at least 100 times more potent than C3a. This 8.3 kDa polypeptide, along with a C5b fragment, is produced by enzymatic cleavage of a C5 precursor during activation of any of the three complement pathways. C5a induces expression of adhesion molecules and chemotactic migration of neutrophils, eosinophils, basophils, and monocytes. It also mediates inflammatory reactions by causing smooth muscle contraction, increasing vascular permeability, inducing basophil and mast cell degranulation, and releasing lysosomal proteases and oxidative free radicals. The anaphylactic and chemotactic effects of C5a are mediated through its interaction with the C5a receptor (C5aR, CD88), a 350-residue GPCR expressed on neutrophils, monocytes, basophils, eosinophils, renal glomerular tissues, lung smooth muscle, and endothelial cells.

Approaches to pharmacological intervention targeting C5aR have focused on two general areas. In the first, targeting of C5aR itself by small molecule, peptide, and anti-C5aR antibodies offers a direct and selective inhibition of C5aR function. In the second approach, binding of antibodies or aptamers to C5 prevents its cleavage to C5a and C5b, indirectly preventing activation of C5aR but also impacting formation of the MAC (Figure 1), which is critical in the clearing of many bacterial infections through MAC-mediated cell lysis. Although inhibition of MAC formation can be a successful strategy for some therapeutic indications (Sections 5 and 6), it is important to note that the two approaches can result in very different biological outcomes, as inhibition of MAC formation can also create some unique risk of infection [1].

2. PEPTIDE AND LARGE MOLECULE AGENTS

Truncated C5a C-terminus hexapeptide derivatives served as the starting point for the discovery of several peptidomimetic antagonists of C5aR. The conformationally constrained cyclic antagonist **1** (PMX-53, 3D53, AcF-[OP-dCha-WR]) was designed after ^{1}H NMR experiments suggested a turned cyclic conformation of prototypical linear hexapeptides [5,6]. A moderate scale (50–100 g) solution phase synthesis of **1** has been reported [7]. Numerous preclinical *in vitro* and *in vivo* studies have been described utilizing **1**, including rat pharmacokinetics [8] and efficacy in animal models of monoarticular arthritis, LPS-induced neutropenia, ulcerative colitis, dermal and peritoneal inflammation, and assorted ischemia/reperfusion injuries [2–4,9].

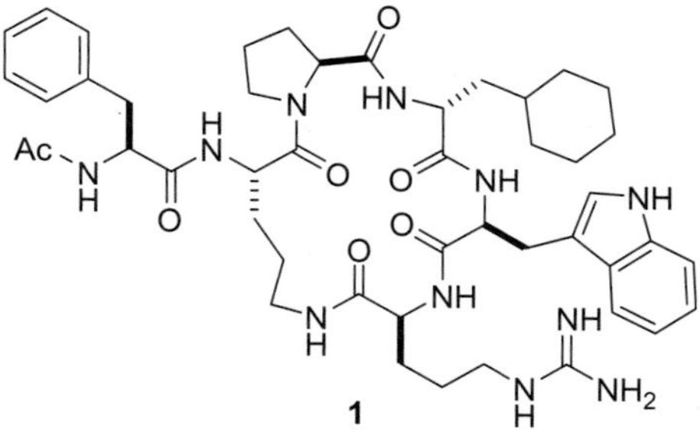

Peptide **1** was shown to be safe and well tolerated in several Phase 1 clinical trials, both in healthy volunteers and in patients with psoriasis and arthritis [10]. A small Phase 1b/2a clinical trial performed with **1** in patients with active RA [11] failed to show a decrease in cell infiltration, synovial inflammation, or changes in key biomarkers associated with clinical efficacy. However, it must be noted that low drug exposure observed in the study, combined with the potential effect of plasma protein binding, may have provided insufficient receptor coverage for efficacy. Compound **1** does not appear to be undergoing any additional clinical activity at the present time.

A peptidomimetic antagonist, JPE1375 (**2**), was derived from the systematic deconstruction of **1** [12]. More specifically, JPE1375 resulted from replacement of the arginine in **1** with the lipophilic phenylalanine, cleavage of the macrocyclic ring, and replacement of the terminal NAc moiety with hydroorotic acid. Compound **2** was reported to have similar functional potency to **1** as measured by inhibition of C5a-induced glucosaminidase release in C5aR-transfected RBL cells ($IC_{50} = 39$ and 29 nM for **2** and **1**, respectively) and in binding to HEK293 cells ($IC_{50} = 111$ and 104 nM for **2** and **1**, respectively). Potency on the mouse receptor was significantly greater for **2** than for **1** as measured by chemotaxis in mouse J774.1 cells ($IC_{50} = 0.42$ vs. 7.1 μM). Stability in human liver microsomes was significantly improved for **2** (80% remaining at 1 h) *versus* **1** (10% remaining after 1 h). The *in vivo* efficacy of **2** was tested in the reverse passive Arthus reaction, a mouse model of immune complex-mediated disease. JPE1375 (dosed at 1 mg/kg, i.v.) significantly reduced the influx of neutrophils into the peritoneum after simultaneous i.v. challenge with OVA peptide and i.p. challenge with an anti-OVA antibody. In a model of tubulointerstitial fibrosis, **2** (dosed at 0.63 mg/kg/day, i.p.) resulted in significant impact on markers of renal fibrosis, including fibronectin protein expression, Sirius Red staining, and PDGF-B mRNA expression after 5 days [13].

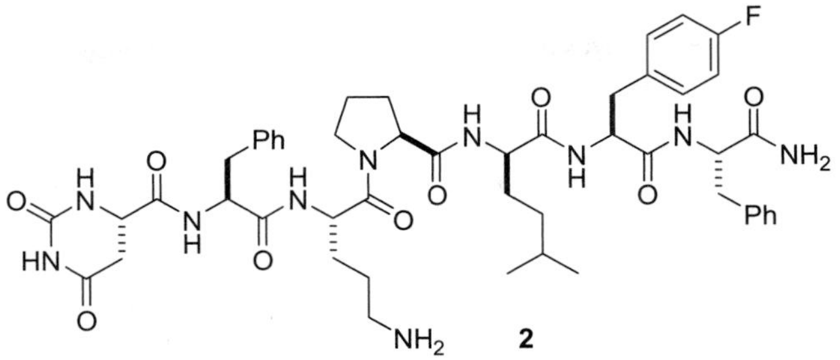

3. MACROMOLECULES

Non-antibody macromolecules have been actively pursued as antagonists of C5aR. The 121-residue immune evasive protein excreted by *Staphylococcus aures*, chemotaxis inhibitory protein of *S. aures* (CHIPS) binds to C5aR. Recombinant $CHIPS_{28-149}$ inhibits the binding of C5a to C5aR with an IC_{50} of 6.7 nM without inhibiting the binding of C5a to the closely related C5L2 receptor [14]. While $CHIPS_{28-149}$ does block binding of C5a-des-Arg to C5L2 with an IC_{50} of 274 nM, it lacks affinity for the closely related ChemR23, FPRL1, or FPRL2 receptors. Although CHIPS is a potent antagonist of C5aR, it is also highly immunogenic, and antibodies to CHIPS have been identified throughout the general population [14], making the therapeutic use of CHIPS itself untenable. Several groups have sought to modify CHIPS by removing or replacing the multiple immunogenic epitopes. In this regard, a 50-residue adapted peptide (CHOPS), designed to maintain binding while reducing the interaction with human IgG, has been reported with micromolar affinity for a model peptide comprising residues 7–28 of the C5aR N-terminus [15].

ADC-1004 is a truncated and mutated form of CHIPS designed by directed evolution [16] to lower the interaction with human IgG and is postulated to inhibit the binding of C5a by interacting with the N-terminal site on C5aR. ADC-1004 binds to, but does not activate, C5aR [17,18]. ADC-1004 has been tested in a porcine ischemia–reperfusion model, reducing myocardial infarction (infarct size) by 21% ($p = 0.007$) when dosed 175 mg *via* i.v. bolus [19]. No clinical activity has been disclosed for ADC-1004 [20].

4. SMALL MOLECULE AGENTS

Many small molecule C5aR antagonists have been disclosed over the past two decades [2–4]. Clinical efficacy with the first generation small molecules was very limited, but some of the second and third generation antagonists have shown considerable promise.

W-54011, **3**, was identified using a high-throughput screen followed by lead optimization [21]. W-54011 is a potent inhibitor of C5a binding to human neutrophils in cell culture buffer with a K_i of 2 nM and is also a functional inhibitor of Ca^{2+} mobilization ($IC_{50} = 3$ nM) and C5a-mediated chemotaxis ($IC_{50} = 3$ nM) in human neutrophils [21]. In the same report, the IC_{50} values for **1**, in Ca^{2+} mobilization and chemotaxis, were 55 and 18 nM, respectively. W-54011 demonstrated potent, and species selective, inhibition of C5a-mediated Ca^{2+} flux in neutrophils from humans, cynomolgus monkeys, and gerbils, but not from mice, rats, guinea pigs, rabbits, or dogs. W-54011 did not block Ca^{2+} mobilization stimulated by fMLP, PAF, or IL-8 and demonstrated good selectivity for C5aR. In a C5a-induced neutropenia gerbil model, W-54011 (dosed p.o.) inhibited C5a-induced neutropenia in a dose-dependent manner, with complete abrogation of the neutropenia at the top 30 mg/kg dose. W-54011 was also investigated for its ability to ameliorate established collagen-induced arthritis in cynomolgus monkeys [22]. Treatment with W-54011 for 15 days (30 mg/kg, p.o.) resulted in significant reduction in joint swelling within 2 days, with continued suppression for the length of the experiment. In the same study, W-54011 also significantly ameliorated radiographic scores of joint destruction. Two recent patent applications have published that disclose a closely related small molecule (**4**) and its associated salts, as well as the preparation of an optically pure intermediate used in its synthesis [23,24]. Although no data for compound **4** has been published in the scientific literature, these applications may indicate a special interest in this specific compound.

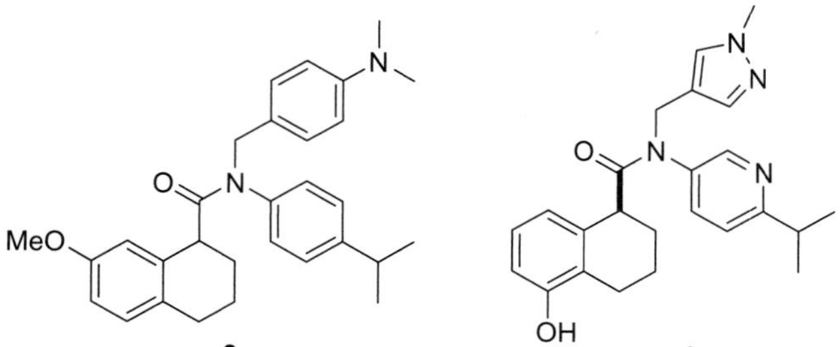

MeO O N N OH O N N N N

3 **4**

NDT9520492 (**5**) has been investigated for activity on the C5a receptor from multiple species [25]. Overall C5aR sequence homology between human and nonhuman primate C5aR is >95%, but it is generally only 65–75% between human and nonprimates. Based on alignment of C5aR sequences, it was shown that a tryptophan residue in the transmembrane domain V is the only transmembrane domain amino acid unique to species (*i.e.*, gerbil, human, and nonhuman primate) that recognize small molecule C5aR antagonists. In binding experiments, NDT9520492 inhibited [^{125}I]-C5a binding with IC_{50} values of 29 nM (human C5aR), 109 nM (gerbil C5aR), and >10,000 nM (mouse C5aR), while in the same experiment, W-54011 and AcF-[OP-dCha-WR] inhibited binding with the same pattern (W-54011: 4, 13, and >10,000 nM, respectively, AcF-[OP-dCha-WR]: 25, 456, and >10,000 nM, respectively). Interestingly, site-directed single-point (L214W) mutagenesis of the mouse receptor to install the tryptophan residue in transmembrane domain V resulted in dramatic increases in potency inhibition of mL214W-C5aR [^{35}S]GTPγS binding for NDT9520492 and W-54011, but not for cyclic peptide AcF-[OP-dCha-WR] [25].

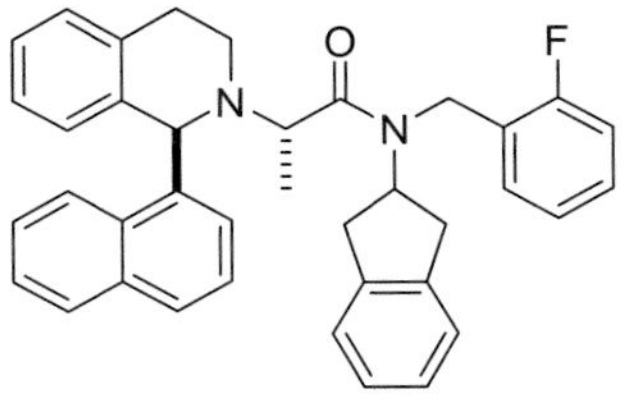

5

NDT9513727 (**6**) has been described as a potent and competitive antagonist of C5aR with properties consistent with inverse agonism [26]. NDT9513727 has been well characterized and displayed an IC_{50} of 11.6 nM in a [^{125}I]-C5a binding assay. NDT9513727 inhibited C5a-stimulated responses, including Ca^{2+} mobilization, oxidative burst, degranulation, chemotaxis, and cell surface CD11b expression in assorted cell types with IC_{50} values from 1 to 9 nM. The compound was selective against C5L2 and in a Cerep screen, and was potent in human, cynomolgus monkey, and gerbil C5aR, but was inactive in the rat, mouse, and dog receptors. NDT9513727 was found to have reasonable PK in cynomolgus monkeys with moderate bioavailability ($F = 26\%$). NDT9513727 was examined in a C5a-induced neutropenia model in cynomolgus monkeys and exhibited 66% inhibition of C5a-stimulated neutropenia at 25 mg/kg dosed orally (plasma concentration = 410 ± 218 nM). In an *ex vivo* assay of C5a-mediated upregulation of CD11b on human blood granulocytes in fresh human whole blood, **6** displayed concentration-dependent inhibition with an IC_{50} of 0.6 μM. The concentration required to reach approximately 50% inhibition was similar in the cyno neutropenia and *ex vivo* human whole blood CD11b expression assays. Modest potency *in vivo* in the cynomolgus neutropenia and in the whole blood CD11b upregulation assay *ex vivo* compared to the potency in the serum-free *in vitro* assays was attributed to the high protein binding (>99%) in human plasma. The use of the physiologically relevant (whole blood, primary cells, presence of plasma proteins) CD11b assay and its correlation with the *in vivo* assay results highlights the importance of the inclusion of these types of effects in assays that hope to predict C5aR receptor blockade *in vivo*.

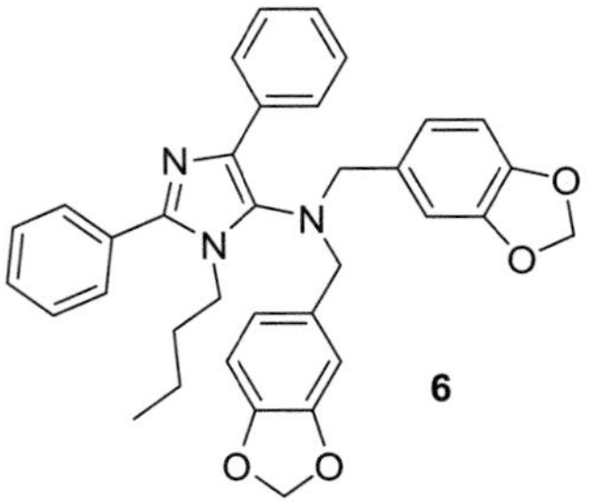

6

NGD 2000-1 (structure undisclosed) has been described as a substituted tetrahydroisoquinoline C5a antagonist [3]. NGD 2000-1 entered Phase 2 clinical trials in both asthma and RA patients. In patients with mild to moderate asthma, NGD 2000-1 did not demonstrate a therapeutic benefit (primary endpoint Forced Expiratory Volume in 1 second, FEV1) [4]. In patients with mild to moderate RA, NGD 2000-1 did not demonstrate an effect in the trial's primary endpoint (changes in C-reactive protein, CRP); however, it did demonstrate a statistically significant change in the Subject Global Assessment of Disease activity at a dose of 100 mg twice daily. Subsequent *post hoc* analysis of the ACR20 response revealed a statistically significant response at the highest dose tested [27]. However, during Phase 1 clinical trials, NGD 2000-1 was found to inhibit cytochrome P450 3A4 [3,28,29], limiting dose levels, which would not allow a sufficient therapeutic window at the doses that were believed to be required for future development in patients with RA [27], and development of NGD 2000-1 has been abandoned [3,28].

SAR studies based on a class of noncompetitive allosteric inhibitors of chemotactic receptors [30] led to **7**, a dual inhibitor of both C5a- and IL-8-mediated human granulocyte chemotaxis [3,31,32]. Descriptions of the potency of **7** against C5a-mediated activity are unclear in the original patent application [31] and in the review literature range from tens of nanomolar [32] to tens of micromolar [3]. Recently, related structures have been described in the patent literature which are selective for inhibition of C5a-mediated human granulocyte chemotaxis and which lack IL-8 activity [33,34], as represented by compound **8**, which showed 60% inhibition of C5a-induced chemotaxis when tested at 10 nM [34].

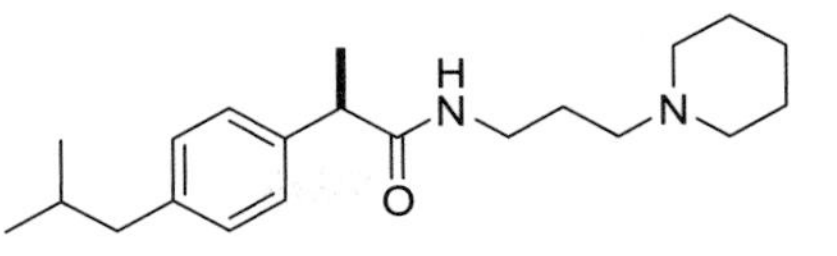

7 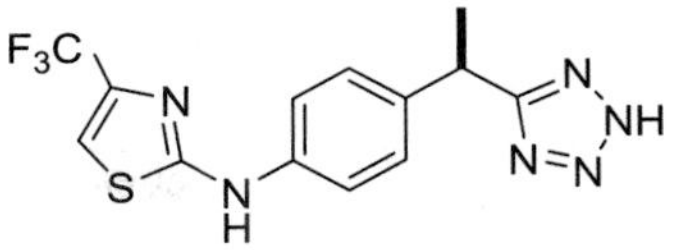 **8**

JSM-7717 is a small molecule C5aR antagonist in preclinical development [35] and belongs to a genus of structures represented by compounds **9–11** [36]. Compound **9** inhibited C5a-induced enzyme release with an IC_{50} of 3 nM, while **11** inhibited [^{125}I]-C5a binding to C5aR in hC5aR-HEK293 cells with an IC_{50} of 43 nM, and inhibited *E. coli*-induced oxidative burst in fresh human whole blood with an IC_{50} of 620 nM. Both **9** (1 mg/kg, i.v.) and **10** (3 mg/kg, i.v.) were examined in a C5a-induced neutropenia model in male gerbils and both demonstrated inhibition in this acute model. Although the patent application does not explicitly disclose the structure of JSM-7717, it may be deduced to be either **9** or **10** based on the identifiers in the neutropenia results (Figure 2 in Ref. 36) as compared with the written description.

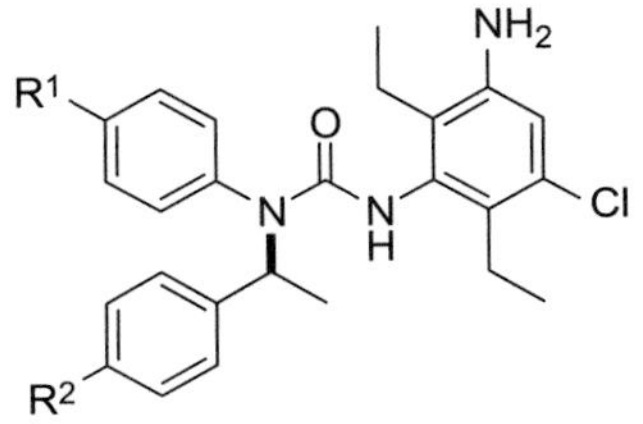

9: R^1 = NHMe, R^2 = Cl
10: R^1 = Et, R^2 = Cl
11: R^1 = NHMe, R^2 = H

A series of 5,6,7,8-tetrahydroquinoline C5a antagonists are exemplified by **12** [37–39]. Compound **12** inhibited C5aR in a Ca^{2+} mobilization assay with an IC_{50} of 7.3 nM, and had a K_i in a [^{125}I]-C5a competition binding assay in the single-digit nanomolar range [37]. Similar compounds showed moderate to medium clearance when dosed i.v. in rat, and moderate to good bioavailability [38,39].

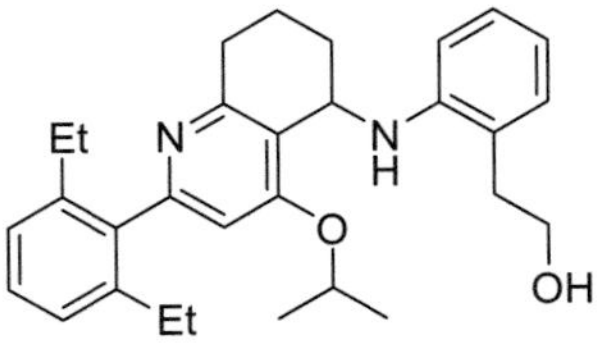

12

A high-throughput screening effort led to the discovery of CP-447,697 [40], which was further elaborated to **13** [41]. **13** inhibited [^{125}I]-C5a binding in U937 cells with an IC_{50} of 27 nM and inhibited C5a-mediated elastase release in human neutrophils with an IC_{50} of 25 nM. In a CD11b upregulation assay in fresh human whole blood, **13** exhibited a potency of 4,800 nM (the concentration required to shift the dose–response curve for the C5a-induced upregulation of CD11b in human whole blood by 10-fold). The investigators postulated that the very high shift in potency in the presence of human whole blood was directly related to the very high human plasma protein binding (99.6%). The high protein binding was attributed to the high lipophilicity of **13**, which led to the abandonment of further lead optimization efforts [41].

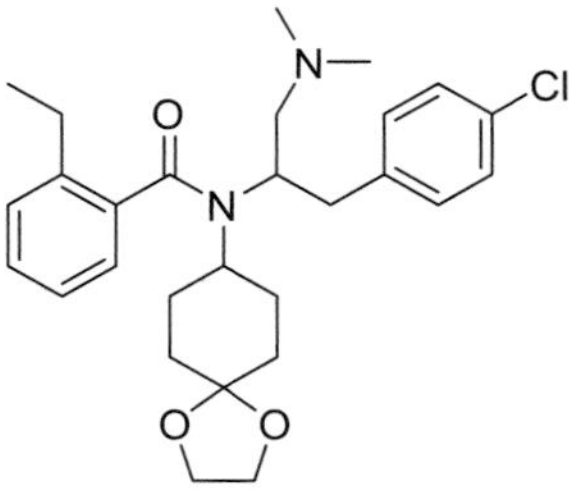

13

High-throughput screening led to the discovery of **14** [42]. Compound **14** inhibited C5a-induced Ca^{2+} mobilization in hC5aR transfected 293 cells with a pIC_{50} of 7.6 and inhibited C5a-stimulated Ca^{2+} mobilization in human neutrophils with a pA_2 of 7.4. The compound was inactive in dog, rat, and mouse receptors. Due to an apparently insurmountable disconnect between potency and metabolic stability, the series was not progressed further into additional lead optimization [42].

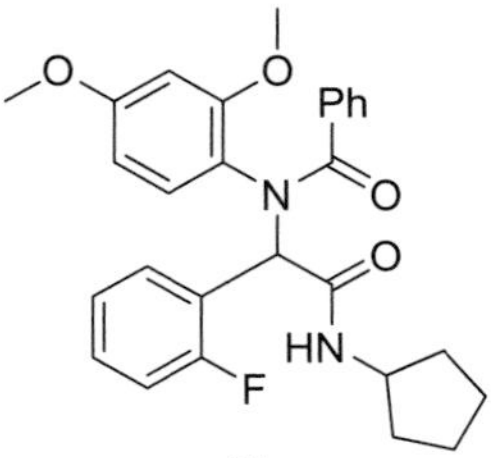

14

CCX168 (structure not disclosed) is a small molecule that potently inhibits [^{125}I]-C5a binding to C5aR with an IC_{50} of 0.62 nM, C5a-mediated chemotaxis with an IC_{50} of 0.25 nM (both in human U937 cells), and C5a-mediated Ca^{2+} mobilization with an IC_{50} of 0.4 nM in human monocytes in cell culture buffer [43]. CCX168 is highly selective with no activity on the closely related C5L2, C3aR, ChemR23, GPR1, and FPR1 receptors. CCX168 remained highly potent under physiologically relevant conditions and inhibited C5aR-mediated chemotaxis in fresh human whole blood with an A_2 of 1.7 nM, and inhibited C5aR-mediated CD11b upregulation on human neutrophils in fresh whole blood with an A_2 of 4 nM. The compound was potent on cynomolgus monkey receptor but inactive on mouse and rat receptors. CCX168 was tested *in vivo* in human C5aR knock-in mice (similar *in vitro* potency) and demonstrated inhibition of C5a-induced leukopenia with an ED_{50} of ~0.03 mg/kg. Oral dosing of CCX168 in a mouse model of vasculitis in humanized mice markedly suppressed the induction of glomerulonephritis by antimyeloperoxidase (anti-MPO) IgG. Daily dosing of CCX168 (30 mg/kg) reduced glomerular crescent formation from 29.3% (vehicle alone) to 3.3% with CCX168 ($p < 0.0001$), and glomerular necrosis was reduced from 8.2% to 1.1% ($p < 0.0001$) [44]. Urine protein, leukocytes and RBCs, and serum BUN and creatinine were reduced as well. A low dose of 0.1 mg/kg/day caused a 30% reduction in crescents. The lowest dose that produced near-maximal therapeutic benefit was 4 mg/kg bid, where plasma levels ranged from 35 to 200 ng/mL throughout the day and C5aR blockade ranged from 95% to 99% based on the potency in whole blood CD11b upregulation [44]. CCX168 is currently in human clinical trials (*vide infra*).

5. CLINICAL UPDATE

ARC1905 is an anti-C5 RNA aptamer that inhibits the cleavage of C5 into C5a and C5b [1], for which detailed characterization data has not been published. ARC1905 is currently in two Phase 1 clinical trials in patients with age-related macular degeneration (AMD). The first trial is to examine safety and tolerability of ARC1905 intravitreous injection in subjects with geographic atrophy secondary to dry AMD [45]. The second trial is intended to evaluate the safety, tolerability, and pharmacokinetics of multiple doses of ARC1905 intravitreous injection when administered in conjunction with multiple doses of Lucentis® 0.5 mg/eye, or with a single induction dose of Lucentis® 0.5 mg/eye in patients with subfoveal neovascularization (CNV) secondary to AMD [46].

CCX168 (structure not disclosed) is a sub-nanomolar C5aR antagonist across many assay formats in cell culture buffer (*vide supra*) that inhibits C5aR-mediated chemotaxis in human whole blood with an A_2 of 1.7 nM and inhibits C5aR-mediated CD11b upregulation on human neutrophils in whole blood with an A_2 of 4 nM [43]. CCX168 has been evaluated in Phase 1 single and multiple dose clinical trials. In an *ex vivo* pharmacodynamic analysis of C5aR receptor coverage performed as part of the Phase 1 trial, CCX168 was found to reduce C5a-induced CD11b upregulation on blood neutrophils 12 h following a single dose of 100 mg. Plasma levels of

CCX168 of 197 ng/mL (~400 nM) were reached after the 100 mg dose, which far exceeds the levels required in an anti-MPO mouse model for near-maximal prevention of glomerulonephritis [44]. At this dose, terminal plasma half-life was ~29 h. On day 7 of the 30 mg CCX168 bid regimen, greater than 90% receptor coverage was maintained throughout the day. The combined pharmacokinetic and pharmacodynamic data for the Phase 1 trials and preclinical studies indicated that 30 mg CCX168 dosed bid in humans should result in greater than 90% C5aR coverage in blood at all times, believed to be optimal for testing C5aR antagonism in Phase 2 trials. Interest has been expressed in further studies to examine CCX168 as a potential therapeutic for antineutrophil cytoplasmic antibody (ANCA) related vasculitis [43,44].

NN8209 is an anti-C5aR antibody currently in a Phase 2 clinical trial in patients with RA to assess the safety, tolerability, and pharmacokinetics of NN8209 in combination with stable doses of methotrexate [47]. In this trial, NN8209 is dosed once weekly subcutaneously over a 3-week period.

MP-435 is a small molecule C5aR antagonist which is currently in a randomized, double blind, placebo controlled Phase 2 efficacy trial in combination with methotrexate in patients with RA [48]. Although the structure of MP-435 has not been disclosed, it is possibly related to **4** (*vide supra*).

Eculizumab (Soliris®) is a recombinant humanized monoclonal $IgG_{2/4}\kappa$ 148 kDa anti-C5 antibody produced by murine myeloma cell culture and approved for use in the United States and European Union in 2007 for the treatment of paroxysmal nocturnal hemoglobinuria (PNH, Section 4). Human pharmacokinetics have been described [49]. Eculizumab has undergone trials in many indications, including psoriasis, RA, SLE, and cardiovascular indications [1,3]. Numerous clinical trials are ongoing or recruiting with eculizumab, including Phase 2 studies in ANCA-associated vasculitis [50], AMD [51], dense deposit disease and C3 nephropathy [52], kidney transplant [53,54], myasthenia gravis [55], and atypical hemolytic uremic syndrome (aHUS). Recent Phase 2 trial results in an interim analysis of 17 patients with aHUS who were resistant to plasma therapy and were treated with eculizumab resulted in a significant ($p < 0.0001$) increase in platelet count observed with treatment compared to baseline [56]. Another recent study summarized results from an interim analysis of 15 patients with aHUS on chronic plasma therapy treated with eculizumab for at least 12 weeks showed a significant 87% (13/15; 95% CI 60–98) number of patients achieved TMA event free status (Thrombotic MicroAngiopathy; defined as stable platelet counts, absence of plasma therapy, and no new dialysis) [57]. A shorter acting variant of eculizumab, pexelizumab, no longer appears to be under active development.

6. MARKETED AGENTS

Eculizumab (Soliris®) is a recombinant humanized monoclonal anti-C5 antibody, approved in 2007 for the treatment of PNH. The pathophysiology of PNH and treatment of the condition with eculizumab in clinical trials have been reviewed [58]. Clinical features of PNH are caused by the MAC attack on erythrocytes, and prevention of MAC formation is believed to protect PNH red blood cells in circulation, and in this context, treatment with eculizumab prevents C5b formation which is necessary to form the MAC. In the treatment of PNH, eculizumab is dosed 600 mg *via* i. v. infusion every 7 days for the first 4 weeks, followed by 900 mg 7 days later, and then 900 mg every 14 days thereafter, for a total of ~25 g of antibody/patient/year. Eculizumab reduces production of C5b (and thus MAC formation), which is expected to lead to higher susceptibility to bacterial infection. As it is also known that people with genetic deficiency for terminal complement proteins (*i.e.*, the MAC) have an increased risk for infection, particularly by *Neisseria meningitides*, patients on eculizumab are vaccinated with a meningococcal vaccine as a prophylactic measure before starting treatment. Sales of eculizumab were US $541M in 2010 [59].

7. CONCLUSION

Over the past decade, significant advances have been made toward the goal of targeting the C5a/C5aR axis, with one agent (eculizumab) approved and five agents currently reported to be undergoing clinical trials. Activity remains split between targeting C5 (anti-C5 antibodies and aptamers) and C5aR itself (small molecule C5aR antagonists and anti-C5aR antibodies). Discovery of small molecule C5aR antagonists with properties appropriate for advancement into the clinic has continued to advance, and the first anti-C5aR antibody has begun clinical trials as well. With two C5aR antagonists in Phase 2 clinical trials for RA, it will be very interesting to see results which follow up on the early, if limited, success of the last C5aR antagonist explored in this indication (NGD 2000-1 in 2003). Ultimately, the future clinical success of C5aR antagonists will depend on their potency under physiologically relevant conditions and robust receptor coverage, safety and tolerability, and the careful choice of appropriate therapeutic indications based on a deep understanding of the biology of C5aR.

REFERENCES

[1] D. Ricklin and J. D. Lambris, *Nat. Biotechnol.*, 2007, **25**, 1265.
[2] A. J. Hutchison and J. E. Krause, *Annu. Rep. Med. Chem.*, 2004, **39**, 139.
[3] L. M. Proctor, T. M. Woodruff and S. M. Taylor, *Expert Opin. Ther. Pat.*, 2006, **16**, 445.

[4] H. Qu, D. Ricklin and J. D. Lambris, *Mol. Immunol.*, 2009, **47**, 185.
[5] A. K. Wong, A. M. Finch, G. K. Pierens, D. J. Craik, S. M. Taylor and D. P. Fairlie, *J. Med. Chem.*, 1998, **41**, 3417.
[6] A. M. Finch, A. K. Wong, N. J. Paczkowski, K. Wadi, D. J. Craik, D. P. Fairlie and S. M. Taylor, *J. Med. Chem.*, 1965, **1999**, 42.
[7] R. C. Reid, G. Abbenante, S. M. Taylor and D. P. Fairlie, *J. Org. Chem.*, 2003, **68**, 59.
[8] M. Morgan, A. C. Bulmer, T. M. Woodruff, L. M. Proctor, H. M. Williams, S. Z. Stocks, S. Pollitt, S. M. Taylor and I. A. Shiels, *Eur. J. Pharm. Sci.*, 2008, **33**, 390.
[9] M. C. H. Holland, D. Morikis and J. D. Lambris, *Curr. Opin. Investig. Drugs*, 2004, **5**, 1164.
[10] J. Kohl, *Curr. Opin. Mol. Ther.*, 2006, **8**, 529.
[11] C. E. Vergunst, D. M. Gerlag, H. Dinant, L. Schultz, M. Vinkenoog, T. J. M. Smeets, M.E. Sanders, K. A. Reedquist and P. P. Tak, *Rheumatology*, 2007, **46**, 1773.
[12] K. Schnatbaum, E. Locardi, D. Scharn, U. Richter, H. Hawlisch, J. Knolle and T. Polakowski, *Bioorg. Med. Chem. Lett.*, 2006, **16**, 5088.
[13] P. Boor, A. Konieczny, L. Villa, A.-L. Schult, E. Bucher, S. Rong, U. Kunter, C. R. C. van Roeyen, T. Polakowski, H. Hawlisch, S. Hillebrandt, F. Lammert, F. Eitner, J. Floege and T. Ostendorf, *J. Am. Soc. Nephrol.*, 2007, **18**, 1508.
[14] A. J. Wright, A. Higgenbottom, D. Philippe, A. Upadhyay, S. Bagby, R. C. Read, P. Monk and L. J. Partridge, *Mol. Immunol.*, 2007, **44**, 2507.
[15] A. Bunschoten, J. H. Ippel, J. A. W. Kruijtzer, L. Feitsma, C. J. C. de Haas, R. M. J. Liskamp and J. Kemmink, *Amino Acids*, 2011, **40**, 731.
[16] E. Gustafsson, A. Rosen, K. Barchan, K. P. P. van Kessel, K. Haraldsson, S. Lindman, C. Forsberg, L. Ljung, K. Bryder, B. Walse, P.-J. Haas, J. A. G. van Strijp and C. Furebring, *Protein Eng. Des. Sel.*, 2010, **23**, 91.
[17] E. Gustafsson, C. Forsberg, S. Lindman, L. Ljung and C. Furebring, *Protein Expr. Purif.*, 2009, **63**, 95.
[18] E. Gustafsson, P.-J. Haas, B. Walse, M. Hijnen, C. Furebring, M. Ohlin, J. A. G. van Strijp and K. P. M. van Kessel, *BMC Immunol.*, 2009, **10**, 13.
[19] J. van der Pals, S. Koul, P. Andersson, M. Gotberg, J. F. A. Ubachs, M. Kanski, H. Arheden, G. K. Olivecrona, B. Larsson and D. Erlinge, *BMC Cardiovasc. Disord.*, 2010, **10**, 45.
[20] Company website, Alligator Bioscience AB, Schleelevagen 19 A, SE-223 70 Lund, Sweden. http://www.alligatorbioscience.com/partnership.aspx.
[21] H. Sumichika, K. Sakata, N. Sato, S. Takeshita, S. Ishibuchi, M. Nakamura, T. Kamahori, S. Ehara, K. Itoh, T. Ohtsuka, T. Ohbora, T. Mishina, H. Komatsu and Y. Naka, *J. Biol. Chem.*, 2002, **277**, 49403.
[22] K. Sakata, H. Sumichika, K. Goto, N. Sato, S. Takeshita, S. Ishibuchi, M. Nakamura, T. Kamahori, N. Kobayashi, T. Takashima, T. Ohtsuka, T. Ohbora, T. Mishina, H. Komatsu and K. Chiba, Immunology 2004; 12th International Congress of Immunology and 4th Annual Conference of FOCIS, Montreal, QC, Canada, July 18–23, 2004.
[23] M. Nakamura, S. Ishibuchi, T. Ohtsuka, H. Sumichika, S. Sekiguchi, T. Ishige and N. Ueda, *Patent Application WO 2006/082975-A1*, 2006.
[24] M. Ishibuchi, K. Nakamura, R. Koya and M. Sano, *Patent Application JP 2005120027-A*, 2005.
[25] S. M. Waters, R. M. Brodbeck, J. Steflik, J. Yu, C. Baltazar, A. E. Peck, D. Severance, L. Y. Zhang, K. Currie, B. L. Chenard, A. J. Hutchinson, G. Maynard and J. E. Krause, *J. Biol. Chem.*, 2005, **280**, 40617.
[26] R. M. Brodbeck, D. N. Cortright, A. P. Kieltyka, J. Yu, C. O. Baltazar, M. E. Buck, R. Meade, G. D. Maynard, A. Thurkauf, D.-S. Chien, A. J. Hutchison and J. E. Krause, *J. Pharmacol. Exp. Ther.*, 2008, **327**, 898.

[27] Neurogen Corporation press release, 2004. http://www.thefreelibrary.com/Neurogen+Reports+Phase+IIa+Clinical+Trial+Results+for+Oral+RA+Drug.-a0132375565.
[28] H. Lee, P. L. Whitfield and C. R. Mackay, *Immunol. Cell Biol.*, 2008, **86**, 153.
[29] Neurogen Corporation 2003 annual report: http://edgar-online.com.
[30] R. Bertini, M. Allegretti, C. Bizzarri, A. Moriconi, M. Locati, G. Zampella, M. N. Cervellera, V. Di Cioccio, M. C. Cesta, E. Galliera, F. O. Martinez, R. Di Bitondo, G. Troiana, V. Sabbatini, G. D'Anniballe, R. Anacardio, J. C. Cutrin, B. Cavalieri, F. Mainiero, R. Strippoli, P. Villa, M. D. Girolamo, F. Martin, M. Gentile, A. Santoni, D. Corda, G. Poli, A. Mantovani, P. Ghezzi and F. Colotta, *Proc. Natl. Acad. Sci. USA*, 2004, **101**, 11791.
[31] M. Allegretti, R. Bertini, V. Berdini, C. Bizzarri, M. C. Cesta, V. Di Cioccio, G. Caselli, F. Colotta and C. Gandolfi, *Patent Application WO 2003/068377-A1*, 2003.
[32] M. Allegretti, A. Moriconi, A. R. Beccari, R. Di Bitondo, C. Biaazrri, R. Bertini and F. Colotta, *Curr. Med. Chem.*, 2005, **12**, 217.
[33] M. Allegretti, A. Moriconi, A. Aramini, M. C. Cesta, A. Beccari and R. Bertini, *Patent Application WO 2007/060215-A2*, 2007.
[34] A. Moriconi and A. Aramini, *Patent Application WO 2009/050258-A1*, 2009.
[35] A. G. Jerini, Invalidenstrasse 130, 10115 Berlin, Germany. Pipeline chart as of 3/2011; http://www.jerini.com.
[36] K. Schnatbaum, D. Scharn, E. Locardi, T. Polkowski, U. Richter, U. Reinkke and G. Hummel, *Patent Application WO 2006/128670-A1*, 2006.
[37] J. K. Barbay, W. He, Y. Gong, J. Li, J. Van Wauwe and M. Buntinx, *Patent Application WO 2009/023669-A1*, 2009.
[38] J. K. Barbay, Y. Gong, M. Buntinx, J. Li, C. Claes, P. J. Hornby, G. Van Lommen, J. Van Wauwe and W. He, *Bioorg. Med. Chem. Lett.*, 2008, **18**, 2544.
[39] Y. Gong, J. K. Barbay, M. Buntinx, J. Li, J. Van Wauwe, C. Claes, G. Van Lommen, P. J. Hornby and W. He, *Bioorg. Med. Chem. Lett.*, 2008, **18**, 3852.
[40] J. Blagg, C. Mowbray, D. C. Pryde, G. Salmon, E. Schmid, D. Fairman and K. Beaumont, *Bioorg. Med. Chem. Lett.*, 2008, **18**, 5601.
[41] J. Blagg, C. Mowbray, D. Pryde, G. Salmon, D. Fairman, E. Schmid and K. Beaumont, *Bioorg. Med. Chem. Lett.*, 2008, **18**, 5605.
[42] H. T. Sanganee, A. Baxter, S. Barber, A. J. H. Brown, D. Grice, F. Hunt, S. King, D. Laughton, G. Pairadeau, B. Thong, R. Weaver and J. Unitt, *Bioorg. Med. Chem. Lett.*, 2009, **19**, 1143.
[43] D. J. Dairaghi, J. C. Jaen, K. Deshayes, D. A. Johnson, M. R. Leleti, S. Miao, J. P. Powers, L. C. Seitz, Y. Wang, T. J. Schall and P. J. Bekker, *Arthritis Rheum.*, 2010, **62**(10S), S850.
[44] H. Xiao, J. C. Jennette, D. J. Dairaghi, L. Ertl, T. Baumgart, S. Miao, J. P. Powers, L. C. Seitz, Y. Wang, P. Hu, R. J. Falk, T. J. Schall and J. C. Jaen, *J. Am. Soc. Nephrol.*, 2010, **21**, 40A.
[45] http://clinicaltrials.gov/ct2/show/NCT00950638.
[46] http://clinicaltrials.gov/ct2/show/NCT00709527.
[47] http://clinicaltrials.gov/ct2/show/NCT01223911.
[48] http://clinicaltrials.gov/ct2/show/NCT01143337.
[49] Eculizumab prescribing information insert: http://www.accessdata.fda.gov/drugsatfda_docs/label/2007/125166lbl.pdf.
[50] http://clinicaltrials.gov/ct2/show/NCT01275287.
[51] http://clinicaltrials.gov/ct2/show/NCT00935883.
[52] http://clinicaltrials.gov/ct2/show/NCT01221181.
[53] http://clinicaltrials.gov/ct2/show/NCT01106027.
[54] http://clinicaltrials.gov/ct2/show/NCT01095887.
[55] http://clinicaltrials.gov/ct2/show/NCT00727194.

[56] C. M. Legendre, S. Babu, R. F. Furman, N. S. Sheerin, D. J. Cohen, A. O. Gaber, F. Eitner, Y. Delmas, C. Loirat and L. A. Greenbaum, *J. Am. Soc. Nephrol.*, 2010, **21**, SA-FC406.
[57] P. Muus, C. M. Legendre, K. Douglas, M. Hourmant, Y. Delmas, B. M. Herthelius, A. Trivelli, C. Loirat, T. H. Goodship and C. Licht, *J. Am. Soc. Nephrol.*, 2010, **21**, F-P01274.
[58] R. Kelly, S. Richards, P. Hillman and A. Hill, *Ther. Clin. Risk Manag.*, 2009, **5**, 911.
[59] Alexion Pharmaceuticals, CT, USA, http://ir.alexionpharm.com/releases.cfm.

PART IV:
Oncology

Editor: Shelli R. McAlpine
School of Chemistry
University of New South Wales
Sydney
Australia

CHAPTER **12**

Inhibition of Translation Initiation as a Novel Paradigm for Cancer Therapy

Bertal H. Aktas*, **Jose A. Halperin***,
Gerhard Wagner** and **Michael Chorev***

Contents

1. INTRODUCTION

The regulation of gene expression at the level of translation initiation is critical for proper control of cell growth, proliferation, differentiation, and apoptosis. Deregulation of translation initiation is frequently observed in tumors and plays an important role in the genesis, progression, and

* Harvard Medical School, One Kendall Sq. Buld. 600, Cambridge, MA 02139, USA
** Harvard Medical School, 240 Longwood Avenue, Boston, MA 02115, USA

Annual Reports in Medicinal Chemistry, Volume 46
ISSN: 0065-7743, DOI: 10.1016/B978-0-12-386009-5.00015-1

maintenance of some cancers. This is because unrestricted translation favors expression of genes that promote cell proliferation, malignant transformation, and cancer progression. Conversely, restricting the translation initiation by molecular and chemical genetic approaches reverts the malignant phenotype because it preferentially reduces translation of mRNAs that code for proteins important for the genesis and progression of cancer. The growing understanding of the structural biology and the mechanistic insight into the translation initiation cascade led to the identification of pharmacological targets for the development of mechanism-specific anticancer agents: a new paradigm for anticancer therapy. This report will not cover the impact of changes in signaling pathways such as PI3K-Akt and Ras-Raf MAPK on translation initiation.

Translation initiation is the process of assembling the translation competent ribosome on the AUG start codon of the *bona fide* open reading frame (ORF). This requires, at a minimum, the assembly of mRNA, 43S preinitiation complex, and 60S ribosomal subunits in a complex such that anticodon of Met-tRNA (Met-tRNAi) is paired with the AUG start codon of the ORF (Figure 1). The translation initiation machinery is made up of a host of translation initiation factors. The ternary complex, which comprises of the Met-tRNAi and the GTP-coupled eukaryotic initiation factor 2 (eIF2)

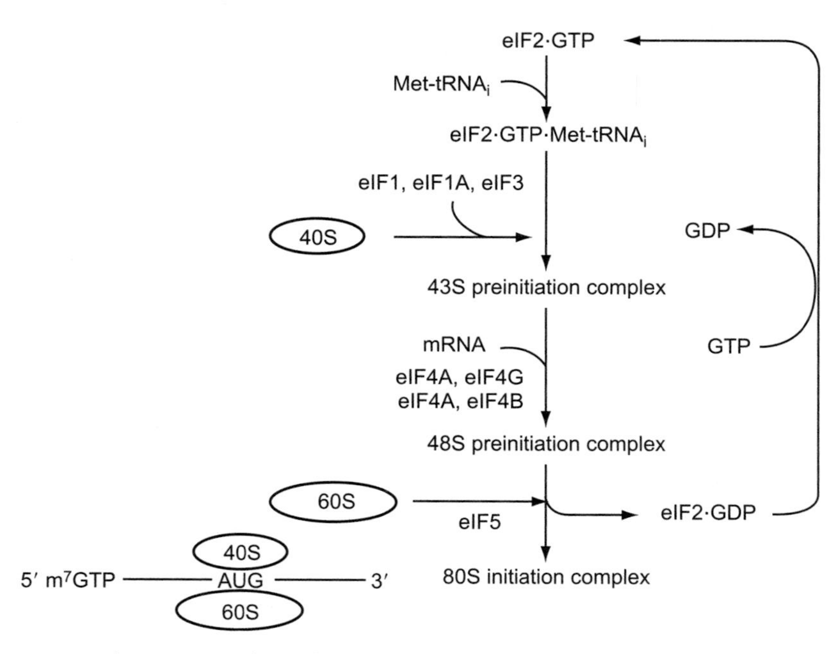

Figure 1 Eukaryotic cap-dependent translation initiation cascade.

binds to the small ribosomal subunit (40S) forming part of the 43S preinitiation complex. This complex also includes other translation initiation factors such as eIF1, eIF1A, and eIF3. The interaction of the 43S preinitiation complex with mRNA is facilitated by the eIF4F complex formed by the eIF4G, the scaffolding protein, eIF4A, the DEAD-box RNA helicase, and the eIF4E, the mRNA cap (the 7-methyl-guanosine 5′-triphosphate) binding protein. The binding of eIF4E to the cap at the 5′-end of mRNA is critical for scanning to locate the AUG initiation codon. The RNA helicase eIF4A in the eIF4F complex unwinds the secondary structure of the mRNA and allows the 43S preinitiation complex to scan the 5′ untranslated region (5′UTR) of the mRNA for the AUG start codon. The scaffolding protein eIF4G also interacts with other translation initiation factors such as eIF3, eIF2, eIF1, and eIF5 and polyadenylate binding protein. The eIF1 in the preinitiation complex plays an important role in the processivity of the 5′UTR scanning and, together with eIF1A and eIF5B, contributes to selection of the *bona fide* start codon. In addition, the hydrolysis of GTP in the ternary complex by eIF5 and eIF5B plays an important role in the binding of the large ribosomal subunit to the 48S preinitiation complex and formation of the 80S initiation complex that is competent for polypeptide synthesis [1–3].

1.1. Role of translation initiation in cancer

Tight regulation of translation ensures that the appropriate quantity and mix of proteins is synthesized. For example, quiescent cells synthesize mostly metabolic and housekeeping proteins. In proliferating cells, not only is the overall rate of protein synthesis increased, but also the mix of newly synthesized proteins is changed to include proteins required for DNA synthesis, chromosome segregation, and cell division. Relieving the physiological restraints on translation initiation induces preferential synthesis of oncogenic proteins and results in malignant transformation.

1.2. Weak and strong mRNAs

The secondary structures in the 5′UTR of mRNAs reduce the processivity of and may cause termination of scanning, while the upstream ORFs (uORFs) cause unproductive initiations. Both these factors lead to reduced translational efficiency. The mRNAs that contain stable secondary structure in the 5′UTR require ATP and the helicase activity of eIF4A to enhance the scanning by the 43S preinitiation complex [1]. Frequent initiation and robust RNA helicase activities are needed to overcome inefficiencies caused by complex secondary structures and uROFs. Interestingly, strong mRNAs coding for housekeeping proteins usually possess a relatively short and simple 5′UTR, while those coding for most pro-proliferation and survival proteins contain a rather long

5′UTR usually burdened by stable secondary structures and/or uORFs. The stable secondary structures and uORFs reduce while strong Kozak consensus sequences [4] increase the efficiency of translation. The stable secondary structures in the 5′UTR, presence of uORFs, and weak Kozak sequences render the mRNA translation highly dependent on the activity of translation initiation factors. This differential dependence of mRNA translatability on the activity of the translation initiation factors forms one of the bases of gene specific regulation at the level of translation initiation [5–7]. Perhaps not surprisingly, malignant transformation is associated with a selective increase in the translation of weak mRNAs that encode for numerous growth factors and oncogenic proteins.

1.3. Expression of translation initiation factors in cancer

Components of the eIF4F complex are overexpressed in many cancers. Levels of eIF4E are elevated in non-Hodgkin's lymphoma, neuroblastomas, and cancers of breast, bladder, colon, prostate, gastrointestinal tract, and lung. Further, in head and neck as well as in breast cancers [8], levels of eIF4E increase during the progression from normal tissue to invasive carcinomas, and this increase correlates with the risk of recurrence after surgical excision [9] and with the cancer-related mortality [10–12]. Levels of eIF4G are elevated in squamous cell lung carcinomas [13–16], and eIF4A is overexpressed in melanomas and primary hepatocellular carcinomas [17,18]. Experimentally, overexpression of eIF4E in mice leads to malignant transformation. However, ectopic expression of eIF4E-binding protein-1 (4E-BP1), which inhibits eIF4E/eIF4G protein–protein interaction, suppresses translation initiation. This results in partial reversal of transformed phenotype and tumorogenesis. These observations support the notion that pharmacological interventions that reduce the activity of eIF4F may offer a new paradigm for anticancer therapy.

Similarly, overexpression of eIF2, the critical component of the ternary complex, and inactivating mutations eIF2α kinases has been reported in human cancers [19–22]. Both these will increase the abundance of the ternary complex rendering translation initiation unrestricted. Experimentally, forced expression of eIF2α-S51A, a nonphosphorylatable eIF2α mutant [23] or of Met-tRNA$_i$, causes transformation of normal cells into malignant cells [24].

Finally, various subunits of eIF3 are overexpressed in some human cancers, and when that occurs, it usually predicts a poor prognosis [25]. Consistently, overexpression of the five subunits of eIF3, either individually or in combination, causes malignant transformation [26].

1.4. Drug targets in the translation initiation cascade

The availability of a high resolution crystal structure of the cap-binding protein eIF4E, which is considered to be rate limiting for the translation initiation process, and its complexes with 7-Me-GDP [27], as well as with the 17 residues from the consensus sequences of 4E-BP1(51–67) and eIF4GII(621–637) [28], made it the first target for the development of anticancer agents. The developments targeted both the eIF4E/eIF4G interface and the cap-binding site on eIF4E. Another obvious target for inhibitors of initiation is eIF4A, the ATP-dependent RNA DEAD-box helicase. Structural studies [29–31] suggest sites for targeting eIF4A either in the domain interface, the ATP binding sites, or interfaces with the eIF4G middle domain. There are additional potential targets in the translation initiation machinery, such as the eIF5, eIF2, eIF1, and eIF1A proteins. However, the impact of inhibition of these translation initiation factors remains to be studied.

2. STATE OF THE ART

2.1. Inhibitors of eIF4F

Direct targeting of eIF4F complex was accomplished by either disrupting the expression of eIF4E or by blocking the interaction between eIF4E and eIF4G (Figure 2). The former was achieved by either developing eIF4E-specific antisense oligonucleotides (ASOs) that trigger RNase H-mediated RNA digestion [32,33] or by using synthetic eIF4E-specific siRNA that binds and cleaves the cognate eIF4A mRNA [34], while the latter was achieved by high-throughput screening campaign of small molecule libraries employing a fluorogenic eIF4GII-derived eIF4E-binding peptide and a transgenically expressed GB1-eIF4E fusion protein [35].

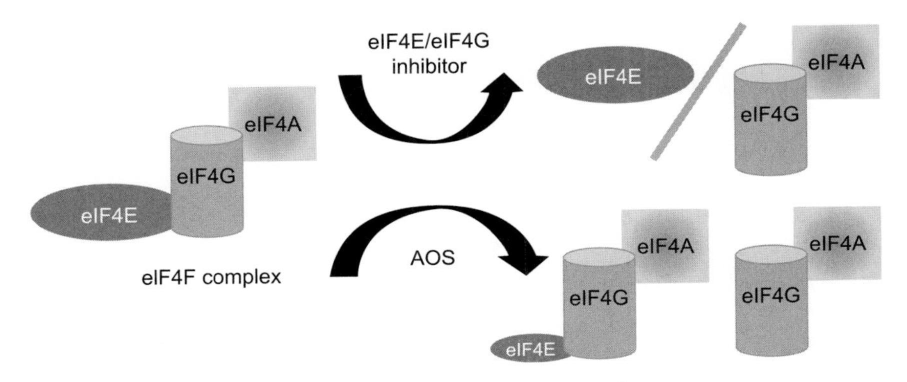

Figure 2 Strategies to inhibit formation of the eIF4F complex.

2.1.1. Targeting eIF4E

2.1.1.1. Antisense Oligonucleotides Early studies demonstrated that many phenotypic changes associated with HeLa cells and *ras*-induced malignant transformation of other cells can be reversed by expression of ASOs complementary to eIF4E mRNA [36,37]. Ectopic expression of eIF4E ASOs in FaDu, a head and neck squamous cell carcinoma (HNSCC) cell line, suppressed tumorigenic and angiogenic properties in a mouse model of human HNSCC [28]. The ASO reduced eIF4E expression in numerous human and murine cell lines and decreased the expression of key malignancy-related proteins such as cyclin D1, VEGF, c-myc, survivin, and BCL-2, but not the expression of housekeeping protein such as β-actin. Moreover, ASO-associated reduction in eIF4E expression leads to apoptosis, reduction in cell viability, and inability of HUVEC to form vessel-like tube structures [38]. Systemic treatment of mice models of human prostate and breast cancers with eIF4E ASO (Figure 3) leads to increased apoptosis, inhibition of tumor growth, and the reduction of eIF4E expression. Immunostaining of VEGF and endothelial specific marker, von Willebrand's factor, suggest that eIF4E ASO treatment reduced tumor vascularity with no apparent toxicity [29]. In a phase I clinical trial with patients who suffer from a variety of cancers, treatment with eIF4E ASO LY2275796 (**1**) induced apoptosis and reduced the tumors' expression of eIF4E and eIF4E-regulated proteins BCL-2 and c-myc [39]. Currently, two phase 1b/2 combination studies of ISIS-EIF4E$_{Rx}$ (LY2275796) with either (a) carboplatin and paclitaxel or (b) docetaxel and prednisone, on non-small cell lung and castrate-resistant prostate cancers, respectively, are underway.

2.1.1.2. Inhibitors of eIF4E/eIF4G Interaction Both eIF4G and 4E-BP1 share a hydrophobic and helical consensus motif, Y(X)$_4$Lϕ, which binds to a conserved and hydrophobic hotspot on the dorsal surface of eIF4E

(A) 5′-*TGTCA*TATTCCTGGA*TCCTT*-3′ eIF4E ASO4 **1**

(B) Base O Me O O O=P ⊖S O Base O O O Me

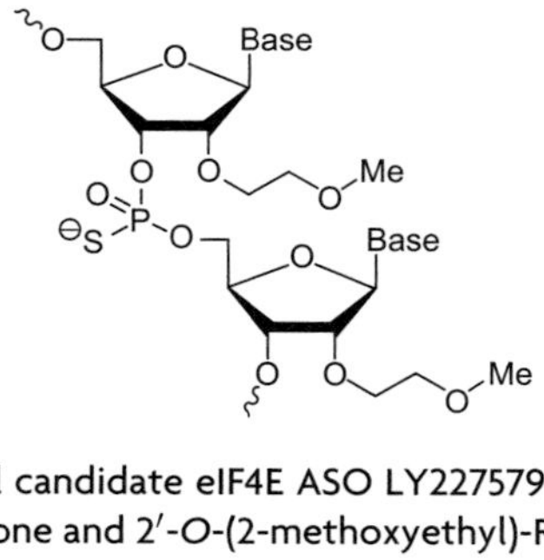

Figure 3 (A) The clinical candidate eIF4E ASO LY2275796 and (B) Schematics of the phosphorothioate backbone and 2′-*O*-(2-methoxyethyl)-RNAs (at the italicized segment in structure A).

(Figure 4) [28]. This motif formed the basis for a high-throughput fluorescence polarization (FP) screening assay, which led to the discovery of a prototypic inhibitor of eIF4E/eIF4G interaction: 4EGI-1 (**2**, Figure 5) [35]. In the cell-free FP assay, 4EGI-1 displayed a $K_d \approx 25\ \mu\text{M}$. In rabbit reticulocyte lysate, 4EGI-1 inhibits binding of full-length eIF4G to eIF4E with a K_d of 20 μM and inhibits cap-dependent translation of a dual-luciferase mRNA reporter construct. Interestingly, 4EGI-1 enhances the binding of 4E-BP1 to eIF4E. NMR titration of GB1-eIF4E fusion protein with 4EGI-1 suggests reversible and tight ligand binding to residues on the convex dorsal surface of eIF4E. Importantly, 4EGI-1 inhibited expression of oncogenic proteins such as c-Myc and Bcl-XL at the translational level, with negligible effects on the translation of housekeeping proteins such as

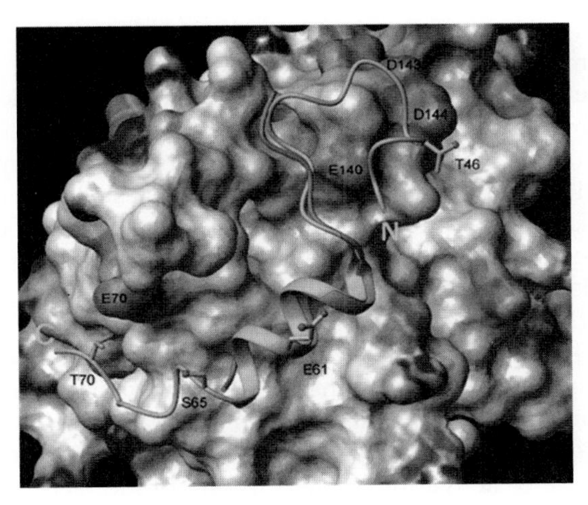

Figure 4 eIF4G- and 4E-BP-derived peptides (cyano and aquamarine, respectively) containing the helical conserved consensus motif, $Y(X)_4L\phi$, bind to the same hydrophobic patch on the convex dorsal surface on eIF4E. (See Color Plate 12.4 in Color Plate Section.)

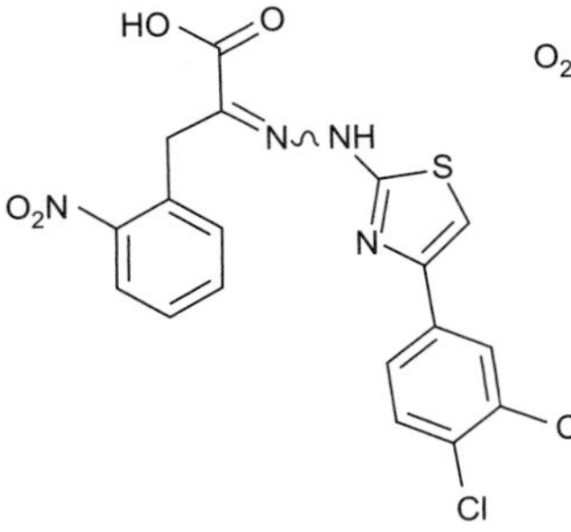

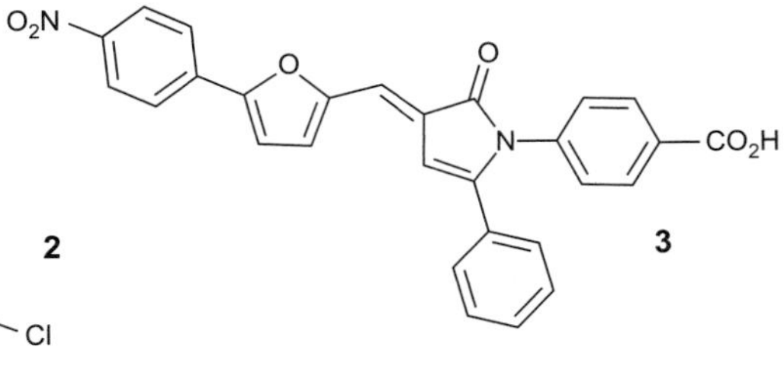

Figure 5 Inhibitors of eIF4E/eIF4G protein–protein interactions: 4EGI-1 (**2**) and 4E1RCat (**3**).

β-actin, tubulin, or ubiquitin. Further, 4EGI-1 inhibits proliferation and/or survival of Jukart cells, various melanoma, breast, lung, and prostate cancer cell lines with IC_{50}s of 1–20 μM, and inhibits preferentially proliferation of p210bcr/abl-transformed Ba/F3 cells compared to vector-transfected immortalized maternal cells (IC_{50} 15 and >40 μM, respectively) [35]. *In vivo*, 4EGI-1 strongly inhibits growth of human CRL-1500 breast and CRL-2813 melanoma cancer xenografts, without any apparent macroscopic- or microscopic-toxicity [40]. As anticipated, in the tumors, treatment with 4EGI-1 led to dissociation of eIF4G from eIF4E and increased the association of 4E-BP1 with eIF4E. The 4EGI-1 treated tumors had a lower expression of proliferating cell nuclear antigen (PCNA) and oncogenic and growth regulatory proteins such as cyclins D1 and E, c-Myc, Bcl-2, and VEGF as compared with the nontreated tumors. Importantly, 4EGI-1 did not increase the fraction of apoptotic cells in the treated tumors as compared with the nontreated ones [40].

The antileukemic potential of 4EGI-1 in acute myeloid leukemia (AML) is demonstrated *ex vivo* by robust reduction of clonogenic growth of AML precursors and a massive induction of blast cell apoptosis [41], with minimal impairment of normal hematopoiesis. Importantly, treatment of AML cells by 4EGI-1, but not the rapalog RAD001, abrogates eIF4F complex formation, reduces translation of c-Myc mRNA, and reduces levels of c-Myc, cyclin D1, and Bcl-xL, proteins implicated in the oncogenic process.

In human lung cancer cells, in addition to the inhibition of cap-dependent translation, 4EGI-1 also augments tumor necrosis factor-related apoptosis-inducing ligand (TRIAL), which mediates apoptosis through induction of death receptor 5 (DR5), and downregulates c-FLIP, which inhibits caspase-8 activation. This pro-apoptotic activity of 4EGI-1 appears to be independent of inhibition of cap-dependent translation [42]. As such, 4EGI-1 contributes in a cooperative manner to the induction of apoptosis by inhibiting cap-dependent translation initiation, while sensitizing human lung cancer cells to TRIAL-induced apoptosis. This could be an exciting demonstration of reinforcing polypharmacology. 4EGI-1 also preferentially suppresses proliferation of malignant pleural mesothelioma cells by inhibiting cap-dependent translation as compared to LP9 normal mesothelial cells and inhibits growth of non-small cell lung cancer by the same mechanism [43,44].

Recently, other screening efforts led to the discovery of 4E1RCat (**3**, Figure 5), which inhibits eIF4E/eIF4G interaction with an $IC_{50} \sim 4$ μM as measured in the time-resolved-FRET assay [45]. It is distinct from **2** because it inhibits not only eIF4E/eIF4G interactions but also the eIF4E/4E-BP1 interaction. In an animal model of Eμ-Myc-driven lymphoma, **3** was effective in reversing chemo-resistance to doxorubicin treatment.

*2.1.1.3. **Ribavirin*** The antiviral drug Ribavirin (**4**, Figure 6) is a nucleotide analog that causes hypermutations in the genome of RNA viruses and disrupts replication of DNA viruses [46]. Ribavirin displays cytostatic activity in mammalian cells and is reportedly useful for treatment of patients with acute myeloid leukemia [47,48]. It is currently being evaluated for treatment of breast cancer patients [49]. Although some reports attributed the anticancer activity of Ribavirin to its mimicry of 7-methyl-guanosine (mRNA cap) and inhibition of translation initiation[50,51], others have challenged this view [52,53]. Specifically, ribavirin was two to four orders of magnitude less potent than 7-Me-GTP *in vitro* in inhibiting cap-dependent translation. Further, unlike 7-Me-GTP, Ribavirin did not preferentially inhibit cap-dependent translation compared to cap-independent translation.

*2.1.1.4. **7-Methylguanidine (Cap) Analogs*** The eIF4E interacts with the mRNA cap, forming the 48S initiation complex, which is critical for translation initiation. Therefore, cap (Me^7GTP **5**) analogs (Figure 7) have been sought after as potential inhibitors of translation initiation and novel anticancer drugs. Several attempts were made over the years to synthesize cap analogs that will be cell permeable, bind to eIF4E with high

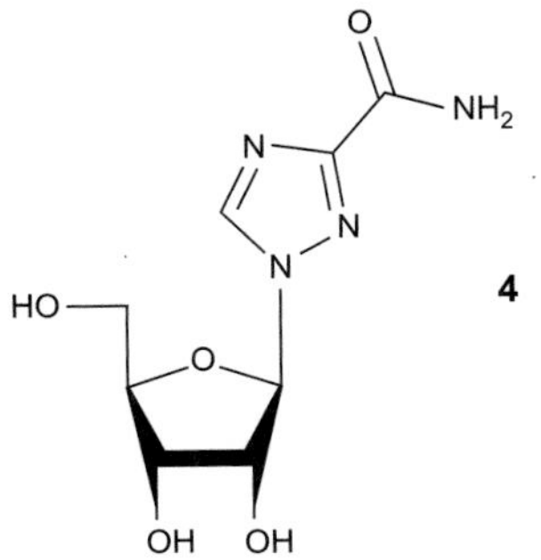

Figure 6 Ribavirin, inhibitor of eIF4E/mRNA cap interaction.

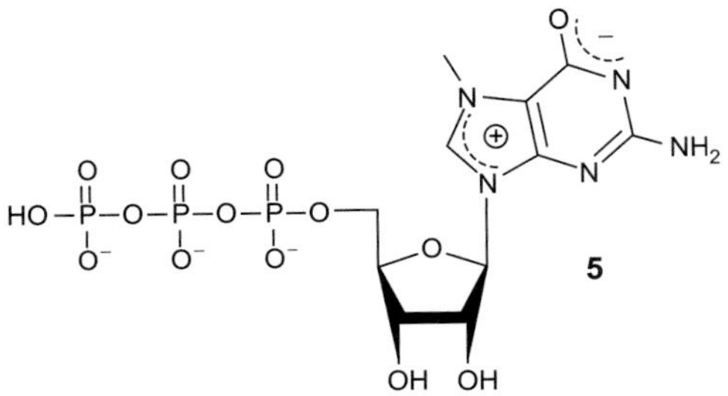

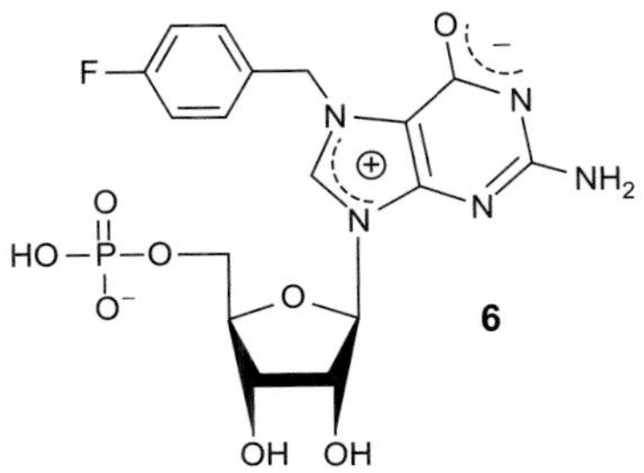

Figure 7 Analogs of the cap, 7-Methylguanidine, Me^7GTP (**5**) and *p*-FBn^{7}GMP (**6**).

affinity, and inhibit effectively cap-dependent translation [54,55]. An interesting series of Bn^7-GMP derivatives yielded one of the most potent cap analogs *p*-FBn^7GMP (**6**, $K_d = 1.96$ μM) [56]. To date, none of these have been employed as anticancer agents *in vivo*.

2.1.1.5. 4E-BP1 and Related Peptides 4E-BP1 is a natural repressor of translation initiation that disrupts eIF4E/eIF4G interaction. Cell permeable peptides that contain the conserved eIF4E-binding motif present in eIF4G and 4E-BP1 will inhibit eIF4E/eIF4G interaction [28,57]. 4E-BP1-derived peptides fused to cell penetrating peptides, such as penetratin [58] and TAT [59], induce apoptosis in cancer cells [58]. More recently, a helix-stabilized eIF4E-binding peptide derived from eIF4G1 fused to TAT was shown to inhibit translation initiation and induce apoptosis in MCF-7 breast cancer cells [59].

In another approach, 4E-BP1-derived eIF4E-binding sequence was conjugated to an agonist of gonadotropin-releasing hormone [D-Lys^6] GnRH, generating a fusion peptide [D-Lys^6]GnRH–4E-BP1 that can be taken up by GnRH receptor-expressing tumor cells in a tissue-specific manner. GnRH in this context serves the dual purpose of facilitating entry of peptide into the cells, and targeting it only to ovarian cancer cells [60]. In a mouse model of intraperitoneally implanted ovarian cancer, this peptide reduced tumor burden compared to saline treatment [60]. Another approach to this peptide based cancer therapy is the delivery of full-length 4E-BP1 into the cells by adenovirus mediated gene therapy [61]. Adenovirus-mediated transfer of 4E-BP1 by itself failed to inhibit the growth of pancreatic or gastric carcinoma cells because hyperphosphorylation of recombinant protein inhibited its binding to eIF4E. Combined treatment with mTOR inhibitors, on the other hand, reduced the phosphorylation of recombinant 4E-BP1 and inhibited proliferation of cancer cell proliferation *in vitro* and tumor growth *in vivo* [61].

2.1.1.6. MAPK-Interacting Kinases Inhibition of the mitogen-activated protein kinase (MAPK) and interacting kinases (Mnk1), which bind to the C-terminal MA3 domain of human eIF4G and phosphorylates Ser209 of eIF4E, has recently been shown to block eIF4E phosphorylation and suppress outgrowth of experimental lung metastases [62]. Screening has identified antifungal agent cercosporamide (**7**, Figure 8) as a potent, selective, and orally bioavailable Mnk inhibitor. Cercosporamide blocks eIF4E phosphorylation *in vitro* and in normal mouse tissue and xenografted tumors within 30 min after oral administration. Cercosporamide reduces tumor growth in HCT116 tumor bearing animals, and suppresses the outgrowth of B16 melanoma lung metastases [62]. Thus, the Mnk/eIF4G complex may be an attractive target for development of antitumor agents.

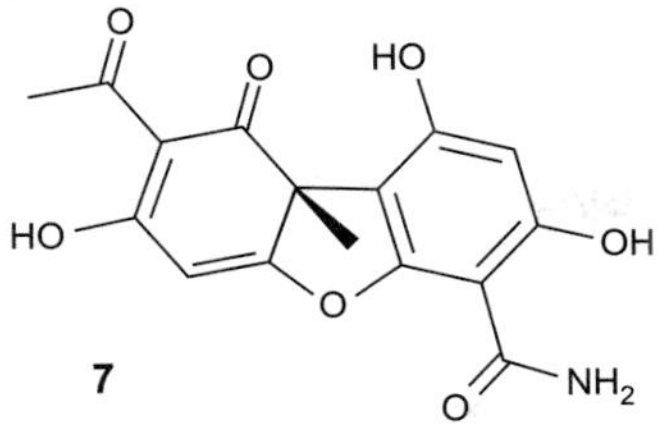

Figure 8 Cercosporamide inhibits Mnk1 and prevents phosphorylation of eIF4E.

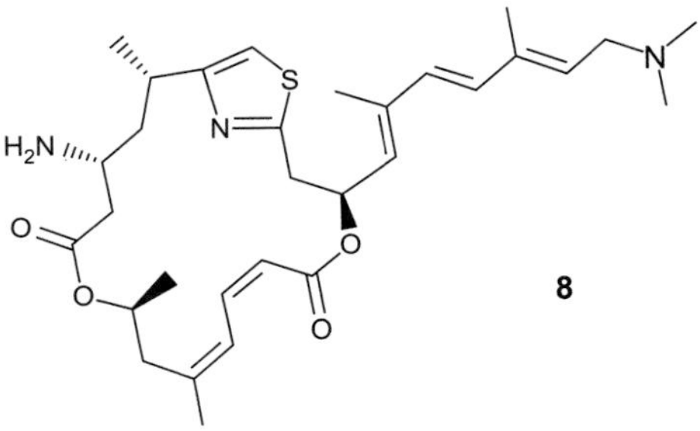

Figure 9 Pateamine A, a thiazole-containing macrolide.

2.1.2. Targeting eIF4A

*2.1.2.1. **Pateamine A*** Pateamine A (**8**, Figure 9), a thiazole-containing macrolide isolated from the marine sponge *Mycale* sp., with well-known antineoplastic activity [63,64] interacts with eIF4AI, disrupts the eIF4A/eIF4G interaction, and inhibits translation initiation [65–68]. In biochemical assays, **8** stimulates rather than inhibits the helicase activity of eIF4A by induction of conformational changes in the structure of eIF4A, but because **8**-bound eIF4A cannot associate with eIF4G, it cannot unwind mRNA secondary structures in the 48S preinitiation complex (Figure 10). Treatment of cells by **8** induces formation of stress granules, presumably by inhibiting translation initiation [69,70]. Pateamine A inhibits the nonsense-mediated mRNA decay by directly targeting eIF4AIII, a core component of the exon junction complex (EJC), independent of inhibition of translation initiation [71]. Simplified **8** analogs have been synthesized and shown to possess potent anticancer activity [72]. Pateamine A causes acute inhibition of DNA synthesis presumably by inhibition of DNA polymerases, indicating that polypharmacology, that is, modulation of several unrelated targets by **8** may contribute to its apparent antineoplastic effects.

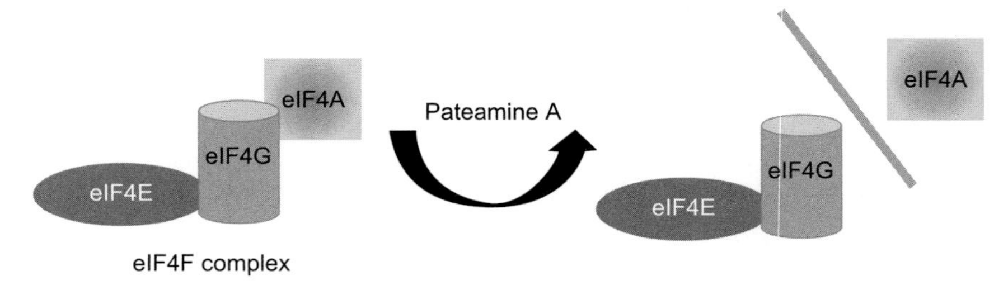

Figure 10 Pateamine A binds to eIF4A and inhibits the assembly of eIF4F complex.

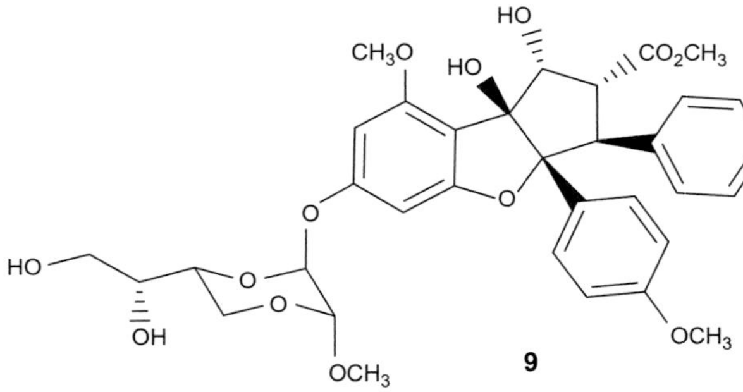

Figure 11 Silvestrol, a natural product and a member of the flavagline family that targets translation initiation.

2.1.2.2. Cyclopenta[b]benzofuran Flavaglines—Silvestrol Cyclopenta[*b*] benzofuran flavaglines (CBFs) are potential anticancer agents originally isolated from the species of *Aglaia* genus of the *Meliaceae* plant family. These agents inhibit protein synthesis and display *in vitro* activity against cancer cells and *in vivo* activity against xenograft tumors [73–78].

Silvestrol (**9**, Figure 11), a CBF isolated from *Aglaia silvestris*, inhibits translation initiation by targeting eIF4A and ribosome loading onto mRNA templates [79]. Silvestrol prevented growth of human Jurkat and LNCaP cells in culture [77,80] and showed weak antitumor potential in xenograft tumor models of human BC1 breast cancer [74]. While ineffective as a single agent against Eμ-myc lymphomas that harbor mutants of tumor-suppressor phosphatase and tensin homolog (PTEN), or overexpress eIF4E [79], Silvestrol inhibited growth of human MDA-MB-231 breast cancer xenografts [81]. Importantly, in combination with doxorubicin, it was effective against *Pten+/−Eμ-myc* and *Eμ-myc/eIF4E* tumors, which were refractory to rapamycin/doxorubicin treatment [79]. Silvestrol preferentially inhibits translation of weak mRNAs coding for proteins such as cyclin D1, survivin, Mcl-1, Bcl-2, and c-Myc [81].

*2.1.2.3. **Hippuristanol*** Hippuristanol (**10**, Figure 12), purified from the coral *Isis hippuris* extract [82], was discovered in a chemical screen for inhibitors of translation initiation [83]. Hippuristanol interferes with binding of eIF4A to RNA and thereby inhibits the ATPase activity, but does not prevent binding of ATP to eIF4A (Figure 13). As such, **10** inhibits RNA dependent helicase activity, resulting in abrogation of cap-dependent translation [66]. Hippuristanol binds to amino acids adjacent to and overlapping with two conserved motifs present in the carboxy-terminal domain of eIF4A that are implicated in interaction with RNA and ATP, and interdomain contacts [30]. Inhibition of cap-dependent translation is thought to underlie inhibition of adult T-cell xenograft tumor growth [84]. Hippuristanol was synthesized by various synthetic approaches and was

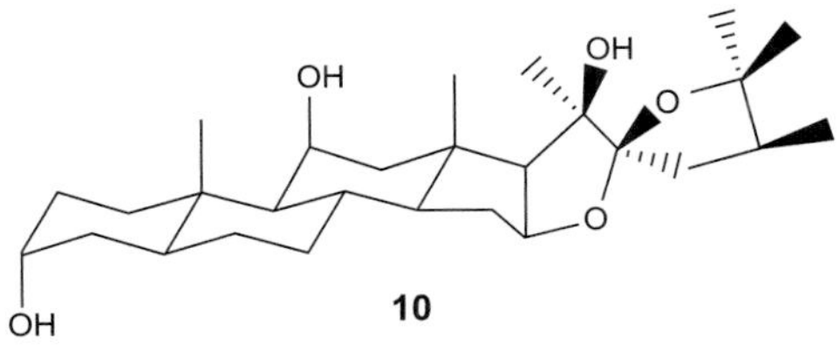

Figure 12 Hippuristanol, a natural product that inhibits RNA dependent helicase activity.

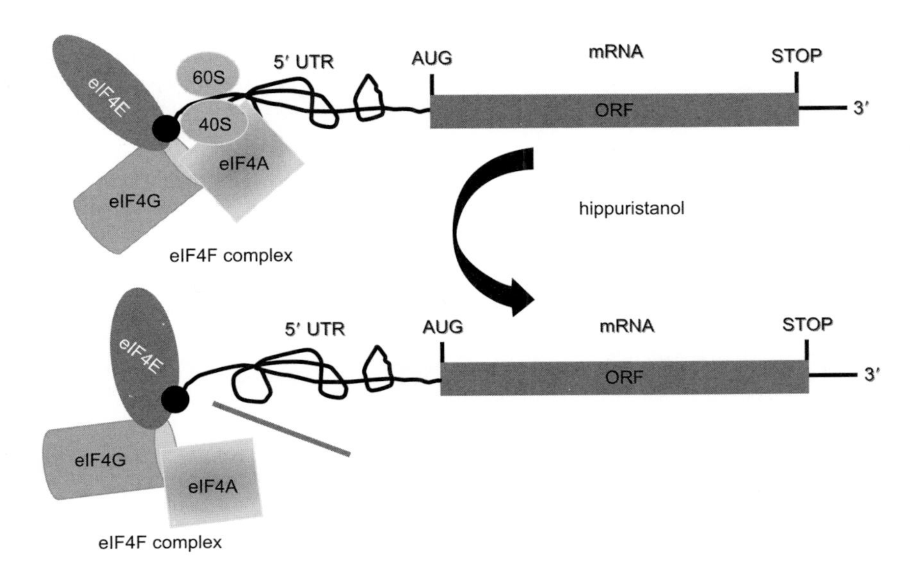

Figure 13 Hippuristanol binds to a subunit of eIF4A and inhibits the helicase activity of eIF4F.

subjected to limited structure activity relationship studies [85–87]. Both the *gem*-dimethyl on the spiro-furan ring F and the size of this ring are important for the antiproliferative activity of **10**.

Overall, these results suggest that inhibition of translation initiation by modulating eIF4A activity is a promising approach to altering drug resistance associated with PI3K/mTOR activation, and for inhibition of tumor growth.

2.2. Inhibitors of ternary complex

2.2.1. Salubrinal

Salubrinal (**11**, Figure 14) was discovered in a screen for agents that protect cells from endoplasmic reticulum stress induced cell death and was shown to cause eIF2α phosphorylation by selectively inhibiting an eIF2α phosphatase [88]. Salubrinal directly interacts with bcl-2 and inhibits induction of apoptosis by DNA damaging agents [89]. Salubrinal inhibits leukemia cell proliferation synergistically with proteasome inhibitors [90] and renders a subpopulation of bortezamib-resistant multiple myeloma cells sensitive to this FDA approved proteasome inhibitor [91].

2.2.2. 15-Deoxyspergualin

15-Deoxyspergualin (**12**, Figure 15), an analog of the immunosuppressive agent spergualin, inhibits cell cycle progression and cytokine production upon naïve T-cell activation [92]. It also displays significant tumoricidal activity [93]. It interacts with the EEVD domain of heat shock protein 70

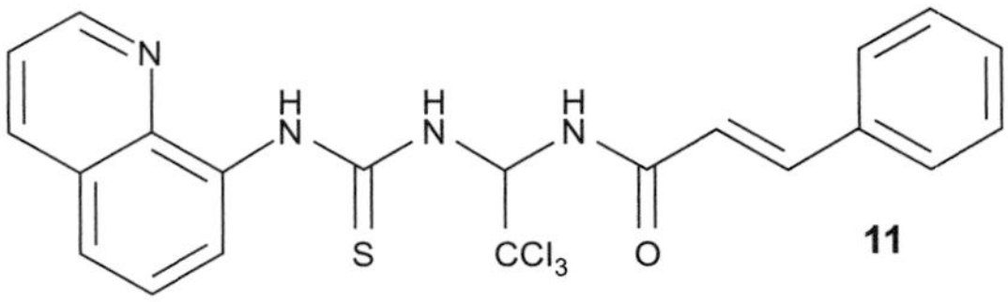

Figure 14 Salubrinal inhibits ternary complex formation by inhibiting eIF2α phosphatase.

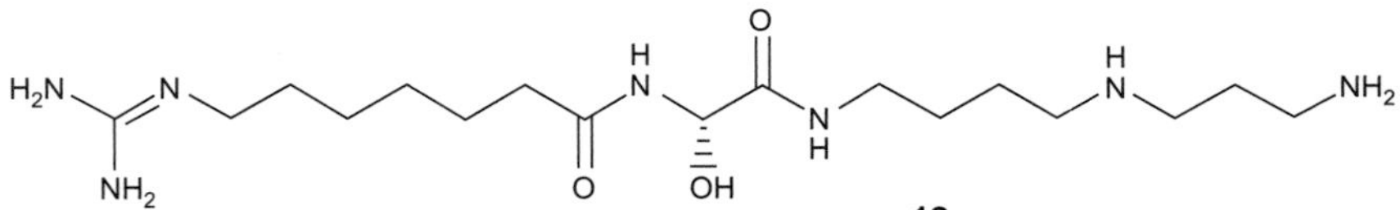

Figure 15 15-Deoxyspergualin inactivates eIF5A by posttranslational hypusination and leads to phosphorylation of eIF2α.

and causes phosphorylation of eIF2α [94]. Activation of eIF5A, an RNA binding protein, involves hypusination of Lys50. Inhibition of eIF5A hypusination by **12** inhibits translation initiation and cell proliferation [95,96]. Consistently, incubation of mouse mammary carcinoma FM3A cells with **12** led to inhibition of cell growth and to the formation of inactive eIF5A.

2.2.3. Thiozolidone-indenones

Thiazolidinediones (TZD) were discovered as peroxisome proliferator activated receptor gamma (PPAR-γ) agonists and developed for the management of lipid and glucose metabolism in type-2 diabetes [97,98]. This class includes drugs such as pioglitazone, rosiglitazone, ciglitazone, and troglitazone (**13**, Figure 16), all of which display some antiproliferative activity against cancer cells *in vitro* and inhibit tumor growth in xenograft models of human cancer [99–102].

Although the activity of TZDs against cancer cell lines or xenograft tumors was thought to be mediated by activation of PPAR-γ, in many instances, it was shown that the antiproliferative activity of these agents is independent of PPAR-γ. These PPAR-γ independent effects are reportedly mediated by: (a) Ca^{2+} store depletion and subsequent phosphorylation of eIF2α and inhibition of translation initiation (Figure 17) [99]; (b) induction of cellular acidosis through inhibition of the Na^+/H^+ exchanger [103]; (c) inhibition of Bcl-xL/Bcl-2 complex [104]; (d) release

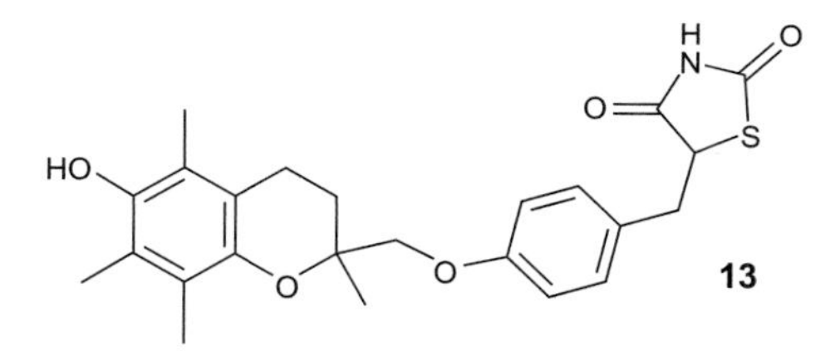

Figure 16 Troglitazone an antiproliferative agent from the TZD family.

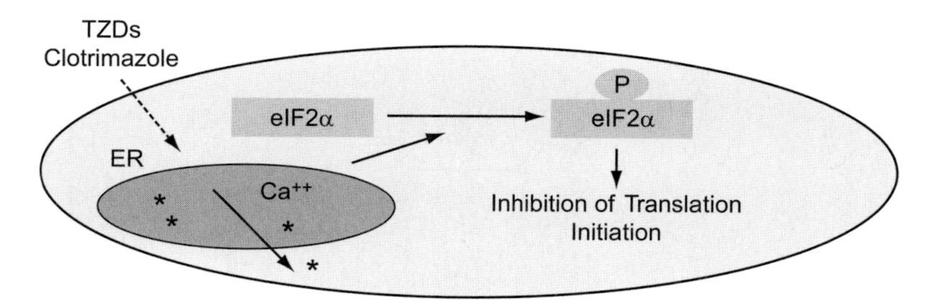

Figure 17 Mechanism of action of antiproliferative agents such as TZDs and clotrimazole that release Ca^{2+} from ER and phosphorylate eIF2α.

of apoptotic factors from the mitochondria through the production of ROS [105]; (e) upregulation of PTEN expression; (f) AMPK phosphorylation; and (g) downregulation of Akt/mTOR/p70S6 signaling cascades [106]. This multitude of seemingly disparate PPAR-γ independent activities of TZDs suggests either that TZDs are very promiscuous agents or that one of the PPAR-γ independent mechanisms underlies most of these activities. We suggest that eIF2α phosphorylation-dependent inhibition of translation initiation may be the unifying mechanism that underlies the antiproliferative effects of TZDs.

2.2.4. Clotrimazole

Clotrimazole (**14**, Figure 18) was developed in 1970s for the treatment of fungal infections [107]. It was subsequently shown to inhibit Ca^{2+} activated potassium channels, release Ca^{2+} from internal stores, inhibit Ca^{2+} release activated Ca^{2+} influx, and inhibit proliferation of cancer cells *in vitro* and tumor growth and metastasis *in vivo* [108]. Further studies demonstrated that partial depletion of Ca^{2+} stores cause phosphorylation of eIF2α and inhibition of translation initiation. This accounts, at least in part, for the anticancer activity of **14** [109]. Further, **14** inhibits angiogenesis [110] and sensitizes glioblastoma cells to radiation therapy [111]. These affects are likely secondary to the inhibition of translation initiation. Nevertheless, like many first generation translation initiation inhibitors, **14** has polypharmacology that may include inhibition of translation initiation and glycolytic supply of ATP that are required for cancer cells proliferation [112].

2.2.5. 3,3-Diaryl oxindoles

3,3-Diaryl oxindoles (Figure 19) were developed through structure activity relationship studies to identify agents that induce sustained-partial depletion of internal Ca^{2+} stores, thereby inducing eIF2α phosphorylation and inhibiting translation initiation [113,114]. Two lead compounds identified through this effort, #1181 (**15**) and #1430 (**16**), inhibit proliferation of a wide variety of cancer cells with low micromolar potencies. *In vitro*

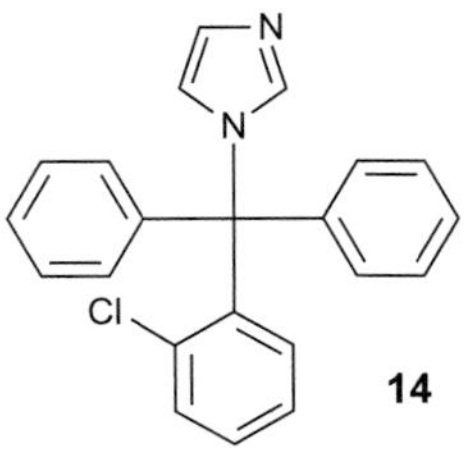

Figure 18 Clotrimazole, an antifungal agents that depletes ternary complex and inhibits cap-dependent translation initiation.

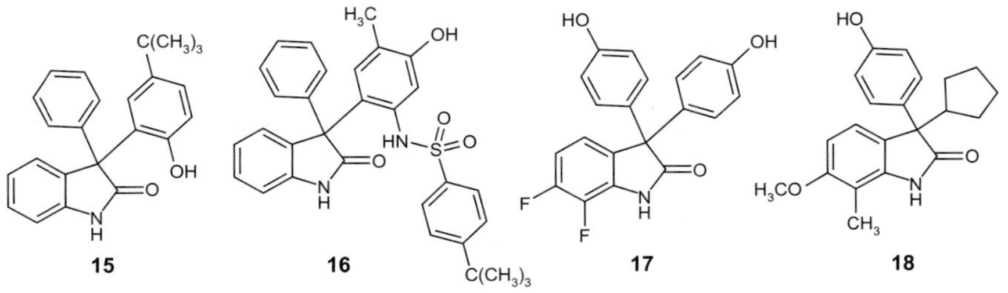

Figure 19 3,3-Diaryl oxindoles with anticancer activity.

studies with compound **15** demonstrate that this agent inhibits translation initiation, and preferentially abrogates the expression of oncogenic and antiapoptotic proteins with negligible effect on the expression of housekeeping proteins. Most, but not all of these effects, are translational. Treatment of mice bearing MCF-7 human breast cancer cell derived tumors (~150 mm^3) with 140 mg/kg/bid of #1181 sc for 3 weeks resulted in ~30% regression in tumor size [40]. This antitumor activity is associated with phosphorylation of eIF2α and inhibition of the expression of oncogenic proteins. These findings provide proof-of-principle that phosphorylation of eIF2α and reduced formation of the ternary complex are pharmacologically viable targets for cancer therapy.

Screening oxyphenistatin analogs for preferential inhibition of human breast cancer cell (MDA-468) proliferation, as compared to a related non-cancer cell, identified 6,7-difluoro-oxyphenistatin (**17**) to have 1000-fold greater antiproliferative activity toward MDA-468 cells (IC_{50} = 20 nM) [115]. In addition, oxindole **17** showed potent antitumor activity in xenograft mice models of human prostate and breast cancers following p.o. or i. v. administration [116]. Further optimization led to the development of 3-aryl,3-cycloalkyl-oxindoles that presented a significant departure from the 3,3-diaryloxindole chemotype [117]. Oxindole **18** showed nanomolar antiproliferative activity toward breast and prostate cancer cells and caused complete tumor regression in a rat xenograft model of human prostate cancer [117]. A very recent publication reports on *N,N′*-diarylureas that activate heme-regulated inhibitor kinase, induce eIF2α phosphorylation, inhibit translation initiation, and display anticancer activity [118].

3. CONCLUSIONS

Prominent targets for anticancer drugs like phosphoinositide 3-kinase (PI3K), AKT kinase, or mammalian target of rapamycin (mTOR) are all upstream of 4E-BP1 phosphorylation, and they trigger inhibition of translation initiation while modulating other downstream targets. Targeting cancer pathways downstream of their cellular signaling networks at the level of translation initiation will be more effective and could offer fewer side effects than inhibiting their upstream targets. The location of translation initiation at the apex of many well-defined oncogenic, pro-apoptotic, and tumor-suppressor pathways, and the plethora of distinct targets make it an attractive field for the development of a new generation of anticancer agents. The diverse nature of these targets includes interaction with classical active sites, as well as challenging targets such as protein–protein interactions. Because restoration of translational control down-regulates oncogenic proteins with minimal effect on housekeeping proteins, it will affect predominantly the addicted cancer cells and spare normal ones. Hence, small molecule inhibitors of translation initiation have excellent potential for achieving a wide therapeutic window. Therefore, this emerging field represents a new and highly promising paradigm in cancer therapy.

REFERENCES

[1] T. V. Pestova and V. G. Kolupaeva, *Genes Dev.*, 2002, **16**, 2906.
[2] T. V. Pestova, J. R. Lorsch and C. U. T. Hellen, in *Translational Control in Biology and Medicine*, (eds. M. Mathews, N. Sonenberg, J. W. B. Hershey), Cold Spring Harbor Laboratory Press, Cold Spring Harbor, NY, 2007, p. 87.
[3] N. Sonenberg and A. G. Hinnebusch, *Cell*, 2009, **136**, 731.
[4] R. Kawaguchi and J. Bailey-Serres, *Nucleic Acids Res.*, 2005, **33**, 955.
[5] H. A. Meijer and A. A. Thomas, *Biochem. J.*, 2002, **367**, 1.
[6] M. Kozak, *Mol. Cell. Biol.*, 1989, **9**, 5134.
[7] M. Kozak, *J. Biol. Chem.*, 1991, **266**, 19867.
[8] M. S. Haydon, J. D. Googe, D. S. Sorrells, G. E. Ghali and B. D. Li, *Cancer*, 2000, **88**, 2803.
[9] C. O. Nathan, N. Amirghahri, C. Rice, F. W. Abreo, R. Shi and F. J. Stucker, *Laryngoscope*, 2002, **112**, 2129.
[10] B. D. Li, J. C. McDonald, R. Nassar and A. De Benedetti, *Ann. Surg.*, 1998, **227**, 756, discussion 761.
[11] B. D. Li, J. S. Gruner, F. Abreo, L. W. Johnson, H. Yu, S. Nawas, J. C. McDonald and A. DeBenedetti, *Ann. Surg.*, 2002, **235**, 732, discussion 738.
[12] Y. Mamane, E. Petroulakis, L. Rong, K. Yoshida, L. W. Ler and N. Sonenberg, *Oncogene*, 2004, **23**, 3172.
[13] R. Pincheira, Q. Chen, Z. Huang, J. T. Zhang and J. Eu, *Eur. J. Cell Biol.*, 2001, **80**, 410.
[14] O. Saramaki, N. Willi, O. Bratt, T. C. Gasser, P. Koivisto, N. N. Nupponen, L. Bubendorf and T. Visakorpi, *Am. J. Pathol.*, 2001, **159**, 2089.
[15] N. N. Nupponen, K. Porkka, L. Kakkola, M. Tanner, K. Persson, A. Borg, J. Isola and T. Visakorpi, *Am. J. Pathol.*, 1999, **154**, 1777.
[16] K. Dua, T. M. Williams and L. Beretta, *Proteomics*, 2001, **1**, 1191.
[17] J. Eberle, K. Krasagakis and C. E. Orfanos, *Int. J. Cancer*, 1997, **71**, 396.
[18] M. Shuda, N. Kondoh, K. Tanaka, A. Ryo, T. Wakatsuki, A. Hada, N. Goseki, T. Igari, K. Hatsuse, T. Aihara, S. Horiuchi, M. Shichita, N. Yamamoto and M. Yamamoto, *Anticancer Res.*, 2000, **20**, 2489.
[19] I. B. Rosenwald, M. J. Hutzler, S. Wang, L. Savas and A. E. Fraire, *Cancer*, 2001, **92**, 2164.

[20] N. Abraham, M. L. Jaramillo, P. I. Duncan, N. Methot, P. L. Icely, D. F. Stojdl, G.N. Barber and J. C. Bell, *Exp. Cell Res.*, 1998, **244**, 394.
[21] I. B. Rosenwald, L. Koifman, L. Savas, J. J. Chen, B. A. Woda and M. E. Kadin, *Hum. Pathol.*, 2008, **39**, 910.
[22] I. B. Rosenwald, S. Wang, L. Savas, B. Woda and J. Pullman, *Cancer*, 2003, **98**, 1080.
[23] O. Donze, R. Jagus, A. E. Koromilas, J. W. Hershey and N. Sonenberg, *EMBO J.*, 1995, **14**, 3828.
[24] L. Marshall, N. S. Kenneth and R. J. White, *Cell*, 2008, **133**, 78.
[25] Z. Dong and J. T. Zhang, *Crit. Rev. Oncol. Hematol.*, 2006, **59**, 169.
[26] L. Zhang, X. Pan and J. W. Hershey, *J. Biol. Chem.*, 2007, **282**, 5790.
[27] J. Marcotrigiano, A. C. Gingras, N. Sonenberg and S. K. Burley, *Cell*, 1997, **89**, 951.
[28] J. Marcotrigiano, A. C. Gingras, N. Sonenberg and S. K. Burley, *Mol. Cell*, 1999, **3**, 707.
[29] J. H. Chang, Y. H. Cho, S. Y. Sohn, J. M. Choi, A. Kim, Y. C. Kim, S. K. Jang and Y. Cho, *Proc. Natl. Acad. Sci. USA*, 2009, **106**, 3148.
[30] L. Lindqvist, M. Oberer, M. Reibarkh, R. Cencic, M. E. Bordeleau, E. Vogt, A. Marintchev, J. Tanaka, F. Fagotto, M. Altmann, G. Wagner and J. Pelletier, *PLoS One*, 2008, **3**, e1583.
[31] P. Schutz, M. Bumann, A. E. Oberholzer, C. Bieniossek, H. Trachsel, M. Altmann and U. Baumann, *Proc. Natl. Acad. Sci. USA*, 2008, **105**, 9564.
[32] R. J. DeFatta, C. O. Nathan and A. De Benedetti, *Laryngoscope*, 2000, **110**, 928.
[33] J. R. Graff, B. W. Konicek, T. M. Vincent, R. L. Lynch, D. Monteith, S. N. Weir, P. Schwier, A. Capen, R. L. Goode, M. S. Dowless, Y. Chen, H. Zhang, S. Sissons, K. Cox, A.M. McNulty, S. H. Parsons, T. Wang, L. Sams, S. Geeganage, L. E. Douglass, B.L. Neubauer, N. M. Dean, K. Blanchard, J. Shou, L. F. Stancato, J. H. Carter and E.G. Marcusson, *J. Clin. Invest.*, 2007, **117**, 2638.
[34] N. Oridate, H. J. Kim, X. Xu and R. Lotan, *Cancer Biol. Ther.*, 2005, **4**, 318.
[35] N. J. Moerke, H. Aktas, H. Chen, S. Cantel, M. Y. Reibarkh, A. Fahmy, J. D. Gross, A. Degterev, J. Yuan, M. Chorev, J. A. Halperin and G. Wagner, *Cell*, 2007, **128**, 257.
[36] A. De Benedetti, S. Joshi-Barve, C. Rinker-Schaeffer and R. E. Rhoads, *Mol. Cell. Biol.*, 1991, **11**, 5435.
[37] C. W. Rinker-Schaeffer, J. R. Graff, A. De Benedetti, S. G. Zimmer and R. E. Rhoads, *Int. J. Cancer*, 1993, **55**, 841.
[38] J. R. Graff, B. W. Konicek, J. H. Carter and E. G. Marcusson, *Cancer Res.*, 2008, **68**, 631.
[39] J. R. Graff, B. W. Konicek, R. L. Lynch, C. A. Dumstorf, M. S. Dowless, A. M. McNulty, S. H. Parsons, L. H. Brail, B. M. Colligan, J. W. Koop, B. M. Hurst, J. A. Deddens, B. L. Neubauer, L. F. Stancato, H. W. Carter, L. E. Douglass and J. H. Carter, *Cancer Res.*, 2009, **69**, 3866.
[40] L. Chen, B. H. Aktas, Y. Wang, X. He, R. Sahoo, N. Zhang, S. Denoyelle, E. Kabha, H. Yang, R. Yefidoff-Freedman, J. G. Supko, M. Chorev, G. Wagner, and J. A. Halperin, Submitted for publication.
[41] J. Tamburini, A. S. Green, V. Bardet, N. Chapuis, S. Park, L. Willems, M. Uzunov, N. Ifrah, F. Dreyfus, C. Lacombe, P. Mayeux and D. Bouscary, *Blood*, 2009, **114**, 1618.
[42] S. Fan, Y. Li, P. Yue, F. R. Khuri and S. Y. Sun, *Neoplasia*, 2010, **12**, 346–356.
[43] A. De, B. A. Jacobson, M. S. Peterson, M. G. Kratzke and R. A. Kratzke, American Association for Cancer Research Annual Meeting (San Diego, California), 2008.
[44] A. De, O. Larsson, B. A. Jacobson, M. Peterson, K. Terai, M. G. Kratzke, A. Z. Dudek, P.B. Bitterman and R. A. Kratzke, American Association for Cancer Research Annual Meeting (San Diego, California), 2008.
[45] R. Cencic, D. R. Hall, F. Robert, Y. Du, J. Min, L. Li, M. Qui, I. Lewis, S. Kurtkaya, R. Dingledine, H. Fu, D. Kozakov, S. Vajda and J. Pelletier, *Proc. Natl. Acad. Sci. USA*, 2011, **108**, 1046.
[46] S. Crotty, C. Cameron and R. Andino, *J. Mol. Med.*, 2002, **80**, 86.

[47] S. Assouline, B. Culjkovic, E. Cocolakis, C. Rousseau, N. Beslu, A. Amri, S. Caplan, B. Leber, D. C. Roy, W. H. Miller Jr. and K. L. B. Borden, *Blood*, 2009, **114**, 257.
[48] K. L. Borden and B. Culjkovic-Kraljacic, *Leuk. Lymphoma*, 1805, **2010**, 51.
[49] http://clinicaltrials.gov/ct2/show/NCT01056757, 2010.
[50] A. Kentsis, I. Topisirovic, B. Culjkovic, L. Shao and K. L. Borden, *Proc. Natl. Acad. Sci. USA*, 2004, **101**, 18105.
[51] A. Kentsis, L. Volpon, I. Topisirovic, C. E. Soll, B. Culjkovic, L. Shao and K. L. Borden, *RNA*, 2005, **11**, 1762.
[52] B. Westman, L. Beeren, E. Grudzien, J. Stepinski, R. Worch, J. Zuberek, J. Jemielity, R. Stolarski, E. Darzynkiewicz, R. E. Rhoads and T. Preiss, *RNA*, 2005, **11**, 1505.
[53] Y. Yan, Y. Svitkin, J. M. Lee, M. Bisaillon and J. Pelletier, *RNA*, 2005, **11**, 1238.
[54] A. Cai, M. Jankowska-Anyszka, A. Centers, L. Chlebicka, J. Stepinski, R. Stolarski, E. Darzynkiewicz and R. E. Rhoads, *Biochemistry*, 1999, **38**, 8538.
[55] J. Kowalska, M. Lukaszewicz, J. Zuberek, M. Ziemniak, E. Darzynkiewicz and J. Jemielity, *Bioorg. Med. Chem. Lett.*, 1921, **2009**, 19.
[56] C. J. Brown, I. McNae, P. M. Fischer and M. D. Walkinshaw, *J. Mol. Biol.*, 2007, **372**, 7.
[57] H. Matsuo, H. Li, A. M. McGuire, C. M. Fletcher, A. C. Gingras, N. Sonenberg and G. Wagner, *Nat. Struct. Biol.*, 1997, **4**, 717.
[58] T. P. Herbert, R. Fahraeus, A. Prescott, D. P. Lane and C. G. Proud, *Curr. Biol.*, 2000, **10**, 793.
[59] C. J. Brown, J. J. Lim, T. Leonard, H. C. Lim, C. S. Chia, C. S. Verma and D. P. Lane, *J. Mol. Biol.*, 2011, **405**, 736.
[60] S. Y. Ko, H. Guo, N. Barengo and H. Naora, *Clin. Cancer Res.*, 2009, **15**, 4336.
[61] R. Mishra, M. Miyamoto, T. Yoshioka, K. Ishikawa, Y. Matsumura, Y. Shoji, K. Ichinokawa, T. Itoh, T. Shichinohe, S. Hirano and S. Kondo, *Int. J. Oncol.*, 2009, **34**, 1231.
[62] B. Konicek, J. Stephens, A. McNulty, N. Robichaud, R. Peery, C. Dumstorf, M. Dowless, P. Iversen, S. Parsons, K. Ellis, D. J. McCann, J. Pelletier, L. Furic, J. M. Yingling, L. F. Stancato, N. Sonenberg and J. R. Graff, *Cancer Res.*, 1849, **2011**, 71.
[63] K. A. Hood, L. M. West, P. T. Northcote, M. V. Berridge and J. H. Miller, *Apoptosis*, 2001, **6**, 207.
[64] P. T. Northcote, J. W. Blunt and M. H. G. Munro, *Tetrahedron Lett.*, 1991, **32**, 6411.
[65] M. E. Bordeleau, J. Matthews, J. M. Wojnar, L. Lindqvist, O. Novac, E. Jankowsky, N. Sonenberg, P. Northcote, P. Teesdale-Spittle and J. Pelletier, *Proc. Natl. Acad. Sci. USA*, 2005, **102**, 10460.
[66] M. E. Bordeleau, R. Cencic, L. Lindqvist, M. Oberer, P. Northcote, G. Wagner and J. Pelletier, *Chem. Biol.*, 2006, **13**, 1287.
[67] W. K. Low, Y. Dang, T. Schneider-Poetsch, Z. Shi, N. S. Choi, W. C. Merrick, D. Romo and J. O. Liu, *Mol. Cell*, 2005, **20**, 709.
[68] W. K. Low, Y. Dang, S. Bhat, D. Romo and J. O. Liu, *Chem. Biol.*, 2007, **14**, 715–727.
[69] R. Mazroui, R. Sukarieh, M. E. Bordeleau, R. J. Kaufman, P. Northcote, J. Tanaka, I. Gallouzi and J. Pelletier, *Mol. Biol. Cell*, 2006, **17**, 4212.
[70] Y. Dang, N. Kedersha, W. K. Low, D. Romo, M. Gorospe, R. Kaufman, P. Anderson and J.O. Liu, *J. Biol. Chem.*, 2006, **281**, 32870.
[71] Y. Dang, W. K. Low, J. Xu, N. H. Gehring, H. C. Dietz, D. Romo and J. O. Liu, *J. Biol. Chem.*, 2009, **284**, 23613.
[72] G. Kuznetsov, Q. Xu, L. Rudolph-Owen, K. Tendyke, J. Liu, M. Towle, N. Zhao, J. Marsh, S. Agoulnik, N. Twine, L. Parent, Z. Chen, J. L. Shie, Y. Jiang, H. Zhang, H. Du, R. Boivin, Y. Wang, D. Romo and B. A. Littlefield, *Mol. Cancer Ther.*, 2009, **8**, 1250.
[73] T. Ohse, S. Ohba, T. Yamamoto, T. Koyano and K. Umezawa, *J. Nat. Prod.*, 1996, **59**, 650.

[74] S. K. Lee, B. Cui, R. R. Mehta, A. D. Kinghorn and J. M. Pezzuto, *Chem. Biol. Interact.*, 1998, **115**, 215.
[75] S. Kim, A. A. Salim, S. M. Swanson and A. D. Kinghorn, *Anticancer Agents Med. Chem.*, 2006, **6**, 319.
[76] B. Y. Hwang, B. N. Su, H. Chai, Q. Mi, L. B. Kardono, J. J. Afriastini, S. Riswan, B. D. Santarsiero, A. D. Mesecar, R. Wild, C. R. Fairchild, G. D. Vite, W. C. Rose, N. R. Farnsworth, G. A. Cordell, J. M. Pezzuto, S. M. Swanson and A. D. Kinghorn, *J. Org. Chem.*, 2004, **69**, 3350.
[77] Q. Mi, B. N. Su, H. Chai, G. A. Cordell, N. R. Farnsworth, A. D. Kinghorn and S. M. Swanson, *Anticancer Res.*, 2006, **26**, 947.
[78] B. Hausott, H. Greger and B. Marian, *Int. J. Cancer*, 2004, **109**, 933.
[79] M. E. Bordeleau, F. Robert, B. Gerard, L. Lindqvist, S. M. Chen, H. G. Wendel, B. Brem, H. Greger, S. W. Lowe, J. A. Porco Jr. and J. Pelletier, *J. Clin. Invest.*, 2008, **118**, 2651.
[80] S. Kim, B. Y. Hwang, B. N. Su, H. Chai, Q. Mi, A. D. Kinghorn, R. Wild and S. M. Swanson, *Anticancer Res.*, 2007, **27**, 2175.
[81] R. Cencic, M. Carrier, G. Galicia-Vazquez, M. E. Bordeleau, R. Sukarieh, A. Bourdeau, B. Brem, J. G. Teodoro, H. Greger, M. L. Tremblay, J. A. Porco Jr. and J. Pelletier, *PLoS One*, 2009, **4**, e5223.
[82] J.-H. Sheu, C.-H. Chao, G.-H. Wang, K.-C. Hung, C.-Y. Duh, M. Y. Chiang, Y.-C. Wud and C.-C. Wud, *Tetrahedron Lett.*, 2004, **45**, 6413.
[83] O. Novac, A. S. Guenier and J. Pelletier, *Nucleic Acids Res.*, 2004, **32**, 902.
[84] T. Tsumuraya, C. Ishikawa, Y. Machijima, S. Nakachi, M. Senba, J. Tanaka and N. Mori, *Biochem. Pharmacol.*, 2011, **81**, 713.
[85] W. Li, Y. Dang, J. O. Liu and B. Yu, *Bioorg. Med. Chem. Lett.*, 2010, **20**, 3112.
[86] W. Li, Y. Dang, J. O. Liu and B. Yu, *Chemistry*, 2009, **15**, 10356.
[87] K. Ravindar, M. S. Reddy, L. Lindqvist, J. Pelletier and P. Deslongchamps, *J. Org. Chem.*, 2011, **76**, 1269.
[88] M. Boyce, K. F. Bryant, C. Jousse, K. Long, H. P. Harding, D. Scheuner, R. J. Kaufman, D. Ma, D. M. Coen, D. Ron and J. Yuan, *Science*, 2005, **307**, 935.
[89] D. Kessel, *Biochem. Biophys. Res. Commun.*, 2006, **346**, 1320.
[90] H. C. Drexler, *PLoS One*, 2009, **4**, e4161.
[91] D. M. Schewe and J. A. Aguirre-Ghiso, *Cancer Res.*, 2009, **69**, 1545.
[92] H. Holcombe, I. Mellman, C. A. Janeway Jr., K. Bottomly and B. N. Dittel, *J. Immunol.*, 2002, **169**, 4982.
[93] W. E. Muller, N. Weissmann, A. Maidhof, M. Bachmann and H. C. Schroder, *J. Antibiot.*, 1987, **40**, 1028.
[94] T. N. Ramya, N. Surolia and A. Surolia, *Biochem. J.*, 2007, **401**, 411.
[95] K. Nishimura, Y. Ohki, T. Fukuchi-Shimogori, K. Sakata, K. Saiga, T. Beppu, A. Shirahata, K. Kashiwagi and K. Igarashi, *Biochem. J.*, 2002, **363**, 761.
[96] M. H. Park, K. Nishimura, C. F. Zanelli and S. R. Valentini, *Amino Acids*, 2010, **38**, 491.
[97] R. Retnakaran and B. Zinman, *Lancet*, 2009, **373**, 2088.
[98] A. Krishnaswami, S. Ravi-Kumar and J. M. Lewis, *Perm. J.*, 2010, **14**, 64.
[99] S. S. Palakurthi, H. Aktas, L. M. Grubissich, R. M. Mortensen and J. A. Halperin, *Cancer Res.*, 2001, **61**, 6213.
[100] Y. Tsubouchi, H. Sano, Y. Kawahito, S. Mukai, R. Yamada, M. Kohno, K. Inoue, T. Hla and M. Kondo, *Biochem. Biophys. Res. Commun.*, 2000, **270**, 400.
[101] N. Suh, Y. Wang, C. R. Williams, R. Risingsong, T. Gilmer, T. M. Willson and M. B. Sporn, *Cancer Res.*, 1999, **59**, 5671.
[102] M. H. Jarrar and A. Baranova, *J. Cell. Mol. Med.*, 2007, **11**, 71.
[103] F. Turturro, E. Friday, R. Fowler, D. Surie and T. Welbourne, *Clin. Cancer Res*, 2004, **10**, 7022.

[104] C. W. Shiau, C. C. Yang, S. K. Kulp, K. F. Chen, C. S. Chen, J. W. Huang and C. S. Chen, *Cancer Res.*, 2005, **65**, 1561.

[105] I. Inoue, S. Katayama, K. Takahashi, K. Negishi, T. Miyazaki, M. Sonoda and T. Komoda, *Biochem. Biophys. Res. Commun.*, 1997, **235**, 113.

[106] S. Han and J. Roman, *Mol. Cancer Ther.*, 2006, **5**, 430.

[107] M. A. Burgess and G. P. Bodey, *Antimicrob. Agents Chemother.*, 1972, **2**, 423.

[108] L. R. Benzaquen, C. Brugnara, H. R. Byers, S. Gatton-Celli and J. A. Halperin, *Nat. Med.*, 1995, **1**, 534.

[109] H. Aktas, R. Fluckiger, J. A. Acosta, J. M. Savage, S. S. Palakurthi and J. A. Halperin, *Proc. Natl. Acad. Sci. USA*, 1998, **95**, 8280.

[110] H. Takahashi, M. Abe, T. Sugawara, K. Tanaka, Y. Saito, S. Fujimura, M. Shibuya and Y. Sato, *Jpn. J. Cancer Res.*, 1998, **89**, 445.

[111] H. Liu, Y. Li and K. P. Raisch, *Anticancer Drugs*, 2010, **21**, 841.

[112] J. Penso and R. Beitner, *Eur. J. Pharmacol.*, 2002, **451**, 227.

[113] A. Natarajan, Y. H. Fan, H. Chen, Y. Guo, J. Iyasere, F. Harbinski, W. J. Christ, H. Aktas and J. A. Halperin, *J. Med. Chem.*, 1882, **2004**, 47.

[114] A. Natarajan, Y. Guo, F. Harbinski, Y. H. Fan, H. Chen, L. Luus, J. Diercks, H. Aktas, M. Chorev and J. A. Halperin, *J. Med. Chem.*, 2004, **47**, 4979–4982.

[115] M. K. Uddin, S. G. Reignier, T. Coulter, C. Montalbetti, C. Granas, S. Butcher, C. Krog-Jensenb and J. Felding, *Bioorg. Med. Chem. Lett.*, 2007, **17**, 2854.

[116] J. Felding, H. C. Pedersen, C. Krog-Jensenb, M. Praestegaard, P. Butcher, V. Linde, T.S. Coulter, C. Montalbetti, M. K. Uddin and S. G. Reignier, *Patent Application WO 2005097107 A2 20051020 CAN 143:405798 AN 2005*, 2007.

[117] M. K. Christensen, K. D. Erichsen, C. Trojel-Hansen, J. Tjornelund, S. J. Nielsen, K. Frydenvang, T. N. Johansen, B. Nielsen, M. Sehested, P. B. Jensen, M. Ikaunieks, A. Zaichenko, E. Loza, I. Kalvinsh and F. Bjorkling, *J. Med. Chem.*, 2010, **53**, 7140.

[118] T. Chen, D. Ozel, Y. Qiao, F. Harbinski, L. Chen, S. Denoyelle, X. He, N. Zvereva, J. G. Supko, M. Chorev, J. A. Halperin and B. H. Aktas, *Nat. Chem. Biol.*, 2011, in press.

CHAPTER 13

The Discovery and Development of Smac Mimetics—Small-Molecule Antagonists of the Inhibitor of Apoptosis Proteins

Stephen M. Condon

Contents

TetraLogic Pharmaceuticals, 343 Phoenixville Pike, Malvern, PA 19355, USA

Annual Reports in Medicinal Chemistry, Volume 46
ISSN: 0065-7743, DOI: 10.1016/B978-0-12-386009-5.00014-X

1. INTRODUCTION

Apoptosis, or programmed cell death, is a critical cellular function for maintenance of normal tissue development and homeostasis as well as for the routine surveillance of cellular dysfunction by the immune system. Apoptosis can be activated by a number of internal and external signals including irradiation, chemotherapy, oxidative stress, and death receptor ligand binding. The key mediators of apoptosis are the caspases, and not surprisingly, the proteolytic activation of caspases as well as the maintenance of activated caspase function is a highly regulated process; the former being partly mediated by the apoptosome, while the latter is controlled by the inhibitor of apoptosis proteins or IAPs [1]. Although originally discovered in baculovirus-infected insect cells, several mammalian IAPs have now been identified. These include the X-linked IAP (XIAP), the cellular IAPs (cIAP-1 and -2), melanoma-linked IAP (ML-IAP), and neuronal IAP (NIAP), among others.

Apoptosis is initiated by either an intracellular (intrinsic) or extracellular (extrinsic) signaling event. Following an apoptotic stimulus, the intrinsic pathway is propagated through cytochrome *c* release from the mitochondria which, together with Apaf-1 and dATP, forms the apoptosome complex (Figure 1). The apoptosome is responsible for converting procaspase-9 to its active form *via* autocatalytic processing. Once activated, caspase-9 serves to cleave procaspase-3 to the effector caspase-3, which then goes on to perform the heavy lifting of apoptosis. The discovery that XIAP could bind to and inhibit activated caspases and that XIAP was present at elevated levels in many cancers led to the hypothesis that tumor resistance to chemotherapy may be the direct result of caspase inhibition by XIAP [3].

A major advance in this field was the discovery of the mitochondrial protein Smac/DIABLO, which is released from the mitochondria upon apoptosis-inducing signals and was shown to neutralize the effects of XIAP on both caspase-9 and caspase-3 inhibition [4]. Thus, the paradigm of apoptosis regulation through caspase inhibition by XIAP—to block apoptosis—and caspase derepression by Smac/DIABLO—to restore apoptosis—was first demonstrated.

2. ORGANIZATION OF INHIBITOR OF APOPTOSIS PROTEINS

IAPs are multidomain and multifunctional proteins (Figure 2) [5]. ML-IAP comprises a single N-terminal baculovirus IAP repeat (BIR) domain and a C-terminal RING domain. XIAP comprises an N-terminal BIR domain (BIR1), two additional BIR domains (BIR2 and BIR3), a ubiquitin (Ub)-associated domain (UBA), and a C-terminal RING domain, which

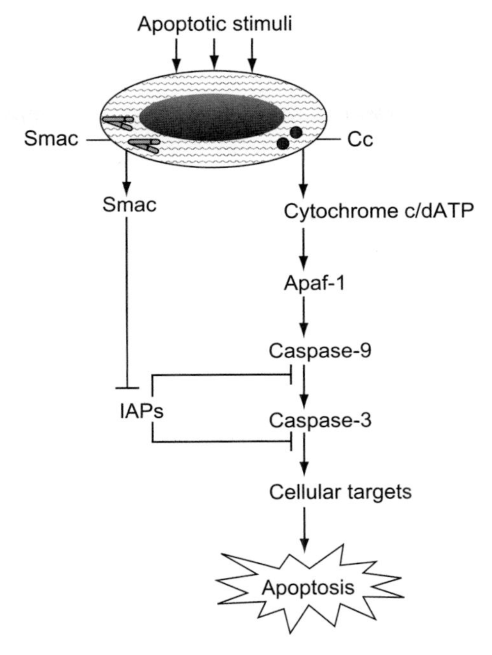

Figure 1 The intrinsic apoptotic cascade (adapted from Ref. [2] with permission from publisher). (See Color Plate 13.1 in Color Plate Section.)

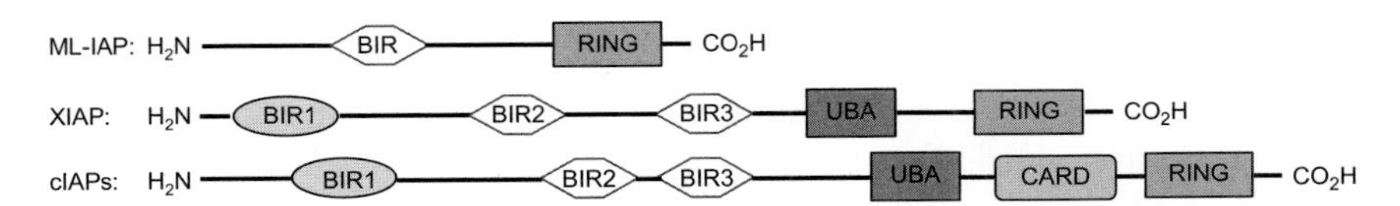

Figure 2 Domain architecture of ML-IAP, XIAP, and the cIAPs. (See Color Plate 13.2 in Color Plate Section.)

functions as a ubiquitin E3 ligase [6]. Caspase-9 inhibition is mediated through the BIR3 domain, while caspase-3 binds to and is inhibited by the linker region proximal to the BIR2 domain. The BIR1 domain accommodates interactions with various adaptor proteins [7,8]. The UBA domain of IAPs is essential for mediating binding to Ub-conjugated proteins and thus linking these binding interactions to downstream biochemical processes [9].

The cIAPs have a similar domain structure but include a caspase-recruitment domain (CARD). Tremendous progress has been made in understanding the mechanistic contribution of each of these domains as well as how they work cooperatively to mediate IAP functions [10]. Several regions of specific IAPs have been crystallized and the structures reported [11].

3. SMAC, XIAP, AND CASPASE-9: STRUCTURE AND MECHANISM

The mature Smac monomer exists as a three-helix bundle, and two monomer units are packed into a homodimeric tertiary structure (Figure 3) [2]. Disruption of the dimer interface through discrete mutation results in a weakened ability of Smac to promote the enzymatic activity of caspase-3. Wild-type Smac interacts independently with the isolated BIR2 and BIR3 domains of XIAP but not with the BIR1 domain despite strong sequence homology. Importantly, deletion of the four N-terminal residues from Smac, Ala-Val-Pro-Ile (AVPI), results in the loss of its ability to promote caspase-3 activity. Conversely, short N-terminal peptides derived from the Smac sequence retain this function. When co-crystallized with XIAP BIR3, the four N-terminal residues of Smac pack into a surface groove of the BIR3 domain *via* a combination of hydrogen bonding and van der Waals interactions [12].

The four N-terminal residues of caspase-9, Ala-Thr-Pro-Phe (ATPF), occupy the same surface groove on XIAP BIR3 as the AVPI sequence of Smac. This caspase-9/BIR3 interaction precludes entry of substrate peptides into the active site of caspase-9. Structurally, therefore, the XIAP/caspase-9/Smac paradigm can be summarized as: (1) activated caspase-9 is inhibited through binding of its N-terminus (ATPF) to XIAP BIR3; (2) Smac competes with caspase-9 for XIAP BIR3 binding thereby releasing active caspase-9 and reconstituting the apoptotic program. The structural details of caspase-3 derepression by Smac are considered to be the result of either homodimeric Smac binding cooperatively to the BIR2 and BIR3 domains of XIAP or *via* steric inhibition of the caspase-3/XIAP interaction by the BIR3-anchored Smac protein. Thus, the protein–ligand interactions defined by the co-crystal structure of Smac/XIAP BIR3 would serve as the starting point for the development of small-molecule antagonists of the Smac/XIAP interaction termed IAP antagonists or Smac mimetics [13].

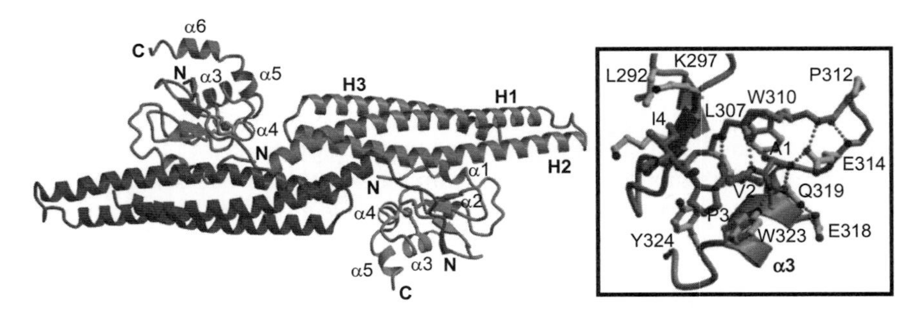

Figure 3 The structures of the Smac homodimer and AVPI bound to the XIAP BIR3 domain (adapted from Ref. [12] with permission from publisher). (See Color Plate 13.3 in Color Plate Section.)

4. STRUCTURE–ACTIVITY RELATIONSHIPS OF SMAC MIMETICS

In general, the N-terminal residue (P_1') is Ala or N(Me)Ala [14]. A basic residue is required to maintain a salt bridge interaction with an acidic residue within the IAP binding pocket. Although both of these residues afford analogs with good binding to BIR3 domains, only N(Me)Ala substitution allows for robust cellular responses. NMR-derived and co-crystal structures of Smac mimetics bound to XIAP BIR3 indicate that the side chain of the P_2' residue is projected away from the protein; therefore, many side chains are tolerated at this position. Large hydrophobic groups like *tert*-leucine or cyclohexylglycine afford analogs with excellent binding, as do other β-branched amino acids. The P_3' residue is typically proline or substituted pyrrolidine, which allows for an important van der Waals interaction between the ligand and Trp323 of the BIR3 binding site. Specific substitutions on the pyrrolidine ring are tolerated, and many constrained compounds that link either the P_2' and P_3' groups or the P_3' and P_4' groups have been prepared. A large, hydrophobic substituent is preferred at the P_4' position. The P_3'–P_4' amide bond and C-terminal acid or amide moieties are dispensable for biological activity.

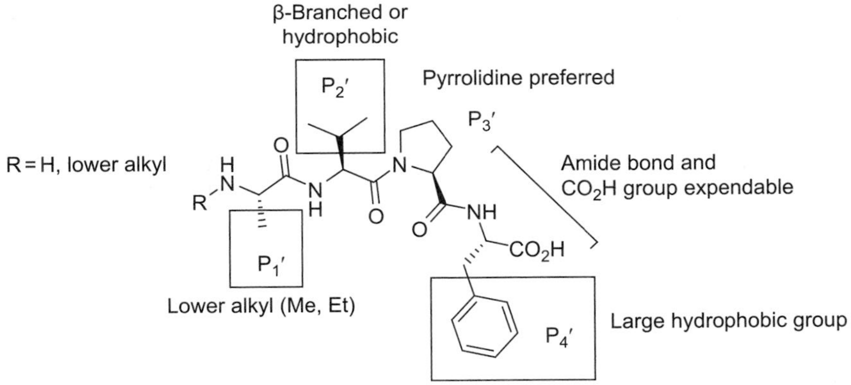

β-Branched or hydrophobic
P_2'
Pyrrolidine preferred
P_3'
R = H, lower alkyl
Amide bond and CO_2H group expendable
P_1'
Lower alkyl (Me, Et)
P_4'
Large hydrophobic group

5. BIVALENCY AND SMAC MIMETIC FUNCTION

5.1. The role of cIAP-1 and death receptors

Since Smac is recognized to act as a dimer, it was expected the bivalent Smac mimetics may have improved properties relative to their monovalent counterparts [2]. This was demonstrated to be the case when an attempt to chemically manipulate the alkyne moiety of **1** afforded the homodimerization product **2** [15]. Although **2** was shown to bind XIAP

BIR3 with comparable affinity to **1**, unexpectedly **2** caused extensive cell death when dosed in combination with the TNFα-related apoptosis-inducing ligand (TRAIL); whereas **1** was much less active under these conditions.

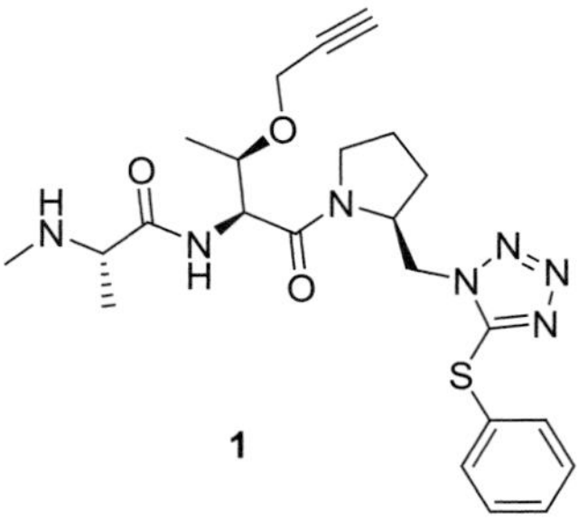

More importantly, however, was the revelation that the cIAP-1 and cIAP-2 were also critical intracellular targets of **2**. Both cIAP-1 and cIAP-2 are present in the TNFα receptor-1 (TNFR1) signaling complex, but it is unclear whether their specific function is inhibition of caspase-8, promotion of NF-κB and JNK signaling, or both. Treatment of HeLa cells with **2** in combination with TNFα resulted in significant apoptosis. Caspase-8 activation was observed within 2 h of **2** treatment and PARP cleavage, a measure of caspase-3 activity, within 4 h. These observations suggested that cIAP-1/2 function to block caspase-8 activation at the TNF receptor level and that antagonism of cIAP-1/2 by a Smac mimetic reverses caspase-8 inhibition leading to apoptosis.

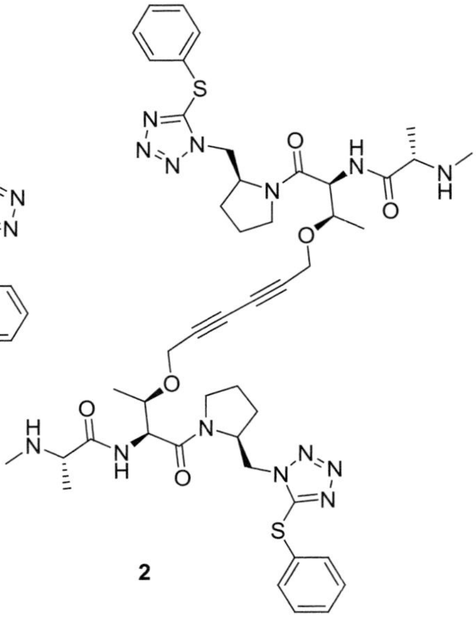

The next piece of the puzzle was provided by four independent laboratories using several chemically unique Smac mimetics [16–19]. In addition to compounds **1** and **2**, bivalent analogs **3** (compound A) and **4** (BV6) as well as monovalent ligands **5** (compound C), **6** (MV1), and **7** (LBW-242) were employed in these studies.

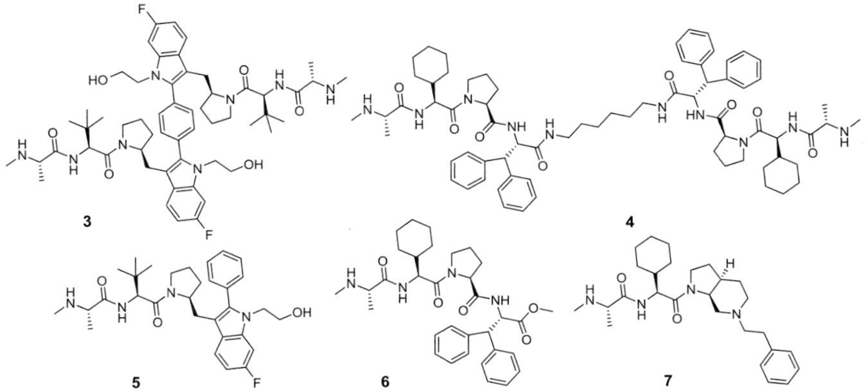

Key observations provided by these studies were as follows: (1) cIAP-1/2 acts to protect cells from TNFα-induced apoptosis; (2) Smac mimetics elicit auto-ubiquitylation and subsequent proteosomal degradation of cIAP-1/2. Smac mimetic-sensitive cell lines undergo apoptosis owing to the presence of an autocrine TNFα loop. In these cell lines, Smac mimetic-induced apoptosis could be blocked upstream by the addition of neutralizing TNFα antibody or downstream by proteasome inhibitors.

cIAP-1 acts primarily as a ubiquitin E3 ligase necessary for maintaining protein levels of the key components of death receptor signaling pathways including the receptor interacting kinase (RIP1) [20], the NF-κ B-inducing kinase (NIK) [21], cIAP-2 [22], and the TNFα receptor-associated factor (TRAF2) [23]. Smac mimetic-induced loss of cIAP-1 results in the dysregulation of death receptor signaling such as activation and/or inhibition of the canonical and noncanonical NF-κB pathways [24]. Thus, a comprehensive understanding of Smac mimetic-induced tumor cell killing revealed a much more complicated mechanism of action (Figure 4).

5.2. The XIAP BIR2–BIR3 binding hypothesis

[8,5]-Bicyclic Smac mimetic **8** (SM-122) binds to both XIAP BIR3 (K_i = 26 nM) and XIAP BIR2 (IC_{50} = 5.6 μM) [25]. Using computational models of **8** independently ligated to the BIR2 and BIR3 domains of XIAP, a suitable position for attachment of a linker group was identified. Bivalent Smac mimetic **9** exhibited excellent binding to an XIAP BIR2-BIR3 protein construct (IC_{50} = 1.39 nM), and a series of gel filtration and NMR experiments provided support for simultaneous occupation of the BIR2 and BIR3 domains. Importantly, these biophysical studies were conducted using IAPs devoid of either the N-terminal BIR1 domain which has

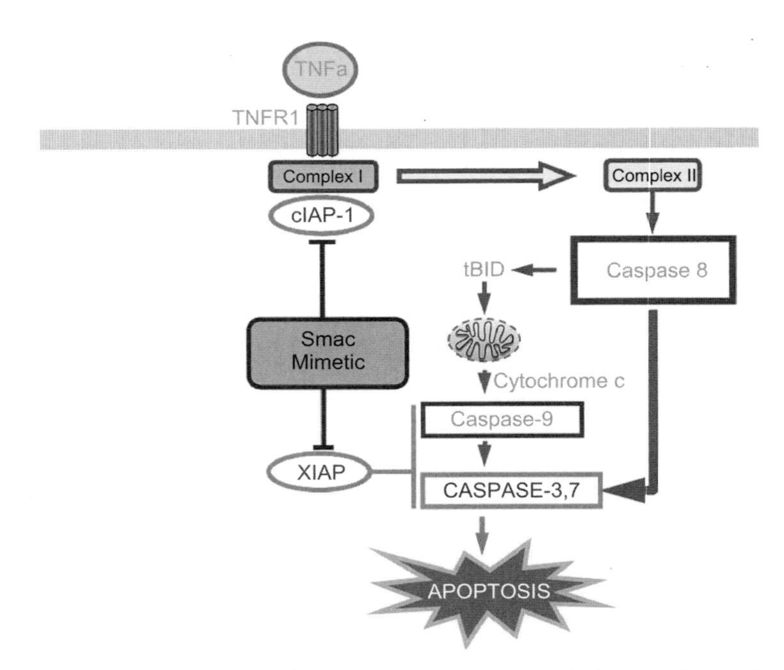

Figure 4 Schematic summary of Smac mimetic mechanism(s) of action. (See Color Plate 13.4 in Color Plate Section.)

been implicated in XIAP/TAB1 [7] and cIAP/TRAF [8] interactions or the C-terminal RING domain which has a key role in protein–protein dimerization and E3 ligase activity [26].

5.3. The cIAP BIR3–BIR3 binding hypothesis

Bivalent Smac mimetic **4** (BV6) was also reported to simultaneously occupy the BIR2 and BIR3 domains of an XIAP BIR2–BIR3 construct [16]. In contrast, however, analytical ultracentrifugation suggested that **4** interacted with cIAP-1 BIR2–BIR3 *via* cross-linking of two BIR3 domains. Recent crystallographic studies using a cIAP BIR3–UBA–CARD–RING construct support the concept that the E3 ligase activity of cIAP-1 is suppressed by BIR3/RING interactions [27]. This assumes that Smac mimetic binding to BIR3 unmasks the RING domain for participation in RING–RING dimerization in advance of ubiquitin transfer. One proposal suggests that the kinetics of Ub transfer is enhanced by a bivalent Smac mimetic-mediated cIAP BIR3–BIR3 cross-link which serves to stabilize the RING–RING dimer [28].

6. RECENTLY DESCRIBED SMAC MIMETICS

6.1. cIAP-selective Smac mimetics

Penicillamine-derived analog **10** is selective for the ML-IAP BIR domain versus XIAP BIR3 ($K_i = 50$ and 770 nM, respectively) [29]. The bias toward ML-IAP BIR binding was attributed to the replacement of Tyr324 in XIAP with Phe148 in ML-IAP—the *para*-hydroxy group of Tyr324 sterically interferes with the pro-R methyl group on **10** thus disfavoring XIAP binding. Since cIAP-1 contains the same Tyr-to-Phe replacement, **10** was shown to maintain good binding to cIAP-1 BIR3 ($K_i = 50$ nM). Antagonist **10** inhibits Smac/ML-IAP binding in a dose-dependent manner, causes single agent cell killing in MDA-MB-231 cells, induces apoptosis in A2058 melanoma cells as measured by caspase-3/7 activation, and demonstrates additivity when administered in combination with doxorubicin.

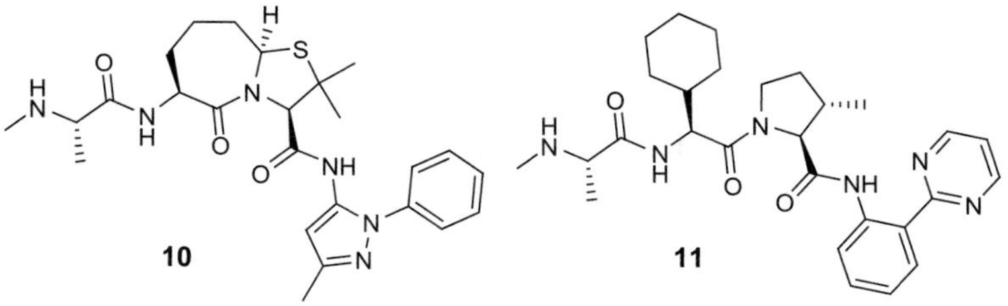

The highly cIAP-1-selective Smac mimetic **11** exploited the capacity of the cIAP-1 BIR3 to tolerate *S*-methyl substitution at the P_3' position, while the 2-pyrimidinyl group made use of both steric and electronic interactions within the P_4' binding site [30]. Together, these SAR elements enhance binding of **11** to cIAP-1 while dramatically reducing its affinity for XIAP BIR3 ($K_i = 0.016$ *vs.* >34 μM, respectively).

6.2. Macrocyclic Smac mimetics

A series of bivalent Smac mimetics culminated in the discovery of macrocyclic Smac mimetic **12** linked *via* the P_2' and P_4' residues [31,32]. Macrocycle **12** binds with subnanomolar affinity to both the XIAP and cIAP-1

BIR3 domains and effects tumor cell killing in the SK-OV-3 ovarian cancer cell line ($CC_{50} = 6$ nM) [33].

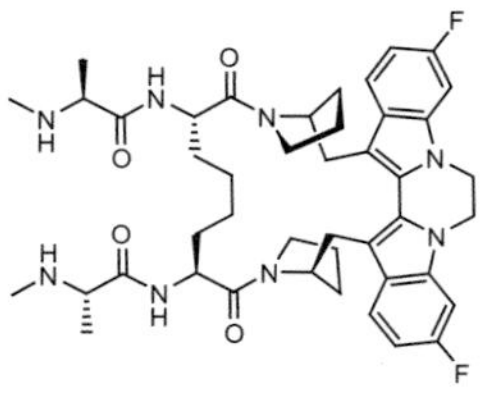

Cyclic peptide **13** binds to BIR2–BIR3-containing XIAP constructs with a biphasic dose–response curve representing two binding sites (IC_{50}s $= 0.5$ and 406 nM) [34]. Gel filtration experiments indicate that **13** forms a 1:1 complex with BIR2–BIR3 XIAP. Co-crystallography of **13** with the XIAP BIR3 domain afforded a 2:1 BIR3:**13** structure, which was utilized to build a model of the BIR2–BIR3/**13** complex.

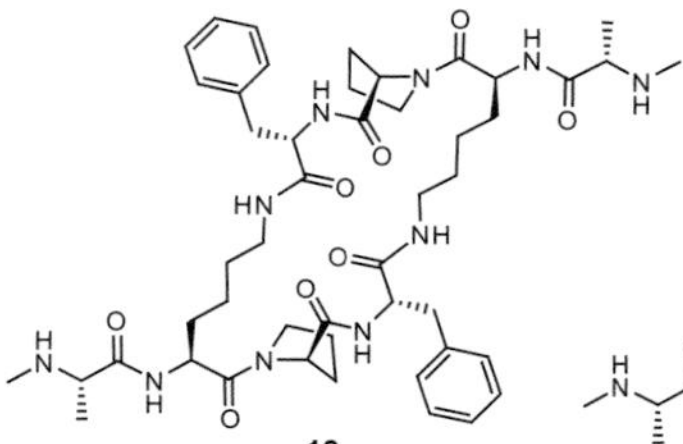

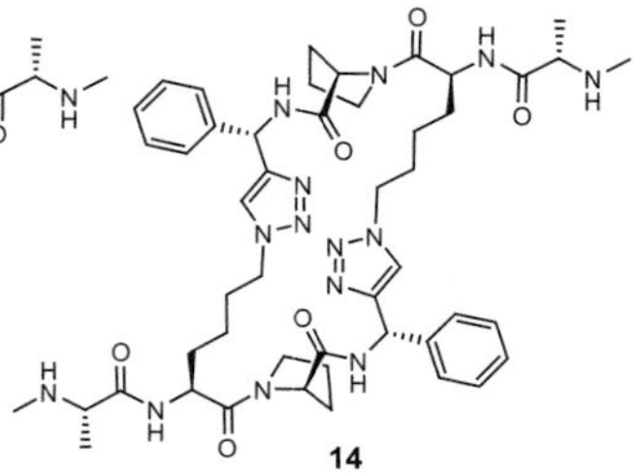

In an effort to reduce the peptide character of **13**, Smac mimetic **14** was prepared [35]. Interestingly, analysis of the binding curve associated with **14** and BIR2–BIR3 XIAP afforded IC_{50} values of 0.43 and 23 nM for the two binding sites. When tested against the BIR2-only XIAP, the IC_{50} value rose sharply to 3.2 μM (compared to 53.9 μM for **13**) suggesting that other protein–ligand interactions may be involved in binding [36].

6.3. Orally bioavailable Smac mimetics

Smac mimetics **15** and **16**, designed using the software program CAVEAT, were shown to have oral bioavailability in mice when dosed as an aqueous hydroxypropyl-β-cyclodextrin/succinic acid solution [37]. Both **15** and **16** were able to bind to the ML-IAP BIR domain with good affinity (K_is $= 82$ and 46 nM, respectively), and the protein–ligand structures were solved by X-ray co-crystallography. Both analogs were advanced into murine xenograft studies, and **16** potently inhibited tumor growth without affecting body weight.

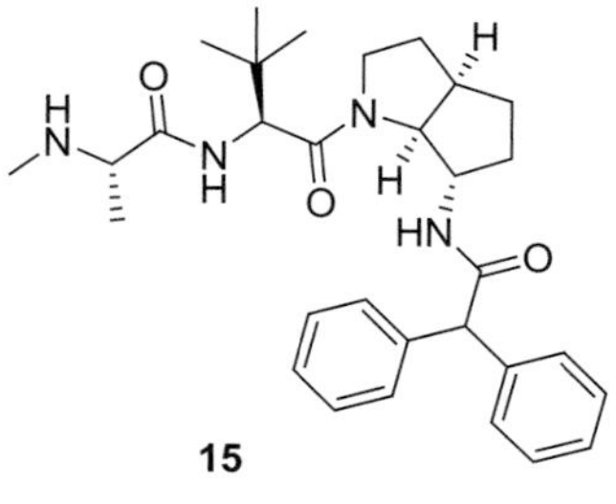

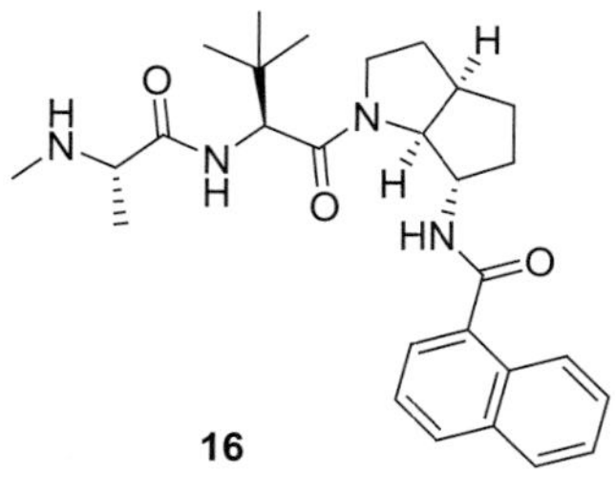

Diazabicycle **17** (SM-337) has nanomolar binding affinity to XIAP, cIAP-1, and cIAP-2 BIR3 domains (K_is = 8.4, 1.5, and 4.1 nM, respectively) and is capable of relieving XIAP-mediated inhibition of caspase-3/7 in a cell-free system [38]; is eight times more potent than **8** at inhibiting MDA-MB-231 cell growth (IC_{50} = 31 *vs.* 259 nM, respectively); and induces proteosomal degradation of cIAP-1 at concentrations as low as 10 nM. At 30 mg/kg, oral dosing of **17** hydrochloride in the rat afforded an AUC of 1985 ± 614 μg/L and a C_{max} which is 17-fold higher than its IC_{50}. Another diazabicycle-based Smac mimetic **18** (AT-406) was selected for clinical evaluation [39]. Smac mimetic **19** (LCL-161) was profiled against a number of ovarian, melanoma, and lung cancer cell lines as a single agent and in combination with taxol® [40] and is administered orally as tablets or in solution in an ongoing clinical trial [41].

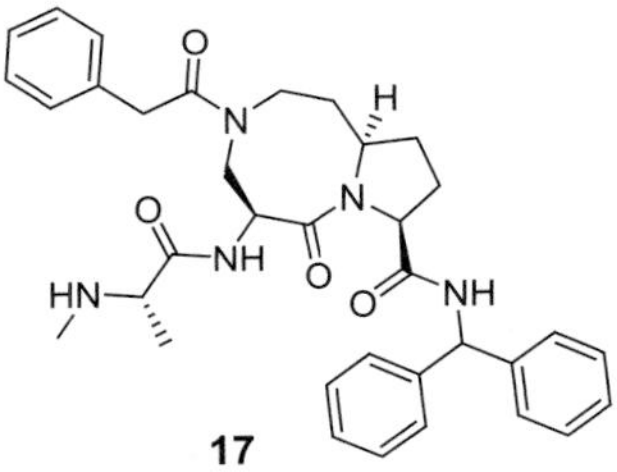

17 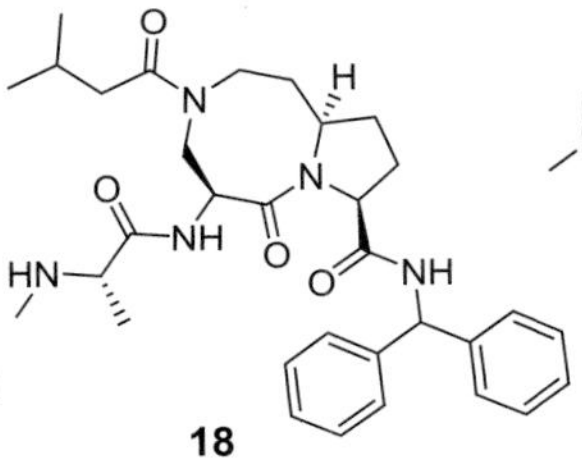 **18** 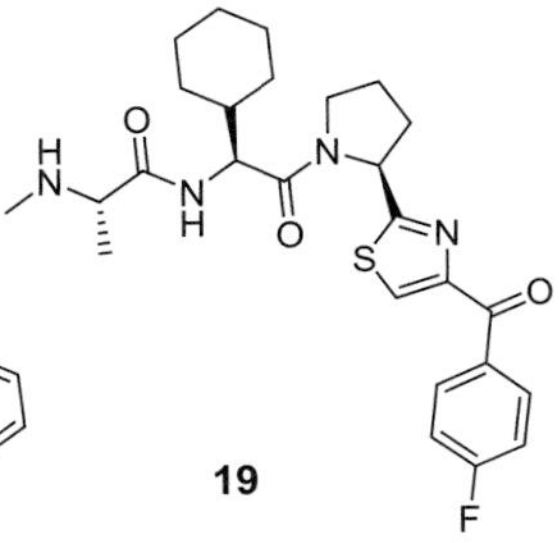 **19**

7. PRECLINICAL AND CLINICAL EVALUATION

To date, six Smac mimetics have advanced to clinical trials: GDC-0152/RG7419, GDC-0917/RG7459, HGS1029/AEG40826·2HCl, LCL-161, TL32711, and AT-406. These compounds represent a panel of monovalent (GDC-0152, GDC-0917, LCL-161, AT-406) and bivalent (HGS1029, TL32711) Smac mimetics, and their routes of administration are either *via* intravenous infusion or *via* oral dosing.

GDC-0152/RG7419 (structure not disclosed), administered by intravenous infusion, was advanced into an open-label Phase 1 dose escalation study for patients with locally advanced or metastatic malignancies. Monocyte chemotactic protein-1 (MCP-1) is expressed during an inflammatory response and has been associated with Smac mimetic treatment in some preclinical species [42]. Although the PK parameters were dose-proportional with moderate variability, no consistent dose-dependent changes in MCP-1 levels were observed in patients following GDC-0152 treatment at the doses studied [43]. This investigation was completed in 2010.

In late 2010, an orally bioavailable Smac mimetic, GDC-0917/RG7459 (structure not disclosed) was introduced into Phase 1 clinical trials for refractory solid tumors or lymphoma.

HGS1029/AEG40826·2HCl (structure not disclosed) is administered as a 15-min infusion provided on days 1, 8, and 15 of a 28-day cycle in patients with relapsed or refractory solid tumors [44]. Of the 27 patients enrolled to date, 1 patient experienced dose-limiting toxicity after the first dose but recovered quickly and no further DLTs were observed. HGS1029 produced a dose-dependent but short-lived lymphocytopenia and, possibly, transient neutrophilia and supraventricular tachycardia. The pharmacokinetics (PK) was linear over the dose range tested and did not vary between single and multiple doses. cIAP-1 loss was observed in peripheral blood mononuclear cells (PBMCs) and was dose dependent. HGS1029 is well tolerated and dose escalation is ongoing. A second Phase 1 trial evaluating HGS1029 in patients with relapsed or refractory lymphoid malignancies was initiated in 2009.

LCL-161 has demonstrated single agent activity in tumor models with high basal TNFα levels and has shown synergistic activity in combination with taxanes. When evaluated in patients with advanced treatment-refractory solid tumors, oral LCL161 exhibited dose-proportional PK up to 1800 mg po [41]. Doses $>$500 mg achieved AUC levels at or greater than those which illicit a tumor response in preclinical models as either a single agent or in combination with standard-of-care agents. Loss of cIAP-1 was observed by Western blot analysis of PBMC-derived cell lysates as well as from a Merkel cell carcinoma biopsy; elevated levels of cleaved cytokeratin-18 were also observed at doses $\geq$320 mg. The high-dose groups (500–1800 mg) displayed increases in both serum IL-8 and MCP-1 levels, suggesting activation of NF-κB pathways. No dose-limiting toxicities have thus far been observed and a combination trial with paclitaxel is planned for 2010.

TL32711 (structure not disclosed) is a bivalent Smac mimetic which selectively antagonizes multiple IAPs and has demonstrated antitumor activity in a number of preclinical models [45]. In animals, TL32711 was well tolerated up to 60 mg/kg, the highest dose tested, on a q3d $\times$ 5 schedule. In the MDA-MB-231-derived xenograft model, the minimally efficacious dose was 1.25 mg/kg (q3d $\times$ 5) and activity was maintained in both large and small tumors. TL32711 was rapidly and extensively distributed into normal and tumor tissue but was more slowly cleared from tumor tissue relative to normal tissue. Rapid reduction in cIAP-1 was observed in tumor tissue where cIAP-1 levels remained suppressed for up to 2 weeks following a single 5 mg/kg dose. A Phase 1a dose escalation safety study in patients with refractory solid tumors and lymphoma began enrolling in 2009 [46]. TL32711 is administered as a 30-min iv infusion once weekly for 3 weeks per repeated 4-week intervals. More recently, a Phase 1b/2a study combining TL32711 with several standard-of-care chemotherapies was initiated.

A similar dose escalation and safety study of AT-406 in patients with solid tumors and lymphomas commenced in 2010. AT-406 is to be administered orally to determine the maximum tolerated dose. Patients will receive **18** on days 1–5, and days 15–19 of a 28-day cycle.

8. CONCLUSION

Based on the unique mechanism of apoptosis induction, Smac mimetics hold promise for treating patients suffering from a wide variety of difficult to treat solid tumors and hematological malignancies [47,48]. As such, they will join a family of other apoptosis-inducing agents under clinical development including Bcl-2 antagonists, antisense XIAP, rhApo2L/TRAIL, and monoclonal antibody DR4/DR5 agonists. Much has been learned about the mechanism of Smac mimetics although it remains unclear as to the specific contributions of cIAP-1, cIAP-2, and XIAP antagonism toward tumor regression [49]. In addition, many details of IAP–ligand interactions remain unresolved such as whether bivalent IAP antagonists bind discretely to individual IAPs (BIR2–BIR3) or to homo- or hetero-IAP complexes *via* BIR3 cross-linking. Treatment with either monovalent or bivalent Smac mimetics results in the proteosomal degradation of cIAP-1, and that loss of cIAP-1 has direct consequences for TNFα and TRAIL signaling. The mechanism by which Smac mimetics synergize with specific chemotherapeutic agents is less well understood. Additionally, cIAP regulation has been implicated in other receptor signaling pathways such as TWEAK/Fn14 [50] and the NOD family of receptors [51,52] as well as direct interaction with the immune system [23,53,54]. The clinical development of both monovalent and bivalent Smac mimetics will provide an advanced look at the therapeutic benefit of this type of cancer treatment. Already, dose escalation studies have determined that many of these agents are well tolerated in patients with advanced disease while showing significant target suppression. The development of specific biomarkers for apoptosis-inducing agents like Smac mimetics will further aid in the development of these novel therapies.

REFERENCES

[1] S. M. Srinivasula and J. D. Ashwell, *Mol. Cell*, 2008, **30**, 123.
[2] J. Chai, C. Du, J.-W. Wu, S. Kyin, X. Wang and Y. Shi, *Nature*, 2000, **406**, 855.
[3] M. Holcik, H. Gibson and R. G. Korneluk, *Apoptosis*, 2001, **6**, 253.
[4] S. M. Srinivasula, R. Hegde, A. Saleh, P. Datta, E. Shiozaki, J. Chai, R.-A. Lee, P. D. Robbins, T. Fernandes-Alnemri, Y. Shi and E. S. Alnemri, *Nature*, 2001, **410**, 112.

[5] M. Gyrd-Hansen and P. Meier, *Nat. Rev. Cancer*, 2010, **10**, 561.
[6] Y. Yang, S. Fang, J. P. Jensen, A. M. Weissman and J. D. Ashwell, *Science*, 2000, **288**, 874.
[7] M. Lu, S.-C. Lin, Y. Huang, Y. J. Kang, R. Rich, Y.-C. Lo, D. Myszka, J. Han and H. Wu, *Mol. Cell*, 2007, **26**, 689.
[8] C. Zheng, V. Kabaleeswaran, Y. Wang, G. Cheng and H. Wu, *Mol. Cell*, 2010, **38**, 101.
[9] M. Gyrd-Hansen, M. Darding, M. Miasari, M. M. Santoro, L. Zender, W. Xue, T. Tenev, P. C. A. da Fonseca, M. Zvelebil, J. M. Bujnicki, S. Lowe, J. Silke and P. Meier, *Nat. Cell Biol.*, 2008, **10**, 1309.
[10] P. D. Mace, S. Shirley and C. L. Day, *Cell Death Differ.*, 2010, **17**, 46.
[11] M. D. Herman, M. Moche, S. Flodin, M. Welin, L. Trésaugues, I. Johansson, M. Nilsson, P. Nordlund and T. Nyman, *Acta Cryst.*, 2009, **F65**, 1091.
[12] G. Wu, J. Chai, T. L. Suber, J. W. Wu, C. Du, X. Wang and Y. Shi, *Nature*, 2000, **408**, 1008.
[13] R. A. Kipp, M. A. Case, A. D. Wist, C. M. Cresson, M. Carrell, E. Griner, A. Wiita, P. A. Albiniak, J. Chai, Y. Shi, M. F. Semmelhack and G. L. McLendon, *Biochemistry*, 2002, **41**, 7344.
[14] S. K. Sharma, C. Straub and L. Zawel, *Int. J. Pept. Res. Ther.*, 2006, **12**, 21.
[15] L. Li, R. M. Thomas, H. Suzuki, J. K. de Brabander, X. Wang and P. G. Harran, *Science*, 2004, **305**, 1471.
[16] E. Varfolomeev, J. W. Blankenship, S. M. Wayson, A. V. Fedorova, N. Kayagaki, P. Garg, K. Zobel, J. N. Dynek, L. O. Elliot, H. J. A. Wallweber, J. A. Flygare, W. J. Fairbrother, K. Deshayes, V. M. Dixit and D. Vucic, *Cell*, 2007, **131**, 669.
[17] J. E. Vince, W. W.-L. Wong, N. Khan, R. Feltham, D. Chau, A. U. Ahmed, C. A. Benetatos, S. K. Chunduru, S. M. Condon, M. McKinlay, R. Brink, M. Leverkus, V. Tergaonkar, P. Schneider, B. A. Callus, F. Koentgen, D. Vaux and J. Silke, *Cell*, 2007, **131**, 682.
[18] A. Gaither, D. Porter, Y. Yao, J. Borawski, G. Yang, J. Donovan, D. Sage, J. Slisz, M. Tran, C. Straub, T. Ramsey, V. Iourgenko, A. Huang, Y. Chen, R. Schlegel, M. Labow, S. Fawell, W. R. Sellers and L. Zawel, *Cancer Res.*, 2007, **67**, 11493.
[19] S. L. Petersen, L. Wang, A. Yalcin-Chin, L. Li, M. Peyton, J. Minna, P. Harran and X. Wang, *Cancer Cell*, 2007, **12**, 445.
[20] M. J. M. Betrand, S. Milutinovic, K. M. Dickson, W. C. Ho, A. Boudreault, J. Durkin, J. W. Gillard, J. B. Jaquith, S. J. Morris and P. A. Barker, *Mol. Cell*, 2008, **30**, 689.
[21] B. J. Zarnegar, Y. Wang, D. J. Mahoney, P. W. Dempsey, H. H. Cheung, J. He, T. Shiba, X. Yang, W.-C. Yeh, T. W. Mak, R. G. Korneluk and G. Cheng, *Nat. Immun.*, 2008, **9**, 1371.
[22] D. B. Conze, L. Albert, D. A. Ferrick, D. V. Goeddel, W.-C. Yeh, T. Mak and J. D. Ashwell, *Mol. Cell. Biol.*, 2005, **25**, 3348.
[23] A. Dupoux, J. Cartier, S. Cathelin, R. Filomenko, E. Solary and L. Dubrez-Daloz, *Blood*, 2009, **113**, 175.
[24] Y. Dai, T. S. Lawrence and L. Xu, *Am. J. Transl. Res.*, 2009, **1**, 1.
[25] H. Sun, Z. Nikolovska-Coleska, J. Lu, J. L. Meagher, C.-Y. Yang, S. Qiu, Y. Tomita, Y. Ueda, S. Jiang, K. Krajewski, P. P. Roller, J. A. Stuckey and S. Wang, *J. Am. Chem. Soc.*, 2007, **129**, 15279.
[26] P. D. Mace, K. Linke, R. Feltham, F.-R. Schumacher, C. A. Smith, D. L. Vaux, J. Silke and C. L. Day, *J. Biol. Chem.*, 2008, **283**, 31633.
[27] E. C. Dueber, A. Schoeffler, A. Lingel, K. Zobel, K. DeShayes, B. Maurer, S. Hymowitz and W. Fairbrother, ESH International Conference on Mechanisms of Cell Death and Disease: Advances in Therapeutic Intervention and Drug Development, Cascais, Portugal, 2010, Poster #15.
[28] R. Feltham, B. Bettjeman, R. Budhidarmo, P. D. Mace, S. Shirley, S. M. Condon, S. K. Chunduru, M. A. McKinlay, D. L. Vaux, J. Silke and Catherine L. Day, *J. Biol. Chem.*, 2011, **286,** 17015.

[29] K. Zobel, L. Wang, E. Varfolomeev, M. C. Franklin, L. O. Elliott, H. J. A. Wallweber, D. C. Okawa, J. A. Flygare, D. Vucic, W. J. Fairbrother and K. Deshayes, *ACS Chem. Biol.*, 2006, **1**, 525.

[30] C. Ndubaku, E. Varfolomeev, L. Wang, K. Zobel, K. Lau, L. O. Elliott, B. Maurer, A. V. Fedorova, J. N. Dynek, M. Koehler, S. G. Hymowitz, V. Tsui, K. Deshayes, W. J. Fairbrother, J. A. Flygare and D. Vucic, *ACS Chem. Biol.*, 2009, **4**, 557.

[31] S. M. Condon, M. G. LaPorte, Y. Deng and S. R. Rippin, *US Patent 7,517,906 B2*, 2009.

[32] S. M. Condon, M. G. LaPorte, Y. Deng, S. R. Rippin, T. P. Kumar, Y.-H. C. Lee, M. Hendi, J. Chou, M. E. Seipel, L. Gu, S. L. Springs, J. M. Burns, C. A. Benetatos, Y. Shi, M. A. McKinlay and S. Chunduru, 232th ACS National Meeting, San Francisco, CA, US, 2006, Abstract MEDI-578.

[33] M. G. LaPorte, Y. Deng, S. R. Rippin, C. A. Benetatos, S. Chunduru, M. A. McKinlay, J. M. Burns, J. Chou, S. L. Springs, M. Hendi, Y.-H. C. Lee, T. P. Kumar and S. M. Condon, 31st National Medicinal Chemistry Symposium, Pittsburgh, PA, USA, 2008, Abstract 234.

[34] Z. Nikolovska-Coleska, J. L. Meagher, S. Jiang, C.-Y. Yang, S. Qiu, P. P. Roller, J. A. Stuckey and S. Wang, *Biochemistry*, 2008, **47**, 9811.

[35] H. Sun, L. Liu, J. Lu, S. Qiu, C.-Y. Yang, H. Yi and S. Wang, *Bioorg. Med. Chem. Lett.*, 2010, **20**, 3043.

[36] K. E. Splan, J. E. Allen and G. L. McLendon, *Biochemistry*, 2007, **46**, 11938.

[37] F. Cohen, B. Alicke, L. O. Elliott, J. A. Flygare, T. Goncharov, S. F. Keteltas, M. C. Franklin, S. Frankovitz, J.-P. Stephan, V. Tsui, D. Vucic, H. Wong and W. J. Fairbrother, *J. Med. Chem.*, 2009, **52**, 1723.

[38] Y. Peng, H. Sun, Z. Nikolovska-Coleska, S. Qiu, C.-Y. Yang, J. Lu, Q. Cai, H. Yi, S. Kang, D. Yang and S. Wang, *J. Med. Chem.*, 2008, **51**, 8158.

[39] Q. Cai, H. Sun, Y. Peng, J. Lu, Z. Nikolovska-Coleska, D. McEachern, L. Liu, S. Qiu, C.-Y. Yang, R. Miller, H. Yi, T. Zhang, D. Sun, S. Kang, M. Guo, L. Leopold, D. Yang and S. Wang, *J. Med. Chem.*, 2011, **54**, 2714.

[40] M. R. Jensen, C. S. Straub, L. Zawel, M. A. Tran and Y. Wang, *Patent Application WO 2007/075525 A2*.

[41] J. R. Infante, E. C. Dees, H. A. Burris III, L. Zawel, J. A. Sager, C. Stevenson, K. Clarke, S. Dhuria, D. Porter, S. K. Sen, E. Zannou, S. Sharma and R. B. Cohen, AACR 101st Annual Meeting, Washington, DC, USA, Abstract 2775, 2010.

[42] H. Wong, N. Budha, K. West, B. Blackwood, J. A. Ware, R. Yu, W. C. Darbonne, S. E. Gould, R. Steigerwalt, R. Erickson, C. E. A. C. Hop, P. LoRusso, S. G. Eckhardt, A. Wagner, I. T. Chan, M. Mamounas, J. Flygare and W. J. Fairbrother, 22nd EORTC-NCI-AACR Meeting, Berlin, GE, 2010, Abstract 82.

[43] P. LoRusso, A. J. Wagner, N. Budha, W. C. Darbonne, Y. Shin, S. Cheeti, H. Wong, R. Yu, W. J. Fairbrother, M. Mamounas, J. Flygare, I. T. Chan, A. Joshi, J. Ware and S. G. Eckhardt, 22nd EORTC-NCI-AACR Meeting, Berlin, GE, Abstract 393, 2010.

[44] S. G. Eckhardt, G. Gallant, B. I. Sikic, D. R. Camidge, H. A. Burriss III, H. A. Wakelee, W. A. Messersmith, S. F. Jones, A. D. Colevas and J. R. Infante, ASCO Annual Meeting, Chicago, IL, USA, 2010, Abstract 2580.

[45] M. M. Moore, V. E. Estrada, F. E. Nieves, Y. Mitsuuchi, J. M. Burns, S. K. Chunduru, S. M. Condon, M. A. Graham, M. A. McKinlay, A. W. Tolcher and M. J. Wick, 21st EORTC-NCI-AACR Meeting, 2009, Abstract B163.

[46] R. K. Amaravadi, R. J. Schilder, G. K. Dy, W. W. Ma, G. J. Fetterly, D. E. Weng, M. A. Graham, J. M. Burns, S. K. Chunduru, S. M. Condon, M. A. McKinlay and A. A. Adjei, 102nd AACR Meeting, Orlando, FL, USA, 2011, Abstract LB-406/4.

[47] D. Chauhan, P. Neri, M. Velankar, K. Podar, T. Hideshima, M. Fulciniti, P. Tassone, N. Raje, C. Mitsiades, N. Mitsiades, P. Richardson, L. Zawel, M. Tran, N. Munshi and K. C. Anderson, *Blood*, 2007, **109**, 1220.

[48] B. Z. Carter, D. H. Mak, W. D. Schober, E. Koller, C. Pinilla, L. T. Vassilev, J. C. Reed and M. Andreeff, *Blood*, 2010, **115**, 306.
[49] H. Kashkar, *Clin. Cancer Res.*, 2010, **16**, 4496.
[50] J. E. Vince, D. Chau, B. Callus, W. W.-L. Wong, C. J. Hawkins, P. Schneider, M. McKinlay, C. A. Benetatos, S. M. Condon, S. K. Chunduru, G. Yeoh, R. Brink, D. L. Vaux and J. Silke, *J. Cell Biol.*, 2008, **182**, 171.
[51] M. J. M. Bertrand, K. Doiron, K. Labbé, R. G. Korneluk, P. A. Barker and M. Saleh, *Immunity*, 2009, **30**, 789.
[52] A. Krieg, R. G. Correa, J. B. Garrison, G. Le Negrate, K. Welsh, Z. Huang, W. T. Knoefel and J. C. Reed, *Proc. Natl Acad. Sci.*, 2009, **106**, 14524.
[53] D. Conte, M. Holcik, C. A. Lefebvre, E. LaCasse, D. J. Picketts, K. E. Wright and R. G. Korneluk, *Mol. Cell. Biol.*, 2006, **26**, 699.
[54] H. Prakash, D. Becker, L. Böhme, L. Albert, M. Witzenrath, S. Rosseau, T. F. Meyer and T. Rudel, *PLoS ONE*, 2009, **4**, e6519, doi:10.1371/journal.pone.0006519.

CHAPTER 14

Case History: Discovery of Eribulin (HALAVEN^{TM}), a Halichondrin B Analogue That Prolongs Overall Survival in Patients with Metastatic Breast Cancer

Melvin J. Yu*, **Wanjun Zheng***, **Boris M. Seletsky***, **Bruce A. Littlefield**** and **Yoshito Kishi**[†]

Contents

* Eisai Inc., 4 Corporate Drive, Andover, MA 01810, USA
** Harvard Medical School, BCMP, 240 Longwood Avenue, Boston, MA 02115, USA
† Department of Chemistry and Chemical Biology, Harvard University, 12 Oxford Street, Cambridge, MA 02138, USA

Annual Reports in Medicinal Chemistry, Volume 46
ISSN: 0065-7743, DOI: 10.1016/B978-0-12-386009-5.00013-8

1. INTRODUCTION

The most recent projections by the National Cancer Institute estimate that one in eight women will be diagnosed with breast cancer at some point in their lifetime [1], making it one of the leading causes of cancer death among women in the United States [2]. Since few therapeutic options exist for patients with advanced or recurrent breast cancer, new and more effective treatment modalities are needed, particularly those that offer improved survival rates over current therapies.

Eribulin mesylate [3,4] (**1**, HALAVEN™, previously E7389, NSC 707389, and ER-086526) is an antitubulin antimitotic agent with distinct microtubule end-binding properties that result in inhibition of microtubule dynamics in ways that differ from those of vinblastine and paclitaxel [5,6] (Figure 1). This agent was recently approved by the U.S. Food and Drug Administration (FDA) for use in patients with metastatic breast cancer who have previously received at least two chemotherapeutic regimens for the treatment of metastatic disease [7]. Prior therapy should have included an anthracycline and a taxane in either the adjuvant or metastatic setting. HALAVEN therefore represents a new and exciting treatment option that for the first time has been shown to improve overall survival in heavily pretreated women with late stage breast cancer. Building on this success, clinical trials are currently ongoing to evaluate activity in additional cancer indications [8].

Although inspired by the structurally complex marine natural product halichondrin B (**2**, HB), eribulin (**1**) is a totally synthetic macrocyclic ketone analogue. As such, the drug substance can be produced in a controlled and predictable manner with a well-defined and reproducible impurity profile, factors that are essential for all marketed drugs irrespective of source. Since material supply was a critical factor that limited preclinical development of the natural product HB by the U.S. National Cancer Institute (NCI), total synthesis of a structurally simplified and optimized derivative represented the best overall solution to access the needed quantities of material in a sustainable and cost-effective fashion.

In the mid-1980s, the Kishi group was considering a number of chemical targets to showcase the synthetic potential of the Nozaki–Hiyama–Kishi reaction in the construction of structurally complex molecules. From among these, the halichondrins were selected due to their chemical architecture and remarkable biological activity. In 1986, HB was reported by Hirata and Uemura [9] to exhibit subnanomolar cell growth inhibitory potency, good physical properties, and outstanding anticancer activity in animal models [10]. Supported by an NCI grant, the Kishi group initiated a synthetic program, which culminated in the first successful total synthesis of HB in 1992 [11].

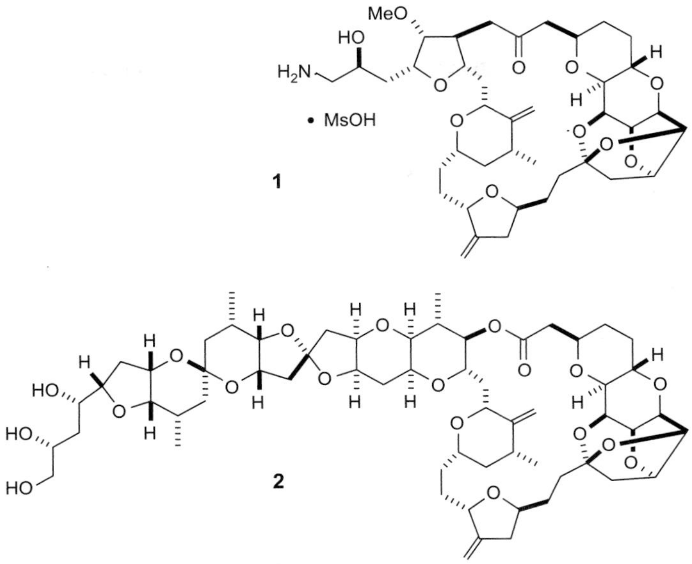

Figure 1 Eribulin mesylate (1) and halichondrin B (2).

That year proved to be a pivotal one as the NCI accepted the natural product HB for preclinical development. That same year, Eisai researchers received a number of synthetic intermediates from the Kishi laboratory and discovered that the right half (RH) C.1–C.38 fragment **6** exhibited potent cell growth inhibitory activity. The technology to synthesize HB was patented by Harvard University [12] and subsequently licensed to Eisai.

2. HALICHONDRIN B

2.1. Material supply

The amount of HB needed to support clinical development was estimated to be around 10 g [13]. Assuming success in clinical trials and approval by regulatory authorities, the amount of HB projected to meet future commercial demand was predicted to be 1–5 kg/year. However, despite even these modest amounts, collection from the wild was determined to be impractical due to the extremely low yield of HB and the low abundance of known HB-producing sponges in the world. In addition, achieving consistent compound purity represented a significant hurdle since material from natural sources must be separated from closely related family members as well as structurally unrelated but highly potent co-isolated bioactive metabolites (*e.g.*, okadaic acid).

Supported in part by an NCI grant, preliminary in-sea aquaculture techniques with the HB-producing sponge *Lissodendoryx* sp. were investigated, but significant challenges remained that rendered future scale-up uncertain [13].

2.2. Total synthesis

A number of chemistry research groups initiated programs directed toward the total synthesis of the halichondrins [14]. The first successful total synthesis was reported by the Kishi group at Harvard University in 1992 [11]. Their synthesis is highly convergent and involves a series of Nozaki–Hiyama–Kishi coupling reactions of advanced intermediates **3–5** to afford RH intermediate **6** (Figure 2). The latter compound, which represents the macrolactone region of the natural product, was then coupled with left half precursor **7** to provide HB. This total synthesis represents a landmark scientific achievement that highlights the power

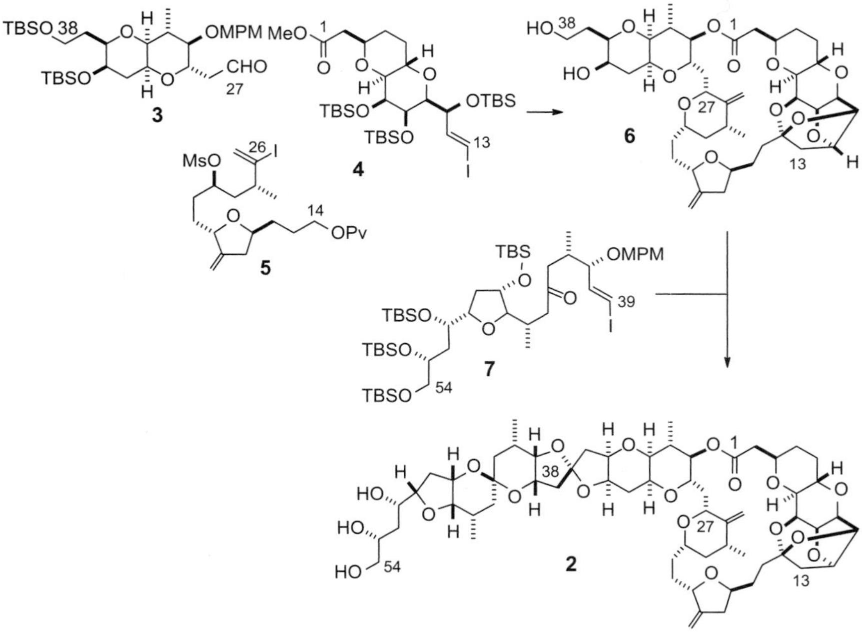

Figure 2 Kishi total synthesis of HB commencing from the C.1–C.13 (**4**), C.14–C.26 (**5**), and C.27–C.38 (**3**) fragments. Coupling of RH (**6**) with the C.39–C.54 left half fragment (**7**) afforded the natural product.

and utility of the Nozaki–Hiyama–Kishi reaction in the construction of structurally complex molecules.

3. ERIBULIN DRUG DISCOVERY PROGRAM

3.1. Medicinal chemistry strategy

Synthetic intermediates from the Kishi laboratory prepared in connection with the total synthesis of HB and norHB were submitted to both the NCI and Eisai for biological evaluation. Interestingly, *in vitro* cell growth inhibitory activity against DLD-1 cells was associated only with the RH macrolactone fragment **6**. None of the left half intermediates submitted for testing was found to be active under the experimental conditions examined. This finding was significant for a number of reasons. First, it established that structurally simplified derivatives of HB, which exhibit potent anticancer activity *in vitro*, could be identified. Second, it reinforced the belief that the material supply problem associated with the halichondrins could be solved in a practical manner by total synthesis. And third, it clearly demonstrated that significant structural modifications could be made to the left half of the molecule without adversely affecting cell growth inhibitory activity, thereby pointing the direction for future optimization.

As outlined in Figure 2, the synthesis of **6** is highly convergent and remarkably modular, allowing each of the three main fragments to be individually modified in a mix and match approach. By stockpiling key fragments, the preparation of new analogues could be transformed from a scientifically complex challenge to that of a technical and material management problem directed by classical medicinal chemistry conventions.

The axiom that formed the basis for Eisai Research Institute's (ERI) medicinal chemistry strategy was to first identify the pharmacophore and then modify the remainder of the structure to increase synthetic accessibility. However, initially, the drug discovery team at ERI (Andover, MA facility) did not have access to ancillary resources that are now generally assumed in today's drug discovery environment. These include (a) ADME studies, (b) a computational chemistry group, (c) a process chemistry department, (d) kilo lab support, and (e) an analytical chemistry unit. Scale-up work and resupply of key intermediates had to be handled entirely within the discovery chemistry team, and questions regarding pharmacokinetics had to be addressed indirectly through hypothesis-directed establishment of new *in vitro* screens.

Often with chemical leads derived from high-throughput screening of small molecule compound libraries, potency is an issue that may be addressed in an additive fashion, for example, structure optimization may involve adding functional groups and substituents to improve affinity of the ligand for the target of interest. In the halichondrin program, however, potency was not a critical path issue. Thus, the optimization campaign was envisioned to proceed in a subtractive mode driven entirely by synthetic accessibility and *in vivo* performance concerns. Using this general design principle, we derived the following initial research objectives:

1. Establish proof of concept (*i.e.*, demonstrate *in vivo* activity for RH)
2. Stockpile key fragments to support analogue synthesis
3. Modify the structure of **6** to identify the minimum pharmacophoric substructure and increase synthetic accessibility
4. Optimize for biological activity

To accomplish the first and second objectives, a scale-up synthesis was initiated where the three key fragments **3–5** were stockpiled. This initial effort afforded approximately 100 mg of **6**, which was more than sufficient to conduct multiple human tumor xenograft experiments. However, in sharp contrast to the natural product, no *in vivo* activity could be observed under the experimental conditions tested. This was clearly a set-back for the program. If **6** rather than HB represented the starting point and chemical lead for the drug discovery program, then in essence the team lost their advanced starting point. Without access to *in vivo* ADME data, it was simply not possible to explain the difference based on pharmacokinetics. Thus, to provide a rational hypothesis-driven path forward a surrogate quantitative cell-based pharmacodynamic assay was needed to prioritize new analogues for *in vivo* evaluation.

As a result, a secondary, orthogonal *in vitro* evaluation system was developed to complement the primary 3- to 4-day cell growth inhibition assay used to evaluate HB and its synthetic intermediates. In the primary screen, cancer cells were continually exposed to the test substance for the duration of the experiment. However, in the human tumor xenograft models, compounds were dosed intermittently, which presumably gave rise to peaks and troughs in free drug concentration. Thus, the hypothesis was generated that the inability to maintain a complete mitotic block (CMB) under drug washout conditions could explain the observed lack of *in vivo* activity for **6**. A new flow cytometric analysis assay using U937 human histiocytic lymphoma cells was therefore established at ERI [15]. In this system, the reversibility ratio was calculated by dividing the test compound concentration found to give a CMB at the 10-h time point by the minimum compound concentration required to induce a CMB at the 0-h time point.

Using this method, an HB concentration of 25 nM (reversibility ratio = 3) was found to be sufficient to induce a CMB at the 10-h time point, whereas the concentration of **6** was found to be >880 nM (reversibility ratio > 30). This result allowed us to reject the null hypothesis and incorporate the reversibility assay into our screening paradigm as a cell-based surrogate for the effects of intermittent dosing.

3.2. Macrolactone

3.2.1. Proof of concept

To demonstrate proof of concept, we needed to identify a RH analogue that exhibited activity in a human tumor xenograft model using a standard intermittent dosing schedule. Since **6** lacks the polyether portion of the natural product and contains a diol, our initial hypothesis was that the C.35–C.38 region needed to more closely resemble that of the natural product. Tetrahydrofuran analogue **8**, which is a direct C.1–C.38 substructure of HB, was therefore evaluated in the mitotic block reversibility assay but found to be highly reversible (Figure 3).

We next hypothesized that the ability of HB to maintain a CMB upon drug washout may be associated with the C.39–C.50 polyether moiety. To test this, a small series of readily accessible "left half surrogate"

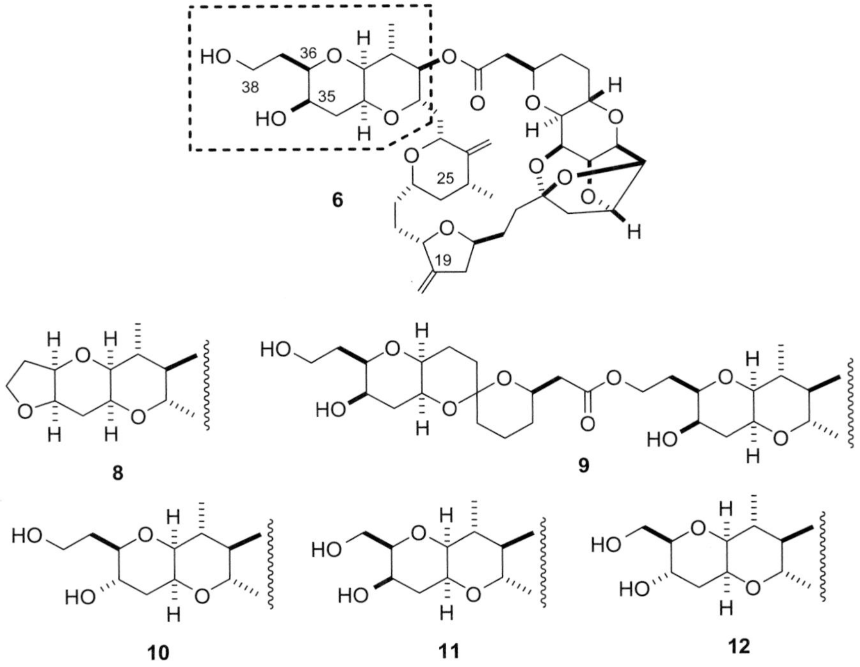

Figure 3 RH macrolactone analogues. Compounds **8–11** were highly reversible in the U937 mitotic block assay. Compound **12** was much less reversible.

derivatives was prepared (*e.g.*, **9**). Unfortunately, these too exhibited unfavorable behavior in the U937 mitotic block reversibility assay, rendering them unsuitable for *in vivo* evaluation.

The breakthrough came when compound **12** was submitted for biological evaluation. At that time, precursors derived from the gluco rather than the galacto series of starting materials were being used to develop the asymmetric Nozaki–Hiyama–Kishi reaction at Harvard. The resulting "norgluco" product **12** contains two structural changes relative to the original RH intermediate **6**—a one carbon truncation at C.37 and inversion of stereochemistry at C.35. Surprisingly, this material exhibited a reversibility ratio of 24, which was the first time that a RH derivative was found to possess an improved reversibility profile. On the basis of the primary and secondary assay results, the synthesis of norgluco RH analogue **12** was scaled up and the material evaluated in the LOX human melanoma xenograft model. Gratifyingly, the compound demonstrated sustained inhibition of *in vivo* tumor growth at 5 mg/kg on a (Q1Dx5) $\times$ 2 dosing schedule, thereby validating the reversibility hypothesis and providing in vivo proof of concept.

Compounds **10** and **11**, which represent single point changes relative to RH, were subsequently prepared at ERI, but both were found to be highly reversible in the secondary *in vitro* screen. Thus, both structural changes present in compound **12** relative to **6** were needed to generate an improved reversibility profile.

The next step was to increase synthetic accessibility by removing or modifying non-pharmacophoric functional groups in a systematic stepwise fashion using the existing stockpile of synthetic intermediates.

3.2.2. C.1–C.13 fragment

Consideration was given to modifying the C.1–C.13 fragment **4**, but there were no clear directions for how that would simplify the chemistry and thus the synthetic accessibility of the compound. Consequently, we decided that this fragment should be left unmodified.

3.2.3. C.14–C.26 fragment

Removal or modification of the C.19 and C.26 exo olefins in fragment **5** was considered next. Since the C.19 center is derived from arabinose, a structure "simplification" would be to retain the asymmetric center already present in the starting material and omit the multistep sequence to introduce the exo olefin. In the event, the resulting C.19 methoxy derivative was active in the primary screen, but less potent than **6** and therefore not pursued further. Complete removal of the exo olefins at C.19 and C.26 similarly afforded less potent compounds.

The C.25 methyl group was then queried. This substituent was introduced by a stereoselective alkylation reaction of an early lactone intermediate. If that could be removed, then the chemistry of the C.14–C.26 fragment would be simplified, albeit incrementally. Disappointingly, however, either deletion or homologation of the methyl group only led to a substantial decrease in activity. Thus, we concluded that the functionality embedded within the C.14–C.26 fragment was critically important and should be retained in the synthesis of future analogues.

3.2.4. C.27–C.38 fragment

Based on the above studies, the decision was made to leave the C.1–C.26 region of the molecule unchanged. Thus, effort focused on the fused C.27–C.38 ring system, which we believed offered the greatest opportunity for structure optimization given the initial observation that HB could be truncated at C.38 and still retain biological activity. Toward that end, formal removal of the C.36 carbon and its attendant hydroxyl group from pivotal compound **12** afforded tetrahydropyran analogue **13** (Figure 4) [16]. Since the C.31 methyl group was introduced *via* a multistep procedure involving a carbohydrate precursor, we envisioned that the existing hydroxyl group in the starting material could instead be converted to a methoxy group in a straightforward manner to ultimately afford **14**. This derivative exhibited good cell growth inhibitory activity and an acceptable level of desired irreversibility in the U937 mitotic block assay [4]. However, despite meeting all of the *in vitro* criteria for compound advancement, it was inactive in the mouse LOX melanoma

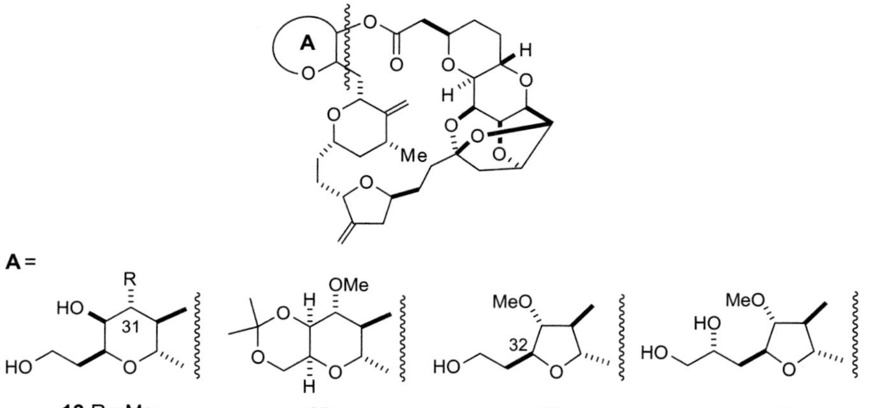

Figure 4 Simplified macrolactone analogues. *In vitro* profile for compounds **13** and **14** was similar. *In vitro* potency for tetrahydrofuran analogues **16** and **17** was superior to that of tetrahydropyran derivative **14**. Compound **17** exhibited superior reversibility characteristics relative to **16**.

xenograft model. Although somewhat puzzling at first, investigation of biological stability *in vitro* revealed that the compound was not stable in the presence of mouse serum [17]. This is in sharp contrast to compound **12**, which was completely stable under the experimental conditions. We hypothesized that the tetrahydropyran ring of **14** was more conformationally flexible than the fused octahydropyrano[3,2-*b*]pyran ring system of **12**, thereby possibly rendering the lactone moiety more susceptible to cleavage by nonspecific mouse serum esterases. As a result, we considered ways to help rigidify the tetrahydropyran ring and indirectly stabilize the macrolactone ring conformation.

One option we pursued was to tie the diol groups into a ketal structure to mimic the fused ring system of **12**. Although **15** was active in the cell growth inhibition assay, it was also highly reversible. As a result, work on this particular approach was abandoned.

The other option that we considered was to reduce the size of the C.29–C.33 six-membered ring to a tetrahydrofuran ring. From earlier studies, we knew that the presence of a C.31 substituent and presumably its conformation relative to the macrocyclic ring were important for cell growth inhibitory activity. Conformational analysis of the norhalichondrin A X-ray crystal structure suggested that the C.31 group in a tetrahydrofuran ring would occupy the appropriate position. Compound **16** (DLD-1 cell growth inhibition $IC_{50} = 0.97$ nM, reversibility ratio $= 14$) was subsequently prepared and found to be both more potent and more irreversible than **14** ($IC_{50} = 2.1$ nM, reversibility ratio >100) [4], making the tetrahydrofuran series considerably more attractive than the tetrahydropyrans. Structure–activity relationship (SAR) studies of *in vitro* cell growth inhibition potency and reversibility relative to C.32-side chain modifications subsequently identified diol **17** as the most interesting. Unfortunately, this compound and all other tetrahydrofuran lactone analogues tested were unstable in the presence of mouse serum. It was not clear why the fused octahydropyrano [3,2-*b*]pyran analogues exemplified by compound **12** were stable in the presence of mouse serum, but the truncated ring derivatives exemplified by **14** and **16** were not. Thus, in the absence of a specific working hypothesis to guide further structure design efforts, hydrolytically stable C.1 lactone bioisosteres appeared to represent the most expedient path forward.

3.3. Macrocyclic ether

One method to stabilize the lactone moiety was to replace it with a simple ether group. A significant stockpile of the C.27–C.35 tetrahydropyran fragment was available at the time and was therefore used for synthesis. Unfortunately, analogue **18** was found to be much less potent than the macrolactone derivatives (Figure 5). Thus, no further work on the macrocyclic ether series was pursued.

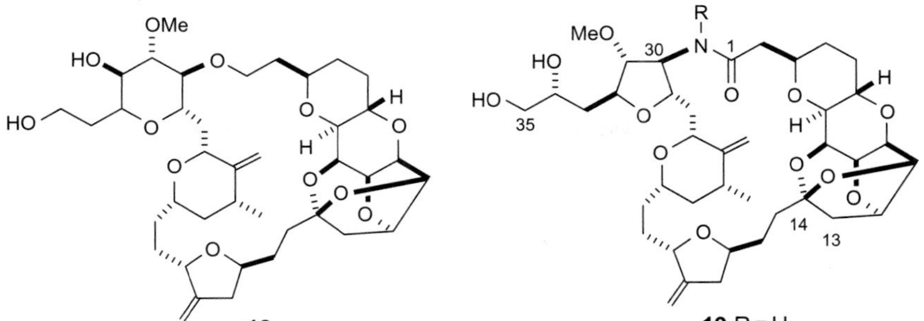

Figure 5 Macrocyclic ether and lactam analogues. All three analogues were much less potent than members of the macrolactone series.

3.4. Macrolactam

Although the lactone to lactam replacement was originally conceived to be a relatively straightforward modification, the existing method to close the macrocyclic ring could not be employed. After considering several possibilities, the macrocyclization precursor was prepared by amide formation between the C.1–C.13 carboxylic acid and an amino derivative of the C.14–C.35 fragment. A Nozaki–Hiyama–Kishi coupling reaction was then used to close the macrocyclic ring at the C.13–C.14 position followed by formation of the ketal "cage" structure. Surprisingly, however, the resulting macrolactam analogues **19** and **20** were approximately two orders of magnitude less potent than the corresponding macrolactone derivatives, suggesting that although a lactam linkage was tolerated, activity was compromised [17].

These results were rationalized on the basis of gas phase molecular dynamics simulations. Over a 600-ps time frame, the lactam dihedral angle exhibited two maxima centered around 0° and 180°, representing the s-*cis* and s-*trans* conformations, respectively. A similar gas phase molecular dynamics simulation placed the lactone dihedral angle maximum at 165° with a population distribution that significantly overlapped that associated with the s-*trans* lactam conformation. The X-ray crystal structure of a norhalichondrin A derivative exhibited a lactone dihedral angle of 163° [9], in good agreement with the gas phase molecular dynamics calculation. If we assume that the solid-state conformation of the norhalichondrin A macrolactone represents the bioactive conformation of HB, then the macrolactam analogues should have been more active than observed. Since this was not the case, we concluded that the bioactive conformation must be one where the dihedral angle lies somewhere in-between 90° and 163°. Molecular dynamics calculations placed the ketone dihedral angle around 90° with a distribution which overlapped that calculated for the lactone, suggesting both the ketone and lactone derivatives could access a common low-energy conformation that would be energetically less accessible by the lactam. On the basis of these modeling studies, we redirected our efforts toward synthesizing a series of C.1 ketone bioisosteric derivatives.

3.5. Macrocyclic ketone

Synthesis of the macrocyclic ketone series proceeded in an analogous manner from the existing stockpile of synthetic intermediates. After modifying the C.27–C.35 tetrahydrofuran fragment to include an appropriately functionalized carbon substituent at C.30, coupling with the C.1–C.13 fragment proceeded smoothly and in high yield. Functional group manipulation and macrocyclic ring formation using the conditions developed for the macrolactam series afforded the desired intermediate that was transformed to the final macrocyclic ketone analogue [18].

In this manner, compound **21** was prepared and found to exhibit excellent potency, good reversibility characteristics, stability in the presence of mouse serum and most importantly, potent inhibition of tumor growth in human cancer xenograft models *in vivo* (Figure 6). Further functional group exploration at C.1 (*e.g.*, exo olefin, alcohol, oxime), C.34 and C.35 (*e.g.*, amides, carbamates, ureas, etc.) led to a series of derivatives, from which eribulin and diol **21** emerged as the most promising [19]. In particular, eribulin exhibited a reversibility ratio of one, indicating irreversible behavior in the mitotic block assay. On the basis of its remarkable biological activity profile, eribulin was nominated and accepted for preclinical development at Eisai. Both eribulin and diol **21** were then submitted to the NCI for further evaluation. Based on a review of all

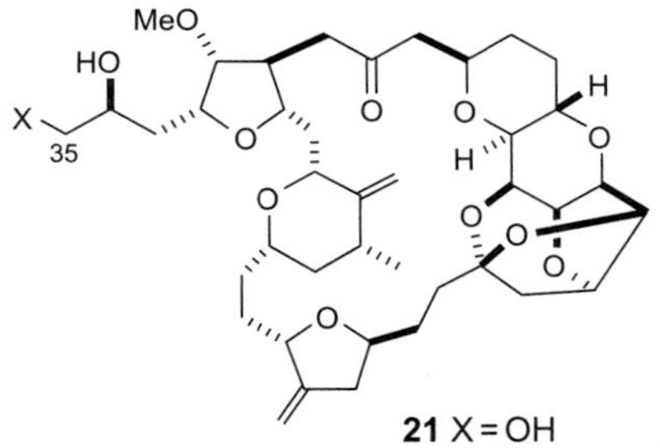

MeO HO X 35 O O O H H O O O O O O O

21 X = OH
1 X = NH_2

Figure 6 Example macrocyclic ketone analogues. Eribulin (**1**) exhibited a reversibility ratio of one, whereas the diol (**21**) exhibited a reversibility ratio of 13.

available data, including cell-based antiproliferative potency, *in vivo* antitumor activity, pharmacokinetic, and stability parameters, eribulin was ultimately selected as the front running candidate [4].

In addition to its remarkable biological and safety profiles, eribulin mesylate exhibited excellent physicochemical properties, which allowed the compound to be formulated in a simple and straightforward manner. Finally, the stage was set for solving the material supply problem with a structurally simplified, optimized, and totally synthetic derivative that not only retained the remarkable biological activity of the natural product that inspired it, but actually surpassed it.

4. FROM DISCOVERY TO DEVELOPMENT

Despite significant simplification relative to HB, the chemical structure of eribulin remains a synthetically challenging target with 19 stereogenic centers. To the best of our knowledge, this molecule exceeds by far the structural complexity of any drug prepared by total synthesis that is either in development or on the market. Not surprisingly, this presented certain difficulties as the program moved from discovery to development. Serious objections were raised regarding the synthetic feasibility and cost of goods for a totally synthetic compound of such unprecedented structural complexity, concerns that nearly terminated the program as it moved through early clinical development. Notwithstanding, the issues were successfully resolved through continued basic scientific contributions from the Kishi group at Harvard [20] and advances from the Eisai chemical development team [21–23]. At 62 steps, manufacturing eribulin in a cost-effective manner on a scale sufficient to meet commercial demand represents a major leap in demonstrating the power of contemporary organic synthesis to solve problems of this magnitude, and one that effectively resets the bar for what is possible.

At the time of candidate nomination, limitations at Eisai precluded moving eribulin into early clinical development. Thus, the best path forward was through a Cooperative Research and Development Agreement (CRADA) in the form of an NCI-sponsored Phase I clinical trial. The early results were promising, eribulin entered full clinical development and Eisai-sponsored trials were initiated. Details of the clinical trial results have recently been published [24]. In particular, the pivotal EMBRACE phase III trial demonstrated an overall survival benefit for eribulin versus single agent therapy of physician's choice in women who were heavily pretreated for metastatic or locally recurrent breast cancer. Several major advantages offered by HALAVEN over TAXOL®, include the option to formulate HALAVEN using ethanol/water without the need for Cremophor® EL, and the ability for the drug to be administered as a 2- to 5-min infusion [25] as compared to several hours for Taxol®. Another advantage suggested by preclinical data is a potentially lower incidence of peripheral neuropathy [26]. On November 15, 2010, the FDA announced the approval of HALAVEN for use in patients with metastatic breast cancer who meet certain criteria.

5. CONCLUSION

The path leading to the discovery and development of eribulin included a close three-way collaboration between the Kishi group at Harvard University, the NCI, and Eisai. At many points along the drug discovery path, setbacks and roadblocks arose that threatened to derail the program. Nevertheless, each of the problems was solved in turn thereby demonstrating that a structurally complex molecule could be optimized through total synthesis to successfully deliver a marketed drug that meets all pharmacological, toxicological, and physicochemical requirements.

Although HALAVEN is currently approved for use in patients with metastatic breast cancer who have previously received at least two chemotherapeutic regimens, clinical trials are ongoing to support additional indications [7]. The journey through the discovery and development pipeline was not a smooth or easy one, but one that was ultimately successful. In the end, the project goals were realized and HALAVEN emerged, bringing with it new hope for cancer patients and their families.

REFERENCES

[1] http://www.cancer.gov/cancertopics/factsheet/Detection/probability-breast-cancer (accessed January 26, 2011).
[2] http://seer.cancer.gov/statfacts/html/breast.html (accessed January 26, 2011).
[3] M. J. Towle, K. A. Salvato, J. Budrow, B. F. Wels, G. Kuznetsov, K. K. Aalfs, S. Welsh, W. Zheng, B. M. Seletsky, M. H. Palme, G. J. Habgood, L. A. Singer, L. V. DiPietro, Y. Wang, J. J. Chen, D. A. Quincy, A. Davis, K. Yoshimatsu, Y. Kishi, M. J. Yu and B. A. Littlefield, *Cancer Res.*, 2001, **61**, 1013.
[4] M. J. Yu, Y. Kishi and B. A. Littlefield, in *Anticancer Agents from Natural Products*, (eds. D. J. Newman, D. G. I. Kingston and G. M. Cragg), Anticancer Agents from Natural Products, Taylor & Francis, Washington, 2005, p. 241.
[5] M. A. Jordan, K. Kamath, T. Manna, T. Okouneva, H. P. Miller, C. Davis, B. A. Littlefield and L. Wilson, *Mol. Cancer Ther.*, 2005, **4**, 1086.
[6] J. A. Smith, L. Wilson, O. Azarenko, X. Zhu, B. M. Lewis, B. A. Littlefield and M. A. Jordan, *Biochemistry*, 2010, **49**, 1331.
[7] http://www.fda.gov/NewsEvents/Newsroom/PressAnnouncements/ucm233863.htm (accessed January 26, 2011).
[8] http://www.cancer.gov/search/ResultsClinicalTrials.aspx?protocolsearchid=8759676 (accessed February 2, 2011).
[9] Y. Hirata and D. Uemura, *Pure Appl. Chem.*, 1986, **58**, 701.

[10] Y. Tsukitani, T. Manda, K. Yoshida and H. Kikuchi, *Japanese Patent 1721245*, 1992.
[11] T. D. Aicher, K. R. Buszek, F. G. Fang, C. J. Forsyth, S. H. Jung, Y. Kishi, M. C. Matelich, P. M. Scola, D. M. Spero and S. K. Yoon, *J. Am. Chem. Soc.*, 1992, **114**, 3162.
[12] Y. Kishi, F. G. Fang, C. J. Forsyth, P. M. Scola and S. K. Yoon, *US Patent 5,338,865*, 1994.
[13] J. B. Hart, R. E. Lill, S. J. H. Hickford, J. W. Blunt and M. H. G. Munro, in *Drugs from the Sea*, (ed. N. Fusetani), Switzerland, 2000, pp. 134.
[14] For a review of halichondrin chemistry, see K. L. Jackson, J. A. Henderson and A. J. Phillips, *Chem. Rev.*, 2009, **109**, 3044.
[15] M. J. Towle, K. A. Salvato, B. F. Wels, K. K. Aalfs, W. Zheng, B. M. Seletsky, X. Zhu, B. M. Lewis, Y. Kishi, M. J. Yu and B. A. Littlefield, *Cancer Res.*, 2011, **71**, 496.
[16] B. M. Seletsky, Y. Wang, L. D. Hawkins, M. H. Palme, G. D. Habgood, L. V. DiPietro, M. T. Towle, K. A. Salvato, B. F. Wels, K. K. Aalfs, Y. Kishi, B. A. Littlefield and M. J. Yu, *Bioorg. Med. Chem. Lett.*, 2004, **14**, 5547.
[17] M. J. Towle, B. F. Wels, H. Cheng, J. Budrow, W. Zheng, B. M. Seletsky, L. D. Hawkins, M. H. Palme, G. J. Habgood, L. A. Singer, L. V. DiPietro, Y. Wang, J. J. Chen, P. J. Lydon, D. A. Quincy, K. Tagami, Y. Kishi, M. J. Yu and B. A. Littlefield, Annual meeting of the American Association for Cancer Research, San Francisco, CA, April 6–10, 2002. Abstract 5721.
[18] B. A. Littlefield, M. H. Palme, B. M. Seletsky, M. J. Towle, M. J. Yu and W. Zheng, *PCT Patent Application WO 9965894-A1*, 1999.
[19] W. Zheng, B. M. Seletsky, M. H. Palme, P. J. Lydon, L. A. Singer, C. E. Chase, C. A. Lemelin, Y. Shen, H. Davis, L. Tremblay, M. J. Towle, K. A. Salvato, B. F. Wels, K. K. Aalfs, Y. Kishi, B. A. Littlefield and M. J. Yu, *Bioorg. Med. Chem. Lett.*, 2004, **14**, 5551.
[20] See, for example: D.-S. Kim, C.-G. Dong, J. T. Kim, H. Guo, J. Huang, P. S. Tiseni and Y. Kishi, *J. Am. Chem. Soc.*, 2009, **131**, 15642 and references cited therein.
[21] B. Austad, C. E. Chase and F. G. Fang, *US Patent Application 2007/0244187-A1*, 2007.
[22] C. Chase, A. Endo, F. G. Fang and J. Li, *US Patent Application 2009/0198074-A1*, 2009.
[23] K. Inanaga, M. Kuboto, A. Kayana and K. Tagami, *US Patent Application 2009/0203771-A1*, 2009.
[24] J. Cortes, J. O'Shaughnessy, D. Loesch, J. L. Blum, L. T. Vahdat, K. Petrakova, P. Chollet, A. Manikas, V. Diéras, T. Delozier, V. Vladimirov, F. Cardoso, H. Koh, P. Bougnoux, C. E. Dutcus, S. Seegobin, D. Mir, N. Meneses, J. Wanders and C. Twelves, *Lancet*, 2011, **377**, 914.
[25] http://www.halaven.com/HALAVEN_Prescribing_Information.pdf (accessed February 8, 2011).
[26] K. M. Wozniak, K. Nomoto, R. G. Lapidus, Y. Wu, V. Carozzi, G. Cavaletti, K. Hayakawa, S. Hosokawa, M. J. Towle, B. A. Littlefield and B. S. Slusher, *Cancer Res.*, 2011, **71**, 3952.

PART V:
Infectious Diseases

Editor: John L. Primeau
AstraZeneca
Waltham
Massachusetts

CHAPTER **15**

Emerging New Therapeutics Against Key Gram-Negative Pathogens

D. Obrecht, F. Bernardini, G. Dale and **K. Dembowsky**

Contents

Polyphor Ltd., Hegenheimermattweg 125, CH-4123 Allschwil, Switzerland

Annual Reports in Medicinal Chemistry, Volume 46
ISSN: 0065-7743, DOI: 10.1016/B978-0-12-386009-5.00012-6

1. INTRODUCTION

Health-care-associated infections (HAIs) are a significant cause of morbidity and mortality and represent a major challenge to patient safety [1–4]. The estimated annual direct cost of treating HAIs in the United States ranges from 28.4 to 45 billion dollars resulting in a heavy burden on the public health system. The management of bacterial infections is becoming increasingly difficult due, in part, to the severity of illness and immune status of patients but more so due to the increased prevalence of multidrug-resistant (MDR) pathogens. Currently, the emergence of MDR Gram-positive and Gram-negative bacteria in both hospital and health-care-acquired infections is occurring [5]. Rice [6] recently described the most dangerous MDR bacteria as the "ESKAPE" pathogens *Enterococcus faecium, Staphylococcus aureus, Klebsiella pneumoniae, Acinetobacter baumannii, Pseudomonas aeruginosa,* and *Enterobacter* species to emphasize that in both the developed and developing world, these pathogens more and more frequently escape the lethal killing action of most antibiotics. Adding to these concerns, the situation has been exacerbated by the stagnation in the development of novel antimicrobial agents to treat these pathogens. Moreover, the number of approved antibacterial drugs has steadily declined in the past decade as pointed out by the Infectious Diseases Society of America (IDSA) in a series of alarming reports [7–9], which also launched the 10×20 initiative for a global effort to develop 10 new antibacterial drugs by 2020 [7].

Antibiotic drug discovery in the past two decades has focused efforts on the development of antibiotics against Gram-positive bacteria and more specifically against multidrug-resistant *S. aureus* (MRSA). In contrast, treatment options against some MDR Gram-negative pathogens have become very limited [5], especially due to the emergence of resistance against the last resort antibiotics, colistin and polymyxin B [10]. There are several reasons which led to this alarming situation. Several major pharmaceutical companies have abandoned the antibacterial field probably frustrated by fading profitability margins [9,11]. Increasing hurdles and uncertainty in clinical trial design for successfully obtaining regulatory approval for novel antibiotics is another reason why companies have stepped out of this therapeutic area. Compounding the problem of limited financial resources is the fact that Gram-negative organisms are inherently difficult to kill. This is due in part to the outer membrane (OM) which is composed by up to 75% by lipopolysaccharides (LPSs) [12,13]. LPS is negatively charged which makes the OM of Gram-negative bacteria highly impermeable for many classes of antibiotics thereby providing the Gram-negative bacteria with a formidable shield to prevent entry of antibacterials. Moreover, these organisms are highly efficient at upregulating, mutating, or acquiring genes that code for mechanisms of antibiotic resistance.

Recent data from the U.S. National Nosocomial Infections Surveillance (NNIS) system reported that in 2003, Gram-negative organisms are associated with hospital-acquired infection in intensive care units and that these pathogens are responsible for 66% of all cases of pneumonia, 24% of bloodstream infections, 42% of surgical site infections and 73 % of urinary tract infections [14].

Mid- to long-term strategies to prevent antimicrobial resistance in the intensive care unit include shorter courses of appropriate antibiotic treatment and narrowing of antimicrobial spectrum based on culture results [15]. A study aiming to assess the influence of broad-spectrum versus narrow-spectrum antibiotic prophylaxis in head-injured patients showed that narrow-spectrum prophylaxis was associated with lower rates of antibiotic resistance among the Gram-negative pathogens isolated from subsequent infections [16].

The focus of this review will be on compounds in late preclinical and clinical development and compounds derived from known and new classes of antibiotics which show promising activity against MDR Gram-negative bacteria. A major emphasis will be on compounds which exhibit a new mechanism of action and on anti-pseudomonal antibiotics. In addition, particular attention will be devoted to strategies which move away from the older model of identifying broad-spectrum antibiotics in favor of more focused, narrow-spectrum approaches.

2. NEW COMPOUNDS FROM KNOWN CLASSES

2.1. New aminoglycoside antibiotics

Since the discovery of streptomycin [17], aminoglycosides (AGs) have played a major role as efficacious broad-spectrum antibiotics. Despite vestibular and auditory toxicities observed as major side effects, streptomycin was remarkably successful and triggered the development of several other natural product-derived, semisynthetic AGs such as neomycin (1949); kanamycin (1957); paromomycin (1959); gentamycin (1963); tobramycin (1963); sisomycin (1970); and amikacin, isepamicin, netilmicin, and arbekacin (**1**) in the 1970s and early 1980s [18,19].

AGs bind to the A-site of ribosomal RNA and interfere with bacterial protein synthesis. Enzymes that inactivate AGs by modifying the molecule by methylation, N-acetylation, O-phosphorylation, or O-adenylation, and bacterial efflux pumps, constitute the two major resistance mechanisms. As an example, 16S rRNA methylases (such as ArmA and RmtC) confer resistance to all AGs [19].

Reported dose-limiting toxicities are neuromuscular blockade, oto- and nephrotoxicity.

The good activity of AGs against Gram-negative pathogens, including *P. aeruginosa*, however, has renewed the interest for developing new AG-derived antibiotics [19,20]. In a recent patent application [21], compounds such as **2** derived from arbekacin showed good broad-spectrum activity against Gram-negative and Gram-positive organisms including MRSA, and a twofold improved activity against *P. aeruginosa* (MIC values <4–$8\ \mu g/ml$). Compound **3** (ACHN-490) was derived from sisomicin [18] and is currently in clinical evaluation for treatment of complicated urinary tract infections and acute pyelonephritis [19,22]. ACHN-490 demonstrated good activity against Gram-positive *Staphylococci* ($MIC_{90} = 2\ \mu g/ml$) and a collection of Gram-negative clinical isolates of *P. aeruginosa*, *Escherichia coli*, *K. pneumoniae*, and *A. baumannii* [19]. In a Phase I clinical trial, ACHN-490 was well tolerated and exhibited linear and dose proportional pharmacokinetics (PK), while mild to moderate adverse events were reported [19]. It is important to note, however, that **3** is also susceptible to resistance *via* the same mechanisms as mentioned above for the class [23,24].

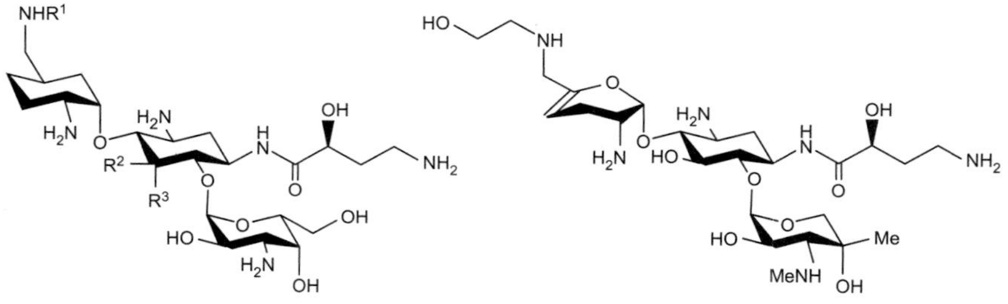

1 ($R^1 = H$; $R^2 = OH$; $R^3 = H$)
2 ($R^1 = Me$; $R^2 = H$; $R^3 = OH$)

3

2.2. New β-lactam antibiotics

Since the discovery and wide therapeutic use of penicillin G, many β-lactam antibiotics have been successfully developed and commercialized [25]. The traditional β-lactam antibiotics comprise the penam (penicillin), penem, carbapenem, cephem (cephalosporins), carbacephem, and monobactam subfamilies. β-Lactams interfere with bacterial cell wall biosynthesis by inhibiting the final transpeptidation step in the synthesis of the peptidoglycan by transpeptidases known as penicillin-binding proteins (PBPs). There are two main mechanisms of resistance to β-lactams: hydrolysis of the β-lactam ring *via* β-lactamases or the synthesis of altered PBPs with decreased sensitivity to β-lactams. The emergence of various classes of β-lactamases has become a serious issue, especially in the fight against Gram-negative bacteria.

In recent years, only a few novel broad-spectrum β-lactam antibiotics with improved activity against Gram-negative bacteria have been discovered. In addition, the combination of β-lactam antibiotics with β-lactamase inhibitors has emerged as a viable strategy to fight MDR Gram-negative bacteria. Among the more recent β-lactam antibiotics that are in late stage clinical development (Phase III) or being marketed are the two anti-MRSA cephalosporins, ceftobiprole and ceftaroline (**4**) [25]. Both compounds are considered fifth generation cephalosporins due to their pronounced anti-MRSA activity, but like the previous generation cephalosporins, they maintain a broad-spectrum activity against Gram-negative bacteria. However, like their predecessors, these cephalosporins do not overcome resistance from Gram-negative bacteria producing extended spectrum β-lactamases (ESBLs) [26,27]. More recently, ceftaroline has been described in combinations with other antibiotics such as amikacin and meropenem to have synergistic activities against cephalosporin-resistant Gram-negative bacteria [28]. Two more recent cephalosporins with enhanced anti-pseudomonal activity include **5** (CXA-101) [29,30] and **6**. CXA-101 has demonstrated potent activity toward *P. aeruginosa* including carbapenem-resistant isolates [31,32]. However, the compound displayed modest activity when tested against cephalosporin-resistant *P. aeruginosa* and against cephalosporin-resistant Enterobacteriaceae and *Bacteroides fragalis* isolates [33]. Compound **6** employs a catechol unit as an iron-chelating group which improves its entry into Gram-negative bacteria, particularly into *P. aeruginosa* [34,35]. CXA-101 has recently completed early clinical trials for safety and efficacy in patients with complicated urinary tract infections and is currently under investigation for treatment of complicated intra-abdominal infections with Gram-negative bacteria.

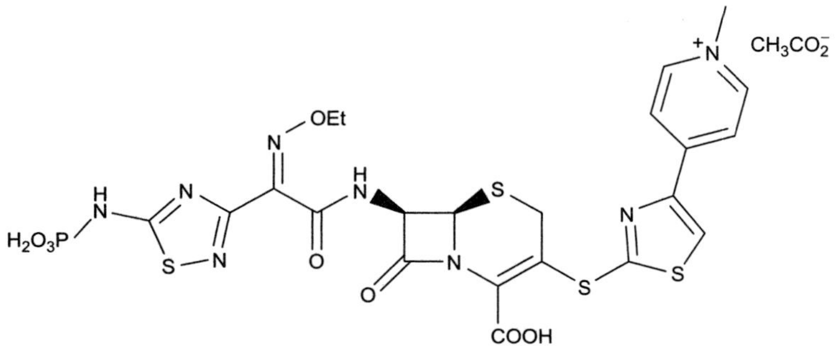

4

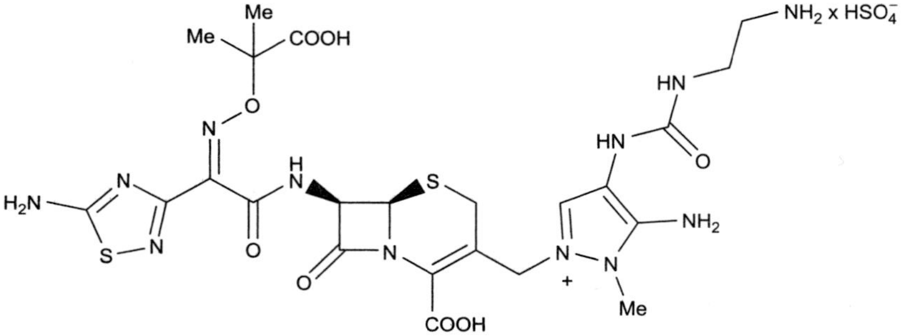

5

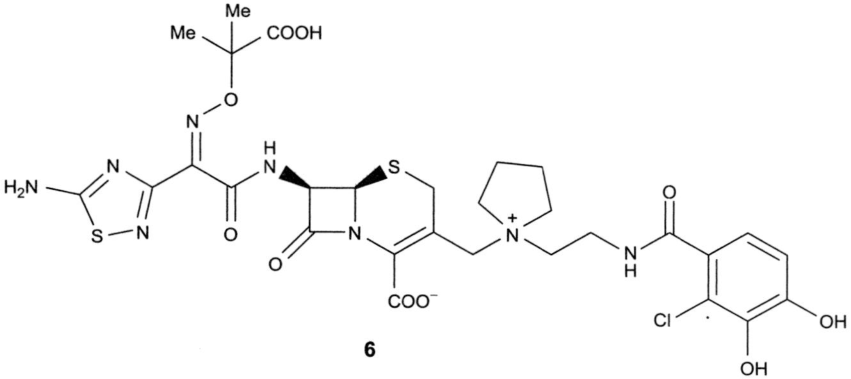

6

New tricyclic carbapenems such as **7** (LK-176) [36] are effective as inhibitors of β-lactamases of class C and also of class A by forming stable acyl enzyme complexes with low turnover rates [37]. Compound **7** demonstrated synergism with cefotaxim and ceftazidime against Enterobactericeae that produce class A and class C β-lactamases [25]. Similar to **6**, some novel agents such as **8** (BAL30072) incorporate siderophores in their structure in order to facilitate the intracellular uptake by iron transport mechanisms [38]. In addition, the siderophore moiety of BAL30072 contributes directly to the good antimicrobial activity and the facilitated uptake appears to contribute significantly to the activity against *A. baumanni*. Compound **8** shows a broad-spectrum antibiotic activity including MDR *P. aeruginosa* and some carbapenem-resistant *Acinetobacter* spp. [38]. In recent years, several successful examples of combinations of β-lactam antibiotics with β-lactamase inhibitors such as **9** (tazobactam) and **10** (NXL104) have been described [25]. CXA-201 is a 2:1 combination of **5** and **9** that is effective against MDR *P. aeruginosa* and other Gram-negative pathogens and is currently in Phase II development [39]. Combinations of **10** with cefazidime and ceftaroline have also been described [25].

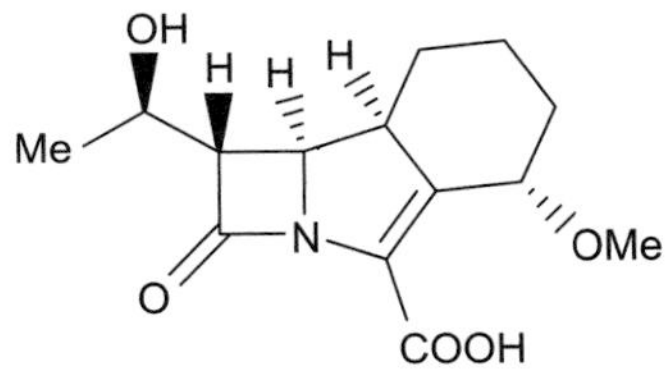

7

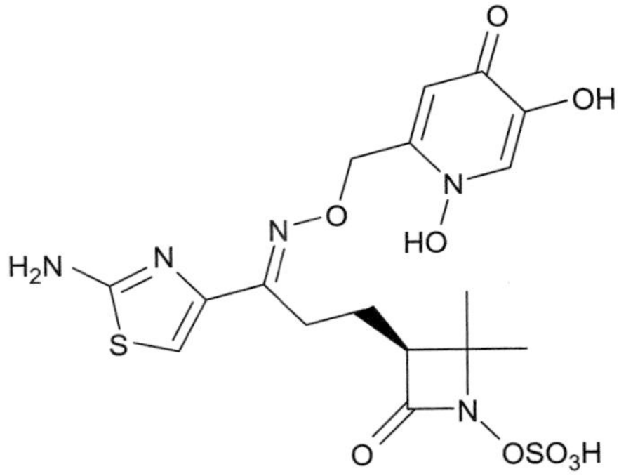

8

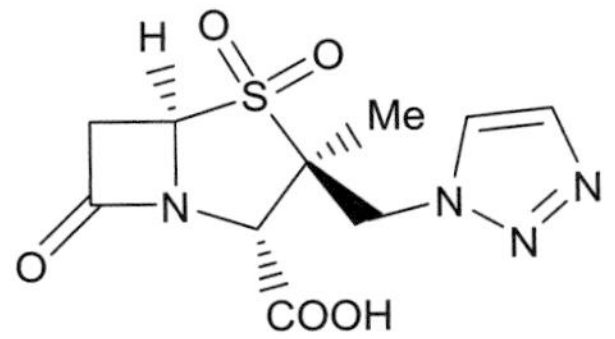

9

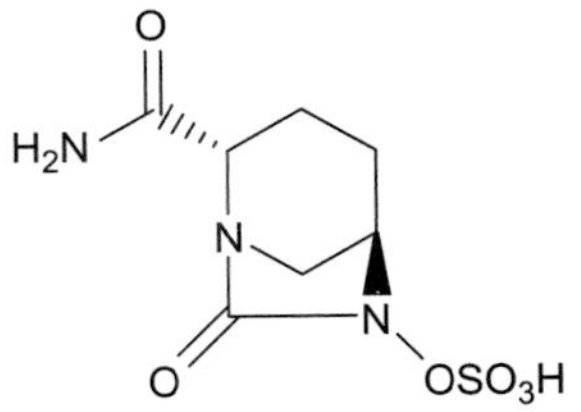

10

2.3. New tetracycline antibiotics

The tetracyclines were discovered in 1945 (chlortetracycline) and belong to one of the important classes of broad-spectrum antibiotics [40]. Two of the more common semisynthetic tetracyclines that have been marketed for decades are doxycycline and minocycline. The tetracyclines inhibit bacterial growth by inhibiting bacterial protein synthesis by preventing the association of aminoacyl-tRNA with bacterial ribosomes [41]. Resistance to tetracyclines is mediated through three main mechanisms: efflux, reduced affinity to the target, as well as enzymatic inactivation. A number of tetracycline analogues have been designed to prevent the development of resistance toward tetracyclines. The glycylcyclines such as **11** (tigecycline) bind with a 5- to 100-fold higher affinity to 30S and 70S ribosomes than tetracyclines, respectively. In 2005, the FDA approved tigecycline for the treatment of complicated skin and soft tissue infections as well as complicated intra-abdominal infections [3,42]. Tigecycline is active against many Gram-positive bacteria, Gram-negative bacteria, and anaerobes—including MRSA, vancomycin-resistant *Enterococci* (VRE), drug-resistant *Streptococcus pneumoniae*, and Gram-negative pathogens of respiratory infections such as *Haemophilus influenzae, Moraxella catarrhalis*, Enterobacteriaceae, and MDR strains of *A. baumannii*. It is also active against *Mycoplasma pneumoniae*. However, tigecycline has limited activity against *Pseudomonas* spp. or *Proteus* spp. [43]. Additional analogues of **11** in clinical development include **12** (PTK-0796; BAY 73-6944), an aminomethylcycline, and more recently **13** (TP-434) [44]. A Phase III study of PTK-0796 in patients with complicated skin and skin structure infection (CSSSI) is planned.

Like tigecycline, these novel compounds overcome tetracycline resistance due to efflux pumps and/or ribosomal protection [45], and show a broad-spectrum activity toward both Gram-positive and Gram-negative pathogens. However, similar to tigecycline, these compounds lack good antibacterial activity against *P. aeruginosa*.

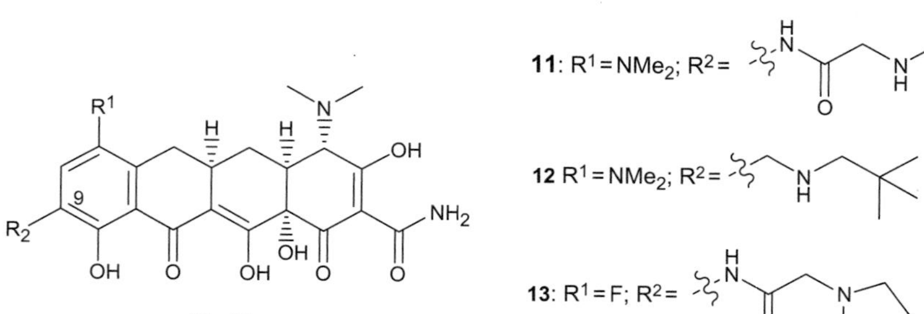

11: $R^1 = NMe_2$; $R^2 =$

12 $R^1 = NMe_2$; $R^2 =$

13: $R^1 = F$; $R^2 =$

11–13

2.4. New quinolone antibiotics

The quinolone class is one of the most important classes of antibiotics identified in the past 50 years [46]. The discovery of fluoroquinolones in the 1980s constituted a breakthrough due to their excellent broad-spectrum activity that includes Gram-negative pathogens. Ciprofloxacin (**14**), levofloxacin, and moxifloxacin have become major pharmaceutical products. Quinolones target the bacterial type II topoisomerases, that is, DNA gyrase, topoisomerase IV [47,48]. These heterotetrameric enzymes manipulate DNA topology by introduction of transient double-stranded breaks in bound DNA. Quinolone antibiotics bind to the enzyme complex and the cut DNA, which stabilizes this so-called cleavage complex and leads to accumulation and eventual release of double-stranded DNA breaks that are ultimately lethal to the cell because bacterial DNA synthesis is inhibited [49] .These enzymes are attractive as antibacterial targets as they are essential, common to all bacteria, and their human homologues are sufficiently different to achieve selectivity. Ciprofloxacin remains the most potent quinolone against Gram-negative bacteria and is effective against many susceptible strains of *P. aeruginosa* and *A. baumannii*. Despite potential toxicity issues and emergence of bacterial resistance, there is still enthusiasm for these antibiotics, especially in view of their activity against Gram-negative bacteria [46]. Quinolones that are in clinical development showing good activity against Gram-negative bacteria are finafloxacin (**15**) and nemonoxacin (**16**). Due to their broad-spectrum activity, quinolones have been tested against various infections. Compound **15** was in Phase II clinical trials for *Helicobacter pylori* and urinary tract infection indications and possesses antibacterial activity against ciprofloxacin-sensitive and -resistant *A. baumannii* strains [50,51]. Phase II trials for community-acquired pneumonia (CAP) and diabetic foot infections were recently completed for quinolone **16** with a favorable safety profile [52]. Among the preclinical compounds, quinolone **17** (WQ-3813) shows interesting activities against nosocomial pathogens MRSA, *E. coli*, and *A. baumannii* as well as major respiratory pathogens including MDR- and quinolone-resistant isolates [48]. It has the potential to be developed for the treatment of respiratory infections.

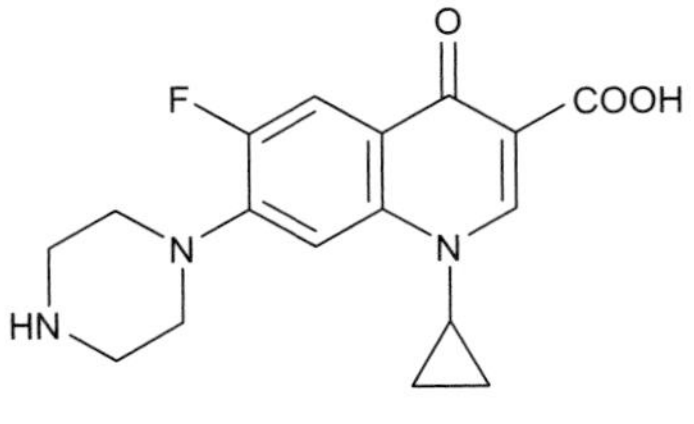

14 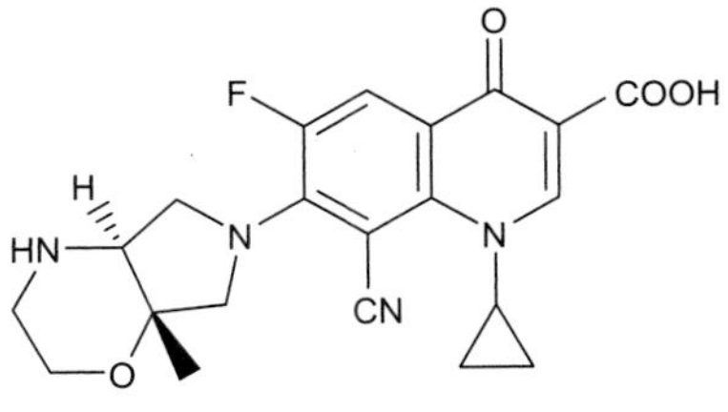 **15**

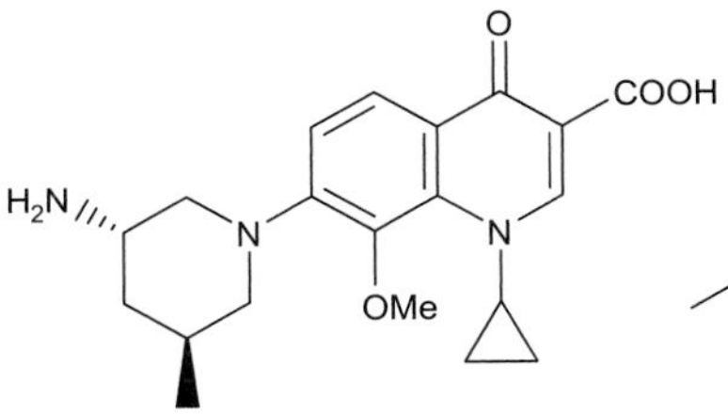

16 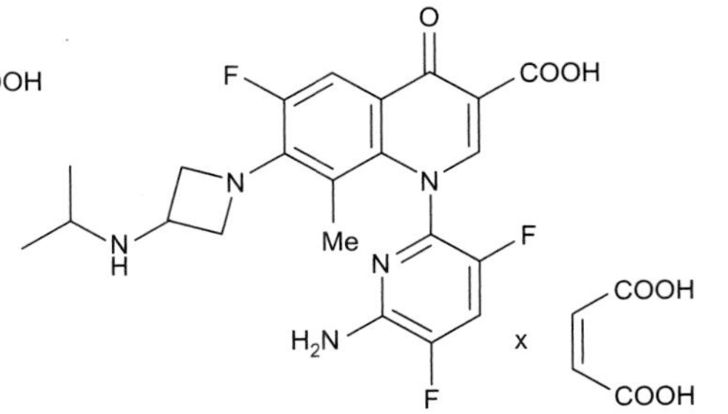 **17**

2.5. New antimicrobial peptides, lipopeptides, and natural product-derived antibiotics

Natural products have been a rich source for the discovery of new antibacterial drugs [53,54]. Among those, antimicrobial peptides (AMPs) [5,55] and lipopeptides [56] hold great promise to the identification and development of new antibiotics against MDR pathogens, including Gram-negative species.

The polymyxin family of antibiotics was discovered in 1947 and comprises polymyxins A–E, where only polymyxins B and E (colistin) have been used clinically [54]. Polymyxins are highly active against a large panel of Gram-negative bacteria, including *P. aeruginosa* and *A. baumannii* [54,57]; however, they were abandoned as systemic therapy in the 1970s because of renal toxicity [10]. Due to the rapid emergence of MDR Gram-negative bacteria, in recent years they have regained importance and are now regarded as last resort antibiotics. However, colistin-resistant *A. baumannii* strains are emerging rapidly in recent years [57]. Hence, new polymyxin derivatives showing an improved toxicity profile and activity against these MDR clinical strains would be highly desirable. Alternatively, the discovery of novel antibiotics with a similar antibacterial profile as polymyxins but with a different mode of action would be desirable to overcome colistin resistance.

Two interesting new polymyxin-derived antibiotics **18** (NAB739) and **19** (NAB7061) have recently been described in preclinical studies [10,58]. Compound **18** shows a similar antimicrobial activity spectrum to polymyxin B or colistin; however, it has an overall net positive charge of +3, compared to +5 for the polymyxins, which could be favorable for reducing renal toxicity. NAB7061, similar to the polymyxin nonapeptide, lacks direct antimicrobial activity but reduces the MICs of many antibiotics when given in combination (*e.g.*, with rifampin and clarithromycin) for *E. coli, K. pneumoniae, K. oxytoca, E. freudii,* and *A. baumannii* [58]. NAB compounds **18** and **19** have decreased affinity in the isolated rat kidney brush border membrane assay in comparison to polymyxin B (~ 15-20%), which might be a predictive assay for kidney toxicity observed for polymyxins and AGs. The relevance of this finding for kidney toxicity *in vivo* remains to be seen [10]. Polymyxin B derivative **20** (CB-182,804) having a net charge of +5 shows activity against Gram-negative pathogens, including ESBLs and carbapenemases (KPCs)-producing strains [59]. Overall, the antibacterial spectrum of activity of CB-182,804 closely resembles that of colistin and polymyxin B [60]. *In vitro* killing kinetics of CB-182,804 showed that the compound had low MICs (0.5–2 μg/ml) and very good killing against 10 drug-susceptible and -resistant Gram-negative species [61].

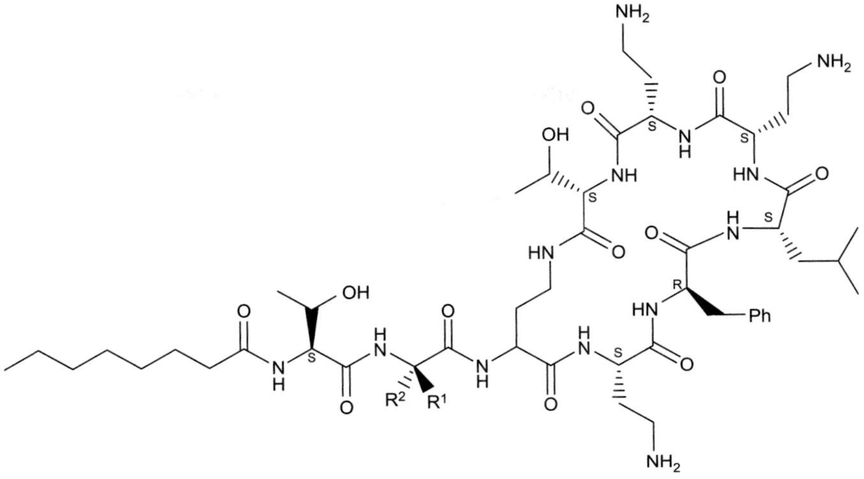

18: $R^1 = CH_2OH$; $R^2 = H$

19: $R^1 = H$; $R^2 = CH_2CH_3$

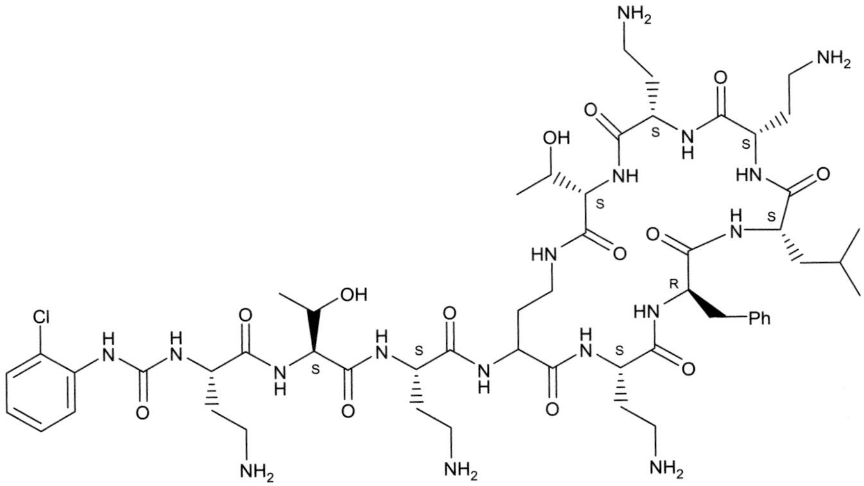

20

3. NOVEL COMPOUND CLASSES

3.1. Oxaboroles

Oxaborole **21** (AN3365) belongs to a novel class of boron-containing antibiotics with good antimicrobial activity against a broad range of Gram-negative bacteria, including MDR *P. aeruginosa* ($MIC_{50/90} = 4/8$ μg/ml), wild-type *Acinetobacter* spp., and *Burkholderia cepacia* [62]. The oxaboroles block protein biosynthesis by inhibiting leucyl-tRNA synthetase *via* a unique mechanism of action [63]. Compound **21** completed a Phase I clinical trial [62] and holds promise for the treatment of Gram-negative bacterial infections against MDR pathogens.

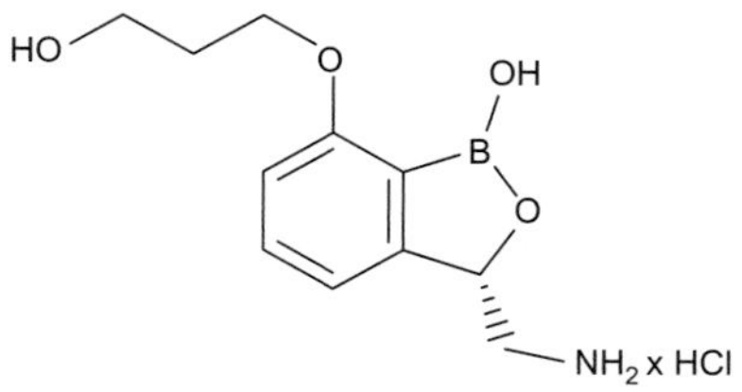

HO O OH B O NH_2 x HCl

21 (AN3365)

3.2. LpxC inhibitors

The zinc-dependent metalloaminase UDP-3-*O*-(*R*-3-hydroxymyristoyl)-N-acetylglucosamine deacylase (LpxC) is a promising target against Gram-negative bacteria as it is highly conserved, and many potent LpxC inhibitors have been described [64]. Among those, compound **22** (CHIR-090) shows antimicrobial activity against *P. aeruginosa* and *E. coli* comparable to tobramycin and ciprofloxacin [64].

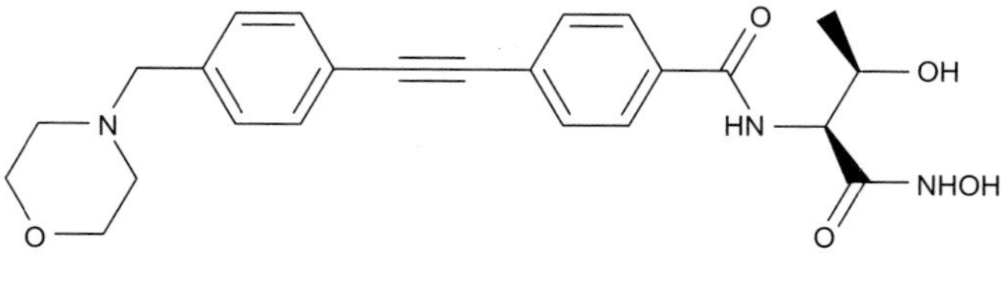

N O O HN OH NHOH O

22 (CHIR-090)

3.3. Ceragenins (CSA-13)

Ceragenins are cholic acid-derived antimicrobial agents that mimic the activity of endogenous AMPs and act *via* membrane depolarization [5,65]. CSA-13 (**23**) is the most potent of the ceragenin class with broad-spectrum antimicrobial activity. MBCs for **23** against *E. coli* (minimal bactericidal concentration (MBC)=1.5 μg/ml), *K. pneumoniae* (MBC = 2.5 μg/ml), and *P. aeruginosa* (MBC = 3.5 μg/ml) were described [65]. In particular, **23** shows also some activity against a MDR *P. aeruginosa* strain [66]. The *in vitro* activities of CSA-13, alone or in combination with colistin, tobramycin, and ciprofloxacin, were investigated using 50 *P. aeruginosa* isolates from CF patients. Synergistic interactions were mostly seen for CSA-13 in combination with colistin. Therefore, CSA-13 seems to be an interesting candidate for the treatment of *P. aeruginosa* strains in CF patients, alone or in combination. However, additional studies on safety, efficacy, and PK are needed [67].

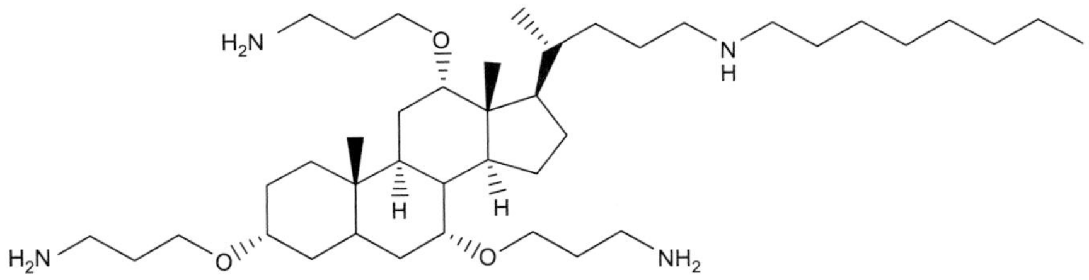

23

3.4. Peptide mimetics: PMX30063

In view of overcoming some unfavorable drug-like properties of AMPs such as low plasma stability and high hemolytic activity, new approaches have been utilized to design and synthesize peptide mimetics. Restrained antimicrobial arylamide foldamers such as **24** have been described in recent years [68].

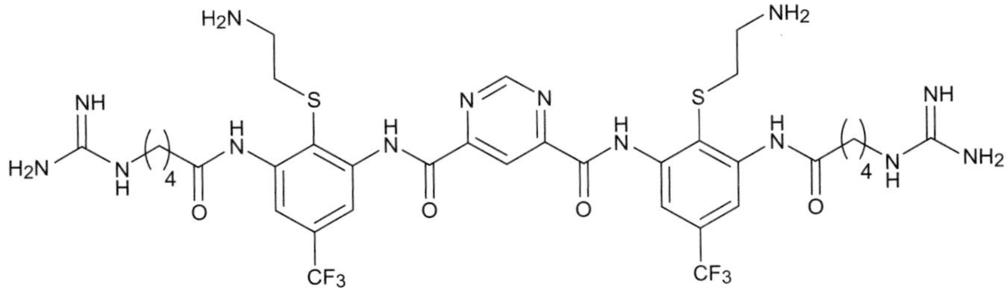

24

Foldamers were designed to mimic amphiphilic structures observed in many α-helical- and β-hairpin-derived AMPs by a non-peptidic aryl amide scaffold. These compounds show a membranolytic mechanism of action which seems, however, to be more specific toward bacterial rather than mammalian membranes resulting in reduced hemolytic activity.

The most active compounds show broad-spectrum activity including activity against some Gram-negative bacteria.

PMX30063 was derived from the foldamer concept and has entered clinical development in 2008 [69,70].

3.5. New β-hairpin mimetics targeting outer-membrane biogenesis of *P. aeruginosa*

AMPs such as protegrin I (**25**) show great potential in large part due to their activity against MDR Gram-negative bacteria [55] and low incidences of bacterial resistance formation. However, their clinical development as systemic drugs has so far been hampered by some unfavorable ADMET properties [5].

By transferring the protegrin I pharmacophore onto a cyclic 14-residue peptide β-hairpin mimetic scaffold [71,72], potent protegrin I analogues could be obtained [72]. Several rounds of optimization aimed at optimizing potency and ADMET properties yielded novel compounds such as **26** (POL7001) [73], which display potent and selective activity against a large panel of *P. aeruginosa* clinical strains, including many isolates from CF patients resistant to one or more classes of antibiotics. MIC_{90} for **26** was 0.25 μg/ml against a broad panel of clinical isolates including several MDR strains. *In vivo* efficacy of **26** was evaluated in a mouse septicemia infection model after dosing subcutaneously at 1 and 5 h after bacterial inoculation with either *P. aeruginosa* ATCC 9027 or ATCC 27853. POL7001 demonstrated excellent activity with ED_{50} in the range of 0.25–0.28 mg/kg as compared to gentamycin (used as a positive control), which showed an ED_{50} of 3.1 and 2.9 mg/kg, respectively [73].

Biochemical and genetic studies showed that these peptidomimetics are bactericidal, however, they act by a non-membranolytic mechanism of action. Instead, through the use of photoaffinity-labeled analogues, the putative target of these novel compounds was identified as the *Pseudomonas* OM protein, LptD [73,74]. LptD is a β-barrel OM protein widely distributed in Gram-negative bacteria that plays a role in the assembly of LPS in the outer leaflet of the outer membrane. LptD represents a novel target in antibiotic drug discovery and development, and its identification emphasizes the essential need to explore novel chemical approaches. One member of the β-hairpin mimetic family (POL7080) is currently under investigation in a Phase I study to assess its safety, tolerability, and pharmacokinetic profile in man for the intravenous administration route.

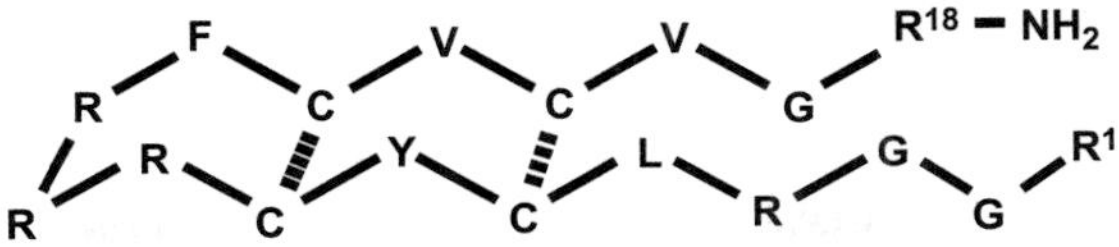

25 (Protegrin I)

C = cystein, F = phenylalanine, G = glycine, L = leucine, R = arginine, Y = tyrosine

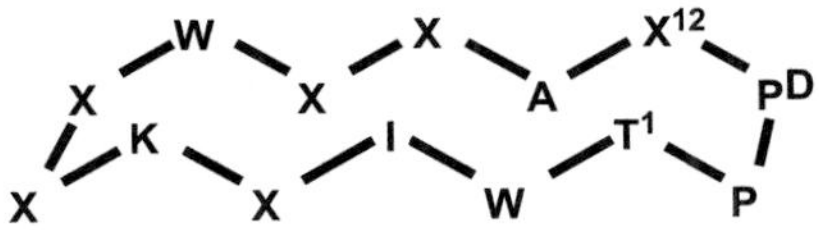

26 (POL7001)

A = alanine, I = isoleucine, K = lysine, P = proline, T = threonine, X: L-2,4-diamino butyric acid

4. CONCLUSION

In this review, we have summarized the most promising antibacterial compounds currently being developed against Gram-negative bacteria. As a major caveat, most of these compounds in preclinical and early development are derived from established classes. Within the established drug classes, novel β-lactam antibiotics containing iron-chelating groups such as **8** seem promising. They use bacterial iron uptake mechanisms as a means to bypass the defense mechanisms of bacteria ("Trojan horse" approach). These compounds show good activity against a range of MDR Gram-negative bacteria. Furthermore, several combinations of established β-lactam antibiotics such as ceftaroline and novel β-lactamase inhibitors such as **10** are emerging. The excellent activity of ciprofloxacin against a broad range of Gram-negative strains has triggered renewed interest in the development of new quinolones. Compound **15** is one of the most promising candidates. However, these new derivatives will encounter the well-established mechanisms which lead to resistance to previous members of the fluoroquinolone class. Furthermore, analogues of the last resort antimicrobial peptide colistin (*e.g.*, **18** and **20**) have been designed and studied by several academic groups and small Biotech companies. Whether these new compounds will show an improved therapeutic window *in vivo* is currently under investigation.

The search for novel antibiotics with a novel mechanism of action has been traditionally very difficult. In that respect, the discovery of the oxaboroles (*e.g.*, **21**) and of a new class of β-hairpin mimetics (*e.g.*, **26**)

constitutes highlights. Both compound classes exhibit novel, bacteria-specific mechanisms of action and may offer solutions to the emerging resistance problems seen with current antibiotics to combat serious and life-threatening infections by Gram-negative bacteria.

REFERENCES

[1] M. Klievens, J. Edwards, C. L. Richards, T. C. Horan, R. P. Gaynes, D. A. Pollock and D. M. Cardo, *Publ. Health Rep.*, 2007, **122**, 160.
[2] L. L. Leape, T. A. Brennan, N. Laird, A. G. Lawthers, A. R. Localio, B. A. Barnes, L. Herbert, J. P. Newhouse, P. C. Weiler and H. Hiatt, *N. Engl. J. Med.*, 1991, **324**, 377.
[3] G. Devasahayam, W. M. Scheld and P. S. Hoffman, *Expert Opin. Invest. Drugs*, 2010, **19**, 215.
[4] A. Y. Peleg and D. C. Hooper, *N. Engl. J. Med.*, 2010, **362**, 1804.
[5] M. Vaara, *Curr. Opin. Pharmacol.*, 2009, **9**, 1.
[6] L. B. Rice, *J. Infect. Dis.*, 2008, **197**, 1079.
[7] D. N. Gilbert, R. J. Guidos, H. W. Boucher, G. H. Talbot, B. Spellberg, J. E. Edwards, M. Scheld, J. S. Bradley and J. G. Bartlett, *Clin. Infect. Dis.*, 2010, **50**, 1081.
[8] G. H. Talbot, J. Bradley, J. E. Edwards Jr., D. Gilbert, M. Scheld and J. G. Bartlett, *Clin. Infect. Dis.*, 2006, **42**, 657.
[9] H. W. Boucher, G. H. Talbot, J. S. Bradley, E. Edwards Jr., D. Gilbert, L. B. Rice, M. Scheid, B. Spellberg and J. Bartlett, *Clin. Infect. Dis.*, 2009, **48**, 1.
[10] M. Vaara, *Curr. Opin. Microbiol.*, 2010, **13**, 574.
[11] M. N. Gwynn, A. Portnoy, S. F. Rittenhouse and D. J. Payne, *Ann. N. Y. Acad. Sci.*, 2010, **1213**, 5.
[12] C. Alexander and E. T. Rietschel, *J. Endotox. Res.*, 2001, **7**, 167.
[13] D. S. Kabanov and I. R. Prokhorenko, *Biochemistry (Moscow)*, 2010, **75**, 383.
[14] R. Gaynes and J. Edwards, *Clin. Infect. Dis.*, 2005, **41**, 848.
[15] M. H. Kollef and S. T. Micek, *Crit. Care Med.*, 2005, **33**, 1845.
[16] A. K. May, S. B. Fleming, R. O. Carpenter, J. J. Diaz JR, O. D. Guillamondegui, S. A. Deppen, R. S. Miller, T. R. Talbot and J. A. Morris, *Surg. Infect.*, 2006, **7**(5), 409.
[17] A. Schatz, E. Bugie and S. A. Waksman, *Proc. Soc. Exp. Biol.*, 1944, **55**, 66.
[18] I. R. Hooper, in *Aminoglycoside Antibiotics*, (eds. H. Umezawa and I. R. Hooper), Springer Verlag, Berlin, 1982.
[19] P. Dozzo and H. E. Moser, *Expert Opin. Ther. Patents*, 2010, **20**, 1321.
[20] J. L. Houghton, K. D. Green and S. Garneau-Tsodikova, *ChemBioChem*, 2010, **11**, 880.
[21] N. Minowa, T. Usui, Y. Akiyama, Y. Hiraiwa, T. Yoneda, T. Hasegawa, K. Maebashi, T. Ida, K. Katsumata, K. Otsuka and D. Ikea, *Patent application WO2005/070945-A1*, 2005.
[22] J. Aggen, A. A. Goldblum, M. Linsell, P. Dozzo, H. E. Moser, D. Hildebrand and M. Gliedt, *Patent application WO2009/067692-A1*, 2009.
[23] T. R. Fritsche, M. Castanheira, G. H. Miller, R. N. Jones and E. S. Armstrong, *Antimicrob. Agents Chemother.*, 2008, **52**, 1843.
[24] S. Mushtaq, M. Warner, J.-C. Zhang, S. Maharjan, M. Doumith, N. Woodford and D. Livermore, *Clin. Microbiol. Infect.*, 2010, **16**(Suppl. 2), 285.
[25] K. Bush and M. J. Macielag, *Expert Opin. Ther. Patents*, 2010, **20**, 1277.
[26] M. G. P. Page, *Curr. Opin. Pharmacol.*, 2006, **6**, 480.
[27] K. M. Amsler, T. A. Davies and W. Shang, et al., *Antimicrob. Agents Chemother.*, 2008, **52**, 3418.
[28] D. Biek, *Patent application WO2010/025328-A1*, 2010.
[29] A. Toda, H. Ohki and T. Yamada, et al., *Bioorg. Med. Chem. Lett.*, 2008, **18**, 4849.

[30] D. M. Livermore, S. Mustaq and Y. Ge, et al., *Int. J. Antimicrob. Agents*, 2009, **34**, 402.
[31] S. Takeda, T. Nakai, Y. Wakai, F. Ikeda and K. Hatano, *Antimicrob. Agents Chemother.*, 2007, **51**, 826.
[32] C. Juan, L. Zamorano, J. L. Perez, Y. Ge and A. Oliver, *Antimicrob. Agents Chemother.*, 2010, **54**, 846.
[33] H. S. Sader, S. D. Putnam and R. N. Jones, 49th ICAAC 2009, F1-1992, 2009.
[34] Y. Nishitani, K. Yamawaki, Y. Takeoka, H. Sugimoto, S. Hisakawa and T. Aoki, *Patent application WO/2010/050468-A1*, 2010.
[35] A. M. Harris, *Curr. Opin. Invest. Drugs*, 1993, **2**, 109.
[36] A. Prezelj, U. Urleb and G. Vilfan, *Patent application WO2009/095387-A1*, 2009.
[37] S. Han, R. P. Zaniewski, E. S. Marr, B. M. Lacey, A. P. Tomaras, A. Evdokimov, J. R. Miller and V. Shanmugasundaram, *PNAS*, 2010, **107**, 22002.
[38] M. G. P. Page and J. Heim, *IDrugs*, 2009, **12**, 651.
[39] http://www.cubist.com/products.
[40] M. O. Griffin, E. Fricovsky, G. Ceballos and F. Villareal, *Am. J. Cell Physiol.*, 2010, **299**, C539.
[41] I. Chopra and M. Roberts, *Microbiol. Mol. Biol. Rev.*, 2001, **65**, 232.
[42] V. Seputienne, J. Povilonis, J. Armalyte, K. Suziedelis, A. Pavilonis and E. Suziedeliene, *Medicina (Kaunas)*, 2010, **46**, 240.
[43] A. C. Gales and R. N. Jones, *Diagn. Microbiol. Infect. Dis.*, 2000, **36**, 19.
[44] J. Sutcliffe, W. O'Brian, C. Achorn, P. Appelbaum, C. Pillar and G. Zurenko, 50th ICAAC, September 12–15, 2010 Boston, Poster 127, F1.2158.
[45] W. Rose and M. Rybak, *Pharmacotherapy*, 2006, **26**(8), 1099.
[46] J. A. Wiles, B. J. Bradbury and M. J. Pucci, *Expert Opin. Ther. Patents*, 2010, **20**, 1295.
[47] G. S. Bisacchi and J. Dumas, *Annu. Rep. Med. Chem.*, 2009, **44**, 379.
[48] B. J. Bradbury and M. J. Pucci, *Curr. Opin. Pharmacol.*, 2008, **8**, 574.
[49] A. Wohlkonig, P. F. Chan, A. P. Fosberry, P. Homes, J. Huang, M. Kranz, V. R. Leydon, T. J. Miles, N. D. Pearson, R. L. Perera, A. J. Shillings, M. N. Gwynn and B. D. Bax, *Nat. Struct. Mol. Biol.*, 2010, **9**, 1152.
[50] S. Vasiliou, M. Vicente and R. Castaner, *Drugs Future*, 2009, **43**, 451.
[51] P. G. Higgins, W. Stubbings, H. Wiplingshoff and H. Seifert, *Antimicrob. Agents Chemother.*, 2010, **54**, 1613.
[52] A. Arjona, *Drugs Future*, 2009, **49**, 196.
[53] F. Von Nussbaum, M. Brands, B. Hinzen, S. Weigand and D. Häbich, *Angew. Chem. Int. Ed. Engl.*, 2006, **45**, 5072.
[54] T. Velkov, P. E. Thompson, R. L. Nation and J. Li, *J. Med. Chem.*, 2010, **53**, 1898.
[55] E. B. Hadley and R. E. W. Hancock, *Curr. Top. Med. Chem.*, 2010, **10**, 1872.
[56] G. Pirri, A. Guiliani, S. F. Nicoletto, L. Pizzuto and A. C. Rinaldi, *Cent. Eur. J. Biol.*, 2009, **4**, 258.
[57] R. L. Nation and J. Li, *Curr. Opin. Infect. Dis.*, 2009, **22**, 535.
[58] M. Vaara, J. Fox, G. Loidl, O. Siikanen, J. Apajalahti, F. Hansen, N. Fridmodt-Moller, J. Nagai, M. Takano and T. Vaara, *Antimicrob. Agents Chemother.*, 2008, **52**, 3229.
[59] M. Castanheira, H. S. Sader and R. N. Jones, *Microb. Drug Resist.*, 2010, **16**, 61.
[60] M. M. Traczewski and S. D. Brown, 50th ICAAC, September 12–15 2010, Boston, F1-1622.
[61] G. Lin, G. Pankuch and P. C. Appelbaum, 50th ICCAC, September 12–15 2010, Boston, F1-1623.
[62] D. J. Biedenbach, R. E. Mendes, M. R. K. Alley, H. S. Sader and R. N. Jones, 50th ICAAC, September 12–15 2010, Boston, F1-1639.
[63] V. Hernandez, T. Akama, M. R. K. Alley, S. Barker, W. Mao, F. Rock, Y. K. Zhang, Y. Zhang, Y. Zhou, T. Crepin, S. Cusack, A. Palencia, J. Nieman, M. Anugla, M. Baek,

C. Diaper, C. Ha, M. Keramane, X. Lu, R. Mohammad, K. Savariraj, R. Sharma, R. Singh, R. Subedi and J. Plattner, 50th ICAAC, September 12–15 2010, Boston, F1-1637.
[64] A. W. Barb and P. Zhou, *Curr. Pharm. Biotechnol.*, 2008, **9**, 9.
[65] R. F. Epand, J. E. Pollard, J. O. Wright, P. B. Savage and R. M. Epand, *Antimicrob. Agents Chemother.*, 2010, **54**, 3708.
[66] J. N. Chin, R. N. Jones, H. S. Sader, P. B. Savage and M. J. Rybak, *J. Antimicrob. Chemother.*, 2008, **61**, 365.
[67] C. Bozkurt Güzel, P. B. Savage and A. A. Gerçeker, 50th ICAAC, September 12–15 2010, Boston, E-2050.
[68] S. Choi, A. Isaacs, D. Clements, D. Liu, H. Kim, R. W. Scott, J. D. Winkler and W. D. DeGrado, *PNAS*, 2009, **106**, 6968.
[69] PolyMedix press release, December 2008.
[70] http://www.polymedix.com/novel_antibiotic.php.
[71] J. A. Robinson, S. DeMarco, F. Gombert, K. Moehle and D. Obrecht, *Drug Discov. Today*, 2008, **13**, 955.
[72] J. A. Robinson, S. C. Shankaramma, P. Jetter, U. Kienzl, R. A. Schwendener, J. W. Vrijbloed and D. Obrecht, *Bioorg. Med. Chem.*, 2005, **13**, 2055.
[73] N. Srinivas, P. Jetter, B. J. Ueberbacher, M. Werneburg, K. Zerbe, J. Steinmann, B. Van der Meijden, F. Bernardini, A. Lederer, R. L. Dias, P. E. Misson, H. Henze, J. Zumbrunn, F. O. Gombert, D. Obrecht, P. Hunziker, S. Schauer, U. Ziegler, A. Käch, L. Eberl, K. Riedl, S. J. DeMarco and J. A. Robinson, *Science*, 2010, **327**, 1010.
[74] This work is supported in part by the European Union FP7-NABATIVI program.

CHAPTER 16

Hepatitis C Virus—Progress Toward Inhibiting the Nonenzymatic Viral Proteins

Nicholas A. Meanwell and **Makonen Belema**

Contents

1. INTRODUCTION

The molecular characterization of the hepatitis C virus by Michael Houghton and colleagues in 1991 catalyzed a significant effort to identity and develop potent and selective virus inhibitors suitable for clinical application [1]. The HCV NS3 protease and NS5B RNA-dependent RNA polymerase were inevitably the initial targets of focus because these enzymes were readily recapitulated functionally using biochemical assays, an effort considerably facilitated by the solving of X-ray

Department of Medicinal Chemistry, Bristol-Myers Squibb Research and Development, 5 Research Parkway, Wallingford, CT 06492, USA

Annual Reports in Medicinal Chemistry, Volume 46
ISSN: 0065-7743, DOI: 10.1016/B978-0-12-386009-5.00011-4

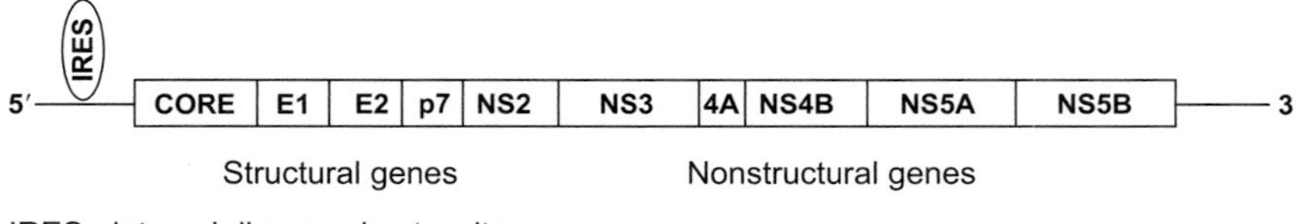

Figure 1 Genomic arrangement of HCV.

crystallographic structures that informed structure-based drug design campaigns [2,3]. In 2011, boceprevir and telaprevir, the first NS3 protease inhibitors to be submitted for marketing authorization, were approved as adjuncts to the current optimal therapy, comprising a combination of pegylated interferon-α and ribavirin, neither of which are specific antiviral agents. While the addition of these agents to current therapy leads to improved sustained viral response rates and/or shorter drug dosing regimens, it is well recognized that additional direct-acting antiviral agents will be needed to be used in conjunction with interferons and ribavirin or in small molecule combinations, if the opportunity to eradicate HCV within 40 years of its discovery is to be realized [4–6]. The advent of subgenomic replicons in 1999 followed in 2005 by replicating virus provided critical tools to identify and assess inhibitors with novel modes of action that may not be amenable to a simple biochemical assay [7,8]. Combined with the maturing clinical portfolio of NS3 and NS5B inhibitors, the advent of these assays stimulated the identification of a number of molecules targeting interesting and effective viral targets beyond the known enzymes [9]. In this chapter, we summarize the key developments that have occurred in this field over the past several years.

The genomic arrangement of HCV is depicted in Figure 1. The viral genome is translated as a single polyprotein that is cleaved by a combination of host cell proteases and the viral proteases NS2 and NS3. All viral proteins are plausible targets, and since many play multiple roles in replication, temporal selectivity based on unique protein function at different stages of the virus life cycle may be possible.

2. HCV CORE (CAPSID) PROTEIN INHIBITORS

The core or capsid protein is essential to virion assembly and structure, interacts with several host cell proteins, and is the most conserved of all of the viral proteins across the six major genotypes [10]. The core protein oligomerizes during the process of viral assembly, and a screen evaluating dimerization, a process mediated by the N-terminal domain, identified small-molecule inhibitors from a library of indoline derivatives

[11–13]. After library expansion, the racemic hexahydroindolo[2,3-*d*][1,8] naphthyridine **1** was identified as the most potent inhibitor of core protein dimerization (IC_{50} = 1.4 μM) that also inhibited genotype 2a virus replication with an EC_{50} = 2.3 μM [10,11]. However, the precise mode by which **1** exerts its antiviral effect remains to be established. More recently, dimeric forms of these compounds tethered *via* the indoline N atom have shown improved potency [13].

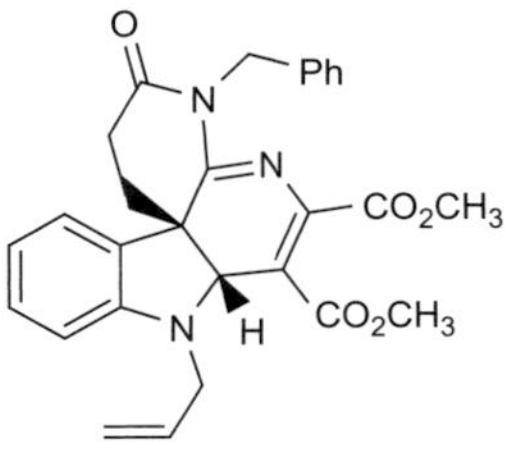

1

3. HCV ENTRY INHIBITORS

The E1 and E2 proteins are expressed on the surface of the viral membrane and associate into heterodimers in the native state, primed to mediate virus attachment and fusion that allows the virus to enter the host cell cytosol. The precise roles and function of E1 and E2 in the entry process and the choreography associated with the complex series of events as HCV engages several host cell proteins to enter hepatocytes have not been elucidated in detail [14–16]. However, association with highly sulfated heparan sulfate is thought to be an important first step in the binding process, with an ensuing interaction between E2 and the large extracellular loop of the tetraspanin CD81 [17]. Several other host proteins have been shown to be important contributors to the entry process including the tight junction protein claudin-1, the scavenger receptor, class B, type I (SR-BI), and occludin, which may play a role in post-binding events [16,18–22]. Endocytosis of the virus into clathrin-coated pits leads to internalization, with the ensuing fusion of viral and host cell membranes occurring in a process dependent on the low pH environment [23,24].

Screening for inhibitors of HCV entry has been mechanistically agnostic, relying upon the use of either pseudoparticles or infectious virus to identify lead compounds, although the determination of a role for SR-BI in virus entry focused interest on compounds known to interfere with the function of this host cell protein [25–27]. ITX-5061 (**2**) has been advanced into clinical trials where it increased HDL levels by 20% in a cohort of hypertriglyceridemic subjects. **2** inhibits soluble HCV E2 binding to hepatoma cells and blocks infection of primary hepatocytes by HCV pseudoparticles expressing envelope proteins from all major genotypes [26]. The EC_{50} for inhibition of a pseudoparticle displaying genotype 1b proteins is $\sim$1 nM, and **2** also inhibits infection by a chimeric genotype 2a virus ($EC_{50} = 0.1$ nM). However, **2** also demonstrates detectable inhibition of p38 mitogen-activated kinase (p38 MAPK) and UDP glucuronosyltransferase 1, polypeptide A1 (UGT1A1), leading to a focus on ITX-7650 (structure not disclosed), which shows enhanced specificity while retaining the antiviral activity, ($EC_{50} = 0.25$ nM) toward the chimeric genotype 2a virus [26].

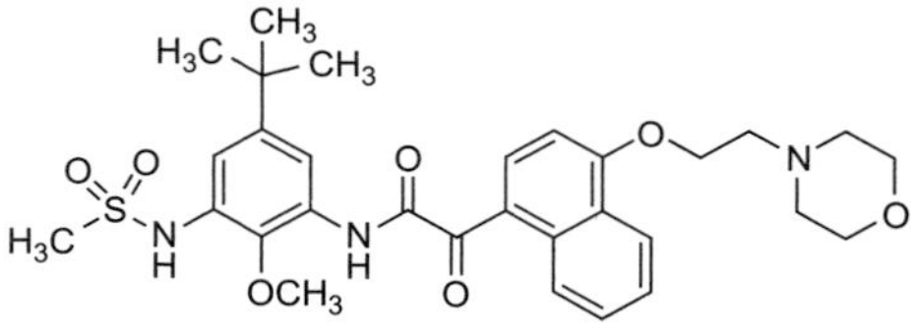

2 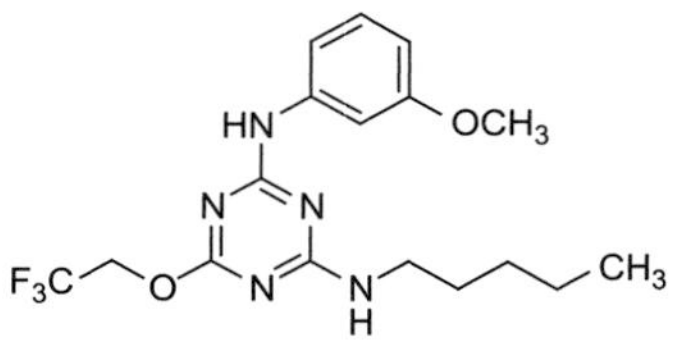 **3**

A series of triazine-based inhibitors of HCV entry discovered by broad screening have been described with **3** exhibiting EC_{50}s of 95 and 54 nM toward HCV pseudoparticles expressing genotype 1a and 1b envelopes, respectively [28–30]. A closely related compound demonstrated potent inhibition of cell culture-adapted infectious HCV, with time-of-addition experiments consistent with an entry inhibiting mechanism. Resistance to this chemotype, which afforded a >50-fold shift in the EC_{50}, mapped to a V719P or V719G substitution in the carboxy terminus region of the E2 protein [30].

The antihistamine terfenadine (**4**) and the broad spectrum antiviral agent arbidol (**5**) have been shown to interfere with HCV entry, with the former identified based on inhibition of the CD-81 large extracellular loop interacting with E2, while the latter associates with phospholipid in membranes and also appears to interact with tryptophan residues in lipopeptides [31–33]. A broad range of fused tricyclic derivatives have been claimed as HCV entry inhibitors, while the lectin cyanovirin-N appears to inhibit HCV entry at low nanomolar concentrations by binding to glycans on the viral envelope proteins [34–36].

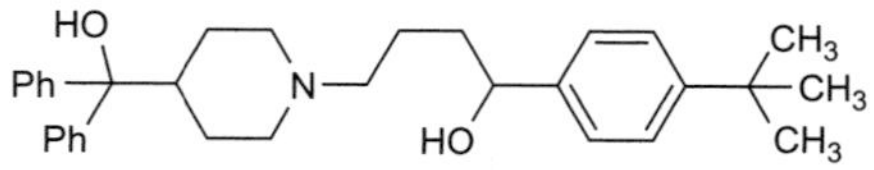

4 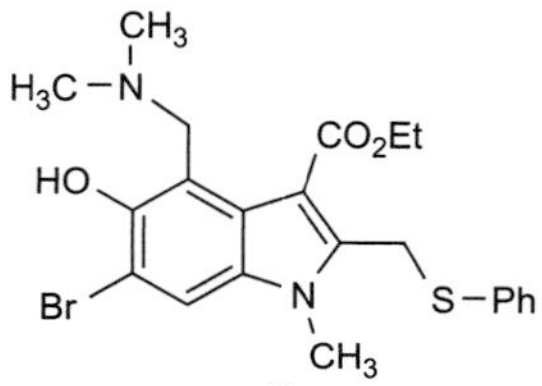 **5**

4. HCV P7 INHIBITORS

The HCV p7 protein is a 63 amino acid, short hydrophobic peptide with two membrane-spanning α-helical domains separated by a loop that faces the cytosol [37–39]. This protein plays an important role in virion assembly and release but not replication, since functional subgenomic replicons do not express p7 [40]. The HCV p7 protein is a member of the viroporin family that includes the influenza M2 and HIV Vpu channels, all of which oligomerize to form hydrophilic cation transporters [37–39]. The functional analogy between p7 and the influenza M2 channel, which extends to p7 being able to substitute for M2 in cell-based assays, led to the observation that amantadine and rimantadine also block p7 ion channel activity and, ultimately, to clinical studies with the two drugs in HCV-infected subjects [38,41,42]. Although amantadine exerts no significant effect on viral load as monotherapy, meta-analyses of early clinical studies suggested increased sustained virological response (SVR) rates when added to a regimen of interferon-α and ribavirin. However, subsequent studies have been less compelling, giving rise to considerable ambiguity with respect to the potential of this compound as adjunct therapy [38,43–45]. The p7 inhibitor BIT-225 (**6**) inhibits the JFH-2a infectious virus *in vitro* and reduced viremia by a modest but significant 0.5 $\log_{10}$ following dosing at 200 mg for 7 days to HCV-infected subjects when compared to placebo [46]. However, **6** is also an inhibitor of the related pestivirus bovine viral diarrhea virus (BVDV), $EC_{50} = 314$ nM, and HIV-1 in cell culture, $EC_{50} = 2.25$ μM, with inhibition occurring post-integration and consistent with inhibition of the accessory protein Vpu which plays a role in virus assembly [47,48].

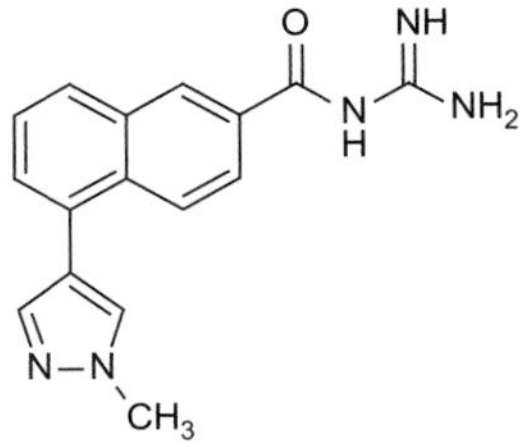

6

5. HCV NS4A INHIBITORS

The HCV NS4A protein is a critical cofactor that associates with the NS3 protease immediately after its proteolytic release, in essence rendering subsequent NS3-mediated polyprotein processing a pseudo-intramolecular event. The thiourea ACH-806 (**7**, GS-9132) is a potent genotype 1b inhibitor ($EC_{50} = 14$ nM) that is thought to bind to the NS4A protein and interfere with assembly of the replication complex [49,50]. **7** interacts in a synergistic fashion with HCV NS3 and NS5B inhibitors *in vitro* and resistance maps to changes at C16S and A39V. In clinical studies, 5 days of monotherapy with **7** reduced viral load by a mean of 0.91 $\log_{10}$ at 300 mg BID. However, kidney toxicity was observed, and more recent patent applications have focused on thiazole derivatives, such as **8**, which presumably gave rise to the second advanced compound ACH-1095 [51–53].

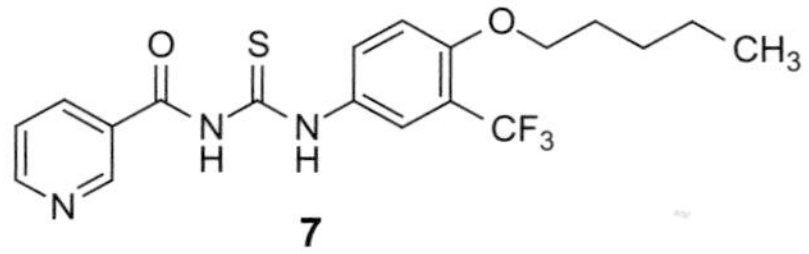

7 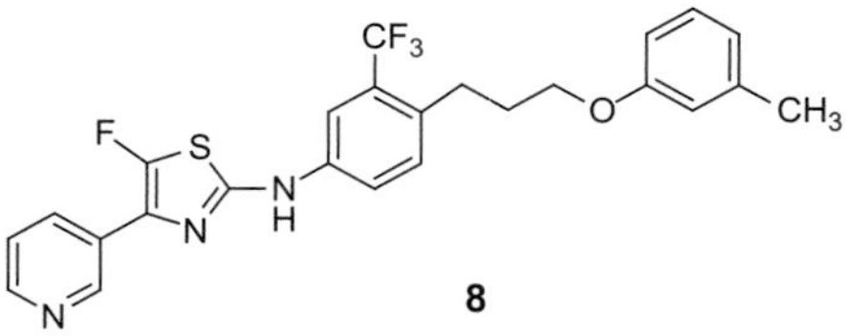 **8**

6. HCV NS4B INHIBITORS

HCV NS4B is a 261 residue hydrophobic protein with at least four domains predicted to be membrane spanning and which is released from the viral polyprotein by the NS3 protease [54–56]. The first 27 residues of the amino terminus contain an amphipathic helix that functions as a membrane anchoring element and associates with the endoplasmic reticulum membrane while oligomerization of NS4B is dependent on N-terminus residues and a second amphipathic helix at residues 43–65, a process facilitated by palmitoylation of the C-terminus [54–57]. While the precise role of NS4B is not well understood, the protein is critical for virus replication, inducing membrane vesicle formation and facilitating the formation of a membranous web that acts as a scaffold for the replication complex. NS4B has been shown to bind RNA and is also reported to possess NTPase activity consistent with the presence of a nucleotide binding domain at residues 129–135. NS4B binds and hydrolyzes both ATP and GTP although this aspect of the function of the protein is not well characterized and remains somewhat controversial [58].

The antihistamine clemizole (**9**) was discovered as an inhibitor of the binding of NS4B to the 3′-end of the negative strand of viral RNA, a potent association with a $K_d = 3.4$ nM, using a high-throughput microfluidic affinity analysis screen designed to overcome issues associated with the fragility of NS4B and preserve the natural folding of the protein [59]. **9** inhibits the NS4B/RNA interaction with an $IC_{50} = 24$ nM and abrogates HCV replication in a cell-based transient assay, $EC_{50} \sim 8$ μM. Resistance mapped to

W55R and R214Q substitutions in HCV NS4B, with the W55R mutant NS4B binding to the 3′-end of the viral RNA with fourfold increased affinity, $K_d = 0.75$ nM. **9** interacted synergistically with telaprevir or boceprevir in a luciferase-based reporter assay and additively with interferon, ribavirin, the nucleoside analog NM-283, or the allosteric NS5B inhibitor HCV-796 [60]. Deuterated derivatives of **9** that demonstrate increased metabolic stability in human liver microsomes have been described [61]. **9** has been evaluated clinically for its effect on viremia in HCV-infected patients, but the results have not been published.

Optimization of **9** has focused on improving the inhibitory potency of the compound toward a genotype 1b replicon and reducing the potential for hERG inhibition [62–64]. The benzimidazole **10** and the indazole **11** were the most potent representatives of their respective chemotypes demonstrating EC_{50}s of 1.1 and 3.3 μM, respectively, although both compounds inhibited the hERG channel with similar potency.

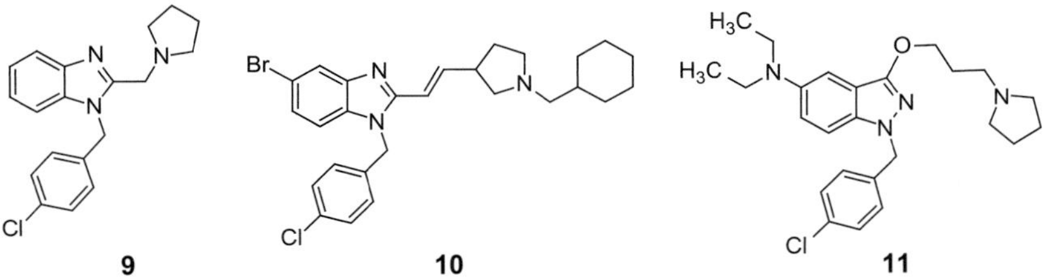

The discovery of a second amphipathic helix in NS4B comprising residues 43–65, designated 4BAH2 to distinguish it from the amphipathic helix in the amino terminus, enabled characterization of this motif as a mediator of protein oligomerization and lipid vesicle aggregation. Using a 384-well assay that monitored the induction of aggregation of fluorescently labeled lipid vesicles upon the addition of the 4BAH2 peptide, several inhibitors were identified, including anguizole (**12**) and the pyrazine **13** [57]. In a transient genotype 1b assay, **12** inhibited replication with an $EC_{50} = 300$ nM but was inactive toward genotype 2a replication, in contrast to **13** which inhibited both viral strains with EC_{50}s $= 2.5$ and 3.7 μM, respectively. Resistance to **12** mapped to an H94R mutation in NS4B, which exhibited a 37-fold shift in the EC_{50} [65].

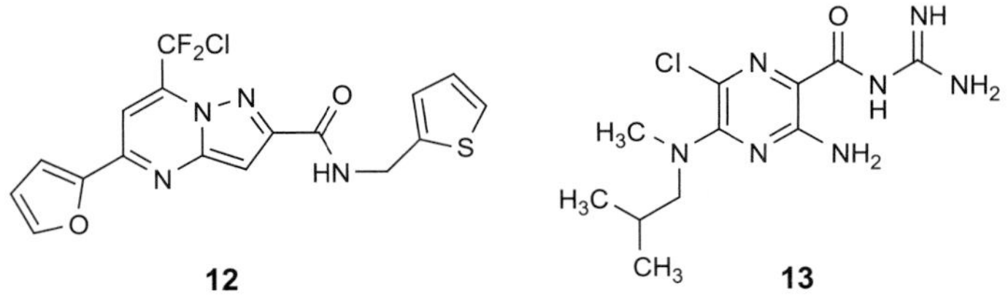

Several patent applications have claimed analogues of **12** including **14**, which completely inhibited HCV replication in a subgenomic replicon at a concentration of 10 μM, while **15** exhibited an $EC_{50} = 7$ nM, and **16** demonstrated balanced inhibition of subgenomic genotype 1a and 1b replicons with EC_{50}s = 10 nM [66–69]. AP-80978 (**17**) inhibited HCV replication *in vitro* with EC_{50}s = 0.9–1.8 μM (genotype 1b) and 7.8 μM (genotype 1a), while a genotype 2a replicon was insensitive, and the enantiomer was essentially inactive, $EC_{50} > 25$ μM [69].

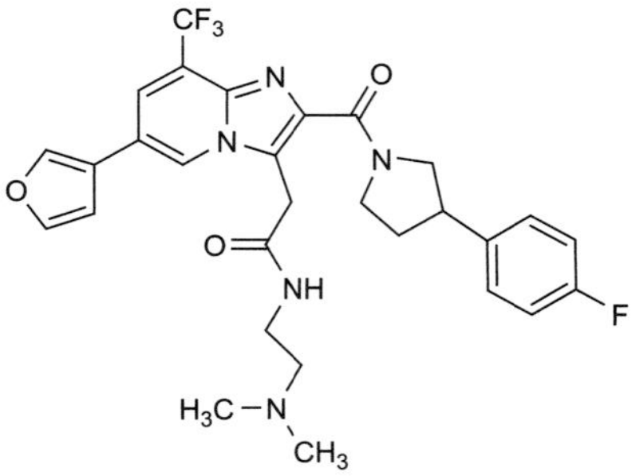

14 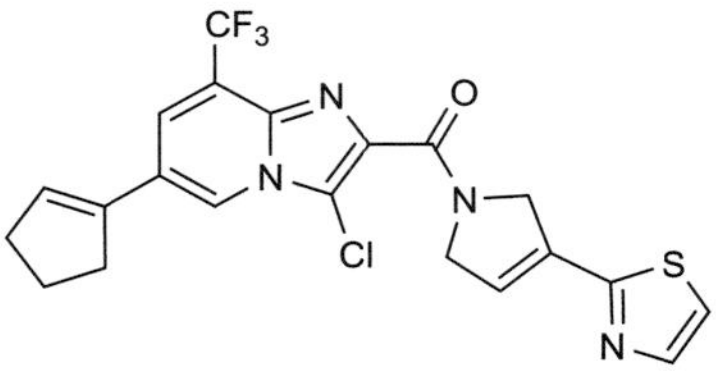 **15**

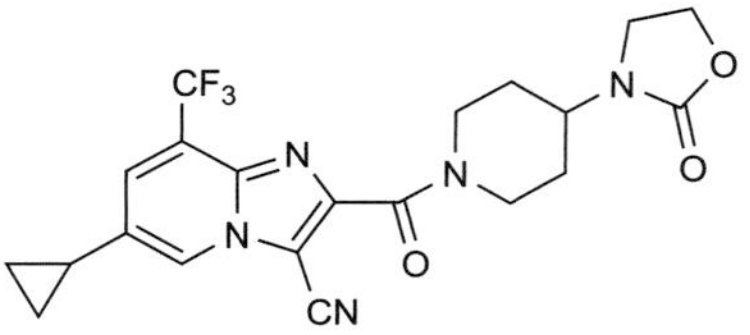

16 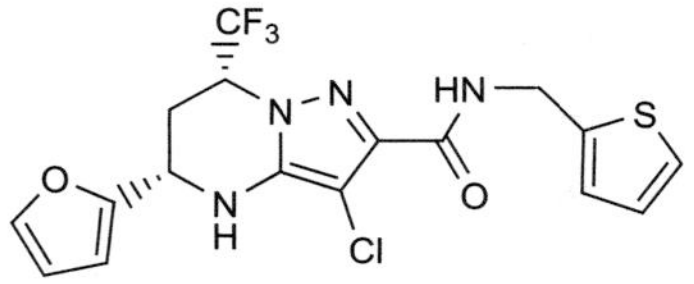 **17**

7. HCV NS5A INHIBITORS

NS5A is a multifunctional protein that plays key roles in viral genome replication, particle assembly, and virus–host interactions. The elucidation of the mechanisms by which NS5A is able to orchestrate such a diverse set of functions in the viral life cycle is an active area of considerable interest and investigation. Despite the lack of a complete understanding of the structure and function of the NS5A protein, considerable progress has been made in identifying molecules that selectively disrupt its function and demonstrate antiviral effects in HCV-infected subjects [9,70].

BMS-790052 (**21**) is a highly potent, first-in-class inhibitor that provided proof-of-concept for the clinical efficacy of NS5A inhibitors when a single 100 mg dose of **21** produced a mean viral load decline of 3.3 $\log_{10}$ at 24 h post-dose in genotype 1-infected subjects [71]. The effort that culminated in its identification began from the screening hit **18** and involved a chemotype evolution based on the discovery of dimeric species **19** that demonstrated potent antiviral activity, leading to the elucidation of the simplified stilbene **20**. Further optimization resulted in **21**, which inhibits HCV genotypes 1–5 with EC_{50}s ranging from 9 to 146 pM [71–74].

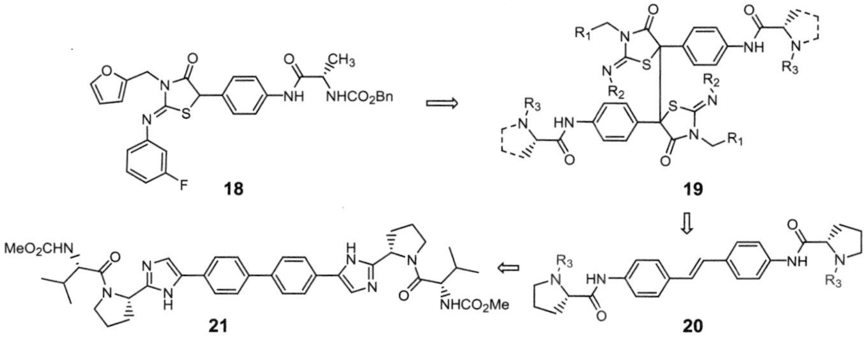

Since monotherapy with **21** leads to the emergence of resistance, Phase-IIa studies have focused on combinations with other agents [75–77]. In a clinical trial of 48 weeks, therapy with **21** (3–60 mg QD) administered with pegylated interferon-α-2a and ribavirin (PEG-IFN/RBV), 92% of the 10 mg group and 83% of the 60 mg group maintained undetectable HCV RNA levels 12 weeks after the end of treatment compared to 25% for the PEG-IFN/RBV control group [77]. In a 24-week study, in which **21** (60 mg QD) was dosed in combination with the NS3 protease inhibitor BMS-650032 (600 mg BID) to null responders with and without PEG-IFN/RBV, a >5 $\log_{10}$ median reduction in viral titer was observed initially in both groups [78]. However, viral breakthrough occurred in some members of the group taking only the direct-acting antiviral agents, but many of these subjects responded to the addition of PEG-IFN/RBV to the dosing regimen. All of the subjects that received the 24 weeks of quadruple therapy had undetectable viral RNA 12 weeks after the end of treatment [78,79].

BMS-824393 (structure not disclosed) is an NS5A inhibitor that, when dosed QD for 3 days, effected a median RNA decline of 2.5–3.9 $\log_{10}$ and 3.2–3.9 $\log_{10}$ in genotype 1a (1–100 mg) and 1b (10–100 mg) infected subjects, respectively [80].

AZD7295 (**23**, A-689), an NS5A inhibitor resulting from optimization of the biarylamide **22**, exhibited EC_{50}s of 1.24 and 0.007 μM toward genotype 1a and 1b replicons, respectively [81,82]. In a 5-day MAD clinical study, **23** at doses of 90 mg TID, 350 mg BID, and 233 mg TID was associated with a 1.2, 1.3, and 2.1 $\log_{10}$ mean reduction in viral load in genotype 1b-infected subjects, respectively, at day 6. However, one-third of the 1b-infected subjects and all subjects infected with genotype 1a or 3 showed no significant reduction in viremia [81,82]. Although **23** and **21** belong to distinct chemotypes, they exhibit partially overlapping resistance profiles that suggest some similarity in their mode of interaction with the NS5A protein.

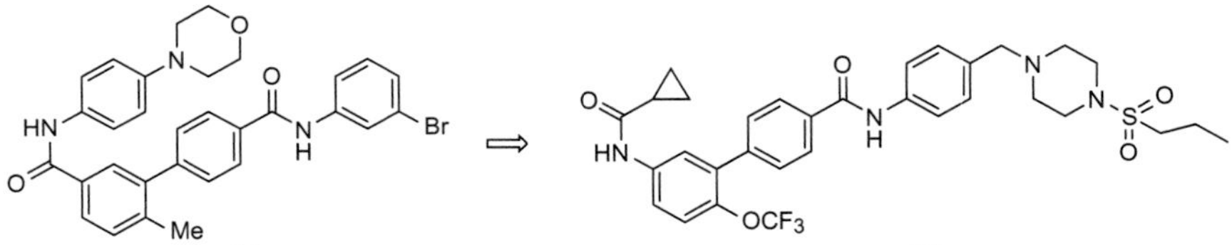

PPI-461 (structure not disclosed) is a potent pan-genotypic HCV inhibitor with EC_{50}s ranging from 0.01 to 9.3 nM that has advanced into Phase 1 clinical studies [83]. In normal healthy volunteers, PPI-461 exhibited a $T_{1/2}$ of 7.9–10.3 h and all subjects receiving doses of 50 mg or higher achieved a C_{24h} concentration greater than the EC_{50} for the least sensitive genotype 3a strain of HCV. A Phase 1b clinical study with PPI-461 has reportedly been initiated [84]. GS-5885 (structure not disclosed) is another potent HCV NS5A inhibitor, EC_{50}s = 0.005–10 nM, that exhibited a dose-proportional increase in C_{max} and AUC and had a long mean $T_{1/2}$ (37–45 h) in normal healthy volunteers [85]. At doses ranging from 3 to 100 mg, GS-5885 achieved C_{24h} plasma concentrations 9- to 366-fold above the genotype 1a protein binding-adjusted EC_{50}. In a MAD study at doses of 1–30 mg administered QD for 3 days conducted in genotype 1-infected subjects, a 3.1–3.3 log_{10} reduction in median viral load was reported [86]. Finally, the NS5A inhibitor ABT-267 (structure not disclosed) has been reported to exhibit PK and safety profiles in healthy volunteers supportive of further development [87].

The disclosure of clinical efficacy associated with **21** stimulated considerable interest in NS5A inhibitors, reflected in patent filings and discussion of advanced preclinical leads, the structures of which have not been disclosed. ITMN-10050 exhibits *in vitro* potency and preclinical PK properties similar to that of **21**; EDP-239 is two- to fourfold more potent than **21** against genotype 1a and 1b replicons and several 1b-resistant mutants; ACH-2928 exhibited EC_{50}s of <103 pM toward a broad range of HCV genotypes, including some with NS5A derived from clinical isolates, and is 20–40% orally bioavailable in the rat and dog; and GSK-2336805 has EC_{50}s of 44, 8, and 54 pM toward genotype 1a, 1b replicons and 2a-JFH virus, respectively [88–93]. PPI-437, PPI-668, and PPI-833 exhibited promising oral bioavailabilities and acceptable animal toxicology profiles in preclinical species, while IDX210 exhibited EC_{50}s of 6–175 pM toward a range of HCV genotypes and a monkey PK profile similar to that of **21** [94,95]. A common structural thread in the majority of the patent applications published to date and, particularly, in the more recent documents, is a dimeric pharmacophore, not necessarily symmetrical, in which variation is focused on a core scaffold that projects two pyrrolidine moieties (see **24**), although some structural variation of the pyrroldine and valine elements has also been described [96–109]. A representative sampling of the chemical diversity that has been surveyed in these patent applications is highlighted below, with the biological data for a selected set of compounds compiled in Table 1.

Finally, a class of NS5A inhibitors exemplified by **25** has been reported that essentially integrates pharmacophoric elements derived from chemotypes that had previously afforded clinical candidates [110].

Table 1 Structure and biological data for select NS5A inhibitors

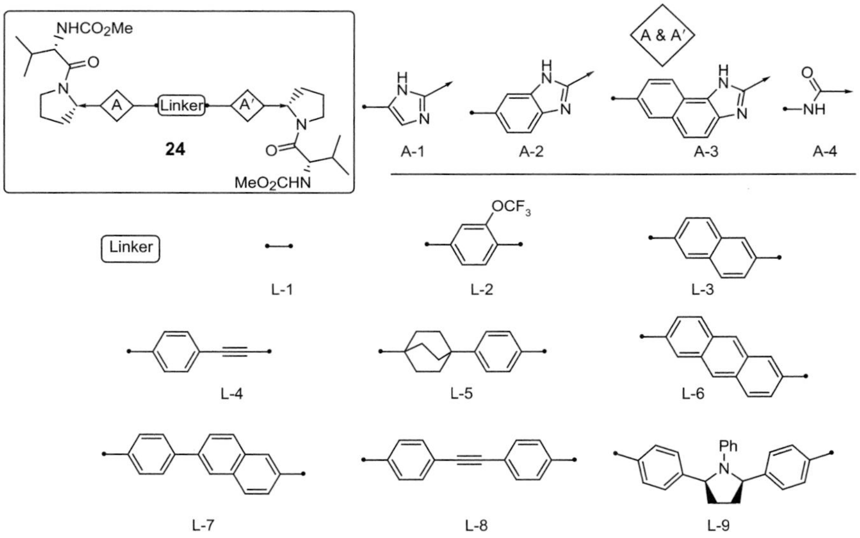

NHCO2Me, Linker, 24, MeO2CHN, A & A′, A-1, A-2, A-3, A-4, OCF3, L-1, L-2, L-3, L-4, L-5, L-6, L-7, L-8, L-9, Ph

Compound	A	Linker	A′	Genotypes and EC_{50} (nM)	Reference
24a	A-1	L-9	A-1	1b < 0.1	[96]
24b	A-2	L-1	A-3	1b < 4.6	[98]
24c	A-1	L-3	A-2	1b = 0.0067	[101]
24d	A-1	L-6	A-1	1a/1b = 0.001/0.0007	[103]
24e	A-2	L-2	A-2	1a/1b = 0.05/0.003	[104]
24f	A-1	L-4	A-2	1a/1b = 0.3/0.016	[105]
24g	A-1	L-7	A-1	1a/1b = 0.022/0.0024	[106]
24h	A-1	L-4	A-3	1b < 1	[107]

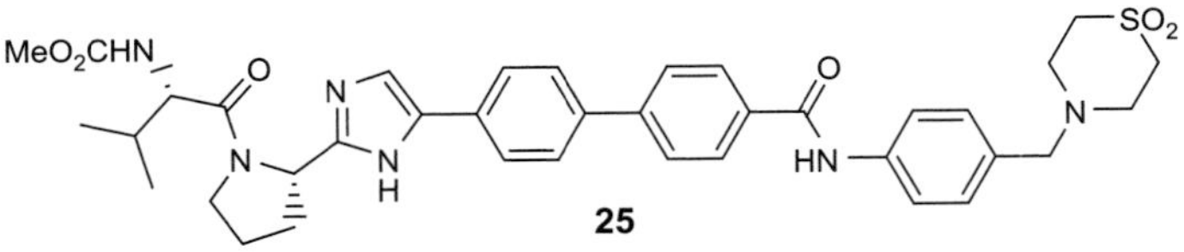

25

1a/1b $EC_{50} = 6.5/0.34$ nM

8. HCV IRES INHIBITORS

HCV uses a highly structured element of the 5 -non-translated region spanning residues 40–372 of the genomic RNA as an internal ribosomal entry site (IRES) to initiate translation [111–113]. The IRES contains two major domains, designated II and III, that contain all of the structural elements that recognize initiation factors and the 40S ribosome with the AUG start codon located in domain IV at residues 342–344. Initial approaches to inhibiting the HCV IRES focused on the application of antisense oligonucleotide (ASO) technology, but small-molecule inhibitors have been described more recently. ISIS-14803 is a 20 unit phosphorothiorate ASO that targets nucleotides 330–349 of the 5′-noncoding region, a stem-loop structure in domain 4 of the IRES that encompasses the AUG start codon [114,115]. ISIS-14803 was evaluated clinically in 24 HCV-infected subjects who were administered doses of 0.5, 1.0, 2.0, or 3.0 mg/kg for 4 weeks, with a modest clinical effect observed. Only two patients receiving 2.0 mg/kg experienced a >1.0 $\log_{10}$ reduction, while viremia in nine additional patients declined by <1.0 $\log_{10}$, an effect difficult to distinguish from natural variation in the assay or patient. However, no clear evidence of the emergence of resistant virus was observed in this study, an observation that encourages optimization and further study of ASOs for the treatment of HCV [114,115].

Using a 20-mer construct derived from domain IIa of the HCV IRES (a region known to be critical to virus translation and replication) to probe for association with small molecules from an HTS library, the benzimidazole **26** was identified as a weak binder by mass spectrometry ($K_D = 100$ μM) that bound poorly to a control 33-mer structured RNA construct [116]. SAR by MS established that the dimethylaminopropyl moiety was critical for recognition, with systematic studies leading to the identification of ISIS-11 (**27**) with a K_D of 1.7 μM for a refined 40mer RNA construct adopted based on an NMR structure of domain II. In a subgenomic replicon assay, ISIS-11 exhibited an EC_{50} of 1.5 μM with no overt cytotoxicity observed at a concentration of 100 μM.

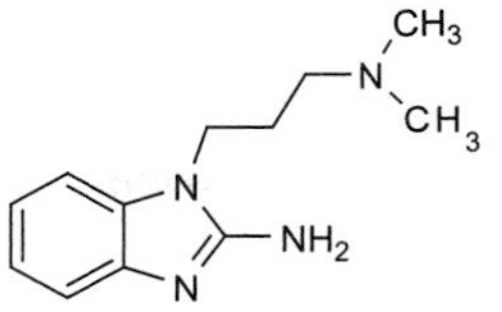

26 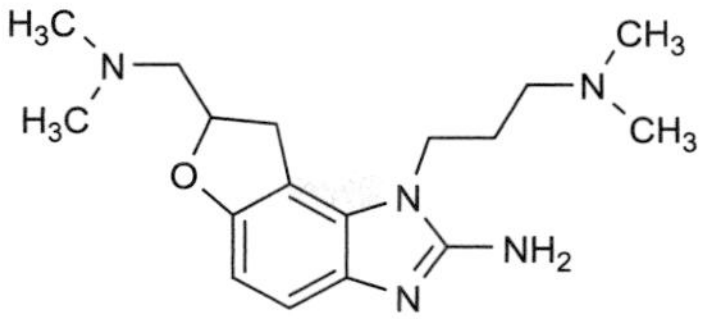 27

Detailed NMR studies confirmed that ISIS-11 bound to the lower bulge region of domain IIa, effecting a significant structural reorganization that led to extrusion of the 5′-residues from the bulge while establishing new stacking interactions on the 3′-side of the bulge that ultimately shifts the apical loop of domain II away from the site where eIF5 binds to the 40S subunit, providing a potential explanation for the antiviral activity [116–119].

Using a similar RNA binding screen, a series of lysine derivatives incorporating additional basic amines were discovered that offer a binding mode distinct to that of the benzimidazole derivatives with **28** characterized in some detail [120]. This class of compound associates with domain IIa with micromolar affinity and appears to compete with Mg^{2+} binding to the RNA structure, with no evidence of a compound-induced conformational change detected by a FRET assay designed to monitor the intrahelical angle between the base-paired stems. Domain IIa incorporates two binding sites for divalent metal ions in both the crystal state and in solution that contribute to the architectural integrity of the RNA by stabilizing a right-angled bend at the intersection of the two helices and providing the correct conformation association with the 40S ribosomal subunit. In a cell-based replicon assay, **28** dose-dependently inhibited HCV IRES-mediated translation of a luciferase reporter at micromolar concentrations [120].

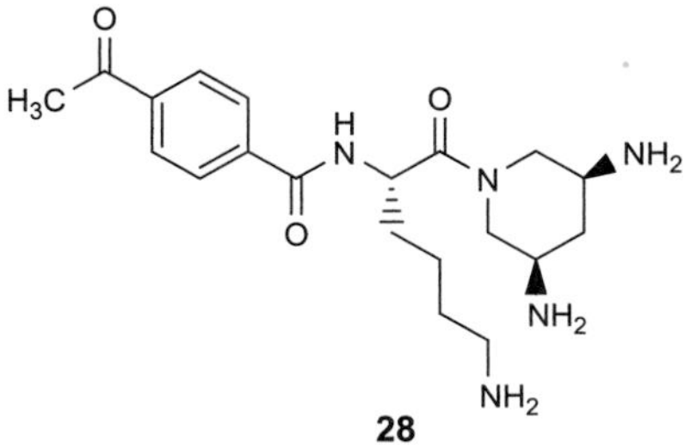

28

A broad range of indole- and azaindole-based HCV IRES inhibitors identified by gene expression modulation by small-molecules (GEMS) technology have been disclosed in a series of patent applications [121–131]. PS-102283 (**29**, SCH-1385145) is one of three compounds specifically identified as valuable in combination with boceprevir. More recently, the azaindole **30** and indole **31** have been shown to interact synergistically with boceprevir.

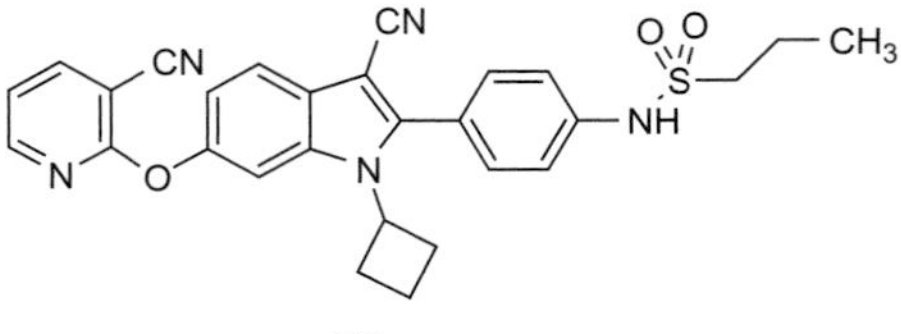

29

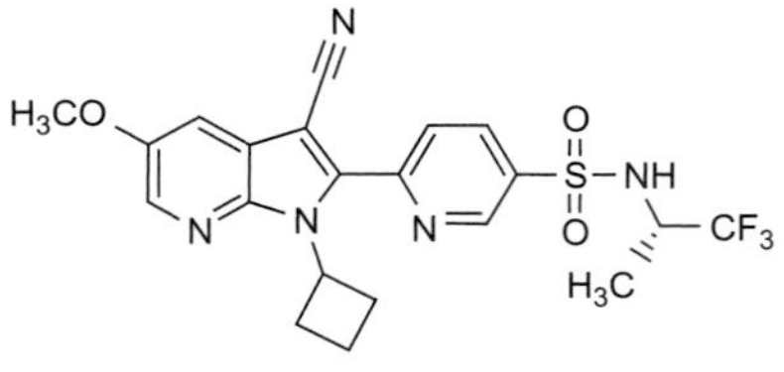

30

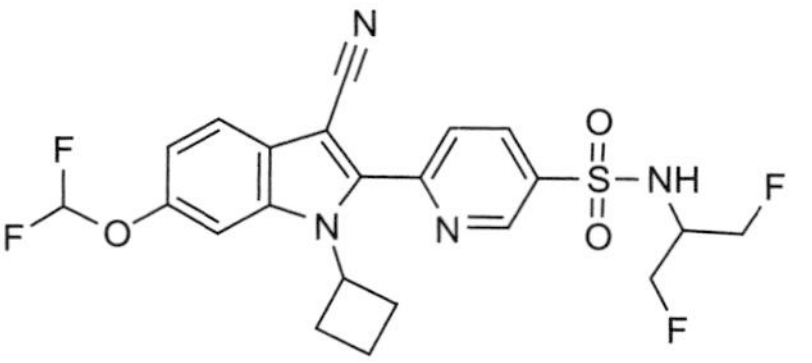

31

9. CONCLUSION

There has been considerable progress made toward the identification of potent and effective inhibitors of HCV that do not target the key enzymes NS3 protease and NS5B polymerase. At this juncture, inhibitors of HCV NS5A are the most clinically advanced and appear to offer promise both as adjunct therapy and in combination with mechanistically complementary direct-acting antiviral agents. Inhibitors of the other nonenzyme targets discussed above remain to be validated clinically, but lead inhibitors have been identified for many of these that have the potential to support drug discovery campaigns.

REFERENCES

[1] M. Houghton, *J. Hepatol.*, 2009, **51**, 939.
[2] A. D. Kwong, L. McNair, I. Jacobson and S. George, *Curr. Opin. Pharmacol.*, 2008, **8**, 522.
[3] T. Asselah, Y. Benhamou and P. Marcellin, *Liver Int.*, 2009, **29**, 57.
[4] S. M. Lemon, J. A. McKeating, T. Pietschmann, D. N. Frick, J. S. Glenn, T. L. Tellinghuisen, J. Symons and P. A. Furman, *Antiviral Res.*, 2010, **86**, 79.
[5] R. F. Schinazi, L. Bassit, C. Gavegnano and J. Viral, *Hepatitis*, 2010, **17**, 77.
[6] J.-M. Pawlotsky, *Clin. Microbiol. Infect.*, 2011, **17**, 105.
[7] R. Bartenschlager, *Curr. Opin. Microbiol.*, 2006, **9**, 416.
[8] N. Appel, T. Schaller, F. Penin and R. Bartenschlager, *J. Biol. Chem.*, 2006, **281**, 9833.
[9] T. P. Holler, T. Parkinson and D. C. Pryde, *Expert Opin. Drug Discov.*, 2009, **4**, 293.

[10] A. D. Strosberg, S. Kota, V. Takahashi, J. K. Snyder and G. Mousseau, *Viruses*, 2010, **2**, 1734.
[11] W. Wei, C. Cai, S. Kota, V. Takahashi, F. Ni, A. D. Strosberg and J. K. Snyder, *Bioorg. Med. Chem. Lett.*, 2009, **19**, 6926.
[12] S. Kota, C. Coito, G. Mousseau, J.-P. Lavergne and A. D. Strosberg, *J. Gen. Virol.*, 2009, **90**, 1319.
[13] F. Ni, S. Kota, V. Takahashi, A. D. Strosberg and J. K. Snyder, *Bioorg. Med. Chem. Lett.*, 2011, **21**, 2198.
[14] D. Moradpour, F. Penin and C. M. Rice, *Nat. Rev. Microbiol.*, 2007, **5**, 453.
[15] M. Donia, B. Cacopardo, M. Libra, G. Scalia, J. A. McCubrey and F. Nicoletti, *Recent Patents Anti-Infect. Drug Discov.*, 2010, **5**, 181.
[16] M. B. Zeisel, I. Fofana, S. Fafi-Kremer and T. F. Baumert, *J. Hepatol.*, 2011, **54**, 566.
[17] M. Brazzoli, A. Bianchi, S. Filippini, A. Weiner, Q. Zhu, M. Pizza and S. Crotta, *J. Virol.*, 2008, **82**, 8316.
[18] M. J. Evans, T. von Hahn, D. M. Tscherne, A. J. Syder, M. Panis, B. Woelk, T. Hatziioannou, J. A. McKeating, P. D. Bieniasz and C. M. Rice, *Nature*, 2007, **446**, 801.
[19] J. Grove, T. Huby, Z. Stamataki, T. Vanwolleghem, P. Meuleman, M. Farquhar, A. Schwarz, M. Moreau, J. S. Owen, G. Leroux-Roels, P. Balfe and J. A. McKeating, *J. Virol.*, 2007, **81**, 3162.
[20] M. B. Zeisel, G. Koutsoudakis, E. K. Schnober, A. Haberstroh, H. E. Blum, F.-L. Cosset, T. Wakita, D. Jaeck, M. Doffoel, C. Royer, E. Soulier, E. Schvoerer, C. Schuster, F. Stoll-Keller, R. Bartenschlager, T. Pietschmann, H. Barth and T. F. Baumert, *Hepatology*, 2007, **46**, 1722.
[21] M. T. Catanese, H. Ansuini, R. Graziani, T. Huby, M. Moreau, J. K. Ball, G. Paonessa, C. M. Rice, R. Cortese, A. Vitelli and A. Nicosia, *J. Virol.*, 2010, **84**, 34.
[22] A. Ploss, M. J. Evans, V. A. Gaysinskaya, M. Panis, H. You, Y. P. de Jong and C. M. Rice, *Nature*, 2009, **457**, 882.
[23] A. Codran, C. Royer, D. Jaeck, M. Bastien-Valle, T. F. Baumert, M. P. Kieny, C. Pereira, M. Augusto and J.-P. Martin, *J. Gen. Virol.*, 2006, **87**, 2583.
[24] E. Blanchard, S. Belouzard, L. Goueslain, T. Wakita, J. Dubuisson, C. Wychowsk and Y. Rouille, *J. Virol.*, 2006, **80**, 6964.
[25] J.-P. Yang, D. Zhou and F. Wang-Staal, Hepatitis C: Methods and Protocols, Vol. 510, 2nd Edition, 2009 p. 295.
[26] A. J. Syder, H. Lee, M. B. Zeisel, J. Grove, E. Soulier, J. MacDonald, S. Chow, J. Chang, T. F. Baumert, J. A. McKeating, J. McKelvy and F. Wong-Staal, *J. Hepatol.*, 2011, **54**, 48.
[27] D. Masson, M. Koseki, M. Ishibashi, C. J. Larson, S. G. Miller, B. D. King and A. R. Tall, *Arterioscler. Thromb. Vasc. Biol.*, 2009, **29**, 2054.
[28] G. Coburn, A. Q. Han, K. Provoncha and Y. Rotshteyn, *Patent Application, WO 2009091388-A2*, 2009.
[29] A. Q. Han, E. Wang, C. Gauss, W. Xie, G. Coburn and J.-M. Demuys, *Patent Application, WO 2010/118367-A2*, 2010.
[30] C. J. Baldick, M. J. Wichroski, A. Pendri, A. W. Walsh, J. Fang, C. E. Mazzucco, K. A. Pokornowski, R. E. Rose, B. J. Eggers, M. Hsu, W. Zhai, G. Zhai, S. W. Gerritz, M. A. Poss, N. A. Meanwell, M. I. Cockett and D. J. Tenney, *PLoS Pathogens*, 2010, **6**, e1001086.
[31] M. Holzer, S. Ziegler, B. Albrecht, B. Kronenberger, A. Kaul, R. Bartenschlager, L. Kattner, C. D. Klein and R. W. Hartmann, *Molecules*, 2008, **13**, 1081.
[32] E.-I. Pecheur, D. Lavillette, F. Alcaras, J. Molle, Y. S. Boriskin, M. Roberts, F.-L. Cosset and S. J. Polyak, *Biochemistry*, 2007, **46**, 6050.
[33] E. Teissier, G. Zandomeneghi, A. Loquet, D. Lavillette, J.-P. Lavergne, R. Montserret, F.-L. Cosset, A. Böckmann, B. H. Meier, F. Penin and E.-I. Pécheur, *PLoS ONE*, 2011, **6**, e15874.

[34] T. J. Cuthbertson, M. Ibanez, C. A. Rijnbrand, A. J. Jackson, G. K. Mittapalli, F. Zhao, J. E. MacDonald and F. Wong-Staal, *Patent Application, WO 2008/021745-A2*, 2008.
[35] P. Gastaminza, C. Whitten-Bauer and F. V. Chisari, *Proc. Natl. Acad. Sci. USA*, 2010, **107**, 291.
[36] F. Helle, C. Wychowski, N. Vu-Dac, K. R. Gustafson, C. Voisset and J. Dubuisson, *J. Biol. Chem.*, 2006, **281**, 25177.
[37] S. Griffin, *Curr. Opin. Invest. Drugs*, 2010, **11**, 175.
[38] S. Khaliq, S. Jahan and S. Hassan, *Liver Int.*, 2011, **31**, 606.
[39] E. Steinman and T. Pietschmann, *Viruses*, 2010, **2**, 2078.
[40] A. L. Wozniak, S. Griffin, D. Rowlands, M. Harris, M.-K. Yi, S. M. Lemon and S. A. Weinman, *PLoS Pathogens*, 2010, **6**, e1001087.
[41] S. D. Griffin, L. P. Beales and D. S. Clark, *FEBS Lett.*, 2003, **535**, 34.
[42] C. StGelais, T. J. Tuthill, D. S. Clarke, D. J. Rowlands, M. Harris and S. Griffin, *Antiviral Res.*, 2007, **76**, 48.
[43] P. Deltenre, J. Henrion, V. Canva, S. Dharancy, F. Texier, A. Louvet, S. De Maeght, J.-C. Paris and P. Mathurin, *J. Hepatol.*, 2004, **40**, 462.
[44] A. Mangia, G. Leandro, B. Heibling, E. L. Renner, M. Tabone, L. Sidoli, S. Caronia, G. R. Foster, S. Zeuzem, T. Berg, V. Di Marco, N. Cino and A. Andriulli, *J. Hepatol.*, 2004, **40**, 478.
[45] M. Maynard, P. Pradat, F. Bailly, F. Rozier, C. Nemoz, S. N. S. Ahmed, P. Adeleine and C. Trépo, and a French Multicenter Group*J. Hepatol.*, 2006, **44**, 484.
[46] S. Riordan, C. A. Luscombe, G. D. Ewart, J. Wilkinson, K. Quan, J. Marjason, B. Leggett, G. Dore, C. Fenn and M. Miller, *Glob. Antivir. J.*, 2009, **5**, HEP DART, Hawaii Dec 6-10th, 2009. Abstract 160. Presentation available at http://www.ihlpress.com/pdf%20files/hepdart09_presentations/late_breaker/11_Luscombe%20HepDART%202009%20Oral%20presentation.pdf.
[47] C. A. Luscombe, Z. Huang, M. G. Murray, M. Miller, J. Wilkinson and G. D. Ewart, *Antiviral Res.*, 2010, **86**, 144.
[48] G. Khoury, G. Ewart, C. Luscombe, M. Miller and J. Wilkinson, *Antimicrob. Agents Chemother.*, 2010, **54**, 835.
[49] W. Yang, Y. Zhao, J. Fabrycki, X. Hou, X. Nie, A. Sanchez, A. Phadke, M. Deshpande, A. Agarwal and M. Huang, *Antimicrob. Agents Chemother.*, 2008, **52**, 2043.
[50] D. L. Wyles, K. A. Kaihara and R. T. Schooley, *Antimicrob. Agents Chemother.*, 2008, **52**, 1862.
[51] S. Zhang, A. Phadke, C. Liu, X. Wang, J. Quinn, D. Chen, V. Gadhachanda, S. Li and M. Deshpande, *Patent Application, WO-2006122011-A2*, 2006.
[52] X. Wang, S. Zhang, V. Gadhachanda, C. Liu, S. Li, J. Quinn, D. Chen, M. Deshpande and A. Phadke, *Patent Application, WO-2008147557-A2*, 2008.
[53] Achillion press release on ACH-1095, available at http://www.achillion.com/PL/pdf/06_ns4a_bg_2.pdf.
[54] H. Dvory-Sobol, P. S. Pang and J. S. Glenn, *Viruses*, 2010, **2**, 2481.
[55] J. Gouttenoire, F. Penin and D. Moradpour, *Rev. Med. Virol.*, 2010, **20**, 117.
[56] R. Rai and J. Deval, *Antiviral Res.*, 2011, **90**, 93.
[57] N.-J. Cho, H. Dvory-Sobol, C. Lee, S.-J. Cho, P. Bryson, M. Masek, M. Elazar, C. W. Frank and J. S. Glenn, *Sci. Transl. Med.*, 2010, **2**, 5ra6.
[58] A. A. Thompson, A. Zou, J. Yan, R. Duggal, W. Hao, D. Molina, C. N. Cronin and P. A. Wells, *Biochemistry*, 2009, **48**, 906.
[59] S. Einav, D. Gerber, P. D. Bryson, E. H. Sklan, M. Elazar, S. J. Maerkl, J. S. Glenn and S. R. Quake, *Nat. Biotechnol.*, 2008, **26**, 1019.
[60] S. Einav, H. Dvory-Sobol, E. Gehrig and J. S. Glenn, *J. Infect. Dis.*, 2010, **202**, 65.
[61] T. Rao and C. Zhang, *Patent Application, WO 2010/118286-A2*, 2010.

[62] S. Einav, J. S. Glenn, R. McDowell and W. Yang, *Patent Application, WO 2009/039248-A2*, 2009.
[63] I. C. Choong, D. Cory, J. S. Glenn and W. Yang, *Patent Application, WO 2010/107739-A2*, 2010.
[64] I. C. Choong, J. S. Glenn and W. Yang, *Patent Application, WO 2010/107742-A2*, 2010.
[65] P. D. Bryson, N.-J. Cho, S. Einav, C. Lee, V. Tai, J. Bechtel, M. Sivaraja, C. Roberts, U. Schmitz and J. S. Glenn, *Antiviral Res.*, 2010, **87**, 1.
[66] F. U. Schmitz, V. W.-F. Tai, R. Roopa, C. D. Roberts, A. D. M. Abadi, S. Baskaran, I. Slobodov, J. Muang and M. L. Neitzel, *Patent Application, WO-2009 023179-A2*, 2009.
[67] S. Baskaran, J. Muang, M. Neitzel, R. Rai, I. Slobodov, V. Tai and H. Zhang, *Patent Application, WO 2010/091409-A1*, 2010.
[68] A. Banka, S. Baskaran, J. Catalano, P. Chong, H. Dickson, J. Fang, J. Maung, M. L. Neitzel, A. Peat, D. Price, R. Rai, C. D. Roberts, B. Shotwell, V. Tai and H. Zhang, *Patent Application*, WO *2010/091411-A1*, 2010.
[69] U. Slomczynska, P. Olivo, J. Beattie, G. Starkey, A. Noueiry and R. Roth, *Patent Application, WO 2010/096115-A1*, 2010.
[70] U. Schmitz and S.-L. Tan, *Recent Patents Anti-Infect. Drug Discov.*, 2008, **3**, 77.
[71] M. Gao, R. E. Nettles, M. Belema, L. B. Snyder, V. N. Nguyen, R. A. Fridell, M. H. Serrano-Wu, D. R. Langley, J.-H. Sun, D. R. O'Boyle II, J. A. Lemm, C. Wang, J. O. Knipe, C. Chien, R. J. Colonno, D. M. Grasela, N. A. Meanwell and L. G. Hamann, *Nature*, 2010, **465**, 96.
[72] J. A. Lemm, D. O'Boyle II, M. Liu, P. T. Nower, R. Colonno, M. S. Deshpande, L. B. Snyder, S. W. Martin, D. R. St. Laurent, M. H. Serrano-Wu, J. L. Romine, N. A. Meanwell and M. Gao, *J. Virol.*, 2010, **84**, 482.
[73] J. L. Romine, D. R. St. Laurent, J. E. Leet, S. W. Martin, M. H. Serrano-Wu, F. Yang, M. Gao, D. R. O'Boyle II, J. A. Lemm, J.-H. Sun, P. T. Nower, X.(.S.). Huang, M. S. Deshpande, N. A. Meanwell and L. B. Snyder, *ACS Med. Chem. Lett.*, 2011, **2**, 224.
[74] R. A. Fridell, D. Qiu, C. Wang, L. Valera and M. Gao, *Antimicrob. Agents Chemother.*, 2010, **54**, 3641.
[75] R. E. Nettles, H. Sevinksy, E. Chung, D. Burt, H. Xiao, T. Marbury, R. Goldwater, M. DeMicco, M. Rodriguez-Torres, E. Fuentes, A. Vutikullird, E. Lawitz, A. Persson, M. Bifano and D. M. Grasela, Abstract P-1881, 61st AASLD, The Liver Meeting, Boston, MA, October, 2010.
[76] M. Gao, C. Wang, J. Sun, D. R. O'Boyle II, R. Hindes, P. D. Yin, S. M. Schnittman, P. T. Nower, L. Valera, J. Lemm, S. Voss, F. McPhee, Y. Fei, J. Kadow, M. Belema, N. A. Meanwell, M. Cockett, R. Nettles, M. Bifano, H. Sevinsky, X. Huang, B. Kienzle, P. Patel, D. Hernandez and R. Fridell, Abstract P-1853, 61st AASLD, The Liver Meeting, Boston, MA, October, 2010.
[77] S. Pol, R. H. Ghalib, V. K. Rustgi, C. Martorell, G. T. Everson, H. A. Tatum, C. Hezode, J. K. Lim, J.-P. Bronowicki, G. A. Abrams, N. Brau, D. W. Morris, P. Thuluvath, R. Reindollar, P. D. Yin, U. Diva, R. Hindes, F. McPhee, M. Gao, A. Thiry, S. Schnittman and E. A. Hughes, Abstract 1373, 46th EASL, Berlin, Germany, March, 2011.
[78] A. Lok, D. Gardiner, E. Lawitz, C. Martorell, G. Everson, R. Ghalib, R. Reindollar, V. Rustgi, P. Wendelburg, K. Zhu, V. Shah, D. Sherman, F. McPhee, M. Wind-Rotolo, M. Bifano, T. Eley, T. Guo, A. Persson, R. Hindes, D. Grasela and C. Pasquinelli, Abstract LB-8, 61st AASLD, The Liver Meeting, Boston, MA, October, 2010.
[79] A. Lok, D. Gardiner, E. Lawitz, C. Martorell, G. Everson, R. Ghalib, R. Reindollar, V. Rustgi, F. McPhee, M. Wind-Rotolo, A. Persson, K. Zhu, D. Dimitrova, T. Eley, T. Guo, D. Grasela and C. Pasquinelli, Abstract 1356, 46th EASL, Berlin, Germany, March, 2011.

[80] R. E. Nettles, X. Wang, S. Quadri, Y. Wu, M. Gao, M. Belema, E. J. Lawitz, R. Goldwater, M. DeMicco, T. Marbury, A. B. Vutikullird, E. Fuentes, A. Persson and D. M. Grasela, Abstract P-1858, 61st AASLD, The Liver Meeting, Boston, MA, October, 2010.
[81] E. Gane, G. R. Foster, J. Cianciara, C. Stedman, S. Ryder, M. Buti, E. Clark and D. Tait, Abstract P-2003, 45th EASL, Vienna, Austria, April, 2010.
[82] M. Carter, B. Baxter, D. Bushnell, S. Cockerill, J. Chapman, S. Fram, E. Goulding, M. Lockyer, N. Mathews, P. Najarro, D. Rupassara, J. Salter, E. Thomas, C. Wheelhouse, J. Borger and K. Powell, *Antiviral Res.*, 2010, **86**, A9 23rd International Conference on Antiviral Research, San Francisco, CA, April, 2010.
[83] R. Colonno, E. Peng, M. Bencsik, N. Huang, M. Zhong, A. Huq, Q. Huang, J. Williams and L. Li, Abstract P-875, 45th EASL, Vienna, Austria, April, 2010.
[84] N. Brown, P. Vig, E. Ruby, A. Muchnik, E. Pottorff, S. Knox, S. Febbraro, B. Wargin, T. Molvadgaard, A. Jones, L. Li and R. Colonno, Abstract LB-12, 61st AASLD, The Liver Meeting, Boston, MA, October, 2010.
[85] J. Link, R. Bannister, L. Beilke, G. Cheng, M. Cornpropst, A. Corsa, E. Dowdy, H. Guo, D. Kato, T. Kirschberg, H. Liu, M. Mitchell, M. Matles, E. Mogalian, E. Mondou, C. Ohmstede, B. Peng, R. Scott, J. Findley, G. Chittick, F. Wang, J. Alianti, J. Sun, J. Taylor, Y. Tian, L. Xu, C. Yang, G. Yuen, K. Wang and G. Eisenburg, Abstract P-1883, 61st AASLD, The Liver Meeting, Boston, MA, October, 2010.
[86] E. Lawitz, D. Gruener, J. Hill, T. Marbury, S. Komjathy, M. DeMicco, A. Murillo, F. Jenkin, K. Kim, J. Simpson, M. Aycock, A. Mathias, C. Yang, E. Dowdy, M. Liles, G. Cheng, E. Mondou, J. Link, D. Brainard, J. McHutchison, C. Ohmstede and R. Bannister, Abstract 1219, 46th EASL, Berlin, Germany, March, 2011.
[87] E. O. Dumas, A. Lawal, R. M. Menon, T. Podsadecki, W. Awni, S. Dutta and L. Williams, Abstract 1204, 46th EASL, Berlin, Germany, March, 2011.
[88] J. B. Nicholas, B. O. Buckman, V. Serebryany, C. J. Schaefer, K. Kossen, D. Ruhrmund, L. Hooi, N. Aleskovski, A. Arfsten, S. R. Lim, L. Pan, L. Huang, R. Rajagopalan, S. Misialek, H. Ramesha and S. Seiwert, Abstract 1228, 46th EASL, Berlin, Germany, March, 2011.
[89] C. M. Owens, B. Brasher, A. J. Polemeropoulos, X. Peng, C. Wang, L. Ying, H. Cao, Y. Qiu, L. Jiang and Y. Or, Abstract P-1863, 61st AASLD, The Liver Meeting, Boston, MA, October, 2010.
[90] L. J. Jiang, S. Liu, T. Phan, C. Owens, B. Brasher, A. Polemeropoulos, X. Luo, K. Hoang, M. Wang, X. Peng, H. Cao, Y. Qiu and Y. S. Or, Abstract 1213, 46th EASL, Berlin, Germany, March, 2011.
[91] Y. Zhao, G. Yang, J. Fabrycki, D. Patel, J. Wiles, X. Wang, A. Phadke, M. Deshpande and M. Huang, Abstract P-1880, 61st AASLD, The Liver Meeting, Boston, MA, October, 2010.
[92] M. Huang, G. Yang, D. Patel, Y. Zhao, J. Fabrycki, C. Marlor, J. Rivera, K. Stauber, V. Gadhachanda, J. Wiles, A. Hashimoto, D. Chen, Q. Wang, G. Pais, X. Wang, M. Deshpande and A. Phadke, Abstract 1212, 46th EASL, Berlin, Germany, March, 2011.
[93] J. Bechtel, R. Crosby, A. Wang, E. Woldu, S. Van Horn, J. Horton, K. Creech, L. H. Caballo, S. You, C. Voitenleitner, J. Vamathevan, M. Duan, A. Spaltenstein, W. Kazmierski, C. Roberts and R. Hamatake, Abstract 764, 46th EASL, Berlin, Germany, March, 2011.
[94] R. Colonno, N. Huang, Q. Huang, E. Peng, A. Hug, M. Lau, M. Bencsik, M. Zhong and L. Li, Abstract 1200, 46th EASL, Berlin, Germany, March, 2011.

[95] C. B. Dousson, C. Charpron, D. Standring, J. P. Bilello, J. McCarville, M. La Colla, M. Seifer, C. Parsy, D. Dukhan, C. Pierra and D. Surleraux, Abstract 815, 46th EASL, Berlin, Germany, March, 2011.
[96] D. A. Degoey, W. M. Kati, C. W. Hutchins, P. L. Donner, A. C. Krueger, C. E. Motter, L. T. Nelson, S. V. Patel, M. A. Matulenko, R. G. Keddy, T. K. Jinkerson, T. N. Soltwedel, D. K. Hutchinson, C. A. Flentge, R. Wagner, C. J. Maring, M. D. Tufano, D. A. Beterbenner, T. W. Rockway, D. Liu, J. K. Pratt, M. J. Lavin, K. Sarris, K. R. Woller, S. H. Wagaw, J. C. Califano and W. Li, *Patent Application WO 2010/144646-A2*, 2010.
[97] J. A. Bender, P. Hewawasam, J. F. Kadow, O. D. Lopez, N. A. Meanwell, J. L. Romine, L. B. Snyder, D. R. Laurent St., G. Wang, N. Xu and M. Belema, *Patent Application, WO 2010/117635-A1*, 2010.
[98] J. L. Romine, D. R. Laurent St., M. Belema, L. B. Snyder, L. G. Hamann, J. F. Kadow, J. Kapur, A. C. Good, O. D. Lopez, R. Lavoie and J. A. Bender, *Patent Application, WO 2010/096302-A1*, 2010.
[99] O. D. Lopez, Q. Chen, M. Belema and L. G. Hamann, *Patent Application, US 2010/0249190-A1*, 2010.
[100] Y.-L. Qiu, W. Ce, X. Peng, L. Ying and S. Y. Or, *Patent Application, WO 2010/099527-A1*, 2010.
[101] H. Guo, D. Kato, T. A. Kirschberg, H. Liu, J. O. Link, M. L. Mitchell, J. P. Parrish, N. Squires, J. Sun, J. Taylor, E. M. Bacon, E. Canales, A. Cho, J. J. Cottell, M. C. Desai, R. L. Halcomb, E. S. Krygowski, S. E. Lazerwith, Q. Liu, R. Mackman, H.-J. Pyun, J. H. Saugier, J. D. Trenkle, W. C. Tse, R. W. Vivian, S. D. Schroeder, W. J. Watkins, L. Xu, Z.-Y. Yang, T. Kellar, X. Sheng, M. O. H. Clarke, C.-H. Chou, M. Graupe, H. Jin, R. McFadden, M. R. Mish, S. E. Metobo, B. W. Phillips and C. Venkataramani, *Patent Application, WO 2010/132601-A1*, 2010.
[102] F. U. Schmitz, R. Rai, C. D. Roberts, W. Kazmierski and R. Grimes, *Patent Application, WO 2010/062821-A1*, 2010.
[103] S. B. Rosenblum, K. X. Chen, J. A. Kozlowski, F. G. Njoroge and C. A. Coburn, *Patent Application, WO 2010/132538-A1*, 2010.
[104] K. X. Chen, G. N. Anilkumar, Q. Zheng, S. B. Rosenblum, J. A. Kozlowski and F. G. Njoroge, *Patent Application, WO 2010/138790-A1*, 2010.
[105] Q. Zheng, K. X. Chen, G. N. Anilkumar, S. B. Rosenblum, J. A. Kozlowski and F. G. Njoroge, *Patent Application, WO 2010/138791-A1*, 2010.
[106] J. B. J. Milbank, D. C. Pryde and T. D. Tran, *Application, WO 2011/004276-A1*, 2011.
[107] L. Li and M. Zhong, *Patent Application, WO 2010/065668-A1*, 2010.
[108] S. D. Demin, D. McGowan, S. J. Last and P. J.-M. B. Raboisson, *Patent Application, WO 2010/122162-A1*, 2010.
[109] J. A. Henderson, J. Maxwell, L. Vaillan-Court, M. Morris, R. Grey Jr., S. Giroux, L. C. C. Kong, S. K. Das, B. Liu, C. Poisson, C. Cadilhac, M. Bubenik, T. J. Reddy, G. Falardeau, C. Yannopoulos, J. Wang, O. Z. Pereira, Y. L. Bennani, A. C. Pierce, G. R. Bhisetti, K. M. Cottrell and V. Marone, *Patent Application, WO 2011/009084-A2*, 2011.
[110] M. C. Carter and N. Mathews, *Patent Application, WO 2010/094977-A1*, 2010.
[111] A. Dasgupta, S. Das, R. Izumi, A. Vemkatesan and B. Barat, *FEMS Microbiol. Lett.*, 2004, **234**, 189.
[112] C. S. Fraser and J. A. Doudna, *Nat. Rev. Microbiol.*, 2007, **5**, 29.
[113] D. R. Davis and P. P. Seth, *Antiviral Chem. Chemother.*, 2010, **21**, 117.
[114] M. Soler, J. G. McHutchison, T. J. Kwoh, F. A. Dorr and J.-M. Pawlotsky, *Antiviral Ther.*, 2004, **9**, 953.
[115] J. G. McHutchison, K. Patel, P. Pockros, L. Nyberg, S. Pianko, R. Z. Yu, F. A. Dorr and T. J. Kwoh, *J. Hepatol.*, 2006, **44**, 88.

[116] P. P. Seth, A. Miyaji, E. A. Jefferson, K. A. Sannes-Lowery, S. A. Osgood, S. S. Propp, R. Ranken, C. Massire, R. Sampath, D. J. Ecker, E. E. Swayze and R. H. Griffey, *J. Med. Chem.*, 2005, **48**, 7099.
[117] R. B. Paulsen, P. P. Seth, E. E. Swayze, R. H. Griffey, J. J. Skalicky, T. E. Cheatham III and D. R. Davis, *Proc. Natl. Acad. Sci. USA*, 2010, **107**, 7263.
[118] J. Parsons, M. P. Castaldi, S. Dutta, S. M. Dibrov, D. L. Wyles and T. Hermann, *Nat. Chem. Biol.*, 2009, **5**, 823.
[119] S. Liu, C. A. Nelson, L. Xiao, L. Lu, P. P. Seth, D. R. Davis and C. H. Hagedorn, *Antiviral Res.*, 2011, **89**, 54.
[120] M. Carnevali, J. Parsons, D. L. Wyles and T. Hermann, *ChemBioChem*, 2010, **11**, 1364.
[121] A. Bhattacharyya, C. R. Trotta and S. W. Peltz, *Drug Discov. Today*, 2007, **12**, 553.
[122] P. S. Hwang, J. Takasugi, H. Ren, R. G. Wilde, A. Turpoff, A. Arefolov, G. M. Karp, G. Chen and J. A. Campbell, *Patent Application, WO 2006/019831-A1*, 2006.
[123] G. M. Karp and G. Chen, *Patent Application, WO 2006/019832-A1*, 2006.
[124] G. M. Karp, *Patent Application, WO 2007/084413-A2*, 2007.
[125] G. M. Karp, P. S. Hwang, J. J. Takasugi, H. Ren, R. G. Wilde, A. Turpoff, A. Arefolov, G. Chen and J. A. Campbell, *Patent Application, WO 2007/084435-A2*, 2007.
[126] F. C. Lahser and G. M. Karp, *Patent Application, WO 2007/106317-A2*, 2007.
[127] M. MacCoss, G. F. Njoroge, A. Nomier, G. Chen, S. X. Huang, R. Kakarla, G. M. Karp, W. J. Lennox, C. Li, R. Liu, Y. Liu, C. Morrill, S. D. Paget, S. W. Smith, J. Takasugi, A. A. Turpoff, H. Ren, N. Zhang, X. Zhang and J. Zhu, *Patent Application, WO 2010/117932-A1*, 2010.
[128] M, MacCoss, G. F. Njoroge, A. Nomier, G, Chen, G. M. Karp, W. J. Lennox, C. Li, C. Morrill, S. D. Paget, H. Ren, N. Zhang and X. Zhang, *Patent Application, WO 2010/117935-A1*, 2010.
[129] F. C. Lahser and Z. Gu, *Patent Application, WO 2010/117936-A1*, 2010.
[130] F. C. Lahser and Z. Gu, *Patent Application, WO 2010/117939-A1*, 2010.
[131] Z. Gu, *Patent Application, WO 2010/118009-A1*, 2010.

CHAPTER **17**

The Emergence of Small-Molecule Inhibitors of Capsid Assembly as Potential Antiviral Therapeutics

Clarence R. Hurt, Vishwanath R. Lingappa and **William J. Hansen**

Contents

Prosetta Antiviral, Inc., 670 5th Street, San Francisco, CA 94107, USA

Annual Reports in Medicinal Chemistry, Volume 46
ISSN: 0065-7743, DOI: 10.1016/B978-0-12-386009-5.00010-2

1. INTRODUCTION

There is a large degree of variation between viruses that cause human disease, but the general life cycle of viruses share common features that are potential targets for antiviral drugs. These are viral attachment, viral entry, uncoating, replication and release. Each of these key processes represents an opportunity to disrupt the viral life cycle and prevent the virus from replicating [1]. Thus, in order for antiviral treatments to be effective, they must either block the entry of the virus into the cell or be active within the infected cell. Because viruses use the host cell's own systems to reproduce, therapeutic approaches can target not only the viral life cycle but also the cellular proteins that are hijacked and manipulated by the virus. Enzyme inhibitors have the unique ability to affect processes within infected cells. They bind to enzymes in a way that disrupts their function, preventing a step in the infectious process from occurring. The classes of enzymes currently targeted by antiviral therapy include reverse transcriptase, protease, integrase, replicase, and neuraminidase [2]. Other approaches that are employed to disrupt the viral life cycle include entry inhibitors, fusion inhibitors, and integrase inhibitors. More recently, attention has focused on the later stages of viral replication and the capsid assembly pathway [3,4].

2. CAPSID ASSEMBLY INHIBITORS

The viral capsid is a structural protein that encloses and protects the genetic material of the virus during the viral replication process. It is a unique protein synthesized by specific genes in the nucleic acid of the virus [5]. Within the viral life cycle, the capsid protein is the most conserved of all the viral structural proteins. Mutations in the sequence and structure of the capsid protein have led to alterations in the binding mode, which prevented dimerization, processing and ultimately diminished viral infectivity [6,7]. Capsid assembly (CA) stands out as a step common to viral replication and infectivity, yet it has not been the subject of extensive drug discovery programs until now. In the past few years, a body of data has opened new opportunities for antiviral research to focus on the modulation of CA processes as a novel approach to the inhibition of viral infections [8]. Most of the early efforts have focused on disrupting the normal interactions of the capsid protein, key protein–protein interactions in the capsid binding region, and modulating the processing of the CA protein N-terminal and C-terminal domains (CA_{NTD}–CA_{CTD}).

2.1. Inhibitors that bind directly to HIV-1 CA

Early validation of this approach was obtained through point mutations in the CA which resulted in diminished levels of processing, dimerization, and infectivity during the viral replication cycle [9,10]. Mutations on CA severely disrupt the cell-cycle independence of HIV. The loss of cell-cycle independence can be cell-type specific, which suggests that a cellular factor affects the ability of HIV to infect nondividing cells. The identification of small molecules capable of disrupting these same interactions could lead to novel approaches to supplement current therapies to combat viral infections.

2.1.1. HIV-1 CA_{NTD} inhibitors

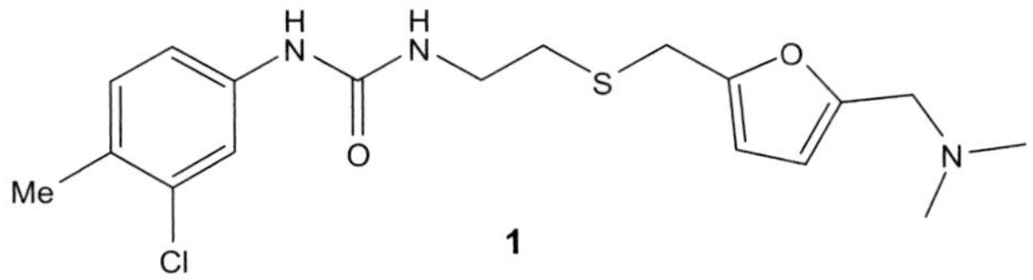

Gag is the major structural protein for the virus particle and is responsible for viral budding. CA is synthesized as a domain within the 55-kDa Gag precursor polyprotein and as thousands of copies of Gag assemble near the plasma membrane, they bud to form an immature, noninfectious viral particle. CA is cleaved through proteolysis, and a conformational shift leads to the formation of the capsid particles. Urea **1** (also known as 1-(3-chloro-4-methylphenyl)-3-(2-((5-((dimethylamino)methyl)furan-2-yl)methylthio)ethyl)urea or CAP-1) was one of the earliest compounds shown to bind to the HIV-1 capsid (CA) protein using nuclear magnetic resonance (NMR) spectroscopy [11,12]. Compound **1** does not inhibit cell growth or virus production, but the particles produced are poorly infectious. The phenotype of aberrant capsid formation was shown to be the direct result of the inhibitor binding to the CA_{NTD} through displacement of Phe 32 residue for the CAP-1 aromatic residues, which disrupts capsid maturation and interferes with Gag–Gag interactions during assembly of immature particles [13].

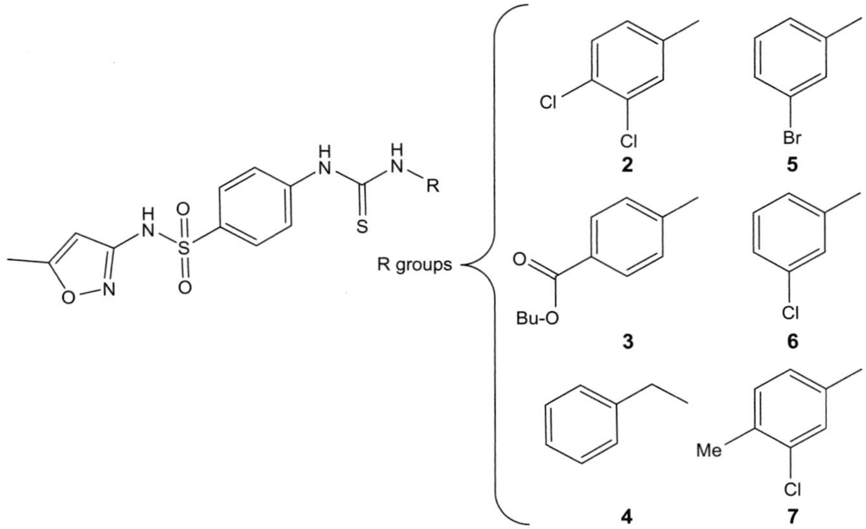

A series of thioureas **2**, **3**, and **4** were shown to act as dual inhibitors of CA and Cyclophilin A (Cyp A) and inhibit HIV-1 in the low micromolar range as measured in an optical density assay for turbidity during the CA process [14]. These compounds also displayed low toxicity for cell proliferation against uninfected CEM cells (TC_{50}) with values greater than 100 μM [14]. Additional analogs **5**, **6**, and **7** were also found to cause 50% inhibition in the submicromolar range [15]. With Cyp A also inhibited by this novel class of thioureas, both capsid disassembly and CA are blocked in the viral replication life cycle [16].

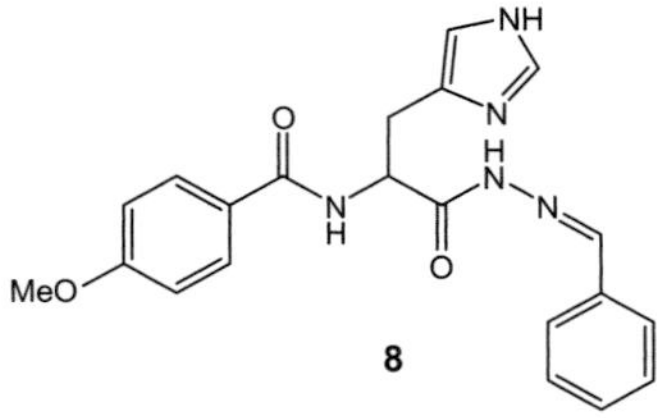

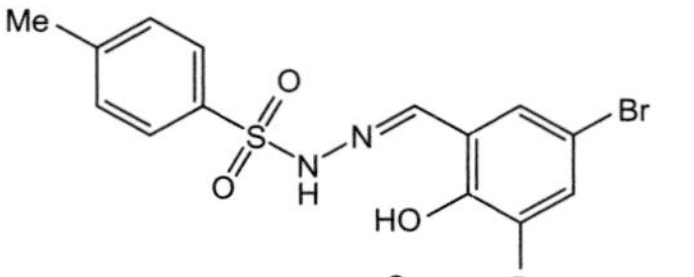

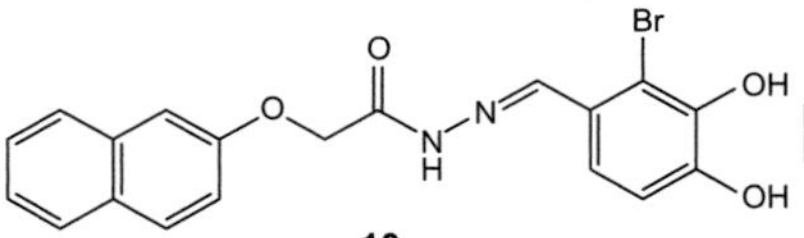

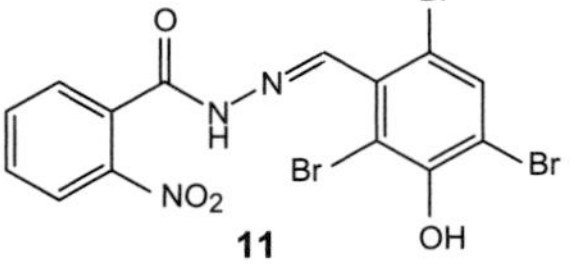

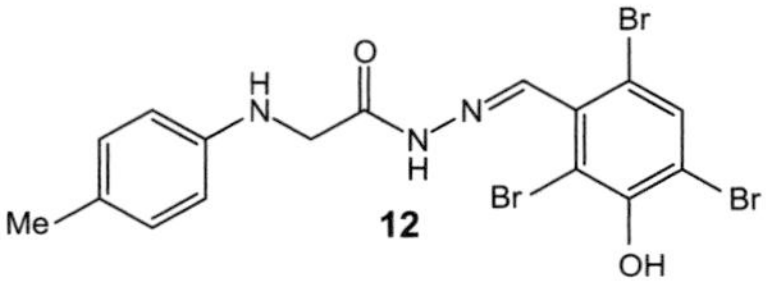

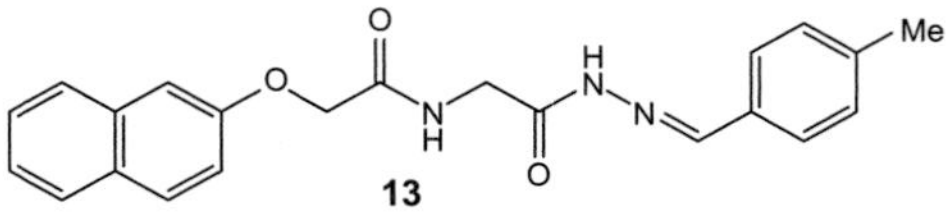

Several groups of acyl hydrazones were also identified as potent inhibitors of Simian Immunodeficiency Virus (SIV) and have been shown to bind to CA_{NTD} in a similar manner as compound **1** [17,18]. Compounds **8** through **12** inhibit SIV with EC_{50} values in the low micromolar range. Compound **13** inhibited SIV with an EC_{50} of 0.47 μM while it was not toxic against uninfected cells at concentrations above 100 μM.

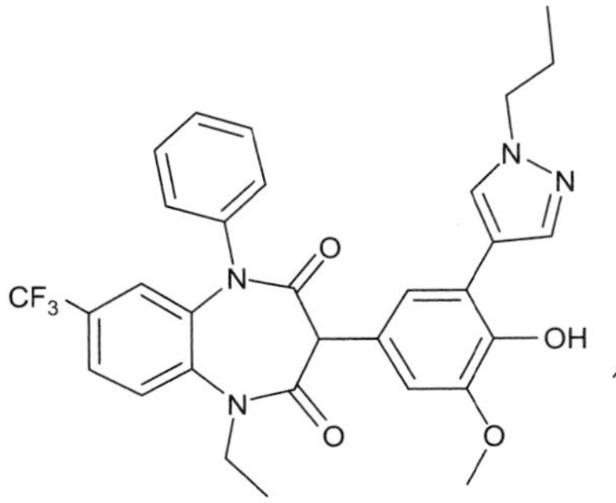

14 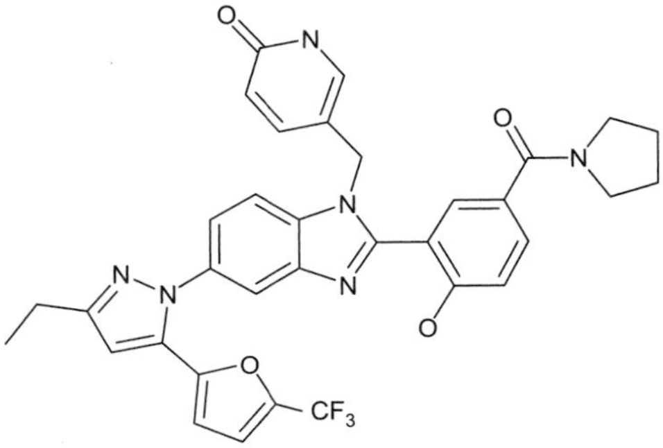 **15**

Two chemical series were identified as binding to CA_{NTD} and inhibiting HIV-1 replication through NMR spectroscopy [19,20]. Compound **14** is a benzodiazepine with $EC_{50} = 70$ nM and low toxicity for a 50% reduction of cell proliferation against uninfected cells (CC_{50}) of 28 μM. Compound **15** is a benzimidazole with $EC_{50} = 62$ nM and a comparable cellular toxicity profile, with CC_{50} greater than 20 μM. The mechanism of action for these compounds was similar to the other CA inhibitors in that they were shown to bind directly to the CA_{NTD} and these compounds do not show cross-resistance with other antiviral mechanisms. Resistance mutants within the CA inhibitor binding pocket map to the CA_{NTD} and these mutations offset the binding and function of the inhibitors. The site of interaction was confirmed by electron microscopy (EM), NMR, and co-crystallization (Co-Xtal) data [19]. Lead optimization for both lead series was stopped due to the inability to reconcile the shift in potency due to protein binding and the development of a complex resistance profile.

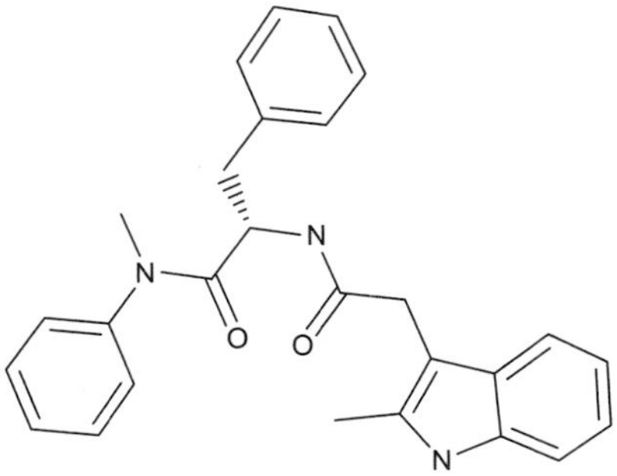

16

Compound **16**, PF74, targets the CA and is active following HIV-1 envelope mediated entry and before reverse transcription [21,22]. It also binds directly to the HIV CA and blocks the formation of infectious virion by triggering premature uncoating of the capsid in target cells. Compound **16** also binds to CA with micromolar affinity and inhibits a wide range of HIV isolates. Selection for drug resistance in culture identified a series of five mutations in CA (Q67H, K70R, H87P, T107N, and L111I) that prevented compound binding, collectively conferring resistance to the molecule.

Ac–NH2–E—Q—A—S—Q—E—V—K—N—W—M—T—E—T—L—L—V—Q—N—A—COOH

17

A 20-mer peptide, **17**, CAC1, was designed to mimic the structural and energetic interactions of the HIV CA_{CTD} [23]. The amino acid sequence corresponds to residues 175–194 of α-helix 2 of the CA_{CTD} of HIV, and **17** interacts specifically with this region. The affinity of **17** for the CA_{CTD} region is approximately 50 μM and may represent an excellent opportunity to initiate the development of peptidomimetic CA inhibitors.

NH_2–I—T—F—E—D—L—L—D—Y—Y—E—P—COOH

18

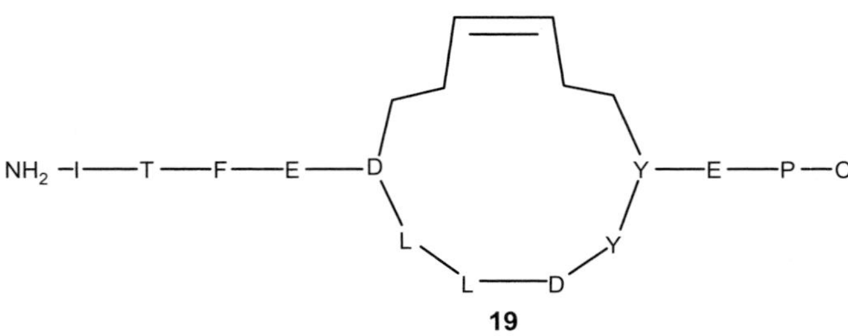

19

A 12-mer peptide, **18**, CAI, that binds the CA domain of Gag was identified through a phage display screen [24–27]. Peptide **18** inhibits assembly of immature and mature HIV capsid particles *in vitro* but not in cell culture due to poor cell penetration. The next generation of these peptide inhibitors, peptide **19**, NYAD-1, displayed activity in cell culture and had 10-fold better potency over CAI [28,29]. Theses peptide inhibitors of HIV-1 block the CA_{NTD}–CA_{CTD} interaction by competing for the natural binding region in the CA_{NTD}. Altering the CA_{NTD}–CA_{CTD} dimer interface that is crucial for connecting the hexameric rings in the CA lattice would weaken the hexamers, impair assembly, and destabilize the assembly cores.

GPG-NH2,20 ALG-NH2,21 RQG-NH2,22

Recently, several tripeptides were identified as inhibitors of HIV-1 CA, with an 80% reduction of HIV-1 replication [30–32]. Tripeptides **20, 21,** and **22** were shown to interact with CA with EC_{50} values in the low- to mid-micromolar range and resulted in the formation of a range of aberrant capsids. The tripepetides were effective against HIV-1 drug-resistant strains, and **22** did not display resistance mutations after 30 passages of the HIV-1 virus. The tripeptides represent a promising new approach to combat HIV starting with molecules of low molecular weight.

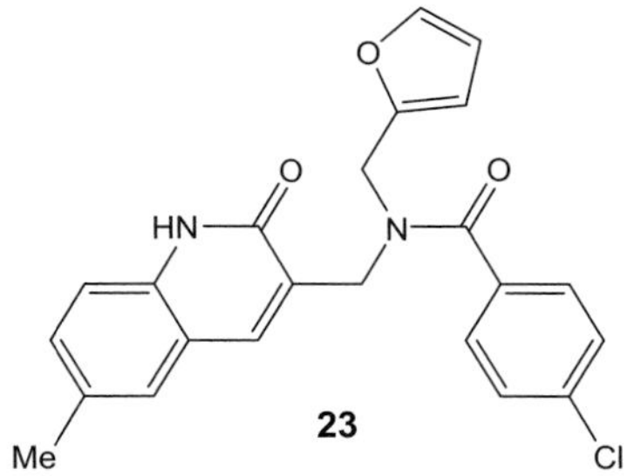

23 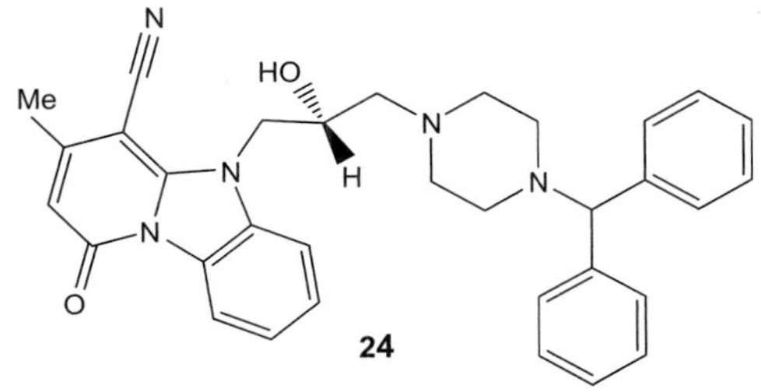 **24**

A docking-based virtual screen of the HIV-1 CA_{CTD} binding pocket occupied by the tripeptide-based inhibitors above leads to the identification of small molecules capable of occupying the same hydrophobic cavity. Compounds **23** and **24** are potential HIV-1 inhibitors with EC_{50} values equal to 1.1 and 1.8 μM, respectively [33]. These compounds are capable of inhibiting the formation of mature viral particles, which was verified with EM studies. Both chemical series show comparable activity against a wide range of HIV-1 laboratory-adapted isolates.

2.2. Maturation inhibitors: Antivirals that block cleavage of HIV-1 CA-SP1

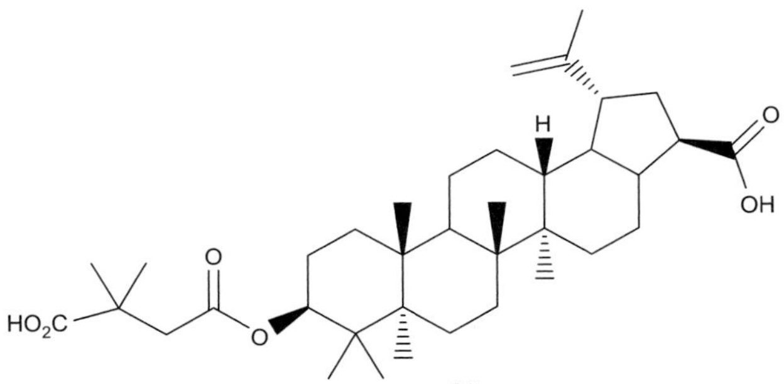

25

Maturation inhibitors are distinct from protease inhibitors (PIs). PIs block Gag and the precursor protein (Gag–Pol) cleavage by aspartyl proteases to the structural capsid proteins, matrix antigen (MA p17), capsid antigen (CA p24), nucleocapsid (NC p7, NC p1), transcriptase (p66/p51), integrase (p32), and other functional proteins (p11/p11) [2]. Maturation inhibitors block the cleavage of capsid precursor (CA-SP1) to mature capsid protein p24 (or CA), as exemplified by compound **25** (BVM, a triterpene derived from betulinic acid) with an $EC_{50} = 1.3$ nM, $CC_{50} = 43$ μM [34–40]. By preventing the cleavage of SP1 from the C-terminal of the CA, there is an accumulation of the CA-SP1 intermediate which results in defective core condensation. The viral particles released were noninfectious. Oral administration of **25** to SCID-hu Thy/Liv mice reduced viral RNA by a factor of 100. A dose-dependent inhibition of capsid-SP1 cleavage in HIV-1-infected human thymocytes obtained from mice was also observed. Maturation inhibitor **25** was advanced to clinical trials for further evaluation of efficacy and drug resistance. The activity of **25** in human patients was found to be dependent on a specific Gag polymorphism. Patients without the mutations Q369, V370, or T371 were more likely to respond. Resistance was also found to be common in treatment-naïve patients as well as patients with previous exposure to an HIV PI. The development of **25** was stopped after completion of a Phase 2b clinical study as the patient population became less responsive.

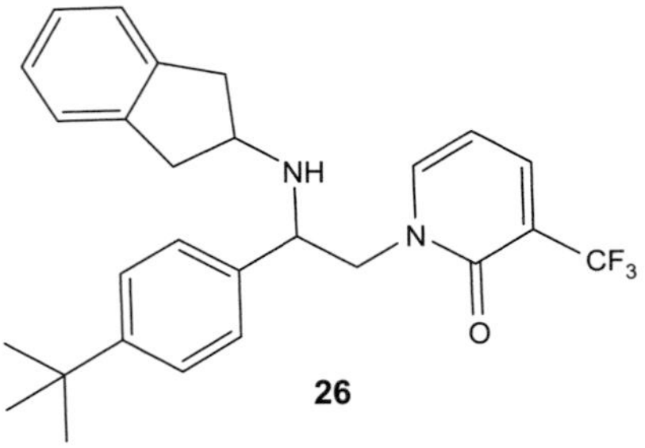

NH N CF_3 O

26

An additional compound **26**, PF-46396, was discovered through a high throughput screen (HTS) for inhibitors of HIV and, subsequently, found to affect viral uncoating and assembly [41]. Compound **26** blocks the processing of HIV-1 in manner similar to BVM and was effective against multiple HIV-1 laboratory strains. Viral resistance to **26** was raised after routine serial passage of the NL4-3 strain of HIV-1. A single amino acid mutation (I201V) conferred resistance to **26** and restored the regular HIV-1 CA-SP1 cleavage, even in the presence of the inhibitor.

2.3. HBV CA inhibitors

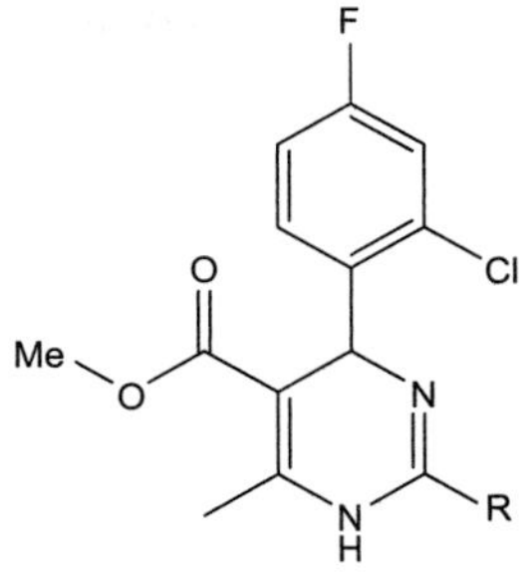

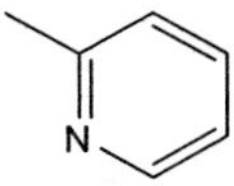

The heteroaryldihydropyrimidines (DHPs) **27, 28, 29,** and **30** represent a class of compounds capable of inhibiting replication of the hepatitis B virus (HBV) in HepG2.2.15 cells and thus suppress the production of viral DNA [42–44]. These compounds have been shown to bind to either HBV capsid or newly synthesized core protein. When the inhibitors were bound to the capsid proteins, the core protein cannot assemble properly and was easily degraded. The inhibition of HBV results from disruption of CA by binding of the compounds to dimers and oligomers of core protein. The mechanism of action was established through extensive co-crystal studies with HAP-1. Other direct binding inhibitors of HBV capsid formation are compounds **31** and **32** [44]. Both bind to the C-terminal region of the HBV.

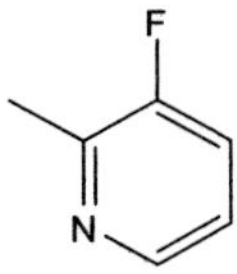

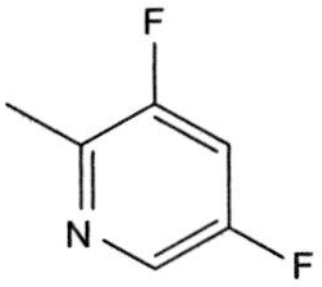

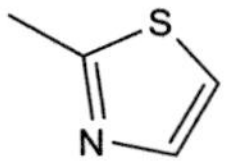

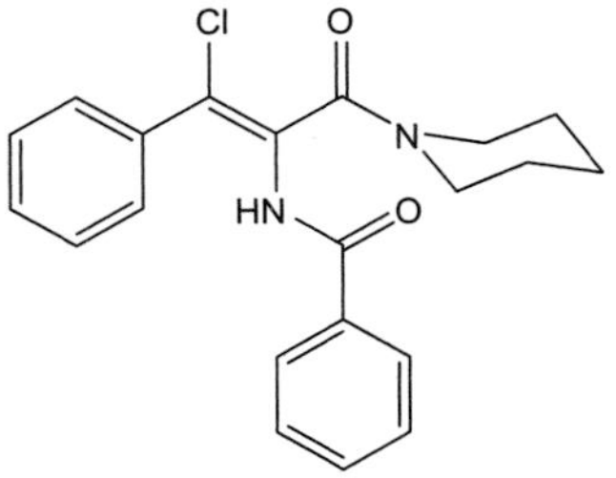

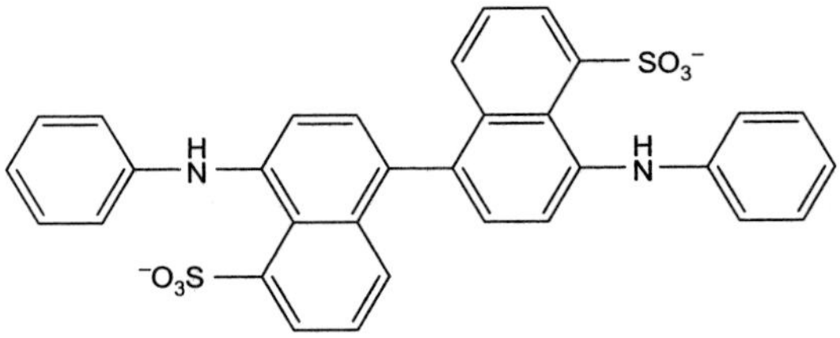

2.4. HCV CA inhibitors

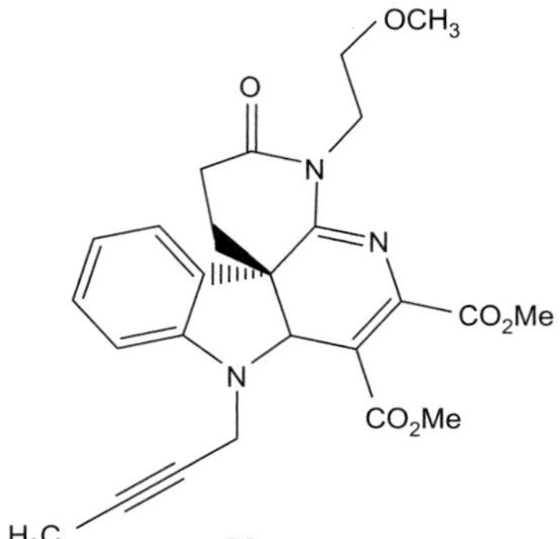

33

Compound **33** was recently identified from an automated screen of indoline alkaloids as a potential inhibitor of HCV [45]. The screen selected for inhibitors of the capsid protein core of the virus, ultimately resulting in the production of a noninfectious virion. The EC_{50} of compound **33** was 2.0 μM with a CC_{50} of greater than 320 μM.

3. CA INHIBITION *VIA* HOST FACTOR MODULATION

Protein–protein interactions are involved in every biological process responsible for cellular function, including the propagation of viral replication. The viral replication process is also subjected to extensive protein–protein interaction during the replication process. In general for viral families, one of the most conserved proteins produced during the replication process is the capsid protein. The capsid protein shares no homology with cellular host proteins even though transcription occurs using host factors and other cellular components. Capitalizing on these differences provides the basis for a novel approach to viral inhibition by focusing on the cellular host factors to disrupt the assembly of the viral capsid.

3.1. Pathway-wide screen using the cell-free screening system

Protein–protein interactions occurring in higher animals have been effectively modeled by homologous interactions occurring in more biochemically tractable systems derived from non-metazoans. The *in vitro* assembly for the *de novo synthesis* of the capsid protein on ribosomes in a cell-free (but cell-derived) extract under physiological conditions was used to direct newly synthesized capsid protein to form structures that were indistinguishable from authentic viral capsids by biochemical and biophysical criteria, and electron microscopic appearance [46]. The assay is one of cell-free translation programmed by viral capsid mRNA in 384-well plates containing small molecules. The compounds that inhibit CA are identified through an antibody detection system.

3.2. Modulation of CA pathway leads to novel antivirals

Compounds identified as potential inhibitors of the CA pathway were evaluated for dose-dependent inhibition and for cellular toxicity against relevant cell lines. Two distinct phenotypes expressed by inhibition of the CA pathway were (1) the blocked release of virus, with accumulation of an intracellular assembly intermediate, presumably the site of action of the host protein drug target [47]; (2) the release of aberrant capsids [48,49], consistent with the demonstration that viral particles are produced upon drug treatment of virus infected cells, leading to the production of noninfectious particles. This novel approach was presented at the 23rd Annual International Conference on Antiviral Research (ICAR) [48–56]. Figure 1 represents a summary of the number of distinct chemical series identified as hits (or pre-leads) for each virus screened with the aforementioned screening format. The EC_{50}s for these antiviral pre-lead hits ranged from the low micromolar to the low nanomolar range. Although numerous distinct chemical classes were identified and described at ICAR (Figure 1), only one chemical class of compounds was presented, shown below as the generic structure **34**. In general, compounds active against one member of a viral family were also found to be comparably active against the other members (and strains) of the same viral family. There was also a fair degree of interfamily antiviral activity, as some compounds

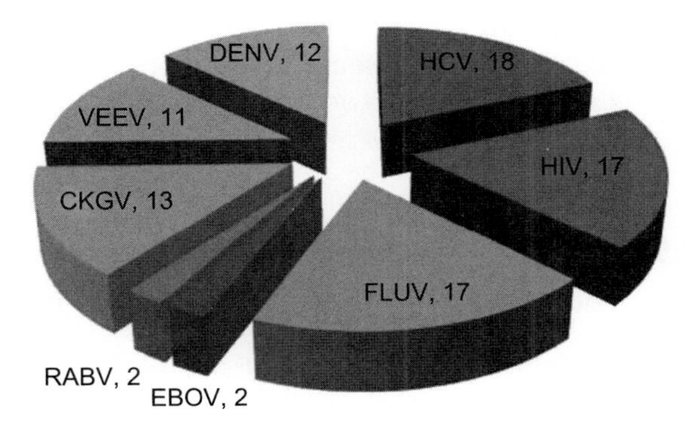

Figure 1 The number of distinct pre-lead chemical series identified from the cell-free screening system. All pre-leads display antiviral activity in the cell culture live virus assays. (See Color Plate 17.1 in Color Plate Section.)

demonstrated activity against multiple viral families. Preliminary results were presented for an animal efficacy study, where mice challenged with the Ebola virus were treated once daily (for 10 days) with a 5-mg/Kg dose intraperitoneal with a compound related to **34** resulting in 100% survival of the animals [56]. Lead optimization is underway for several therapeutic programs along with elucidation of the mechanism of action.

34

4. CONCLUSION

Over the past three decades, great strides have been made to advance antiviral research and to develop new antiviral drugs, with the majority of these approaches focusing on different stages of the viral life cycle. With the frequent development of resistance and toxicity from drug treatment, interest has increased for research on novel approaches to antiviral therapies. Research that targets the later stages of viral replication, such as the CA process, offers a new opportunity to develop therapeutic approaches to treat viral infections. Antivirals that interfere with the normal assembly or processing of the viral capsid are gaining momentum as potential therapies because the integrity of the viral capsid protein is crucial for infectivity and the replication of the virus. The early success of BVM, which demonstrated efficacy in the HIV-1 mouse model and progressed to Phase II in the clinic, provides encouragement that future endeavors along these lines of research may achieve comparable success. Antiviral programs which feature capsid assembly inhibitors will need to address the issues of drug resistance and toxicity from drug treatment in order to be successful.

REFERENCES

[1] N. E. Davey, G. Trave and T. J. Gibson, *Trends Biochem. Sci.*, 2010, **159**, 1.
[2] E. De Clercq, *Nat. Rev. Drug Discov.*, 2007, **6**, 1001.
[3] W. C. Greene, Z. Debyser, Y. Ikeda, E. O. Freed, E. Stephens, W. Yonemoto, R. W. Buckheit, J. A. Este and T. Cihlar, *Antiviral Res.*, 2008, **80**, 251.
[4] Y. Jiang, X. Liu and E. De Clercg, *Curr. Med. Chem.*, 2011, **18**, 16.
[5] Y. Morikawa, *Curr. HIV Res.*, 2003, **1**, 1.
[6] J. Kafaie, M. Dolatshahi, L. Ajamian, R. Song, A. J. Mouland, I. Rouiller and M. Laughrea, *Virology*, 2009, **385**(1), 233.

[7] I. Scholz, B. Arvidsan, D. Huseby and E. Barklis, *J. Virol.*, 2005, **79**, 1470.
[8] E. Barklis, A. Alfadhi, C. McQuaw, S. Yalamuri, A. Still, R. L. Barklis, B. Kukull and C. S. Lopez, *J. Mol. Biol.*, 2009, **387**, 376.
[9] S. Tang, T. Murakami, B. F. Agresta, S. Campbell, E. O. Freed and J. G. Levin, *J. Virol.*, 2001, **75**, 9357.
[10] M. Yamashita, O. Perez, T. J. Hope and M. Emerman, *PLos Pathogens*, 2007, **3**, e156.
[11] M. F. Summers, C. Tang and M. Huang, *Patent Application WO03089615-A2*, 2003.
[12] C. Tang, E. Loeliger, I. Kinde, S. Kyere, K. Mayo, E. Barklis, Y. Sun, M. Huang and M. F. Summers, *J. Mol. Biol.*, 2003, **327**, 1013.
[13] B. N. Kelly, S. Kyere, I. Kinde, C. Tang, B. R. Howard, H. Robinson, W. I. Sundquist, M. F. Summers and C. P. Hill, *J. Mol. Biol.*, 2007, **373**, 355.
[14] K. Chen, Z. Tan, M. He, J. Li, S. Tang, I. Hewlett, F. Yu, Y. Jin and M. Yang, *Chem. Biol. Drug Des.*, 2010, **76**, 25.
[15] Z. Tan, J. Li, R. Pang, S. He, M. He, S. Tang, I. Hewlett and M. Yang, *Med. Chem. Res.*, 2010, **20**, 314.
[16] J. Li, Z. Tan, S. Tang, I. Hewlett, R. Pang, M. He, S. He, B. Tian, K. Chen and M. Yang, *Bioorg. Med. Chem.*, 2009, **17**, 3177.
[17] P. Prevelige, *Patent Application WO 2007/048042-A2*, 2007.
[18] B. Tian, M. He, Z. Tan, S. Tang, I. Hewlett, S. Chen, Y. Jin and M. Yang, *Chem. Biol, Drug Des.*, 2011, **77**, 189.
[19] L. D. Fader, R. Bethell, P. Bonneau, M. Bos, Y. Bousquet, M. G. Cordingley, R. Coulombe, P. Derby, A. M. Faucher, A. Gagnon, N. Goudreau, C. G. Maitre, I. Guse, O. Hucke, S. H. Kawai, J. E. Lacoste, S. Landry, C. T. Lemke, E. Malenfant, S. Mason, S. Morin, J. O'Meara, B. Simoneau, S. Titolo and C. Yoakim, *Bioorg. Med. Chem. Lett.*, 2011, **21**, 398.
[20] S. Titolo, J.-F. Mercie and E. Wardrop, et al., *17th Conference on Retroviruses & Opportunistic Infections (CROI 2010)*. San Francisco, February 16–19, 2010, Abstract 50.
[21] W. S. Blair, C. Pickford, S. L. Irving, D. G. Brown, M. Anderson, R. Bazin, J. Cao, G. Claramella, J. Isaacson, L. Jackson, R. Hunt, A. Kjerrstrom, J. A. Nieman, A. K. Patick, M. Perros, A. D. Scott, K. Whitby, H. Wu and S. L. Butler, *PLos Pathogen*, 2010, **6**, 1.
[22] J. Shi, J. Zhou, V. B. Shah, C. Aiken and K. Whitby, *J. Virol.*, 2011, **85**, 542.
[23] M. T. Garzon, M. C. Lidon-Moya, F. N. Barrera, A. Prieto, J. Gomez, M. G. Mateu and J. L. Neira, *Protein Sci.*, 2004, **13**, 1512.
[24] E. Barklis, A. Alfadhli, C. McQuaw, S. Yalamuri, A. Still, R. L. Barklis, B. Kukull and C. S. Lopez, *J. Mol. Biol.*, 2009, **387**, 376.
[25] F. Ternois, J. Sticht, S. Duquerroy, H.-G. Krausslich and F. A. Rey, *Nat. Struct. Mol. Biol.*, 2005, **12**, 678.
[26] K. Braun, M. Frank, R. Pipkorn, J. Reed, H. Spring, J. Debus, B. Didinger, C.-W. von der Lieth, M. Wiessler and W. Waldeck, *Int. J. Med. Sci.*, 2008, **5**, 230.
[27] J. Sticht, M. Humbert, S. Findlow, J. Boden, B. Muller, U. Dietrich, J. Werner and H. -G. Krausslich, *Nat. Struct. Mol. Biol.*, 2005, **12**, 671.
[28] J. L. Neira, *FEBS*, 2009, **276**, 6110.
[29] H. Zhang, Q. Zhao, S. Bhattacharya, A. A. Waheed, X. Tong, A. Hong, S. Heck, F. Curreli, M. Goger, D. Cowburn, E. O. Freed and A. K. Debnath, *J. Mol. Biol.*, 2008, **378**, 565.
[30] S. Abdurahman, A. Vegvari, M. Levi, S. Hoglund, M. Hogberg, W. Tong, I. Romero, J. Balzarini and A. Vahlne, *Retrovirology*, 2009, **7**, 34.
[31] A. Jejcic, S. Hoglund and A. Vahlne, *Retrovirology*, 2010, **7**, 20.
[32] P. Na Nakorn, W. Treesuwan, K. Choowonkomon, S. Hannongbua and N. Boonyalai, *J. Theor. Biol.*, 2011, **270**, 88.
[33] F. Curreli, H. Zhang, X. Zhang, I. Pyatkin, Z. Victor, A. Altieri and A. K. Debnath, *Bioorg. Med. Chem.*, 2011, **19**, 77.

[34] K. Qian, R. Y. Kuo, C. H. Chen, L. Huang, S. L. Morris-Natschke and K. H. Lee, *J. Med. Chem.*, 2010, **53**, 3133.
[35] C. A. Stoddart, P. Joshi, B. Sloan, J. C. Bare, P. C. Smith, G. P. Allaway, C. T. Wild and D. E. Martin, *PLos One*, 2007, **11**, 1251.
[36] C. A. Adamson, S. D. Ablan, I. Boeras, R. Goila-Gaur, F. Soheilian, K. Nagashima, F. Li, K. Salzwedel, M. Sakalian, C. T. Wild and E. O. Freed, *J. Virol.*, 2006, **80**, 10957.
[37] V. M. Vogt, *Nat. Struct. Mol. Biol.*, 2005, **12**, 638.
[38] K. Salzwedel, D. E. Martin and M. Sakalian, *AIDS Rev.*, 2007, **9**, 162.
[39] C. Aiken and C. H. Chen, *Trends Mol. Med.*, 2005, **11**, 31.
[40] C. S. Adamson, M. Sakalian, K. Salzwedel and E. O. Freed, *Retrovirology*, 2010, **7**, 36.
[41] W. S. Blair, J. Cao, J. Fok-Seang, P. Griffin, J. Isaacson, R. L. Jackson, E. Murray, A. K. Patick, Q. Peng, M. Perros, C. Pickford, H. Wu and S. L. Butler, *Antimicrob. Agents Chemother.*, 2009, **53**, 5080.
[42] S. J. Stray, C. R. Bourne, S. Punna, W. G. Lewis, M. G. Finn and A. Zlotnick, *PNAS*, 2005, **102**, 8138.
[43] K. Deres, C. H. Schoder, A. Paessens, S. Goldmann, H. J. Kacker, O. Weber, T. Kramer, U. Niewohner, U. Pleiss, J. Stoltefuss, E. Graef, D. Koletzki, R. N. A. Masantschek, A. Reimann, R. Jaeger, R. Grob, B. Beckermann, K.-H. Schlemmer, D. Haebich and H. R. Waigmann, *Science*, 2003, **299**, 893.
[44] I.-G. Choi and Y. G. Yu, *Infect. Disorders Drug Targets*, 2007, **7**, 251.
[45] W. Wei, C. Cai, S. Kota, V. Takahashi, F. Ni, A. D. Strosberg and J. K. Syder, *Bioorg. Med. Chem. Lett.*, 2009, **19**, 6926.
[46] V. R. Lingappa, E. Harrell, K. Copeland, M. D. Prasad, J. R. Lingappa, C. Kelleher, C. R. Hurt and W. Hansen, Antiviral Research, ICAR 23rd Annual Meeting Poster Session 2010 Poster 143.
[47] Q. Li, A. L. Brass, A. Ng, Z. Hu, R. J. Xavier, T. J. Liang and S. J. Elledge, A genome-wide genetic screen for host factors required for hepatitis C virus propagation, *Proc. Natl. Acad. Sci. USA*, 2009, **106**(38), 16410.
[48] B. Petsch, C. R. Hurt, B. Freeman, E. Zirdum, A. Ganesh, A. Schörg, A. Kitaygorodskyy, Y. Marwidi, O. Ducoudret, C. Kelleher, W. Hansen, V. R. Lingappa, C. Essrich and L. Stitz, Antiviral Research, ICAR 23rd Annual Meeting Poster Session, 2010, Vol. **86**, Poster 88.
[49] M. Baker-Wagner, N. Wolcott, Y. Marwidi, S. F. Yu, D. Dey, B. Onisko, K. Barlow, S. Potluri, C. Sahlman, A. Calayag, V. R. Lingappa, P. Glass, M. Farmer, C. H. Hurt and W. Hansen, Antiviral Research, ICAR 23rd Annual Meeting Poster Session, 2010, Vol. **86**, Poster 36.
[50] B. Kelley-Clarke, P. J. Glass, C. R. Hurt, B. K. Molter, B. K. Thielen, B. Freeman, A. Kitaygorodskyy, C. Kelleher, M. Corpuz, E. Harrell, W. J. Hansen, J. S. Lee, K. C. Klein, V. R. Lingappa and J. R. Lingappa, A drug that inhibits alphavirus replication identified in a high-throughput cell-free assembly pathway screen. Submitted to Cell Host & Microbe.
[51] J. Gentzsch, C. R. Hurt, V. R. Lingappa and T. Pietschmann, Antiviral Research, ICAR 23rd Annual Meeting Poster Session, 2010, Vol. **86**, Poster 14.
[52] K. Copeland, W. Hansen, V. Asundi, S. Hong, J. Chamberlin, D. Dey, S. Broce, H. Himmel, C. Decloutte, S. Ram, I. Steffen, S. Pöhlmann, J. R. Lingappa, R. Jaisri, C. R. Hurt and V. R. Lingappa, Antiviral Research, ICAR 23rd Annual Meeting Poster Session, 2010, Vol. **86**, Poster 21.
[53] V. R. Lingappa, J. Gentzsch, K. Copeland, I. Jaing, M. Corpuz, P. Glass, J. R. Lingappa, B. Kelley-Clarke, D. Dey, C. Kelleher, A. Atuegbu, A. Anderson, J. Lehrer-Graiwer, T. Pietschmann, C. R. Hurt and W. Hansen, Antiviral Research, ICAR 23rd Annual Meeting Poster Session, 2010, Vol. **86**, Poster 75.

[54] M. Karpuj, D. Smith, B. Kelley-Clarke, A. Stossel, A. Honko, S. Broce, N. Van Loan, E. Harrell, C. Kelleher, J. R. Lingappa, W. Hansen, C. R. Hurt, L. Hensley and V. R. Lingappa, Antiviral Research, ICAR 23rd Annual Meeting Poster Session, 2010, Vol. **86**, Poster 134.
[55] V. R. Lingappa, U. Lingappa, E. Borst, J. Pajda, I. Brown, S. Long, B. Rami, A. Nalca, W. I. Lipkin, C. Rupprecht, M. Messerle, C. R. Hurt and W. Hansen, Antiviral Research, ICAR 23rd Annual Meeting Poster Session, 2010, Vol. **86**, Poster 142.
[56] J. Francis, W. Kalina, T. Warren, K. Edwards, I. Jaing, A. P. Abola, A. Nissar, A Kitaygorodskyy, C. R. Hurt, S. Bavari, W. Hansen and V. R. Lingappa, Antiviral Research, ICAR 23rd Annual Meeting Poster Session, 2010, Vol. **86**, Poster 123.

PART VI:
Topics in Biology

Editor: John Lowe
JL3Pharma LLC
Stonington
Connecticut

CHAPTER **18**

Molecular Mechanism of Action (MMoA) in Drug Discovery

David C. Swinney

Contents

Institute for Rare and Neglected Diseases Drug Discovery Belmont, CA, USA

Annual Reports in Medicinal Chemistry, Volume 46
ISSN: 0065-7743, DOI: 10.1016/B978-0-12-386009-5.00009-6

ABBREVIATIONS

ADME	absorption distribution metabolism excretion
BE	biochemical efficiency
CB	cannabinoid
EGF	epidermal growth factor
GABA	γ-aminobutyric acid
GR	glucocorticoid receptor
inhA	enoyl-ACP reductase of *Mycobacterium tuberculosis*
LBD	ligand binding domain
MK2	mitogen-activated protein kinase-activated protein kinase 2
MMoA	molecular mechanism of action
NA	neuroaminidase
NCoR	nuclear receptor co-repressor
NMDA	*N*-methyl-D-aspartate
NNRT	nonnucleotide reverse transcriptase
NSAIDs	nonsteroidal anti-inflammatory drugs
PDGFR	platelet-derived growth factor receptors
PPARγ	peroxisome proliferator-activated receptor
VEGFR	vascular endothelial growth factor receptor

1. INTRODUCTION-MOLECULAR MECHANISM OF ACTION

The molecular mechanism of action (MMoA) of a medicine is the connection of the molecular interactions between the therapeutic treatment and the biological target (*e.g.*, receptor, enzyme, etc.) that yields the physiological response. Pharmacological action begins with the interaction between these molecules as noted by Paul Ehlrich in 1913 that a substance will not work unless it is bound *corpora non agunt nisi fixata* [1]. However, the binding of the "medicine" to the target is not always sufficient for a substance to communicate the desired message to physiology. There are many facets of this interaction that ultimately result in the desired therapeutic outcome. For example, the site of interaction (allosteric *vs.* orthosteric), molecular descriptors of the binding interaction (affinity, binding kinetics), the functional impact (agonist, modulation, antagonism), and the specificity of the functional outcome (activation of specific signaling pathways) all contribute to the MMoA and impact the ultimate pharmacological response.

1.1. A significant factor in optimizing MMoA is understanding the potential for mechanism-based toxicity

Mechanism-based toxicity (on-target toxicity) is a potential concern with most human targets. MMoAs provide the means to minimize the mechanism-based toxicity while retaining the desired response [2–4]. Competition with endogenous effector (surmountable with fast on and off rates), uncompetitive inhibition, partial agonism, functional selectivity, and allosteric partial antagonism are mechanisms in which the physiological environment can help to shape the dose–response curves to minimize mechanism-based toxicity while retaining sufficient drug efficacy (see Section 4.1 for examples).

However, a drug such as an anti-infective agent with no potential for mechanism-based toxicity will maximize its therapeutic index *via* mechanisms in which drug binding is efficiently coupled to efficacy. Kinetic mechanisms that are irreversible, insurmountable, noncompetitive, full agonistic or slow dissociating will be safer because lower drug concentrations are required for efficacy. Lower drug concentrations minimize off-target toxicity and result in an increased therapeutic index. This is a principle driver for the desire for slow dissociation rates, long residence times, and irreversible inhibition.

1.2. Details of MMoA are not fully captured by IC_{50} and K_I

A challenge for drug discovery is that many of the molecular features important to an optimal MMoA are not captured by IC_{50} and K_I values in target-based screening assays. This is exemplified in a recent report on the molecular features that differentiate binding between functionally different ligands for the β_2-adrenergic receptor (β_2AR), a member of the G-protein-coupled receptor superfamily [5]. Wacker and colleagues found no discernable structural difference in binding between inverse agonist and antagonist. β_2AR bound to pharmacologically distinct ligands (antagonists and inverse agonists) has virtually identical backbone conformations in the crystal structures suggesting that the conformational changes capable of modifying signaling properties are very small, beyond the resolution of the obtained data. Alternatively, the major effect of inverse agonists and antagonists on β_2AR, which may be extrapolated to agonists, is not on modifying a specific conformation with large conformational changes, but on minor structural changes and a significantly larger contribution from receptor dynamics.

Another example of MMoA not captured by K_I demonstrates the contribution of association rate for the binding of benzimidazol-2-one derivatives (**1**) to HIV-1 reverse transcriptase [6]. The lysine 103 to asparagine mutant (K103N) has a minimal influence on the bound conformation of NNRTIs, while it significantly affects the kinetics of the inhibitor-binding process. A detailed enzymatic analysis elucidating the molecular mechanism of interaction between benzimidazol-2-one derivatives and the K103N mutant of HIV RT showed that the loss of potency of these molecules toward the K103N RT was specifically due to a reduction of their association rate to the enzyme. Unexpectedly, these compounds showed a strongly reduced dissociation rate from the K103N mutant, as compared to the wild-type enzyme, suggesting that, once occupied by the drug, the mutated binding site could achieve a more stable interaction with these molecules. The K103N mutant has a minimal influence on the bound conformation of NNRTIs, while it significantly affects the kinetics of the inhibitor-binding process. Available structural and biochemical data strongly support the view that the mutated N103 residue slows the k_{on} rate for inhibitor binding through the formation of a hydrogen bond with the Y188 side chain.

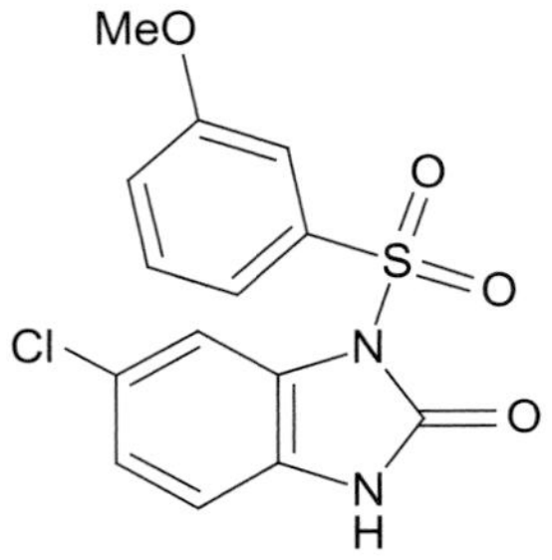

1

These important nuances of MMoA that are not captured by IC_{50} and K_I in binding assays highlight the gap between the molecular and physiological approaches to drug discovery. MMoA provides an opportunity to bridge the gap by mapping molecular changes to physiological outcomes. Molecular descriptors such as binding kinetics combined with metrics such as biochemical efficiency (BE) and discovery strategies based on increased understanding of the MMoA will help to bridge this gap, as described in the following sections.

2. MOLECULAR DESCRIPTORS

2.1. Equilibrium dissociation constant, K_I

The equilibrium dissociation constant measures the propensity for the bound drug/target complex to dissociate to free drug and target. Equilibrium dissociation constants are related to kinetic rate constants by the relationship $K_I = k_{off}/k_{on}$.

2.2. Reversible kinetic rates, k_{on}, k_{off}

The concentration at which the rate of dissociation equals the rate of association is equal to the equilibrium dissociation constant (K_I for inhibitors/antagonists, K_d for ligands). Association rate is concentration dependent in units of $M^{-1}\ s^{-1}$, and the dissociation rate is concentration independent with units in reciprocal time.

2.2.1. Association rates

There is a renewed interest in binding kinetics with an emphasis on residence time and dissociation rates [3,7,8]. Binding is not always diffusion controlled, and association rates can change within a series in a manner such that k_{off} does not correlate with K_I. Diffusion-controlled association rates for reactions in a cellular environment are estimated to be in the range of 10^5–$10^6\ M^{-1}\ s^{-1}$. The binding of all bimolecular reactions must initiate with a diffusion-controlled interaction. Transition to a stable ground state complex can involve local or global molecular movements that will slow the association rate of complex formation. When the timescale is very slow, this is termed slow binding. It can be observed by a lag in the approach to equilibrium as well as a shift in IC_{50} with preincubation time. It is straightforward to estimate if the association rate for the reaction is slower that diffusion controlled from the relationship $K_I = k_{off}/k_{on}$. A drug with a 1-nM K_I and diffusion-controlled reaction rate of $10^6\ M^{-1}\ s^{-1}$ will have a k_{off} rate of $0.001\ s^{-1}$ ($t_{1/2} = 693$ s from the relationship $t_{1/2} = 0.693/k$). Association rates of drugs with nanomolar K_I values and dissociation half-lives less than 10 min are most likely diffusion controlled, while the association rates of drugs with longer half-lives must involve a rate limiting conformational movement.

When binding is diffusion controlled within a structural series for a specific target, k_{off} will correlate with K_I. However, if the binding is more complex then K_I does not always correlate with k_{off}. Ligand binding to the M3 muscarinic receptor involves kinetically distinct conformational states. Sykes *et al.* [9] observed that the change in affinity for a series of antagonists better correlated with association rate as opposed to dissociation rate.

A recent report by Schrieber and coworkers describes some insights into the interactions involved in the association process [10]. They point out that there are two major classes of protein–protein interactions that describe most protein–protein association processes. One class is regulated primarily by electrostatic forces that provide both a long-range steering function and dominate the overall binding energy. The second class is more predominant and involves molecular contact interfaces having neutral or weakly charged surfaces in which the binding energy is governed primarily by hydrophobic forces. Studies based on weakly charged interfaces by Schreiber and coworkers indicate the individual interface side chains have little or no effect on association rates [11,12]. For weakly charged interfaces, Horn and coworkers, using human growth hormone receptor as a model system, concluded that precise matching of surface terrains of the two molecules defines the competent transition state. Interestingly, they also concluded that final fine-tuning of the structure after the transition state contributes significantly to the thermodynamic character of the interaction [13].

2.2.2. Dissociation rates

Residence time and dissociation rates are important features of a medicine's action [3,7,8]. The rational chemical design of molecules with a desired residence time remains a great challenge. Efforts to improve activity almost invariably focus on maximizing binding through stabilizing the energy of the drug–target complex. However, achieving long residence time requires modulation of the energy between both ground states and transition states on the reaction coordinate. An excellent recent report by Lu and Tonge shows the relative contribution of both ground state and transition state to residence time [14]. Analysis of the relative contributions for many different molecules emphasizes the point that transition state energy can play a major role in modulating residence time. They also noted that compounds with long residence times such as efavirenz (4.1 h) (**2**) can have relatively low thermodynamic affinity for their targets (5 μM), stressing the potential disconnect between their thermodynamic stability of a drug–target complex and the lifetime of that complex.

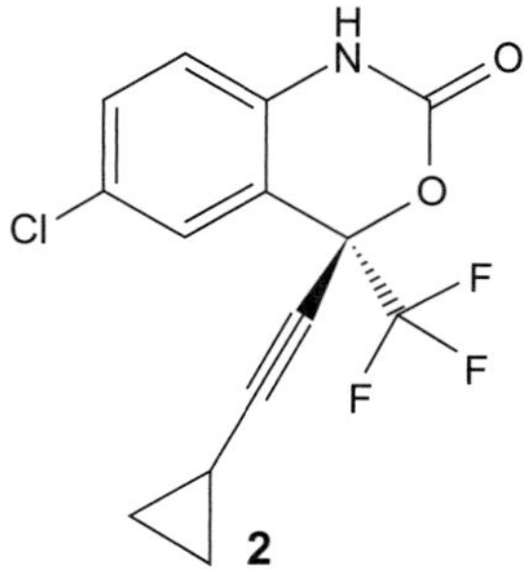

Recently, there were a number of reports of compounds with long residence times and long-lasting pharmacodynamic action attributed to slow dissociation rates including AZ12491187 (**3**) binding to the CB1 receptor [15], TAK-593 (**4**) a novel vascular endothelial growth factor receptor/platelet-derived growth factor receptors (VEGFR/PDGFR) inhibitor with a long residence time ($t_{1/2}$ from VEGFR2 with activity based assay was 5100 min) [16], and indacaterol (**5**) as an ultralong-acting inhaled β2-adrenoreceptor agonist [17].

In another significant contribution, Lu and coworkers showed that the rate of breakdown of the enzyme–inhibitor complex of fatty acid synthase-II enoyl reductase from *Francisella rularensis* is a better predictor of *in vivo* activity than the overall thermodynamic stability of the complex [18].

2.3. IC_{50}

The IC_{50} is the concentration of drug required for 50% inhibition. IC_{50} is an operational term dependent on the assay conditions. IC_{90} or IC_{99} is sometimes used when complete inhibition is required. Calculation of the fractional occupancy shows that IC_{90} concentration is approximately 10-fold greater than the IC_{50} concentration assuming one-site binding at equilibrium with a Hill coefficient of 1. The IC_{99} concentration is approximately 100-fold the IC_{50} concentration. The percent bound (also referred to as fractional occupancy) comes from the relationship percent bound $= [I]/[I] + K_I$, where K_I is the equilibrium dissociation constant for I and [I] is the concentration of inhibitor.

How do binding kinetics relate to the half maximal inhibitory concentration (IC_{50})? As noted above, IC_{50} is an operational term dependent on the assay conditions, whereas k_{on}, k_{off}, and K_I are intrinsic to a bimolecular interaction. For example, the IC_{50} for a competitive inhibitor will increase with increasing substrate concentrations even though the K_I, k_{on}, and k_{off} are unchanged. The change in IC_{50} due to competition is mathematically described by the Cheng–Prusoff relationship: $IC_{50}/K_I = 1 + [S]/K_m$, where [S] is the substrate concentration and K_m is the Michaelis constant. The IC_{50} is equal to K_I for noncompetitive inhibition.

2.4. Quantitation of irreversible inhibition

The effect of irreversible inhibition on a response is time dependent, and the rate of covalent modification is measured as the k_{inact}/K_I ($M^{-1}\ s^{-1}$), where k_{inact} is the kinetic rate of the covalent irreversible modification expressed in time and K_I is the equilibrium dissociation constant for initial binding of the inhibitor to the protein.

2.5. Conformational changes

There are no quantitative molecular descriptors for conformational changes. The effect of ligand-specific conformational changes is measured empirically in functional assays.

3. METRIC, BIOCHEMICAL EFFICIENCY

BE is defined as how effectively the binding of an inhibitor to the target provides the desired pharmacological response [3,4]. Quantitatively, BE is the ratio of the K_I obtained in a binding or enzyme assay to the IC_{50} in a physiologically relevant functional assay (BE $= K_I/IC_{50}$). A ratio of one indicates that binding is efficiently coupled to the physiological response. BE will directly influence the therapeutic index *via* the impact on the physiological drug concentrations required for pharmacodynamic response.

There are many factors that can influence the shift in dose–response curves between binding and functional assays, including

- Pharmacokinetics and ADME properties (solubility, cell penetration, efflux, metabolism, and protein binding)
- Assay relevance
    - Is the functional assay appropriate for the target?
    - Are the assays technically accurate?
- The target must be involved in the functional readout and biology.
- Molecular mechanisms of action

While all these can and do contribute to the relationship between binding affinity and the functional consequence, the role of MMoA is not always considered. The concept of biochemical efficiency was introduced to quantitative this possibility [3,4]. When using biochemical efficiency as a measure of an optimal MMoA, it is important that the other factors are also considered. For example, when evaluating for biochemical efficiency the assays must be run in the absence of serum (or plasma) to eliminate the shift in IC_{50} due to serum protein binding.

Confirmation of the importance of MMoA and the use of biochemical efficiency comes from the observation that most drugs are competitive with endogenous ligands (80%) [4] and also have good biochemical efficiency [3]. At first glance this is surprising, as endogenous competitive effectors will shift a dose–response curve to higher doses. Higher concentrations of drug will be required for equivalent occupancy and to achieve the desired pharmacological response, resulting in reduced biochemical efficiency. Further analysis of the MMoAs revealed that the medicines have mechanistic features that minimize the negative effects of competition with endogenous effector.

A recent publication by Yun and coworkers on the epidermal growth factor (EGF) receptor exemplifies the physiological consequences of biochemical efficiency [19]. The diminished ATP affinity of the oncogenic EGF receptor mutants opens a "therapeutic window (index)," which renders them more easily inhibited relative to the wild-type EGF receptor and other kinases. The authors describe how resistance mutations in the ATP binding site restore the affinity of the kinase for ATP to wild-type levels. The mutations enable ATP to compete more effectively with the inhibitor. The work also provides an explanation for why EGF receptor covalent irreversible inhibitors are insensitive to the mutations. Due to a lack of competition with ATP, the dose–response curves for the irreversible inhibitors are not shifted to higher concentrations.

The challenge to overcome poor BE can make a specific target intractable by a specific mechanistic approach. Mourey *et al.* [20] described the pharmacologic properties of a benzothiophene MK2 inhibitor, PF-3644022, with good activity but poor BE (**6**). PF-3644022 is a potent freely reversible ATP-competitive compound that inhibits MK2 activity ($K_I = 3$ nM) with good selectivity when profiled against 200 human kinases. They noted that of the MK2 inhibitor chemotypes reported, few have submicromolar potency at inhibiting TNFα production in cells, perhaps because of poor physiochemical properties, poor cell permeability, poor BE, or inadequate enzyme potency. Of several MK2 chemotypes investigated, only the benzothiophenes have cellular IC_{50} values less than 500 nM. PF-3644022 is a highly permeable and potent MK2 inhibitor ($K_I =$ 3 nM), yet it exhibits poor BE with at least 30-fold weaker activity at inhibiting TNFα production in cells. Given that PF-3644022 is an ATP-competitive inhibitor, the shift in cellular potency may be caused by competition with high cellular concentrations of ATP ($\sim$5 mM). The binding constant of MgATP for nonactive MK2 is 30 μM. The authors believe that the best K_I values achievable with MK2 are low nanomolar because they were unable to achieve further potency even after gaining additional interactions in the ATP pocket. They also developed several irreversible MK2 inhibitors as tool compounds that did in fact exhibit BEs near one but had insufficient selectivity to explore as drug leads. Although the MK2 knockout mouse validated MK2 as a very attractive target for TNFα inhibition, the very low BE suggests a low probability of success developing MK2 inhibitors as drugs.

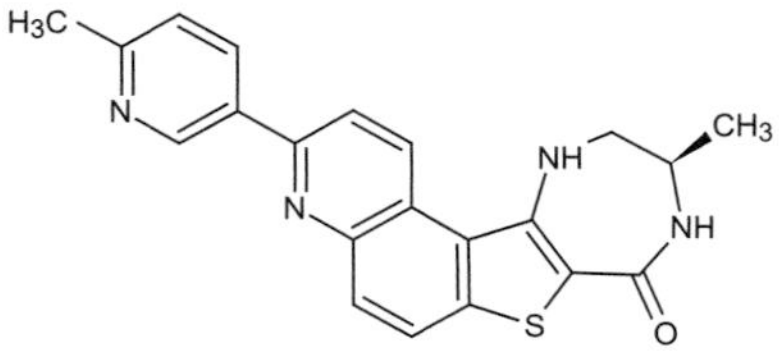

6

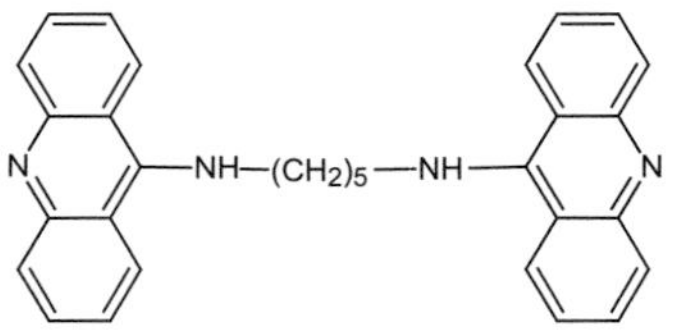

7

4. STRATEGIES FOR AN OPTIMAL MMOA

The strategy for an optimal MMoA is dependent upon the potential for mechanism-based toxicity.

4.1. Strategies for minimizing mechanism-based toxicity

Fast kinetics can minimize mechanism-based toxicity by enabling equilibrium competition that may be advantageous if sufficient efficacy can be maintained. This is exemplified by a number of medicines including atypical antipsychotics working through the D_2 receptor [21], NSAIDs such as ibuprofen [22], and *N*-methyl-D-aspartate (NMDA) antagonists [23]. Effective and safe NMDA receptor antagonists for the treatment of neurodegenerative disorders are required to leave the NMDA channel quickly when the membrane is depolarized under physiological conditions. To meet this criterion, the antagonist is expected to inhibit the channel at negative membrane potentials, then leave the channel quickly and have no effect at depolarized membrane potentials. Luo and coworkers [24] recently reported that bis(propyl)-cognitin (**7**), a novel dimeric acetylcholinesterase inhibitor and γ-aminobutyric acid subtype A receptor antagonist, is an uncompetitive NMDA receptor antagonist with a fast off-rate (UFO). In cultured rat hippocampal neurons, they demonstrated that **7** voltage-dependently, selectively, and moderately inhibited NMDA-activated currents. The inhibitory effects of **7** increased with the rise in NMDA and glycine concentrations. Kinetics analysis showed that the inhibition was of fast onset and offset with an off-rate time constant of 1.9 s.

Drugs that act as allosteric modulators can offer significant advantages over classical agonists and antagonists. For example, the benzodiazepines exhibit large therapeutic indexes, probably because they enhance the action of endogenous γ-aminobutyric acid (GABA) without activating the receptor directly. The low toxicity of the benzodiazepines has contributed to their usefulness in a wide variety of clinical contexts, ranging from anxiety to epilepsy. [25].

Receptor ligands can induce or stabilize ligand-specific conformations that couple to specific functional responses and minimize mechanism-based toxicity. Selective estrogen receptor modulators (SERMs) are the prototype for this MMoA [26]. It is now clear that many receptor modulators bind to a target and stabilize specific conformations that drive ligand-specific and selective functional responses. This new understanding has changed the pharmacological dogma from a linear one ligand/one target/one response paradigm into one in which multiple ligands can interact with one target to produce ligand-specific responses [27,28].

The molecular details that drive this diversity are not as yet fully understood. For example, mifepristone is known to induce mixed passive antagonism, active antagonism, and agonism effects *via* the glucocorticoid receptor (GR) pathway. Schoch and coworkers [29] report the crystal structure of a ternary complex of the GR ligand binding domain (GR-LBD) with mifepristone and a receptor-interacting motif of NCoR. The structures of three different conformations of the GR-LBD mifepristone complex show how the 11β substituent in mifepristone triggers the helix 12 molecular switch to reshape the coactivator site into the co-repressor site. Two observed conformations exemplify the active antagonist state of GR with NCoR bound. In another conformation, helix 12 completely blocks the coregulator binding site and explains the passive antagonistic effect of mifepristone on GR.

4.2. Strategies to maximize efficacy

When there is minimal risk due to mechanism-based toxicity, a MMoA that maximizes efficacy is desirable. A nonequilibrium reversible inhibitor with long residence times will reduce a competition dependent shift in dose–response curves (see Figure 1, left curves). This phenomena has been called insurmountable and pseudo-irreversible. The consequence of this increased BE is a maximized therapeutic index due to the requirement for lower drug concentrations to achieve the efficacious response. This is exemplified in a report by Behm and coworkers [30] with GSK1562590 (**8**), a slowly dissociating urotensin-II receptor antagonist. While optimizing HTS hits, they identified two compounds, GSK1440115 (**9**) and GSK1562590 which exhibited differential characteristics consistent with rapidly and slowly reversible modes of action, respectively. Both compounds were high-affinity ligands; however, GSK1440115 was a

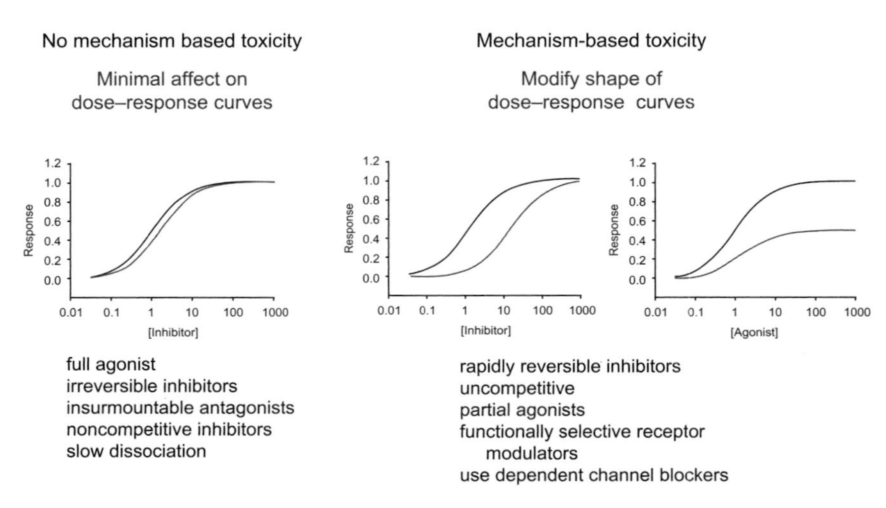

Figure 1 A primary driver of the impact of binding mechanism on the therapeutic index is the potential for mechanism-based toxicity. The curves show the relationship of concentration to binding (black) versus function (red). When there is no mechanism-based toxicity the binding should be efficiently coupled to function and the concentration–response curves will optimally be overlapping (left). When there is potential for mechanism-based toxicity, the functional curves may be shifted to the higher concentrations to limit mechanism-based toxicity (center); this is what would be expected with rapidly reversible competitive inhibitors at equilibrium. A decrease in maximal response as seen with partial agonists is another mechanism to minimize mechanism-based toxicity (right). Source: Swinney [2] reproduced with permission from Wolters Kluwer© 2008. (See Color Plate 18.1 in Color Plate Section.)

surmountable antagonist of urotensin in arteries in all species tested, whereas GSK1562590 was an insurmountable antagonist in all species with the exception of monkey. GSK1562590 was more than two orders of magnitude more potent than GSK1440155 in suppression of the maximal contractile response for human urotensin-II in rat isolated aorta. In addition, a minimal difference in potency between binding and functional assays (BE) was observed for GSK1562590 as compared to ~10-fold for GSK1440155. The authors hypothesize that the slow receptor off-rate might be important for urotensin antagonists to counteract the pseudo-irreversible binding characteristics of urotensin-II.

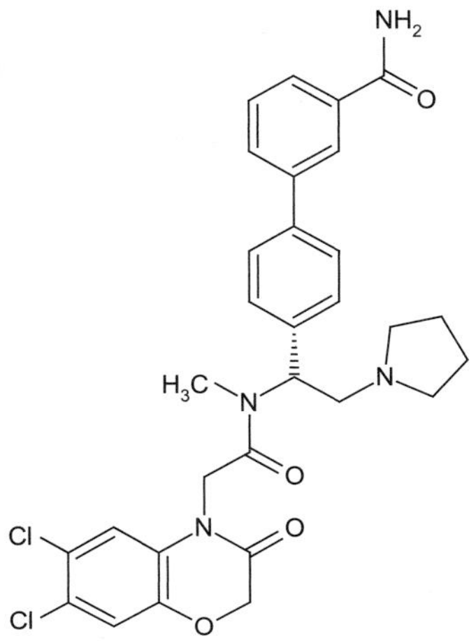

8 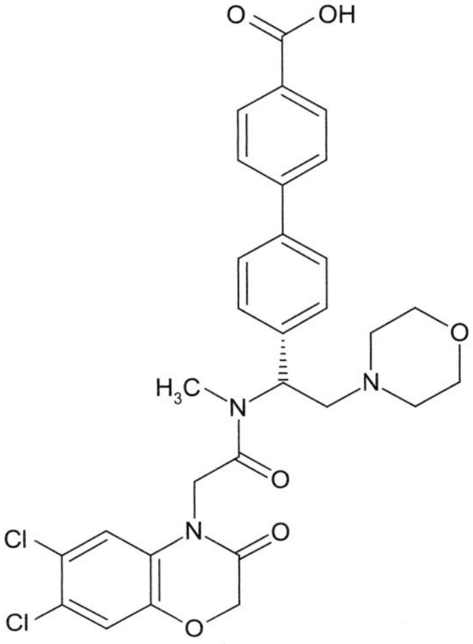 **9**

Irreversible inhibition is another strategy to limit competition and increase BE. Novel irreversible inhibitors have been reported for EGFR [31], peroxisome proliferator-activated receptor γ (PPARγ) [32], and VEGFR [33] tyrosine kinases. Carmi and coworkers [31] reported irreversible EGFR kinase inhibitors can circumvent acquired resistance to first-generation reversible, ATP-competitive inhibitors in the treatment of non-small-cell lung cancer. They contained both a driver group, which assures target recognition, and a warhead, generally an acrylamide or propargylamide fragment that binds covalently to Cys797 within the kinase domain of EGFR. They performed a systematic exploration of the role for the warhead group introducing different cysteine-trapping fragments at the 6-position of a traditional 4-anilinoquinazoline scaffold. They identified different groups that were able to irreversibly bind to EGFR through nucleophilic addition (epoxides), nucleophilic substitution (phenoxyacetamides), carbamoylation (carbamate), Pinner reaction (nitrile), and disulfide bond formation (isothiazolinone, benzisothiazolinone, thiadiazole). They also highlighted another interesting warhead represented by phenoxyacetamides which in principle can release a leaving group on nucleophilic substitution. The lead molecule from this work (**10**) showed promising biological result with efficacy at lower doses than gefitinib.

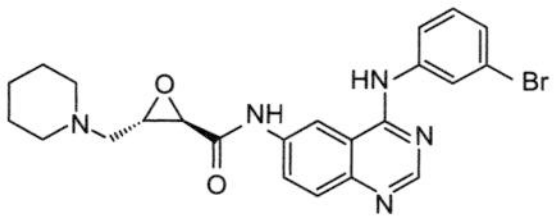

In other reports of irreversible inhibitors, Shearer and coworkers [32] identified a series of trifluoro-methyl-2-pyridylsulfones (**11**) as covalent binders of PPARγ and Barluenga *et al.* [33] reported modifications of the natural product hypothemycin (**12**) and related resorcylic acid lactones bearing a *cis*-enone moiety as irreversible VEGFR kinase inhibitors and with *in vivo* efficacy.

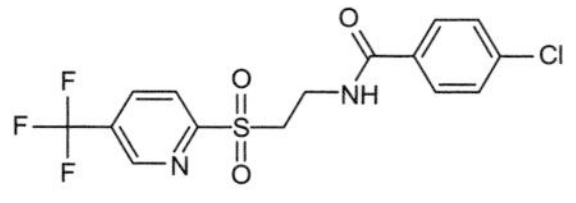

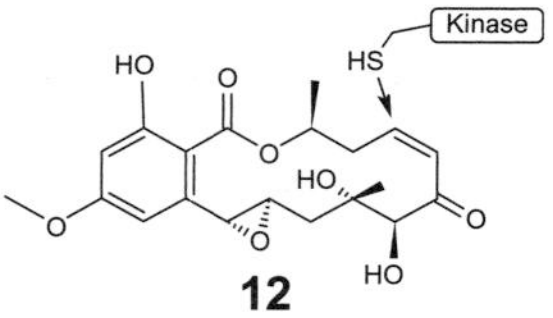

Positive allosteric modulators provide another MMoA to maximize efficacy. Allosteric modulators are ligands that bind to allosteric sites to alter the biological properties of the endogenous orthosteric ligand *via* changing its affinity (generally through the dissociation rate), its efficacy, or both [34,35]. A recent finding by Valant and coworkers identifies a novel and largely unappreciated mechanism of "directed efficacy," whereby functional selectivity may be engendered in a GPCR (muscarinic receptor) by utilizing an allosteric ligand to direct the signaling of an orthosteric ligand encoded within the same molecule [36].

5. CHEMISTRY OF BINDING KINETICS

A challenge for medicinal chemists is that many of the molecular interactions that contribute to binding kinetics and MMoA are dynamic involving structural movements and conformational rearrangements. As described below, some insights are provided by studies with influenza B neuraminidase, which reveals rotations in residues important for binding [37] and InhA, the enoyl-ACP reductase of *Mycobacterium tuberculosis*, which shows a loop ordering involved in the slow kinetics [38].

A mechanism of slow-binding inhibition for oseltamivir, **13**, is revealed by a mutation in neuroaminidase (NA) from influenza B virus. Loss of slow binding is generally associated with mutations in the NA active site, leading to NA inhibitor resistance. Upon binding, residue E275 of B/PerthNA fails to rotate to allow binding of the sec-pentyl moiety to the aliphatic portion of this residue as observed in the equivalent residue (E276) in N1 and N9 NAs. The authors conclude that the rotation of residue E275 needed for high-affinity binding of oseltamivir also does not occur in the current strains of influenza B wild-type NAs, which may possibly lead to decreased clinical efficacy of oseltamivir in children [37].

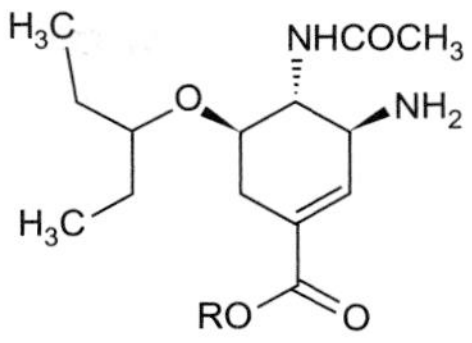

13 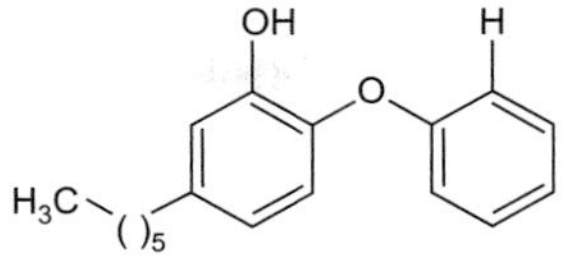 **14**

Luckner and coworkers [38] have used structure-based design to develop a slow onset inhibitor that directly targets InhA. Previous work resulted in the development of a series of alkyl-diphenyl ethers that are nanomolar inhibitors of InhA. The best inhibitor of this series, 8PP (**14**), is active against drug-sensitive and drug-resistant strains of MTB. Although these inhibitors have high affinity for InhA, they are still rapid reversible inhibitors. Luckner and coworkers rationally modified the alkyl-diphenyl ethers to promote interactions between the inhibitor and the loop that becomes ordered during slow onset inhibition. The ordered substrate-binding loop covers the entrance to the binding pocket and thereby locks the inhibitor into the cavity and increases its residence time. It is conceivable that the conformational change of the loop is responsible for the slow step observed in the binding studies.

6. CONCLUSIONS

MMoA is a feature of drug action that bridges the gap between specific molecular interactions and pharmacological activity. Progress is being made to better understand and use binding kinetics, and ligand-specific conformational changes to help design and discover molecules with an increased chance to become therapeutically useful medicines.

ACKNOWLEDGMENT

I wish to thank Dr. Marc Labelle for insightful discussions, critical review of the chapter, and preparation of the structures for the chapter.

REFERENCES

[1] P. Ehrlich, *Lancet*, 1915, **182**, 445.
[2] D. C. Swinney, *Pharm. Med.*, 2008, **22**, 23.
[3] D. C. Swinney, *Nat. Rev. Drug Discov.*, 2004, **3**, 801.
[4] D. C. Swinney, *Curr. Top. Med. Chem.*, 2006, **6**, 461.

[5] D. Wacker, G. Fenalti, M. A. Brown, V. Katritch, R. Abagyan, V. Cherezov and R. C. Stevens, *J. Am. Chem. Soc.*, 2010, **132**, 11443.
[6] A. Samuele, E. Crespan, S. Viterllaro, A. M. Monforte, P. Logoteta, A. Chimirri and G. Maga, *Antiviral Res.*, 2010, **86**, 268.
[7] R. A. Copeland, D. L. Pompliano and T. D. Meek, *Nat. Rev. Drug Discov.*, 2006, **5**, 730.
[8] P. J. Timmino and R. A. Copeland, *Biochemistry*, 2008, **47**, 5481.
[9] D. A. Sykes, M. R. Dowling and S. J. Charlton, *Mol. Pharmacol.*, 2009, **76**, 543.
[10] G. Schreiber, G. Haran and H.-X. Zhou, *Chem. Rev.*, 2009, **109**, 839.
[11] G. Schreiber and A. R. Fersht, *J. Mol. Biol.*, 1995, **248**, 478.
[12] M. Vijayakumar, K. Y. Wong, G. Schrieber, A. R. Fersht, A. Szabo and H. X. Zhou, *J. Mol. Biol.*, 1998, **278**, 1015.
[13] J. R. Horn, T. R. Sosnick and A. A. Kossiakoff, *Proc. Natl. Acad. Sci. USA*, 2009, **106**, 2559.
[14] H. Lu and P. J. Tonge, *Curr. Opin. Chem. Biol.*, 2010, **14**, 1.
[15] H. Lu, K. England, C. am Ende, J. J. Truglio, S. Luckner, B. G. Reddy, N. L. Marlenee, S. E. Knudson, D. L. Knudson, R. A. Bowen, C. Kisker, R. A. Slayden and P. J. Tonge, *ACS Chem. Biol.*, 2009, **4**, 221.
[16] M. Wennerberg, L. Cheng, S. Hjorth, J. C. Clapham, A. Balendran and G. Vauquelin, *Fund. Clin. Pharmacol.*, 2011, **25**, 200.
[17] H. Iwata, S. Imamura, A. Hori, M. S. Hixon, H. Kimura and H. Miki, *Biochemistry*, 2011, **19**, 738.
[18] F. Baur, D. Beattie, D. Beer, D. Bentley, M. Bradley, I. Bruce, S. J. Carlton, B. Cuenoud, R. Ernst, R. A. Fairhurst, B. Faller, D. Farr, T. Keller, J. R. Fozard, J. Fullerton, S. Garman, J. Hatto, C. Hayden, H. He, C. Howes, D. Janus, Z. Jiang, C. Lewis, F. Loeuillet-Ritzler, H. Moser, J. Reilly, A. Steward, D. Sykes, L. Tedaldi, A. Trifilieff, M. Tweed, S. Watson, E. Wessler and D. Wyss, *J. Med. Chem.*, 2010, **53**, 3675.
[19] C. H. Yun, K. E. Mengwasser, A. V. Toms, M. S. Woo, H. Greulich, K. K. Wong, M. Meyerson and M. J. Eck, *Proc. Natl. Acad. Sci. USA*, 2008, **105**, 2070.
[20] R. J. Mourey, B. L. Burnette, S. J. Brustkern, J. S. Daniels, J. L. Hirsch, W. F. Hood, M. J. Meyers, S. J. Mnich, B. S. Pierce, M. J. Saabye, J. F. Schindler, S. A. South, E. G. Webb, J. Zhang and D. R. Anderson, *J. Pharmacol. Exp. Ther.*, 2010, **333**, 797.
[21] S. Kapur and P. Seeman, *Am. J. Psychiatry*, 2001, **158**, 360.
[22] S. A. Lipton, *Nat. Rev. Drug Discov.*, 2006, **5**, 160.
[23] D. C. Swinney, *Lett. Drug Des. Discov.*, 2006, **3**, 569.
[24] J. Luo, W. Li, Y. Zhao, H. Fu, D. L. Ma, J. T. C. Li, R. W. Peoples, F. Li, Q. Wang, P. Huang, J. Xia, Y. Pang and Y. Han, *J. Biol. Chem.*, 2010, **285**, 19947.
[25] D. W. Choi, D. H. Farb and G. D. Fischbach, *Nature*, 1977, **269**, 342.
[26] A. M. Brzozowski, A. C. Pike, Z. Dauter, R. E. Hubbard, T. Bonn, O. Engström, L. Ohman, G. L. Greene, J. A. Gustafsson and M. Carlquist, *Nature*, 1997, **389**, 753.
[27] J. D. Urban, W. P. Clarke, M. von Zastrow, D. E. Nichols, B. Kobilka, H. Weinstein, J. A. Javitch, B. L. Roth, A. Christopoulos, P. M. Sexton, K. J. Miller, M. Spedding and R. B. Mailman, *J. Pharmacol. Exp. Ther.*, 2007, **320**, 1.
[28] T. Kenakin, *J. Pharmacol. Exp. Ther.*, 2011, **336**, 296.
[29] G. A. Schoch, B. D'Arcy, M. Stihle, D. Burger, D. Bar, J. Benz, R. Thoma and A. J. Ruf, *J. Mol. Biol.*, 2010, **395**, 568.
[30] D. J. Behm, N. V. Aiyar, A. R. Olzinski, J. J. McAtee, M. A. Hilfiker, J. W. Dodson, S. E. Dowdell, G. Z. Wang, K. B. Goodman, C. A. Sehon, M. R. Harpel, R. N. Willette, M. J. Neeb, C. A. Leach and S. A. Douglas, *Br. J. Pharmacol.*, 2010, **161**, 207.
[31] C. Carmi, A. Cavazzoni, S. Vezzosi, F. Bordi, F. Vacondio, C. Silva, S. Riviara, A. Lodola, R. R. Alfieri, S. La Monica, M. Galetti, A. Ardizzoni, P. G. Petronini and M. Mor, *J. Med. Chem.*, 2010, **53**, 2038.

[32] B. G. Shearer, R. W. Wiethe, A. Ashe, A. N. Billin, J. M. Way, T. B. Stanley, C. D. Wagner, R. X. Xu, L. M. Leesnitzer, R. V. Merruhew, T. W. Shearer, M. R. Jeune, J. C. Ulrich and T. M. Willson, *J. Med. Chem.*, 2010, **53**, 1857.
[33] S. Barluenga, R. Jogireedy, G. K. Koripelly and N. Winssinger, *Chem. Biol. Chem.*, 2010, **11**, 1.
[34] C. J. Langmead and A. Christopoulos, *Trends Pharmacol. Sci.*, 2006, **27**, 475.
[35] L. T. May, K. Leach, P. M. Sexton and A. Christopoulos, *Annu. Rev. Pharmacol. Toxicol.*, 2007, **28**, 382.
[36] C. Valant, K. J. Gregory, N. E. Hall, P. J. Scammells, M. J. Lew, P. M. Sexton and A. Christopoulos, *J. Biol. Chem.*, 2008, **283**, 29312.
[37] A. J. Oakley, S. Barrett, T. S. Peat, J. Newman, V. A. Streltsov, L. Waddington, T. Saito, M. Tashiro and J. L. McKimm-Breschkin, *J. Med. Chem.*, 2010, **53**, 6421.
[38] S. R. Luckner, N. Liu, C. W. am Ende, P. J. Tonge and C. Kisker, *J. Biol. Chem.*, 2010, **285**, 14330.

CHAPTER 19

Aryl Hydrocarbon Receptor (AhR) Activation: An Emerging Immunology Target?

Peter G. Klimko

Contents

Alcon Laboratories, 6201 South Freeway, Fort Worth, TX 76134, USA

Annual Reports in Medicinal Chemistry, Volume 46
ISSN: 0065-7743, DOI: 10.1016/B978-0-12-386009-5.00008-4

ABBREVIATIONS

AhR	aryl hydrocarbon receptor
ARNT	aryl hydrocarbon nuclear translocator
BaP	benzo[*a*]pyrene
bHLH	basic helix-loop-helix
BMDCs	bone marrow-derived dendritic cells
cDNA	complementary deoxyribonucleic acid
CYP	cytochrome P450
DCs	dendritic cells
DNA	deoxyribonucleic acid
DRE	dioxin response element
EAE	experimental autoimmune encephalomyelitis
ER	estrogen receptor
FICZ	6-formylindolo[3,2-*b*]carbazole
GR	glucocorticoid receptor
H/W	Hans–Wistar
hsp	heat shock protein
IC_{50}	half maximal inhibitory concentration
IDO	indoleamine dioxygenase
IL	interleukin
ITE	2-(1′H-indole-3′-carbonyl)-thiazole-4-carboxylic acid methyl ester
L/E	Long–Evans
LPS	lipopolysaccharide
mRNA	messenger ribonucleic acid
MS	multiple sclerosis
NF-κB	nuclear factor kappa-light-chain-enhancer of activated B cells
NHR	nuclear hormone receptor
NLS	nuclear localization sequence
NOD	non-obese diabetic
PAH	polycyclic aromatic hydrocarbon
PAS	period clock, aryl hydrocarbon nuclear translocator, and single-minded
PGE_2	prostaglandin E_2
RNA	ribonucleic acid
ROR-γt	retinoic acid receptor-related orphan receptor-γt
Saa	serum amyloid A
SAhRM	selective aryl hydrocarbon receptor modulator
TAD	transcriptional activation domain
TCDD	2,3,7,8-tetrachlorodibenzo-*p*-dioxin
TGF-β	transforming growth factor-β

TNF-α	tumor necrosis factor-α
Treg	regulatory T cell
XRE	xenobiotic response element

1. INTRODUCTION

The aryl hydrocarbon receptor (AhR) was originally characterized as a nanomolar-affinity receptor for the polychlorinated aromatic species dioxin (**1**; also called TCDD for 2,3,7,8-tetrachlorodibenzo-*p*-dioxin), responsible for the toxic effects—such as thymic involution and tumor promotion—observed in animals following exposure to this industrial pollutant. This fashioned the view that the primary role of the AhR was in sensing and orchestrating the response to xenobiotic exposure. However, within the past decade, this view has dramatically changed. This review will highlight research on the key role that the AhR plays in the immune system.

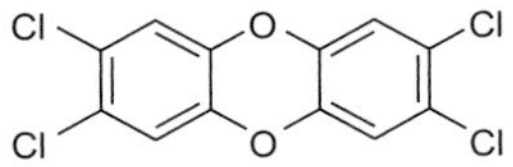

Cl Cl O O Cl Cl

1

2. OVERVIEW OF THE AhR

2.1. Description

The AhR is a ligand-activated transcription factor that resides in the cytoplasm in its unactivated state. The N-terminal portion contains a nuclear localization signal (NLS), a bHLH (basic helix-loop-helix) domain, and two PAS (period clock, aryl hydrocarbon nuclear translocator, and single-minded) subdomains. These regions are necessary for heterodimerization with the aryl hydrocarbon nuclear translocator (ARNT) and DNA binding (*via* the bHLH region). The ligand binding site spans the more C-terminal of the two PAS subdomains. The C-terminal motif contains the transcriptional activation domain (TAD) [1].

Constitutive tissue expression in adults generally appears to be highest in the liver, with significant expression also in the lung, placenta, thymus, and cornea [2–4]. Induced expression of the receptor has been reported in several types of immune cells when they are exposed to particular stimuli (see below for examples).

Although the AhR belongs to the bHLH and not the nuclear hormone receptor (NHR) superfamily, its canonical signaling pathway (see below) is reminiscent of that for type 1 NHRs, such as the glucocorticoid and estrogen receptors (GR and ER, respectively) [5]. The AhR is an orphan receptor, since an endogenous agonist has not yet been conclusively demonstrated.

2.2. Canonical signaling pathway

The AhR in its resting state is a cytoplasmic complex with the repressor proteins hsp90, p23 (also known as prostaglandin E synthase 3), and hepatitis B virus X-associated protein 2. The canonical signaling pathway begins with agonist binding to the receptor, which causes a conformational change in the protein that induces dissociation of the repressor proteins. Subsequent unmasking of an NLS is followed by nuclear translocation and heterodimerization with ARNT to form the active transcription factor. The ligated AhR–ARNT complex binds to a consensus nucleotide sequence containing a core recognition element, which has been dubbed the dioxin response element (DRE), or, alternatively, the xenobiotic response element (XRE). This nomenclature reflects the initial characterization of the AhR as the cognate endogenous dioxin receptor, which directs the response to xenobiotic challenge. Up- or downregulation of protein production follows, depending on whether the DRE is contained within the associated gene promoter or repressor [1].

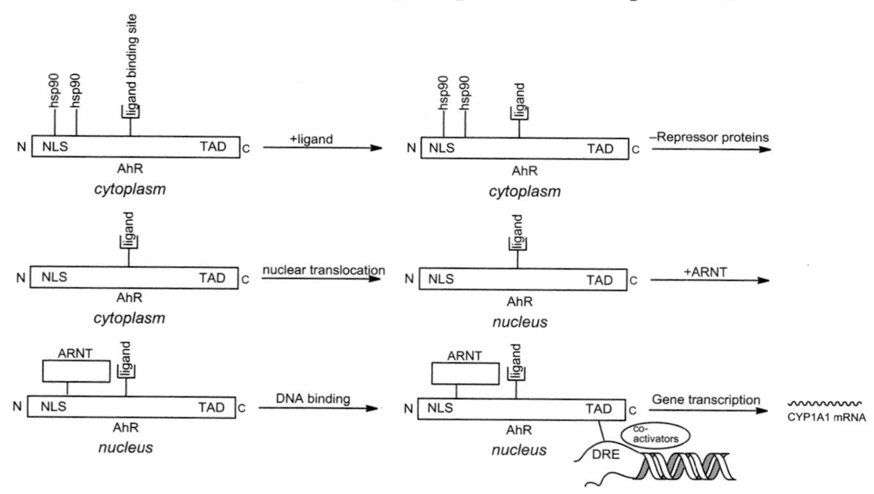

AhR canonical signaling pathway schematic

The canonical signaling result is upregulation of the cytochrome P450 (CYP) enzyme subtypes CYP1A1 and CYP1B1 [6]. Dioxin itself is a poor substrate for enzymatic oxidation, as evidenced by plasma half-lives of weeks (in rodents) to years (in humans) following systemic exposure [7]. Dioxin biopersistence thus leads to sustained CYP enzyme expression elevation, which may lead to tumor promotion through the constant generation of DNA-damaging reactive oxygen species and electrophilic compounds. Additionally, selective AhR agonists of the polycyclic aromatic hydrocarbon (PAH) class are themselves transformed into DNA-damaging species *via* CYP-catalyzed oxidation, as for example in the conversion of benzo[*a*]pyrene (BaP; **2**) into the mutagenic epoxide **3**.

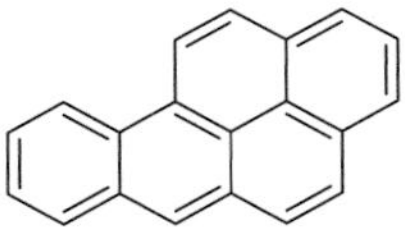

2.3. Noncanonical signaling pathways

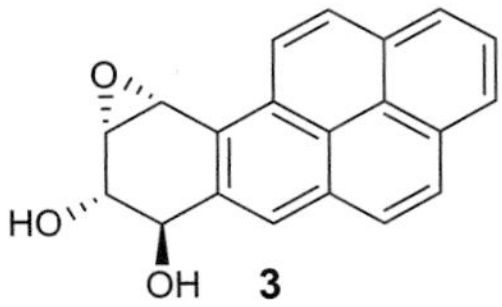

The most immunologically significant noncanonical signaling pathway involves the NF-kB protein family. Physical interaction has been observed between the AhR and members of the NF-kB family, such as RelB, RelA/p65, and p50 [8–10], and different consequences of these interactions have been reported. Interestingly, a binding site in the promoter region of the *Il-8* gene for the AhR/RelB complex has been described, with dioxin causing upregulation of IL-8 protein in macrophages [8]. This implies that NF-kB-linked noncanonical activation is proinflammatory.

In contrast, dioxin and several other AhR agonists inhibited cytokine-induced transcription of the acute phase response genes *Saa1-3* in several cell types [11]. This effect was dependent on AhR nuclear translocation and complexation with ARNT but not DRE binding, thus implicating noncanonical signaling. Furthermore, AhR-knockout mice were more susceptible to lipopolysaccharide (LPS)-induced lethality and cigarette smoke-induced lung neutrophil influx, with coincident loss of nuclear RelB in lung cells and increase in macrophage production of TNF-α and IL-6. Thus, the literature in general suggests that activation of noncanonical NF-κB-linked signaling is anti-inflammatory [12,13].

2.4. Pharmacology tools

The isolation of soluble rodent AhR protein is typically accomplished from the liver, which is the richest source tissue. Use of this technique in transgenic mice with conditional, liver-specific protein knock-in, has been reported for human AhR isolation [14]. The placenta has also been used as an alternative source for the human receptor [15].

Alternatively, a robust procedure for routine scalable isolation of soluble rat and human receptor consists of cloning human or rat AhR cDNA into recombinant baculovirus, followed by infection of insect (*Spodoptera frugiperda*) cells. Cell lysis followed by standard cytosol isolation affords protein with a 20-fold higher ligand binding capacity compared to standard techniques [16].

2.4.1. Receptor binding and functional activation

Receptor binding assays have commonly used [^{3}H]-dioxin (for scintillation-based detection) or the photoaffinity ligand 2-azido-3-[^{125}I]-7,8-dibromodibenzo-*p*-dioxin (for autoradiography-based detection) as the high-affinity radioactive ligand [16,17]. Receptor functional activation frequently has been measured in an electrophoretic mobility shift assay detecting the binding of the AhR–ARNT heterodimer with a DRE-containing oligonucleotide, or by measuring upregulation of CYP1A1 mRNA or protein expression [18]. A higher-throughput method for detecting receptor activation comprises measurement of luciferase activity in a cell line containing a DRE-driven luciferase reporter gene, after test agent treatment [19].

2.4.2. Benchmark ligands

For historical reasons and due to its high binding affinity, dioxin has been the key comparator agonist since the initial characterization of the receptor. A large number of toxicology studies in rodents, and a smaller but burgeoning number of reports from more discrete rodent disease models, have used dioxin in this role.

However, this presents at least two problems. First, dioxin's exceptionally long half-life implies sustained pharmacodynamics far in excess of other well-characterized signaling systems. This limits the extrapolation of these results to the effects of shorter-lived agonists, especially potential endogenous ligands.

Second, it is well established that the receptor binding affinity of dioxin varies substantially between humans and mice and within different mouse strains. Thus, the routinely used C57/Bl6 mouse strain harbors an AhR with a 10-fold higher dioxin binding affinity than that for the human and the DBA/2 mouse strain receptor [17,20]. Dioxin *in vivo* effects are also strain dependent for both mice and rats (see below).

β-Naphtoflavone (**4**) and the tryptophan photoproduct 6-formylindolo [3,2-*b*]carbazole (FICZ; **5**) are reported dioxin-alternative comparator AhR agonists [21,22]. Their main advantage is an increased metabolic liability leading to a shorter plasma half-life. Consequently, their pharmacodynamic effects are more easily interpreted. Also, FICZ is a candidate endogenous AhR agonist, as it is formed in a pharmacologically-relevant concentration in skin. Their main disadvantage is a much smaller database of published studies.

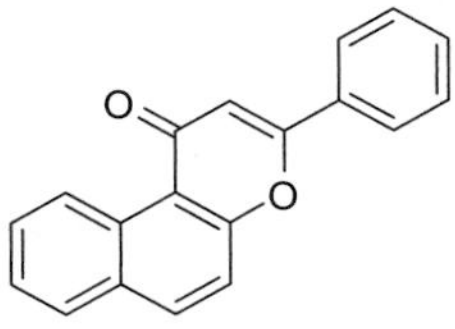

4 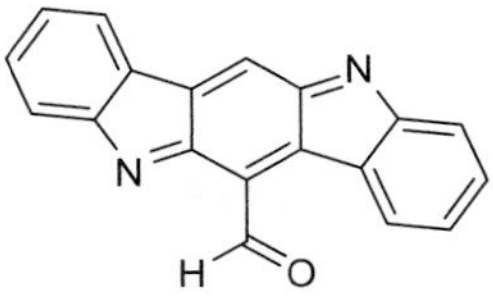 5

Although there are a number of small molecules claimed in the literature to be AhR antagonists, only a few have been commonly used as pharmacology tools, with α-naphthoflavone (**6**) being a popular example [23]. The pyrazole **CH-223191** (**7**) has emerged as an alternative, based on its potent antagonism of dioxin-induced AhR signaling in mice *in vitro* and *in vivo* [24]. Unlike most other reported antagonists, **7** has no detectable agonism at either the AhR or the ER. However, **7** does not antagonize signaling by β-naphthoflavone [25].

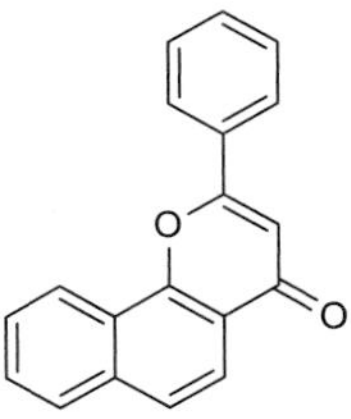

6 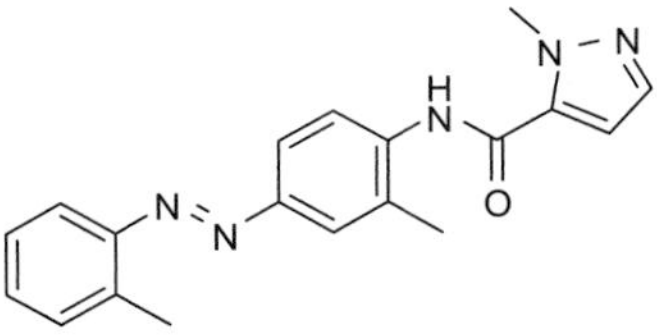 7

Analogous to other hormone receptors such as the GR and ER, there exists for the AhR a third class of ligands that elicit a subset of agonist effects reported for dioxin *in vitro*. Thus, the indazoles **WAY169916** (**8**) (itself a pathway-selective ERα agonist) and **SGA360** (**9**) and the α-naphthoflavone analog **10** all displayed submicromolar binding affinity to the human AhR. These compounds antagonized dioxin-induced AhR-dependent effects *in vitro* (such as *Cyp1a1* message upregulation) yet exhibited AhR-dependent anti-inflammatory effects *in vitro* and *in vivo* in mice [26–28]. These have been dubbed selective AhR modulators (SAhRMs).

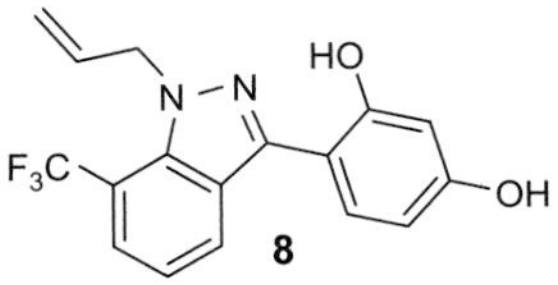

8 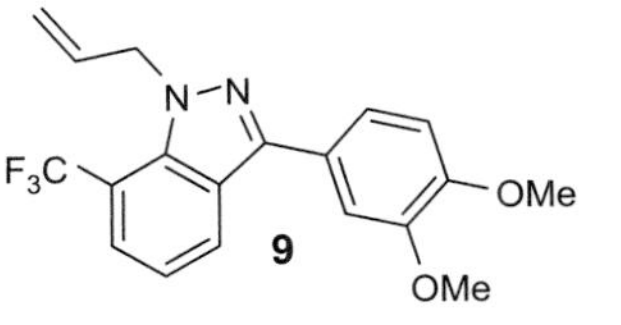 9 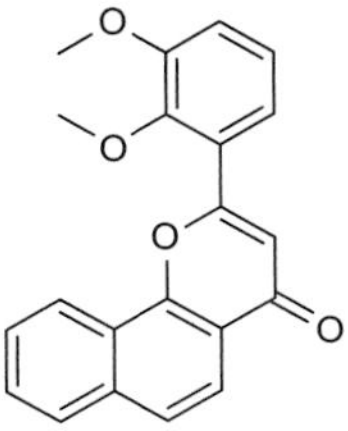 10

2.4.3. Rodent AhR

Wild-type mice can be separated into two groups: those with high ($IC_{50} \leq 2$ nM) and those with low ($IC_{50} \geq 20$ nM) dioxin binding affinities for their respective AhRs [20]. High-affinity strains are classified as having AhR^b alleles, while low-affinity strains are designated as having AhR^d alleles. The former group includes the C57/Bl6 and Balb/c strains, while the latter group includes the DBA/2 and non-obese diabetic (NOD) strains. The human AhR is more similar to the murine AhR^d (*i.e.*, lower affinity) allele. However, even between these two proteins, there are significant differences in binding affinities for other AhR ligands [17]. Additionally, dioxin gene expression effects were substantially different in hepatocytes from wild-type C57/Bl6 mice versus those from mice with liver-specific human AhR knock-in [29].

In rats, different susceptibilities to dioxin toxicity are well documented. In particular, an at least ~1000-fold increased dioxin LD_{50} dose was observed for the Hans–Wistar (H/W) versus the Long–Evans (L/E) rat. Although the AhR for the resistant H/W strain has a ~40 amino acid deletion in its TAD compared to the L/E strain, no difference was observed between the two strains in either dioxin binding affinity or dioxin potency and efficacy in hepatocyte *Cyp1a1* message and enzymatic activity induction [30].

AhR-knockout mice have been the most well studied. Although the specific phenotype depends on the animal genetic background and the method of knockout generation, widespread observations include an expected loss of sensitivity to dioxin toxicity, development of several anatomic abnormalities such as increased liver fibrosis, and reduced fertility [1]. Immunologic-based phenotypic dysfunction observed in these animals (besides the increased inflammatory response to LPS and cigarette smoke mentioned earlier) includes increased lethality upon *Toxoplasma gondii* infection [31]. Most of the effects of AhR knockout on cell function have been described in immune cells, as discussed below.

Other AhR mutations useful for analyzing the contribution of the discrete sequences of AhR signaling include the $AhR^{nls/nls}$ genotype, in which the protein lacks a functional NLS, and the $AhR^{dbd/dbd}$ genotype, where the mutant AhR–ARNT dimer is unable to recognize DRE-containing DNA [32].

3. EFFECTS ON IMMUNE CELL FUNCTION

3.1. Dendritic cells

Deactivation of dendritic cells (DCs) plays an important role in the immunosuppressive effects of dioxin. There are several mechanisms by which this occurs. Dioxin activated naïve murine splenic DCs *in vitro*, increased expression of co-stimulatory molecules like CD40 and proinflammatory cytokines like IL-12, and induced proliferation of cocultured allogenic T cells. This antigen-independent activation caused DC death, as evidenced by a lower splenic DC count in dioxin-treated mice [33].

In contrast, LPS upregulated AhR protein expression in bone marrow-derived DCs (BMDCs) *in vitro*, affording a several hundred-fold increase in the secretion of the anti-inflammatory cytokine IL-10. *In vitro*, this effect which was greatly diminished in $AhR^{-/-}$-derived BMDCs [34]. Additionally, dioxin upregulated the enzyme indoleamine dioxgenase (IDO) in DCs [35]. IDO catalyzes the biosynthesis of kynurenine (**11**) *via* oxidation of tryptophan and has long been known to be immunosuppressive. In a further twist, kynurenine itself was characterized as an AhR agonist in murine hepatocytes, with a potency consistent with its physiological concentration [36]. Thus kynurenine may function endogenously as both a downstream and an upstream effector of AhR activation-induced, DC-mediated immune tolerance.

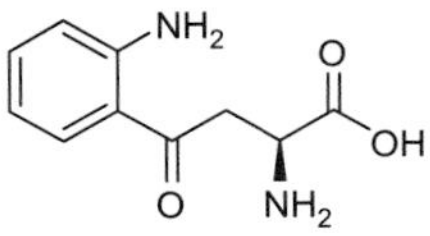

NH_2 O OH O NH_2

11

Furthermore, AhR activation-induced, DC-mediated immunosuppression likely occurs largely *via* DC-evoked conversion of naïve T cells to regulatory T cells (Tregs), a deactivated phenotype important in immune tolerance, as discussed below. For example, adoptive transfer of AhR agonist-treated, antigen-stimulated DCs induced $CD4^+Foxp3^+$ Tregs in host animals and reduced pathology in animal autoimmune disease models [37,38].

3.2. $CD4^+$ T cells

Research to date on the effects of AhR activation on T cell function has concentrated on naïve $CD4^+$ T cell differentiation to either T_H17 cells or Tregs. T_H17 cells are important in host defense against extracellular microorganisms and are important in the pathology of several autoimmune diseases. They are defined by their expression of the transcription factor retinoic acid receptor-related orphan receptor-γt (ROR-γt), are characterized by their secretion of the cytokines IL-17A and IL-22, and in mice, are induced by treatment of naïve $CD4^+$ T cells with TGF-β + IL-6.

$CD4^+$ Tregs are important in maintaining tolerance to self-antigens. The most commonly used molecular biomarker is Foxp3 protein expression. However, Treg definition is based more on phenotypic behavior, such as anergy to antigen stimulation and deactivation of neighboring effector T cells, than on Foxp3 expression. They are classically generated by treatment of naïve $CD4^+$ T cells with TGF-β.

Thus these two T cell types have opposite functions. An increased T_H17/Treg ratio from homeostatic values can lead to loss of immune self-tolerance and is implicated as a causative factor in several autoimmune diseases.

There is some uncertainty in the literature regarding whether there is a general effect, and if so in which direction, of AhR activation on T_H17/Treg differentiation. Part of this confusion may be due to definitions of protein expression in these cell types that are not absolute. For example, a population of murine $CD4^+$ T cells secreting substantial IL-17A but not expressing ROR-γt has been characterized [39]. $CD4^+Foxp3^-$, dioxin-inducible functional Tregs have been described in mice and humans [40,41]. Moreover in human $CD4^+$ T cells, Foxp3 expression below a threshold level was insufficient for Treg function [42].

However, the main reason for the controversy is that in mice, the alternative AhR agonist FICZ has been reported to polarize naïve $CD4^+$ T cells to a population with a high T_H17/Treg ratio *in vitro*, while the classical agonist dioxin has been reported to produce the opposite effect. In support of the T_H17-inducing hypothesis, AhR protein expression was upregulated by several hundred fold when naïve murine $CD4^+$ T cells were exposed to T_H17-polarizing conditions [43,44]. Murine Tregs exhibited a much lower relative expression of the AhR. Differentiation of murine naïve T cells to T_H17 cells *in vitro* was inhibited unless natural AhR agonists like tryptophan were present in culture [45]. The anomalous result with dioxin is typically attributed to an endogenously irrelevant, sustained AhR activation due to dioxin biopersistence. Alternatively, dioxin may selectively kill T_H17 cells (which express the highest AhR levels for any helper T cell subset) *via* receptor overstimulation, by analogy to dioxin-induced DC dropout mentioned earlier [46].

In support of a general T_H17-inhibiting/Treg-inducing effect of AhR activation, a number of AhR agonists besides dioxin, such as kynurenine, tryptophan-derived 2-(1′H-indole-3′-carbonyl)-thiazole-4-carboxylic acid methyl ester (ITE; **12**), and the aminopyrimidine **VAF347** (**13**), induced Treg polarization *in vitro* [36–38]. Differentiation of TGF-β-treated naïve murine $CD4^+$ T cells to Foxp3+ Tregs was depressed up to 70% in AhR-knockout cells [36]. Knockdown of *Ahr* message by siRNA inhibited *Il-10* message by 50% in a murine $Foxp3^-$, IL-10-secreting Treg population [47]. In contrast to its effects in murine T cells, FICZ inhibited human T_H17 cell generation *in vitro* [48]. Thus the literature on balance suggests that AhR activation induces Treg phenotype and immune tolerance in humans; however, the general applicability of this hypothesis awaits confirmation with a wider variety of agonists.

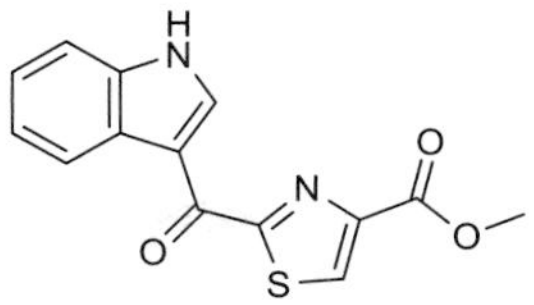

With respect to other murine CD4$^+$ T helper cell subtypes, polarization to T_H2 cells was inhibited in an AhR-dependent manner by the benzimidazole **M50354 (14)** *via* suppression of the T_H2 master transcription factor *Gata-3* mRNA. T_H2 polarization was also inhibited by expression of a constitutively active AhR in naïve CD4$^+$ T cells [49].

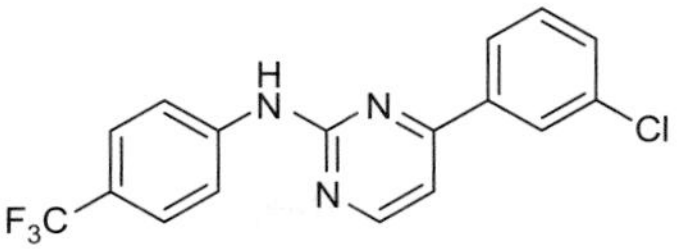

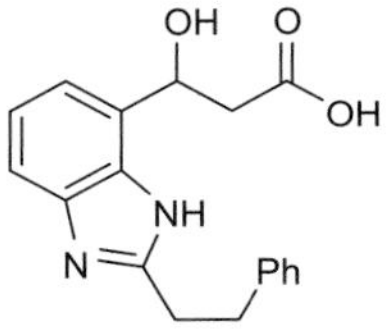

4. THERAPEUTIC EFFECTS IN ANIMAL MODELS OF HUMAN DISEASES

4.1. Experimental autoimmune encephalomyelitis (EAE)

Multiple sclerosis (MS) is a chronic, usually progressive neurological disease with symptoms including abnormal sensations, muscle weakness, and paralysis. Disease pathology is believed to be due to attack of effector T cells and macrophages on cells that form the myelin sheath of myelinated nerves in the central nervous system. The standard of care during acute attacks is treatment with immunosuppressants, such as steroids and interferon-β-1. Consensus opinion is that these treatments are only palliative.

Rodent EAE is a widely-used model of human MS and is typically induced by immunization with a myelin oligodendrocyte glycoprotein. Disease progression from minimal locomotor effects to complete paralysis is represented by increased numerical scores. Of the three AhR agonists with published data in murine EAE models, ITE and dioxin decreased, while FICZ increased, disease scores [37,43,44]. Dioxin and ITE therapeutic effects were not observed in animals with the AhR^d (low affinity) allele. In the absence of agent treatment, these mice also exhibited exacerbated pathology compared to AhR^b (high affinity) animals [37,44].

4.2. Colitis

Colitis (inflammation of the bowel) due to Crohn's disease is characterized by chronic diarrhea and abdominal pain. Affected patients are at increased risk of malnutrition due to nutrient malabsorption, and intestinal cancer due to persistent inflammation. The disease is thought to be caused by autoimmune attack on the gastrointestinal tract and is typified histologically by neutrophil infiltration into and granuloma formation within the bowel. Similar to the situation with MS, the standard of care is treatment with anti-inflammatory/immunosuppressant agents, such as steroids and TNF-α blockers.

Colitis is typically induced experimentally in mice by oral treatment with dextran sodium sulfate or 2,4,6-trinitrobenezenesulfonic acid. Observed effects of clinical relevance to humans include weight loss and colonic ulceration. In mouse models of colitis, dioxin inhibited weight loss and decreased colon damage with concomitant reduction in colonic TNF-α and increase in cytoprotective PGE_2 and Foxp3$^+$ Tregs [50,51]. Seemingly contradictory, in another disclosure dextran sodium sulfate-treated, AhR$^{+/-}$ mice had reduced damage and decreased colonic TNF-α expression compared to wild-type controls, while human inflammatory bowel disease patients had higher colonic AhR expression compared with normals [52]. Thus the effect of AhR activation on inflammatory bowel disease pathology awaits clarification *via* further research.

4.3. Allergic asthma

Asthma is an airway disease in which abnormal bronchial constriction and mucus hypersecretion lead to insufficient oxygen reaching the lungs. Acute attacks are often triggered by exposure to an allergen. The standard of care for acute attacks is inhalation of a β-adrenoceptor agonist like albuterol (or in emergency situations, injection of the endogenous β-adrenoceptor agonist epinephrine). Management of chronic disease is typically achieved with a combination of a long-acting β-adrenoceptor agonist as a bronchial smooth muscle relaxant and a glucocorticoid as an anti-inflammatory. Reducing lung influx of professional inflammatory cells, especially eosinophils, is likely an important effect of the anti-inflammatory component.

In a Balb/c mouse model of ovalbumin-induced allergic asthma, prodrugs of the aforementioned AhR agonists **13** and **14**, namely **VAG539** (**15**) and **M50367** (**16**), reduced several pathological hallmarks, such as serum IgE levels, acetylcholine-induced bronchoconstriction, and pulmonary eosinophil influx. The *in vivo* and *in vitro* anti-allergic effects of these compounds were not observed in AhR-knockout animals [49,53,54].

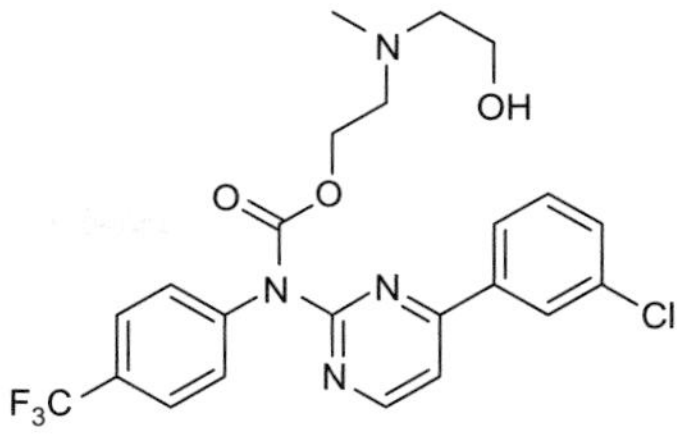

15 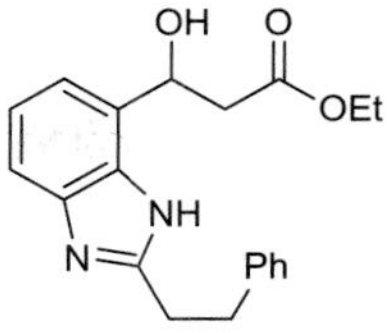 **16**

4.4. Type 1 diabetes

The NOD mouse is a spontaneous model of human type 1 diabetes. As is believed to be true for the human disease, the NOD mouse model is characterized by pancreatic infiltration of autoreactive T cells. Effector T cell-mediated destruction of insulin-secreting beta cells leads to a large increase in blood glucose concentration. The standard of care is daily injection of exogenous insulin.

Treatment of female NOD mice, which harbor the low dioxin-affinity AhR^d allele, with dioxin prevented diabetes (as defined by blood glucose concentration) in all of the treated animals after 23 weeks, as compared to a $>70\%$ diabetes incidence in vehicle-treated animals [55]. This was accompanied by a 50% increase in the absolute and relative number of $CD4^+CD25^+Foxp3^+$ Tregs in pancreatic lymph nodes. This result is interesting in light of the substantial literature implicating dioxin exposure to insulin resistance and the risk of type 2 diabetes—a form of the disease where secreted insulin is insufficient to reduce blood glucose concentration—in humans [56].

4.5. Graft versus host disease

The rejection of allogenic transplanted tissue is the primary complication for transplant recipients. It is frequently described as a graft versus host disease, since it characterized by infiltration of host effector T cells into the allogenic tissue. The standard of care is treatment with immunosuppressants such as rapamycin, with attendant increased risk of infection.

In a mouse model of graft versus host disease, streptozotocin-induced diabetic Balb/c mice were transplanted with pancreatic islet allografts from nondiabetic C57/Bl6 mice. Transplanted mice pretreated with **15** had a 70% graft survival rate after 30 days, compared to a 0% rate for vehicle-treated animals. Harvesting of DCs from the allograft-surviving cohort 60 days after transplantation and transfer to a group of freshly transplanted mice afforded a similar 70% allograft survival rate after 30 days. Both the **15** directly treated group and the DC adoptive transfer group had a higher frequency of splenic $CD4^+Foxp3^+$ Tregs as compared to vehicle- and naïve DC-treated animals [38].

4.6. Experimental autoimmune uveitis (EAU)

Noninfectious uveitis is a frank inflammatory intraocular disease. The autoimmune form of the disease is distinguished by influx of autoreactive effector T cells into the posterior chamber of the eye and is frequently comorbid with an extraocular autoimmune disease, such as ankylosing spondylitis. If the affected tissue is not restored to its normally immune-privileged state, blindness typically results. The standard of care is treatment with steroids, frequently in combination with anti-rheumatoid arthritis drugs like methotrexate.

EAU can be induced in rodents by immunization with a retina-specific antigen, such as interphotoreceptor retinoid-binding protein. Subsequent ocular pathology includes retinal hemorrhage, retinal influx of $CD4^+$ effector T cells, and disruption of retinal cellular organization. In a mouse EAU model, a single 1 μg dose of dioxin given 1 day before disease induction completely inhibited development of disease after 21 days. Compared to vehicle-treated mice, dioxin-treated animals also had a 50% increase in splenocyte $CD4^+Foxp3^+$ Treg frequency, with a concomitant 90% reduction in antigen-stimulated IL-17A secretion *in vitro* [57].

4.7. Miscellaneous

The arachidonic acid metabolite lipoxin A_4 (**17**) has been characterized as a 300 nM potency AhR agonist in mouse liver cells [58]. In wild-type mice challenged with *T. gondii* soluble extract, lipoxin A_4 decreased spontaneous IL-12p40 secretion from splenocyte DCs by 75%, as compared to vehicle-treated animals. A 50% loss of lipoxin inhibitory efficacy was observed in AhR-knockout mice [59].

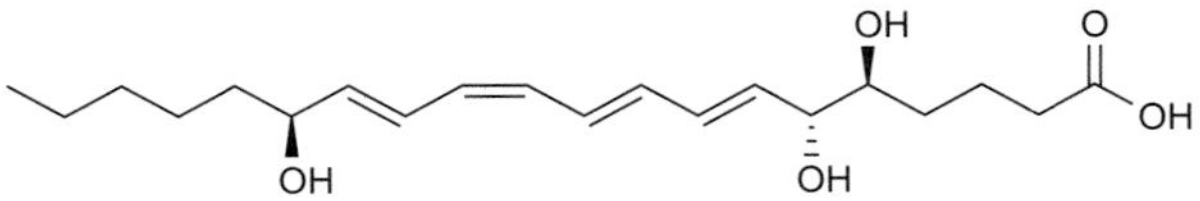

OH O OH OH OH

17

In wild-type mice, topical dosing to the ear of the previously mentioned SAhRM **9** inhibited phorbol ester-induced ear edema and mRNA induction for several inflammatory proteins, such as COX-2 and IL-6. These therapeutic effects were abrogated in AhR-knockout animals [27].

5. CONCLUSION

Extensive rodent studies reporting dioxin-induced, AhR-dependent wasting, immunosuppression, and carcinogenesis have suggested that AhR activation is a pathological event. However, recent evidence that dioxin at nontoxic doses (''the dose makes the poison'') [60] has therapeutic activity in rodent models of autoimmune diseases, with concomitant induction of Tregs and suppression of T_H17 development, has kindled interest in AhR agonism as a therapeutic target. This is supported by key findings that AhR activation-induced CYP protein induction is not inexorably toxic (as demonstrated in H/W rats); that several drug-like compounds (such as **13** and **14**) have AhR-dependent therapeutic activity in *in vivo* animal models of human disease; and that CYP induction can be decoupled from AhR-dependent therapeutic effects (as for **9**).

REFERENCES

[1] J. Abel and T. Haarmann-Stemmann, *Biol. Chem.*, 2010, **391**, 1235.
[2] A. E. Vickers, T. C. Sloop and G. W. Lucier, *Environ. Health Perspect.*, 1985, **59**, 121.
[3] W. Li, S. Donat, O. Döhr, K. Unfried and J. Abel, *Arch. Biochem. Biophys.*, 1994, **315**, 279.
[4] T. Sugamo, K. Nakamura and H. Tamura, *J. Health Sci.*, 2009, **55**, 923.
[5] R. J. Kewley, M. L. Whitelaw and A. Chapman-Smith, *Int. J. Biochem. Cell Biol.*, 2004, **36**, 189.
[6] K. Kawajiri and Y. Fujii-Kuriyama, *Arch. Biochem. Biophys.*, 2007, **464**, 207.
[7] J. E. Huff, A. G. Salmon, N. K. Hooper and L. Zeise, *Cell Biol. Toxicol.*, 1991, **7**, 67.
[8] C. F. Vogel, E. Sciullo, W. Li, P. Wong, G. Lazennec and F. Matsumura, *Mol. Endocrinol.*, 2007, **21**, 2941.
[9] Y. Tian, S. Ke, M. S. Denison, A. B. Rabson and M. A. Gallo, *J. Biol. Chem.*, 1999, **274**, 510.
[10] A. Kimura, T. Naka, T. Nakahama, I. Chinen, K. Masuda, K. Nohara, Y. Fujii-Kuriyama and T. Kishimoto, *J. Exp. Med.*, 2009, **2027**, 206.
[11] R. D. Pate, I. A. Murray, C. A. Flaveny, A. Kusnadi and G. H. Perdew, *Lab. Invest.*, 2009, **89**, 695.
[12] H. Sekine, J. Mimura, M. Oshima, H. Okawa, J. Kanno, K. Igarashi, F. J. Gonzalez, T. Ikuta, K. Kawajiri and Y. Fujii-Kuriyama, *Mol. Cell. Biol.*, 2009, **29**, 6391.
[13] T. H. Thatcher, S. B. Maggirwar, C. J. Baglole, H. F. Lakatos, T. A. Gasiewicz, R. P. Phipps and P. J. Sime, *Am. J. Pathol.*, 2007, **170**, 855.
[14] C. A. Flaveny and G. H. Perdew, *Mol. Cell. Pharmacol.*, 2009, **1**, 119.
[15] D. K. Manchester, S. K. Gordon, C. L. Golas, E. A. Roberts and A. B. Okey, *Cancer Res.*, 1987, **47**, 4861.
[16] M. Q. Fan, A. R. Bell, D. R. Bell, S. Clode, A. Fernandes, P. M. Foster, J. R. Fry, T. Jiang, G. Loizou, A. MacNicoll, B. G. Miller, M. Rose, O. Shaikh-Omar, L. Tran and S. White, *Anal. Biochem.*, 2009, **384**, 279.
[17] P. Ramadoss and G. H. Perdew, *Mol. Pharmacol.*, 2004, **66**, 129.
[18] C. A. Flaveny, I. A. Murray, C. R. Chiaro and G. H. Perdew, *Mol. Pharmacol.*, 2009, **75**, 1412.
[19] J. L. Raucy and J. M. Lasker, *Curr. Drug Metab.*, 2010, **11**, 806.
[20] A. B. Okey, L. M. Vella and P. A. Harper, *Mol. Pharmacol.*, 1989, **35**, 823.

[21] Y. Fujita, M. Yonehara, M. Tetsuhashi, T. Noguchi-Yachide, Y. Hashimoto and M. Ishikawa, *Bioorg. Med. Chem. Lett.*, 2010, **18**, 1194.
[22] M. E. Jönsso, D. G. Frank, B. R. Woodin, M. J. Jenny, R. A. Garrick, L. Behrendt, M. E. Hahn and J. J. Stegeman, *Chem. Biol. Interact.*, 2009, **181**, 447.
[23] T. A. Gasiewicz and G. Rucci, *Mol. Pharmacol.*, 1991, **40**, 607.
[24] S. H. Kim, E. C. Henry, D. K. Kim, Y. H. Kim, K. J. Shin, M. S. Han, T. G. Lee, J. K. Kang, T. A. Gasiewicz, S. H. Ryu and P. G. Suh, *Mol. Pharmacol.*, 1871, **2006**, 69.
[25] B. Zhao, D. E. Degroot, A. Hayashi, G. He and M. S. Denison, *Toxicol. Sci.*, 2010, **117**, 393.
[26] I. A. Murray, J. L. Morales, C. A. Flaveny, B. C. Dinatale, C. Chiaro, K. Gowdahalli, S. Amin and G. H. Perdew, *Mol. Pharmacol.*, 2010, **77**, 247.
[27] I. A. Murray, G. Krishnegowda, B. C. DiNatale, C. Flaveny, C. Chiaro, J. M. Lin, A. K. Sharma, S. Amin and G. H. Perdew, *Chem. Res. Toxicol.*, 2010, **23**, 955.
[28] I. A. Murray, C. A. Flaveny, C. R. Chiaro, A. K. Sharma, R. S. Tanos, J. C. Schroeder, S. G. Amin, W. H. Bisson, S. K. Kolluri and G. H. Perdew, *Mol. Pharmacol.*, 2011, **79**, 508.
[29] C. A. Flaveny, I. A. Murray and G. H. Perdew, *Toxicol. Sci.*, 2010, **114**, 217.
[30] A. B. Okey, M. A. Franc, I. D. Moffat, N. Tijet, P. C. Boutros, M. Korkalainen, J. Tuomisto and R. Pohjanvirta, *Toxicol. Appl. Pharmacol.*, 2005, **207**, 43.
[31] Y. Sanchez, J. de D. Rosado, L. Vega, G. Elizondo, E. Estrada-Muñiz, R. Saavedra, I. Juárez and M. Rodríguez-Sosa, *J. Biomed. Biotechnol.*, 2010, **2010**, 505694. http://www.hindawi.com/journals/jbb/2010/505694.html (accessed February 18, 2011).
[32] J. Bankoti, B. Rase, T. Simones and D. M. Shepherd, *Toxicol. Appl. Pharmacol.*, 2010, **246**, 18.
[33] B. A. Vorderstrasse and N. I. Kerkvliet, *Toxicol. Appl. Pharmacol.*, 2001, **171**, 117.
[34] N. T. Nguyen, A. Kimura, T. Nakahama, I. Chinen, K. Masuda, K. Nohara, Y. Fujii Kuriyama and T. Kishimoto, *Proc. Natl. Acad. Sci. USA*, 2010, **107**, 19961.
[35] C. F. Vogel, S. R. Goth, B. Dong, I. N. Pessah and F. Matsumura, *Biochem. Biophys. Res. Commun.*, 2008, **375**, 331.
[36] J. D. Mezrich, J. H. Fechner, X. Zhang, B. P. Johnson, W. J. Burlingham and C. A. Bradfield, *J. Immunol.*, 2010, **185**, 3190.
[37] F. J. Quintana, G. Murugaiyan, M. F. Farez, M. Mitsdoerffer, A. M. Tukpah, E. J. Burns and H. L. Weiner, *Proc. Natl. Acad. Sci. USA*, 2010, **107**, 20768.
[38] E. Hauben, S. Gregori, E. Draghici, B. Migliavacca, S. Olivieri, M. Woisetschläger and M. G. Roncarolo, *Blood*, 2008, **112**, 1214.
[39] A. Kimura, T. Naka, K. Nohara, Y. Fujii-Kuriyama and T. Kishimoto, *Proc. Natl. Acad. Sci. USA*, 2008, **105**, 9721.
[40] N. B. Marshall, W. R. Vorachek, L. B. Steppan, D. V. Mourich and N. I. Kerkvliet, *J. Immunol.*, 2008, **181**, 2382.
[41] R. Gandhi, D. Kumar, E. J. Burns, M. Nadeau, B. Dake, A. Laroni, D. Kozoriz, H. L. Weiner and F. J. Quintana, *Nat. Immunol.*, 2010, **11**, 846.
[42] S. E. Allan, G. X. Song-Zhao, T. Abraham, A. N. McMurchy and M. K. Levings, *Eur. J. Immunol.*, 2008, **38**, 3282.
[43] M. Veldhoen, K. Hirota, A. M. Westendorf, J. Buer, L. Dumoutier, J. C. Renauld and B. Stockinger, *Nature*, 2008, **453**, 106.
[44] F. J. Quintana, A. S. Basso, A. H. Iglesias, T. Korn, M. F. Farez, E. Bettelli, M. Caccamo, M. Oukka and H. L. Weiner, *Nature*, 2008, **453**, 65.
[45] M. Veldhoen, K. Hirota, J. Christensen, A. O'Garra and B. Stockinger, *J. Exp. Med.*, 2009, **206**, 43.
[46] B. Stockinger, *J. Biol.*, 2009, **8**, 61. http://jbiol.com/content/pdf/jbiol170.pdf (accessed February 18, 2011).
[47] L. Apetoh, F. J. Quintana, C. Pot, N. Joller, S. Xiao, D. Kumar, E. J. Burns, D. H. Sherr, H. L. Weiner and V. K. Kuchroo, *Nat. Immunol.*, 2010, **11**(9), 854.

[48] J. M. Ramirez, N. C. Brembilla, O. Sorg, R. Chicheportiche, T. Matthes, J. M. Dayer, J. H. Saurat, E. Roosnek and C. Chizzolini, *Eur. J. Immunol.*, 2010, **40**, 2450.
[49] T. Negishi, Y. Kato, O. Ooneda, J. Mimura, T. Takada, H. Mochizuki, M. Yamamoto, Y. Fujii-Kuriyama and S. Furusako, *J. Immunol.*, 2005, **175**, 7348.
[50] J. M. Benson and D. M. Shepherd, *Toxicol. Sci.*, 2011, **120**, 68.
[51] T. Takamura, D. Harama, S. Matsuoka, N. Shimokawa, Y. Nakamura, K. Okumura, H. Ogawa, M. Kitamura and A. Nakao, *Immunol. Cell Biol.*, 2010, **88**, 685.
[52] R. Arsenescu, V. Arsenescu, J. Zhong, M. Nasser, R. Melinte, R. W. Dingle, H. Swanson and W. J. de Villiers, *Inflamm. Bowel Dis.*, 2010, **17**, 1149.
[53] B. P. Lawrence, M. S. Denison, H. Novak, B. A. Vorderstrasse, N. Harrer, W. Neruda, C. Reichel and M. Woisetschläger, *Blood*, 2008, **112**, 1158.
[54] Y. Kato, T. Manabe, Y. Tanaka and H. Mochizuki, *J. Immunol.*, 1999, **162**, 7470.
[55] N. I. Kerkvliet, L. B. Steppan, W. Vorachek, S. Oda, D. Farrer, C. P. Wong, D. Pham and D. V. Mourich, *Immunotherapy*, 2009, **1**, 539.
[56] D. O. Carpenter, *Rev. Environ. Health*, 2008, **23**, 59.
[57] L. Zhang, J. Ma, M. Takeuchi, Y. Usui, T. Hattori, Y. Okunuki, N. Yamakawa, T. Kezuka, M. Kuroda and H. Goto, *Invest. Ophthalmol. Vis. Sci.*, 2010, **51**, 2109.
[58] C. M. Schaldach, J. Riby and L. F. Bjeldanes, *Biochemistry*, 1999, **38**, 7594.
[59] F. S. Machado, J. E. Johndrow, L. Esper, A. Dias, A. Bafica, C. N. Serhan and J. Aliberti, *Nat. Med.*, 2006, **12**, 330.
[60] Paraphrase of an English translation of a German statement attributed to the 16th century German scientist Paracelsus; see also B. N. Ames and L. S. Gold, *Mutat. Res.*, 2000, **447**, 3.

CHAPTER **20**

Peptidyl Prolyl Isomerase Inhibitors

Patrick T. Flaherty and **Prashi Jain**

Contents

ABBREVIATIONS

AD	Alzheimer's disease
CNS	central nervous system
CsA	cyclosporine A
FKPB	FK506 binding protein
HCV	hepatitis C virus
HIV	human immunodeficiency virus
Hsp	heat-shock protein
HTS	high-throughput screen
ITC	isothermal calorimetry
LSF	late SV40 factor

Department of Medicinal Chemistry, School of Pharmacy, Duquesne University, 410A Mellon Hall, 600 Forbes Ave, Pittsburgh, PA 15282, USA

Annual Reports in Medicinal Chemistry, Volume 46
ISSN: 0065-7743, DOI: 10.1016/B978-0-12-386009-5.00007-2

NFT	neurofibrillary tangle
PPI	peptidyl prolyl isomerase
pSer	phosphoserine
pThr	phosphothreonine
siRNA	small interfering RNA
TOSCY	total correlation spectroscopy
TRPC1	transient receptor potential cation channel 1
TS	thymidylate synthase
Vpr	multifunctional viral protein R

1. INTRODUCTION

Posttranslational modifications of newly synthesized proteins include proper folding of proteins and covalent modification of side-chain residues with auxiliary groups. Properly folded proteins possess unique physical and biological properties and exhibit unique capacities to undergo subsequent covalent posttranslational modification. *cis–trans* peptidyl prolyl isomerases (PPIs) catalyze the conversion of nascent protein strands containing *cis*-proline residues to the more stable *trans*-proline conformation. The less common interconversion from *trans* to the thermodynamically unfavored *cis*-proline form is catalyzed by *trans–cis* PPIs. Although PPIs [EC 5.2.1.8] all catalyze the *cis–trans* interconversion of proline residues, they can be divided into three enzyme families on the basis of structure and mechanism. Each PPI family exhibits selective binding of both ligands and inhibitors.

cis ⇄ trans

2. CATEGORIES OF PEPTIDYL PROLYL ISOMERASES

Review of the biochemistry of the PPIs is well represented in the literature with an emphasis on overall biological relevance [1–4], foundational biochemistry [5–12], mechanism [7,8,10,13], inhibitors [14–19], and possible therapeutic utility [5,13,17].

2.1. Cyclophilins

Cyclophilins were the first PPIs discovered and currently are represented by at least 60 enzymes [20]. There have been 17 human isoforms of cyclophilin identified [21]. Of the three families, cyclophilins display the greatest sequence homology within their PPI family. The current nomenclature is to assign a letter suffix. Older nomenclature employs a numeric suffix derived from the enzyme weight in kilodaltons. As a result, a discrete cyclophilin may have multiple names: for example, cyclophilin A is identical to cyclophilin 18. Cyclophilins all contain a *cis–trans* PPI domain with an eight-strand β-bundle and two associated α-helices; this site binds a portion of the macrocyclic immunosuppressive drugs represented by cyclosporine A (CsA) **1**. This dimeric complex then binds a third partner, such as calcineurin. The identity of this third binding partner and the ensuing formation of an active or inactive ternary complex are dependent on the identity of bound ligand; this has been referred to as ligand specific protein binding and activation [22].

Pathological states associated with cyclophilin activity include viral infection [23,24], chemotaxis during inflammation, cancer [25,26], and mitochondrial stress, during which cyclophilin inhibitors are protective [27]. The most widely studied cyclophilins are A, B, and D.

The ternary complex of cyclophilin A/CsA/calcineurin generates active pSer (pThr) 2B protein phosphatase which, in part, mediates the immunosuppressive properties of CsA.

Cyclophilin A also catalyzes a *cis/trans* interconversion at a requisite Pro35 residue of multifunctional viral protein R (Vpr), an essential component for HIV virion assembly [28]. This is consistent with a chaperonin role for cyclophilin A facilitating virion assembly and HIV virulence. Cyclophilin A has also been identified as essential for the assembly of NS5B into the virion replication complex of HCV [23,28–31]. The role of cyclophilin B in HIV and HCV continues to be examined [11,32,33], and although siRNA strategies [24] offer better isoform selective analysis, the exact role of cyclophilin B is not clear. Additional interactions of cyclophilin A, B, and **1** with the membrane bound immunological glycoprotein CD147 have been reviewed [34] with an emphasis on neutrophil chemotaxis and migration.

X-ray crystal structures have recently been obtained with cyclophilin B and calnexin, calreticulin, or calmegin [35], providing a structural basis for the folding of *N*-glycosylated proteins in the calnexin cycle. Possible roles for cyclophilin B in the prolyl-3-hydroxylase complex have also been examined with an emphasis on osteogenesis imperfecta [36,37].

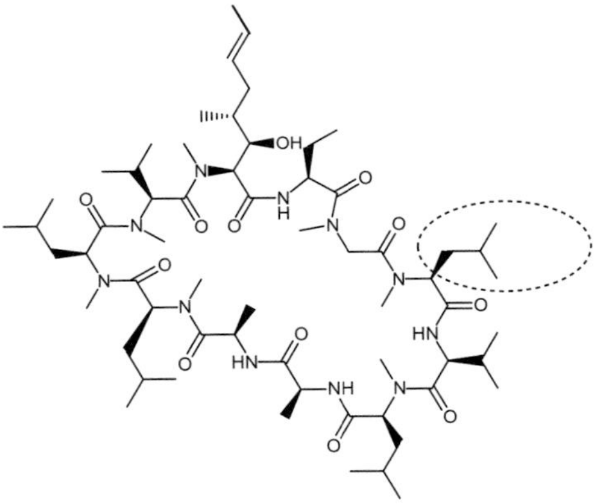

Cyclophilin D has also been shown to activate the mitochondrial permeability transition pore complex in the inner mitochondrial membrane [38]. This pore regulates mitochondrial Ca^{2+} stores, and inhibition of pore activation has been postulated to provide a significant protective effect in multiple models of neurodegeneration [3,39]. Cyclophilin D also associates with the ATP synthase complex in the inner mitochondrial membrane, although the specific nature of this interaction requires further characterization [40]. Cyclophilin D knockout mice, however, display cardiac hypertrophy and less tolerance to physical stress consistent with less metabolic flexibility in the myocardium and a shift from fatty acid to glucose utilization [41]. The conversion of cyclophilin D from a PPI to a pore-activating complex has been proposed as arising from oxidative formation of a disulfide bond between Cys203 and Cys157 [27] and a resultant change in its biological function.

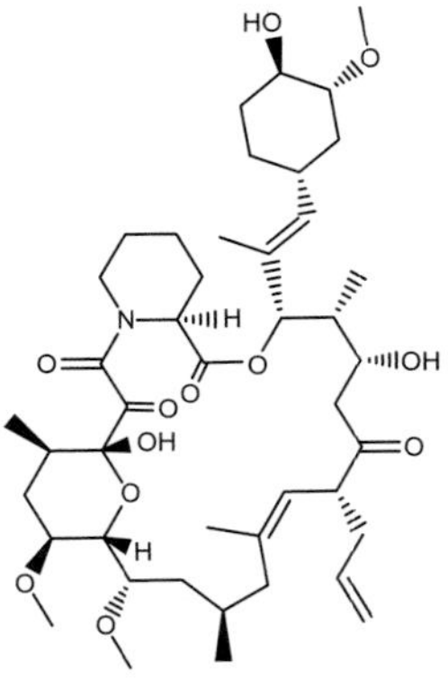

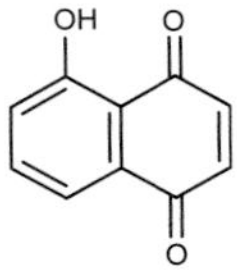

Cyclophilins contain additional protein binding domains and have been demonstrated to form complex multi-protein assemblies [21].

2.2. FK506 binding proteins

PPIs that display preferential binding to the immunosuppressive drug FK506 (Tacrolimus), **2**, [ARMC 29, p. 347] are categorized as FK506 binding proteins (FKBPs). The ternary complex of FKBP, FK506, and calcineurin negatively regulates calcineurin activity in a manner analogous to the ternary complex of cyclophilin A, CsA, and calcineurin. At least 50 FKBPs have been identified and exhibit potent biological activities including signal transduction and cellular adaption to increased metabolic demand and oxidative stress. Both cyclophilins and FKBPs are commonly referred to as immunophilins as they bind immunosuppressive drugs [22]. There are 15 principle members of the human FKPB family. FKPBs have six antiparallel β-strands, one α-helix, several loop regions near the α-helix, and a deep hydrophobic pocket facilitating proline binding. FKBPs are 10–50 times elevated in the CNS (depending on isoform) compared to the periphery. Four of these isoforms exist primarily in the brain: FKBP12, FKBP38, FKBP52, and FKBP65. The role of FKBPs in the brain has been difficult to study due to limited CsA partitioning into the CNS. Nonimmunological pathological disease states associated with FKBPs include neurodegeneration and depression. Emphasis has been placed on FKBP/tau/microtubule interactions [42], neurite outgrowth [43], protein fibrilization [44], and FKBP/glucocorticoid receptor [45,46] interactions [18].

FK506 reduced α-synuclein fibrilization in SHSY5Y neuroblastoma cells in a dose-dependent manner and reduced apoptotic cell death. siRNA or stable knockdown of FKBP12 or FKBP52 (to a lesser extent) produced similar effects [47]. Subsequently, extensive tertiary structure analysis with NMR and modeling has been conducted on FKBP12 [44,48]. Modification of α-synuclein aggregation may find relevance in understanding the pathology of Parkinson's disease or Lewy body dementia; both of these neurodegenerative disorders present Lewy bodies composed of fibrilized α-synuclein.

FKBP51 and 52 assist in stabilizing the microtubule-associated protein tau and stabilize microtubules. The proposed mechanism is salvage of hyperphosphorylated tau from ubquitination through stepwise sequestration by heat-shock protein (Hsp) 90, tau dephosphorylation, and finally FKBP-mediated refolding of tau to its original microtubule stabilizing form [42,49]. Additionally, FKBP52 may interact with the TRPC1 channel assisting with neuron growth cone steering and direction of neurite outgrowth [46]. These processes may underlie synaptic remodeling events relevant to memory and deposition of hyperphosphorylated tau protein in Alzheimer's disease (AD).

Both FKBP51 and 52 interact with the glucocorticoid receptor. FKPB51 effects selective downregulation of receptor activity, although the nature of these interactions is not currently well characterized [45]. Interaction of FKBP5 with the glucocorticoid receptor has been postulated as mediating unipolar depression driven by elevated cortisol levels [46,50].

2.3. Parvulins

Parvulins are PPIs that preferentially bind 5-hydroxy-1,4-naphthoquinone (juglone) **3** rather than immunosuppressive drugs **1** and **2**. Consequently, parvulins are not classified as immunophilins. Parvulins are present in many species, but there are only three human isoforms of the parvulins. Pin1 is the most widely studied isoform. Parvulins effect a phosphoserine (pSer) or a phosphothreonine (pThr) directed *cis–trans* isomerization of the prolyl bond. The enzyme has two distinct domains: a phosphate-recognizing domain that associates with the pSer or pThr and the catalytic PPI domain. The PPI domain of parvulins consists of four antiparallel β-sheets, an α-helix, and convergent loops [51]. Pin1 inhibitors have been the focus of extensive research [15,52] and only recent developments will be discussed here. Pathological states associated with parvulins include AD, cancer, and immunological response.

Many studies have examined the role of Pin1 in tau hyperphosphorylation as a contributing factor in Alzheimer's and other neurodegenerative diseases [51,53,54]. Pin1 has also been colocalized in NFT deposits. Its current role in AD requires additional characterization. Overexpression of Pin1 in hepatic carcinomas suggests Pin1 inhibition as a plausible therapeutic strategy. Pin1 binds the thymidylate synthase (TS) regulating transcription factor late SV40 factor (LSF) at pThr329Pro330, resulting in the *trans*Pro330 product. This resultant pThr329 *trans*Pro330 form of LSF is then dephosphorylated at Ser291 and Ser301 permitting TS expression and successful navigation of the G1/S cell cycle transition [55]. Additionally, Pin1 phosphorylates and activates Notch, which subsequently undergoes γ-secretase-mediated proteolysis. Notch upregulation was correlated with increased growth of breast tumor cells; Pin1 levels were also increased [56].

Pin1 activity has been proposed as essential for eosinophil survival. Pin1 generates pThr167 *trans*Pro168 Bax that is refractory to calpain hydrolysis and the subsequent apoptosis-inducing caspase cascade. As a consequence, Pin1 inhibition has been proposed as a drug target for asthma and other eosinophilic diseases [57]. However, Pin1 knockout mice display cell proliferation deficiency phenotypes [58] and display deficient telomere maintenance [59].

3. SMALL-MOLECULE INHIBITORS

Although small-molecule inhibitors of PPIs were identified over 15 years ago, initially in the context of FK506 work, it has only recently been shown that small molecules can differentiate between isoforms of cyclophilins, specifically cyclophilins A and B. These results have emanated primarily from the laboratories of Fischer [14,60,61] and Li [62–64]. Recent structural analysis of the 17 known human cyclophilins utilizing both X-ray crystal structures and homology modeling assuming an invariant Arg (Arg55 in the case of cyclophilin A) has identified two specific locations near the catalytic site likely to confer isoform selectivity: the S2 pocket and the S1′ pocket. Unique residues on the S2 pocket exist for each isoform suggesting the potential for rationally developing selective inhibitors. Analysis of *p*-nitrophenyl-tagged tetra-peptide-based ligands with isothermal calorimetry (ITC) and $^1H/^1H$ TOSCY experiments was consistent with the hypothesis of isoform variation in the S2 pocket contributing to ligand selectivity for a given cyclophilin [21].

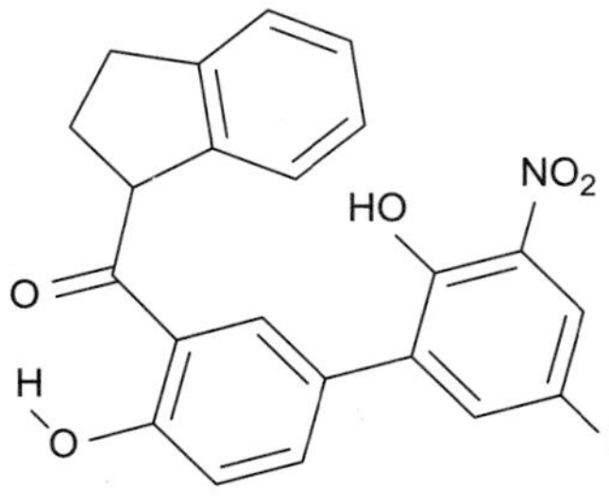

4 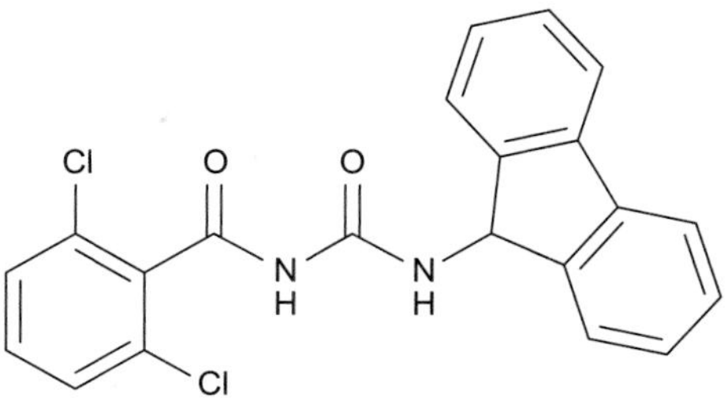 **5**

Compound **4** from the Fisher lab represents a parvulin series compound modified for selective cyclophilin A inhibition [14,65]. The requisite *ortho*-hydroxyl and the dihydroindane ring has been proposed to mimic the prolyl acyl group, and the 3′-nitro biphenyl system increases potency. As a result, **4** displays 520 nM inhibition of cyclophilin A as determined with a fluorescence quench assay. This compound shows no inhibition at the B or C isoforms and is four times less active against cyclophilin D. ITC studies are consistent with the biphenyl group making an entropic contribution that surpasses hydrogen bonding contributions from the acyl group. This is consistent with Dugave's model of a hydrophobic pocket in cyclophilin A comprising residues from Ile57 to Phe60 [7]; this would correspond to the S1′ pocket [21]. Reduction of the 3′-*meta*-nitro group of **4** to the corresponding amine and subsequent testing gave a compound with a K_i of 300 nM for cyclophilin A and a reduction of 40-fold for cyclophilin A versus cyclophilin B selectivity. The differential activity of the 3′-*meta*-nitro and 3′-*meta*-amino, the symmetry of inhibition curves presented, and the concentration-dependent displacement of CsA in the ITC experiments are consistent with a competitive binding in the prolyl isomerase site rather than artifacts arising from bulk hydrophobic properties of aggregates [66,67]. Compound **5** is reported to have near equipotent activity in cyclophilin A inhibition compared to CsA [64]. Selectivity for cyclophilin A over other PPIs was not presented. These observations build upon prior observations from a series of thiourea and related derivatives by the same research group [68,69]. These have been proposed to bind the catalytic site of cyclophilin A. Docking simulations indicate a 90° angle between the plane of the 2,6-dichloro phenyl ring and the amide carbonyl of **5**. Additionally, the aromatic ring may exhibit hydrophobic interactions with the aryl ring of Phe60 of the S1′ pocket of cyclophilin A [70]. Also, the amide acyl is proposed to interact with the catalytic Arg55 residue. A similar series of thiourea derivatives, represented by **6**, have also been explored as dual inhibitors of cyclophilin A and viral capsid assembly in the design of anti-HIV therapeutic agents [71]. The cyclophilin A portion of the design strategy utilized an ionic interaction of the sulfonamide isoxazole with the catalytic Arg55 and extension of the thiourea portion into the lipophilic portion of the S2 pocket [71].

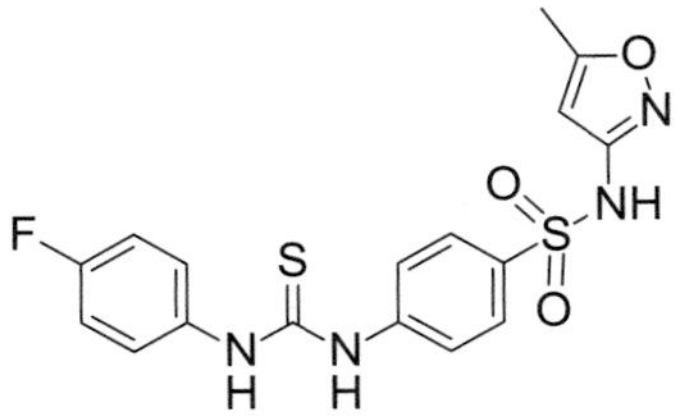

6

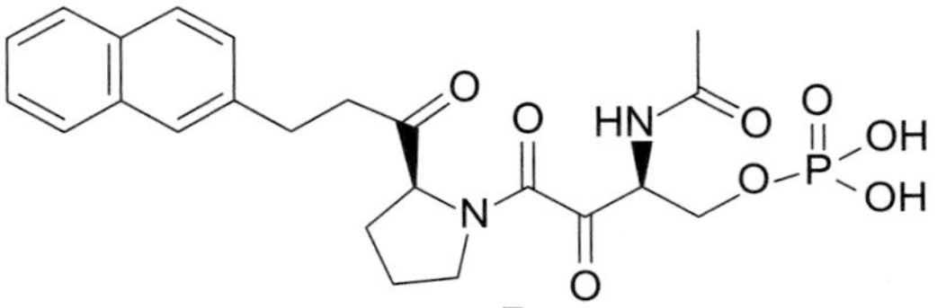

7 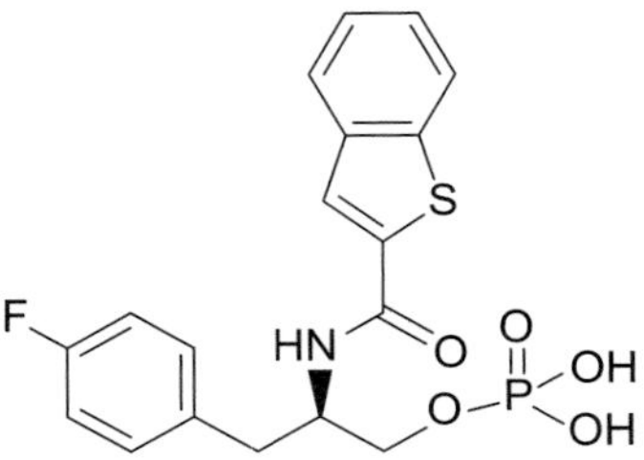 **8**

Recent small-molecule inhibitors of Pin1 have also been reported. Starting with a known core for FKBP inhibition, the 2-amido phosphate moiety, structural variations were examined to optimize inhibition at Pin1 in terms of both potency and selectivity.

Mechanistic probes including **7** were synthesized to evaluate competing enzymatic proposals for Pin1 prolyl bond catalysis [72]. Two broad categories of catalysis by PPIs are frequently invoked: a twisted amide versus a Cys-mediated tetrahedral intermediate. Compound **7** used the α-keto amide functional group, a motif that has led to potent FKPB inhibitors and is consistent with a twisted amide transition state. Although this compound was subsequently prepared in diastereomerically pure form from enantiopure *S*-proline, both compounds were modestly active in the high millimolar range. These results indicate that either the enzymatic mechanism(s) employed by parvulins differ from FKBP or that the structural characteristics of **7** are not yet sufficiently optimized.

A series of single and double-digit nanomolar amido phosphate inhibitors of Pin1 represented by **8** were identified using a structure-based design strategy [73]. A phosphate group was identified (PDB ID: 3IK8) as interacting with an anionic binding domain (Arg68 and Arg69) of Pin1 originally identified in the X-ray crystal structure of 3IK8. Several prolyl acyl bioisosteres were examined with the amide group presented as optimal. Additional exploration of the proline binding catalytic binding pocket identified hydrophobic regions that were explored with aryl and heterocyclic aryl groups resulting in the 6 nM Pin1 inhibitor **8**. Hydrophobic interactions of **8** with Pin1 not accessed by traditional proline mimics include Phe134 and Leu122. Data for cross-inhibition with other parvulins were not presented, and inactivity in cellular assays was attributed to poor transport into the cell due to the phosphate group.

4. MACROCYCLIC INHIBITORS

Despite the imposing structural components presented by **1** and **2**, both are used clinically and have been the subject of multiple total syntheses [65]. Another related compound DEBIO-025, **9**, is in clinical trials as a potential treatment for HCV [74].

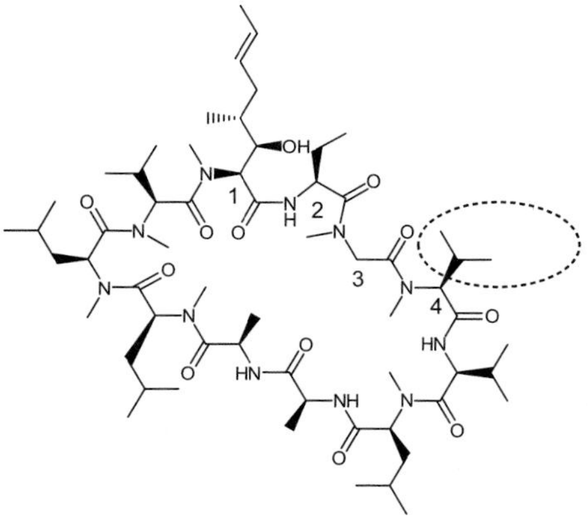

9 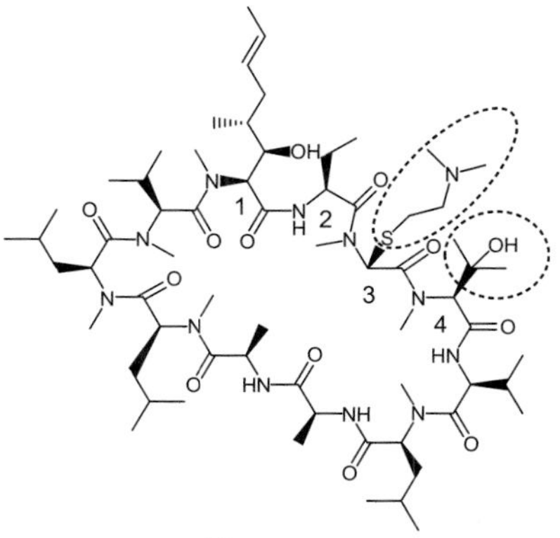 **10**

DEBIO-025, a semisynthetic derivative differing from **1** by a single amino acid side chain, is predominantly a cyclophilin inhibitor [75,76] and is under examination for treatment of HCV particularly in individuals coinfected with both HCV and HIV. DEBIO-025 has been shown to reduce HIV-1 infection and replication *in vitro* and has also been shown to reduce infection and replication against the more challenging HIV-2 genotype, but only when used in combination with existing HIV therapeutic agents. Unlike CsA, **9** does not activate calcineurin and as a result has reduced immunosuppressive effects. SCY-635, **10**, exhibits equipotent binding to CsA and like DEBIO-025 does not activate calcineurin [77]. DEBIO-025 and SCY-635 deviate from the structure of CsA by variation on the three and four amino acid residues; this portion of the macrocycle is directed away from the prolyl isomerase active site toward calcineurin. Three significant points should be emphasized: (1) immunosuppressive and antiviral effects of CsA are possibly independent, (2) inhibition of cyclophilin is a potentially useful strategy for reducing HIV infectivity and replication *in vivo*, and (3) this antiviral effect may be synergistic with currently used therapies.

5. CONCLUSION

The understanding of PPIs on a biological level continues to advance with a current emphasis on mechanism and relevance to disease states. Drug design strategies from one family of PPI are finding application in other PPIs as evidenced by compounds **4**, **8**, DEBIO-025, and SCY-635. Several structural classes of molecules have demonstrated the ability to inhibit cyclophilin with the possibility of isoform selectivity. Additionally, small molecules have been demonstrated to inhibit the active catalytic site without a direct proline mimic. The presentation of a comprehensive structure-based analysis of all cyclophilins suggests that the development of rationally designed selective inhibitors of PPIs is a reasonable goal. The development of HTS assays should facilitate efficient identification of PPI inhibitors [44,78,79]. However, the complexity of mechanism of activity evidenced by Pin1 and the possibility of significant modification of enzymatic activity in response to cellular metabolism evidenced by cyclophilin D suggest that *in vitro* analysis of PPI activity should be supported with relevant whole animal models of relevant human disease states.

REFERENCES

[1] T. Aumuller, G. Jahreis, G. Fischer and C. Schiene-Fischer, *Biochemistry*, 2010, **49**, 1042.
[2] G. Fischer, *Protein Folding Handb.*, 2005, **3**, 377.
[3] V. Giorgio, M. E. Soriano, E. Basso, E. Bisetto, G. Lippe, M. A. Forte and P. Bernardi, *Biochim. Biophys. Acta Bioenerg.*, 2010, **1797**, 1113.
[4] J. Lee, *Arch. Pharmacol Res.*, 2010, **33**, 181.
[5] C. Dugave (ed.), in *cis-trans Isomerization in Biochemistry*, Wiley-VCH, Weinheim, Germany, 2006, 261.
[6] C. Dugave (ed.), in *cis-trans Isomerization in Biochemistry*, Wiley-VCH, Weinheim, Germany, 2006, 7.
[7] C. Dugave and L. Demange, *Chem. Rev.*, 2003, **103**, 2475.
[8] D. Hamelberg and J. A. McCammon, *J. Am. Chem. Soc.*, 2009, **131**, 147.
[9] M. A. Sahai, M. Szoeri, B. Viskolcz, E. F. Pai and I. G. Csizmadia, *J. Phys. Chem. A*, 2007, **111**, 8384.
[10] S. Wawra and G. Fischer, in *cis-trans Isomerization in Biochemistry* (ed. C. Dugave), Wiley-VCH, Weinheim, Germany, 2006, 167.
[11] S. Barik, *Cell. Mol. Life Sci.*, 2006, **63**, 2889.
[12] G. Fischer, in *cis-trans Isomerization in Biochemistry* (ed. C. Dugave), Wiley-VCH, Weinheim, Germany, 2006, 195.

[13] F. Edlich and G. Fischer, *Handb. Exp. Pharmacol.*, 2006, 359.
[14] S. Daum, M. Schumann, S. Mathea, T. Aumueller, M. A. Balsley, S. L. Constant, B. Feaux de Lacroix, F. Kruska, M. Braun and C. Schiene-Fischer, *Biochemistry*, 2009, **48**, 6268.
[15] X. J. Wang and F. A. Etzkorn, *Biopolymers*, 2006, **84**, 125.
[16] J. Dornan, P. Taylor and M. D. Walkinshaw, *Curr. Top. Med. Chem.*, 2003, **3**, 1392.
[17] G. Finn and K. P. Lu, *Curr. Cancer Drug Targets*, 2008, **8**, 223.
[18] G. S. Hamilton and C. Thomas, *Adv. Med. Chem.*, 2000, **5**, 1.
[19] P. C. Waldmeier, K. Zimmermann, T. Qian, M. Tintelnot-Blomley and J. J. Lemasters, *Curr. Med. Chem.*, 2003, **10**, 1485.
[20] G. Fischer and H. Bang, *Biochim. Biophys. Acta Protein Struct. Mol. Enzymol.*, 1985, **828**, 39.
[21] T. L. Davis, J. R. Walker, V. Campagna-Slater, P. J. Finerty Jr., R. Paramanathan, G. Bernstein, F. MacKenzie, W. Tempel, H. Ouyang, W. H. Lee, E. Z. Eisenmesser and S. Dhe-Paganon, *PLoS Biol.*, 2010, **8**, e1000439.
[22] M. T. G. Ivery, *Med. Res. Rev.*, 2000, **20**, 452.
[23] G. Fischer, P. Gallay and S. Hopkins, *Curr. Opin. Invest. Drugs*, 2010, **11**, 911.
[24] P. D. Nagy, R. Y. Wang, J. Pogany, A. Hafren and K. Makinen, *Virology*, 2011, **411**, 374.
[25] N. Baum, C. Schiene-Fischer, M. Frost, M. Schumann, K. Sabapathy, O. Ohlenschlaeger, F. Grosse and B. Schlott, *Oncogene*, 2009, **28**, 3915.
[26] F. T. Bane, J. H. Bannon, S. R. Pennington, G. Campaini, D. C. Williams, D. M. Zisterer and M. M. McGee, *J. Pharmacol. Exp. Ther.*, 2009, **329**, 38.
[27] D. Linard, A. Kandlbinder, H. Degand, P. Morsomme, K. J. Dietz and B. Knoops, *Arch. Biochem. Biophys.*, 2009, **491**, 39.
[28] Z. Liu, F. Yang, J. M. Robotham and H. Tang, *J. Virol.*, 2009, **83**, 6554.
[29] F. Yang, J. M. Robotham, H. B. Nelson, A. Irsigler, R. Kenworthy and H. Tang, *J. Virol.*, 2008, **82**, 5269.
[30] U. Chatterji, M. Bobardt, S. Selvarajah, F. Yang, H. Tang, N. Sakamoto, G. Vuagniaux, T. Parkinson and P. Gallay, *J. Biol. Chem.*, 2009, **284**, 16998.
[31] J. A. Heck, X. Meng and D. N. Frick, *Biochem. Pharmacol.*, 2009, **77**, 1173.
[32] A. Galat and J. Bua, Cell, *Mol. Life Sci.*, 2010, **67**, 3467.
[33] K. Watashi, M. Khan, V. R. K. Yedavalli, M. L. Yeung, K. Strebel and K. T. Jeang, *J. Virol.*, 2008, **82**, 9928.
[34] V. Yurchenko, S. Constant, E. Eisenmesser and M. Bukrinsky, *Clin. Exp. Immunol.*, 2010, **160**, 305.
[35] G. Kozlov, S. Bastos-Aristizabal, P. Maeaettaenen, A. Rosenauer, F. Zheng, A. Killikelly, J. F. Trempe, D. Y. Thomas and K. Gehring, *J. Biol. Chem.*, 2010, **285**, 35551.
[36] J. W. Choi, S. L. Sutor, L. Lindquist, G. L. Evans, B. J. Madden, H. Robert Bergen III, T. E. Hefferan, M. J. Yaszemski and R. J. Bram, *PLoS Genet.*, 2009, **5**, e1000750.
[37] A. M. Barnes, E. M. Carter, W. A. Cabral, M. Weis, W. Chang, E. Makareeva, S. Leikin, C. N. Rotimi, D. R. Eyre, C. L. Raggio and J. C. Marini, *N. Engl. J. Med.*, 2010, **362**, 521.
[38] C. P. Baines, R. A. Kaiser, N. H. Purcell, N. S. Blair, H. Osinska, M. A. Hambleton, E. W. Brunskill, M. R. Sayen, R. A. Gottlieb, G. W. Dorn, J. Robbins and J. D. Molkentin, *Nature*, 2005, **434**, 658.
[39] H. Du, L. Guo, F. Fang, D. Chen, A. A. Sosunov, G. M. McKhann, Y. Yan, C. Wang, H. Zhang, J. D. Molkentin, F. J. Gunn-Moore, J. P. Vonsattel, O. Arancio, J. X. Chen and S. D. Yan, *Nat. Med.*, 2008, **14**, 1097.
[40] V. Giorgio, E. Bisetto, M. E. Soriano, F. Dabbeni-Sala, E. Basso, V. Petronilli, M. A. Forte, P. Bernardi and G. Lippe, *J. Biol. Chem.*, 2009, **284**, 33982.
[41] J. W. Elrod, R. Wong, S. Mishra, R. J. Vagnozzi, B. Sakthievel, S. A. Goonasekera, J. Karch, S. Gabel, J. Farber, T. Force, J. H. Brown, E. Murphy and J. D. Molkentin, *J. Clin. Invest.*, 2010, **120**, 3680.
[42] U. K. Jinwal, J. Koren III, S. I. Borysov, A. B. Schmid, J. F. Abisambra, L. J. Blair, A. G. Johnson, J. R. Jones, C. L. Shults, J. C. O'Leary III, Y. Jin, J. Buchner, M. B. Cox and C. A. Dickey, *J. Neurosci.*, 2010, **30**, 591.

[43] S. Shim, J. P. Yuan, J. Y. Kim, W. Zeng, G. Huang, A. Milshteyn, D. Kern, S. Muallem, G.-L. Ming and P. F. Worley, *Neuron*, 2009, **64**, 471.
[44] G. Zoldak, T. Aumueller, C. Luecke, J. Hritz, C. Oostenbrink, G. Fischer and F. X. Schmid, *Biochemistry*, 2009, **48**, 10423.
[45] I. M. Wolf, S. Periyasamy, T. Hinds, W. Yong, W. Shou and E. R. Sanchez, *J. Steroid Biochem. Mol. Biol.*, 2009, **113**, 36.
[46] A. Zobel and W. Maier, *Eur. Arch. Psychiatry Clin. Neurosci.*, 2010, **260**, 407.
[47] M. Gerard, A. Deleersnijder, V. Daniels, S. Schreurs, S. Munck, V. Reumers, H. Pottel, Y. Engelborghs, d. H. C. Van, J.-M. Taymans, J.-M. Taymans, Z. Debyser and V. Baekelandt, *J. Neurosci.*, 2010, **30**, 2454.
[48] U. Brath and M. Akke, *J. Mol. Biol.*, 2009, **387**, 233.
[49] A. Salminen, J. Ojala, K. Kaarniranta, M. Hiltunen and H. Soininen, *Prog. Neurobiol.*, 2011, **93**, 99.
[50] A. Zobel, A. Schuhmacher, F. Jessen, S. Hoefels, W. O. von, M. Metten, U. Pfeiffer, C. Hanses, T. Becker, M. Rietschel, L. Scheef, W. Block, H. H. Schild, W. Maier and S. G. Schwab, *Int. J. Neuropsychopharmacol.*, 2010, **13**, 649.
[51] P. Rudrabhatla and H. C. Pant, *J. Alzheimer's Dis.*, 2009, **19**, 389.
[52] G. G. Xu and F. A. Etzkorn, *Drug News Perspect.*, 2009, **22**, 399.
[53] A. Bulbarelli, E. Lonati, E. Cazzaniga, M. Gregori and M. Masserini, *Mol. Cell. Neurosci.*, 2009, **42**, 75.
[54] P. Rudrabhatla, W. Albers and H. C. Pant, *J. Neurosci.*, 2009, **29**, 14869.
[55] U. H. Saxena, L. Owens, J. R. Graham, G. M. Cooper and U. Hansen, *J. Biol. Chem.*, 2010, **285**, 31139.
[56] A. Rustighi, L. Tiberi, A. Soldano, M. Napoli, P. Nuciforo, A. Rosato, F. Kaplan, A. Capobianco, S. Pece, P. P. Di Fiore and G. Del Sal, *Nat. Cell Biol.*, 2009, **11**, 133.
[57] Z. J. Shen, S. Esnault, A. Schinzel, C. Borner and J. S. Malter, *Nat. Immunol.*, 2009, **10**, 257.
[58] Y. C. Liou, A. Ryo, H. K. Huang, P. J. Lu, R. Bronson, F. Fujimori, T. Uchida, T. Hunter and K. P. Lu, *Proc. Natl. Acad. Sci. USA*, 2002, **99**, 1335.
[59] T. H. Lee, A. Tun-Kyi, R. Shi, J. Lim, C. Soohoo, G. Finn, M. Balastik, L. Pastorino, G. Wulf, X. Z. Zhou and K. P. Lu, *Nat. Cell Biol.*, 2009, **11**, 97.
[60] S. Daum, F. Erdmann, G. Fischer, B. Feaux de Lacroix, A. Hessamian-Alinejad, S. Houben, W. Frank and M. Braun, *Angew. Chem. Int. Ed.*, 2006, **45**, 7454.
[61] R. Golbik, C. Yu, E. Weyher-Stingl, R. Huber, L. Moroder, N. Budisa and C. Schiene-Fischer, *Biochemistry*, 2005, **44**, 16026.
[62] J. Li, J. Chen, C. Gui, L. Zhang, Y. Qin, Q. Xu, J. Zhang, H. Liu, X. Shen and H. Jiang, *Bioorg. Med. Chem.*, 2006, **14**, 2209.
[63] J. Li, J. Chen, L. Zhang, F. Wang, C. Gui, Y. Qin, Q. Xu, H. Liu, F. Nan, J. Shen, D. Bai, K. Chen, X. Shen and H. Jiang, *Bioorg. Med. Chem.*, 2006, **14**, 5527.
[64] S. Ni, Y. Yuan, J. Huang, X. Mao, M. Lv, J. Zhu, X. Shen, J. Pei, L. Lai, H. Jiang and J. Li, *J. Med. Chem.*, 2009, **52**, 5295.
[65] M. L. Maddess, M. N. Tackett and S. V. Ley, *Prog. Drug Res.*, 2008, **66**(13), 15.
[66] K. E. D. Coan and B. K. Shoichet, *J. Am. Chem. Soc.*, 2008, **130**, 9606.
[67] S. L. McGovern, B. T. Helfand, B. Feng and B. K. Shoichet, *J. Med. Chem.*, 2003, **46**, 4265.
[68] Z. Tan, J. Li, R. Pang, S. He, M. He, S. Tang, I. Hewlett and M. Yang, *Med. Chem. Res.*, 2011, **20**(3), 314.
[69] F. Fan, J. Zhu, S. Ni, J. Cheng, Y. Tang, C. Kang, J. Li and H. Jiang, *QSAR Comb. Sci.*, 2009, **28**, 183.
[70] S. V. Sambasivarao and O. Acevedo, *J. Chem. Inf. Model.*, 2011, **51**, 475.
[71] K. Chen, Z. Tan, M. He, J. Li, S. Tang, I. Hewlett, F. Yu, Y. Jin and M. Yang, *Chem. Biol. Drug Des.*, 2010, **76**, 25.
[72] G. G. Xu and F. A. Etzkorn, *Org. Lett.*, 2010, **12**, 696.
[73] C. Guo, X. Hou, L. Dong, E. Dagostino, S. Greasley, R. Ferre, J. Marakovits, M. C. Johnson, D. Matthews, B. Mroczkowski, H. Parge, T. Van Arsdale, I. Popoff,

J. Piraino, S. Margosiak, J. Thomson, G. Los and B. W. Murray, *Bioorg. Med. Chem. Lett.*, 2009, **19**, 5613.

[74] C. Campas and R. Castaner, *Drugs Future*, 2008, **33**, 1012.

[75] S. Ciesek, E. Steinmann, H. Wedemeyer, M. P. Manns, J. Neyts, N. Tautz, V. Madan, R. Bartenschlager, T. von Hahn and T. Pietschmann, *Hepatology*, 2009, **50**, 1638.

[76] R. G. Ptak, P. A. Gallay, D. Jochmans, A. P. Halestrap, U. T. Ruegg, L. A. Pallansch, M. D. Bobardt, M.-P. de Bethune, J. Neyts, E. De Clercq, J.-M. Dumont, P. Scalfaro, K. Besseghir, R. M. Wenger and B. Rosenwirth, *Antimicrob. Agents Chemother.*, 2008, **52**, 1302.

[77] S. Hopkins, B. Scorneaux, Z. Huang, M. G. Murray, S. Wring, C. Smitley, R. Harris, F. Erdmann, G. Fischer and Y. Ribeill, *Antimicrob. Agents Chemother.*, 2010, **54**, 660.

[78] T. Mori, S. Itami, T. Yanagi, Y. Tatara, M. Takamiya and T. Uchida, *J. Biomol. Screen.*, 2009, **14**, 419.

[79] T. Mori and T. Uchida, *Curr. Enzyme Inhib.*, 2010, **6**, 46.

CHAPTER 21

MicroRNAs—Basic Biology and Therapeutic Potential

A. Katrina Loomis* and Graham J. Brock**

Contents

ABBREVIATIONS

3′UTR	3′untranslated region
Ago	Argonaute
BMBC	brain metastatic breast cancer
CLIA	Clinical Laboratory Improvement Amendments
HCV	hepatitis C virus
HITS-	high-throughput sequencing of RNAs isolated by
CLIP	cross-linking immunoprecipitation
LNA	locked nucleic acid
mRNA	messenger RNA

* Human Genetics and Epigenetics, PharmaTx Precision Medicine, Pfizer, Inc., Groton, Connecticut, USA
** Molecular Profiling, Department of Drug Discovery, Biogen Idec, Inc., Cambridge, Massachusetts, USA

Annual Reports in Medicinal Chemistry, Volume 46
ISSN: 0065-7743, DOI: 10.1016/B978-0-12-386009-5.00006-0

NGS	next-generation sequencing
nt	nucleotide
PARE	parallel analysis of RNA ends
PK	pharmacokinetic
qRT-PCR	quantitative reverse transcriptase-polymerase chain reaction
RISC	RNA induced silencing complex
RNA	ribonucleic acid
RNA Pol II	RNA polymerase II
TARBP	transactivation-responsive RNA binding protein 2
T_m	melting temperature

1. INTRODUCTION

MicroRNAs are single-stranded RNA molecules, approximately 22 nucleotides (nt) long that play a significant role in the post-transcriptional regulation of gene expression. This chapter outlines what is currently known about microRNAs, recent advances in our understanding, their potential therapeutic uses, and future directions for investigation.

1.1. Background

MicroRNAs were originally discovered in *Caenorhadbitis elegans* [1], and due to their high sequence conservation, their subsequent discovery in many other organisms including mammals was facilitated. MicroRNA genes are frequently located in intronic regions of protein-coding genes but may also be found in intergenic regions of the genome and can occur singly or in clusters. When microRNAs are located in introns, their expression is thought to be co-regulated with that of their host gene. MicroRNA genes are transcribed by RNA polymerase II (RNA Pol II) to form primary microRNAs (pri-miRNAs) which are then capped and polyadenylated [2]. These primary transcripts are subsequently processed into ~70 nt precursor microRNAs (pre-miRNAs) by Drosha, an RNase III endonuclease [3] and exported from the nucleus by Exportin-5 [4] (see Figure 1). Significantly, the ~70 nt pre-microRNA folds into a distinct hairpin conformation, and many microRNA sequences are highly conserved across multiple species. Consequently, to identify other putative microRNAs, many computational methods utilize algorithms to scan the genome for sequences capable of forming these characteristic hairpin structures and to identify sequences that are highly similar if not identical to microRNA sequences from other species. As of 2011, the microRNA registry contained predictions of over 15,000 microRNAs, over a thousand of which are in the human genome [5]. Experimental methods are

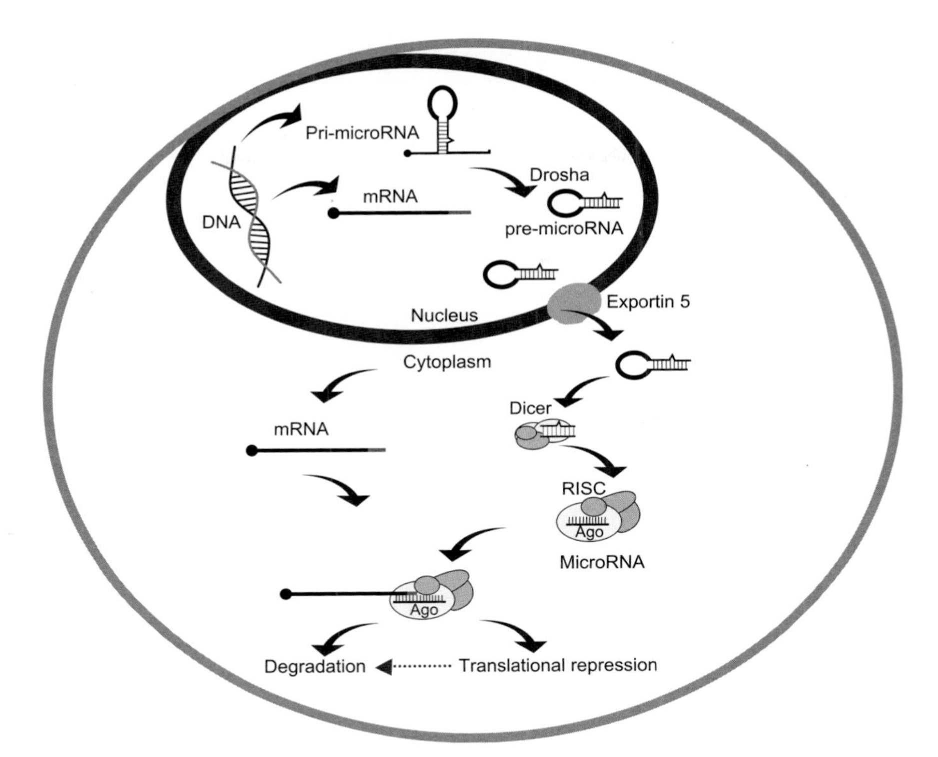

Figure 1 Schematic outline of the main steps in microRNA biogenesis. MicroRNAs are initially synthesized as pri-microRNA transcripts by RNA Pol II in the nucleus. Processing by Drosha results in ~70 nt hairpin pre-microRNA molecules which are exported to the cytoplasm by Exportin 5. Dicer processing leads to the formation of mature microRNAs which are then loaded into the RISC. The passenger strand microRNA is cleaved and degraded, while the guide strand microRNA is retained and used by RISC to identify its target mRNA. Regulation of gene expression is through either translational repression or an mRNA cleavage/degradation mechanism.

needed to confirm the existence of these predicted microRNAs. As with some protein-coding genes, some microRNAs belong to large microRNA families with members differing in sequence by 1 or 2 nt (Table 1).

Once in the cytoplasm, the double-stranded pre-microRNAs are processed by Dicer, another member of the RNase III endonuclease family, into mature ~22 nt microRNAs [3] (Figure 1). This mature double-stranded microRNA species is made up of the guide strand (designated as miR) and its complementary passenger strand (designated as miR*). The regulation of mRNAs by a mature microRNA requires incorporation into the RNA-induced silencing complex (RISC), which comprises multiple proteins including Argonaute, Dicer, and TAR RNA binding proteins (TRBP) (reviewed in Ref. [6]). Following incorporation, the passenger

Table 1 The hsa-let-7 family of microRNAs showing high sequence similarity

MicroRNA	Sequence 5′ to 3′
hsa-let-7a	UGAGGUAGUAGGUUGUAUAGUU
hsa-let-7b	UGAGGUAGUAGGUUGU**G**U**G**GUU
hsa-let-7c	UGAGGUAGUAGGUUGUAU**G**GUU
hsa-let-7d	**A**GAGGUAGUAGGUUG**C**AUAGUU
hsa-let-7e	UGAGGUAG**G**AGGUUGUAUAGUU
hsa-let-7f	UGAGGUAGUAG**A**UUGUAUAGUU
hsa-let-7g	UGAGGUAGUAG**U**UUGUA**C**AGUU
hsa-let-7i	UGAGGUAGUAG**U**UUGU**GCU**GUU

The seed region (nts 2–7) of this microRNA family is underlined and nucleotides that differ from hsa-let-7a are shown in bold.

microRNA strand (miR*) is degraded and released from the RISC. The relative thermodynamic stability of each end of the double-stranded microRNA species plays a large role in determining which strand becomes the guide strand and is retained, and which strand is degraded [7,8]. The recognition of mRNA targets by the RISC occurs through the guide strand microRNA and is thought to be based on the perfect (or almost perfect) sequence complementarity to the ''seed region,'' defined as nucleotides 2–7 at the 5′end of the microRNA. Previously, suppression of gene expression was thought to result either from cleavage of the targeted mRNA followed by its degradation or from translational repression of the target [9–11]. For the cleavage/degradation mechanism to occur, extensive complementarity between the microRNA and its target is required, whereas translational repression is facilitated primarily through complementarity with the seed region. As the majority of microRNAs do not have extensive matches with their predicted mRNA targets, this suggests translational repression as the main mode of action of microRNAs. However, it has recently been reported that degradation of the mRNA target frequently follows translational repression [12].

The limited complementarity between a miRNA and its target mRNAs (with complementarity focusing on the seed region) has hindered the identification of mRNA targets of microRNA action. Consequently, the function of large numbers of microRNAs remains unknown. Nonetheless, current understanding of their basic mechanism of action has established them as an important class of regulatory molecules, adding a new level of eukaryotic gene regulation at the posttranscriptional level [13,14]. Furthermore, microRNA expression patterns may be altered during the progression of many diseases. Therefore, their potential utility as biomarkers in human diseases has been an area of intense investigation, and a database of reported microRNA disease associations is available [15].

To date, studies of altered microRNA expression patterns have mainly focused on oncology, but recent reports have indicated that microRNA expression can be disrupted in other human diseases as well. Questions surrounding the feasibility of using such molecules as biomarkers relate to the technical challenges inherent in discriminating between such highly similar sequences. Despite these challenges, microRNA signatures have proven to be more robust prognostic markers in oncology than mRNA signatures [16,17].

1.2. Technical considerations when studying microRNA expression

The initial step in a microRNA analysis is usually to undertake genome-wide expression profiling to determine expression levels in various cell types or disease states. This will identify changes in the expression of subsets or even individual microRNAs that can then be evaluated as biomarkers. However, as noted, the relatively short length and highly similar sequence characteristics of microRNAs make profiling technically demanding. Nevertheless, several commercially available platforms with varying degress of sensitivity, specificity, and coverage are available. Recently, the advent of next-generation sequencing (NGS) techniques has offered the potential for validating predicted microRNA sequences [18] as well as quantitating microRNA levels in various cell types [19]. In addition, such sequencing methodologies allow *de novo* identification of microRNAs not previously predicted. However, technical issues remain, such as the potential for bias during construction of the sequencing library [20]. Prior to the advent of NGS, genome-wide analysis methods involved microRNA arrays with hybridization to hundreds or even thousands of probes generated using conventional deoxyribonucleotide or locked nucleic acid (LNA) chemistries [21–24]. Alternatively, although less comprehensive in their coverage, methods that use multiple qPCR primer sets, such as the TaqMan Human MicroRNA Array or a similar system from Roche/Exiqon that uses LNA primers, have been widely used. While the numbers of interrogated microRNAs are currently smaller than most hybridization arrays, these qRT-PCR arrays have increased sensitivity, require considerably less starting material, and/or do not require a preamplification or purification step. Regardless of the method chosen, these techniques all have to address the highly varied melting temperatures (T_m, 45–74 °C) and short sequences (~22 nt) differing by only 1–2 nt found in currently known mature microRNAs. The issue of sequence similarity requires that some array platforms truncate the probes to obtain equivalent T_m between the microRNAs represented on an array. As an alternative, LNA-based arrays add modified nucleotides that increase the thermostability of the RNA/DNA duplexes.

Depending on the number of nucleotides modified, the T_m can be altered by 2–10 °C allowing for a more uniform hybridization temperature across the array. The alternative TaqMan and other qRT-PCR-based methods also have to account for the mature microRNA being only slightly longer than the primer sequence used for amplification. The TaqMan system uses a probe to increase specificity, whereas the Roche/Exiqon method uses shorter LNA-based primer sequences while allowing products to be cloned and verified if needed. Finally, in addition to the issue of similarity, there is the added complication of cross-hybridizing with precursor microRNA species. In the case of arrays, this can be minimized through additional steps to enrich for mature microRNAs while excluding larger pri- and pre-microRNAs.

Recent studies have examined array platforms and compared the results with qRT-PCR and/or NGS and they revealed good intra-platform reproducibility for all technologies [25,26]. However, when differential expression of a subset of microRNAs included in all platforms was examined there was little or no concordance [25]. That there are several potential explanations for these discrepancies (bias due to hybridization or amplification, for example) is perhaps illustrative of the relatively novel nature of the technology and the inherent technical difficulties.

1.3. Mode of action, mRNA target recognition, and outcome

The first microRNA families examined in detail were reported to interact with the 3′UTRs of the mRNAs they regulated [27,28]. In a manner analogous to the development of algorithms for pri-microRNA hairpin identification, algorithms have been developed to identify potential mRNA targets of microRNA interaction [27,29]. These algorithms had to account for the fact that the majority of animal microRNAs (unlike plant microRNAs) lack perfect base-pairing to their mRNA targets. In most cases, the regulation of an mRNA is thought to require an interaction between the 6–8 nt of the microRNA "seed region" and the target mRNA sequence, referred to as the "seed match." However, other factors such as the secondary structure of the target mRNA's 3′UTR or synergistic action involving other microRNAs may also be important. Consequently, bioinformatic analysis and prediction of microRNA binding sites may produce high numbers of false-positive predictions. Although various prediction algorithms are predicated upon similar principles, it is not uncommon to find poor agreement in the results obtained from different prediction algorithms (reviewed in Ref. [30]). Such prediction programs may ultimately be improved as a consequence of experimental investigation and a better understanding of microRNA–mRNA interactions. This will enable more accurate identification of multiple microRNA target interactions, critical to the advancement of the field.

Historically, validation of predicted microRNA–mRNA interaction has involved cell-based experimental techniques, including cotransfection of the microRNA and a reporter construct, which may not accurately reflect biological mechanisms. The validation of these individual putative microRNA–mRNA interactions is time consuming and usually involves luciferase reporter constructs containing the 3′UTR of interest for each target [31]. In many cases, a microRNA can have several hundred predicted mRNA targets. These reporter constructs, which aim to provide evidence that a particular microRNA can downregulate protein synthesis through interaction with the target mRNA, are not definitive. Indeed in some cases, artificially high levels of the transfected microRNA and the use of a fragment of 3′UTR potentially lacking secondary structure may call into question the validity of the results.

The majority of studies involving microRNA function have focused on the 3′UTRs of mRNAs. However, there is emerging evidence that microRNAs may also target coding regions and introns of an mRNA [32,33]. Another dogma for predicting microRNA binding sites, the importance of cross-species conservation, has also been challenged. Recently, several mouse microRNA binding sites have not been found to be conserved in other organisms [32].

Additional evidence suggesting the importance of non-3′UTR microRNA binding sites comes from a study identifying microRNA–mRNA interactions *in vivo*, which found that 25% of the mRNA binding sequences identified in the mouse brain were in coding regions of the mRNA [34]. This study used an alternative method to the widely used reporter construct assays for identification of microRNA–mRNA interactions. The technique is called high-throughput sequencing of RNAs isolated by cross-linking immunoprecipitation (HITS-CLIP) [34]. In this method, Argonaute (an important component of the RISC) is covalently cross-linked to mRNA and immunoprecipitation of these complexes is followed by high-throughput sequencing of the cross-linked RNA species. Bioinformatic analysis is then used to identify microRNAs as well as potential seed matches in mRNAs. While this method does not identify one-to-one relationships for microRNA–mRNA interactions, it is a significant improvement over existing prediction programs. It greatly reduces the number of false positives and restricts potential mRNA target sites to the 45–60 bases of the Argonaute interaction [34].

In cases where microRNAs mediate endonucleolytic cleavage of their target mRNAs, another experimental approach for identifying microRNA–mRNA interactions has been developed. Called Degradome-Seq or parallel analysis of RNA ends (PARE) [35], this technique utilizes high-throughput sequencing and bioinformatics tools to create and analyze libraries that contain 3′ cleavage products of mRNAs. These cleavage products are a result of Argonaute activity and are unique in that they have a ligation-competent mRNA end with a 5′-phosphate which is not found in other mRNA species. This difference is exploited to create specific libraries of cleavage products. The sequences of these cleavage products are then mapped back to the genome and used to identify matches to known or potential microRNAs [35]. Other methods that identify microRNA–mRNA interactions have also been described using variations of this experimental approach [36–39]. As an alternative strategy to "Argonaute immunoprecipitation," labeled synthetic microRNAs have also been used in pull-down assays to elucidate the mRNA targets of each microRNA [40–42]. Finally, in the tandem affinity purification of microRNA target mRNAs (TAP-Tar) method [43], a two-step procedure is used to isolate microRNA–mRNA complexes from cells expressing a flag-tag version of Argonaute that have been transfected with biotinylated microRNA. The microRNA–mRNA complexes are pulled down with anti-FLAG antibodies and then purified on streptavidin beads. The authors claim that this method reduces the background often seen with single-step pull-down methods.

Recently, an integrated database was developed to facilitate the annotation and analysis of microRNA–mRNA targets that are being identified by some of these techniques. Called starBase (sRNA target Base), this database will document and integrate HITS-CLIP and Degradome-Seq data from several organisms, including humans [44]. At the time of writing, the methods listed above are relatively novel, and rigorous technological evaluation is still required to determine any drawbacks and to identify the most appropriate use for each technique.

2. MICRORNAS IN HUMAN DISEASE

Given the critical role microRNAs are thought to play in the regulation of cell development, division, and survival, it is not surprising that their dysregulation has been associated with human diseases. Genome-wide microRNA and targeted expression profiling experiments have both identified microRNAs with potential involvement in disease. To date, most reports have focused on cancer, for example, leukemias [45], hepatocellular carcinomas [46], melanoma [47], and ovarian cancers [48] (also reviewed in Ref. [49]). The microRNA profiles generated in these studies have been reported to be able to identify a tumor's origin, classify its subtype, and predict survival or response to specific therapies [49,50]. The majority of these studies simply show associations between microRNA levels and diseases, but there are examples where the function of a microRNA has been elucidated and is reported to be involved in the development or progression of a disease. One such example is miR-95. Its upregulation has been associated with colorectal cancer, and it reportedly functions as an oncogene by increasing cell proliferation via targeting, and thus downregulating, expression of Sorting Nexin 1 [51].

In addition to oncology, recent reports have focused on other areas including diabetes [52], cardiovascular [53,54], and autoimmune diseases. Reports have, for example, suggested a role for miR-375 in type 2 diabetes [55–57], and its expression profile may ultimately serve as a biomarker of this disease [57]. Many studies have reported identifying microRNA profiles associated with various types of cardiac disease [58–61]. Cardiac-expressed microRNAs are sensitive to changes in clinical cardiac status, and studies indicate that these changes may be measurable in peripheral blood, making them potentially valuable biomarkers of heart disease (reviewed in Ref. [53]).

Emerging evidence also indicates a role for microRNAs in a diverse range of autoimmune disorders as well as in the maintenance of normal immune function [62]. Recent reviews have covered developments in lupus [63] and rheumatoid arthritis [64]. The level of miR-181a, which is reportedly important in the regulation of B cell and T cell function [62], is downregulated in pediatric systemic lupus erythematosus patients in comparison to control subjects [65]. Finally, the significance and therapeutic potential of disrupting microRNAs reported to regulate Toll-like receptors (TLRs) was also recently reviewed [66].

One advantage of investigating microRNA profiles as biomarkers or for diagnostic applications is their reported stability in both fresh and archived serum and plasma [67]. The use of microRNA profiles in the field of diagnostics is also arguably preferable to the use of mRNAs due to this stability, especially in samples that are difficult to process. For example, a study has reported that microRNA profiles were superior to mRNA profiles when comparing formalin-fixed, paraffin-embedded (FFPE) samples to fresh malignant melanoma samples [68]. In another analysis, classification of poorly differentiated tumors by the comparatively less complex microRNA profile was reported to be more accurate than classification using mRNA profiles (generated using standard mRNA microarray techniques) when applied to the same samples [69]. Another report claimed that tumor-derived microRNAs in serum or plasma are very stable and may be used as biomarkers in blood-based detection tests for cancer [70]. Indeed, the potential for using circulating microRNAs in cancer detection has resulted in a commercial CLIA (Clinical Laboratory Improvement Amendments) laboratory test offered by Exiqon/Oncotech. This test is reported to be capable of classifying stage II colon cancer patients who may be at significantly higher risk for recurrence and for whom adjuvant chemotherapy may be warranted. This feature of circulating microRNAs and their potential applications is an area of current intense investigation and is further reviewed in Ref. [67].

3. MICRORNAS AS POTENTIAL THERAPEUTICS

Because of the strong association of microRNAs with cancer progression and prognosis as well as the regulation of cell proliferation, they have been investigated as potential therapeutics in cancer treatment (reviewed in Ref. [71]). However, from a therapeutic intervention point of view, there is a need to first understand the mechanism by which these microRNAs could be contributing to the disease etiology. For example, the upregulation of microRNAs could result in inhibition of a tumor suppressor gene or a microRNA itself that functions as a tumor suppressor could be downregulated. This gain or loss of microRNA expression could be attributed to gene amplification, transcriptional deregulation, or disruption of epigenetic mechanisms. Any intervention strategy would then have to be tailored to the effect of the microRNA, that is, inhibition of tumor promoters or enhancement of tumor suppressors. Several studies have attempted to restore the expression of downregulated microRNAs. For example, stable expression of miR-1258 in brain metastatic breast cancer (BMBC) cells inhibited the *in vitro* activity of heparanase, a prometastatic enzyme that is overexpressed in BMBC cells and is associated with a highly aggressive cancer phenotype [72]. In another example, expression of miR-16 was reported to be significantly reduced in most prostate tumors compared to normal prostate tissue, and this reduced expression was associated with rapid tumor growth [73]. Systemic delivery of miR-16 using atelocollagen (a highly purified low immunogenicity form of collagen) as a delivery vehicle resulted in restoration of miR-16 expression in a mouse model of bone-metastatic human prostate cancer [73]. Finally, since most cancers reportedly show a decrease in global microRNA expression, a recent study used the fluoroquinolone antibacterial compound enoxacin, **1**, to enhance microRNA biogenesis. The restoration of global microRNA expression was shown to inhibit the growth of several cancer cell types and is supposedly a consequence of enoxacin binding to the TARBP (TAR RNA binding protein 2) [74].

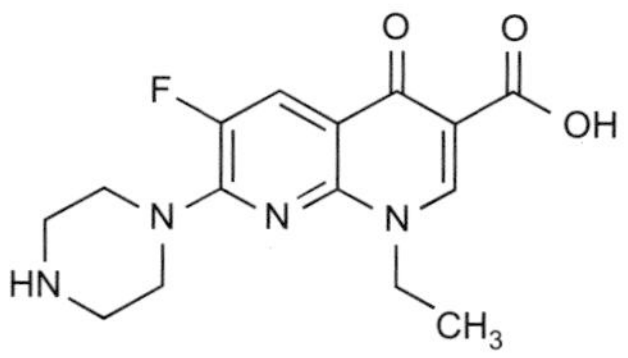

1

Approaches for disrupting the function of aberrantly upregulated microRNAs involve the use of antisense oligonucleotides that are complementary in sequence to the mature microRNAs they are intended to target and inhibit [75,76]. Called antagomirs, these single-stranded oligonucleotides typically incorporate modified nucleosides such as 2′-methoxyribonucleosides or LNAs, **2** and **3**, respectively (below). These chemical modifications increase the thermal stability of the antagomir–microRNA duplex. They also help protect the antagomir from nuclease degradation. Since a single microRNA can potentially have multiple target mRNAs, an antagomir against a single microRNA could potentially increase the expression of multiple genes and increase the activities of multiple associated pathways. Thus, in addition to the issue of antagomir delivery *in vivo*, the potential for producing undesired effects will need to be addressed if antagomirs are to be a safe and effective therapeutic.

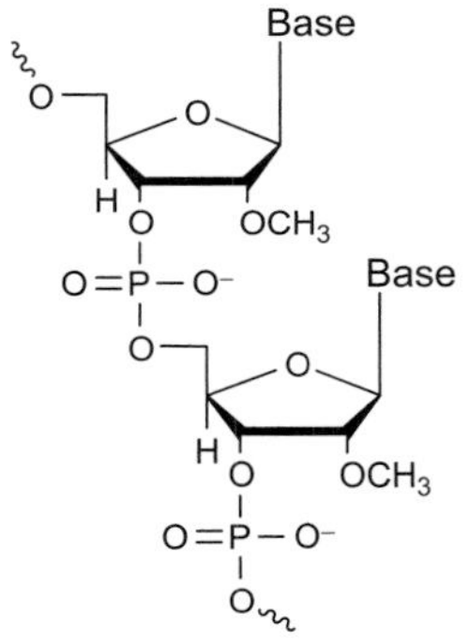

2 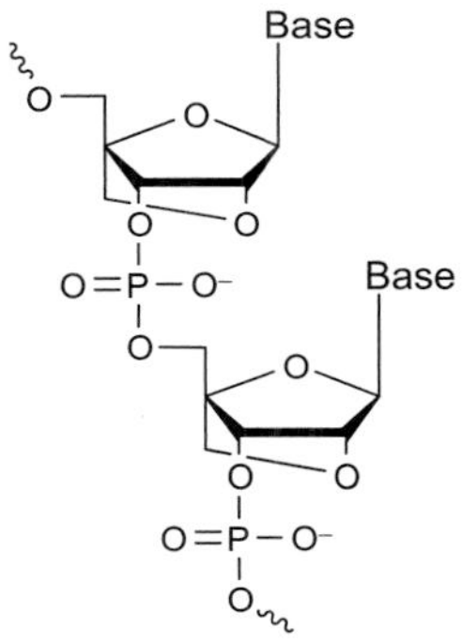 **3**

Antagomirs that are complementary to the entire 22 nt microRNA can potentially partially hybridize to multiple microRNAs. Incorporation of LNAs can help address this specificity issue. Since LNA oligonucleotides have reduced conformational flexibility and hybridize to target microRNAs with increased thermal stability [40], LNA-containing antagomirs as short as 15 nt in length are capable of inhibiting microRNA function. Recently, such a 15-nt LNA-containing antagomir targeted against miR-122 was shown to successfully suppress viremia in chimpanzees chronically infected with hepatitis C virus (HCV), demonstrating the feasibility of using this type of microRNA inhibitor as a therapeutic [77]. Since miR-122 is a host microRNA which appears to be essential for HCV infection, it may be an ideal therapeutic target in that viral escape through adaptive mutations is unlikely. The miR-122 LNA antagomir exhibited good pharmacokinetic (PK) and safety profiles in primates in this study [77]. In Phase I clinical studies, the antagomir was well tolerated and demonstrated dose-dependent pharmacology. It is currently being evaluated in HCV infected patients in Phase II studies.

Another approach reported to inhibit microRNA function in cell lines and transgenic organisms uses microRNA "sponges." These are transcripts expressed from strong promoters that contain multiple tandem binding sites specific for the seed region of the microRNA of interest [78,79]. Sponges can be designed with binding sites that are complementary to the seed region. Alternatively, these binding sites may contain mismatches in the middle of the seed match. Sponge designs with mismatches are thought to be more effective because they reportedly form a more stable interaction with the microRNA [79]. Testing microRNA sponges with 6–18 microRNA binding sites indicated that increasing the number of binding sites above 6 produces a marginal increase in activity [78]. While the approach has potential, it should be noted that because the inhibition is based on the seed match, this method may inhibit activity of whole families of microRNAs which share a common seed region. This could significantly increase any off target effects by disrupting other pathways and/or mRNAs.

4. CONCLUSION

MicroRNAs have a significant regulatory role in post-transcriptional gene expression. While some microRNAs are only required during development, other microRNAs appear to be important for normal cell homeostasis and as such, their dysregulation may contribute to human disease [80–83]. A deeper understanding of their ability to act as regulatory switches which control individual mRNAs and associated pathways is beginning to emerge. However, it should also be noted that this is a relatively novel and rapidly developing field and one for which the rules are still being defined.

The identification of aberrant expression of an individual or small group of microRNAs suggests the potential of microRNAs to serve as biomarkers. MicroRNAs are relatively stable, their expression may not fluctuate as much as that of mRNA molecules, and as they are fewer in number, they present a possibly less complex profile. However, current laboratory techniques for accurately quantitating and discriminating between microRNAs, particularly closely related family members, may need to be improved.

While the importance of microRNAs as a class of regulatory molecules is undisputed, the function of most individual microRNAs remains to be determined. Current techniques for determining microRNA function can be slow and labor intensive. In addition, methods for identifying their putative mRNA targets are fraught with many difficulties and challenges.

Fortunately, these areas are currently the subjects of intense scientific focus, and new techniques such as HITS-CLIP and PARE have been developed. Continued research in these areas will undoubtedly lead to a better understanding of the role of microRNA dysregulation in human disease and facilitate the identification of new gene targets and therapeutic modalities for their treatment.

REFERENCES

[1] R. C. Lee, R. L. Feinbaum and V. Ambros, *Cell*, 1993, **75**, 843.
[2] X. Cai, C. H. Hagedorn and B. R. Cullen, *RNA*, 2004, **10**, 1957.
[3] Y. Lee, C. Ahn, J. Han, H. Choi, J. Kim, J. Yim, J. Lee, P. Provost, O. Radmark, S. Kim and V. N. Kim, *Nature*, 2003, **425**, 415.
[4] R. Yi, Y. Qin, I. G. Macara and B. R. Cullen, *Genes Dev.*, 2003, **17**, 3011.
[5] S. Griffiths-Jones, *Nucleic Acids Res.*, 2004, **32**, 109.
[6] V. N. Kim, J. Han and M. C. Siomi, *Nat. Rev. Mol. Cell Biol.*, 2009, **10**, 126.
[7] A. Khvorova, A. Reynolds and S. D. Jayasena, *Cell*, 2003, **115**, 209.
[8] D. S. Schwarz, G. Hutvagner, T. Du, Z. Xu, N. Aronin and P. D. Zamore, *Cell*, 2003, **115**, 199.
[9] J. G. Doench and P. A. Sharp, *Genes Dev.*, 2004, **18**, 504.
[10] P. H. Olsen and V. Ambros, *Dev. Biol.*, 1999, **216**, 671.
[11] S. M. Hammond, E. Bernstein, D. Beach and G. J. Hannon, *Nature*, 2000, **404**, 293.
[12] H. Guo, N. T. Ingolia, J. S. Weissman and D. P. Bartel, *Nature*, 2010, **466**, 835.
[13] V. Ambros, *Cell*, 2001, **107**, 823.
[14] M. Lagos-Quintana, R. Rauhut, W. Lendeckel and T. Tuschl, *Science*, 2001, **294**, 853.
[15] Q. Jiang, Y. Wang, Y. Hao, L. Juan, M. Teng, X. Zhang, M. Li, G. Wang and Y. Liu, *Nucleic Acids Res.*, 2009, **37**, 98.
[16] M. Jung, A. Schaefer, I. Steiner, C. Kempkensteffen, C. Stephan, A. Erbersdobler and K. Jung, *Clin. Chem.*, 2010, **56**, 998.
[17] Y. Xi, G. Nakajima, E. Gavin, C. G. Morris, K. Kudo, K. Hayashi and J. Ju, *RNA*, 2007, **13**, 1668.
[18] R. D. Morin, M. D. O'Connor, M. Griffith, F. Kuchenbauer, A. Delaney, A. L. Prabhu, Y. Zhao, H. McDonald, T. Zeng, M. Hirst, C. J. Eaves and M. A. Marra, *Genome Res.*, 2008, **18**, 610.
[19] P. Landgraf, M. Rusu, R. Sheridan, A. Sewer, N. Iovino, A. Aravin, S. Pfeffer, A. Rice, A. O. Kamphorst, M. Landthaler, C. Lin, N. D. Socci, L. Hermida, V. Fulci, S. Chiaretti, R. Foa, J. Schliwka, U. Fuchs, A. Novosel, R. U. Muller, B. Schermer, U. Bissels, J. Inman, Q. Phan, M. Chien, D. B. Weir, R. Choksi, G. De Vita, D. Frezzetti, H. I. Trompeter, V. Hornung, G. Teng, G. Hartmann, M. Palkovits, R. Di Lauro, P. Wernet, G. Macino, C. E. Rogler, J. W. Nagle, J. Ju, F. N. Papavasiliou, T. Benzing, P. Lichter, W. Tam, M. J. Brownstein, A. Bosio, A. Borkhardt, J. J. Russo, C. Sander, M. Zavolan and T. Tuschl, *Cell*, 2007, **129**, 1401.
[20] G. Tian, X. Yin, H. Luo, X. Xu, L. Bolund and X. Zhang, *BMC Biotechnol.*, 2010, **10**, 64.
[21] C. G. Liu, G. A. Calin, S. Volinia and C. M. Croce, *Nat. Protoc.*, 2008, **3**, 563.
[22] M. Castoldi, S. Schmidt, V. Benes, M. W. Hentze and M. U. Muckenthaler, *Nat. Protoc.*, 2008, **3**, 321.
[23] J. Q. Yin, R. C. Zhao and K. V. Morris, *Trends Biotechnol.*, 2008, **26**, 70.
[24] P. T. Nelson, D. A. Baldwin, L. M. Scearce, J. C. Oberholtzer, J. W. Tobias and Z. Mourelatos, *Nat. Methods*, 2004, **1**, 155.

[25] A. Git, H. Dvinge, M. Salmon-Divon, M. Osborne, C. Kutter, J. Hadfield, P. Bertone and C. Caldas, *RNA*, 2010, **16**, 991.
[26] Y. Chen, J. A. Gelfond, L. M. McManus and P. K. Shireman, *BMC Genomics*, 2009, **10**, 407.
[27] B. P. Lewis, I. H. Shih, M. W. Jones-Rhoades, D. P. Bartel and C. B. Burge, *Cell*, 2003, **115**, 787.
[28] M. C. Vella, K. Reinert and F. J. Slack, *Chem. Biol.*, 2004, **11**, 1619.
[29] M. Kiriakidou, P. T. Nelson, A. Kouranov, P. Fitziev, C. Bouyioukos, Z. Mourelatos and A. Hatzigeorgiou, *Genes Dev.*, 2004, **18**, 1165.
[30] P. Alexiou, M. Maragkakis, G. L. Papadopoulos, M. Reczko and A. G. Hatzigeorgiou, *Bioinformatics*, 2009, **25**, 3049.
[31] J. L. Clancy, M. Nousch, D. T. Humphreys, B. J. Westman, T. H. Beilharz and T. Preiss, *Methods Enzymol.*, 2007, **431**, 83.
[32] Y. Tay, J. Zhang, A. M. Thomson, B. Lim and I. Rigoutsos, *Nature*, 2008, **455**, 1124.
[33] J. R. Lytle, T. A. Yario and J. A. Steitz, *Proc. Natl. Acad. Sci. USA*, 2007, **104**, 9667.
[34] S. W. Chi, J. B. Zang, A. Mele and R. B. Darnell, *Nature*, 2009, **460**, 479.
[35] M. A. German, M. Pillay, D. H. Jeong, A. Hetawal, S. Luo, P. Janardhanan, V. Kannan, L. A. Rymarquis, K. Nobuta, R. German, E. De Paoli, C. Lu, G. Schroth, B. C. Meyers and P. J. Green, *Nat. Biotechnol.*, 2008, **26**, 941.
[36] R. Jain, T. Devine, A. D. George, S. V. Chittur, T. E. Baroni, L. O. Penalva and S. A. Tenenbaum, *Methods Mol. Biol.*, 2011, **703**, 247.
[37] G. Malterer, L. Dolken and J. Haas, *RNA Biol.*, 2011 8.
[38] L. P. Tan, E. Seinen, G. Duns, D. de Jong, O. C. Sibon, S. Poppema, B. J. Kroesen, K. Kok and A. van den Berg, *Nucleic Acids Res.*, 2009, **37**, 137.
[39] F. V. Karginov, C. Conaco, Z. Xuan, B. H. Schmidt, J. S. Parker, G. Mandel and G. J. Hannon, *Proc. Natl. Acad. Sci. USA*, 2007, **104**, 19291.
[40] U. A. Orom and A. H. Lund, *Methods*, 2007, **43**, 162.
[41] W. Zheng, H. W. Zou, Y. G. Tan and W. S. Cai, *Mol. Biotechnol.*, 2011, **47**, 200.
[42] R. J. Hsu, H. J. Yang and H. J. Tsai, *Nucleic Acids Res.*, 2009, **37**, 77.
[43] N. Nonne, M. Ameyar-Zazoua, M. Souidi and A. Harel-Bellan, *Nucleic Acids Res.*, 2010, **38**, 20.
[44] J. H. Yang, J. H. Li, P. Shao, H. Zhou, Y. Q. Chen and L. H. Qu, *Nucleic Acids Res.*, 2011, **39**, 202.
[45] G. A. Calin, C. G. Liu, C. Sevignani, M. Ferracin, N. Felli, C. D. Dumitru, M. Shimizu, A. Cimmino, S. Zupo, M. Dono, M. L. Dell'Aquila, H. Alder, L. Rassenti, T. J. Kipps, F. Bullrich, M. Negrini and C. M. Croce, *Proc. Natl. Acad. Sci. USA*, 2004, **101**, 11755.
[46] Y. Murakami, T. Yasuda, K. Saigo, T. Urashima, H. Toyoda, T. Okanoue and K. Shimotohno, *Oncogene*, 2006, **25**, 2537.
[47] D. W. Mueller and A. K. Bosserhoff, *Br. J. Cancer*, 2009, **101**, 551.
[48] R. Eitan, M. Kushnir, G. Lithwick-Yanai, M. B. David, M. Hoshen, M. Glezerman, M. Hod, G. Sabah, S. Rosenwald and H. Levavi, *Gynecol. Oncol.*, 2009, **114**, 253.
[49] E. Chan, D. E. Prado and J. B. Weidhaas, *Trends Mol. Med.*, 2011, **17**, 235.
[50] C. A. Andorfer, B. M. Necela, E. A. Thompson and E. A. Perez, *Trends Mol. Med.*, 2011, **17**, 313.
[51] Z. Huang, S. Huang, Q. Wang, L. Liang, S. Ni, L. Wang, W. Sheng, X. He and X. Du, *Cancer Res.*, 2011, **71**, 2582.
[52] D. Ferland-McCollough, S. E. Ozanne, K. Siddle, A. E. Willis and M. Bushell, *Biochem. Soc. Trans.*, 2010, **38**, 1565.
[53] G. W. Dorn 2nd, *Curr. Cardiol. Rep.*, 2010, **12**, 209.
[54] J. Ai, R. Zhang, Y. Li, J. Pu, Y. Lu, J. Jiao, K. Li, B. Yu, Z. Li, R. Wang, L. Wang, Q. Li, N. Wang, H. Shan and B. Yang, *Biochem. Biophys. Res. Commun.*, 2010, **391**, 73.
[55] M. N. Poy, J. Hausser, M. Trajkovski, M. Braun, S. Collins, P. Rorsman, M. Zavolan and M. Stoffel, *Proc. Natl. Acad. Sci. USA*, 2009, **106**, 5813.

[56] Y. Li, X. Xu, Y. Liang, S. Liu, H. Xiao, F. Li, H. Cheng and Z. Fu, *Int. J. Clin. Exp. Pathol.*, 2010, **3**, 254.
[57] H. Zhao, J. Guan, H. M. Lee, Y. Sui, L. He, J. J. Siu, P. P. Tse, P. C. Tong, F. M. Lai and J. C. Chan, *Pancreas*, 2010, **39**, 843.
[58] S. Ikeda, S. W. Kong, J. Lu, E. Bisping, H. Zhang, P. D. Allen, T. R. Golub, B. Pieske and W. T. Pu, *Physiol. Genomics*, 2007, **31**, 367.
[59] C. Sucharov, M. R. Bristow and J. D. Port, *J. Mol. Cell. Cardiol.*, 2008, **45**, 185.
[60] S. V. Naga Prasad, Z. H. Duan, M. K. Gupta, V. S. Surampudi, S. Volinia, G. A. Calin, C. G. Liu, A. Kotwal, C. S. Moravec, R. C. Starling, D. M. Perez, S. Sen, Q. Wu, E. F. Plow, C. M. Croce and S. Karnik, *J. Biol. Chem.*, 2009, **284**, 27487.
[61] S. J. Matkovich, D. J. Van Booven, K. A. Youker, G. Torre-Amione, A. Diwan, W. H. Eschenbacher, L. E. Dorn, M. A. Watson, K. B. Margulies and G. W. Dorn 2nd, *Circulation*, 2009, **119**, 1263.
[62] E. Sonkoly, M. Stahle and A. Pivarcsi, *Semin. Cancer Biol.*, 2008, **18**, 131.
[63] R. J. Rigby and C. G. Vinuesa, *Curr. Opin. Rheumatol.*, 2008, **20**, 526.
[64] V. Furer, J. D. Greenberg, M. Attur, S. B. Abramson and M. H. Pillinger, *Clin. Immunol.*, 2010, **136**, 1.
[65] Y. A. Lashine, A. M. Seoudi, S. Salah and A. I. Abdelaziz, *Clin. Exp. Rheumatol.*, 2011, **10**, 41.
[66] E. J. Hennessy, A. E. Parker and L. A. O'Neill, *Nat. Rev. Drug Discov.*, 2010, **9**, 293.
[67] G. Reid, M. B. Kirschner and N. van Zandwijk, *Crit. Rev. Oncol. Hematol.*, Epub Dec, 2010, in press.
[68] A. Liu, M. T. Tetzlaff, P. Vanbelle, D. Elder, M. Feldman, J. W. Tobias, A. R. Sepulveda and X. Xu, *Int. J. Clin. Exp. Pathol.*, 2009, **2**, 519.
[69] J. Lu, G. Getz, E. A. Miska, E. Alvarez-Saavedra, J. Lamb, D. Peck, A. Sweet-Cordero, B. L. Ebert, R. H. Mak, A. A. Ferrando, J. R. Downing, T. Jacks, H. R. Horvitz and T. R. Golub, *Nature*, 2005, **435**, 834.
[70] P. S. Mitchell, R. K. Parkin, E. M. Kroh, B. R. Fritz, S. K. Wyman, E. L. Pogosova-Agadjanyan, A. Peterson, J. Noteboom, K. C. O'Briant, A. Allen, D. W. Lin, N. Urban, C. W. Drescher, B. S. Knudsen, D. L. Stirewalt, R. Gentleman, R. L. Vessella, P. S. Nelson, D. B. Martin and M. Tewari, *Proc. Natl. Acad. Sci. USA*, 2008, **105**, 10513.
[71] M. Angelica Cortez, C. Ivan, P. Zhou, X. Wu, M. Ivan and G. A. Calin, *Adv. Cancer Res.*, 2010, **108**, 113.
[72] L. Zhang, P. S. Sullivan, J. C. Goodman, P. H. Gunaratne and D. Marchetti, *Cancer Res.*, 2011, **71**, 645.
[73] F. Takeshita, L. Patrawala, M. Osaki, R. U. Takahashi, Y. Yamamoto, N. Kosaka, M. Kawamata, K. Kelnar, A. G. Bader, D. Brown and T. Ochiya, *Mol. Ther.*, 2010, **18**, 181.
[74] S. Melo, A. Villanueva, C. Moutinho, V. Davalos, R. Spizzo, C. Ivan, S. Rossi, F. Setien, O. Casanovas, L. Simo-Riudalbas, J. Carmona, J. Carrere, A. Vidal, A. Aytes, S. Puertas, S. Ropero, R. Kalluri, C. M. Croce, G. A. Calin and M. Esteller, *Proc. Natl. Acad. Sci. USA*, 2011, **108**, 4394.
[75] J. Krutzfeldt, N. Rajewsky, R. Braich, K. G. Rajeev, T. Tuschl, M. Manoharan and M. Stoffel, *Nature*, 2005, **438**, 685.
[76] M. Yang and J. Mattes, *Pharmacol. Ther.*, 2008, **117**, 94.
[77] R. E. Lanford, E. S. Hildebrandt-Eriksen, A. Petri, R. Persson, M. Lindow, M. E. Munk, S. Kauppinen and H. Orum, *Science*, 2010, **327**, 198.
[78] M. S. Ebert, J. R. Neilson and P. A. Sharp, *Nat. Methods*, 2007, **4**, 721.
[79] M. S. Ebert and P. A. Sharp, *RNA*, 2010, **16**, 2043.
[80] J. T. Mendell, *Cell Cycle*, 2005, **4**, 1179.
[81] J. S. Mattick and I. V. Makunin, *Hum. Mol. Genet.*, 2005, **14**(Spec. No. 1), 121.
[82] A. M. Krichevsky, K. S. King, C. P. Donahue, K. Khrapko and K. S. Kosik, *RNA*, 2003, **9**, 1274.
[83] I. Alvarez-Garcia and E. A. Miska, *Development*, 2005, **132**, 4653.

PART VII:
Topics in Drug Design and Discovery

Editor: Manoj C. Desai
Medicinal Chemistry, Gilead Sciences, Inc.
Foster City
California

CHAPTER 22

Induced Pluripotent Stem Cells as Human Disease Models

John T. Dimos, Irene Griswold-Prenner, Marica Grskovic, Stefan Irion, Charles Johnson and **Eugeni Vaisberg**

Contents

1. INTRODUCTION

Recent advances in reprogramming technologies allow conversion of adult somatic cells into induced pluripotent stem cells (iPSC), permitting generation of disease- and patient-specific stem cell lines. Like embryonic

iPierian, 951 Gateway Boulevard, South San Francisco, CA 94080, USA

Annual Reports in Medicinal Chemistry, Volume 46
ISSN: 0065-7743, DOI: 10.1016/B978-0-12-386009-5.00005-9

stem cells, iPSC potentially can be expanded without limits and differentiated into any somatic cell type. This technology potentially allows for any human cell type to be generated at a scale impossible to obtain from primary sources. Previously inaccessible human cell types (*e.g.*, neurons) can now be generated for investigating basic biological and pathological processes. Differentiated cells derived from patients' iPSC are being used to generate disease-specific models that can then be applied in drug discovery assays, drug development applications, toxicology screening and biomarker discovery. The iPSC technology provides human pharmacological and disease-relevant models to increase the biological and pathological context of these translational applications. We review how introducing increased human biological content and context (*i.e.*, patient-derived disease-relevant cells) may improve the probability of success in identifying and developing novel disease modifying drugs.

2. iPSC TECHNOLOGY

In a breakthrough advance several years ago, Shinya Yamanaka and colleagues at Kyoto University demonstrated that mouse somatic cells could be reprogrammed to an embryonic-like, pluripotent state by the enforced expression of a defined set of factors [1]. They tested ectopic expression in somatic cells of 24 genes active in embryonic stem cells and found that four factors, Oct4, Sox2, Klf4 and c-Myc together, converted a small percentage of cells to a pluripotent state. The result was cell colonies with morphology and growth characteristics of embryonic stem cells. These cells were named iPSC to reflect that pluripotency was induced in what had been differentiated somatic cells. The iPSC that were selected for the activation of essential pluripotency factors *Nanog* and *Oct4*, or based on morphology alone, were remarkably similar to embryonic stem cells yielding live chimeric mice, some with germ line contribution [2,3]. The most compelling evidence for iPSC having full developmental potential came from tetraploid complementation studies [4–6]. In these experiments, tetraploid blastocysts, which are incapable of progressing through embryonic development, are injected with iPSC to complement or rescue the defective early embryo-like structure. Fertile mice were generated entirely derived from the injected cells. These data show that iPSC can generate adult mice; thus, iPSC are functionally indistinguishable from embryonic stem cells.

The potential of iPSC was recognized immediately, and the technology was rapidly and successfully applied to human cells [7,8]. To date, iPSC derived from a number of patients with specific clinical conditions have been differentiated into disease-relevant cell types, creating "disease in a dish" models to facilitate drug development. The availability of nearly limitless amounts of disease-relevant cell types from patients will likely have huge benefits for drug discovery [9].

3. iPSC DERIVATION AND PRODUCTION

Deriving high-quality, comprehensively characterized iPSC in a scalable process is crucial for their use in translational applications. A typical process for iPSC production starts with acquiring a small skin biopsy from an appropriately selected patient. The skin biopsy is used to generate a fibroblast cell line, into which reprogramming factors are introduced (see below). After several weeks in culture, iPSC colonies emerge, are manually selected, purged from any background cells and a cell line is derived. The process is followed by a thorough characterization of iPSC lines, their expansion and storage in a biobank.

To date, human iPSC have been generated from a number of human tissues, including keratinocytes [10,11], hepatocytes [12], adipose-derived stem cells [13,14], neural stem cells [15], astrocytes [16], cord blood [17–19] and amniotic cells [20,21], illustrating the robustness of current reprogramming methods. Dermal fibroblasts remain the chief source of human iPSC and have demonstrated reliability and relatively high efficiency of reprogramming [8,22–26]. Due to its accessibility, peripheral blood is also an attractive source of donor cells [27,28]. It is likely that practical issues (*e.g.*, cell accessibility and limited discomfort of the patient) will determine the choice of starting somatic cell types.

Due to its reliability and relatively high efficiency, retrovirus-mediated transduction remains the most widely used method for delivering reprogramming factors. Genes coding for reprogramming factors delivered in this way are randomly and stably integrated into the genome and could affect the process of reprogramming, the differentiation of iPSC into mature cell types [29] and variability between different iPSC lines from a single patient. Newer methods have been designed to overcome these problems, including excisable vectors [30–32], nonintegrating vectors [33,34], transient plasmid transfections [34,35], direct protein transduction [36], RNA-based Sendai viruses [37–39] and mRNA-based transcription factor delivery [40]. Many considerations such as availability, efficiency, reliability, cost, time and convenience will determine which method will become widely used for a particular application.

The screening of small molecule collections to find compounds that either enhance reprogramming or replace the transcription factors Oct4, Sox2, Klf4 and c-Myc has shown some success [41,42]. Some compounds that have enhanced reprogramming act on chromatin remodeling. Examples are valproic acid (HDAC inhibitor), 5-azacytidine, *N*-phthalyltryptophan (DNA methyltransferase inhibitors) and BIX-01294 (G9a histone methyl transferase inhibitor) [43–45]. Additional reprogramming enhancers have been found and are believed to inhibit GSK3, TGF-β, Alk5, MEK or others targets [46–48]. Reprogramming transcription factors have been replaced with combinations of small molecules [49–51]. Adult human keratinocytes after transduction with only OCT4 were reprogrammed by treatment with the small molecules shown in Figure 1 [52]. Importantly, reprogramming of adult somatic cells solely with a cocktail of small molecules has not been accomplished.

4. DIFFERENTIATION—PROBLEMS AND PROMISE

To use iPSC in drug discovery and human disease modeling, mature differentiated cell type(s) of interest must be generated reliably and consistently [53,54]. The most successful differentiation approaches have been based on models that try to mimic normal embryonic development. A selected set of signaling pathways, including WNT [55], BMP/activin [56], FGF [57] and Notch [58,59], play a major role during this process and, their exact dosage and precise timing allow for the hundreds of individual cell types to be specified. For example, during the development of the human liver, BMP and FGF signaling from nascent cardiac mesoderm directs uncommitted endodermal progenitors into the hepatic lineage [60], while inhibition of the BMP signaling pathway directs these cells into the pancreatic lineage [61–63]. To generate populations of differentiated cell types, the natural process is mimicked as closely as possible under defined conditions. In addition, screening small molecule libraries led to the discovery of many compounds that influence lineage commitment [64–67]. While these approaches show promising results, the maturity of the cells generated has yet to be addressed. Often, differentiated cells will only show an adult phenotype when grown for extended periods of time *in vitro* [68] or *in vivo* [62], and their ability to engraft functionally into animal models is variable between cell types and based on desired maturation/integration endpoints [69,70]. For certain applications, an immature cell type might be sufficient in disease modeling, especially to investigate a developmental disease, or if the physiological phenotype manifests in immature cells. Developing methods to produce and culture differentiated cell types remains an active and productive line of basic investigation.

In principle, iPSC can give rise to all somatic cell types, but in practice, *in vitro* differentiation protocols to date have been developed for only a subset of specific cell types [23,54,71,72]. In many cases, the differentiation process is inefficient and produces cultures with mixed cell types. Developing efficient assays to evaluate such cultures may require either additional cell type sorting or selection, which is often limited by availability

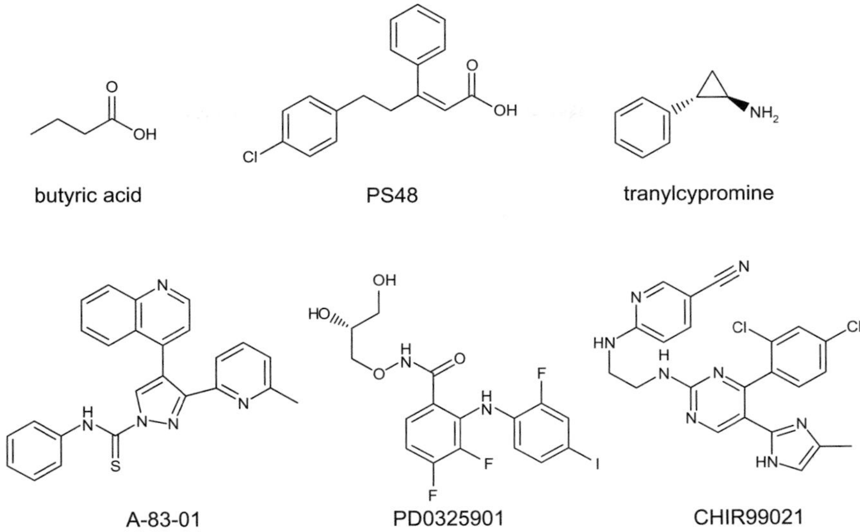

Figure 1 Small molecules used to enhance reprogramming of human keratinocytes transduced with OCT4 [52].

of selective surface markers or by introduction of reporter systems to facilitate cell sorting or selection. The extent to which iPSC differentiated cell cultures can be or should be homogeneous remains unclear and likely highly situation dependent.

5. LEADS FOR DRUG DISCOVERY AND DEVELOPMENT

The identification of a relevant disease phenotype—a molecular or functional difference in the patient-derived differentiated iPSC compared to cells from healthy control individuals—remains a challenge for using iPSC-based models in drug discovery. Identification of a disease-relevant "phenotype," or *in vitro* disease-correlate, provides a cellular model of the pathology in which disease mechanisms can be investigated, and in which agents with therapeutic potential can be identified and tested. These disease models can identify and validate targets for drug discovery and development.

iPSC have been derived from patients with a variety of conditions [22,23,73]. iPSC based disease phenotypes have been identified mainly for monogenic diseases, including spinal muscular atrophy (SMA) [25], fragile X syndrome [74], Hutchinson Gilford Progeria [75], familial dysautonomia (FD) [24], LEOPARD syndrome [26], Rett syndrome (RTT) [76,77], Long-QT syndrome [78] and multiple liver diseases (familial hypercholesterolemia, glycogen storage disease type 1a and alpha1-antitrypsin deficiency) [73]. More recently, disease phenotypes have been identified in neurons from iPSC from a familial Parkinson's disease patient [79] and in patients with schizophrenia [80], demonstrating that phenotypes in multifactorial diseases may be possible to identify as well. These studies show that disease-relevant cells can be generated from iPSC. Further, these cells can manifest disease phenotypes and are beginning to demonstrate benchmark compound responsiveness under limited testing conditions (see below).

Use of iPSC in drug discovery under development	
Phenotypic screen	Identify hits that alter disease phenotype in human patient affected cells
Target ID and target validation	RNAi screens to identify targets that alter disease phenotype in human patient cells
Lead optimization	Select between compounds for whole cell activity and compound tracking with disease phenotype modulation
Candidate selection	Select between compounds with various mechanisms of action across a panel of patients
Tox screening	Test in cardiomyocytes, hepatocytes and neurons for toxic effects
Biomarker discovery	Identify biomarkers in human patient affected cells that track with disease and/or compound efficacy
Mechanism-based safety	Tests for target activity regulation in nontarget cells
Trial cohort selection ("*in vitro* clinical trial")	Tests PDC in a panel of patient cells to choose potential patient responders
IND enabling studies	Supplement or replace animal models of efficacy
Personalized medicine	Companion diagnostic

An example of the application of iPSC in neurodegeneration studies is for SMA, a motor neuron degenerative disorder and the most common cause of infant death by a heritable disease. It is caused by a decrease in survival of motor neuron 1 (SMN1) protein due to deletions in the *SMN1* gene. Although SMN protein is expressed ubiquitously, motor neurons seem most vulnerable in patients, suggesting a specific or additional role of SMN in motor neurons. Lacking patient motor neurons, researchers have screened for compounds that elevate SMN levels in engineered cell lines and SMA patient fibroblasts. However, mechanisms that control SMN protein expression in fibroblasts may be different than in motor neurons. In the first step in disease model establishment, Ebert and colleagues created iPSC lines from SMA patients [25]. Target protein-rich structures called Gems were detected in iPSC from both healthy and SMA patients, but the number of Gems in SMA iPSC was lower than in healthy iPSC. Two compounds, valproic acid, an HDAC inhibitor, and tobramycin, an aminoglycoside antibiotic, that had both been shown to increase SMN Gems and protein levels in patient fibroblasts [81] were tested in SMA iPSC to determine if reprogramming changed their responsiveness to these compounds. Ebert *et al.* showed that both valproic acid and tobramycin increase Gem numbers and SMN levels in iPSC derived from an SMA patient. They further differentiated iPSC into motor neurons and documented that SMA-iPSC cultures have a decreased number of Gems and motor neurons, indicating that disease-relevant cellular phenotypes can be recapitulated in patient-derived motor neurons. It will be interesting to determine if valproic acid and/or tobramycin also increase SMN levels in their SMA patient-derived motor neurons or if altered mechanisms regulate SMN levels in motor neurons. This work provides additional validation that SMN deficiency is retained after reprogramming and differentiation, and that a disease-relevant phenotype can be identified in neurons from patient-derived iPSC.

Although seeing expected disease phenotypes in differentiated cells from patient-derived iPSC is encouraging, the next challenge will be discovering phenotypes in more complex or idiopathic diseases such as amyotrophic lateral sclerosis (ALS), Alzheimer's disease, or type 2 diabetes mellitus. Successful demonstration of cellular pathology *in vitro*, such as TDP-43 displacement from nucleus to cytoplasm in ALS, an increase in hyperphosphorylated microtubule-associated Tau protein in Alzheimer's disease [82], or an insulin resistance phenotype in type 2 diabetes [83], will provide novel and potentially revolutionary opportunities for drug discovery. The use of patient-derived cells is expected to be highly complementary to current drug discovery methodologies, especially where the introduction of high-throughput human pharmacology is needed [84].

6. STEM CELL MODULATORS

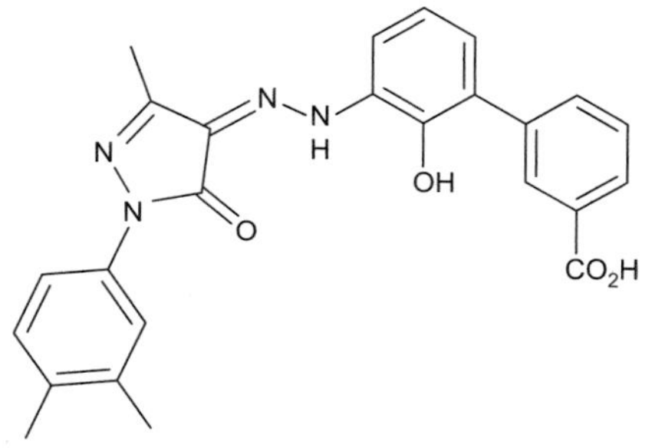

eltrombopag 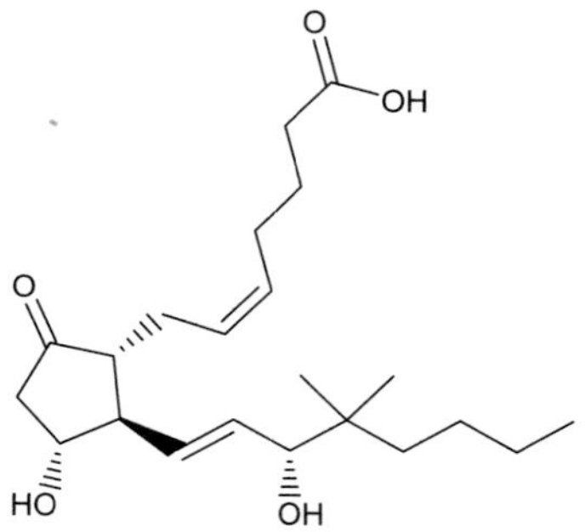 16,16-dimethylprostaglandin E2

Drugs acting on endogenous stem cells already have shown great success as therapies. For example, the protein erythropoietin (Epogen®) has blockbuster sales and enhances red blood cell production *via* stimulation of blood stem cell (colony-forming unit-erythroid [CFU-E]) conversion. The use of iPSC technology has several possibilities for aiding discovery of drugs that act on endogenous stem cell populations. Such agents have been called stem cell modulators and could have a therapeutic effect by increasing or decreasing stem cell proliferation or by promoting differentiation of endogenous stem cells to a particular mature cell type [85]. Therapeutically targeting endogenous stem cells *in vivo* is greatly enabled by having the ability to produce stem cells from patient-derived iPSC. In principle, endogenous stem cells could be produced from iPSC in the same way that fully differentiated cells are produced, thereby enabling the identification of small molecules for control of stem cell fate. The discovery of eltrombopag shows that manipulating stem cells *in vivo* can be achieved by small molecules. Eltrombopag is an orally bioavailable, nonpeptide agonist of the thrombopoietin receptor [86]. It increases platelet production by stimulating proliferation and differentiation of megakaryocytes from endogenous blood stem cells leading to a therapeutic effect. Eltrombopag activity is specific to human and chimpanzee, underscoring the importance of using human cells in drug discovery.

Another stem cell modulator, 16,16-dimethylprostaglandin E2 (Ft1050), has entered clinical trials for optimizing transplantation of human hematopoietic stem cells from umbilical cord blood [87]. Adult or umbilical cord hematopoietic stem cells (HSC) are routinely transplanted into patients following myeloablative chemotherapy. The number of HSC in cord blood is low, and adults require the blood from two cords for potentially successful reconstitution of the blood system. The clinical trial underway briefly treats cord blood before transplantation with 16,16-dimethylprostaglandin E2 to activate the endogenous blood stem cells. Since the efficiency of reconstitution is limited, improved homing, engraftment or proliferation could require less material in transplantation.

7. PREDICTIVE TOXICOLOGY WITH iPSC

The use of *in vitro* generated, patient-derived mature cells offers the opportunity to establish predictive models for cardio-, hepato- and neurotoxicology [88–90]. Using iPSC technology, cardiomyocytes, hepatocytes and neurons can be produced at scale from distinct patient populations and used in standard toxicology assays. Using such a renewable source should result in lower variability than preparations derived from human cadavers [91,92]. The increased use of human cells in discovery and development is expected to predict toxic effects and efficacy early in the discovery process. For example, human pooled primary liver microsomes are used for metabolic stability prediction [93]. Primary cells are difficult to obtain; however, iPSC-derived hepatic cells are a renewable source that will accelerate evaluation in human cells. Human cell toxicology is more predictive of clinical responses than transformed or primary animal cells [94]. An unlimited source of cells from various tissues with known genotypes will likely accelerate human cell-based assays. Now that iPSC-derived cardiomyocytes have become commercially available, the performance of these cells can be compared to current standards. In addition, the opportunity to obtain cardiomyocytes, hepatocytes, and neuronal cell types from the same individual offers a new angle on drug toxicity evaluation. The pharmaceutical industry must fully vet this technology before it can be reviewed by regulatory agencies as a complement or alternative to long-used and established *in vitro* toxicology systems [95].

8. *IN VITRO* CLINICAL TRIAL

Drug candidates are identified and optimized using nonclinical models, which generally encompass a small number of human or animal cell lines, animal efficacy and toxicology studies. This paradigm, for all its success, poses a significant liability: drug candidates remain isolated from the diversity and heterogeneity of human systems and populations until tested in a traditional clinical trial. In other words, potential drugs and human pharmacology do not collide until a clinical trial. This pipeline structure, regardless of thoughtful milestones, allows a large number of drug candidates to fail after many millions of dollars have been spent on their optimization, development and clinical testing. Most trials that fail do so for one of two main reasons: (1) lack of efficacy in the selected patient cohort or (2) adverse effects and safety concerns. A preclinical, disease-relevant human pharmacology model that identifies, optimizes, and selects drug candidates could mitigate these risks. The iPSC technology may enable "*in vitro* clinical trials."

An *in vitro* clinical trial would allow a development candidate to be tested across a broad and diverse cohort of patient samples for activity. Cell-based assays would assess a compound's ability to modify a disease phenotype *in vitro* in patient-derived disease-relevant cells. This *in vitro* clinical trial could test a single or small number of candidate compounds at multiple concentrations, including vehicle only, across a panel of cell lines designed to represent potential patient cohorts for an eventual clinical trial. This design reduces the number of samples needed as all testing is internally controlled and longitudinal, and has potential to predict clinical efficacy years before costly actual clinical trials are conducted. As importantly, this design would allow for refinement of the target patient population, and clinical trials could be better focused on patients with disease subtypes more likely to respond to treatment.

Patient populations differ not only with respect to disease subtype but also in general genetic background. Estimates from the international Haplotype Mapping Project suggest that there are about 10 million single nucleotide polymorphisms in the human population, in addition to copy number variations, deletions, insertions, inversions and epigenetic differences. Regulatory agency-approved nonclinical toxicology provides good prediction of possible clinical toxicity, but iPSC technology may provide an expanded toxicology evaluation using a human pharmacological model. In addition to testing a compound's activity on a broad sampling of patient-derived disease-relevant cells, cells important for drug metabolism could also be generated. For example, a carefully selected cohort of patients representing the major p450 isotypes could be used to generate hepatocytes for metabolic and toxicity profiling. As with testing compound activity for disease modification, this iPSC patient focused approach allows for testing on a broad range of patient cells for toxicological problems from drug–drug interactions years before they would normally be uncovered in a costly clinical trial.

9. PERSONALIZED MEDICINE: PATIENT PROFILING FOR OPTIMAL DRUG EFFICACY

Widespread industrialization of iPSC could enable this technology in a personalized medicine context. Specific differentiated cells could be used to screen for drug toxicity, efficacy and drug–drug interaction on that patient's own cells. If performed routinely, especially for drugs that show fatal adverse reactions, this *in vitro* prescreening might mitigate these risks. Additionally, the most effective or least toxic drug could be selected from a panel of available treatments prior to trial-and-error testing in patients. This scenario requires a dramatic reduction in the time it takes to generate such a screening panel. In addition, the cost of such an individualized approach might be overwhelming and a panel of iPSC lines that covers the majority of a given patient population might be a suitable alternative. Regardless of how this eventually might be implemented, iPSC technology potentially can allow for the discovery, development and selection of the right drug for the right patient.

10. CONCLUSION AND THE ROLE OF SMALL MOLECULE CHEMISTRY

Small molecule medicinal chemistry will likely be impacted and perhaps be changed by iPSC technology. The search is already underway to find small molecules to enhance or replace genetic-based factor delivery for cellular reprogramming. Chemically reprogrammed cells could be used with fewer concerns of genomic integration. These cells may be safe for human transplantation and useful for regenerative medicine applications. Development of small molecule compounds for reprogramming, stabilization of the pluripotent state and differentiation will benefit iPSC technology. For differentiation of iPSC, the stability, cost and specificity of small molecules have an advantage. Many cell types have yet to be produced by *in vitro* differentiation: small molecules could have a large impact in expanding access to more cell types. For traditional small molecule drug discovery, the ability to execute HTS in disease-specific human cells could be transformative. With sufficient development, iPSC-derived panels of patient cells should improve selection of efficacious compounds for clinical trials, as well as selecting those patients who would benefit from specific drugs. For toxicology screening, human cardiomyocytes and other cell types involved in safety assessment of new compounds are available today. In the near future, large panels of many types of cells from many types of patients could assist in more rapid toxicology assessment of new chemical entities. The promise of iPSC technology is to use human cell phenotypic assays to conduct more efficient and reliable small molecule medicinal chemistry due to increased biological and disease input.

ACKNOWLEDGMENT

The authors would like to thank Berta Strulovici and Michael Venuti for their editorial help and advice.

REFERENCES

[1] K. Takahashi and S. Yamanaka, *Cell*, 2006, **126**, 663.
[2] K. Okita, T. Ichisaka and S. Yamanaka, *Nature*, 2007, **448**, 313.
[3] A. Meissner, M. Wernig and R. Jaenisch, *Nat. Biotechnol.*, 2007, **25**, 1177.
[4] M. J. Boland, J. L. Hazen, K. L. Nazor, A. R. Rodriguez, W. Gifford, G. Martin, S. Kupriyanov and K. K. Baldwin, *Nature*, 2009, **461**, 91.
[5] L. Kang, J. Wang, Y. Zhang, Z. Kou and S. Gao, *Cell Stem Cell*, 2009, **5**, 135.
[6] X. Y. Zhao, W. Li, Z. Lv, L. Liu, M. Tong, T. Hai, J. Hao, C. L. Guo, Q. W. Ma, L. Wang, F. Zeng and Q. Zhou, *Nature*, 2009, **461**, 86.
[7] K. Takahashi, K. Tanabe, M. Ohnuki, M. Narita, T. Ichisaka, K. Tomoda and S. Yamanaka, *Cell*, 2007, **131**, 861.
[8] J. Yu, M. A. Vodyanik, K. Smuga-Otto, J. Antosiewicz-Bourget, J. L. Frane, S. Tian, J. Nie, G. A. Jonsdottir, V. Ruotti, R. Stewart, I. I. Slukvin and J. A. Thomson, *Science*, 2007, **318**, 1917.
[9] G. P. Nolan, *Nat. Chem. Biol.*, 2007, **3**, 187.
[10] T. Aasen, A. Raya, M. J. Barrero, E. Garreta, A. Consiglio, F. Gonzalez, R. Vassena, J. Bilic, V. Pekarik, G. Tiscornia, M. Edel, S. Boue and J. C. Izpisua Belmonte, *Nat. Biotechnol.*, 2008, **26**, 1276.
[11] B. W. Carey, S. Markoulaki, J. Hanna, K. Saha, Q. Gao, M. Mitalipova and R. Jaenisch, *Proc. Natl. Acad. Sci. USA*, 2009, **106**, 157.
[12] H. Liu, Z. Ye, Y. Kim, S. Sharkis and Y. Y. Jang, *Hepatology*, 2010, **51**, 1810.
[13] N. Sun, N. J. Panetta, D. M. Gupta, K. D. Wilson, A. Lee, F. Jia, S. Hu, A. M. Cherry, R. C. Robbins, M. T. Longaker and J. C. Wu, *Proc. Natl. Acad. Sci. USA*, 2009, **106**, 15720.
[14] T. Aoki, H. Ohnishi, Y. Oda, M. Tadokoro, M. Sasao, H. Kato, K. Hattori and H. Ohgushi, *Tissue Eng. A*, 2010, **16**, 2197.
[15] J. B. Kim, B. Greber, M. J. Arauzo-Bravo, J. Meyer, K. I. Park, H. Zaehres and H. R. Scholer, *Nature*, 2009, **461**, 649.
[16] S. Ruiz, K. Brennand, A. D. Panopoulos, A. Herrerias, F. H. Gage and J. C. Izpisua-Belmonte, *PLoS One*, 2010, **5**, e15526.
[17] A. Haase, R. Olmer, K. Schwanke, S. Wunderlich, S. Merkert, C. Hess, R. Zweigerdt, I. Gruh, J. Meyer, S. Wagner, L. S. Maier, D. W. Han, S. Glage, K. Miller, P. Fischer, H. R. Scholer and U. Martin, *Cell Stem Cell*, 2009, **5**, 434.
[18] A. Giorgetti, N. Montserrat, T. Aasen, F. Gonzalez, I. Rodriguez-Piza, R. Vassena, A. Raya, S. Boue, M. J. Barrero, B. A. Corbella, M. Torrabadella, A. Veiga and J. C. Izpisua Belmonte, *Cell Stem Cell*, 2009, **5**, 353.
[19] S. Eminli, A. Foudi, M. Stadtfeld, N. Maherali, T. Ahfeldt, G. Mostoslavsky, H. Hock and K. Hochedlinger, *Nat. Genet.*, 2009, **41**, 968.
[20] C. Li, J. Zhou, G. Shi, Y. Ma, Y. Yang, J. Gu, H. Yu, S. Jin, Z. Wei, F. Chen and Y. Jin, *Hum. Mol. Genet.*, 2009, **18**, 4340.
[21] H. X. Zhao, Y. Li, H. F. Jin, L. Xie, C. Liu, F. Jiang, Y. N. Luo, G. W. Yin, Y. Li, J. Wang, L. S. Li, Y. Q. Yao and X. H. Wang, *Differentiation*, 2010, **80**, 123.
[22] I. H. Park, N. Arora, H. Huo, N. Maherali, T. Ahfeldt, A. Shimamura, M. W. Lensch, C. Cowan, K. Hochedlinger and G. Q. Daley, *Cell*, 2008, **134**, 877.
[23] J. T. Dimos, K. T. Rodolfa, K. K. Niakan, L. M. Weisenthal, H. Mitsumoto, W. Chung, G. F. Croft, G. Saphier, R. Leibel, R. Goland, H. Wichterle, C. E. Henderson and K. Eggan, *Science*, 2008, **321**, 1218.
[24] G. Lee, E. P. Papapetrou, H. Kim, S. M. Chambers, M. J. Tomishima, C. A. Fasano, Y. M. Ganat, J. Menon, F. Shimizu, A. Viale, V. Tabar, M. Sadelain and L. Studer, *Nature*, 2009, **461**, 402.
[25] A. D. Ebert, J. Yu, F. F. Rose Jr., V. B. Mattis, C. L. Lorson, J. A. Thomson and C. N. Svendsen, *Nature*, 2009, **457**, 277.

[26] X. Carvajal-Vergara, A. Sevilla, S. L. D'Souza, Y. S. Ang, C. Schaniel, D. F. Lee, L. Yang, A. D. Kaplan, E. D. Adler, R. Rozov, Y. Ge, N. Cohen, L. J. Edelmann, B. Chang, A. Waghray, J. Su, S. Pardo, K. D. Lichtenbelt, M. Tartaglia, B. D. Gelb and I. R. Lemischka, *Nature*, 2010, **465**, 808.
[27] Y. H. Loh, S. Agarwal, I. H. Park, A. Urbach, H. Huo, G. C. Heffner, K. Kim, J. D. Miller, K. Ng and G. Q. Daley, *Blood*, 2009, **113**, 5476.
[28] Y. H. Loh, O. Hartung, H. Li, C. Guo, J. M. Sahalie, P. D. Manos, A. Urbach, G. C. Heffner, M. Grskovic, F. Vigneault, M. W. Lensch, I. H. Park, S. Agarwal, G. M. Church, J. J. Collins, S. Irion and G. Q. Daley, *Cell Stem Cell*, 2010, **7**, 15.
[29] C. A. Sommer, A. G. Sommer, T. A. Longmire, C. Christodoulou, D. D. Thomas, M. Gostissa, F. W. Alt, G. J. Murphy, D. N. Kotton and G. Mostoslavsky, *Stem Cells*, 2010, **28**, 64.
[30] K. Kaji, K. Norrby, A. Paca, M. Mileikovsky, P. Mohseni and K. Woltjen, *Nature*, 2009, **458**, 771.
[31] A. Lacoste, F. Berenshteyn and A. H. Brivanlou, *Cell Stem Cell*, 2009, **5**, 332.
[32] K. Woltjen, I. P. Michael, P. Mohseni, R. Desai, M. Mileikovsky, R. Hamalainen, R. Cowling, W. Wang, P. Liu, M. Gertsenstein, K. Kaji, H. K. Sung and A. Nagy, *Nature*, 2009, **458**, 766.
[33] M. Stadtfeld, M. Nagaya, J. Utikal, G. Weir and K. Hochedlinger, *Science*, 2008, **322**, 945.
[34] J. Yu, K. Hu, K. Smuga-Otto, S. Tian, R. Stewart, I. I. Slukvin and J. A. Thomson, *Science*, 2009, **324**, 797.
[35] F. Jia, K. D. Wilson, N. Sun, D. M. Gupta, M. Huang, Z. Li, N. J. Panetta, Z. Y. Chen, R. C. Robbins, M. A. Kay, M. T. Longaker and J. C. Wu, *Nat. Methods*, 2010, **7**, 197.
[36] D. Kim, C. H. Kim, J. I. Moon, Y. G. Chung, M. Y. Chang, B. S. Han, S. Ko, E. Yang, K. Y. Cha, R. Lanza and K. S. Kim, *Cell Stem Cell*, 2009, **4**, 472.
[37] N. Fusaki, H. Ban, A. Nishiyama, K. Saeki and M. Hasegawa, *Proc. Jpn. Acad. Ser. B Phys. Biol. Sci.*, 2009, **85**, 348.
[38] T. Seki, S. Yuasa, M. Oda, T. Egashira, K. Yae, D. Kusumoto, H. Nakata, S. Tohyama, H. Hashimoto, M. Kodaira, Y. Okada, H. Seimiya, N. Fusaki, M. Hasegawa and K. Fukuda, *Cell Stem Cell*, 2010, **7**, 11.
[39] K. Nishimura, M. Sano, M. Ohtaka, B. Furuta, Y. Umemura, Y. Nakajima, Y. Ikehara, T. Kobayashi, H. Segawa, S. Takayasu, H. Sato, K. Motomura, E. Uchida, T. Kanayasu-Toyoda, M. Asashima, H. Nakauchi, T. Yamaguchi and M. Nakanishi, *J. Biol. Chem.*, 2011, **286**, 4760.
[40] L. Warren, P. D. Manos, T. Ahfeldt, Y. H. Loh, H. Li, F. Lau, W. Ebina, P. K. Mandal, Z. D. Smith, A. Meissner, G. Q. Daley, A. S. Brack, J. J. Collins, C. Cowan, T. M. Schlaeger and D. J. Rossi, *Cell Stem Cell*, 2010, **7**, 618.
[41] M. Stadtfeld and K. Hochedlinger, *Genes Dev.*, 2010, **24**, 2239.
[42] A. Kochegarov, *Expert. Opin. Ther. Pat*, 2009, **19**, 275.
[43] T. S. Mikkelsen, J. Hanna, X. Zhang, M. Ku, M. Wernig, P. Schorderet, B. E. Bernstein, R. Jaenisch, E. S. Lander and A. Meissner, *Nature*, 2008, **454**, 49.
[44] W. Li, H. Zhou, R. Abujarour, S. Zhu, J. J. Young, T. Lin, E. Hao, H. R. Scholer, A. Hayek and S. Ding, *Stem Cells*, 2009, **27**, 2992.
[45] Y. Shi, J. T. Do, C. Desponts, H. S. Hahm, H. R. Scholer and S. Ding, *Cell Stem Cell*, 2008, **2**, 525.
[46] J. Yu, K. F. Chau, M. A. Vodyanik, J. Jiang and Y. Jiang, *PLoS One*, 2011, **6**, e17557.
[47] J. K. Ichida, J. Blanchard, K. Lam, E. Y. Son, J. E. Chung, D. Egli, K. M. Loh, A. C. Carter, F. P. Di Giorgio, K. Koszka, D. Huangfu, H. Akutsu, D. R. Liu, L. L. Rubin and K. Eggan, *Cell Stem Cell*, 2009, **5**, 491.
[48] T. Lin, R. Ambasudhan, X. Yuan, W. Li, S. Hilcove, R. Abujarour, X. Lin, H. S. Hahm, E. Hao, A. Hayek and S. Ding, *Nat. Methods*, 2009, **6**, 805.

[49] C. A. Lyssiotis, R. K. Foreman, J. Staerk, M. Garcia, D. Mathur, S. Markoulaki, J. Hanna, L. L. Lairson, B. D. Charette, L. C. Bouchez, M. Bollong, C. Kunick, A. Brinker, C. Y. Cho, P. G. Schultz and R. Jaenisch, *Proc. Natl. Acad. Sci. USA*, 2009, **106**, 8912.
[50] W. Li, W. Wei, S. Zhu, J. Zhu, Y. Shi, T. Lin, E. Hao, A. Hayek, H. Deng and S. Ding, *Cell Stem Cell*, 2009, **4**, 16.
[51] Y. Shi, C. Desponts, J. T. Do, H. S. Hahm, H. R. Scholer and S. Ding, *Cell Stem Cell*, 2008, **3**, 568.
[52] S. Zhu, W. Li, H. Zhou, W. Wei, R. Ambasudhan, T. Lin, J. Kim, K. Zhang and S. Ding, *Cell Stem Cell*, 2010, **7**, 651.
[53] S. Irion, M. C. Nostro, S. J. Kattman and G. M. Keller, *Cold Spring Harb. Symp. Quant. Biol.*, 2008, **73**, 101.
[54] G. Keller, *Genes Dev.*, 2005, **19**, 1129.
[55] C. Y. Logan and R. Nusse, *Annu. Rev. Cell Dev. Biol.*, 2004, **20**, 781.
[56] A. F. Schier, *Annu. Rev. Cell Dev. Biol.*, 2003, **19**, 589.
[57] L. Niswander and G. R. Martin, *Development*, 1992, **114**, 755.
[58] A. Aulehla and O. Pourquie, *Curr. Opin. Cell Biol.*, 2008, **20**, 632.
[59] J. Liu, C. Sato, M. Cerletti and A. Wagers, *Curr. Top. Dev. Biol.*, 2010, **92**, 367.
[60] V. Gouon-Evans, L. Boussemart, P. Gadue, D. Nierhoff, C. I. Koehler, A. Kubo, D. A. Shafritz and G. Keller, *Nat. Biotechnol.*, 2006, **24**, 1402.
[61] K. S. Zaret and M. Grompe, *Science*, 2008, **322**, 1490.
[62] E. Kroon, L. A. Martinson, K. Kadoya, A. G. Bang, O. G. Kelly, S. Eliazer, H. Young, M. Richardson, N. G. Smart, J. Cunningham, A. D. Agulnick, K. A. D'Amour, M. K. Carpenter and E. E. Baetge, *Nat. Biotechnol.*, 2008, **26**, 443.
[63] M. C. Nostro, F. Sarangi, S. Ogawa, A. Holtzinger, B. Corneo, X. Li, S. J. Micallef, I. H. Park, C. Basford, M. B. Wheeler, G. Q. Daley, A. G. Elefanty, E. G. Stanley and G. Keller, *Development*, 2011, **138**, 861.
[64] A. I. Lukaszewicz, M. K. McMillan and M. Kahn, *J. Med. Chem.*, 2010, **53**, 3439.
[65] Y. Xu, Y. Shi and S. Ding, *Nature*, 2008, **453**, 338.
[66] C. A. Lyssiotis, L. L. Lairson, A. E. Boitano, H. Wurdak, S. Zhu and P. G. Schultz, *Angew. Chem. Int. Ed. Engl.*, 2011, **50**, 200.
[67] A. J. Firestone and J. K. Chen, *ACS Chem. Biol.*, 2010, **5**, 15.
[68] C. Qiu, E. N. Olivier, M. Velho and E. E. Bouhassira, *Blood*, 2008, **111**, 2400.
[69] M. Rubart and L. J. Field, *Nat. Biotechnol.*, 2007, **25**, 993.
[70] D. James, H. S. Nam, M. Seandel, D. Nolan, T. Janovitz, M. Tomishima, L. Studer, G. Lee, D. Lyden, R. Benezra, N. Zaninovic, Z. Rosenwaks, S. Y. Rabbany and S. Rafii, *Nat. Biotechnol.*, 2010, **28**, 161.
[71] S. M. Chambers, C. A. Fasano, E. P. Papapetrou, M. Tomishima, M. Sadelain and L. Studer, *Nat. Biotechnol.*, 2009, **27**, 275.
[72] F. Soldner, D. Hockemeyer, C. Beard, Q. Gao, G. W. Bell, E. G. Cook, G. Hargus, A. Blak, O. Cooper, M. Mitalipova, O. Isacson and R. Jaenisch, *Cell*, 2009, **136**, 964.
[73] S. T. Rashid, S. Corbineau, N. Hannan, S. J. Marciniak, E. Miranda, G. Alexander, I. Huang-Doran, J. Griffin, L. Ahrlund-Richter, J. Skepper, R. Semple, A. Weber, D. A. Lomas and L. Vallier, *J. Clin. Invest.*, 2010, **120**, 3127.
[74] A. Urbach, O. Bar-Nur, G. Q. Daley and N. Benvenisty, *Cell Stem Cell*, 2010, **6**, 407.
[75] J. Zhang, Q. Lian, G. Zhu, F. Zhou, L. Sui, C. Tan, R. A. Mutalif, R. Navasankari, Y. Zhang, H. F. Tse, C. L. Stewart and A. Colman, *Cell Stem Cell*, 2011, **8**, 31.
[76] M. C. Marchetto, C. Carromeu, A. Acab, D. Yu, G. W. Yeo, Y. Mu, G. Chen, F. H. Gage and A. R. Muotri, *Cell*, 2010, **143**, 527.
[77] A. R. Muotri, M. C. Marchetto, N. G. Coufal, R. Oefner, G. Yeo, K. Nakashima and F. H. Gage, *Nature*, 2010, **468**, 443.

[78] A. Moretti, M. Bellin, A. Welling, C. B. Jung, J. T. Lam, L. Bott-Flugel, T. Dorn, A. Goedel, C. Hohnke, F. Hofmann, M. Seyfarth, D. Sinnecker, A. Schomig and K. L. Laugwitz, *N. Engl. J. Med.*, 2010, **363**, 1397.
[79] H. N. Nguyen, B. Byers, B. Cord, A. Shcheglovitov, J. Byrne, P. Gujar, K. Kee, B. Schule, R. E. Dolmetsch, W. Langston, T. D. Palmer and R. R. Pera, *Cell Stem Cell*, 2011, **8**, 267.
[80] K. J. Brennand, A. Simone, J. Jou, C. Gelboin-Burkhart, N. Tran, S. Sangar, Y. Li, Y. Mu, G. Chen, D. Yu, S. McCarthy, J. Sebat and F. H. Gage, *Nature*, 2011, **473**, 221.
[81] E. C. Wolstencroft, V. Mattis, A. A. Bajer, P. J. Young and C. L. Lorson, *Hum. Mol. Genet.*, 2005, **14**, 1199.
[82] I. Grundke-Iqbal, K. Iqbal, Y. C. Tung, M. Quinlan, H. M. Wisniewski and L. I. Binder, *Proc. Natl. Acad. Sci. USA*, 1986, **83**, 4913.
[83] K. F. Petersen, S. Dufour, D. B. Savage, S. Bilz, G. Solomon, S. Yonemitsu, G. W. Cline, D. Befroy, L. Zemany, B. B. Kahn, X. Papademetris, D. L. Rothman and G. I. Shulman, *Proc. Natl. Acad. Sci. USA*, 2007, **104**, 12587.
[84] S. M. Paul, D. S. Mytelka, C. T. Dunwiddie, C. C. Persinger, B. H. Munos, S. R. Lindborg and A. L. Schacht, *Nat. Rev. Drug Discov.*, 2010, **9**, 203.
[85] P. Grayson, J. Mendlein, S. Thies and J. Yingling, *Drug Discov. Today*, 2009, **6**, 141.
[86] C. L. Erickson-Miller, E. Delorme, S. S. Tian, C. B. Hopson, A. J. Landis, E. I. Valoret, T. S. Sellers, J. Rosen, S. G. Miller, J. I. Luengo, K. J. Duffy and J. M. Jenkins, *Stem Cells*, 2009, **27**, 424.
[87] T. E. North, W. Goessling, C. R. Walkley, C. Lengerke, K. R. Kopani, A. M. Lord, G. J. Weber, T. V. Bowman, I. H. Jang, T. Grosser, G. A. FitzGerald, G. Q. Daley, S. H. Orkin and L. I. Zon, *Nature*, 2007, **447**, 1007.
[88] S. J. Kattman, C. H. Koonce, B. J. Swanson and B. D. Anson, *J. Cardiovasc. Transl. Res.*, 2011, **4**, 66.
[89] R. Pal, M. K. Mamidi, D. A. Kumar and R. Bhonde, *J. Cell Physiol.*, 2011, **226**, 1583.
[90] S. Greenhough, C. N. Medine and D. C. Hay, *Toxicology*, 2010, **278**, 250.
[91] A. Guillouzo, *Environ. Health Perspect.*, 1998, **106**(Suppl. 2), 511.
[92] R. Gebhardt, J. G. Hengstler, D. Muller, R. Glockner, P. Buenning, B. Laube, E. Schmelzer, M. Ullrich, D. Utesch, N. Hewitt, M. Ringel, B. R. Hilz, A. Bader, A. Langsch, T. Koose, H. J. Burger, J. Maas and F. Oesch, *Drug Metab. Rev.*, 2003, **35**, 145.
[93] D. Dalvie, R. S. Obach, P. Kang, C. Prakash, C. M. Loi, S. Hurst, A. Nedderman, L. Goulet, E. Smith, H. Z. Bu and D. A. Smith, *Chem. Res. Toxicol.*, 2009, **22**, 357.
[94] A. Trevisan, A. Nicolli and F. Chiara, *Expert Opin. Drug Metab Toxicol.*, 2010, **6**, 1451.
[95] K. Vojnits and S. Bremer, *Toxicology*, 2010, **270**, 10.

CHAPTER 23

The Future of Drug Repositioning: Old Drugs, New Opportunities

Trinh L. Doan*, Michael Pollastri, Michael A. Walters*** and **Gunda I. Georg***

Contents

* College of Pharmacy, University of Minnesota, Minneapolis, MN, USA
** Department of Chemistry and Chemical Biology, Northeastern University, Boston, MA, USA

Annual Reports in Medicinal Chemistry, Volume 46
ISSN: 0065-7743, DOI: 10.1016/B978-0-12-386009-5.00004-7

1. INTRODUCTION

Drug repositioning is a promising field in drug discovery that identifies new therapeutic opportunities for existing drugs. In order to circumvent some of the most expensive drug discovery processes, companies pursue this strategy to increase their productivity (new drugs to market) by reducing the discovery and development timeline. This decreases the overall cost of bringing the drug to market because the safety and pharmacokinetic profiles of the repositioned candidates are already established.

The term "drug repositioning" has been used interchangeably with "drug repurposing" or "drug reprofiling." All these expressions are relatively synonymous for describing the process that seeks to discover new applications for an existing drug that were not previously referenced and not currently prescribed or investigated (Table 1). For consistency, this review will refer to all research that explores the multiple therapeutic applications of drugs as drug repositioning.

Several comprehensive reviews on the strategy and advantages of drug repositioning have been published [1–3,12–14]. This review will summarize novel methods being used to accelerate the discovery of old drugs that could potentially treat new indications, either *via* the established mechanism of action or by identification of new ones. Representative case studies of these approaches to therapeutics discovery will also be highlighted. Researchers have previously identified repositioned drugs by serendipity [1], novel insights, or target searching. The innovative strategies directed toward drug repositioning discussed in this review are phenotypic, high throughput, and *in silico* screening of commercial, public, and pharmaceutical compound libraries, the prospective mining of drug/activity databases, the exchange of compound information in collaborative networks, and data collection from the internet and social networks (Figure 2).

2. PERSPECTIVES OF DRUG REPOSITIONING

During the past two decades, the drug repositioning mindset has led to the discovery of several important and profitable drugs. Such an approach often allows pharmaceutical companies to extend a drug's patent life and to reduce the cost of drug development.

2.1. Previous case studies

Sildenafil (**1**; Figure 1), also known as Viagra®, is one of the most notable drugs that has been repositioned. It is a phosphodiesterase (PDE) type 5 inhibitor that was originally developed to treat angina, but it was

Table 1 Descriptions of various terms for drug repositioning

Term	Description
Drug repositioning	Finding new uses outside the scope of the original medical indication for existing drugs [1] or developing new indications for existing drugs or biologics [2]
Drug repurposing	Identifying, developing, and commercializing new uses for existing or abandoned drugs [3]
Drug reprofiling	Reducing the risks and costs associated with drug development with the advantage that the drug has already undergone preclinical and clinical testing [4]
Drug rediscovery	Investigating new uses for currently prescribed drugs [5]
Drug redirecting	Described as drug repositioning [1]
Drug reformulating	Finding ways to modify a formulation to allow a drug to enter a new market [6]
Therapeutic switching	Opening up new possibilities for old medicines that were not appreciated at the time of original discovery and can be made therapeutically different through new formulations [7]
Indication switching	Exploiting established drugs that have already been approved for treatment [8]
Indications discovery	Identifying new indications for clinical candidates that have been discontinued for their primary indications for reasons other than safety [9]; or research units that aim to find new uses for compounds that had failed in clinical trials or still in development [10,11]

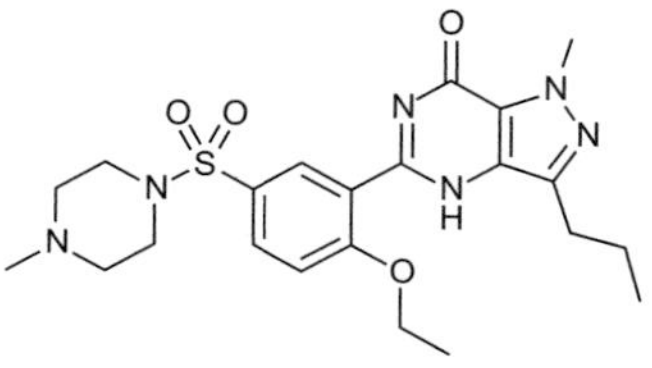

Figure 1 Repositioned drugs of relevance to this review.

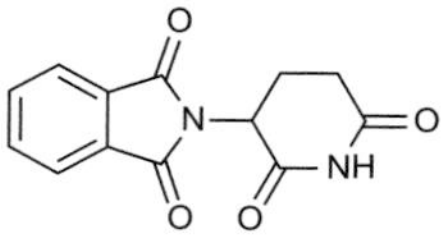

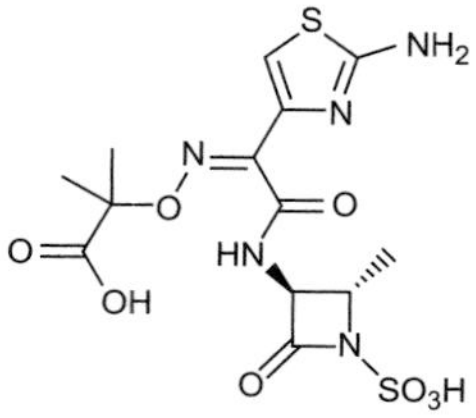

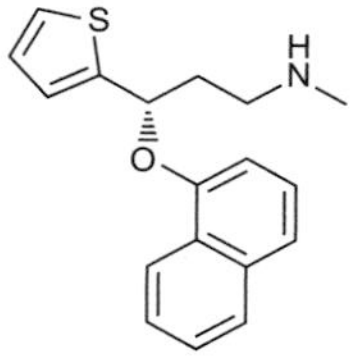

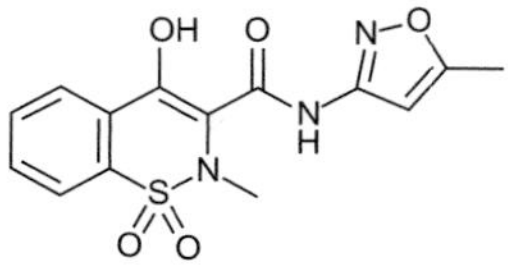

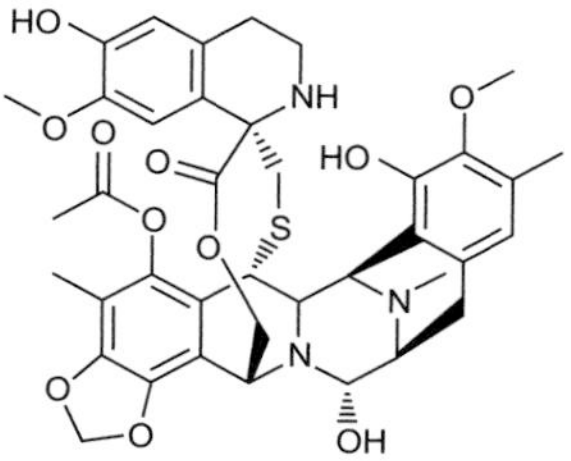

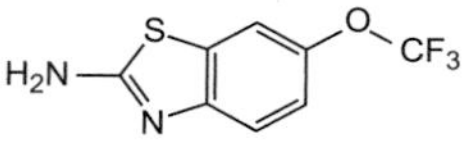

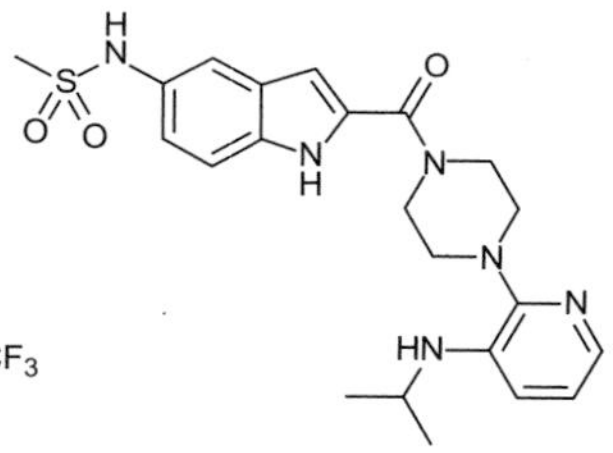

repositioned during clinical testing to treat male erectile dysfunction in order to fully take advantage of its therapeutically relevant side effect profile [15]. As a result of this serendipitous repositioning, sildenafil citrate has grossed over $15 billion in revenue since its release in 1998 [16].

Another notable example is thalidomide (**2**), a sedative that received much notoriety in the early 1960s due to its severe teratogenic effects [17]. Despite its infamy, thalidomide was repositioned to treat erythema nodosum leprosum and was approved for the treatment of this form of leprosy by the Food and Drug Administration (FDA) in 1998 (Thalomid®) with the caveat of its teratogenic effects [18]. Moreover, thalidomide was repositioned again to treat multiple myeloma and FDA approved in 2006 in combination with dexamethasone for newly diagnosed multiple myeloma patients [19].

2.2. Recent case studies

Some recent examples of drug repositioning include aztreonam (**3**) and duloxetine (**4**). Aztreonam is a monobactam antibiotic that was approved in 1986 for the treatment of Gram-negative bacterial infections *via* intravenous or intramuscular injection (Azactam®) [20]. This drug was marketed to treat bacterial infections and is safe for patients with penicillin allergies [21]. Previously, aztreonam was only approved for intravenous or intramuscular injection but not for inhalation because the solution for aztreonam contains an arginine buffer. Arginine is a substrate for nitric oxide production in many organs, and the generation of nitric oxide can lead to tissue injury [22]. Previous inhalation studies in patients with cystic fibrosis demonstrated that long-term use of inhaled arginine was associated with airway inflammation and deterioration of symptoms [23]. Aztreonam was later repositioned as Cayston® for inhalation by reformulating the solution to contain lysine rather than arginine [24]. In February 2010, the FDA approved aztreonam for inhalation *via* an ultrasonic nebulizer for the treatment of pulmonary *Pseudomonas aeruginosa* infections in cystic fibrosis patients [25].

Duloxetine is a serotonin and norepinephrine reuptake inhibitor which is marketed for major depressive disorder (Cymbalta®) in the United States and for stress urinary incontinence (Yentreve®) in Europe [1]. Trial studies revealed that duloxetine reduced painful physical symptoms in depressed patients [26], which led to its FDA approval for the management of diabetic peripheral neuropathic pain (DPNP) in 2004 [27] and fibromyalgia in 2008 [28]. Due to its effectiveness at relieving pain in DPNP and fibromyalgia patients, its effect on chronic lower back pain (CLBP) was evaluated in patients with CLBP. Double-blind studies on the analgesic effects of duloxetine showed significant improvement in patients with CLBP with a maintained analgesic effect for over 40 weeks [29,30]. In November 2010, the FDA approved the use of duloxetine to treat chronic musculoskeletal pain, including discomfort from osteoarthritis and CLBP [31]. Duloxetine has earned over $9 billion in revenue since its release in 2004 for all of its indications [32].

3. NEW STRATEGIES TOWARD DRUG REPOSITIONING

This section reviews the novel approaches toward drug repositioning being developed in both academic institutions and the pharmaceutical industry (Figure 2). These strategies utilize advances in modern technology to explore possible indications for drugs that have entered clinical trials or are clinically approved. The methods described herein could discover drugs that could be directly repositioned, find lead compounds

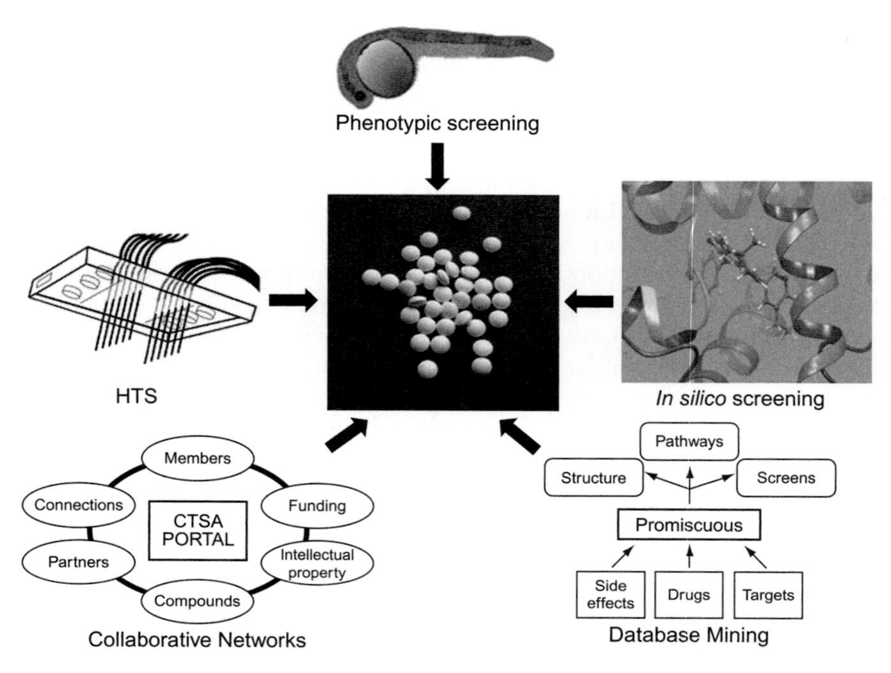

Figure 2 New strategies for drug repositioning.

that could suggest the screening of other drugs in their class, or identify drugs that would require minor structural changes for optimization against their new target. It should be made clear that the strategy of drug repositioning, while holding out the promises of being less costly and taking less time than other methods of drug discovery and development, is never a trivial exercise and may eventually prove to be as costly and time consuming as other drug development strategies. Moreover, repositioning approaches that yield only nonoptimal starting compounds will most likely offer only slight advantages over more standard routes to new drugs.

3.1. Phenotypic screening

The use of zebrafish (*Danio rerio*) as an animal model for developmental research allows quick and economical testing of the efficacy and safety of hundreds of compounds, often in parallel [33]. Zebrafish are small freshwater tropical vertebrates whose embryos develop *ex utero* within 2–3 days [34] and are ideal organisms for high-throughput phenotyping because their embryos are optically transparent. This allows for visual detection of functional and morphological changes without sacrificing the organism. Moreover, females are able to produce up to 300 eggs at a time, and these embryos are less than a millimeter in diameter. Due to their small size and large quantities, numerous embryos can be screened simultaneously [33]. A drug repositioning screen using larval zebrafish allows for rapid *in vivo* screening and is a cost-effective way of determining potential candidates for further development. Recent improvements in automated high-throughput chemical assays involving zebrafish have increased productivity and aided the data assessment of these high content screens [34].

3.2. HTS methods

The high-throughout screening of libraries of known drugs has become a popular method of discovering compounds for repositioning. Advances in technology have engendered innovations in automation, imaging software, and liquid-handling robots to support high-throughput screening (HTS). These advances allow researchers to test a multitude of compounds against targets of interest or in various assays. Additionally, new software has been developed to interpret, calculate, or generate data from these screens [14,35,36]. Flow cytometry is a sensitive and quantitative platform for the measurement of particle fluorescence [37] and has been employed in the drug repositioning efforts toward HIV combination therapy [38]. Microfluidic chip technology allows the miniaturization, integration, automation, and parallelization of chemical or biochemical assays on a silicon or glass chip [39,40]. This lab-on-a-chip technology is cost-effective, and the chips offer increased sensitivity and high throughput by implementing parallel sample processing and miniaturization of integrated on-chip components [41].

3.3. *In silico* screening

A prospective approach for drug repositioning using molecular modeling, cheminformatics and virtual screening (VS) can be effective and efficient because of its inherently low cost and rapid testing of multiple hypotheses. VS can be broadly grouped into two categories: ligand-based VS (based on the similarity of ligands) and structure- or target-based VS (based on the predicted interactions of ligands with enzymes or receptors). This latter category is generally considered to include computational docking of agents into experimentally determined target structures, or into computationally predicted structures. A recent review categorized the types of protein families pursued by VS approaches and compared of the successes of ligand- versus target-based VS [42]. While there is more literature precedence for structure-based VS efforts versus those that are ligand-based, the latter can generate more potent hits, particularly when both two- and three-dimensional approaches are implemented. Thus, developments in algorithms for VS-based drug discovery can support new target repositioning programs. Though VS of targets against *in silico* libraries of approved drugs has been successful for initiating drug repositioning campaigns [43], VS of much larger, target-agnostic compound libraries is far more common. Nonetheless, *in silico* screens of large libraries that contain approved agents can also ultimately identify established drugs.

3.4. Database mining

The field of systems biology offers unprecedented opportunities to mine data bases for drug repositioning [44]. Systems biology examines the relationships between biological targets and pathways to formulate models as frameworks to integrate and interpret multiple data sources for drug discovery and development [45]. The goal of this approach is to understand the physiology of the disease from the perspective of the whole organism through the use of computational and informatics tools, rather than only focusing on one or two biological targets [44].

For example, PROMISCUOUS is a publicly available, network-focused program that combines three different types of data: drugs, proteins, and side effects [46]. This resource provides data on protein–protein and protein–drug [47] interactions along with side effects [48] and structural information and enables the investigation of off-target effects that may be useful in drug repositioning. The network consists of 25,000 drugs, 12,000 proteins, 104,000 associated protein–protein interactions, and 21,500 drug–protein relationships acquired from public databases [46]. With PROMISCUOUS, researchers can identify potential candidates for drug repositioning by examining the side effect data and the different interactions available from the databases.

3.5. Collaborative networks

An important new direction in compound repositioning is the development of collaborative networks that bridge the gap between pharmaceutical companies, biotechnology firms, and academic institutions. One new effort in this area is the Clinical and Translational Science Award (CTSA) Pharmaceutical Assets Portal [49,50]. This collaborative network is constructed on the foundation of the CTSA consortium which is currently comprised of 55 research institutions and was launched in 2006 by the NIH [51]. The primary goal of the consortium is to speed the translation of medical research into treatment and strategies for patients. The Portal is headquartered at the University of California-Davis' Clinical and Translational Science Center and was established in 2008 to collect discontinued, late-stage compounds of pharmaceutical companies and allow academic and nonprofit researchers to investigate the repositioning of these compounds to treat disease [52]. The Portal is sponsored jointly by Pfizer and the National Center for Research Resources (NCRR). It contains many of the core functions that would be necessary to make the Portal successful: Foci-of-Expertise, Partnership for Cures (a nonprofit organization that funds "Rediscovery Research" project), University-Industry Demonstration Project, and the Center for World Health and Medicine (CWHM) [53]. The Portal plans to house, maintain, and distribute a repository of discontinued compounds (compounds that are no longer preclinical or clinical candidates). In early 2011, the Portal issued a request to the pharmaceutical industry to establish a consortium with the goal of collecting discontinued compounds for this central repository. In an example of how such a Portal might work, Pfizer signed a collaborative agreement with Washington University in St. Louis, MO in 2010, allowing researchers access to information about discontinued compounds from the Pfizer collection [11]. Another collaborative network is sponsored by Collaborative Drug Discovery (CDD) [54]. CDD aggregates and hosts public access data relevant to drug discovery that is deposited with them by leading research groups worldwide. These data could open the door to the discovery of new uses for old drugs, primarily in the area of neglected diseases.

4. CASE STUDIES OF DRUG REPOSITIONING STRATEGIES

The following case studies provide glimpses into current efforts toward drug repositioning employing the aforementioned strategies. Further research and development efforts could bring these already approved drugs or lead compounds to the public for use in new indications.

4.1. Phenotypic screening

Phenotypic screening with zebrafish larvae offers the advantage of allowing the drug response in whole organism to be monitored with the added advantage that zebrafish have been shown to have a high degree of conservation of drug responses with humans [55]. This approach was utilized to identify potential candidates to treat multiple sclerosis (MS) [56]. The aim was to identify a potential candidate for therapeutic remyelination enhancement *via* a phenotypic screen with zebrafish [4]. For this study, a library of 1170 compounds that contained marketed drugs and known bioactive substances was screened. Dorsally migrated *olig2*$^{+}$ cells in the zebrafish were monitored in a three-screen cascade: the first screen blindly counted the number of migrated *olig2*$^{+}$ cells in response to compound exposure; the second screen identified oligodendrocyte differentiation and myelination by active compounds from the first screen and compounds of interest from the literature; then, selected compounds were subject to a series of tertiary screens [4]. The researchers were able to identify several compounds from the first two screens that were known to have effects on oligodendrocytes or myelination, which validates their screening approach. The study also uncovered potential remyelination-promoting compounds, including the cyclooxygenase (COX-2) inhibitor isoxicam, **5**, that were not previously tested for this indication [57].

4.2. HTS methods

The National Institutes of Health's Chemical Genomics Center (NCGC) has created a collection of approved and investigational drugs for HTS purposes. This allows researchers to screen these drugs against new targets in an effort to reposition them for other indications, such as rare and neglected diseases [58]. With its quantitative HTS approach, the NCGC has screened their collection against more than 200 cell-based models of disease and characterized the pharmacology of each compound. The NCGC Pharmaceutical Collection (NPC) is publicly accessible *via* PubChem and serves as a resource for validating new models of disease and understanding the molecular basis of disease pathology and intervention [59]. In addition to repositioning the drugs, the NCGC will also screen the collection in their Tox21 system, which is a high-speed robotic screening system that evaluates potential compound toxicities [60]. Other public and private collections of clinically approved drugs are also available for HTS [11,14,35,61].

A quantitative HTS format was utilized to assess approximately 2800 clinically approved drugs from the NPC at 15 different concentrations in search of small-molecule inhibitors of NF-κB signaling [62]. Regulation of the NF-κB pathway is implicated in a myriad of important physiological processes, while dysregulation of this transcription factor is associated with autoimmune diseases and cancer [63]. A NF-κB mediated β-lactamase reporter gene assay was used to determine inhibition of the NF-κB signaling pathway, which is activated by tumor necrosis factor alpha (TNFα), interleukin-1, and bacterial lipopolysaccharides [64]. By adding TNFα as a positive control, researchers were able to perform fluorescence resonance energy transfer (FRET) analysis to determine β-lactamase expression. From the initial screen of 2816 NPC compounds, 55 compounds exhibited activity, and 19 compounds were further analyzed for their mechanism of inhibition, effect on apoptosis, and cytotoxicity [62]. Overall, many agents were identified that were previously approved for other clinical uses, including several anticancer drugs not previously known to inhibit the NF-κB pathway. The most potent compound for

NF-κB pathway inhibition was found to be ectinascidin 743 (**6**), which is approved for the treatment of ovarian cancer [65].

4.3. *In silico* screening

The simultaneous inhibition of pteridine reductase (PTR1) and dihydrofolate reductase (DHFR) in *Leishmania major* represents an innovative approach to the development of new antiparasitic drugs [66]. In a search for non-folate inhibitors of *L. major* PTR1, an initial virtual screen of ~350,000 compounds from the Available Chemicals Directory (ACD) database was performed against the published three-dimensional crystal structure of LmjPTR1 [67]. A comprehensive summary of the *in silico* screen is outlined in Figure 3 [68]. Filtering of the 21,394 docked structures was performed on the basis of the quality of the docked molecules' interactions with the LmDHFR binding site, followed by visual assessment of the specific ligand–protein interactions. This second assessment identified unique interactions between ligands and LmDHFR that were not present in the human DHFR (hDHFR). From this, a set of 53 molecules was selected for screening and compound **9** was found to be the most potent inhibitor of LmjPTR1.

From aminothiadiazole **9**, design of analogs was performed by iterative analysis of active compounds docked into the drug binding site, with cross-docking against hDHFR to eliminate those likely to bind to the human enzyme. Taken together, these docking experiments led to the extension of the thiadiazole ring of **9** into an adjacent lipophilic pocket, providing benzothiazole **10**, which displayed a threefold improvement in potency and retained selectivity over hDHFR. Further explorations of the core provided **7** (riluzole) [69], an established CNS agent that is currently utilized for amyotrophic lateral sclerosis. Importantly, though the authors

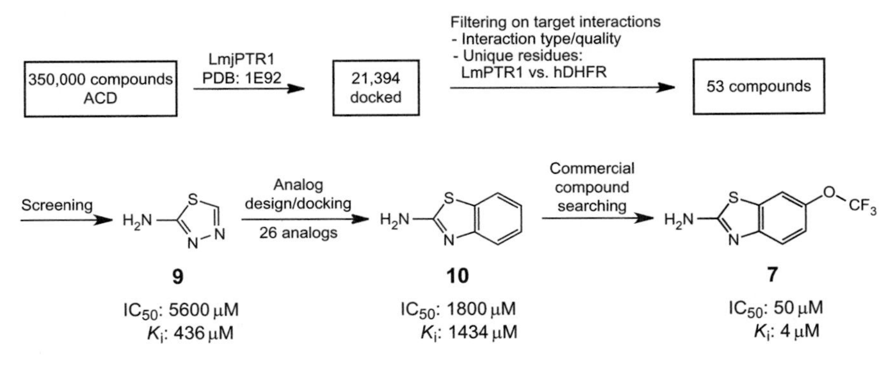

Figure 3 *In silico* identification of anti-leishmanial agents.

initially intended to identify novel LmDHFR inhibitors from a commercial compound set, they identified an established drug that has potential as an anti-leishmanial therapy. This result suggests this type of *in silico* screening approach could be more directly applied to drug repositioning by selection of agents from libraries of approved drugs (rather than from large, random libraries).

5. FUTURE DIRECTIONS OF DRUG REPOSITIONING

5.1. Predicting new targets for known drugs

In this age of target-based drug discovery, the notion of unselective drugs is commonly considered to be undesirable, as off-target effects can be a source of toxicity or drug side effects. However, "selectivity" is often an illusion; while it is conceivable to screen compounds for cross-reactivity against all *known* off-targets, there are always emerging targets and pathways that could not possibly be considered during a drug discovery program. Indeed, for a number of drugs, polypharmacology is desirable for the intended effects, as the inhibition of parallel or redundant pathways may be required [70]. In addition, it would be attractive to utilize cross-reactivity between targets to generate new therapeutic approaches.

A method for identifying likely molecular targets for drugs based on ligand similarity properties has been described [71]. This approach could detect new targets for established drugs that may be pursued for drug repositioning and could help reveal the source of adverse side effects. This was achieved by comparing drugs and investigational agents to a database of established ligand/target combinations. While it is tempting to focus on target sequence/structure similarity, this study showed that *ligand similarity* is in many cases better able to provide compelling predictions. Using a similarity ensemble approach [72], the authors were able to make a prospective prediction of pharmacology based only on ligand similarities, and to suggest new pharmacology of known agents. In a striking example, the HIV protease inhibitor delavirdine (**8**) was predicted and confirmed to bind to the histamine H_4 receptor (which does not bear structural similarity to HIV protease) with a K_i of 5.3 μM. While the potency difference between target and off-target is high in this case, the K_i against H_4 is within the steady-state plasma concentrations for the drug (15 μM), and thus the H_4 activity can help explain the known side-effect profile of the drug. While this is a demonstration of the impact that this cheminformatic approach can have on predicting toxicology and side-effect profiles, these methods can also provide a new approach towards drug repositioning by providing initial starting points for assessment of known agents against targets of putative importance to disease.

5.2. Pharmacovigilance 2.0

The future of drug repositioning may be driven by proactive analysis collections of side-effect data with new methods of data mining and new sources of side-effect information. The tracking of side effects, or adverse drug reactions (ADR), is commonly referred to as pharmacovigilance [73]. Pharmacovigilance has been accepted as the so-called clinical Phase IV, or postmarketing surveillance, of compound development and is undertaken to detect rare or long-term adverse events that were not detectable in Phase I–III trials. The potential of enhanced pharmacovigilance in drug repurposing has recently been highlighted by "type 2" pharmacovigilance [74] or pharmacovigilance 2.0 (PV2.0), which will use data from the internet and exploit the distributed knowledge or interests of large groups to the collection and analysis of potentially useful drug side effects (Figure 4).

Pharmacovigilance typically involves the passive monitoring of ADR by telephone systems like RADAR [75] (Research on Adverse Drug events And Reports) at Northwestern's Feinberg School of Medicine or the FDA's MedWatch [76]. While retrospective data mining of electronic health records (EHR) and other databases has been a staple of pharmacovigilance for some time [77,78], barriers to this process have recently been lowered by the use of ASTER, a spontaneous triggered electronic reporting system that collects data from physicians as they enter it into the EHR [79]. Rather than using electronic records, surveying physicians, or mining governmental databases, PV2.0 strategies will feature the active

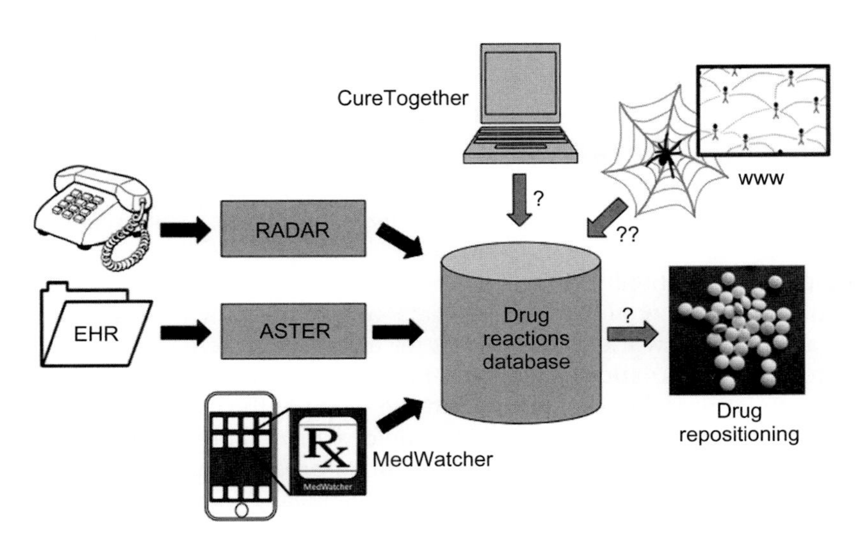

Figure 4 Potential sources of drug side effect data in pharmacovigilance and PV2.0.

mining of the web and social networking information to collect DRs that could potentially be used to reposition known drugs. Pharmacovigilance has already entered the internet age in both passive and active forms. In the passive mode, data on drug side effects are collected using data mining of information gathered from web search virtual robots. For example, passive data mining has been used to obtain patient comments on the effects and side effects of antidepressants [80]. The active mode involves the collection of data through smart phone applications, like MedWatcher (a program that allows the reporting and distribution of adverse events) [81], or affinity communities on the web, like CureTogether (a sharing website for patients to offer views on their medications) [82]. A mix of passive and active data gathering has already been employed to track patient views on the deployment of the H1N1 vaccine in Canada [83].

In work that is suggestive of how PV2.0 could be deployed to a broader audience, drug anticounterfeiting efforts linked by cell phone applications have been developed [84]. This process features scratch-off numeric codes on the pharmaceutical product that the purchaser sends as a text message to the analytics company to determine the product's authenticity. During this cell phone transaction, the analytics company can also survey the individual. It is reasonable to assume that similar feedback systems could be used to extract beneficial DRs from consumers that could be used in repositioning efforts. With the technology of text analytics and smart phone applications becoming more readily available, data mining of web sites and active engagement of social media networks could soon be more widely applied to the repositioning of drugs.

6. CONCLUSION

With economic and financial demands on the pharmaceutical industry to produce more drugs for its pipeline, methods that accelerate drug repositioning efforts are becoming increasingly important. This review presented a wide variety of innovative approaches currently being used to jumpstart repositioning efforts. These strategies include phenotypic, high throughput, and *in silico* screening, database mining, and the formation of collaborative networks. Case studies for the different screening strategies were provided to add insight into the use of modern technology in the field and approaches that go beyond the scope of the current drug repositioning efforts were addressed. The field of drug repositioning appears to be moving toward a broad, consortium approach. This trend is exemplified by groups like the CTSA Pharmaceutical Assets Portal and the development of more collaborative academic–industrial partnerships. In addition, there is the potential for drug repositioning to capitalize on the popularity of social media with the emergence of strategies like PV2.0. Presumably, consumers could eventually be directly involved in the discovery of new therapeutic uses for existing drugs.

REFERENCES

[1] T. T. Ashburn and K. B. Thor, *Nat. Rev. Drug Discov.*, 2004, **3**, 673.
[2] E. L. Tobinick, *Drug News Perspect.*, 2009, **22**, 119.
[3] D. W. Carley, *IDrugs*, 2005, **8**, 306.
[4] C. E. Buckley, A. Marguerie, A. G. Roach, P. Goldsmith, A. Fleming, W. K. Alderton and R.J. Franklin, *Neuropharmacology*, 2010, **59**, 149.
[5] H. J. Ting and F. Khasawneh, *J. Invest. Med.*, 2010, **58**, 208.
[6] D. Bradley, *Nat. Rev. Drug Discov.*, 2007, **6**, 423.
[7] D. Cavalla, *Nat. Rev. Drug Discov.*, 2009, **8**, 849.
[8] A. Duenas-Gonzalez, P. Garcia-Lopez, L. A. Herrera, J. L. Medina-Franco, A. Gonzalez-Fierro and M. Candelaria, *Mol. Cancer*, 2008, **7**, 82.
[9] P. F. Dimond, *Genet. Eng. Biotech. News*, 2010 30.
[10] A. Hopkins, J. Lanfear, C. Lipinski and L. Beeley, *Annu. Rep. Med. Chem.*, 2005, **40**, 339.
[11] L. M. Jarvis, *Chem. Eng. News*, 2010, **88**, 14.
[12] D. W. Carley, *IDrugs*, 2005, **8**, 310.
[13] S. H. Sleigh and C. L. Barton, *Pharm. Med.*, 2010, **24**, 151.
[14] C. R. Chong and D. J. Sullivan Jr., *Nature*, 2007, **448**, 645.
[15] I. Goldstein, T. F. Lue, H. Padma-Nathan, R. C. Rosen, W. D. Steers and P. A. Wicker, *N. Engl. J. Med.*, 1998, **338**, 1397.
[16] Pfizer Financial Reports, http://www.pfizer.com/investors/financial_reports/financial_reports.jsp.
[17] P. Richardson, T. Hideshima and K. Anderson, *Biomed. Pharmacother.*, 2002, **56**, 115.
[18] M. C. Okafor, *Pharmacotherapy*, 2003, **23**, 481.
[19] FDA Approves Thalomid (thalidomide) to Treat Multiple Myeloma, http://www.fda.gov/AboutFDA/CentersOffices/CDER/ucm095651.htm.
[20] R. N. Brogden and R. C. Heel, *Drugs*, 1986, **31**, 96.
[21] B. A. Cunha, *Urology*, 1993, **41**, 249.
[22] M. S. Mulligan, J. M. Hevel, M. A. Marletta and P. A. Ward, *Proc. Natl. Acad. Sci. USA*, 1991, **88**, 6338.
[23] H. J. Dietzsch, B. Gottschalk, K. Heyne, W. Leupoid and P. Wunderlich, *Pediatrics*, 1975, **55**, 96.
[24] G. L. Plosker, *Drugs*, 2010, **70**, 1843.
[25] Y. Waknine, Medscape Medical News, http://www.medscape.com/viewarticle/717553, 2010.
[26] D. J. Goldstein, Y. Lu, M. J. Detke, J. Hudson, S. Iyengar and M. A. Demitrack, *Psychosomatics*, 2004, **45**, 17.
[27] FDA Approves Drug for Neuropathic Pain Associated With Diabetes, http://www.fda.gov/NewsEvents/Newsroom/PressAnnouncements/2004/ucm108349.htm.
[28] Y. Waknine, *Medscape Med. News*, June 2008.
[29] V. Skljarevski, D. Desaiah, H. Liu-Seifert, Q. Zhang, A. S. Chappell, M. J. Detke, S. Iyengar, J. H. Atkinson and M. Backonja, *Spine*, 2010, **35**, E578.
[30] V. Skljarevski, S. Zhang, A. S. Chappell, D. J. Walker, I. Murray and M. Backonja, *Pain Med.*, 2010, **11**, 648.
[31] FDA Approves Cymbalta to Treat Chronic Musculoskeletal Pain, http://www.drugs.com/newdrugs/fda-clears-cymbalta-chronic-musculoskeletal-pain-2398.html.

[32] Eli Lilly Annual Reports, http://investor.lilly.com/annuals.cfm.
[33] L. I. Zon and R. T. Peterson, *Nat. Rev. Drug Discov.*, 2005, **4**, 35.
[34] M. Tsang, *Birth Defects Res. C Embryo Today Rev.*, 2010, **90**, 185.
[35] C. R. Chong, X. Chen, L. Shi, J. O. Liu and D. J. Sullivan Jr., *Nat. Chem. Biol.*, 2006, **2**, 415.
[36] J. C. Engel, K. K. Ang, S. Chen, M. R. Arkin, J. H. McKerrow and P. S. Doyle, *Antimicrob. Agents Chemother.*, 2010, **54**, 3326.
[37] B. S. Edwards, T. Oprea, E. R. Prossnitz and L. A. Sklar, *Curr. Opin. Chem. Biol.*, 2004, **8**, 392.
[38] C. L. Clouser, S. E. Patterson and L. M. Mansky, *J. Virol.*, 2010, **84**, 9301.
[39] S. Haeberle and R. Zengerle, *Lab Chip*, 2007, **7**, 1094.
[40] D. Mark, S. Haeberle, G. Roth, F. von Stetten and R. Zengerle, *Chem. Soc. Rev.*, 2010, **39**, 1153.
[41] D. Wlodkowic and Z. Darzynkiewicz, *World J. Clin. Oncol.*, 2010, **1**, 18.
[42] P. Ripphausen, B. Nisius, L. Peltason and J. Bajorath, *J. Med. Chem.*, 2010, **53**, 8461.
[43] W. H. Bisson, A. V. Cheltsov, N. Bruey-Sedano, B. Lin, J. Chen, N. Goldberger, L. T. May, A. Christopoulos, J. T. Dalton, P. M. Sexton, X. K. Zhang and R. Abagyan, *Proc. Natl. Acad. Sci. USA*, 2007, **104**, 11927.
[44] E. C. Butcher, E. L. Berg and E. J. Kunkel, *Nat. Biotechnol.*, 2004, **22**, 1253.
[45] P. Y. Lum, J. M. Derry and E. E. Schadt, *Pharmacogenomics*, 2009, **10**, 203.
[46] J. von Eichborn, M. S. Murgueitio, M. Dunkel, S. Koerner, P. E. Bourne and R. Preissner, *Nucleic Acids Res.*, 2010, **39**, D1060.
[47] A. F. Fliri, W. T. Loging and R. A. Volkmann, *J. Med. Chem.*, 2009, **52**, 8038.
[48] C. G. Wermuth, *Drug Discov. Today*, 2006, **11**, 160.
[49] CTSA Pharmaceutical Assets Portal, http://www.ctsapharmaportal.org.
[50] T. Gower, Proto, http://protomag.com/assets/drug-repositioning, 2009.
[51] CTSA Clinical Translational Science Awards: Translating Discoveries to Medical Practice, http://www.ctsaweb.org/.
[52] C. Schubert, *Nat. Med.*, 2010, **16**, 7.
[53] Center for World Health & Medicine at Saint Louis University, http://www.cwhm.org/.
[54] Collaborative Drug Discovery, http://www.collaborativedrug.com/.
[55] D. J. Milan, T. A. Peterson, J. N. Ruskin, R. T. Peterson and C. A. MacRae, *Circulation*, 2003, **107**, 1355.
[56] A. Compston and A. Coles, *Lancet*, 2008, **372**, 1502.
[57] M. Ouellet, J. P. Falgueyret and M. D. Percival, *Biochem. J.*, 2004, **377**, 675.
[58] NIH researchers create comprehensive collection of approved drugs to identify new therapies for rare and neglected diseases, http://www.genome.gov/27544241.
[59] The NCGC Pharmaceutical Collection, http://tripod.nih.gov/npc/.
[60] New robot system to test 10,000 chemicals for toxicity, http://www.genome.gov/27543708.
[61] The Johns Hopkins Clinical Compound Screening Initiative, www.jhccsi.org.
[62] S. C. Miller, R. Huang, S. Sakamuru, S. J. Shukla, M. S. Attene-Ramos, P. Shinn, D. Van Leer, W. Leister, C. P. Austin and M. Xia, *Biochem. Pharmacol.*, 2010, **79**, 1272.
[63] A. Kumar, Y. Takada, A. M. Boriek and B. B. Aggarwal, *J. Mol. Med.*, 2004, **82**, 434.
[64] M. Karin, Y. Yamamoto and Q. M. Wang, *Nat. Rev. Drug Discov.*, 2004, **3**, 17.
[65] P. Schoffski, H. Dumez, P. Wolter, C. Stefan, A. Wozniak, J. Jimeno and A. T. Van Oosterom, *Expert Opin. Pharmacother.*, 2008, **9**, 1609.
[66] B. Nare, J. Luba, L. W. Hardy and S. Beverley, *Parasitology*, 1997, **114**(Suppl.), S101.
[67] D. G. Gourley, A. W. Schuttelkopf, G. A. Leonard, J. Luba, L. W. Hardy, S. M. Beverley and W. N. Hunter, *Nat. Struct. Mol. Biol.*, 2001, **8**, 521.

[68] S. Ferrari, F. Morandi, D. Motiejunas, E. Nerini, S. Henrich, R. Luciani, A. Venturelli, S. Lazzari, S. Calo, S. Gupta, V. Hannaert, P. A. Michels, R. C. Wade and M. P. Costi, *J. Med. Chem.*, 2011, **54**, 211.
[69] P. Jimonet, F. Audiau, M. Barreau, J. C. Blanchard, A. Boireau, Y. Bour, M. A. Coleno, A. Doble, G. Doerflinger, C. D. Huu, M. H. Donat, J. M. Duchesne, P. Ganil, C. Gueremy, E. Honor, B. Just, R. Kerphirique, S. Gontier, P. Hubert, P. M. Laduron, J. Le Blevec, M. Meunier, J. M. Miquet, C. Nemecek and S. Mignani, et al., *J. Med. Chem.*, 1999, **42**, 2828.
[70] A. L. Hopkins, *Nat. Biotechnol.*, 2007, **25**, 1110.
[71] M. J. Keiser, V. Setola, J. J. Irwin, C. Laggner, A. I. Abbas, S. J. Hufeisen, N. H. Jensen, M. B. Kuijer, R. C. Matos, T. B. Tran, R. Whaley, R. A. Glennon, J. Hert, K. L. Thomas, D. D. Edwards, B. K. Shoichet and B. L. Roth, *Nature*, 2009, **462**, 175.
[72] A. Schuffenhauer, J. Zimmermann, R. Stoop, J.-J. van der Vyver, S. Lecchini and E. Jacoby, *J. Chem. Inf. Comput. Sci.*, 2002, **42**, 947.
[73] The Importance of Pharmacovigilance—Safety Monitoring of Medicinal Products, http://apps.who.int/medicinedocs/en/d/Js4893e/1.html.
[74] M. S. Boguski, K. D. Mandl and V. P. Sukhatme, *Science*, 2009, **324**, 1394.
[75] RADAR (Research on Adverse Drug events And Reports), http://www.cancer.northwestern.edu/research/research_programs/radar/index.cfm.
[76] MedWatch: The FDA Safety Information and Adverse Event Reporting Program, http://www.fda.gov/Safety/MedWatch/default.htm.
[77] J. Almenoff, J. M. Tonning, A. L. Gould, A. Szarfman, M. Hauben, R. Ouellet-Hellstrom, R. Ball, K. Hornbuckle, L. Walsh, C. Yee, S. T. Sacks, N. Yuen, V. Patadia, M. Blum, M. Johnston, C. Gerrits, H. Seifert and K. LaCroix, *Drug Saf.*, 2005, **28**, 981.
[78] M. Hauben and A. Bate, *Pharm. Data Min.*, 2010 341.
[79] X. Wang, G. Hripcsak, M. Markatou and C. Friedman, *J. Am. Med. Inf. Assoc.*, 2009, **16**, 328.
[80] C. Rizo, A. Deshpande, A. Ing and N. Seeman, *J. Affect. Disord.*, 2011, **130**, 290.
[81] New iPhone app tracks drug safety, http://www.boston.com/news/health/blog/2010/09/new_iphone_app.html.
[82] Find the best treatments, feel better faster, http://www.curetogether.com/.
[83] N. Seeman, A. Ing and C. Rizo, *Healthc. Q.*, 2010, **13**(Spec. No.), 8.
[84] Sproxil, Inc., http://sproxil.com/.

CHAPTER 24

Deuterium in Drug Discovery and Development

Scott L. Harbeson and **Roger D. Tung**

Contents

Concert Pharmaceuticals, Inc., 99 Hayden Avenue, Suite 500, Lexington, MA 02421, USA

Annual Reports in Medicinal Chemistry, Volume 46
ISSN: 0065-7743, DOI: 10.1016/B978-0-12-386009-5.00003-5

ABBREVIATIONS

E	free enzyme
EOS	oxygenating enzyme species with substrate bound
EOS_D	active oxygenating enzyme species for C–D bond oxidation
EOS_H	active oxygenating enzyme species for C–H bond oxidation
EP	enzyme–product complex
ES	enzyme–substrate complex
P	product
P_D	product derived from C–D bond oxidation
P_H	product derived from C–H bond oxidation
S	substrate
S_D	substrate with oxidation of C–D bond
S_H	substrate with oxidation of C–H bond

1. INTRODUCTION

Deuterated compounds have been widely studied in nonclinical settings and have seen broad application as metabolic or pharmacokinetic (PK) probes both *in vitro* and *in vivo* [1,2]. Depending upon a given compound's route of metabolism and the location of the deuterium, deuteration can be metabolically silent, enabling utility as a PK tracer, or it can alter the compound's metabolism allowing use as a mechanistic probe. It is difficult to predict *a priori* which effect deuterium may have on a drug's metabolism. In spite of the potential to alter a compound's metabolic fate, deuterium-containing compounds have rarely been clinically explored in the context of creating new therapeutic agents [3,4]. To date, no deuterated compound has advanced beyond Phase 2 clinical evaluation [5]. The incorporation of deuterium into pharmacologically active agents offers potential benefits such as improved exposure profiles and decreased production of toxic metabolites [6,7], which could yield improvements in efficacy, tolerability, or safety. As noted in a recent review [8], there has been a resurgence of interest in the application of deuterium in medicinal chemistry as evidenced by the emergence of several new companies largely or solely focused on this technology. This chapter will provide a brief review of the use of deuterium to alter the metabolic properties of compounds and will discuss past and current development of potential deuterium-containing drugs.

2. DEUTERIUM BACKGROUND

Deuterium is a naturally occurring, stable, nonradioactive isotope of hydrogen discovered in 1932 [9]. Hydrogen consists of one electron and one proton and has a mass of 1.008 atomic mass units (AMU), whereas deuterium also contains a neutron, which results in a mass of 2.014 AMU. Deuterium occurs at a natural abundance of approximately 1 part in 6400 or 0.015%, which allows large quantities of deuterium to be isolated as heavy water (D_2O) in very high isotopic purity [10]. D_2O can then serve as a direct or indirect source of deuterium for a wide range of chemical reagents and building blocks for preparing deuterated drugs.

2.1. Primary deuterium isotope effect

Due to the greater atomic mass of deuterium, a deuterium–carbon bond has a lower vibrational frequency and, therefore, a lower zero-point energy than a corresponding hydrogen–carbon bond [11]. In contrast, the frequencies associated with scission of C–H and the corresponding C–D bonds in the transition states are similar. Therefore, the lower zero-point energy translates to higher activation energy for C–D bond cleavage and a slower reaction rate (represented by rate constant k). This effect on rate is known as the primary deuterium isotope effect (DIE) and is expressed as k_H/k_D, the ratio of the reaction rate constants of C–H versus C–D bond cleavage with a theoretical limit of about 9 at 37°C in the absence of tunneling effects [12,13]. In principle, the DIE has the potential to affect the biological fate of many drugs that are metabolized by pathways involving hydrogen–carbon bond scission. In practice, the observed DIE, $(k_H/k_D)_{obs}$ for a metabolic reaction is often "masked", which means that it is smaller than k_H/k_D or, in some cases, entirely absent [14]. There are also literature examples of inverse DIEs in which $(k_H/k_D)_{obs} < 1$ [15].

A very large number of studies have appeared in the literature over the years reporting DIEs for a great many enzyme-catalyzed reactions. The most important enzymes in drug metabolism are the cytochrome P450s (CYPs), which are responsible for the Phase I metabolism of most drugs. The structures and mechanisms of CYPs have been reviewed in addition to the application of DIEs to the study of CYP-catalyzed reactions [12,16]. Other enzymes in drug metabolism include monoamine oxidase, alcohol dehydrogenase, and aldehyde oxidase. The emerging importance of aldehyde oxidase was recently reviewed, and the potential use of deuterium as a blocking group at aldehyde oxidase metabolic sites was noted [17].

The complexity of biological systems and the number of competing effects that can mask the DIE have made the application of deuterium to drug discovery highly unpredictable and challenging [13]. In order to observe a DIE, it is necessary that the C–H bond cleavage step is at least partially rate limiting. Enzyme-catalyzed reactions are complex, and catalysis is often dependent upon several partially rate-limiting steps; therefore, the observed DIEs, if any, for deuterium–carbon bond cleavage are often significantly less than the theoretical limit [18].

An irreversible step prior to the C–H bond cleavage in the catalytic sequence can also mask the DIE. As shown for CYP-catalyzed oxidation in Equation (1), an irreversible step, k_3—formation of EOS, the active oxygenating species—occurs prior to the isotopically sensitive C–H bond-cleaving step. In the absence of an alternate pathway, even though the DIE results in $k_{5D} < k_{5H}$, the concentration of EOS_D increases, which compensates for the DIE rate reduction as shown in Eq. (2). This can result in little or no change in the catalytic turnover of the hydrogen versus deuterated substrate [13]. Another mechanism for masking the DIE is "metabolic switching" [19], which is illustrated in Eq. (3). In this case, the site of metabolism switches from the site of deuteration (EOS_D) to another site in the molecule (EOS_H). Under these conditions, the rate of substrate turnover may not change, but there will be a decrease in P_D and an increase in P_H. CYP-catalyzed aromatic oxidation generally does not involve direct C–H bond cleavage. In this case, an arene-epoxide intermediate is formed, followed by an intramolecular hydride (deuteride) shift, known as the NIH shift [20], which results in a different DIE from what is observed for aliphatic hydroxylation. As a result of these complexities, the observed magnitude and even direction of the DIEs are unpredictable; DIEs appear to depend on the CYP enzyme involved, the compound being substituted, and the specific substitution pattern.

$$\mathrm{E} + \mathrm{S} \underset{k_2}{\overset{k_1}{\rightleftharpoons}} \mathrm{ES} \overset{k_3}{\rightarrow} \mathrm{EOS} \overset{k_5}{\rightarrow} \mathrm{EP} \overset{k_7}{\rightarrow} \mathrm{E} + \mathrm{P} \tag{1}$$

$$k_{5H} \cdot [EOS_H] \approx k_{5D} \cdot [EOS_D] \tag{2}$$

$$\mathrm{E + S} \rightleftharpoons \mathrm{ES} \rightleftharpoons \begin{cases} \mathrm{EOS_D} \longrightarrow \mathrm{E + P_D} \\ \mathrm{EOS_H} \longrightarrow \mathrm{E + P_H} \end{cases} \tag{3}$$

2.2. Sourcing and properties of deuterium

Deuterium can be sourced through a multistep distillation process that concentrates naturally occurring deuterium from bulk water to produce highly enriched D_2O. Because D_2O is used as a moderator in nuclear reactors, multiton quantities are commercially available [10]. Depending on the desired sites of deuteration, in some cases, deuterium from D_2O can be exchanged directly into finished drug compounds, or into reagents that are useful for synthesizing drug molecules [21,22]. Deuterium gas, available through electrolysis of D_2O, can also be used for incorporating deuterium into molecules. Catalytic deuteration of olefinic and acetylenic bonds provides a rapid route to incorporation of deuterium [23]. Metal catalysts (*e.g.*, Pd, Pt, and Rh) in the presence of deuterium gas can be used to directly exchange deuterium for hydrogen in functional group containing hydrocarbons [24]. A wide variety of deuterated reagents and synthetic building blocks are now commercially available.

When deuterium is incorporated into molecules in place of hydrogen, in most respects, the deuterated compound is very similar to the all-hydrogen compound. Since the electron clouds of its component atoms define the shape of a molecule, deuterated compounds have shapes and sizes that are essentially indistinguishable from their all-hydrogen analogs [25]. Small physical property changes have been detected in partially or fully deuterated compounds, including reduced hydrophobicity [26,27], decreased acidity of carboxylic acids and phenols [28], and increased basicity of amines [29]. These differences tend to be quite small, and the authors are aware of only one report: deuterated analogs of sildenafil (**1**), in which deuteration of a noncovalent drug appears to change its biochemical potency or selectivity to relevant pharmacological targets [30].

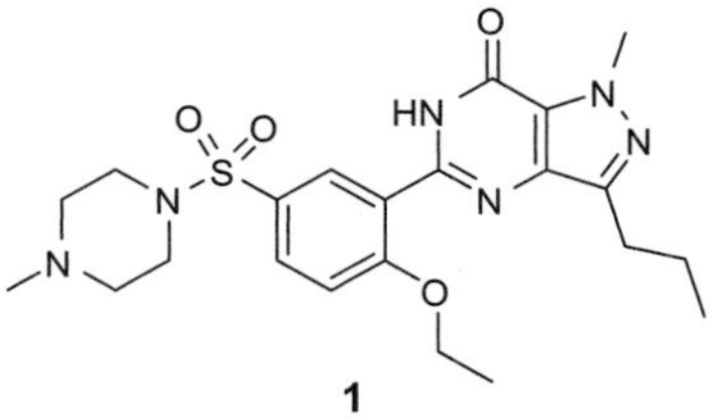

1

3. DEUTERIUM SAFETY AND PHARMACOLOGY

3.1. Deuterium exposure in organisms

The availability of sufficient quantities of deuterium-containing compounds has facilitated exploration of the effects of deuterium. In general, deuterium exposure in the form of D_2O has remarkably low systemic toxicity. Single-celled organisms can often be grown in conditions of full deuteration [31]. Lower organisms including fish and tadpoles reportedly survive at least 30% D_2O. Mice and dogs do not display visible effects from long-term replacement of at least 10–15% of body-fluid hydrogen with deuterium, although concentrations above 25% are broadly toxic to those species [6,32].

Humans can tolerate significant levels of deuterium in body fluids. Acute exposure levels of 15–23% deuterium replacement in whole body plasma levels have been reported with no evident adverse effects [33]. Deuterium (in the form of D_2O or HDO) is excreted by humans *via* the urine with a half-life of about 10 days, which is slightly shorter than the ~14-day half-life of water [34].

3.2. Effects upon metabolism

As discussed previously, deuterium substitution can result in reduced rates of metabolism and/or metabolic switching in which there is a change in the ratio of metabolites formed [35]. In spite of the ability of deuterium to alter metabolism patterns, the authors are not aware of any reports of deuteration resulting in the formation of unique metabolites that were not also observed for the all-hydrogen analog.

3.3. Receptor interactions

As cited previously [30], deuterated analogs of sildenafil were tested for their inhibitory activity versus phosphodiesterases I–VI. As shown in Table 1, compound **2** was a slightly more potent inhibitor of phosphodiesterase V than sildenafil by 28%. Compound **2** was also reported to be twice as selective for phosphodiesterase V versus phosphodiesterase VI. All three compounds were more potent than sildenafil in an *in vitro* functional assay. Although this appears to be the only example of a binding isotope effect on the pharmacology of a reversible-binding drug with its target enzyme, binding isotope effects are well known and have been recently reviewed [36–38]. Although binding isotope effects have been previously considered negligible, these more recent data support that they are unpredictable and can be insignificant or contribute positively or negatively to measured DIEs.

4. DEUTERIUM-CONTAINING DRUGS

4.1. Deuterated tramadol

Deuterated analogs (**5**, **6**, **7**) of tramadol have been evaluated in both *in vitro* and *in vivo* systems [39]. The compounds showed *in vitro* pharmacological activity (mu opioid binding and inhibition of 5-hydroxytryptamine and norepinephrine reuptake) similar to tramadol. When each of the three analogs was incubated in the presence of either human hepatocytes

Table 1 Activities of sildenafil and deuterated analogs

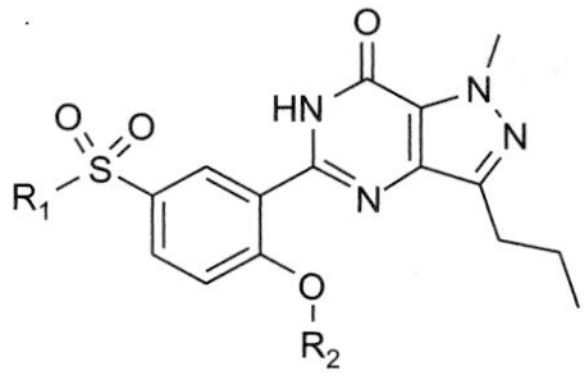

Compounds	R_1	R_2	PDE V IC_{50} (nM)[a]	PDE VI IC_{50} (nM)[a]	ED_{50} (nM)[b]
1	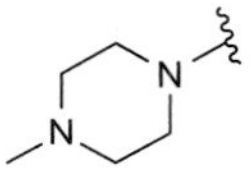		8.73	37.0	245
2	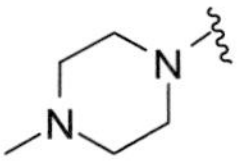	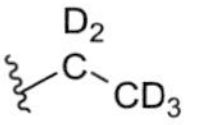 $C(D_2)CD_3$	6.31	46.4	85
3	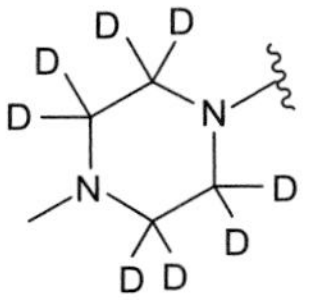		12.3	38.2	91
4	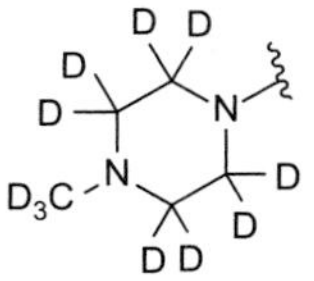		10.0	43.5	121

[a] Inhibition of $[^3H]$cGMP conversion to $[^3H]$guanosine.
[b] Relaxation of phenylephrine-precontracted corpus cavernosum strips (rabbit).

or human liver microsomes (HLM) and the loss of parent was measured, only **7** had a significantly longer half-life ($t_{1/2}$) than tramadol. When metabolic stability versus tramadol was measured by the formation of the *O*-demethylated metabolite, compounds **5** and **7** showed an approximately fivefold reduction in the formation of the metabolite. However, the increased *in vitro* stability of **7** did not translate to enhanced *in vivo* activity since **7** was no better than tramadol with respect to either potency or duration of analgesia in the rat tail-flick latency model.

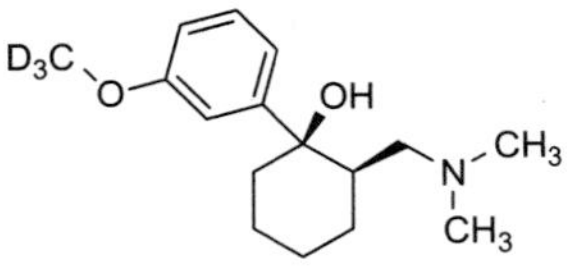

5 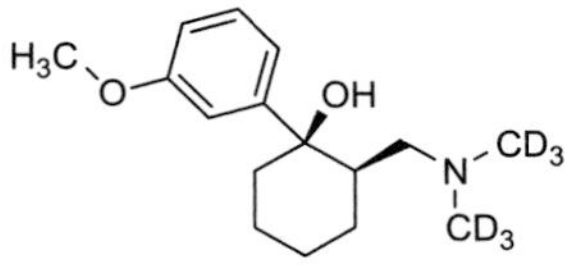 **6** 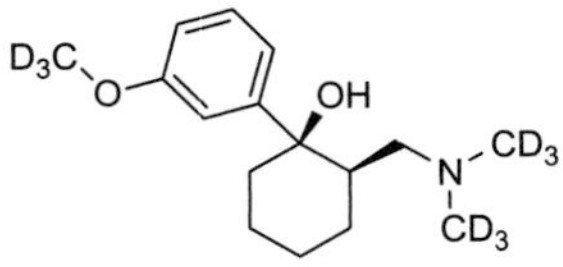 **7**

4.2. Deuterated rofecoxib

Improved PK in rats has been reported for **8**, a deuterated analog of rofecoxib [40]. The activities of rofecoxib and **8** for inhibition of cyclooxygenase-1 (COX1) in human platelets were assessed and showed similar IC_{50} values of 169 and 173 nM, respectively. When **8** and rofecoxib were orally dosed individually to rats, the PK parameters C_{max} (peak plasma concentration) and AUC (area under the curve, plasma exposure) for **8** versus rofecoxib were increased by 1.6- and 1.5-fold, respectively; however, **8** showed no increase in $t_{1/2}$.

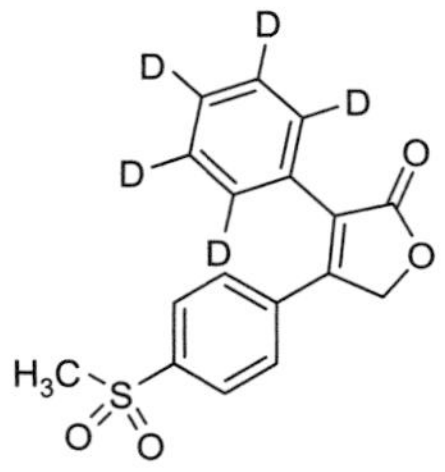

8 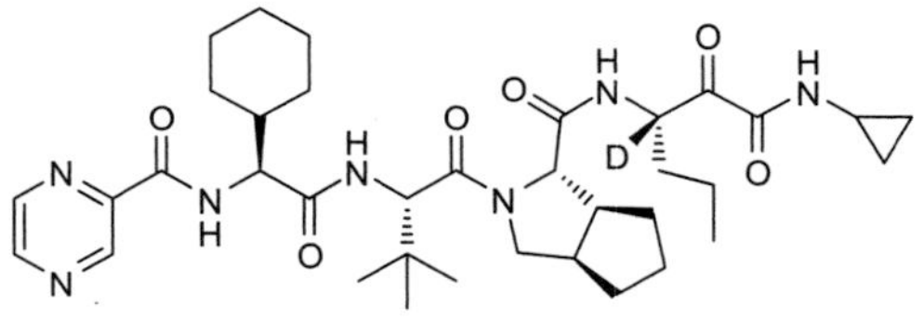 **9**

4.3. Deuterated telaprevir

Telaprevir, an inhibitor of hepatitis C viral NS3-4A protease, was recently approved by the FDA for the treatment of HCV and is marketed as Incivek™. Compound **9**, a deuterated analog of telaprevir in which the hydrogen at the chiral center adjacent to the ketoamide moiety was replaced with deuterium, has been reported [41]. The chirality for this center is (*S*) in the active diastereomer of telaprevir, whereas the compound with (*R*) chirality is approximately 30-fold weaker as a protease inhibitor. The deuterium substitution was intended to increase *in vivo* exposure of the active (*S*) compound by decreasing the rate of epimerization at that chiral center. The DIEs observed for the epimerization of **9** versus telaprevir in plasma (dog, rat, and human) ranged from 4 to 7. Upon oral codosing (1:1) of **9** and telaprevir in rats, a modest 13% increase in AUC for **9** was observed. Compound **9** was equipotent to telaprevir in both a NS3-4A protease inhibition assay and a viral replication assay.

4.4. Deuterated nevirapine

Compound **10**, a deuterated analog of nevirapine, is an example in which the deuterated compound appears to be more rapidly cleared in rats relative to nevirapine [42]. This anomalous observation was attributed to metabolic switching away from a reactive metabolite that inactivates CYP450, resulting in less CYP inhibition by **10** and faster *in vivo* clearance versus nevirapine. Support for this explanation was provided when rats were codosed with compound **10** and a CYP inhibitor and similar plasma levels of **10** and nevirapine were measured.

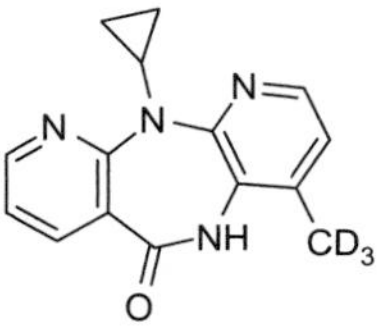

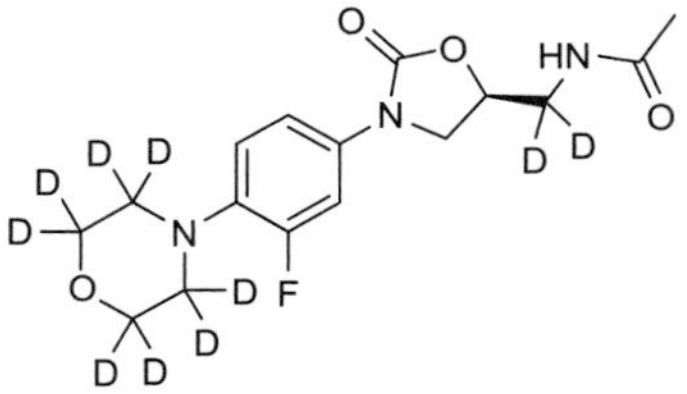

4.5. Deuterated linezolid

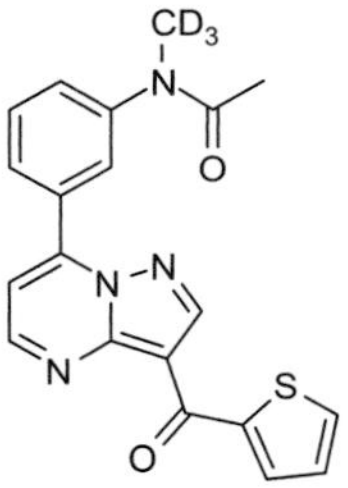

Compound **11** is a deuterated version of linezolid that shows improved *in vivo* PK in primates [43]. When **11** was codosed with linezolid (IV; 1:1), the $t_{1/2}$s were 6.3 versus 4.5 h, respectively. Compound **11** and linezolid were codosed (1:1) orally to primates and **11** showed enhanced exposure versus linezolid, $AUC_{24h} = 16.4$ versus 11.3 μg-h/mL, respectively. The improved oral PK translated to improved trough levels as shown by the C_{24h} ratios of **11** to linezolid: 2.99 (female primate) and 2.20 (male primate). The improved PK observed in primates supports the potential use of deuterated linezolid as a once-daily agent in humans [44].

4.6. Deuterated indiplon

Deuteration of the $GABA_A$-agonist sleep agent indiplon [45] provided compound **12** in which replacement of the N–CH_3 with N–CD_3 resulted in decreased *in vitro* metabolism in both rat and HLM [46]. This *in vitro* result was predictive of the *in vivo* PK in rat. Individual oral dosing of indiplon or **12** in rats showed a distinct PK advantage for the deuterated molecule since it had both a longer $t_{1/2}$ (~2×) and increased AUC (>2×). The binding of **12** to the central benzodiazepine receptor is reported to be similar to indiplon indicating that deuteration did not alter the pharmacology. Since the $t_{1/2}$ for indiplon in humans is approximately 1.3 h, the doubling of $t_{1/2}$ observed in rat suggests that **12** may have potential as an improved agent for sleep maintenance in humans.

5. DEUTERATED DRUGS AS CLINICAL AGENTS

5.1. Fludalanine

Fludalanine (**13**) appears to be the most extensively studied deuterated-drug candidate to have entered clinical trials. Fludalanine, combined with cycloserine, displays broad and potent antibacterial activity [47]. Its all-hydrogen analog is also a highly effective anti-infective agent; however, preclinical studies demonstrated that it was metabolized by D-aminoacid oxidase to 3-fluoropyruvate followed by L-lactate dehydrogenase reduction to 3-fluorolactate [48]. A recent letter reported that the fluorolactate metabolite caused brain vacuolization in rats and primates [5]. Fludalanine showed a reduced rate of metabolism and produced levels of fluorolactate deemed acceptable in healthy volunteers. The first trial in patients, however, showed higher levels of fluorolactate than expected, and studies on fludalanine were discontinued at Phase 2b. This unexpected result was attributed to metabolic changes associated with the disease state—a not uncommon observation when transitioning from healthy volunteers to patients.

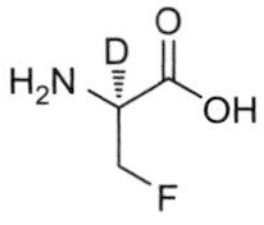

H_2N D O OH F D O H N D_3C D CH_3 D D D_3C N CD_3 OH H D_3C O

13 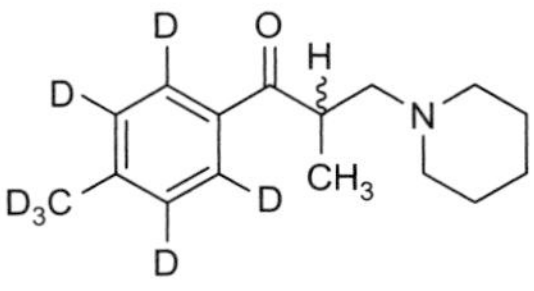 **14** 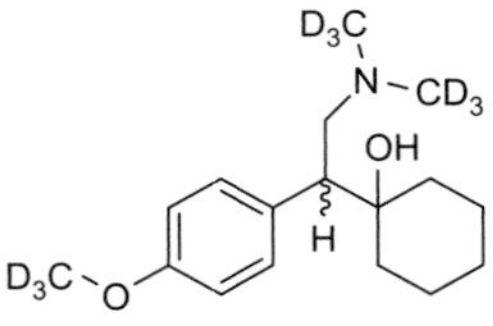 **15**

5.2. BDD-10103, deuterated tolperisone

BDD-10103, a deuterated analog of the muscle relaxant tolperisone, has reportedly been dosed in humans [49]; however, neither the structure of BDD-10103 nor the human PK data have been published. The major metabolic path for tolperisone is hydroxylation of the aryl methyl, predominantly *via* CYP2D6 and CYP2C19 [50]. Although the structure of

BDD-10103 has not been disclosed, compound **14** has been disclosed [51] and is reported to show a 10-fold decrease in aryl methyl hydroxylation versus tolperisone by both CYP2D6 and CYP2C19. The current clinical status of BDD-10103 is unclear.

5.3. SD-254, deuterated venlafaxine

A selectively deuterated analog of venlafaxine, SD-254 (**15**), has also advanced into clinical development. Although little data are available, the compound has been dosed in a Phase 1 healthy-volunteer study. It was metabolized more slowly than venlafaxine, and maintained an efficacious exposure for a longer time. Development in neuropathic pain is reportedly planned for SD-254 [8,52].

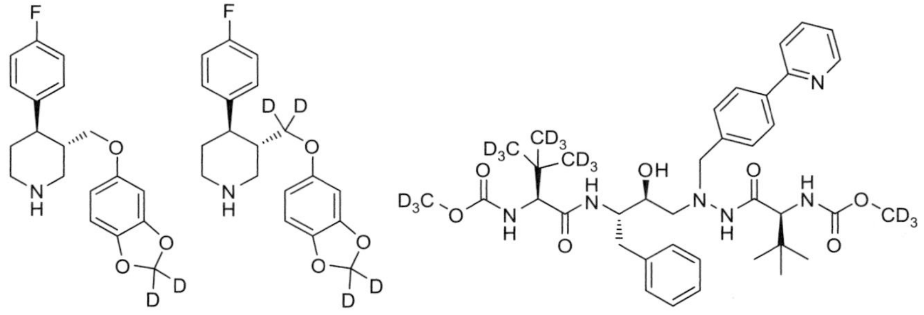

16 **17** **18**

5.4. CTP-347, deuterated paroxetine

CTP-347, a selectively deuterated analog of paroxetine for the treatment of hot flashes has completed Phase 1 clinical evaluation [53]. While it showed inhibition of serotonin and norepinephrine reuptake similar to paroxetine [54], CTP-347 was reported to be the first example of the use of deuterium to mitigate CYP2D6 inactivation in a clinical setting [7,55]. The structure of CTP-347 has not been disclosed; however, compounds **16** and **17** have been disclosed in the patent literature with data showing *increased* metabolism versus paroxetine in HLM [56]. The percent of parent remaining after 60 min incubation in HLM for paroxetine, **16**, and **17** was 51%, 26%, and 10%, respectively. The greater stability of paroxetine is due to its irreversible inhibition of CYP2D6, whereas the decreased stability of the deuterated compounds results from reduced CYP2D6 inhibition. Metabolism experiments in HLM confirmed that while paroxetine was a mechanism-based inactivator of CYP2D6 ($K_{inact} = 0.08\ min^{-1}$), CTP-347 showed little or no CYP2D6 inactivation [7,57]. This was possibly due to metabolic shunting that prevents the formation of a reactive metabolite that forms an irreversible complex at the active site of CYP2D6 [58]. A Phase 1 study was conducted to assess the effect of CTP-347 on mechanism-based inhibition of CYP2D6 in healthy women [55]. Subjects receiving CTP-347 retained a greater ability to metabolize dextromethorphan (CYP2D6 substrate) than has been reported previously for paroxetine [59,60]. Minor CYP2D6 inhibition was observed at higher CTP-347 doses, which is consistent with the reversible, competitive inhibition seen *in vitro*.

5.5. CTP-518, deuterated atazanavir

CTP-518, a deuterated version of the HIV-protease inhibitor atazanavir, has entered Phase 1 clinical studies [8] with a goal of developing a drug that retains the antiviral potency of atazanavir without the need for ritonavir or another PK-boosting agent [52]. The structure of CTP-518 has not been disclosed. Compound **18** has been disclosed in a patent [61] with data showing increased stability versus atazanavir both *in vitro* (HLM) and *in vivo* (codose studies in primates). The $t_{1/2}$ in HLM for **18** was increased 51% versus atazanavir, which translated to an average 52% increase in $t_{1/2}$ versus atazanavir following an IV codose (1:1) in primates. Atazanavir and **18** showed similar HIV antiviral activity in an *in vitro* viral assay.

6. PATENTABILITY OF DEUTERATED DRUGS

As might be expected from the increase in preclinical and clinical work on deuterated drugs over the past 5 years, there have been a growing number of published patent applications and issued patents in this area. A search from 2005 to 2010 for the terms "deuterium" or "deuterated" in the claims section and "pharmaceutical" in all fields of published U.S. patent applications and issued U.S. patents returns 761 and 142 hits, respectively. For the years 2001–2005, only 77 patents were issued. Figure 1 shows the breakdown for each of the past 6 years. There has been a large overall increase in published U.S. applications for 2008 through 2010 with a smaller but steady increase for issued patents over the same period. This trend shows that patent protection may be obtained for deuterated drugs. As more applications enter prosecution at the USPTO, it will be of great interest to see if the number of issued patents continues to grow and what data may be needed to support patentability.

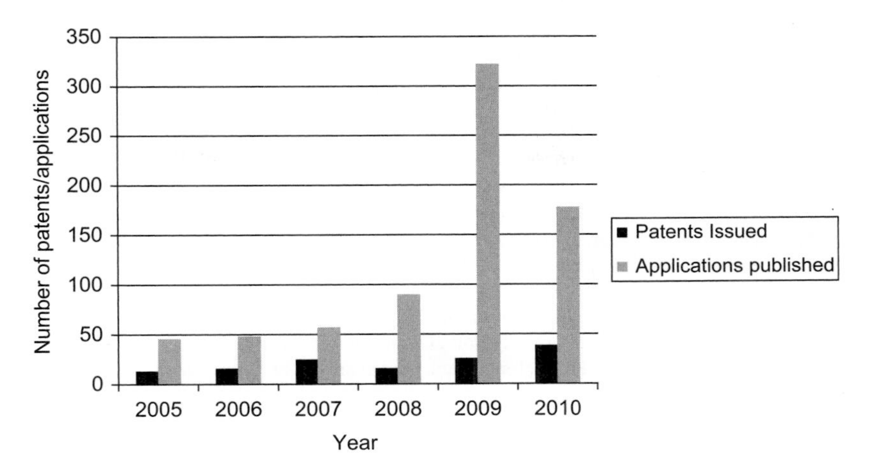

Figure 1 Issued patents and published applications containing "deuterium" or "deuterated" in the claims and "pharmaceutical" in any field for 2005–2010.

7. CONCLUSIONS

Deuterium can be a powerful medicinal chemistry tool that has, until recently, received little attention in the context of new drugs. A majority of the examples that have appeared in the recent literature are deuterated versions of known, well-characterized drugs with established therapeutic utilities. Most deuterated compounds reported-to-date appear to retain full biochemical potency and selectivity. However, in select cases, deuterated drugs show a differentiated PK profile versus the hydrogen-only compounds. In contrast to earlier reviews that were pessimistic about deuterium as an effective strategy for creating new drugs, recent reports suggest the opposite. More scientists in the pharmaceutical industry are now using deuterium as a tool to improve drug properties [7,8,52]. With more deuterium-containing compounds entering clinical evaluation, it appears increasingly likely that the approach will succeed in producing important new medicines if preclinical indications of improvements in safety, tolerability, and/or efficacy are recapitulated in humans.

REFERENCES

[1] R. Schoenheimer and D. Rittenberg, *Physiol. Rev.*, 1940, **20**, 218.
[2] T. A. Baillie, *Pharmacol. Rev.*, 1981, **33**, 81.
[3] M. I. Blake, H. L. Crespi and J. J. Katz, *J. Pharm. Sci.*, 1975, **64**, 367.
[4] A. B. Foster, *Adv. Drug Res.*, 1985, **14**, 1.
[5] F. M. Kahan, *Chem. Eng. News*, 2009, **87**, 4.
[6] D. J. Kushner, A. Baker and T. G. Dunstall, *Can. J. Physiol. Pharmacol.*, 1999, **77**, 79.

[7] R. Tung, *Innov. Pharm. Technol.*, 2010, **32**, 24.
[8] L. Shao and M. C. Hewitt, *Drug News Perspect.*, 2010, **23**, 398.
[9] H. C. Urey, F. G. Brickwedde and G. M. Murphy, *Phys. Rev.*, 1932, **39**, 164.
[10] W. L. Marter, D. W. Hayes and D. W. Jones, (eds. J. C. McKetta and W. A. Cunningham), in *Encyclopedia of Chemical Processing and Design, Issue* 15, Marcel Dekker, Inc., New York, NY, 1982, p. 308.
[11] K. B. Wiberg, *Chem. Rev.*, 1955, **55**, 713.
[12] S. D. Nelson and W. F. Trager, *Drug Metab. Dispos.*, 2003, **31**, 1481.
[13] M. B. Fisher, K. R. Henne and J. Boer, *Curr. Opin. Drug Discov. Dev.*, 2006, **9**, 101.
[14] W. P. Jenks, *Catalysis in Chemistry and Enzymology*, Dover Publications Inc., Mineola, NY, 1987 p. 243.
[15] K. J. Ling and R. P. Hanzlik, *Biochem. Biophys. Res. Commun.*, 1989, **169**, 844.
[16] B. Meunier, S. P. de Visser and S. Shaik, *Chem. Rev.*, 2004, **104**, 3947.
[17] D. C. Pryd, D. Dalvie, Q. Hu, P. Jones, R. S. Obach and T.-D. Tran, *J. Med. Chem.*, 2010, **53**, 8441.
[18] D. B. Northrop, *Biochemistry*, 1975, **14**, 2644.
[19] M. G. Horning, K. D. Haegele, K. R. Sommer, J. Nowlin, M. Stafford and J. P. Thenot, *Proc. Int. Conf. Stable Isotopes, 2nd NTIS*, Springfield, VA, CON-751027, 41, 1976.
[20] G. Guroff, J. W. Daly, D. M. Jerina, J. Renson, B. Witkop and S. Udenfriend, *Science*, 1967, **157**, 1524.
[21] H. Esaki, N. Ito, S. Sakai, T. Maegawa, Y. Monguchi and H. Sajiki, *Tetrahedron*, 2006, **62**, 10954.
[22] H. Esaki, F. Aoki, M. Umemura, M. Kato, T. Maegawa, Y. Monguchi and H. Sajiki, *Chem. Eur. J.*, 2007, **13**, 4052.
[23] H. J. Leis, G. Fauler and W. Windischhofer, *Curr. Org. Chem.*, 1998, **2**, 131.
[24] J. G. Atkinson and M. O. Luke, *US Patent 3,966,781*, 1976.
[25] L. Di Costanzo, M. Moulin, M. Haertlein, F. Meilleur and D. W. Christianson, *Arch. Biochem. Biophys.*, 2007, **465**, 82.
[26] N. E. Tayar, H.v.d Waterbeemd, M. Gryllaki, B. Testa and W. F. Trager, *Int. J. Pharm.*, 1984, **19**, 271.
[27] M. Turowski, N. Yamakawa, J. Meller, K. Kimata, T. Ikegami, K. Hosoya, N. Tanaka and E. R. Thornton, *J. Am. Chem. Soc.*, 2003, **125**, 13836.
[28] C. L. Perrin and Y. Dong, *J. Am. Chem. Soc.*, 2007, **129**, 4490.
[29] C. L. Perrin, B. K. Ohta, J. Liberman and M. Erdelyi, *J. Am. Chem. Soc.*, 2005, **127**, 9641.
[30] F. Schneider, E. Mattern-Dogru, M. Hillgenberg and R.-G. Alken, *Arzneim. Forsch./Drug Res.*, 2007, **57**, 293.
[31] E. Flaumenhaft, S. Bose, H. L. Crespi and J. J. Katz, *Int. Rev. Cytol.*, 1965, **18**, 313.
[32] P. J. H. Jones and S. T. Leatherdale, *Clin. Sci.*, 1991, **80**, 277.
[33] N. Blagojevic, G. Storr, J. B. Allen, H. Hatanaka and H. Nakagawa, (eds. R. Zamenhof, G. Solares and O. Harling), in *Dosimetry & Treatment Planning for Neutron Capture Therapy*, Advanced Medical Publishing, Madison, WI, 1994, p. 125.
[34] P. R. Schloerb, B. J. Friis-Hansen, I. S. Edelman, A. K. Solomon and F. D. Moore, *J. Clin. Invest.*, 1950, **29**, 1296.
[35] A. E. Mutlib, R. J. Gerson, P. C. Meunier, P. J. Haley, H. Chen, L. S. Gan, M. H. Davies, B. Gemzik, D. D. Christ, D. F. Krahn, J. A. Markwalder, S. P. Seitz, R. T. Robertson and G. T. Miwa, *Toxicol. Appl. Pharmacol.*, 2000, **169**, 102.
[36] M. W. Ruszczycky and V. E. Anderson, *J. Theor. Biol.*, 2006, **243**, 328.
[37] B. E. Lewis and V. L. Schramm, (eds. A. Kohen and H.-H. Limbach), in *Isotope Effects in Chemistry and Biology*, CRC Taylor & Francis, Boca Raton, FL, 2006, p. 1019.
[38] V. L. Schramm, *Curr. Opin. Chem. Biol.*, 2007, **11**, 529.
[39] L. Shao, C. Abolin, C. M. C. Hewitt, P. Koch and M. Varney, *Bioorg. Med. Chem. Lett.*, 2006, **16**, 691.

[40] F. Schneider, M. Hillgenberg, R. Koytchev and R.-G. Alken, *Arzneim. Forsch/Drug Res.*, 2006, **56**, 295.
[41] F. Maltais, Y. C. Jung, M. Chen, J. Tanoury, R. B. Perni, M. Mani, L. Laitinen, H. Huang, S. Liao, H. Gao, H. Tsao, E. Block, C. Ma, R. S. Shawgo, C. Town, C. L. Brummel, D. Howe, S. Pazhanisamy, S. Raybuck, M. Namchuk and Y. L. Bennani, *J. Med. Chem.*, 2009, **52**, 7993.
[42] J. Chen, B. M. Mannargudi, L. Xu and J. Uetrecht, *Chem. Res. Toxicol.*, 2008, **21**, 1862.
[43] R. Tung and S. Harbeson, *Patent Application US 2008/0139563-A1*, 2008.
[44] A. Morales, D. Wells, R. Tung, R. Zelle, C. Cheng and S. Harbeson, Abstract A-990, *48th Annual Interscience Conference on Antimicrobial Agents and Chemotherapy (ICAAC) and the 46th Annual Meeting of the Infectious Diseases Society of America (IDSA)*, Washington, DC, 2008.
[45] R. Rosenberg, T. Roth, M. Scharf, D. A. Lankford and R. Farber, *J. Clin. Sleep Med.*, 2007, **3**, 374.
[46] A. J. Morales, R. Gallegos, V. Uttamsingh, C. Cheng, S. Harbeson, R. Zelle, R. Tung and D. Wells, Abstract 285, *The 15th North American Meeting of the International Society of Xenobiotics*, San Diego, CA, 2008.
[47] R. Wise and J. M. Andrews, *Antimicrob. Agents Chemother.*, 1984, **25**, 612.
[48] G. K. Darland, R. Hajdu, H. Kropp, F. M. Kahan, R. W. Walker and W. J. Vandenheuvel, *Drug Metab. Dispos.*, 1986, **14**, 668.
[49] www.birdspharma.com/projects/bdd_10103.
[50] B. Dalmadi, J. Leibinger, S. Szeberenyi, T. Borbas, S. Farkas, Z. Szombathelyi and K. Tihanyi, *Drug Metab. Dispos.*, 2003, **31**, 631.
[51] R.-G. Alken, *Patent Application US 2004/0186136-A1*, 2004.
[52] A. Yarnell, *Chem. Eng. News*, 2009, **87**, 36.
[53] V. Uttamsingh, D. Wells, D. Soergel and R. Zelle, Abstract 11552, *The 38th American College of Clinical Pharmacology*, San Antonio, TX, 2009.
[54] D. S. Wells, R. Gallegos, V. Uttamsingh, C. Cheng, G. Bridson, C. Masse, S. L. Harbeson, P. B. Graham, R. Zelle, R. Tung and A. J. Morales, Abstract P-4, *19th Annual Meeting of the North American Menopause Society (NAMS)*, 2008.
[55] D. Wells, D. Soergel and R. Zelle, Abstract 164, *38th American College of Clinical Pharmacology Annual Meeting*, 2009.
[56] R. Tung, *US Patent 7,687,914*, 2010.
[57] R. Gallegos, V. Uttamsingh, A. Morales, C. Cheng, G. Bridson, J. F. Liu, R. Tung, R. Zelle, S. Harbeson and D. S. Wells, Abstract 188, *The 16th North American Meeting of the International Society of Xenobiotics*, 2009.
[58] M. Murray, *Curr. Drug Metab.*, 2000, **1**, 67.
[59] H. L. Liston, C. L. DeVane, D. W. Boulton, S. C. Risch, J. S. Markowitz and J. Goldman, *J. Clin. Psychopharmacol.*, 2002, **22**, 169.
[60] C. L. Alfaro, Y. M. Lam, J. Simpson and L. Ereshefsky, *J. Clin. Pharmacol.*, 2000, **40**, 58.
[61] S. L. Harbeson and R. D. Tung, *Patent Application US 2009/0036357-A1*, 2009.

CHAPTER 25

Drug-Induced Phospholipidosis

Peter R. Bernstein*, Paul Ciaccio** and
James Morelli†

Contents

1. INTRODUCTION

1.1. What is drug-induced phospholipidosis?

Drug-induced phospholipidosis (PLD) is a condition in which drugs cause the excessive accumulation of phospholipids in cells [1]. It is characterized by the accumulation of phospholipid–drug complexes as intracellular concentric lamellar bodies that are visible within lysosomes by electron microscopy [2]. In most instances, the drugs that induce this

* PhaRmaB LLC, 14 Forest View Road, Rose Valley, PA 19086, USA
** AstraZeneca Pharmaceuticals, 35 Gatehouse Drive, B2.85, Waltham, MA 02451, USA
† AstraZeneca Pharmaceuticals, 1800 Concord Pike, Wilmington, DE 19850-5437, USA

Annual Reports in Medicinal Chemistry, Volume 46
ISSN: 0065-7743, DOI: 10.1016/B978-0-12-386009-5.00001-1

condition are those that have both a hydrophilic domain, with one or more positively charged nitrogen groups, and a hydrophobic domain that frequently contains an aryl group. Generally such compounds are called cationic amphiphilic drugs (CADs).

1.2. What is the mechanism of PLD?

There is no single causative mechanism for induction of PLD, as diverse drugs seem to induce it in different ways [3] and others may involve multiple causative mechanisms [4]. Some CADs are attracted to negatively charged phospholipids and accumulate in membranes, changing the susceptibility of the membranes to breakdown. Others cause phospholipid buildup by directly inhibiting the action of phospholipases, and some appear to increase phospholipid synthesis [5].

There are numerous PLD-responding cell types and tissues. Cells of the macrophage system (*e.g.*, bronchiolar macrophages and Kupffer cells) are prominent responders that act to scavenge lamellar bodies and facilitate the lamellar body/drug complex removal. Such cells are particularly lysosome rich, and the CADs ionize and partition into the more acidic environment of the lysosome which serves to protect the cell from cytotoxicity. PLD induction seems to be a dose-dependent process with manifestation of PLD being related to accumulation of the CADs.

1.3. What are the toxicological implications?

Numerous (>50) marketed drugs are known to induce PLD in preclinical species at doses that are significantly higher (>10-fold) than are routinely prescribed in the clinic [6]. PLD is generally considered to be reversible, but the rate of reversibility varies widely within drugs, tissues, and species. It appears more commonly preclinically than clinically, and in one important study, CNS PLD did not yield electrophysiological functional deficits [7]. Viewed this way, PLD would not be considered adverse or high risk for drug development.

However, PLD has been circumstantially associated with toxicities *in vivo* both preclinically and clinically [8]. Although there are no comprehensive published reviews of pharmaceutical industry data to sort these risks, various published reports include: PLD observed *sometimes* concurrently with cataracts and corneal opacities, pneumonitis, myodegeneration, neurodegeneration and neuropathies, and liver toxicity [8]. Additionally, the physicochemical properties of phospholipogenics *may* predispose them to additional toxicities apart from direct PLD, such as mitochondrial toxicity and increased risk of QT prolongation.

2. EVOLVING REGULATORY AND INDUSTRY VIEWS

Since the linkage between toxicological responses and PLD remains unclear, there are currently no regulatory guidelines. In an effort to address this issue, the Food and Drug Administration (FDA) announced that guidance on PLD was being developed and established a "FDA Phospholipidosis Working Group" [9]. Although a formal report has not been issued, the challenges are documented in presentations given at a recent FDA advisory committee meeting [10]. These include the questions: Why is it that only a small number of compounds that show PLD also show toxicity and why is it that PLD seen in animals is not predictive of clinical results?

Multiple reports reveal that PLD-inducing compounds have either a higher incidence of histological findings in toxicity studies than PLD negative (non-phospholipogenics) compounds [11] or a strong correlative relationship between PLD and preclinical toxicities (toxicity lowest effect level [LOEL] *vs.* PLD LOEL) [12]. A high proportion of drugs known to cause PLD in preclinical animal models are known to cause QT prolongation clinically [13,14]. The mechanistic basis for this increased QT risk is unproven since (a) the intrinsic PLD potencies and hERG blockade potencies of agents that induce Torsades des Pointes [Torsadogens] do not appear to correlate [11] and (b) the risk is not directly correlated with hERG potency as QT effects are found with PLD positive agents at much higher hERG IC_{50} values than those found for PLD negative ones [15]. Specifically, induction of Torsades des Pointes is usually associated with potent hERG blockers; however, if the Torsadogen is also a known PLD inducer, then the cardiac effects manifest themselves even if the direct effect at hERG is very weak [12]. While it is not known why this occurs and although only a small number of phospholipogenics also cause cardiac PLD, a potential hypothesis for the increased risk of QT prolongration is pharmacokinetic, that is, amphiphilic drugs may partition into cardiac tissue, reaching concentrations that exceed plasma levels and hERG IC_{50} values.

Given these considerations, discovery of PLD during preclinical *in vivo* toxicology has the potential to cause significant delays and increased expenditures in the drug development process. On many occasions, drug development of potentially beneficial compounds showing PLD has been terminated [16]. These decisions are made not because of a proven toxicological problem but because of the uncertainty a PLD finding introduces and the increased hurdle/cost of developing such molecules [17]. The possibility of discovering PLD late in the drug discovery process has caused significant effort to be spent on strategies to minimize the potential for compounds to induce PLD as evidenced by numerous recent publications [6,18].

It is incumbent upon drug makers to challenge views that assume a low risk and adaptive nature for PLD, or assume a low PLD risk for compounds that are CADs. They must be prepared to support risk assessments after obtaining experimental evidence of PLD-inducing potential prior to pivotal toxicology study support [19]. Some companies may profile phospholipogenic compound safety risks on a case-by-case basis. Others are developing more sophisticated risk avoidance and management strategies that include: (a) *in silico* and *in vitro* screening of compounds in discovery; (b) identification of both *in vitro* and *in vivo* intrinsic PLD potency and associated tissue degeneration *in vivo* before pivotal studies are conducted; (c) confirmatory *in vivo* tools (electron microscopy, measurement of phospholipid content, Lamp2 [lysosome-associated membrane protein 2] immunohistochemistry, and gene expression profiling), as well as mechanistic investigations by which PLD affects cell function (*e.g.*, inhibition of autophagy and impact on lysosomal protein degradation and release, fusion abilities and endocytosis; apoptosis; lysosomal *vs.* mitochondrial role in cytotoxicity); (d) drug accumulation and PK association with responses over time; and (e) functional response evaluations (*e.g.*, QT, supporting clinical need for thorough evaluations, and neurologic effects).

In their comprehensive review of an internal Pfizer risk assessment strategy, Chatman *et al.* [20] emphasized the importance of demonstrating reversibility in pivotal studies and used a tiered testing approach, starting in discovery with *in silico* and *in vitro* tools, followed by thorough characterization of responses *in vivo*. Pharmaceutical sponsors might also consider additional points, including (1) understanding the correlation of response severity to intrinsic phospholipogenic potency or tissue levels, (2) examining the presence of supraproportionate tissue levels relative to plasma exposure, and (3) investigating the time dependency of proposed biomarkers to discriminate toxic effects from PLD-only responses. A particularly promising noninvasive PLD biomarker is bis(monacylglycerol)phosphate (BMP), a lysosomal membrane phospholipid that marks membranes for degradation, a feature that distinguishes it mechanistically from perimeter membrane phospholipids [8].

3. SCREENING METHODS

3.1. Introduction

In an effort to reduce the risk of drug attrition due to toxicity associated with physicochemical parameters, the drug industry has been introducing numerous methods that will allow companies to select compounds that are less likely to fail for reasons other than lack of efficacy [21].

To reduce the risk of failure *in vivo* due to PLD, the methodology has included *in silico* models and *in vitro* assays.

3.2. *In silico* techniques

Although many computational models have been disclosed, a common shortcoming is that many models are proprietary and built with limited datasets [22,23]. An openly accessible, effective *in silico*, model that yielded rapid and accurate predictions of PLD would be of great value to the drug discovery industry. A recent review provides a detailed description of most *in silico* approaches [18]. Amphiphilicity of drug molecules is linked to their ability to induce PLD. This is because binding with both the polar phosphate region and the lipophilic core of phospholipid membranes requires specific spatial distribution of the molecule's lipophilic and charged groups. Recognizing this fact, a group at Roche developed an *in silico* program (CAFCA) that calculated the free energy of amphiphilicity for molecules [24]. Picking the correct conformations for these calculations was important for this approach to successfully provide good agreement with experimentally determined values. Each conformation produces a different value and although the reason was not clear, "in all cases the lowest calculated free energies of amphiphilicity came closest to the measured values." In an early attempt to produce an *in silico* model, a group at Organon published a method for predicting PLD-inducing potential that was based on ClogP and calculated pK_a [25]. The result was a prediction that cationic amphiphilic compounds would be phospholipogenic provided that $pK_a > 8$ and $\text{ClogP} > 1$ and also that Eq. (1) is satisfied.

$$(\text{ClogP})^2 + (\text{calculated}\, pK_a)^2 > 90 \qquad (1)$$

The limitations of this approach were discussed in a published analysis by a Pfizer group that also described modifications to the Organon model that significantly improved the percentage of compounds that were correctly diagnosed as not inducing PLD [26]. QSAR models based on two commercially available software programs MC4PC-QSAR and MDL-QSAR have been described by an FDA-based team [27]. Although the statistics for the single MC4PC-QSAR model were modest, addition of the second model and seeking consensus prediction by the two models significantly increased the accuracy of the tool. Another approach to increasing the predictability of *in silico* models based on ClogP and pK_a was illustrated by a Bristol-Myers Squibb team, that showed the addition of volume of distribution (V_d) resulted in improved concordance with PLD [28]. Most recently, a team at the University of St. Andrew's reviewed the status of predicting PLD using machine learning [23].

The Pfizer, FDA, and St. Andrew's team all cited improved accuracy as the size of the datasets they used to build and validate their models were increased.

3.3. *In vitro* techniques

An early *in vitro* screen for PLD was described in 1991 using human foreskin fibroblasts treated with a fluorescently labeled surrogate phospholipid, NBD-PC [29]. NBD-PC consists of phosphotidylcholine (PC) linked to the fluorescent moiety, nitrobenzoxadiazole (NBD). Comparison of control to treated samples using a fluorescent spectrophotometer following cell disruption yielded an insufficient signal-to-noise ratio for screening purposes. Despite this limitation, the methodology laid the ground work for later assay development by several groups using other methods, including high content biology, which was not available at the time.

In 2001, a team at Abbott described an assay, using rat and human hepatocytes and phosphoethanolamine (NBD-PE), combined with laser scanning confocal microscopy, as part of an effort to find alternatives for matrix metalloproteinase inhibitors that caused PLD [30]. The authors underscored the importance of assaying metabolites of the parent drug, as initial attempts to detect PLD using hepatocytes with the parent drug alone yielded negative results, but use of the amine metabolite resulted in strong detection of NBD-PE accumulation.

Kasahara *et al.* found that treatment of the CHO-K1 cell line with NBD-PC and drug combined with fluorescent spectrophotometer readings gave good results; it correctly identified a large set of known positive and negative phospholipogenic compounds with high specificity and sensitivity [31].

In 2006, a team at AstraZeneca described the use of a mouse splenic macrophage cell line, NBD-PE, and the Cellomics Arrayscan High Content Screening (HCS) instrument to develop an assay with high dynamic range, sensitivity, and specificity [32]. Elimination of individual dead cells from quantitation using Sytox Orange Dye allowed for treatment of cells through a wide range of drug concentrations (typically up to 300 μM), while measuring for effects only on live cells.

In 2009, a team at JNJ developed a flow cytometric-based method for detecting PLD in a high-throughput 96-well format [33]. This method uses the human monocyte cell line THP-1, Invitrogen's LipidTox reagent for phospholipid staining, and a live-dead cell stain combination for assessment of cell toxicity. The group claims that one significant advantage of the method is its sensitivity, exemplified by its ability to identify compounds which cause only a twofold increase in LipidTox retention over controls.

PLD assays based on toxicogenomics have also been described. In 2005, Sawada analyzed RNA from treated HepG2 cells using Affymetrix microarray technology [4]. Analysis of the microarray studies yielded 12 marker genes which could be used as signatures to predict the potential for PLD. Shortly after another team also published a gene expression-based assay [34]. While these assays provide important contributions to understanding the mechanisms occurring during PLD, their broader use is restricted due to high cost and low throughput.

In conclusion, many methods for assaying compounds for the potential to cause PLD exist. Each has advantages and disadvantages in terms of cost, required instrumentation, and desired throughput. Despite these diverse options, *in vitro* methods have many limitations. *In vitro* assays cannot predict multiorgan PLD, which tissues will be affected, which species will be affected, or even if there will be a human response to the drug. For this, additional knowledge gathering is needed.

4. EXAMPLES OF PROJECT RESPONSES TO FINDING PLD

The impact on drug development of finding drug-induced PLD during toxicological studies has been documented in many disease areas and projects. In most cases, the induction of PLD has been traced to a direct effect of the parent molecule. Following the discovery of PLD, teams have used a variety of approaches to develop new compounds with different physicochemical properties that retain potency at the pharmacologic target but which have reduced probability of inducing PLD. Several examples follow and additional description and examples may be found in a recent review [18].

At Roche, a team working to develop DPP-IV inhibitors found that compound **1** had many favorable properties but induced PLD in a concentration-dependent manner in cultured fibroblasts (at 2.5–20 μM) [35]. Knowing that the "CAFCA" program [25] shows reduced potential for inducing PLD for compounds with a calculated free energy of amphiphilicity ($\Delta\Delta G_{AM}$) greater than -6 kJ/mol and that **1** had $\Delta\Delta G_{AM} = -6.6$ they prepared a set of compounds that were predicted to be less amphiphilic. This effort led to compound **2**, which matched the DPP-IV potency of **1** but which had $\Delta\Delta G_{AM} = -5.6$ and which did not induce PLD in the fibroblast assay at its highest tested concentration [20 μM].

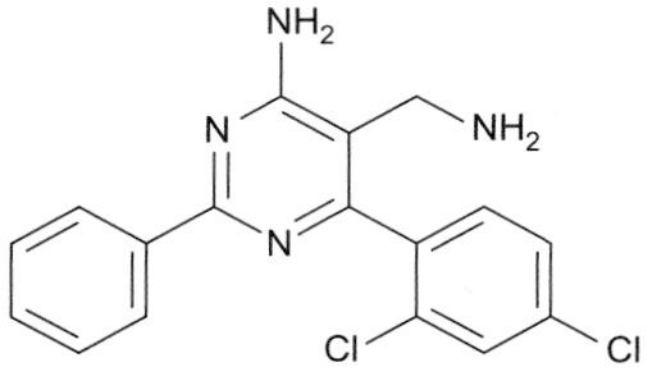

1 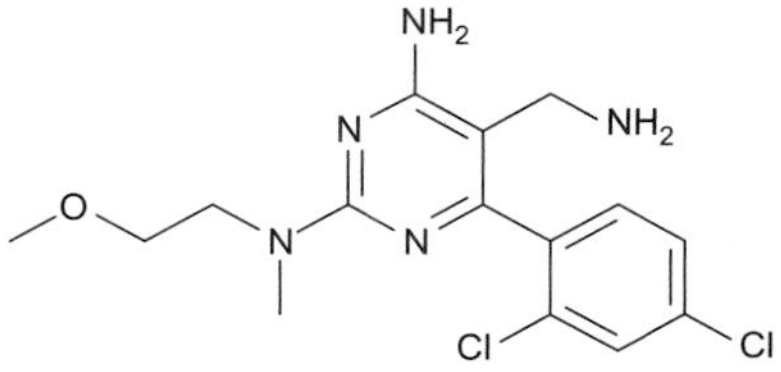 **2**

A team at AstraZeneca disclosed that a series of 5-HT_{1B} antagonists, exemplified by AZD8129 (**3**, Table 1), had exhibited PLD in preclinical rat and dog toxicology studies [38]. They used computational modeling and an *in vitro* assay [32] to develop a follow-up compound (AZD3783, **4**) that had improved 5-HT_{1B} potency to PLD ratios, both *in vitro* and *in vivo*, compared to **3** (as well as many reference drugs, including amiodarone and fluoxetine). A comparison of the *in vivo* dorsal root ganglia and sciatic nerve responses to these two candidate drugs is shown in Table 1. The apparent improvement of **4** over **3** illustrates early success in this program [36]. Dorsal root ganglia [DRG] are a structural unit that is vulnerable to toxicants, mainly due to fenestrated endothelium and discontinuous capillary basement membranes that enable higher exposures than surrounding tissues. DRG are linked functionally to the sciatic nerve. At lower exposures in a 1-month repeat dose rat study, **3** administration yielded greater dorsal root ganglia PLD, sciatic neurodegeneration and intrinsic PLD potency than **4** even as animals were exposed to higher amounts of **4** than **3**.

Unfortunately, AZD3783 was also withdrawn from development due to other toxicities potentially linked to buildup of compound and PLD (data not shown). So, the team designed **5**, a picomolar 5-HT_{1B} antagonist [37]. Since **5** no longer contained the basic piperazine found in **4**, it was no longer a CAD and was predicted to not be phospholipidogenic. This was confirmed by its inactivity at 300 μM in the *in vitro* PLD assay at which concentration **4** showed 81% inhibition.

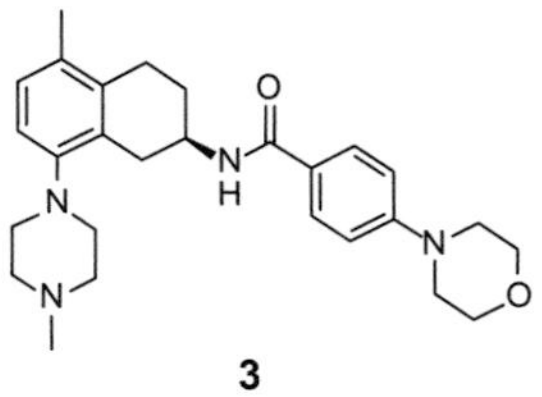

3 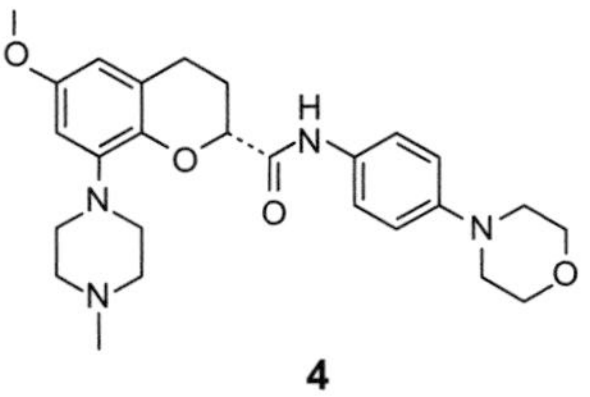 **4** 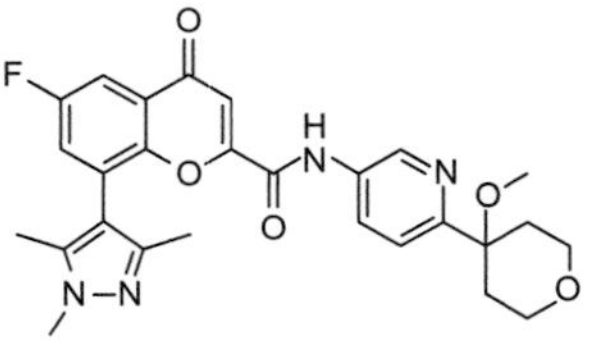 **5**

Table 1 AUC levels and *in vitro/in vivo* PLD potency of 5-HT_{1B} antagonists [36,37]

Compound	Total AUC exposures μmol*h	*In vitro* PLD potency (EC_{50}) μM	Rat dorsal root ganglia PLD response severity(Incidence)[a]	Rat sciatic nerve degeneration severity (Incidence)
3	145	10	+++ (5/6)	+ (2/7)
4	764	164	++ (5/8)	– (0/8)

[a] Severity (–, no incidence: +, slight; ++ minimal; +++, moderate; ++++, severe)/incidence (# rats responding/# rats tested).

Teams at several companies developing H3 antagonists have described needing to overcome PLD. A group at Roche discovered a naphthalene based series exemplified by **6** [39]. While this compound had many positive attributes, it was strongly positive in an *in vitro* PLD assay at 5 μM and above. Analysis *via* CAFCA showed that this compound was highly amphiphilic and had $\Delta\Delta G_{AM} = -10.4$. They explored replacements for all three regions of the molecule: piperidinyl propyl ether, amide, and naphthyl core. The new compounds were designed to decrease amphiphilicity but would still fit an H3-pharmacophore model. This led to the discovery of **7**, a compound which retained most of the H3 potency of **6**, and its other positive features but which had $\Delta\Delta G_{AM} = -4.27$ and did not trigger a PLD flag. Efforts to further reduce $\Delta\Delta G_{AM}$ by replacing the naphthalene core with quinolone led to unacceptable loses in H3 potency.

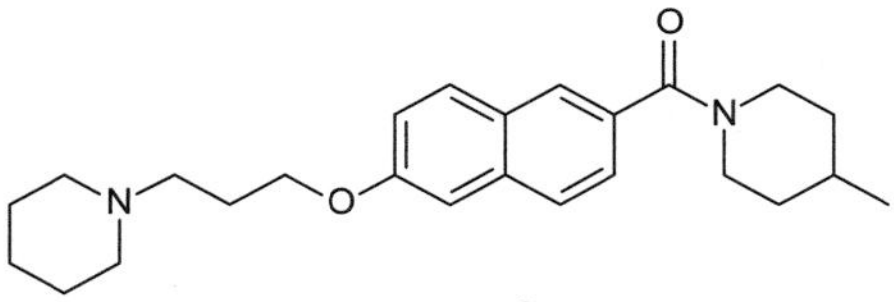

6 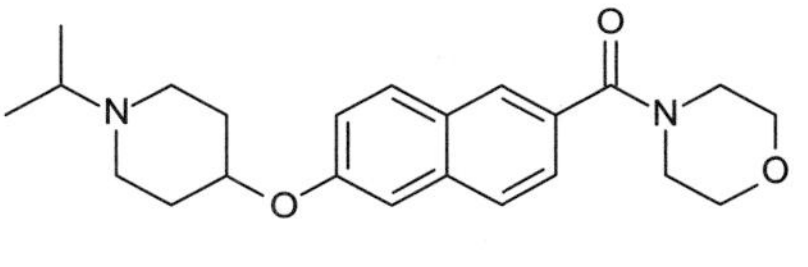 **7**

At Abbott a team working on backups to the H3-antagonist ABT-239 (**8**) disclosed **9**, an analog with a basic side chain that was more potent than amiodarone in a PLD assay in cultured rat hepatocytes [40]. This was shown to be due to the inhibition of phospholipases upon the binding of dibasic amines with membranes [41]. By either reducing the basicity of the second amine by arylating it (*e.g.*, **10**) or removing the amine side chain, they were able to prepare potent H3 antagonists with reduced PLD-inducing potential.

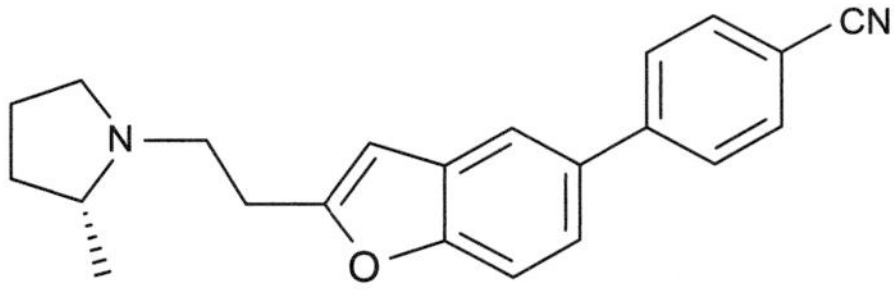

8

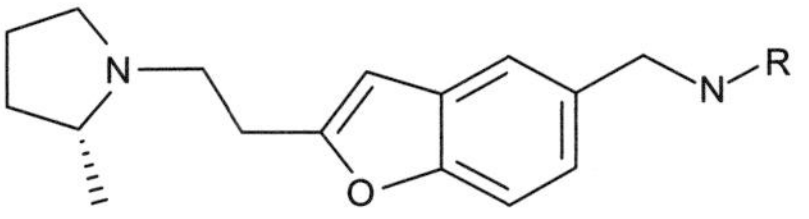

9 R = H
10 R = pyrimidin-2-yl

Similarly, JNJ researchers disclosed that JNJ-5207852 (**11**), a potent selective H3 antagonist, produced PLD following prolonged high exposure in rodents [42]. Since **11** was a dibasic CAD with a very high volume of distribution at steady state (Vdss) and a long $t_{1/2}$ (multiple days in brain), the approach used to produce compounds with an improved safety profile included reducing both the ClogP and the pK_a of the two amines [43]. This initially led to JNJ-10181457 (**12**), a compound in which one of the two piperidine groups was replaced with a morpholine. It had a much shorter $t_{1/2}$, with no detectable brain levels remaining after 24 h. Replacement of the second piperidine and increasing polarity, *via* incorporating an amide linker, led to compounds (**13**) and (**14**). These were still active *in vivo* and had reduced Vdss, improved clearance (in rats), and did not exhibit PLD [44].

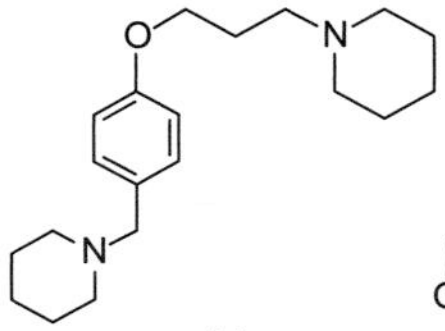

11 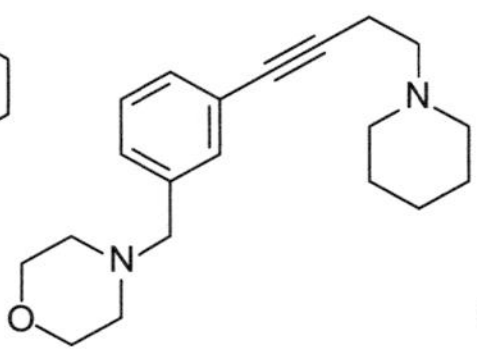 **12** 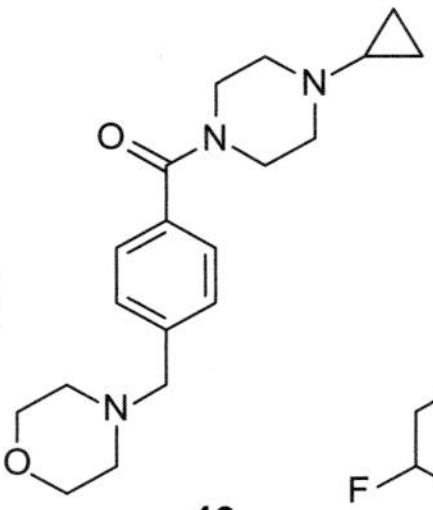 **13** 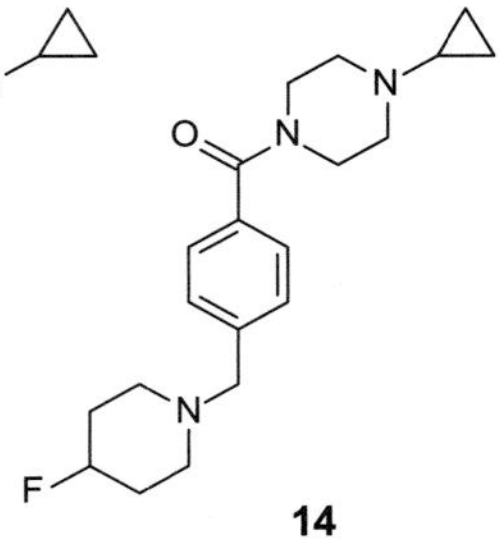 **14**

An additional complication that teams face is that in rare cases PLD has been found during *in vivo* studies even though the compound had not been predicted to show PLD by earlier assays [19]. In several of these cases, it has been possible to show that circulating metabolites do profile *in vitro* (or *in silico*) as probable inducers of PLD, thereby providing a link.

5. CONCLUSION

The original view that PLD is just an adaptive/reversible response has been replaced by a cautionary one that considers it a sign of a potential toxicological response. Unfortunately, there is no clear understanding why some PLD responses, and CAD chemistries, are linked with concurrent toxicities and others are not. Until clarification of this occurs, it is incumbent upon the drug discovery and development community to minimize the potential for PLD to manifest.

This presents a problem that could potentially be solved by avoiding all CADs. However, given the number of marketed CADs that is safe and effective, this option could hinder access to many potentially valuable treatments. One approach that is illustrated in this review is to use *in silico* models and *in vitro* assays to maximize the target potency/PLD induction ratio, then confirm with *in vivo* studies that PLD has been minimized or eliminated, and that more importantly, related concurrent toxicities are not occurring. An alternative approach, which targets the same receptors/disease but avoids CADs and is the subject of much current research in the field of 7TM-GPCR receptors, is to develop allosteric modulators [45].

ACKNOWLEDGMENTS

The authors gratefully acknowledge Anna-Lena Berg (AstraZeneca Pharmaceuticals) for her seminal pathology contributions to the AstraZeneca 5-HT_{1B} project and for directly influencing our general understanding of the phospholipidotic response.

REFERENCES

[1] W. H. Halliwell, *Toxicol. Pathol*, 1997, **25**, 53.
[2] F. A. de la Iglesia, G. Feuer, E. J. McGuire and A. Takada, *Toxicol. Appl. Pharmacol.*, 1977, **34**, 28.
[3] U. P. Kodavanti and H. M. Mehendale, *Pharmacol. Rev.*, 1990, **42**, 327.
[4] H. Sawada, K. Takami and S. Asahi, *Toxicol. Sci.*, 2005, **83**, 282.
[5] B. D. Wilson, C. E. Clarkson and M. L. Lippman, *Am. Rev. Resp. Dis.*, 1991, **143**, 1110.
[6] M. J. Reasor, K. L. Hastings and R. G. Ulrich, *Expert Opin. Drug Saf.*, 2006, **5**, 567.
[7] M. E. Cartwright, J. Petruska, J. Arezzo, D. Frank, M. Litwak, R. E. Morrissey, J. MacDonald and T. E. Davis, *Toxicol. Pathol.*, 2009, **37**, 902.
[8] E. A. Tengstrand, G. T. Miwa and F. Y. Hsieh, *Expert Opin. Drug Metab. Toxicol.*, 2010, **6**, 555.
[9] Pink Sheet: August 16, 2004 | Volume 66| Number 033 |p. 24: FDA Phospholipidosis Guidance Being Developed.
[10] For slides presented at the FDA PS&CP Advisory Committee meeting on April 14th 2010, see: http://www.fda.gov/downloads/AdvisoryCommittees/CommitteesMeetingMaterials/Drugs/AdvisoryCommitteeforPharmaceuticalScienceandClinicalPharmacology/UCM210798.pdf.
[11] L. Barone, S. Boyer, J. Damewood, J. Fikes, S. Matis, G. Gipson and P. Ciaccio, Are Phospholipogenic Compounds Associated with More Toxicities than Non-Phospholipogenic Compounds?, Society of Toxicology Meeting, Baltimore, MD, March, 2009.
[12] J. M. Willard, FDA PLWG Activities on Phospholipidosis: Data Mining, Modeling and Laboratory Research, *Phospholipidosis Working Group and Division of Cardiovascular and Renal Products—CDER. FDA ADVISORY MEETING*, April 14, 2010. see Ref.[10] for slide set.
[13] B. G. Small, L. R. Barone, J. P. Valentine, C. Pollard and P. J. Ciaccio, *J. Pharmacol. Toxicol. Methods*, 2007, **56**(2), e70.
[14] L. R. Barone, P. J. Ciaccio, B. G. Small, J.-P. Valentin and C. E. Pollard, Society of Toxicology Meeting, Charlotte, NC, March, 2007.
[15] See Slides 77 and 78 in: J. M. Willard, Ref.[12], FDA ACPS meeting, April 14, 2010.
[16] M. J. Reasor, Overview of Drug Induced Phospholipidosis, *Phospholipidosis Working Group and Division of Cardiovascular and Renal Products—CDER. FDA Advisory Meeting*, April 14, 2010, see Ref. [10] for slide set.
[17] B. R. Berridge, L. A. Chatman, M. Odin, A. E. Schultze, P. E. Losco, J. T. Meehan, T. Peters and S. L. Vonderfecht, *Toxicol. Pathol.*, 2007, **35**, 325.
[18] A. J. Ratcliffe, *Curr. Med. Chem.*, 2009, **16**, 2816.
[19] K. L. Hastings, Phospholipidosis Industry Perspective, *Phospholipidosis Working Group and Division of Cardiovascular and Renal Products—CDER. FDA Advisory Meeting*, April 14, 2010. see Ref. [10] for slide set.
[20] L. A. Chatman, D. Morton, T. O. Johnson and S. D. Anway, *Toxicol. Pathol.*, 2009, **37**, 997.
[21] P. D. Leeson and J. R. Empfield, *Ann. Rep. Med. Chem.*, 2010, **45**, 393.
[22] D. K. Monteith, R. E. Morgan and B. Halstead, *Expert Opin. Drug. Metab. Toxicol.*, 2006, **2**, 687.
[23] R. Lowe, R. C. Glen and J. B. O. Mitchell, *Mol. Pharm.*, 2010, **7**, 1708.

[24] H. Fisher, M. Kansy and D. Bur, *Chimia*, 2000, **54**, 640.
[25] J.-P. H. T. M. Ploemen, J. Kelder, T. Hafmans, H. van de Sandt, J. A. van Burgsteden, P. J. M. Salemink and E. van Esch, *Exp. Toxicol. Pathol.*, 2004, **55**, 347.
[26] D. J. Pelletier, D. Gehlhaar, A. Tilloy-Ellul, T. O. Johnson and N. Greene, *J. Chem. Inf. Model.*, 2007, **47**, 1196.
[27] N. L. Kruhlak, S. S. Choi, J. F. Contrera, J. L. Weaver, J. M. Willard, K. L. Hastings and L. F. Sancilio, *Toxicol. Mech. Methods*, 2008, **18**, 217.
[28] U. M. Hanumegowda, R. Yordanova, G. Wenke, J. P. Corradi, A. Regueiro-Ren and S. P. Adams, *Chem. Res. Toxicol.*, 2010, **23**, 749.
[29] R. G. Ulrich, K. S. Kilgore, E. L. Sun, C. T. Cramer and L. C. Ginsberg, *Toxicol. Methods*, 1991, **1**, 89.
[30] R. J. Gum, D. Hickman, J. A. Fagerland, M. A. Heindel, G. D. Gagne, J. M. Schmidt, M. R. Michaelides, S. K. Davidsen and R. G. Ulrich, *Biochem. Pharmacol.*, 2001, **62**, 1661.
[31] T. Kasahara, K. Tomita, H. Murano, T. Harada, K. Tsubakimoto, T. Ogihara, S. Ohnishi and C. Kakinuma, *Toxicol. Sci.*, 2006, **90**, 133.
[32] J. K. Morelli, M. Buehrle, F. Pognan, L. R. Barone, W. Fieles and P. J. Ciaccio, *Cell Biol. Toxicol.*, 2006, **22**, 15.
[33] M. Natalie, S. Margino, H. Erik, P. Annelieke, V. Geert and V. Philippe, *Toxicol. In Vitro*, 2009, **23**, 217.
[34] P. Nioi, B. K. Perry, E.-J. Wang, Y.-Z. Gu and R. D. Snyder, *Toxicol. Sci.*, 2007, **99**, 162.
[35] J. U. Peters, D. Hunziker, H. Fischer, M. Kansy, S. Weber, S. Kritter, A. Muller, A. Wallier, F. Ricklin, M. Boehringer, S. M. Poli, M. Csato and B.-M. Loeffler, *Bioorg. Med. Chem. Lett.*, 2004, **14**, 3575.
[36] A. L. Berg and P. Ciaccio, Compounds Causing Phospholipidosis, Including Effects on the Central Nervous System, Classic examples in Toxicologic Pathology XVI. Seminar for the European Society of Toxicologic Pathology and the Graduate School for Biomedical Sciences at the University of Veterinary Medicine in Hannover, February 27–28, 2009.
[37] D. A. Nugiel, J. R. Krumrine, D. Hill, J. R. Damewood, P. R. Bernstein, C. D. Sobotka-Briner, J. Liu, A. Zacco and M. E. Pierson, *J. Med. Chem.*, 1876, **2010**, 53.
[38] P. R. Bernstein, 3rd RSC/SCI Symposium on GPCRs in Medicinal Chemistry, MSD, Oss, The Netherlands, September 20–22, 2010.
[39] R. M. R. Sarmiento, M. T. Nettekoven, S. Taylor, J. M. Plancher, H. Richter and O. Roche, *Bioorg. Med. Chem. Lett.*, 2009, **19**, 4495.
[40] M. Sun, C. Zhao, G. A. Gfesser, C. Thiffault, T. R. Miller, K. Marsh, J. Wetter, M. Curtis, R. Faghih, T. A. Esbenshade, A. A. Hancock and M. Cowart, *J. Med. Chem.*, 2005, **48**, 6482.
[41] J. P. Montenez, F. Van Bambeke, J. Piret, R. Brasseur, P. M. Tulkens and M. P. Mingeot-Leclercq, *Toxicol. Appl. Pharmacol.*, 1999, **156**, 129.
[42] P. Bonaventure, M. Letavic, C. Dugovic, S. Wilson, L. Aluisio, C. Pudiak, B. Lord, C. Mazur, F. Kamme, S. Nishino, N. Carruthers and T. Lovenberg, *Biochem. Pharmacol.*, 2007, **73**, 1084.
[43] See discussion pages 114–116: S. Celanire, F. Lebon and H. Stark, in *The Third Histamine Receptor: Selective Ligands as Potential Therapeutic Agents in CSN Disorders* (eds. D. Vohora), CRC Press, Boca Raton, 2009, pp. 103–166.
[44] N. I. Carruthers, (1-[4-(3-Piperidin-1-ylpropoxy)benzyl]piperidine): A Template for the Design of Potent and Selective Non-Imidazole Histamine H3 Receptor Antagonists, 33rd Meeting of the European Histamine Research Society, Cologne, Germany, April 28th–May 2nd, 2004.
[45] P. J. Conn, A. Christopoulos and C. W. Lindsley, *Nat. Rev. Drug Discov.*, 2009, **8**, 41.

PART VIII:
Trends and Perspectives

CHAPTER 26

To Market, To Market—2010

Joanne Bronson*, **Murali Dhar****,
William Ewing[†] and **Nils Lonberg**[‡]

Contents

* Bristol-Myers Squibb Company, Wallingford, CT 06492, USA
** Bristol-Myers Squibb Company, Princeton, NJ 08543, USA
† Bristol-Myers Squibb Company, Pennington, NJ 08534, USA
‡ Bristol-Myers Squibb Company, Milpitas, CA 95035, USA

Annual Reports in Medicinal Chemistry, Volume 46
ISSN: 0065-7743, DOI: 10.1016/B978-0-12-386009-5.00002-3

OVERVIEW

Approvals of new molecular entities (NMEs) were somewhat lower in 2010 than in previous years. For example, 21 new drugs were approved in the United States in 2010, compared with 25 in 2009 and 24 in 2008 [1]. Seven of the 21 U.S. approved NMEs had previously been approved outside of the United States. Approvals continue to include first-in-class agents, along with drugs for established targets that offer significant new benefits. This year's To-Market-To-Market chapter provides summaries for 24 NMEs that have been approved for the first time worldwide, of which 19 are small molecules. Twelve of the summaries are for NMEs that were first approved in the United States and seven are for NMEs first approved in the European Union. The remaining NMEs are from Canada, Japan, and Russia. Treatments for cancer dominate the list, with five drug approvals for small molecules and approval of the first vaccine for hormone-refractory prostate cancer. The infectious disease and cardiovascular disease areas are next in number of new drug approvals, with three each. Two new drugs were approved for treatment of multiple sclerosis (MS) and associated symptoms. Other therapeutic categories saw approval of one to two new drugs each. For completeness, a few marketed drugs that were approved in 2009 and were not covered in Volume 45 of *ARMC* are summarized in this chapter. The following overview is organized by therapeutic area, starting with drugs that are covered in the detailed summaries and followed by approvals that are not covered in the detail but are of significant interest, including new combinations, new indications for previously approved drugs, new vaccines, and new enzymes.

In the anticancer area, five new small molecules were approved along with a novel therapeutic vaccine for the treatment of hormone-refractory prostate cancer. Of the small molecules, four are related to natural products and one is a cyclic peptide. Jevtana® (cabazitaxel), a microtubule inhibitor, is a semisynthetic analog of the natural product taxol® that was approved by the Food and Drug Administration (FDA) in combination with prednisone for the treatment of metastatic castration-resistant prostate cancer (mCRPC) in patients who were previously treated with a docetaxel-containing regimen for late-stage disease. Prostate cancer usually occurs in older men and is the second most common cancer among men in the United States, behind skin cancer. Cabazitaxel is dosed intravenously at 25 mg/m^2 over 1 h every 3 weeks along with 10 mg of oral prednisone administered daily throughout cabazitaxel treatment. The median overall survival rate for patients on cabazitaxel treatment was ~2.4 months longer than current alternatives. Halaven™ (eribulin), another microtubule inhibitor, is a synthetic analog of the marine natural product halichondrin B that was approved for the treatment of metastatic breast cancer (MBC) in patients who previously received at least two chemotherapeutic regimens for late-stage disease. Breast cancer is the second leading cause of cancer-related deaths among women in the United States after lung cancer, according to the National Cancer Institute. Eribulin mesylate is dosed intravenously at 1.4 mg/m^2 over 2–5 min on days 1 and 8 of a 21-day cycle. The median overall survival rate for patients on eribulin mesylate treatment was found to be ~2.5 months longer than current alternatives. Eribulin has 19 stereocenters, and the commercial route to its synthesis is purported to involve 62 steps. Junovan® (mifamurtide) is a liposomal formulation of muramyl tripeptide phosphatidylethanolamine that has been approved by the European Commission (EC) for the treatment of osteosarcoma. Phase III results with mifamurtide clearly demonstrated ~30% decrease in the risk for death, with ~78% of patients surviving through the sixth year of treatment. Mifamurtide is thought to act by stimulating the innate immune system to release proinflammatory cytokines, leading to tumoricidal activity. Mifamurtide is dosed intravenously at 2 mg/m^2 administered as adjuvant therapy over 1 h twice weekly for an initial 12 weeks. Istodax® (romidepsin) is a natural product isolated from *Chromobacterium violaceum* that was approved by the FDA as a single-agent therapy for the treatment of cutaneous T cell lymphoma (CTCL) in patients who previously received at least one systemic therapy. Romidepsin is a cyclic peptide derivative and is an inhibitor of histone deacetylases (HDACs) with modest selectivity for class I HDACs. Romidepsin is dosed intravenously at 14 mg/m^2 over 4 h on days 1, 8, and 15 of a 28-day cycle. In clinical trials, the overall objective disease response (ODR) and complete response (CR) rates were found to be ~34% and 6%, respectively. Javlor® (vinflunine) is a fluorinated, semisynthetic analog of the natural *vinca* alkaloids vinblastine and vincristine that has been approved by EMEA for the treatment of bladder cancer for patients who were previously treated with a first-line platinum-containing regimen. The recommended dose of vinflunine ditartrate is 320 mg/m^2 administered intravenously over 20 min once every 3 weeks. In a Phase III study, vinflunine treated patients showed improvement in the median overall survival of 6.9 months compared to 4.3 months for patients who were treated with best supportive care (BSC). Provenge® (sipuleucel-T) is an autologous vaccine for the treatment of asymptomatic or minimally symptomatic metastatic hormone-refractory prostate cancer. The vaccine is made from individual patient's blood cells that have been cultured with a recombinant fusion protein comprising the prostate tumor antigen human prostatic acid phosphatase fused to granulocyte macrophage colony-stimulating factor (GM-CSF). It is administered as three separate infusions given at 2-week intervals. Sipuleucel-T improved median survival by 4.5 months. In addition to new single-agent approvals, several new indications for previously approved drugs for anticancer treatments are noteworthy. Two oral kinase inhibitors received extended approval for the treatment of a rare type of leukemia. Tasigna® (nilotinib) from Novartis and Sprycel® (dasatinib) from Bristol-Myers Squibb both received approval for the additional indication of treatment of Philadelphia chromosome positive chronic phase chronic myeloid leukemia (Ph+ CP-CML), a slowly progressing blood and bone marrow disease linked to a genetic abnormality. Tasigna® was previously approved for treatment of chronic myeloid leukemia. Sprycel® was previously approved for treatment of chronic myelocytic leukemia and acute lymphocytic leukemia. The FDA extended the indications for the monoclonal antibody Rituxan® (rituximab) with its approval for treatment of chronic lymphocytic leukemia (CLL), a slowly progressing blood and bone marrow cancer. Rituxan® is being developed by Biogen Idec and Genentech (Roche) and is intended for patients with CLL who are beginning chemotherapy for the first time and for those who have not responded to other cancer drugs for CLL. Rituxan® was previously approved for treatment of B cell lymphoma and rheumatoid arthritis. Herceptin® (trastuzumab), a monoclonal antibody from Genentech that selectively binds with high affinity to the extracellular domain of the human epidermal growth factor receptor 2 protein (HER2), was approved for treatment of gastric cancer. Herceptin® is specifically indicated in combination with cisplatin and capecitabine or 5-fluorouracil for the treatment of patients with HER2 overexpressing metastatic gastric or gastroesophageal junction adenocarcinoma, who have not received prior treatment for metastatic disease. Herceptin® was previously approved for treatment of HER2 overexpressing breast cancer.

The area of infectious disease treatments saw approvals for one antibacterial drug and two anti-influenza drugs, along with two new vaccines. Teflaro® (ceftaroline fosamil), a novel cephalosporin prodrug, was approved as an injectable antibiotic to treat adults with community-acquired bacterial pneumonia (CABP) and acute bacterial skin and skin structure infections (ABSSSI), including those caused by methicillin-resistant *Staphylococcus aureus* (MRSA). Infections due to drug-resistant gram-positive bacteria, particularly MRSA, are a continuing concern worldwide. Four Phase III trials were used to demonstrate the efficacy of ceftaroline fosamil in ABSSSI and CABP. In ABSSSI, clinical cure rates were similar and high ($>90\%$) for both the ceftaroline and vancomycin plus aztreonam-treated groups. In CABP, clinical cure rates were higher for ceftaroline fosamil than for ceftriaxone for gram-positive organisms and were similar for gram-negative organisms. Like most cephalosporins, ceftaroline fosamil was safe and well tolerated. Two new neuraminidase (NA) inhibitors, Inavir® (laninamivir octanoate) and Rapiacta (peramivir), were approved in Japan in 2010 for treatment of influenza infection, joining previously approved NA inhibitors Relenza® (zanamivir) and Tamiflu® (oseltamivir). Influenza is a global health concern, with both seasonal epidemics and unpredictable pandemics resulting in significant morbidity and mortality, particularly for patients at high risk for influenza-associated complications. Laninamivir octanoate, an ester prodrug form of the active drug laninamivir, is given by intranasal administration at a 20 or 40 mg dose. Laninamivir octanoate has a long half-life in humans such that efficacy can be achieved after only a single dose. Peramivir is the only NA inhibitor available for intravenous (IV) use. Peramivir is given as a single 300 mg IV dose for adult and pediatric uncomplicated seasonal influenza infection, and as single and multiple 600 mg IV doses for patients at high risk for complications associated with influenza virus infection. Two vaccines were approved in the infectious diseases arena. Prevnar 13® from Pfizer is a pneumococcal 13-valent conjugate vaccine that was approved in the United States for infants and young children ages 6 weeks through 5 years for the prevention of invasive disease caused by 13 different serotypes of the bacterium *Streptococcus pneumoniae*. It is also approved for the prevention of otitis media caused by the seven serotypes shared with a previously approved vaccine, Prevnar®. *S. pneumoniae* can cause pneumonia as well as infections of the blood, middle ear, and the covering of the brain and spinal cord. Prevnar 13 will be the successor to Prevnar®, extending protection to six additional types of disease-causing bacteria. Menveo® from Novartis was approved as a vaccine to prevent meningococcal disease. Meningococcal disease is a leading cause of bacterial meningitis, an infection of the membrane around the brain and spinal cord, and sepsis, an often life-threatening blood infection. The vaccine is indicated for the active immunization of people between 2 and 55 years old in the prevention of invasive meningococcal disease caused by *Neisseria meningitidis* serogroups A, C, Y, and W-135.

Within the cardiovascular disease arena, three new agents were approved, along with several new drug combinations. Kalbitor® (ecallantide) is an analog of tissue factor pathway inhibitor (TFPI) and is a first-in-class drug that potently inhibits plasma kallikrein and effectively treats the symptoms and frequency of hereditary angioedema (HAE), an autosomal disease that affects between 1:10,000 and 1:50,000 people. Patients with HAE go through a cycle of flare-ups with symptoms of pain and swelling of cutaneous and mucosal tissues. The disease results from a genetic deficiency of the C1-esterase inhibitor protein, which in turn leads to the excessive formation of bradykinin resulting in edema. In Phase III trials, ecallantide was found to be effective in relieving HAE symptoms and reducing the frequency and duration of HAE attacks. The drug was effective in moderate and severe HAE patients and was found to have a 24-h duration of action. Ecallantide is given as three 10 mg subcutaneous doses. Brilique™ (ticagrelor), an antagonist of the $P2Y_{12}$ receptor that effectively blocks platelet aggregation, was approved for the prevention of atherothrombotic events in patients with acute coronary syndrome (ACS). It is the first reversible $P2Y_{12}$ receptor antagonist to be approved. ACS and related events are the leading cause of mortality in the United States. The rupture of an atherosclerosis plaque in coronary arteries initiates a series of biochemical events that leads to platelet aggregation and thrombus formation with serious medical consequences. Stimulation of the $P2Y_{12}$ receptor is a powerful initiation step in the GPIIb/IIIa pathway of platelet activation and aggregation. As a $P2Y_{12}$ antagonist, ticagrelor is effective in treatment of events associated with ACS, including myocardial infarction, stroke, death from vascular complications, ischemia, and stent thrombosis. Brinavess™ (vernakalant) has been approved as an IV agent for the treatment of patients experiencing short duration or recent onset atrial fibrillation (AF), a condition in which signals from the atria of the heart are discoordinated with the ventricle signals, resulting in irregular heart beat. Episodes can last from a few minutes to days to weeks and years. The symptoms include chest pains, palpitations, and weakness. The condition is associated with congestive heart failure (CHF). Vernakalant is a selective class III antiarrhythmic that has more of its effects in atrial tissues than ventricular tissues. It has activity for cardiac $Na+$ and $K+$ channels and also for the atrial-selective Kv1.5 channel. Two cardiovascular drugs that were previously approved outside the U.S. received FDA approval in 2010: Pradaxa® (dabigatran) for prevention of stroke and Lumizyme® (alglucosidase alfa) for treatment of Pompe disease, a rare, inherited and often fatal disorder that disables the heart and muscles. These were reviewed in To-Market-To-Market in previous years. For combination agents, there were three new approvals for the treatment of hypertension. It is estimated that about one billion people globally have high blood pressure, and many of these remain either untreated, or even if treated, are not at their ideal blood pressure target. Tekamlo® from Novartis is a single-pill combination of the renin inhibitor Tekturna® (aliskiren) and the calcium-channel blocker, amlodipine. The FDA approval of Tekamlo® for high blood pressure was based on clinical trial data involving more than 5000 patients with mild-to-moderate high blood pressure. Tekamlo® gave greater reductions in systolic and diastolic blood pressure compared with either agent alone. The aliskiren component of Tekamlo® inhibits the activity of the renin angiotensin aldosterone system (RAAS), an important regulator of blood pressure, while the calcium-channel blocker, amlodipine, lowers blood pressure by relaxing the blood vessel walls through the inhibition of calcium influx. Amturnide™ is a single oral tablet from Novartis combining Tekturna® and amlodipine with the diuretic hydrochlorothiazide (HCTZ). Approval of Amturnide™ was based on a double-blind, active-controlled study in 1181 patients with moderately or severely elevated blood pressure. Both patient populations achieved greater systolic and diastolic blood pressure reductions with Amturnide™ compared to the dual combinations. Amturnide™ is indicated for the treatment of hypertension in patients not adequately controlled with any two of the following: aliskiren, dihydropyridine calcium-channel blockers, and thiazide diuretics. Tribenzor™ from Daiichi-Sankyo, a combination of three drugs (amlodipine, HCTZ, and an angiotensin II receptor antagonist olmesartan medoxomil), was approved as an oral agent for the treatment of hypertension. The FDA approval of Tribenzor™ was based on a double-blind, active-controlled study in 2492 hypertensive patients, with subjects receiving Tribenzor™ or one of three dual therapies. After 8 weeks of treatment, the triple combination therapy produced greater reductions in both systolic and diastolic blood pressures compared to each of the three dual combination therapies. Tribenzor™ is the second three-drug combination pill approved by the FDA, with Exforge HCT® (Novartis; amlodipine/valsartan/HCTZ) receiving approval in 2009.

Approvals for endocrine disease therapeutics included drugs for treating diabetes, agents for infertility and contraception, and a combination drug for treating benign prostatic hyperplasia. Diabetes is a disease with increasing worldwide prevalence that has a significant impact on human mortality and the cost of healthcare. Nesina® (alogliptin) is an inhibitor of dipeptidyl peptidase-4 (DPP-4) that increases the concentration and half-life of the incretin GLP-1, a master regulator of glucose homeostasis. Alogliptin is the fourth DPP-4 inhibitor approved for the treatment of diabetes. Victoza® (liraglutide), a GLP-1 analog, was approved in the United States for diabetes treatment. This drug was previously approved in several countries outside the United States and was summarized in To-Market-To-Market in 2009. Kombiglyze XR™ from Bristol-Myers Squibb and Astra-Zeneca is a new combination tablet for the treatment of type 2 diabetes consisting of metformin and saxagliptin, a DPP-4 inhibitor. The FDA approved once-a-day Kombiglyze XR™ based on two Phase III clinical trials and bioequivalence studies. Kombiglyze XR™ offers strong glycemic control and is indicated as an adjunct to diet and exercise to improve glycemic control in adults with type 2 diabetes mellitus when treatment with both saxagliptin and metformin is appropriate. In the area of reproductive health, one new drug was approved for infertility and a new combination was approved for contraception. Elonva® (corifollitropin alfa) is a long-acting follicle-stimulating hormone (FSH) analog that was approved for increasing the success rate for conception. Corifollitropin alfa is first-in-class for sustained follicle stimulants (SFS). One injection of corifollitropin alfa replaces seven daily injections of recombinant FSH (rFSH) in a controlled ovarian stimulation cycle. Natazia® from Bayer Healthcare Pharmaceuticals is a combination hormonal tablet approved for use as an oral contraceptive. Natazia® contains two female hormones, an estrogen (estradiol valerate) and a progestin (dienogest), and is the first oral contraceptive marketed in the United States to vary the doses of progestin and estrogen four times throughout each 28-day treatment cycle. The safety and efficacy of Natazia® were evaluated in two multicenter Phase III clinical trials involving 1867 women and nearly 30,000 28-day treatment cycles. Natazia® was found to be effective as a hormonal contraceptive in both studies. The synthetic steroid ella® (ulipristal acetate) was approved in the United States for emergency contraception when taken orally within 120 h (5 days) after a contraceptive failure or unprotected intercourse. Ulipristal was previously approved in the European Union and was covered in *To-Market-To-Market* in 2009. Jalyn™ is a new combination drug from Glaxo that received approval for treatment of benign prostatic hyperplasia (BPH). BPH is one of the most common prostate disorders, affecting nearly half of all men 50 years of age or older in the United States. Jalyn™ combines the synthetic steroid dutasteride, a 5-alpha reductase inhibitor, with tamsulosin, an alpha1A adrenoreceptor blocker. Dutasteride reduces symptoms and shrinks the prostate and is already approved as a single agent (Avodart®) for treatment of BPH. Tamsulosin acts by relaxing the muscles in the bladder and prostate. Approval of the combination was based on 2-year results from the CombAT (Combination of Avodart® and Tamsulosin) study, one of the largest clinical trials to date of men with BPH.

Approvals in the metabolic disease category include drugs for osteoporosis, lipid disorders, and gout. Osteoporosis is a disease in which the bones become weak and are more likely to break. The vast majority of people with osteoporosis are women; one of every two women over age 50 will break a bone in their lifetime due to osteoporosis. Prolia®/Xgeva® (denosumab) is a fully human monoclonal antibody that inhibits bone resorption by blocking the ligand for the osteoclast receptor RANK. It was approved in two different dosage forms for treatment of postmenopausal women with osteoporosis at high risk for fracture, and for the prevention of skeletal-related events in patients with bone metastases from solid tumors. Denosumab is given as a 60-mg subcutaneous injection every 6 months for osteoporosis and 120 mg every 4 weeks for patients with bone metastases. For osteoporosis patients, the drug provided an approximately threefold reduction in fractures over placebo; for cancer patients, bone mineral density, progression free survival, overall survival, and adverse events were comparable to bisphosphonate therapy. In the area of lipid disorders, Egrifta™ (tesamorelin) is the first FDA-approved treatment for HIV patients with lipodystrophy, a condition in which excess fat develops in different areas of the body, most notably around the liver, stomach, and other abdominal organs. The use of antiretroviral agents for the treatment of HIV infection has been widespread since the mid-1990s because of their effectiveness in improving the symptoms and halting progression of the virus. One of the side effects observed with antiretroviral treatment (ART) is the accumulation of fat in some patients. Tesamorelin is a growth hormone-releasing factor (GRF) drug that is administered in a once-daily injection. VPRIV® (velaglucerase alfa) from Shire Human Genetic Therapies has been approved as a long-term, injectable enzyme replacement therapy for the treatment of pediatric and adult patients with type 1 Gaucher disease, a rare genetic disorder in which lipid accumulates in cells and certain organs due to a hereditary deficiency of the enzyme glucocerebrosidase. About 1:50,000 to 1:100,000 people in the general population have Gaucher disease. The safety and effectiveness of VPRIV® was assessed in three clinical studies involving 82 patients with type 1 Gaucher disease ages 4 years and older. The studies included patients who switched to VPRIV® after being treated with Cerezyme, a previously approved enzyme replacement therapy. The primary endpoint, hemoglobin concentration, either improved or stabilized upon treatment with VPRIV®. Krystexxa® (pegloticase) from Savient Pharmaceuticals is a polyethylene glycol conjugate of a recombinant mammalian uricase enzyme that was approved for IV treatment of chronic gout in adult patients refractory to conventional therapy. Gout occurs due to buildup of uric acid, which is eventually deposited as needle-like crystals in the joints or in soft tissue and causes the severe pain associated with the disease. Krystexxa® catalyzes the oxidation of uric acid to allantoin, a water-soluble metabolite that is readily eliminated, thereby lowering uric acid levels. Approval of Krystexxa® was based on two replicate, multicenter, randomized, double-blind, placebo-controlled 6-month studies in 212 patients. The difference between Krystexxa® and placebo was statistically significant for an every 2-week dosing regimen, but not for an every 4-week dosing regimen. Carbaglu® (carglumic acid) was approved in the United States for treatment of hyperammonemia due to *N*-acetylglutamate synthase (NAGS) deficiency, an extremely rare, genetic disorder that results in too much ammonia in the blood. Hyperammonemia can be fatal if it is not detected and treated early. Carbaglu® was previously approved in Europe in 2003.

For diseases of the central nervous system (CNS), new agents were approved for the treatment of schizophrenia and for multiple sclerosis (MS). The atypical antipsychotic Latuda® (lurasidone hydrochloride) was approved as a once-daily, oral agent for the treatment of patients with schizophrenia. Schizophrenia is a debilitating mental disorder that affects

1% of the population worldwide. Like other atypical antipsychotic agents, lurasidone has potent antagonist activity at D_2 and 5-HT_{2A} receptors. It is also a 5-HT_{1A} receptor partial agonist and a potent 5-HT_7 receptor antagonist. The efficacy of lurasidone for the treatment of schizophrenia was established in four 6-week, placebo-controlled studies in adult patients. The recommended starting dose for lurasidone is 40 mg/day. Unlike many atypical antipsychotics, lurasidone has a neutral effect on weight gain. Two new small molecule drugs were approved for people with MS, a chronic and disabling disease that affects the brain, spinal cord, and optic nerves. According to the National Multiple Sclerosis Society, there are about 400,000 people in the United States and 2.1 million people worldwide with MS. Ampyra® (dalfampridine) is a voltage-gated potassium channel blocker that is the first oral therapy approved by the FDA to improve walking in MS patients with existing gait impairment. The extended release formulation of dalfampridine improves pharmacokinetic parameters and minimizes the side effects of dalfampridine. A drug with a similar structure and mechanism of action, Firdapse™ (amifampridine) from BioMarin Pharmaceuticals, was approved in 2009 for the treatment of Lambert-Eaton myasthenic syndrome, a rare disorder of neuromuscular transmission. Gilenya® (fingolimod), a sphingosine-1-receptor agonist, is the first approved oral therapy for the relapsing–remitting form of multiple sclerosis (RRMS). Fingolimod is a prodrug and is converted *in vivo* to the active drug, (*S*)-fingolimod phosphate, primarily by sphingosine kinase-2. (*S*)-Fingolimod phosphate acts on sphingosine 1-phosphate receptors and blocks lymphocyte egress from lymph nodes and prevents their recirculation and entry into the CNS. The U.S. FDA approved fingolimod for RRMS at a dose of 0.5 mg once daily.

Two new drugs were approved for treatment of ophthalmologic diseases. Diquas (diquafosol) was approved in Japan in 2010 as a 3% ophthalmic solution for treatment of dry eye disease. Diquafosol is a $P2Y_2$ purinergic receptor agonist with the ability to activate the receptor on the ocular surface and stimulate water, lipid, and mucin secretion. The 3% solution was reported to improve dry eye symptoms by promoting secretion of mucin and water, thereby bringing the tear film closer to a normal state. Lastacaft™ (alcaftadine) was approved as a once-daily, 0.25% ophthalmic solution for the prevention of itching and redness associated with allergic conjunctivitis. Alcaftadine blocks histamine receptors and thus prevents the inflammatory effects of histamine.

In the respiratory diseases area, two new drugs were approved in 2010. Daxas® (roflumilast) is a selective, orally active phosphodiesterase (PDE) 4 inhibitor that was approved as an add-on to bronchodilator treatment for maintenance therapy of severe chronic obstructive pulmonary disease (COPD) associated with chronic bronchitis in adult patients with a history of frequent exacerbations. COPD is a chronic inflammatory disease and is the sixth most common cause of death worldwide and a major cause of morbidity. Roflumilast is a potent and competitive inhibitor of PDE4 and is equipotent against PDE4A, 4B, and 4D but is inactive against PDE4C and other members of the PDE family. Despite its inhibition of PDE4D, roflumilast shows the lowest incidence of nausea (3–5%) of the PDE4 inhibitors investigated to date. In clinical trials, patients with COPD having more severe airway obstruction showed a significant reduction in exacerbation frequency with roflumilast. In patients with chronic bronchitis, a 500-μg daily oral dose of roflumilast also gave a significant improvement in forced expiratory volume in 1 s (FEV_1) and a reduction in exacerbation rate. Bilaxten™ (bilastine) is an orally bioavailable, selective inhibitor of the histamine H_1 receptor that has a rapid biological onset and long duration of action. A total of 28 clinical trials were run with bilastine in 5000 patients. Bilastine effectively treats allergic rhinitis, shows no effect on cardiovascular parameters, and has no sedation side effects.

For treatment of pain, one new drug was approved in 2010. Civanex® (zucapsaicin) is a topical analgesic that was approved for use in conjunction with oral cyclooxygenase-2 (COX-2) inhibitors or nonsteroidal anti-inflammatory drugs (NSAIDs) to relieve severe pain in adults with osteoarthritis of the knee. Zucapsaicin is the *cis*-isomer of the natural product capsaicin. Its analgesic action is mediated through the transient receptor potential vanilloid type 1 (TRPV1) channel, a ligand-gated ion channel that is expressed in the spinal cord and brain and is localized on neurons in sensory ganglia. Initial constant activation of TRPV1 is followed by desensitization to a variety of noxious stimuli. In clinical trials, 0.075% zucapsaicin cream was efficacious in patients experiencing severe pain. Application site burning sensation was the most frequently reported adverse event and was predominantly mild to moderate. A new indication was approved for Eli Lilly's Cymbalta® (duloxetine), namely for the treatment of chronic musculoskeletal pain, including discomfort from osteoarthritis and chronic lower back pain. Cymbalta® was first used to treat major depressive disorder. It has also been previously approved for the treatment of diabetic peripheral neuropathy, generalized anxiety disorder, maintenance treatment of major depression, and fibromyalgia. The efficacy of Cymbalta® in chronic low back pain and osteoarthritis was assessed in four double-blind, placebo-controlled, randomized clinical trials. At the end of the study period, patients taking Cymbalta had a significantly greater pain reduction compared with placebo.

Additional approvals in 2010 include an agent for the treatment of a debilitating hand disease and one for removal of varicose veins. Xiaflex® (collagenase clostridium histolyticum) from Auxilium Pharmaceuticals was approved in the United States for treatment of a hand disease known as Dupuytren's contracture, in which buildup of collagen in the palm leads to formation of rope-like cords of tissue that prevent normal function of the fingers. Xiaflex® is a collagenase that acts by breaking down excessive collagen that builds up in the disease. In one 66-patient study, 44% of those injected with Xiaflex® were treated successfully, compared to 5% for patients who received a placebo. In a separate 306-patient study, 64% of patients given Xiaflex® were treated successfully, compared to only 7% of patients receiving the placebo. Asclera® (polidocanol) from Merz Aesthetics and Chemische Fabrik Kreussler was approved in the United States for removal of varicose veins. Polidocanol is a dodecyl ether derivative of a nine-unit polyethylene glycol. It works by damaging the cell lining of blood vessels, causing them to close and eventually be replaced by other types of tissue.

1. ALCAFTADINE (0.25%) (OPHTHALMOLOGIC, ALLERGIC CONJUNCTIVITIS) [2–6]

Class	Antihistamine
Country of origin	United States
Originator	Janssen Research Foundation
First introduction	United States
Introduced by	Vistakon Pharmaceuticals, LLC
Trade name	Lastacaft™
CAS registry no.	147084-10-4
Molecular weight	307.4

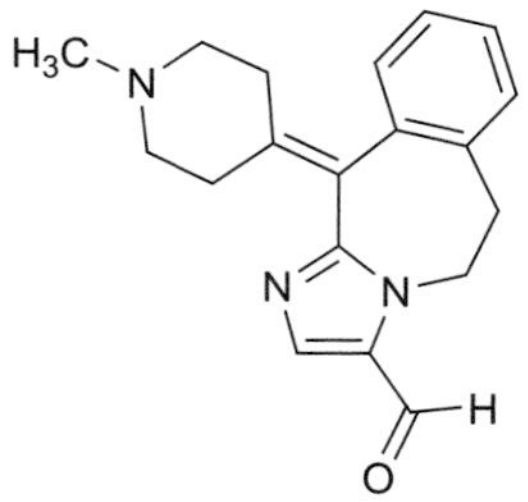

H_3C N N N H O

Alcaftadine, a histamine H_1/H_2 receptor antagonist, was approved in the United States in 2010 for the prevention of itching and redness associated with allergic conjunctivitis. Seasonal and perennial allergic conjunctivitis affects up to 40% of the population worldwide. There are numerous treatment options, with topical antihistamines being an effective therapy. Some of the primary symptoms and signs of allergic conjunctivitis are ocular itching and conjunctival redness. The pharmaceutical market for conjunctivitis is substantial and steadily increasing. For example, the market for allergic conjunctivitis in the United States increased from $6 million in 1993 to $2 billion in 2008. The market in Europe has experienced similar growth. In a healthy eye, tight junctions form a barrier that prevents foreign agents from penetrating the conjunctival epithelium and allows for the paracellular passage of nutrients and water. The tight junctions comprise cell membrane protein complexes that provide stability and adhesion by linking the cytoskeleton of adjacent epithelial cells. In patients with seasonal allergic conjunctivitis, epithelial cell adhesion proteins and cytoskeletal elements are found to be downregulated. Some of the common contributors to this degradation process are proteases from dust mites and peptidases from pollens. Once the protective element of the ocular surface is compromised, allergens penetrate and cause an allergic reaction. The degree of severity depends on a number of factors including the allergen load, the degree of allergen dilution within the tear film, and the expression of epithelial cell molecules. Allergen proteins cross-link with IgE to initiate an allergic cascade resulting in mast cell degranulation and in turn the release of allergen molecules including principally histamine. Ocular itching has been shown to result from histamine's stimulation of the H_4 receptors. Histamine's actions on H_1 and H_2 receptors result in the redness of the conjunctiva. Chemosis and lid swelling also result from H_1 receptor activation [2,3]. Alcaftadine is a potent antagonist of the H_1 receptor ($K_i = 3.1$ nM) and the H_2 receptor ($K_i = 58$ nM) and, additionally, is an antagonist of the H_4 receptor ($K_i = 2900$ nM). Clinical pharmacological studies identified one active metabolite of significance, the corresponding carboxylic acid of the aldehyde group found in alcaftadine. The metabolite forms through non-CYP-mediated enzymatic processes [3]. In an allergic mouse model, alcaftadine was compared to olopatadine where alcaftadine showed differentiation under the experimental protocol and doses chosen. Alcaftadine was found to significantly inhibit eosinophil recruitment when compared to olopatadine and the vehicle control group. The effects of eosinophils have been ascribed to H_4 receptor stimulation. Alcaftadine was found to protect epithelial tight junction proteins, assessed by measuring zonula occludens (ZO-1) and E-cadherin expression levels. Alcaftadine ophthalmic topical solution was assessed in clinical studies where it was found to be safe and effective in preventing ocular itching and redness of the conjunctiva. In a dose escalation study, a 0.25% alcaftadine ophthalmic solution was found to be the most effective at relieving ocular itching and conjunctive redness. Alcaftadine (0.25%) ophthalmic solution was found to have a fast onset of action (3–15 min) and a long duration of action (16 h) making it suitable for use as a once-a-day therapy. In Phase III studies using the conjunctival allergen challenge (CAC) model, alcaftadine (0.25%) ophthalmic solution showed statistically significant lower scores for conjunctival redness versus placebo at 7, 15, and 20 min after an allergen challenge with results persisting for 16 h.

Alcaftadine (0.25%) ophthalmic solution was found to be effective at preventing ocular itching 3, 5, and 7 min after an allergen challenge, demonstrating a fast onset of action. From the Phase III studies, alcaftadine (0.25%) ophthalmic solution was found to prevent ocular itching, reduce conjunctival redness and almost all other allergic signs and symptoms at 15 min and 16 h postdose [4–6]. In July 2010, the FDA approved alcaftadine (0.25%) ophthalmic solution for use in patients with allergic conjunctivitis to prevent itching. The once-daily ophthalmic solution, containing 2.5 mgs of alcaftadine, is marketed by Vistakon, a subsidiary of Johnson and Johnson, under the brand name Lastacaft™.

2. ALOGLIPTIN (ANTIDIABETIC) [7–15]

Class	DPP-4 inhibitor
Country of origin	Japan
Originator	Syrrx Inc. (now Takeda San Diego)
First introduction	Japan
Introduced by	Takeda Pharmaceuticals, Furiex Pharmaceuticals
Trade names	Nesina®
CAS registry no.	850649-61-5
	850649-62-6 (benzoate)
Molecular weight	339.4

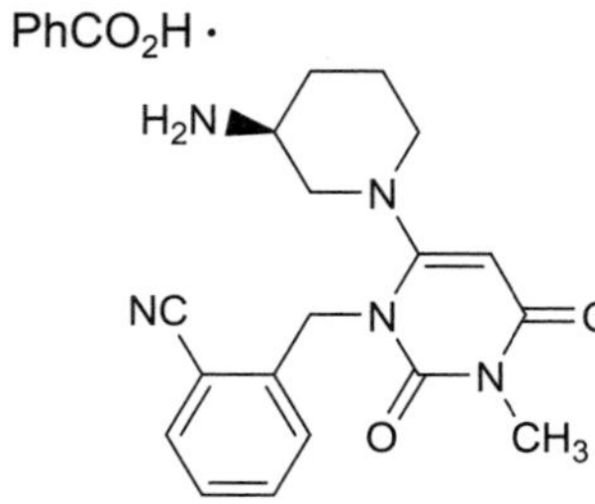

Alogliptin is a dipeptidyl-peptidase IV (DPP-4) inhibitor that was approved in Japan in 2010 for treatment of type 2 diabetes, a disease in which insulin resistance and β-cell dysfunction lead to hyperglycemia. According to the American Diabetic Association, diabetes is the seventh leading cause of death and increases the risk of heart disease and stroke by two to four times. Macro- and microvascular complications result from the progression of the severity of diabetes. The prevalence of diabetes continues to increase worldwide with an estimated 370 million people projected to be affected by 2030. The current number of cases in the United States (8% of the population) is predicted to double by 2050. In the United States, the economic impact of diabetes was estimated at $176 billion in 2007, with $116 billion attributed medical expenditures [7,8]. As a diabetic patient's metabolic control deteriorates, there is a need to escalate therapeutic intervention by increasing diabetic drug doses and then prescribing combination therapy. The progression of diabetes is attributed to several factors. The process of being in a hyperglycemic state with increases in free fatty acids, cytokines, adipokines, and associated metabolites leads to the loss of β-cell function and β-cell mass in islets [9]. As islet function is lost, the severity of insulin resistance increases. The introduction of DPP-4 inhibitors has brought a novel class of insulinotropic agents for the treatment options available to type 2 diabetic patients. The therapeutic potential of glucagon-like peptide 1 (GLP-1), an incretin peptide, for the treatment of type 2 diabetes was realized in the 1990s. The insulinotropic effects of GLP-1 depend closely on glucose concentrations providing the possibility of glucose normalization without the risk of hypoglycemia. GLP-1 has other non-insulinotropic physiological actions that are advantageous. It suppresses glucagon secretion from α cells and slows gastric emptying, which contributes to satiety and to a slower passage and reabsorption of carbohydrates. GLP-1 also contributes to satiety *via* a central mechanism as a neurotransmitter with effects on the hypothalamus. GLP-1 stimulates β-cell formation from precursor cells and inhibits their apoptosis leading to an increase in β-cell mass and to an improvement in β-cell function. GLP-1 reduces inappropriate glucagon secretion from the pancreas. The observation that oral intake of glucose gives a greater insulin response than IV glucose administration is termed the incretin effect; in type 2 diabetes, the incretin effect is impaired. In patients with type 2 diabetes, infusion of GLP-1 has been found to increase insulin secretion and to normalize both fasting and post-prandial blood glucose. GLP-1 is a 30-amino acid peptide that is inactivated by cleavage of the N-terminal dipeptide sequence (His-Ala) through the peptidase action of DPP-4. This inactivation occurs rapidly, with the half-life of circulating GLP-1 being <2 min. DPP-4 has several other substrates including another beneficial incretin peptide, gastric inhibitory peptide (GIP). Inhibitors of DPP-4 have been shown in man to increase GLP-1 and GIP levels two- to threefold. DPP-4 inhibitors are the first class of agents to utilize the pharmacology of GLP-1 and thus offer a novel way to increase endogenous incretin peptide concentrations. Because insulin secretion *via* the actions of GLP-1 occurs only in response to rising glucose levels, the risk of hypoglycemia is low, resulting in the wide acceptance of DPP-4 inhibitors into clinical practice. DPP-4 inhibitors are primarily once-a-day, weight-neutral drugs with a favorable adverse-effect profile. As shown by animal studies, the class can decrease β-cell apoptosis and increase β-cell survival. In animal models, DPP-4 inhibitors increase the number of insulin positive β-cells in islets. Islet insulin content is found to be increased and glucose-stimulated insulin secretion in isolated islets is improved [8–10].

Alogliptin (SYR-322) has been described as a potent, highly selective DPP-4 inhibitor. The discovery of alogliptin arose out of a designed series of quinazolinone-based inhibitors of DPP-4. The lead quinazolinone was found to inhibit CYP3A4 and to have micromolar affinity for the hERG channel. Replacement of the quinazolinone ring with a pyrimidinedione addressed both of these issues with none of the unwanted off-target activities observed. This discovery resulted in alogliptin (SYR-322). Alogliptin inhibits DPP-4 with an $IC_{50} < 10$ nM and is highly selective >10,000-fold) against DPP-8 and -9 as well as over other endopeptidases. The synthesis of alogliptin described in the literature starts with chloropyrimidinedione. One of the nitrogen atoms is selectively benzylated followed by methylation of the remaining nitrogen. This is followed by chloro displacement using 3-aminopiperidine to give alogliptin [11]. In animal models, alogliptin rapidly inhibits DPP-4 activity (15 min post-dose); plasma DPP-4 inhibition is sustained for 12 h in rats and dogs and for 24 h in monkeys. The animal pharmacokinetic profile was found to be supportive for once-daily dosing of alogliptin in man. The reported efficacy of alogliptin in multiple animal models of diabetes has been reported. In a 4-week ob/ob mouse model of type 2 diabetes, alogliptin dose dependently reduced plasma DPP-4 activity and increased GLP-1 levels. In addition, plasma insulin levels were increased, plasma glucose levels were decreased and HbA1c levels were reduced. Glucagon and triglyceride levels were also decreased [12,13]. In clinical studies, alogliptin has been found to be safe and well tolerated. In dose escalation studies, the pharmacokinetic parameters were linear over the 25–800 mg dose range. In man, alogliptin is rapidly absorbed and eliminated primarily intact with 40% hepatic and 60% renal elimination. There are two metabolites formed at very low levels. An active metabolite (M1) resulting from CYP 2D6 N-demethylation is formed as <2% of parent drug concentration. A second inactive metabolite resulting from the N-acetylation of M1 is <6% of total parent drug concentration. The half-life of alogliptin determined over six Phase III studies is 10.9–21.8 h. Alogliptin has been studied as monotherapy and in combination with pioglitazone and as an add-on with metformin, glyburide, pioglitazone, and insulin. As monotherapy, alogliptin significantly improved HbA1c levels from baseline as both the 12.5 and 25 mg doses, which are now the marketed doses [14,15]. Alogliptin benzoate, Nesina®, a highly selective DPP-4 inhibitor for the treatment of type 2 diabetes, received regulatory approval from the Japanese Ministry of Health, Labor and Welfare in April 2010. Nesina® is the fourth marketed inhibitor of DPP-4. Nesina® is being codeveloped by Takeda Pharmaceutical Company and Furiex Pharmaceuticals, Inc.

3. BILASTINE (ANTIALLERGY) [16–23]

Class	Selective histamine H_1 receptor antagonist
Country of origin	Spain
Originator	FAES FARMA, S.A.
First introduction	European Union
Introduced by	FAES FARMA, S.A., Menarini, Pierre Fabre Medicament, Merck-Serano
Trade names	Bilaxten™
CAS registry no.	202189-78-4
Molecular weight	463.6

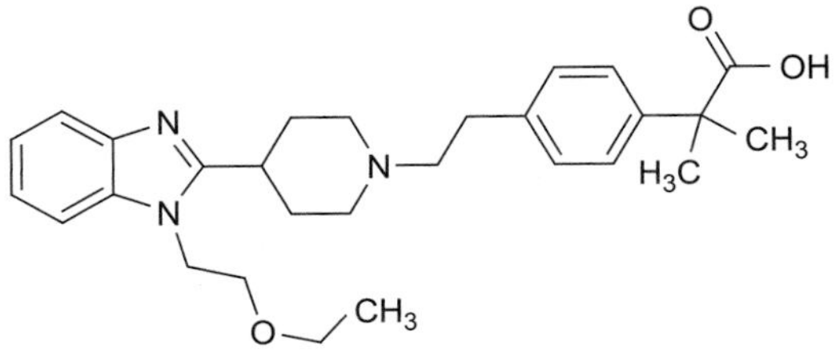

Bilastine, a potent and selective histamine H_1 receptor antagonist, was approved in Europe in 2010 for the treatment of allergic rhinoconjunctivitis (AR) and urticaria (hives or skin rash). The prevalence of AR is estimated to be >20% worldwide with some studies suggesting that 42% of the worldwide population is affected. In the United States, the allergies in America survey found that 14.2% of the population had been diagnosed with AR. The total worldwide market for antihistamines is >4.2 billion euros with one quarter of the market share in Europe. AR affects quality of life. Sufferers have symptoms of sneezing, nasal congestion, rhinorrhea, itching, headaches, and associated ocular conditions. In many AR patients, sleep and social functioning are affected. Thus, in addition to medical costs, there are indirect costs of AR such as decreased productivity and absenteeism from work or school. AR is a major risk factor for the development of asthma [16]. The inflammatory response in AR is triggered by an IgE-mediated reaction to allergens. One of the major mediators of this response is histamine. In response to stimuli, histamine is released from mast cells and basophilic granulocytes and interacts with the H_1, H_2, and H_4 histamine receptors. The resulting biological processes that are triggered include stimulation of sensory nerve and cough receptors, increased vascular permeability, and smooth muscle contraction. The main pharmacological agents to intervene with these processes are histamine H_1 antagonists. Bilastine is a potent and selective histamine H_1 receptor antagonist with a $K_i = 44$ nM (guinea pig ileum) and no activity (>100 μM) for the H_2 and H_3 receptors. In *in vitro* experiments conducted in guinea pig and rat tissues, bilastine showed no activity against serotonin, bradykinin, leukotriene (LTD4), muscarinic (M3), and adrenoreceptors. When tested in a panel screen (MDS Pharma Services), no activity was observed for the 30 assays which included adenosine, adrenoreceptors, dopamine, muscarinic, opiate, serotonin, and steroidal receptors. The functional antagonist activity of bilastine was demonstrated by its ability to block histamine and interleukin IL-4 release from human mast cells and peripheral blood granulocytes [17,18]. In animal studies, orally administered bilastine was found to have a rapid onset of action and a long duration of action supporting once-daily dosing in man. Bilastine was effective in rat models used to assess its antihistamine activity as measured by the reduction in the increase of histamine-mediated capillary permeability. In guinea pigs, bilastine was shown to be effective at reducing the microvascular extravasation induced by histamine in the trachea. As well, the antiallergic effects of bilastine were demonstrated in several other animal models. Based on the preclinical data, bilastine was advanced into clinical trials [19,20]. The original synthesis of bilastine [21] involves alkylation of 2-piperidinyl-1H-benzimidazole with a phenethyltosylate, the *para* position of which is substituted with a dimethyloxazoline moiety serving as a masked carboxylic acid group. Alkylation of the benzimidazole nitrogen with 2-chloroethyl ethyl ether followed by unmasking of the oxazoline moiety with sulfuric acid provided bilastine. An alternate synthesis of bilastine has since been reported [22].

In healthy male volunteers, bilastine was found to be rapidly absorbed with a T_{max} of 1 h. Bilastine showed linear pharmacokinetics over the dose range of 10–100 mg given once daily. The half-life was found to be shorter on day 1 when compared to day 14 (4.7 *vs.* 9.6 h for the 20 mg dose). After a single dose, bilastine is excreted in the feces (67%) and in the urine (33%). In Phase III clinical studies, a once-daily oral dose of 20 mg of bilastine effectively relieved the symptoms of allergic rhinitis. In addition, bilastine was effective at improving quality of life as measured by reduction of sleep disturbances and parameters of general discomfort. Studies in healthy volunteers and patients have demonstrated that bilastine is safe, lacking the unwanted sedative and cardiotoxic effects associated with some antihistamine drugs. In two major clinical trials, bilastine was effective at relieving allergic rhinitis as assessed by measuring the severity of nasal (obstruction, rhinorrhea, itching, sneezing) and nonnasal (ocular itching, tearing, ocular redness, itching of ears, and/or palate) symptoms. A total of 28 clinical trials were conducted with bilastine in 5000 patients [23]. In September 2010, the German health agency (BfARm) approved bilastine for treatment of AR and urticaria. Bilastine will be sold under the trade name Bilaxten™ in Spain as once-daily oral dosage of 20 mg.

4. CABAZITAXEL (ANTICANCER) [24–30]

Class	Tubulin inhibitor
Country of origin	France
Originator	Sanofi-Aventis
First introduction	United States
Introduced by	Sanofi-Aventis
Trade name	Jevtana®
CAS registry no.	183133-96-2
Molecular weight	835.9

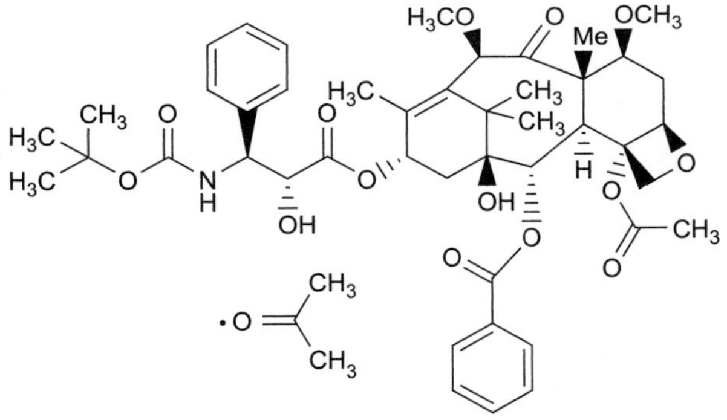

In June 2010, the U.S. FDA approved cabazitaxel (also referred to as XRP6258 and RPR 116258A) in combination with the steroid prednisone for the treatment of metastatic Castration-Resistant Prostate Cancer (mCRPC) for patients who were previously treated with a docetaxel-containing regimen for late-stage disease. The 2010 statistics from the U.S. National Cancer Institute are that ~220,000 men will be diagnosed with prostate cancer and ~32,000 men will die of the disease. Depending on the stage of the disease, prostate cancer symptoms may include urinary problems, impotence, blood in the urine or semen, and pain in the lower back, hip, or upper thighs. Chemotherapeutic standard of care for mCRPC usually involves treatment with the anticancer drug docetaxel in combination with prednisone. Cabazitaxel is a semisynthetic analog of the natural product taxol®, which is isolated from the bark of the yew tree. Cabazitaxel is a microtubule inhibitor that binds to the taxol-binding site of tubulin. Similar to other tubulin inhibitors of the taxol class, cabazitaxel inhibits microtubule disassembly resulting in mitotic blockade and cell death. Docetaxel, also a semisynthetic taxol analog, was approved by the FDA for the treatment of mCRPC in 2004. However, docetaxel is a substrate for P-gp, which is thought to contribute to the constitutive and acquired resistance of cancer cells to taxanes. Cabazitaxel has poor affinity for P-gp and showed antitumor activity in preclinical *in vitro* studies and *in vivo* tumor models that overexpress this protein. Cabazitaxel is synthesized on a commercial scale from 10-deacetylbaccatin [24]. Preclinical *in vitro* studies suggested that the antitumor activity of cabazitaxel is comparable to docetaxel with IC_{50}s ranging from ~4 to 35 nM. Moreover, cabazitaxel is active in cancer cell lines that are resistant to docetaxel with resistance factor ratios ranging from 1.8 to 10 for cabazitaxel versus 4.8–59 for docetaxel. In preclinical human tumor xenograft models, cabazitaxel showed complete regression rates in colon, lung, pancreatic, head and neck, kidney, and prostate carcinoma cell lines, when dosed intravenously [25,26].

Cabazitaxel exhibited dose-proportional pharmacokinetics over a dose range of 10–25 mg/m^2 following a 1-h IV infusion every 3 weeks to patients ($n = 25$) with solid tumors in a Phase I clinical trial [27]. The C_{max} and AUC increased in a dose-proportional fashion with a mean terminal half-life of ~77 h. The clearance and volume of distribution were high with mean values of 53.5 L/h and 2034 L/m^2, respectively. Cabazitaxel is extensively metabolized (>95% in liver) predominantly by CYP3A4/5 and to a minor extent by CYP2C8. The safety and efficacy of cabazitaxel were established in an open-labeled, randomized, Phase III study comprising 755 men with hormone-refractory metastatic cancer who were previously treated with a regimen that contained docetaxel [28]. Of the 755 patients, 378 were treated with cabazitaxel at 25 mg/m^2 dosed intravenously every 3 weeks for a maximum of 10 cycles along with 10 mg of prednisone orally and 377 were treated with 12 mg/m^2 of mitoxantrone dosed intravenously every 3 weeks for a maximum of 10 cycles along with 10 mg of prednisone dosed orally. The Phase III study clearly demonstrated that cabazitaxel-treated patients showed improvement in the median overall survival of 15.1 months compared to 12.7 months for patients in the mitoxantrone group. Most common serious adverse events (>5%) associated with cabazitaxel treatment were neutropenia, febrile neutropenia, leukopenia, anemia, diarrhea, fatigue, hypersensitivity, and asthenia. The recommended dose of cabazitaxel is 25 mg/m^2 administered intravenously over 1-h every 3 weeks along with 10 mg of oral prednisone administered daily throughout cabazitaxel treatment [29,30].

5. CEFTAROLINE FOSAMIL (ANTIBACTERIAL) [31–38]

Class	Bacterial cell wall synthesis inhibitor
Country of origin	Japan
Originator	Takeda
First introduction	United States
Introduced by	Forest Laboratories Inc.
Trade name	Teflaro®
CAS registry no.	402741-13-3
Molecular weight	684.7

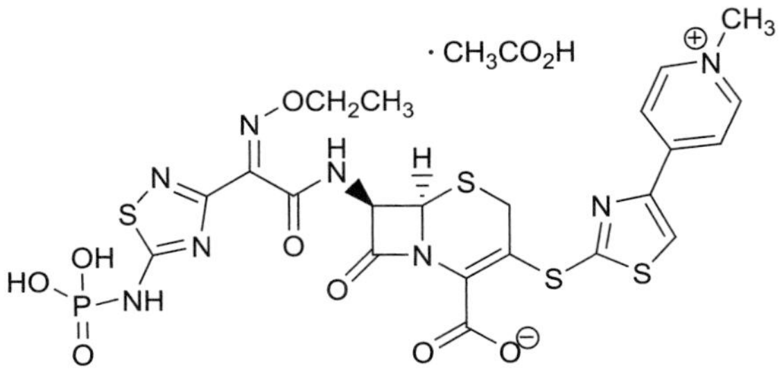

Ceftaroline fosamil, also referred to as TAK-599, is a cephalosporin antibacterial agent that was approved in the United States in October 2010 for the IV treatment of acute bacterial skin and skin structure infections (ABSSSI) and community-acquired bacterial pneumonia (CABP). Ceftaroline fosamil is the water-soluble, *N*-phosphono prodrug of ceftaroline (T-91825), a broad-spectrum, bactericidal agent with potent activity against methicillin-resistant *Staphylococcus aureus* (MRSA) strains, multidrug resistant *S. pneumonia*, and common gram-negative organisms [31–33]. Infections due to drug-resistant gram-positive bacteria, particularly MRSA, are a continuing and growing concern worldwide. For example, in 2005 more than 100,000 people in the United States developed serious MRSA infection and nearly 19,000 people died during a hospital stay due to serious MRSA infections [34]. About 85% of all invasive MRSA infections in the United States were associated with healthcare, with two-thirds occurring outside of the hospital and one-third occurring during hospitalization. In the European Union, a study published in 2010 showed that more than 150,000 patients within the healthcare setting have MRSA infections, with the proportion of MRSA infections ranging from 1% to 50%, depending on the country [35]. New MRSA strains have also emerged as community and livestock-associated human pathogens. Like other β-lactam antibiotics, ceftaroline acts by inhibiting the essential transpeptidase activity of penicillin-binding proteins (PBPs), which leads to inhibition of bacterial cell wall synthesis and, ultimately, bacterial cell death. MRSA resistance to β-lactams arises from an alteration in PBP to a new form, PBP2a, which retains function in cell wall synthesis, but has lower affinity for β-lactam antibiotics. Ceftaroline binds to PBP2a [36] as well as other PBPs with high affinity and, as a result, retains potent activity. Ceftaroline exhibits activity against most gram-positive pathogens, including β-lactam-susceptible and -resistant *S. aureus*, vancomycin-resistant *S. aureus*, and resistant and susceptible forms of *S. pneumoniae* but has weak activity against *Enterococcus* sp. The gram-negative antibacterial activity of ceftaroline is limited mainly to respiratory pathogens such as *Moraxella catarrhalis* and *Haemophilus influenzae.* Ceftaroline has demonstrated *in vivo* efficacy in a number of preclinical infection models [31–33], including murine pneumonia, murine soft tissue infection, rat and rabbit endocarditis, and rabbit osteomyelitis models. Ceftaroline fosamil is synthesized by coupling 4-(pyridin-4-yl)thiazol-2-amine with a protected 7-amino-3-methanesulfonyloxy cephalosporinic acid ester derivative. Quaternization of the pyridine nitrogen with methyl iodide followed by deprotection of the 7-amino and 4-carboxylic acid groups provides the amino cephalosporin intermediate [37,38]. Reaction of 2-(5-amino-1,2,4-thiadiazol-3-yl)-2-(ethoxyimino)acetic acid with phosphorous pentachloride provides the requisite C-7 side-chain intermediate. Coupling of the two fragments, followed by acidic hydrolysis, affords ceftaroline fosamil, which is converted to its acetate salt. The aqueous solubility of ceftaroline fosamil is >100 mg/mL compared with 2.3 mg/mL for ceftaroline. Ceftaroline fosamil is available as 600 or 400 mg of sterile powder in single-use 20 mL vials. The powder is dissolved in 20 mL of sterile water and then further diluted with ≥ 250 mL of an appropriate vehicle prior to administration. The recommended dosage is 600 mg administered every 12 h by IV infusion over 1 h.

Ceftaroline fosamil is rapidly converted into bioactive ceftaroline by plasma phosphatases. After a single 600 mg dose of ceftaroline fosamil as a 1-h infusion in healthy volunteers, the ceftaroline pharmacokinetic parameters were $C_{max} = 19$ μg/mL; $T_{max} \sim 1$ h; AUC $= 57$ μg h/mL; half-life $= 1.6$ h. Pharmacokinetic parameters were similar for single and multiple dose IV administration of ceftaroline fosamil. The average binding of ceftaroline to human plasma proteins is $\sim$20%. The steady-state volume of distribution was similar to extracellular fluid volume. As with other β-lactam antimicrobial agents, the time that unbound plasma concentrations of ceftaroline exceed the minimum inhibitory concentration (MIC) of the infecting organism correlates with efficacy in preclinical infection models with *S. aureus* and *S. pneumoniae*. Ceftaroline is not a substrate or an inhibitor of hepatic CYP450 enzymes. Hydrolysis of the β-lactam ring occurs to form the microbiologically inactive, ring-opened metabolite M-1 to an extent of $\sim$30%. Ceftaroline and its metabolites were primarily eliminated by the kidneys: there was ~90% recovery of radioactivity in urine and 6% in feces within 48 h of a 600 mg IV dose. Of the radioactivity recovered in urine, ~64% was excreted as ceftaroline and ~2% as M-1. The renal clearance of ceftaroline was 5.56 L/h, suggesting that ceftaroline is predominantly eliminated by glomerular filtration. Four Phase III trials were used to demonstrate the efficacy of ceftaroline fosamil in ABSSSI and CABP. In two ABSSSI trials, half of the ~1500 patients received ceftaroline fosamil (600 mg, 1-h IV infusion) and half received vancomycin plus aztreonam. These studies included patients with infections caused by both susceptible and resistant organisms. Clinical cure rates were similar and high (>90%) for both the ceftaroline and vancomycin plus aztreonam groups. For CABP, a total of ~1200 adults were enrolled in two randomized, double-blind, non-inferiority trials comparing ceftaroline fosamil (600 mg, 1-h IV infusion every 12 h) with ceftriaxone (1 g, 30-min IV infusion every 24 h). One trial included oral clarithromycin as adjunctive therapy. Patients with known or suspected MRSA were excluded from both trials. Treatment duration was 5–7 days. Clinical cure rates by pathogen were higher for ceftaroline fosamil than for ceftriaxone for gram-positive organisms and were similar for gram-negative organisms. Like most cephalosporins, ceftaroline fosamil was safe and well tolerated. The most common adverse events occurring in >2% of patients were diarrhea, nausea, and rash. No single adverse event occurred in >5% of patients. In the four pooled Phase III clinical trials, treatment discontinuation due to adverse events occurred in 2.7% of patients receiving ceftaroline fosamil and 3.7% of patients receiving comparator drugs. In a randomized, positive- and placebo-controlled crossover thorough QTc study, no significant effect on QTc interval was detected at peak plasma concentration or at any other time in healthy subjects given a 1500-mg single dose by IV infusion over 1 h. Ceftaroline was approved in the United States in October, 2010, and is marketed under the trade name Teflaro®.

6. CORIFOLLITROPIN ALFA (INFERTILITY) [39–43]

Class	Follicle-stimulating hormone receptor agonist
Country of origin	Netherlands
Originator	N.V. Organon
First introduction	European Union
Introduced by	Merck
Weight	47 kDa
Trade names	Elonva®
CAS registry no.	195962-23-3

Corifollitropin is a follicle-stimulating hormone (FSH) receptor agonist that was approved as an infertility treatment in the European Union in 2010. Infertility is a recognized worldwide public health issue by the World Health Organization (WHO). The prevalence in more- and less-developed countries is similar; however, the reasons contributing to infertility may differ. There is a steady rise in age-related infertility in more-developed countries, whereas in less well-developed countries, infections play a large contributing role to infertility. Worldwide on average, 9% of women of child bearing years are infertile. Infertility has social consequences as a large percentage of couples identify having children as one of their life-long goals. However, only 50% of infertile couples have been reported to seek fertility services [39]. Nearly all of the assisted reproductive techniques (ART) used to increase the chances for conception can be categorized as *in vitro* fertilization (IVF), intracytoplasmic sperm injection (ICSI), and controlled ovarian stimulation (COS). Multifollicular development resulting from COS is essential to ART and is primarily governed by circulating levels of FSH [39,40]. FSH belongs to the gonadotropin family of glycoproteins that are excreted from the pituitary gland by gonadotrope cells. Other members include thyroid-stimulating hormone (TSH), luteinizing hormone (LH), and human chorionic gonadotropin (hCG). TSH and LH are produced in the pituitary gland, whereas hCG is produced in the placenta. The gonadotropins all have a dimeric structure in which the α-subunit is a shared, nearly identical, substructure, and the glycosylated β-subunit is varied and is responsible for receptor selectivity. Variants of FSH have been used in therapies to enhance conception since the 1950s. Since the 1990s, rFSH has been employed. The dosing regimen of FSH requires several daily injections over a 9-day period and has the drawbacks of low patient compliance and injection discomfort. Several approaches have been applied to increase the half-life of FSH *via* formulation or chemical modification [41–43]. A unique approach developed by Boime and his research group involved using the carboxy peptide terminus (CPT) of the β-chain of hCG to increase the half-life of FSH, while maintaining the β-chain of FSH to confer selectivity and potency for the FSH receptor. These hybrid glycopeptides were produced using recombinant DNA methods in Chinese Hamster Ovary (CHO) cells. One of the analogs produced from this effort was corifollitropin alfa. Corifollitropin has been compared to rFSH both *in vitro* and *in vivo*. Corifollitropin binds to the FSH receptor with equal potency compared to rFSH. In an FSH receptor transactivation assay, corifollitropin was 1.8-fold less active than rFSH but remained a potent activator with an $EC_{50} = 5$ pM. Corifollitropin alfa lacks activity for the LH and TSH receptors. In a rat PK model, corifollitropin alfa has two times the C_{max} and 1.6 times the half-life of rFSH. A similar result was reported in beagle dogs, with the half-life of corifollitropin alfa being 47 h compared to 23 h for rFSH. In a rat model of fertility, the extended half-life of corifollitropin resulted in better efficacy in increasing ovarian weight when compared to rFHS when both were dosed in combination with hCG. In another rat efficacy model, the number of ova per rat was found to be increased to higher levels with corifollitropin alfa when compared to rFSH [41–43]. In humans, the mean half-life from Phase I and Phase II studies is 65 h, compared with 35 h for rFSH, and the AUC is dose proportional over the dose range of 60–240 μg. Peak levels of corifollitropin are reached between 25 and 45 h after injection. In human efficacy studies, a single dose of corifollitropin alfa is able to sustain follicular development for 1 week. The exposure of corifollitropin alfa was found to have some variability with regard to body weight. Based on this finding, in Phase III studies, subjects weighing more than 60 kg were given a 150-μg subcutaneous dose and subjects weighing less than 60 kg were given a 100-μg subcutaneous dose. In the largest double-blind fertility agent trial ever performed, the ENGAGE trial studied 1506 patients. The ENGAGE trial (Phase III) compared a single dose (150 μg) of corifollitropin alfa to rFSH (200 IU/day) over 7 days to induce multifollicular growth in subjects weighing more than 60 kg who were undergoing COS as part of IVF or ICSI. The primary endpoint of the study was pregnancy rate as assessed 10 weeks after placental implant. After the 7 days, subjects were then treated with daily rFSH up to the day that the subjects were given hCG. The maximal duration of drug stimulation was 19 days. Patients also received ganirelex (GnRH antagonist) starting on day 5. The rate of pregnancy and number of oocytes retrieved were comparable between the corifollitropin alfa and rFSH groups. The ENSURE trial (Phase III) compared corifollitropin alfa to rFSH in patients weighing less than 60 kg undergoing COS as part of IVF or ICSI. The primary endpoint of this study was the number of oocytes retrieved. In this study, patients were given a single dose of corifollitropin (100 μg) or daily injections of rFSH (150 IU/day) for 7 days to induce multifollicular growth. From day 8 onward, subjects received daily injections of rFSH dose adjusted up to 200 IU until the day that hCG was given to cause oocyte maturation. The maximal total duration of stimulation was 19 days. Patients received ganirelex on day 5. The number of oocytes and ongoing pregnancy rate were comparable between the corifollitropin alfa and rFSH groups [41–43]. In January 2010, the EC approved corifollitropin alfa injection for COS in combination with a GnRH antagonist for the development of multiple follicles in women participating in an assisted reproductive technology (ART) program. Merck and Co., Inc. market corifollitropin alfa injection under the brand name Elonva®. Elonva® is the first in a class of sustained follicle stimulants (SFS).

7. DALFAMPRIDINE (MULTIPLE SCLEROSIS) [44–50]

Class	Potassium channel blocker
Country of origin	United States
Originator	Rush University Medical Center
First introduction	United States
Introduced by	Acorda Therapeutics Inc.
Trade name	Ampyra®
CAS registry no.	504-24-5
Molecular weight	94.1

NH_2
N

Dalfampridine (also referred to as 4-AP) is the first drug approved by the FDA to improve walking in patients with multiple sclerosis (MS). MS is an autoimmune disease that affects the brain and spinal cord. MS is caused by inflammation-mediated damage to the myelin sheath, the insulating layer that surrounds the core of a nerve fiber or axon and facilitates the transmission of nerve impulses. Repeated episodes of inflammation can occur along any area of the brain, optic nerve, and spinal cord. Among the many symptoms that negatively affect the quality of life in MS patients, fatigability of strength (defined as the decrease in strength that occurs with repetitive movements) is the most common impairment that is not addressed by currently available therapies. For example, in a survey of 1011 people conducted by the United States National MS society, 64% reported difficulty with walking; of this group, 78% reported that this impacted their ability to work [44]. Dalfampridine addresses this issue *via* a novel mechanism of action. In MS patients, the normal pattern of nerve conduction is slowed down because of damage to myelinated fibers, which is manifested clinically as weakness and fatigability of strength. Dalfampridine is a voltage-gated potassium channel blocker that readily penetrates the CNS and increases the conduction and duration of action potential across nerve fibers resulting in enhanced functionality as observed in the walking speed of MS patients [45]. 4-AP has been widely used in clinical practice based on trials with immediate-release formulations that showed improvement in motor and visual functions in some patients. However, fluctuations in peak drug levels leading to unintentional overdose [46] or inadequate serum levels (requiring frequent dosing) necessitated the development of an extended-release formulation to improve pharmacokinetic parameters and minimize side effects of 4-AP.

A 10 mg dose of extended-release dalfampridine administered to healthy volunteers or patients with MS gave peak concentrations of ~20 ng/mL over 3–4 h post-administration [47]. In contrast, administration of 10 mg dose of immediate-release dalfampridine led to a C_{max} of ~43 ng/mL with a T_{max} of ~1.3 h. Extended-release dalfampridine tablet has a relative bioavailability of 96% when compared to an aqueous oral solution (immediate-release formulation). Dalfampridine shows very little binding to human plasma proteins (1–3%) and has a volume of distribution of ~2.6 L/kg. After two Phase II trials showed some level of efficacy, extended-release dalfampridine was evaluated in two Phase III studies. In the first randomized, multicenter, double-blind, placebo-controlled Phase III trials, 301 patients were enrolled [48]. The second trial involved 240 patients [49]. Walking speed as measured by the Timed 25-foot Walk (T25FW) was the primary measure of efficacy in both trials. Results from both clinical studies showed that extended-release dalfampridine (10 mg, twice daily) significantly improved walking speed and strength in a majority of the MS patients compared to placebo (T25FW responder rate in Trial 1: 34.8% *vs.* 8.3% for placebo; Trial 2: 42.9% *vs.* 9.3% for placebo). From a safety perspective, seizure was the most important adverse event from various dalfampridine clinical trials. Therefore, dalfampridine is contraindicated in patients with a history of seizures. Since dalfampridine is primarily excreted by the kidney as unchanged drug, it is also contraindicated in patients with moderate to severe renal impairment. Assessing the overall benefit–risk profile from various clinical trials, the FDA approved dalfampridine (daily oral dose of 10 mg, b.i.d.) to improve walking in MS patients with existing gait impairment [50].

8. DENOSUMAB (OSTEOPOROSIS AND METASTATIC BONE DISEASE) [51–59]

Class	Recombinant monoclonal antibody
Country of origin	United States
Originator	Amgen
First introduction	United States
Type	Fully human IgG2, anti-RANKL
Introduced by	Amgen
Weight	~147 kDa
Trade names	Prolia®/Xgeva®
Expression system	Rodent CHO-cell line
CAS registry no.	615258-40-7

Denosumab, which was approved in the United States in 2010, is a fully human sequence IgG2 monoclonal antibody that inhibits bone resorption by blocking the activity of receptor activator of nuclear factor-κB ligand (RANKL). RANKL is a TNF family protein that is expressed in both secreted and cell surface forms by a variety of bone marrow cell types and mediates bone resorption through its receptor (RANK), which is found on osteoclasts and osteoclast precursors [51]. Denosumab was discovered using Xenomouse™ transgenic mice comprising human immunoglobulin genes [52,53]. The antibody is approved for treatment of postmenopausal women with osteoporosis at high risk for fracture, and for the prevention of skeletal-related events in patients with bone metastases from solid tumors. Denosumab competes directly with bisphosphonates such as alendronic acid in postmenopausal osteoporosis and with zoledronic acid in both of these indications. It has been shown to have comparable efficacy and safety to bisphosphonates with some tolerability and patient acceptability advantages. The drug is formulated as a solution for subcutaneous injection. The recommended dosage for treatment of postmenopausal women with osteoporosis is 60 mg administered every 6 months. At this dose, the observed mean C_{max} was 6.75 mcg/mL followed by serum concentration decline with a mean half-life of 25.4 days. The recommended dosage for patients with bone metastases is 120 mg administered every 4 weeks. With multiple subcutaneous doses, up to a 2.8-fold accumulation in serum denosumab concentrations was observed and steady state was achieved by 6 months. At steady state, the mean serum trough concentration was 20.5 mcg/mL, and the mean elimination half-life was 28 days. For both regimens, administration of vitamin D and calcium supplements is recommended to treat or prevent hypocalcemia. Denosumab was tested in three randomized Phase III trials in postmenopausal osteoporosis and three randomized Phase III trials in cancer patients with bone metastases. In all six of these trials, subjects were randomized 1:1 to either denosumab treatment or a control regimen. The largest of these trials, the pivotal FREEDOM trial [54], enrolled 7868 women who had low bone mineral density. Subjects were randomly assigned to receive either 60 mg of denosumab or placebo subcutaneously every 6 months for 36 months. Denosumab reduced the risk of new radiographic vertebral fracture (a cumulative incidence of 2.3% *vs.* 7.2% in the placebo group, $p < 0.001$), hip fracture (cumulative incidence of 0.7% *vs.* 1.2% in the placebo group, $p = 0.04$), and nonvertebral fracture (a cumulative incidence of 6.5% *vs.* 8.0% in the placebo group, $p = 0.01$). There was no increase in the risk of cancer, infection, cardiovascular disease, delayed fracture healing, or hypocalcemia. The two supportive, non-pivotal, Phase III trials in postmenopausal osteoporosis patients, the STAND [55] and DECIDE [56] trials, enrolled 504 and 1189 women, respectively, and directly compared denosumab to alendronic acid. The STAND trial examined the effect of transitioning from bisphosphonate therapy to denosumab. Subjects received open-label alendronate 70 mg once weekly for 1 month and then were randomly assigned to either continued weekly alendronate therapy or subcutaneous denosumab 60 mg every 6 months and were followed for 12 months. In subjects treated with denosumab, total hip bone mineral density increased by 1.90% at month 12 compared with a 1.05% increase in subjects on alendronate ($p < 0.0001$). Bone mineral density at the lumbar spine, femoral neck, and one-third radius was also greater with denosumab at 12 months (all $p < 0.0125$). In the DECIDE trial, patients were treated with either denosumab or alendronate without necessarily transitioning from prior alendronate therapy (~12% had received prior bisphosphonate therapy). Subjects were randomized to receive subcutaneous denosumab injections (60 mg every 6 months) plus oral placebo weekly or oral alendronate weekly (70 mg) with subcutaneous placebo injections. Denosumab treatment increased total hip bone mineral density compared with alendronate at month 12 (3.5% *vs.* 2.6%; $p < 0.0001$). Greater increases in bone mineral density were also observed with denosumab treatment at all measured skeletal sites (12-month treatment difference: 0.6%, femoral neck; 1.0%, trochanter; 1.1%, lumbar spine; 0.6%, one-third radius; $p \leq 0.0002$ all sites). In both the STAND and DECIDE trials, adverse events and serious adverse events occurred with similar types and frequencies in the denosumab and alendronate treatment groups. Prevention of skeletal-related events in patients with bone metastases from solid tumors was demonstrated in three randomized, double-blind trials comparing denosumab with zoledronic acid in 2046 breast cancer patients [57], 1776 solid tumor (excluding breast cancer and prostate cancer) and multiple myeloma patients [58], and 1901 prostate cancer patients [59]. In all three trials, patients were randomized to receive 120 mg denosumab subcutaneously every 4 weeks or 4 mg zoledronic acid intravenously every 4 weeks (dose adjusted for reduced renal function). In each trial, the main outcome measure was demonstration of non-inferiority of time to first skeletal-related event as compared to zoledronic acid. The percentage of patients having skeletal-related events in the denosumab-treated groups for each of the three trials was 30.7%, 31.4%, and 35.9%, respectively. The percentages for the zoledronic acid-treated groups were 36.5%, 36.3%, and 40.6%. Median time to first skeletal-related event was not reached in the denosumab arm of the breast cancer trial. It was 26.4 months for the zoledronic acid arm. For the other two trials, the denosumab-treated group had median first event times of 20.5 and 20.7 months, compared to 16.3 and 17.1 months, respectively for zoledronic acid treatment. The p values for non-inferiority for denosumab over zoledronic acid were <0.001 for all three trials. The p values for superiority were 0.01, 0.06, and 0.008, respectively. Overall survival and progression-free survival were similar between arms in all three trials; however, mortality was higher with denosumab in a subgroup analysis of 180 patients with multiple myeloma (hazard ratio of 2.26; 95% CI: 1.13, 4.50). For this reason, denosumab is not approved for prevention of skeletal-related events in patients with multiple myeloma. Adverse events and serious adverse events (including new malignancies and serious infections) occurred with similar types and frequencies in the denosumab and zoledronic acid treatment groups for all three trials. Cumulative incidence of osteonecrosis of the jaw across the three trials was 1.9% for the denosumab-treated patients and 1.3% for the zoledronic acid-treated patients. The incidence of adverse events associated with renal toxicity was higher with zoledronic acid, despite dose adjustments for patients with reduced renal function. Acute phase reactions were also more frequent in the zoledronic acid cohorts. Overall experience with denosumab shows that the drug is relatively nonimmunogenic. Cumulative data from 8115 postmenopausal osteoporosis patients and 2758 cancer patients treated with denosumab showed that less than 1% of the patients tested positive for antidrug antibodies and none of the 10,873 patients tested positive for neutralizing antibodies.

9. DIQUAFOSOL (OPHTHALMOLOGIC, DRY EYE) [60–67]

Class	$P2Y_2$ purinergic receptor agonist
Country of origin	United States
Originator	Inspire Pharmaceuticals
First introduction	Japan
Introduced by	Santen
Trade name	Diquas
CAS registry no.	211427-08-6 (tetrasodium salt) 59985-21-6 (acid)
Molecular weight	878.2 (tetrasodium salt)

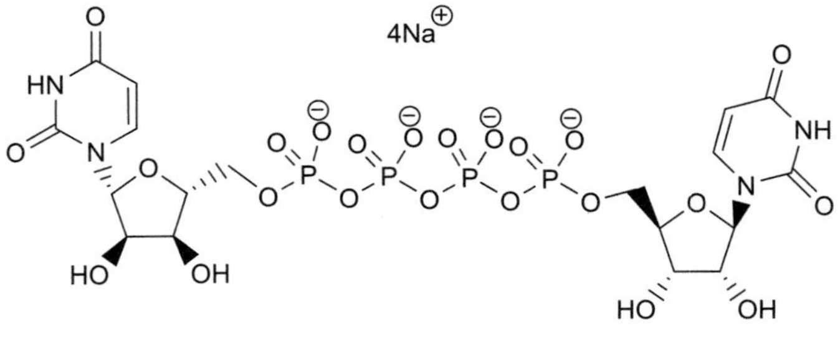

4Na⊕
O HN O N O HO OH O⊖ O O⊖ O O⊖ O O⊖ O P O P O P O P O O O O NH O N O HO OH

Diquafosol (INS-365) was approved in Japan in 2010 as a 3% ophthalmic solution for treatment of dry eye disease [60]. Dry eye disease is a common condition that begins with symptoms of ocular discomfort such as burning, stinging, or a sandy/gritty sensation. The disease is characterized by a lack of tear volume and/or an improper tear composition and damage to the ocular surface. Dry eye is a highly prevalent condition, affecting 14–33% of the population worldwide [61]. Clinical diagnosis of dry eye is difficult because the condition presents a variety of symptoms. Treatment options include tear supplements (lubricants), anti-inflammatory drugs (*e.g.*, cyclosporine eye drops or steroid eye drops), and tear retention devices. Diquafosol is a unique agent for the treatment of dry eye in that it acts as a $P2Y_2$ purinergic receptor agonist with the ability to activate this receptor on the ocular surface and stimulate water, lipid, and mucin secretion [62]. These are the three main processes that are needed to produce an appropriate tear. Diquafosol is a full agonist at the $P2Y_2$ receptor ($EC_{50} = 0.10$ μM) and is nearly as potent as the native agonist uridine diphosphate (UDP; $EC_{50} = 0.03$ μM) [63,64]. It is also a full agonist at the $P2Y_4$ receptor ($EC_{50} = 0.4$ μM) and a weak agonist at the $P2Y_6$ receptor ($EC_{50} = 20$ μM). Diquafosol is prepared by treatment of uridine triphosphate (UTP) with DCC, followed by condensation with uridine monophosphate (UMP) to give diquafosol in $\sim$30% yield [64,65]. Alternatively, the compound could be prepared by activation of UDP with CDI, followed by the addition of a second molecule of UDP. Diquafosol is administered topically, which limits exposure to the local site of action. In 25 dry eye patients monitored over 6 months of treatment, there were no detectable systemic concentrations of diquafosol or its metabolites. Diquafosol is metabolized by phosphodiesterases to UTP, UDP, UMP, and uridine. Diquafosol has been well tolerated in clinical trials, with side effects being local to the ocular surface. In one Phase III trial, burning and stinging were reported in 7% of diquafosol subjects compared with 2% on placebo. For approval of a new prescription drug for the treatment of dry eye disease, the FDA requires demonstration of efficacy for both a sign (an objective measure such as corneal staining) and a symptom (a subjective measure such as sensation of a foreign body in the eye) of the disease [66]. There have been five Phase III trials with 2% solutions of diquafosol conducted. Results have been mixed with some endpoints being met, but with failure to meet others. Three additional Phase III trials have been carried out in Japan using a 3% ophthalmic solution of diquafosol [67]. The 3% solution was reported to improve dry eye symptoms by promoting secretion of mucin and water, thereby bringing the tear film closer to a normal state. In addition, no serious ocular or systemic adverse drug reactions were found during the clinical trials [60]. The 3% solution of diquafosol is the approved treatment for dry eye in Japan.

10. ECALLANTIDE (ANGIOEDEMA, HEREDITARY) [68–72]

Class	Plasma kallikrein inhibitor
Country of origin	United States
Originator	Dyax Corp.
First introduction	United States
Introduced by	Dyax Corp.
Trade name	Kalbitor®
CAS registry No.	460738-38-9
Weight	7053.8
Type	Modified tissue factor pathway inhibitor, 60 amino acid recombinant protein

[Glu^{20},Ala^{21},Arg^{36},Ala^{38},His^{39},Pro^{40},Trp^{42}] tissue factor pathway inhibitor (human)-(20-79)-peptide

Ecallantide, also known as DX-88, was approved in 2009 in the United States for treatment of hereditary angioedema (HAE), a condition characterized by episodic attacks of localized edema in cutaneous and mucosal tissues. The pain and swelling of face, genitalia, extremities, and abdomen affects quality of life with swelling of the pharynx and larynx being life threatening. Patients with HAE can have one to three episodes per month with each episode lasting between 2 and 5 days. HAE is a rare autosomal disease affecting between 1:10,000 and 1:50,000 people in the United States [68]. There is no difference in prevalence with regard to sex or ethnic group, but women have more severe clinical symptoms than men. HAE results from deficiencies or disorders of C1-esterase inhibitor protein (C1-1NH). Mutation of the gene that encodes C1-1NH causes the lack or altered activity of the serine protease, C1-1NH. C1-1NH regulates the kallikrein–kinin (contact activation) and complement cascade systems. C1-1NH inhibits the complement system by binding C1 and prevents the formation of the C1 complex. Overactivation of the kinin–kallikrein system leads to the overproduction of plasma kallikrein. Left unchecked, plasma kallikrein produces high levels of bradykinin and stimulates release of C5a, further activating the complement cascade system. Activation of the complement system causes the release of cytokines, TNF-α and interleukins [68,69]. Bradykinin is a potent vasodilator, and the increased vascular permeability leads to the accumulation of fluids that is seen in HAE patients. There are three manifestations of the disease. Type I results from low serum levels of C1-1NH and represents 85% of the cases. In Type II HAE, C1-1NH levels are normal to elevated, but dysfunctional, affecting 15% of patients. A third type of HAE, primarily affecting women, has been characterized and is caused by mutations in the factor XII gene [69,70]. Ecallantide (DX-88) was designed to inhibit the action of plasma kallikrein. Ecallantide is a potent and selective inhibitor of plasma kallikrein with a $K_i = 25$ pM. The discovery program that identified ecallantide used phage display technology and a library of designed variants of the first Kunitz domain of TFPI. Ecallantide, a 60-amino acid peptide, with 3-disulfide bonds, differs from TFPI by 7-amino acids. Ecallantide has been developed as a subcutaneous administered formulation [70,71]. In Phase I studies with healthy volunteers, ecallantide was found to have a high volume of distribution (26.4 L $\pm$ 7.81) with a mean elimination half-life of 2.0 $\pm$ 0.5 h. The bioavailability was found to be 90%. Ecallantide, in Phase III EDEMA3 and EDEMA4 trials, was found to be effective in relieving HAE symptoms and the frequency and duration of HAE attacks. The drug was effective in moderate and severe HAE patients and was found to have a 24-h duration of action in the EDEMA trials. The most serious adverse events reported for ecallantide are hypersensitivity and anaphylaxis, with incidences of 3.9% (10 patients) and 2.7% (5 patients), respectively. In all cases, the condition resolved with treatment and without further complications. Ecallantide is given as three 10 mg subcutaneous doses and is a first-in-class drug targeting plasma kallikrein to treat HAE [71,72]. In December 2009, the U.S. FDA granted approval for Kalbitor® (ecallantide) for the treatment of acute attacks of HAEin patients 16 years of age and older.

11. ERIBULIN MESYLATE (ANTICANCER) [73–79]

Class	Tubulin inhibitor
Country of origin	United States
Originator	Eisai
First introduction	United States
Introduced by	Eisai
Trade name	Halaven™
CAS registry no.	253128-41-5 (free base)
	441045-17-6
Molecular weight	729.9 (free base)

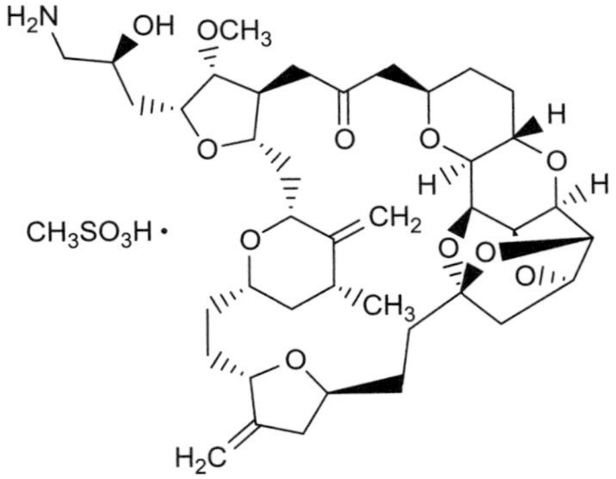

The U.S. FDA approved eribulin mesylate (also referred to as E7389) in November 2010 for the treatment of metastatic breast cancer (MBC) for patients who previously received at least two chemotherapeutic regimens for late-stage disease. Although the overall incidence of early stage breast cancer has been declining in recent years, one-third of the women in this group will develop MBC within 5 years of their initial diagnosis. Despite significant advances in the understanding of tumor biology, MBC is still incurable and accounted for over 40,000 deaths in the United States in 2009. A number of cytotoxic agents are currently available for the treatment of breast cancer, but their use in a metastatic setting is more limited. For example, taxane-based therapies are restricted by cumulative neurotoxicity and tumor progression on earlier taxane-based therapy. Therefore, development of novel cytotoxic agents is still desirable. Eribulin is a synthetic analog of the marine natural product halichondrin B, which is isolated from the sea sponge *Halichondria okadai*. Eribulin retains most of the structural elements that constitute the right hand side of halichondrin B; structure–activity relationship (SAR) studies suggested that the antitumor activity of halichondrin B resides in that part of the molecule [73]. Eribulin is a microtubule inhibitor that binds close to the *vinca*-binding site of tubulin. Unlike most tubulin inhibitors like taxanes, epothilones, and *vinca* alkaloids that inhibit microtubule dynamic instability by changing tubulin addition and loss parameters, eribulin's effects on dynamic instability are novel in that eribulin inhibits the growth phase of microtubules without affecting the shortening phase by binding to microtubule plus ends [74]. The net effect is blockage of cell cycle progression at the G2–M phase leading to apoptotic cell death after prolonged mitotic blockage. Eribulin has 19 stereocenters and protocols for its synthesis are outlined in the patent and open literature [75–77]. Among the key steps involved in the reported synthetic approaches to eribulin are catalytic asymmetric Ni–Cr-mediated coupling reactions and a Ni–Cr-mediated macrocyclization. Eisai's process route for the commercial production of eribulin is purported to involve 62 steps. Preclinical *in vitro* studies demonstrated the cytotoxic ability of eribulin to inhibit breast, prostrate, colon, and melanoma cancer cell lines with IC_{50} values ranging from 0.09 to 9.5 nM. The cytotoxic activity of eribulin was maintained in taxol-resistant® cell lines including those with mutations in β-tubulin. Cytotoxicity was not observed up to a concentration of 1 μM in IMR-90 human fibroblasts as well as in *in vivo* studies (*e.g.*, body weight loss and water consumption). Eribulin demonstrated *in vivo* activity in various mouse xenograft models (MDA-MB-435, COLO 205, LOX, OVCAR-3) when given intravenously or intraperitoneally at doses ranging from 0.05 to 1 mg/kg. In addition to showing a wider therapeutic window (four- to fivefold) in these studies compared to paclitaxel (twofold), tumor regrowth was suppressed to a significant extent after treatment cessation with eribulin when compared to paclitaxel.

In a Phase I clinical trial, eribulin showed linear kinetics over a dose range of 0.25–4.0 mg/m^2 following IV administration to patients with solid tumors. The C_{max} (0.086–0.231 μg/mL) and AUC (0.171–0.563 μg h/mL) increased in a dose-proportional fashion with a mean elimination half-life of ~41 h. The mean clearance and volume of distribution were 2 L/h and 72 L/m^2, respectively. Eribulin is not extensively metabolized (metabolite concentrations <0.6% of parent) and is primarily eliminated in feces. Eribulin competitively inhibits CYP3A4-mediated testosterone and midazolam hydroxylation of human liver microsomal preparations with an apparent $K_i = 20$ μM. Eribulin does not inhibit CYP1A2, CYP2C9, CYP2C19, and CYP2D6 in human liver microsomes up to a concentration of 5 μM. The human plasma protein binding of eribulin ranges from 49% to 65% at concentrations of 100–1000 ng/mL, respectively. The safety and efficacy of eribulin mesylate were established in an open-labeled, randomized, multicenter Phase III EMBRACE study, comprising 762 women with MBC who were previously treated with at least two different chemotherapy regimens for late-stage disease [78]. Of the 762 women, 508 were treated with eribulin mesylate and 254 were treated with an agent of their physician's choice. The EMBRACE study indicated that eribulin mesylate-treated patients showed improvement in the median overall survival of 13.12 compared to 10.65 months for patients who were treated with a single-agent therapy chosen by their physician. Most common serious adverse events (≥25%) associated with eribulin treatment were neutropenia (although it did not translate into a high rate of febrile neutropenia), anemia, hair loss, weakness, peripheral neuropathy, nausea, and constipation. In addition, ECG monitoring is recommended in patients with CHF and bradyarrhythmias. The recommended dose of eribulin mesylate is 1.4 mg/m^2 administered intravenously over 2–5 min on days 1 and 8 of a 21-day cycle [79].

12. FINGOLIMOD HYDROCHLORIDE (MULTIPLE SCLEROSIS) [80–87]

Class	Sphingosine-1-phosphate (S1P) receptor agonist
Country of origin	United States
Originator	Yoshitomi Pharmaceutical Industries (now Mitsubishi Tanabe Pharma)
First introduction	Russia
Introduced by	Novartis
Trade name	Gilenya®
CAS registry no.	162359-56-0 162359-55-9 (free base)
Molecular weight	307.5 (free base) 343.9 (hydrochloride salt)

HO OH NH_2 $CH_3(CH_2)_7$ ·HCl

Approved by the U.S. FDA in September 2010, fingolimod (also referred to as FTY720) is the first approved oral therapy for the relapsing-remitting form of multiple sclerosis (RRMS). MS is a chronic autoimmune demyelinating disease of the CNS that affects over 2.5 million individuals worldwide. The relapsing–remitting form of MS, which occurs in approximately 85% of newly diagnosed patients, is characterized by recurrent acute exacerbations (relapses) of neurological dysfunction, followed by recovery. Six drugs are currently approved by the FDA for the treatment of patients with MS: an interferon beta-1b (IFNβ-1b) product, two IFNβ-1a formulations, glatiramer acetate (GA), mitoxantrone (an antineoplastic anthracenedione), and natalizumab, a recombinant monoclonal antibody. However, these drugs are not orally administered and have other limitations. For example, IFNβ-1a is administered subcutaneously and causes side effects that include influenza-like symptoms. Natalizumab has been associated with hypersensitivity reactions and progressive multifocal leukoencephalopathy (PML). Because of these limitations, there has been a significant effort in the industry and academia to identify novel treatments for the treatment of MS that are effective and orally bioavailable. Fingolimod was first described as an immunosuppressant based on SAR studies around the fungal metabolite myriocin or ISP-1 isolated from *Isaria sinclairii* [80]. Because of fingolimod's structural resemblance to sphingosine, a metabolite of sphingolipids that constitutes the cell membrane of all eukaryotic cells, it was hypothesized that fingolimod may be affecting sphingolipid metabolism in cells. A series of elegant *in vitro* and *in vivo* studies [81,82] confirmed that fingolimod is converted to (*S*)-fingolimod phosphate primarily by sphingosine kinase-2 and that (*S*)-fingolimod phosphate mediates multiple biological processes by binding to novel GPCR's referred to as sphingosine-1-phosphate (S1P) receptors. S1P receptors are divided into five subtypes, $S1P_{1-5}$, which have varying tissue and cellular distribution. $S1P_{1-3}$ receptors are ubiquitously expressed in the immune, cardiovascular, and central nervous systems, $S1P_4$ is restricted to the hematopoietic system, and $S1P_5$ is mostly localized in the white matter of CNS. $S1P_{1-3}$ receptors play important roles in endothelial barrier function, maintaining vascular tone, regulating heart rate and allowing for lymphocyte egress from secondary lymphoid organs. The functional role of $S1P_4$ is unknown, while the $S1P_5$ receptor is thought to be involved in natural killer cell trafficking and oligodendrocyte function. *In vitro* binding and functional studies revealed that (*S*)-fingolimod phosphate is a full agonist at $S1P_1$, $S1P_4$, and $S1P_5$ receptors ($EC_{50} = 0.3$–0.6 nM); a partial agonist at $S1P_3$ receptor ($EC_{50} \sim 3$ nM); and inactive at the $S1P_2$ receptor. Fingolimod itself is significantly less potent at the $S1P_1$ and $S1P_5$ receptors ($IC_{50} \sim 300$ and 2600 nM, respectively) and inactive at $S1P_{2-4}$ receptors ($IC_{50} > 5000$ nM). Binding of (*S*)-fingolimod phosphate to lymphocytic $S1P_1$ leads to internalization of the receptor and sequestration of naïve T cells and self-reactive central memory T cells in lymph nodes, thereby preventing their recirculation and entry into the CNS. Since naïve T cells and self-reactive central memory T cells are believed to be responsible for the inflammation and neural damage found in MS, their retention in lymph nodes is believed to be responsible for the beneficial effects seen with fingolimod in MS patients. Effector memory T cells are not affected during fingolimod treatment, and lymphocytes are not destroyed; therefore, immune functions such as activation and proliferation and effector functions of T and B cells remain unaffected during treatment. The synthesis of fingolimod has been described in a number of publications and patents [83a–c]. One approach to the synthesis of fingolimod employs a Friedel–Crafts acylation followed by reduction to install the C8-side chain and alkylation of an amino malonate unit followed by reduction to install the amino diol head piece [83a,b]. An alternate approach to the synthesis of the amino diol headpiece employs a Petasis reaction of an appropriately substituted arylvinylboronic acid, benzyl amine, and dihydroxyacetone in a key step [83c].

Fingolimod was shown to be efficacious in preclinical models of MS (rat experimental autoimmune encephalomyelitis) at doses ranging from 0.3 (prophylactic administration) to 12–28 mg/kg (therapeutic administration). Phase I multiple dose pharmacokinetic studies with fingolimod indicated that at a dose of 1.25 mg/day, the mean C_{max} observed was 5.0 ± 1.0 ng/mL and at a dose of 5 mg/day, the mean C_{max} was 18.2 ± 4.1 ng/mL. The median T_{max} was $\sim$12 h for both doses. Fingolimod has a half-life of ~9 days partly because of a large volume of distribution (>1000 L); pharmacokinetic steady state is achieved after ~2 months. Fingolimod is highly protein bound (>99.8%) and is predominantly metabolized by CYP4F2. A dose-dependent decrease in peripheral lymphocyte count is observed at both the 1.25 mg (77%) and 0.5 mg (73%) doses of fingolimod. The cell counts remained stable for the entire treatment period and returned to normal range within 6 weeks after stopping fingolimod treatment. After an initial 6-month placebo-controlled Phase II study that showed benefit in RRMS patients, fingolimod was evaluated in two Phase III studies. In the 2- year double-blind FREEDOMS study [84], involving 1272 patients with RRMS, fingolimod was compared with placebo, and in the 1 year, double-blind TRANSFORMS study [85] involving 1292 patients with RRMS, fingolimod was compared with IFN-β1a. Both studies used 0.5 mg and 1.25 mg once-daily doses of fingolimod. In the FREEDOMS trial, patients on fingolimod had an annualized relapse rate significantly lower than in patients who received placebo [0.18 (0.5 mg); 0.16 (1.25 mg) *vs.* 0.40 (placebo); $p < 0.001$]. Disability progression was significantly delayed, compared to placebo, with both doses of fingolimod. In the TRANSFORMS study, patients on fingolimod had an annualized relapse rate significantly lower than in patients who received IFN-β1a [0.16 (0.5 mg); 0.20 (1.25 mg) *vs.* 0.33 (IFN-β1a); $p < 0.001$]. From a safety perspective, bradycardia and atrioventricular block were observed on first dose of fingolimod treatment. Macular edema was seen in four patients in the 1.25 mg group (1%) and two patients in the 0.5 mg group (0.5%). Other serious adverse events include basal cell carcinoma and a 2–3% reduction in the mean FEV_1. The most common adverse events occurring in 10–20% of fingolimod-treated patients were fatigue, influenza, lower respiratory tract or lung infection, back pain, diarrhea, cough, and elevations in liver-enzyme levels. Assessing the overall benefit–risk profile, the FDA approved fingolimod for RRMS at 0.5 mg once daily [86,87].

13. LANINAMIVIR OCTANOATE (ANTIVIRAL) [88–97]

Class	Neuraminidase inhibitor
Country of origin	Japan
Originator	Sankyo Co., Ltd.
First introduction	Japan
Introduced by	Daiichi-Sankyo Co., Ltd.
Trade name	Inavir®
CAS registry no.	203120-17-6
Molecular weight	346.34

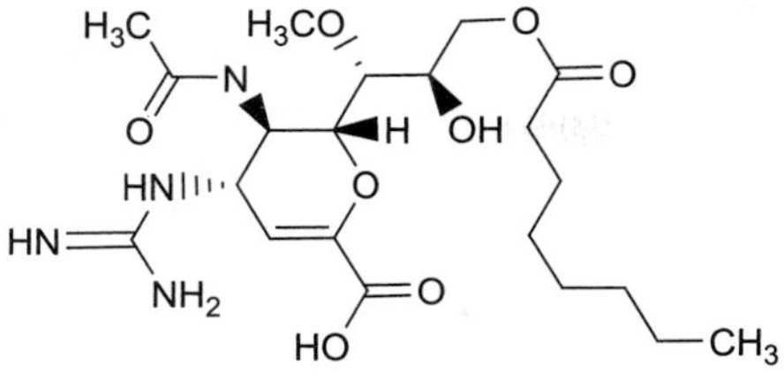

Laninamivir octanoate (CS-8958) [88,89], an ester prodrug form of the neuraminidase (NA) inhibitor laninamivir (R-125489), was approved in Japan in 2010 for treatment of influenza virus infections. Laninamivir octanoate is given by intranasal administration at a 20 mg or 40 mg dose. It has a long half-life in humans such that efficacy can be achieved after only a single dose. It is the second NA inhibitor approved in 2010, the first being peramivir (*vide infra*). Influenza is a serious and contagious respiratory illness that is caused by influenza A and B viruses. Influenza is a global health concern, with both seasonal epidemics and unpredictable pandemics resulting in significant morbidity and mortality, particularly for patients at high risk for influenza-associated complications. In addition to vaccines for immunoprophylaxis, antiviral drugs play an essential role in the treatment of influenza virus infections [90]. Two viral proteins have been targeted for therapeutic intervention: the M2 ion channel and NA. The M2 ion channel is blocked by drugs such as amantadine and rimantidine, which inhibit viral replication at the stage of viral entry and viral release. These agents have seen declining use due to widespread resistance and lack of activity against influenza B. Zanamivir (intranasal) and oseltamivir (oral) are the two approved drugs in the NA inhibitor class. These agents act by inhibiting the release of newly formed virus particles from infected cells by blocking the cleavage of the terminal sialic acid residues from glycoconjugates. The frequency of resistance to NA inhibitors is much lower than for M2 inhibitors; however, resistance has been documented, particularly to oseltamivir. For example, in a study of H1N1 strain of influenza A viruses from the 2009 pandemic, 0.7% of viruses were resistant to oseltamivir but remained sensitive to zanamivir and laninamivir [91]. More than 99% were resistant to amantadine. Laninamivir differs from zanamivir by the replacement of the 1′-hydroxyl group on the side chain at the 2-position of the 3,4-dihydro-2*H*-pyran core with a methoxy group. Laninamivir octanoate is prepared starting from a neuraminic acid precursor [88,92]. The route from 2,3-didehydroneuramic acid entails a multistep sequence to protect the acid and hydroxyl groups at the 4-, 2′-, and 3′-positions. Methylation of the remaining 1′-hydroxyl by treatment with dimethylsulfate and NaH is followed by conversion of the 4-hydroxyl to an amine.

Cleavage of the 2′,3′-dihydroxy protecting group, conversion of the 4-NH_2 to the guanidine, and acylation of the 3′-OH group afford laninamivir octanoate. This three-step sequence can be reordered such that the guanidine is introduced first, followed by deprotection of the 2′,3′-diOH groups and acylation. An alternative sequence involves a Boc-protected guanidine intermediate, which is converted in a four-step sequence (deprotection of the acid and 2′,3′-hydroxyl groups, reprotection of the acid as its diphenylmethyl ether, acylation of the 3′-OH and deprotection of the guanidine group) to laninamivir octanoate [92]. Laninamivir can also be synthesized from the α-methyl glycoside of *N*-acetylneuramic acid methyl ester by an analogous route. Laninamivir is a less potent NA inhibitor than zanamivir or oseltamivir against many influenza A strains, but it has superior potency for inhibition of viral replication (two- to fivefold depending on the virus strain) [92,93]. It is a more potent inhibitor of NA from the highly pathogenic avian influenza A H5N1 strains. The prodrug laninamivir octanoate is inactive. Laninamivir octanoate shows efficacy in a variety of preclinical influenza models, including a model of prophylaxis where a single dose given intranasally 7 days before infection was effective in prolonging survival time in infected mice [93]. Zanamivir was not efficacious under these conditions. It is believed that long retention of the compound in the lungs may contribute to the long-lasting activity. Laninamivir octanoate was also active in *in vivo* infection models with the avian H5N1 influenza A virus [94].

In Phase 1 studies, healthy male volunteers received single intranasal doses from 5 to 120 or 20 or 40 mg doses twice daily for 3 days [95]. The T_{max} for laninamivir octanoate was 0.5–1 h, while the T_{max} for laninamivir parent drug was 4 h. The half-life for laninamivir octanoate was 1.8 and 70–80 h for the parent drug. AUC and C_{max} increased proportionally with dose. The cumulative excretion in urine over 144 h was 2.3–3.6% for the prodrug and 10.7–14.6% for laninamivir. Plasma protein binding is 67% for laninamivir octanoate and $<0.1\%$ for parent drug. There were no adverse events related to test drug in these studies. Laninamivir octanoate was evaluated in a double-blind, randomized, non-inferiority clinical trial in ~1000 patients in comparison with oseltamivir [96]. Most patients were infected with influenza A virus, of which ~65% were H1N1 and ~35% were H3N2 strains. The H1N1 strains were oseltamivir resistant (H274Y mutation, mean $IC_{50} = 690$ nmol/L) and laninamivir-sensitive (mean $IC_{50} = 1.7$ nmol/L). The H3N2 strain was sensitive to both drugs. Laninamivir octanoate inhaled once at 20 or 40 mg was compared with 75 mg of oseltamivir given orally twice a day for 5 days. In H1N1 infected patients, there was little difference in the median times to illness alleviation for the 40-mg laninamivir octanoate group (73 h) compared with the oseltamivir group

(74 h). However, the proportion of patients shedding virus at day 3 in the 40-mg laninamivir octanoate group was significantly lower than in the oseltamivir group, which is consistent with the greater potency. Time to illness alleviation and reduction in viral shedding were similar for the two drugs in H3N2-infected patients. The 20-mg laninamivir octanoate dose group had a longer median time to illness alleviation (86 h). Both drugs were well tolerated with the most common adverse events being gastrointestinal events. In a similar trial in pediatric patients (median 9 years of age) infected with oseltamivir-resistant H1N1 influenza A, single inhalation doses of 20 or 40 mg laninamivir alleviated influenza illness more rapidly than oseltamivir given twice daily for 5 days at 2 mg/kg body weight [97]. Overall, laninamivir has proven to be effective, long-lasting, and well tolerated for the treatment of influenza infection.

14. LURASIDONE (ANTIPSYCHOTIC) [98–104]

Class	D_2, 5-HT_{2A}, and 5-HT_7 receptor antagonist; 5-HT_{1A} partial agonist
Country of origin	Japan
Originator	Dainippon Sumitomo Pharma
First introduction	United States
Introduced by	Dainippon Sumitomo Pharma
Trade name	Latuda®
CAS registry no.	367514-87-2 (free base) 367514-88-3 (HCl salt)
Molecular weight	492.3 (free base)

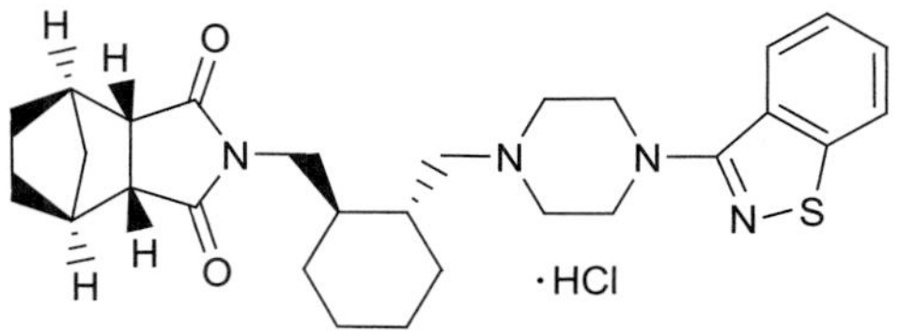

·HCl

The atypical antipsychotic lurasidone (also known as SM-13496) was approved in the United States in 2010 as an oral agent for the treatment of patients with schizophrenia [98,99]. Schizophrenia is a debilitating mental disorder that affects 1% of the population worldwide. The disease is characterized by three symptom domains: positive (psychotic) symptoms, such as delusions and hallucinations; negative symptoms, such as emotional flatness and a lack of motivation for daily activities; and cognitive symptoms, including difficulties with memory and concentration, and an inability to organize thoughts. Available drug treatments for schizophrenia are primarily aimed at decreasing dopaminergic transmission. First-generation antipsychotic agents such as haloperidol are dopamine D_2 receptor antagonists that are effective at treating positive symptoms but have little impact on negative and cognitive symptoms and are associated with significant side effects, particularly movement disorders (extrapyramidal side effects, EPS). Second-generation, or atypical, antipsychotics act at both D_2 and 5-HT_{2A} receptors and treat positive symptoms as well as some negative symptoms. While the atypical agents have less EPS, other issues such as weight gain, prolactin and glucose elevation, and sedation are observed with these agents, primarily as a result of off-target activities. The overall efficacy/side-effect profile of these agents is related to their complex, multitarget, pharmacological profile and leads to their differentiation in the clinical setting. Lurasidone has potent affinity for D_2 ($K_i = 1.7$ nM) and 5-HT_{2A} ($K_i = 2.0$ nM) receptors and acts as an antagonist at both receptors [100]. It is also a partial agonist at the 5-HT_{1A} receptor and, unlike other atypical agents, is a potent antagonist at the 5-HT_7 receptor; both of these activities are thought to confer beneficial cognitive properties. Lurasidone is further differentiated by its lack of affinity for muscarinic and histamine H1 receptors and its weak affinity for the 5-HT_{2C} receptor. Antagonism at H_1 and 5-HT_{2C} receptors has been implicated in weight gain associated with atypical agents, while muscarinic receptor antagonism is associated with cognitive deficits. Preclinical behavioral studies have shown lurasidone to be efficacious in models of psychosis, depression, and anxiety [100]. In cognition models, lurasidone has shown efficacy in a MK-801-induced impairment models of learning and memory, while other atypical antipsychotic agents were inactive [101,102]. At doses up to 1000 mg/kg, lurasidone did not induce catalepsy in rats or mice. It showed weak activity in additional rodent models for motoric side effects, indicating a lower potential for EPS [100]. The compound is synthesized by treatment of 1-(1,2-benzoisothiazol-3-yl)piperazine with the di-mesylate of (*R*,*R*)-cyclohexane-1,2-diyldimethanol under basic conditions to give an intermediate spiropiperazinium salt that undergoes reaction with bicyclo[2.2.1]heptane-2,3-dicarboximide to provide lurasidone [103].

In healthy volunteers and patients [98,103,104], lurasidone showed dose-proportional pharmacokinetics over a range of 20–160 mg, with 9–19% of drug being absorbed. Lurasidone reached peak levels 1–3 h after oral administration and had an elimination half-life of 18 h after a single dose, extending to 36 h at steady state. The clearance rate was found to be 3.9 L/min. The compound is highly protein bound ($\sim$99%) in serum. Lurasidone is not a significant inhibitor of P450 isozymes. The metabolism of lurasidone is mainly mediated by CYP3A4, with elimination occurring primarily *via* the liver ($\sim$80%) and to a lesser extent *via* the kidney ($\sim$10%). Use of lurasidone in combination with CYP3A4 inhibitors (*e.g.*, ketoconazole) or inducers (*e.g.*, rifampin) is contraindicated as a result of observations in drug–drug interaction trials. The major metabolic pathways for lurasidone are N-dealkylation, oxidation of the [2.2.1]-bicycloheptane ring, and oxidation of the benzoisothiazole sulfur atom. In a food effect study, exposures of lurasidone were three times higher for C_{max} and two times higher for AUC in subjects given food compared to those who were fasted. Therefore, it is recommended that lurasidone be taken with a meal. Positive emission tomography (PET) studies with [^{11}C]-raclopride following single oral doses lurasidone showed $\sim$40% D_2 receptor occupancy at a 10 mg dose, increasing to 75–85% at 60 mg and 70–80% at 80 mg. The efficacy of lurasidone for the treatment of schizophrenia was established in four 6-week, placebo-controlled studies in adult patients (mean age of 38.8 years, range 18–72) who met diagnostic criteria for schizophrenia. In a 6-week, placebo-controlled trial with 145 subjects, doses of 40 or 120 mg per day of lurasidone were superior to placebo on the Brief Psychiatric Rating Scale-derived (BPRSd) total score and the Clinical Global Impression severity scale (CGI-S). In a similar trial with 180 subjects, treatment with an 80-mg/day dose of lurasidone was superior to placebo using the same rating scales. In a 6-week, placebo- and active-controlled trial with 473 subjects with 40 or 120 mg/day of lurasidone, both lurasidone doses and the active control olanzapine were superior to placebo on the Positive and Negative Syndrome Scale (PANSS) total score and the CGI-S. In a 6-week, placebo-controlled trial ($N = 489$) with 40, 80, or 120 mg/day of lurasidone, only the 80 mg/day dose was superior to placebo on the PANSS total score and the CGI-S. Based on these results, the recommended starting dose for lurasidone is 40 mg/day and the maximum recommended dose is 80 mg/day. The 120 mg/day dose did not provide additional benefit and was associated with an increase in some adverse events. Commonly observed adverse events (incidence $\geq$ 5% and at least twice the rate for placebo) included somnolence, akathisia, nausea, parkinsonism, and agitation, all of which appeared dose-related. The overall discontinuation rate due to adverse events was 8% for lurasidone versus 4% for placebo. Lurasidone showed a neutral effect on weight and was not associated with significant QTc prolongation. Like all atypical antipsychotics, lurasidone carries a black box warning for increased mortality in elderly patients with dementia-related psychosis. Lurasidone is supplied as an HCl salt in 40 and 80 mg tablets for oral administration.

15. MIFAMURTIDE (ANTICANCER) [105–112]

Class	Activator of monocytes and macrophages
Country of origin	Switzerland
Originator	Novartis
First introduction	Austria
Introduced by	Takeda
Trade name	Junovan®
CAS registry no.	838853-48-8 (sodium salt)
	83461-56-7 (anhydrous, free acid)
Molecular weight	1277.5 (sodium salt)

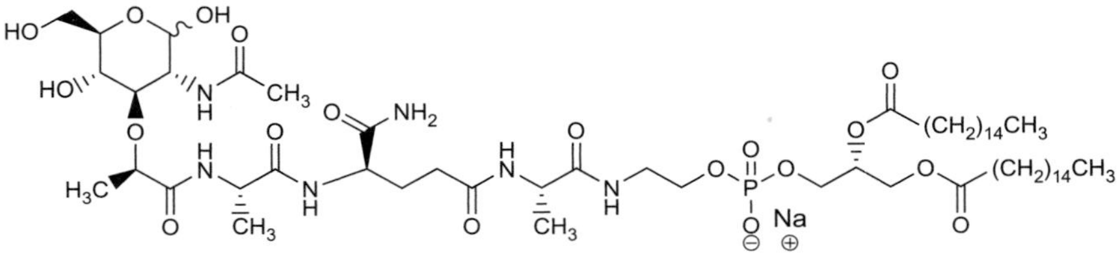

Mifamurtide (also referred to as MTP-PE) was approved by the EC in 2009 for the treatment of high-grade nonmetastatic osteosarcoma patients between the ages 2 and 30 in combination with postoperative multiagent chemotherapy. Osteosarcoma, which is a primary malignant bone tumor, is diagnosed in over 1000 new patients each year in North America and Europe. The standard of care for osteosarcoma is surgical removal of the tumor followed by chemotherapy. Doxorubicin, cisplatin, high-dose methotrexate with leucovorin rescue, and ifosfamide are the approved chemotherapeutic agents for osteosarcoma. These agents increase the event-free survival (EFS) of patients at 3–5 years by 60–70% in nonmetastatic conditions. Despite these advances, there is a need to improve the EFS of patients with nonmetastatic osteosarcoma. Mifamurtide is the first new drug approved for the treatment of osteosarcoma in over 20 years. The active component of mifamurtide is muramyl dipeptide (MDP), a component of bacterial cell walls, which is linked *via* an alanine moiety to phosphatidyl ethanolamine (PE) to form the tripeptide MTP-PE. Mifamurtide is combined with a liposome formulation consisting of synthetic phospholipids dioleoyl phosphatidyl serine and 1-palmitoyl-2-oleoyl phosphatidyl choline in a ratio of 3:7. This liposome formulation (L-MTP-PE) allows for the specific *in vivo* targeting of monocytes and macrophages in the liver, spleen, and lungs by mifamurtide [105]. *In vitro* and *in vivo* studies have shown that activation of monocytes and macrophages by mifamurtide leads to the production of proinflammatory cytokines such as TNF-α, IL-1, IL-6, IL-8, NO, prostaglandin E2 and PGD_2, and adhesion molecules such as LFA-1 and ICAM-1 [106]. MTP-PE is a ligand for nucleotide-binding oligomerization domain (Nod)-2 receptor, and it has been postulated that the activation of monocytes and macrophages is mediated by Nod-2 following phagocytosis of L-MTP-PE and liberation of MDP [105,107]. *In vitro*, human monocytes treated with MTP-PE had tumoricidal activity on allogeneic and autologous tumor cells but were nontoxic toward normal cells. The liberation of proinflammatory cytokines following activation of monocytes and macrophages and the tumoricidal activity of MTP-PE are not clearly understood. Mifamurtide is synthesized by DCC coupling of (a) *N*-acetylmuramyl-L-alanyl-D-isoglutaminyl-L-alanine with 2-aminoethyl-2,3-dipalmitoylglyceryl-phosphoric acid [108] or (b) *N*-acetylmuramyl-L-alanyl-D-isoglutamine and alanyl-2-aminoethyl-2,3-dipalmitoylglyceryl-phosphoric acid [109]. In a Phase I pharmacokinetic study in 21 healthy subjects, mifamurtide dosed at 4 mg intravenously over 30 min showed a C_{max} of ~16 nM, an AUC of ~17 nM h, and a mean half-life of ~2 h. The safety and efficacy of mifamurtide were established in a randomized, Phase III study comprising 662 patients with nonmetastatic resectable osteosarcoma [110]. All patients received similar doses of cisplatin, doxorubicin, and methotrexate. Patients were then randomly assigned to receive or not to receive ifosfamide and/or mifamurtide. There was a clear trend in the mifamurtide group toward better EFS ($p = 0.039$) and overall survival ($p = 0.030$) compared to the ifosfamide group. These results demonstrated that the addition of mifamurtide to chemotherapy improved the 6-year overall survival from 70% to 78% ($p = 0.03$). Most common serious adverse events associated with mifamurtide treatment were fever, chills, and tachycardia. The recommended dose of mifamurtide is 2 mg/m^2 administered as adjuvant therapy intravenously over 1-h twice weekly for an initial 12 weeks followed by once weekly for an additional 24 weeks for a total of 48 infusions in 36 weeks [111,112].

16. PERAMIVIR (ANTIVIRAL) [113–120]

Class	Neuraminidase inhibitor
Country of origin	United States
Originator	BioCryst Pharmaceuticals Inc.
First introduction	Japan

Introduced by	BioCryst Pharmaceuticals Inc.
Trade name	Rapiacta, PeramiFlu
CAS registry no.	330600-85-6
Molecular weight	328.4

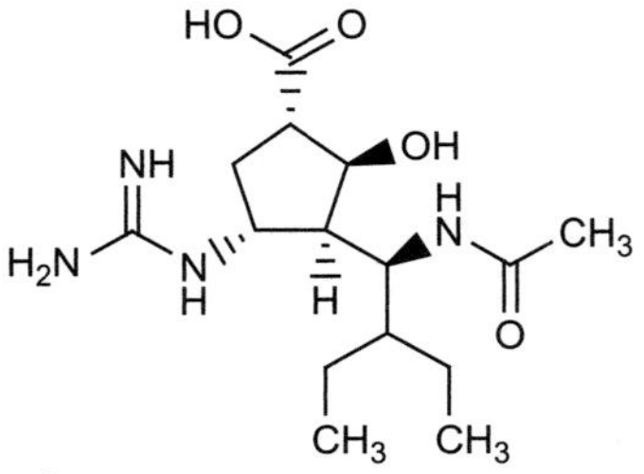

HO O NH OH H N CH3 H2N N H H O CH3 CH3

Peramivir is a neuraminidase (NA) inhibitor that was approved in Japan in 2010 for treatment of patients with influenza. It is the only NA inhibitor available for IV use and is the first of two NA inhibitors approved in 2010, the second being the inhaled drug laninamivir octanoate (*vide supra*). Peramivir is given as a 300-mg single dose for adult and pediatric uncomplicated seasonal influenza infection, and as single- and multiple 600 mg dose for patients at high risk for complications associated with influenza virus infection. From October 2009 to June 2010, peramivir was given Emergency Use Authorization (EUA) in the United States for treatment of certain hospitalized patients with suspected or confirmed cases of H1N1 influenza virus infection [113]; this is the first time an EUA has been granted for an unapproved drug [114]. Influenza is an infectious respiratory tract disease that annually affects approximately 10% of the world's population [115]. The virus that causes the disease is divided into three main types: A, B, and C. Influenza A viruses are responsible for seasonal flu, including the pandemics in 1918, 1957, 1968, and, most recently, in 2009 with the pandemic caused by the H1N1 strain of influenza A. While influenza is usually a self-limiting disease, there is a risk of complications and death, often in high-risk populations such as the very young or the elderly, but also in healthy individuals infected with highly pathogenic strains such as H5N1 (avian influenza A). In addition, influenza is highly communicable and causes a significant economic burden, for example, due to lost work time. Vaccines are highly effective in prophylaxis and controlling a flu epidemic; however, limitations in vaccine efficacy and delay in strain-specific production make antiviral drugs important prophylactic and treatment options [90,116]. Two classes of antiviral drugs have been approved for treatment of influenza: M2-ion channel inhibitors (amantadine, rimantidine) and NA inhibitors (zanamivir, oseltamivir). A major concern with antiviral agents is the development of resistance. Indeed, resistance to M2 inhibitors is widespread, although some strains remain susceptible. Resistance to NA inhibitors has also emerged, although thus far with lower frequency. Zanamivir is approved for intranasal use, while oseltamivir is an oral agent. Peramivir is the only NA inhibitor approved for IV use, which gives it a unique place in influenza treatment for seriously ill patients. Peramivir was discovered using structure-based drug design and is synthesized in six steps from Boc-protected methyl (*1S,4R*)-4-amino-cyclopent-2-enecarboxylate, which is prepared from 2-azabicyclo[2.2.1]hept-5-en-3-one [117–119]. Cycloaddition of the cyclopentene olefin with a nitrile oxide provided an intermediate fused cyclopentane-dihydroisoxazole. Hydrogenolysis and acetylation set up a fully functionalized cyclopentane with all four stereocenters established. Deprotection of the amine and acid groups was followed by installation of the guanidine moiety to provide peramivir. Like zanamivir and oseltamivir, peramivir is a potent inhibitor of influenza virus A and B NA [strain A(H1N1) IC_{50} = 0.34 nM; strain A(H3N2) IC_{50} = 0.60 nM; strain B IC_{50} = 1.36 nM]. However, peramivir is less potent against oseltamivir-resistant viruses that have the H275Y NA mutation. These viruses remain sensitive to zanamivir. Peramivir is active against influenza A and B viruses and has a low enzymatic off-rate, suggesting that it could inhibit NA activity for a prolonged period and allow lower frequency of dosing. Peramivir has proven efficacious in preclinical animal models of influenza infection [90].

The recommended dose of peramivir in adults is 600 mg IV over 30 min once daily for 5–10 days. Based on several Phase I trials in adults [114], there is a linear relationship between the IV dose of peramivir and the C_{max}. The half-life was 7.7–20.8 h. The AUC following a single 600 mg IV dose of peramivir was 80 μg·h/mL. Peramivir is excreted unchanged by the kidneys after IV dosing, with renal clearance of unchanged parent drug accounting for ~90% of total clearance. Peramivir clearance was 7.58 L/h/70 kg in adults with influenza and was 6.19 L/h/70 kg in healthy adults. There was no accumulation following multiple dose administration. The most common adverse events related to peramivir include diarrhea, nausea, vomiting, and neutropenia [114]. Similar rates of gastrointestinal events were observed in patients treated with peramivir and placebo. In a Phase II clinical study, 200 or 400 mg of peramivir proved beneficial when given IV to patients hospitalized within 72 h of the onset of symptoms [90]. In a Phase III clinical trial [115], previously healthy adult subjects were recruited within 48 h of the onset of influenza symptoms and randomized to single IV infusion of 300 mg peramivir, 600 mg peramivir, or matching placebo. Peramivir significantly reduced the time to alleviation of symptoms at both the 300 and 600 mg doses compared with placebo. No serious adverse events were reported. In another Phase III trial [90], peramivir when given IV over multiple days at 300 or 600 mg per day alleviated symptoms in all patients. In a limited study under an emergency IND, severely ill patients in the United States with H1N1 viral pneumonia and with progressing disease despite oseltamivir treatment were treated with IV peramivir for 1–14 days [120]. The drug was associated with recovery in most patients. Overall, peramivir has shown consistent efficacy in patients and has been safe and well tolerated. Peramivir was approved in Japan in January 2010 and is marketed under the trade name Rapiacta.

17. ROFLUMILAST (CHRONIC OBSTRUCTIVE PULMONARY DISORDER) [121–126]

Class	PDE4 inhibitor
Country of origin	Germany
Originator	BYK Gulden Lomberg Chemische Fabrik GmbH
First introduction	Germany
Introduced by	Nycomed (Altana)
Trade name	Daxas®
CAS registry no.	162401-32-3
Molecular weight	403.2

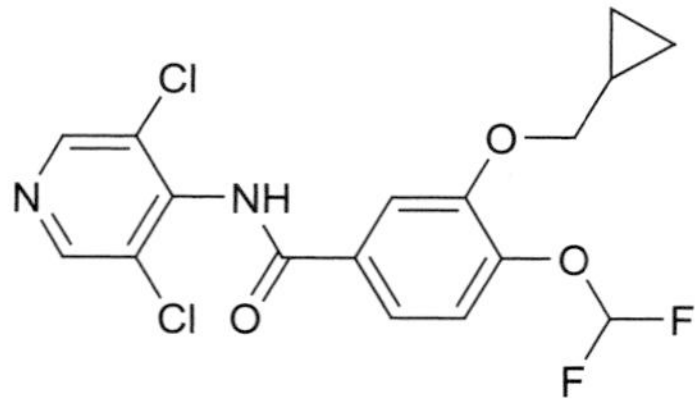

Cl
N
NH
Cl O
O
O
F
F

Roflumilast is a selective, orally active PDE4 inhibitor that was approved in Germany in July 2010 as an add-on to bronchodilator treatment for maintenance therapy of severe chronic obstructive pulmonary disorder (COPD) associated with chronic bronchitis in adult patients with a history of frequent exacerbations [121,122]. COPD is a chronic inflammatory disease that is characterized by an increase in neutrophils, macrophages, and $CD8^+$ T-lymphocytes in airways, lungs, and pulmonary vasculature. It is the sixth most common cause of death worldwide and is a major cause of morbidity. Disease manifestations include lung damage, progressive airway obstruction, chronic cough, and mucus hypersecretion. Acute exacerbations are accompanied by deterioration of lung function and worsening disability. PDEs are enzymes responsible for hydrolysis of cyclic nucleotide phosphates. PDE4 is expressed in several tissue types involved in diseases of the airway and is known to play a role in inflammation [123]. Inhibition of PDE4 blocks the hydrolysis of cAMP, leading to elevated levels of cAMP and regulation of inflammatory cells. There are four known subtypes of the PDE4 family, PDEs A–D. Since PDE4D inhibition has been associated with nausea and vomiting, a challenge in the development of PDE4 inhibitors as therapeutic agents has been subtype selectivity. A classic example is rolipram, which showed promising antidepressant efficacy in early clinical trials, but development was discontinued due to drug-induced nausea. Roflumilast and its primary metabolite roflumilast *N*-oxide are potent and competitive inhibitors of PDE4 and are equipotent against PDE4A, B, and D but inactive against PDE4C and the other ten members of the PDE family (PDEs 1–3, 5–11). Despite its inhibition of PDE4D ($IC_{50} = 0.80$ nM, *N*-oxide $IC_{50} = 2.0$ nM), roflumilast shows the lowest incidence of nausea (3–5%) among the PDE4 inhibitors investigated in clinical trials. Anti-inflammatory effects of roflumilast have been demonstrated in preclinical cellular and animal models [122,124,125]. Roflumilast is synthesized in four steps from 3-(cyclopropylmethoxy)-4-hydroxybenzaldehyde [126]. The difluoromethyl ether is introduced by alkylation of the free phenolic group with chlorodifluoromethane and base. The aldehyde moiety is oxidized to the benzoic acid, which is then converted to an acid chloride and coupled with 3,5-dichloro-4-aminopyridine.

Roflumilast is rapidly absorbed and metabolized to its active metabolite, roflumilast *N*-oxide [121,122]. Metabolism is mediated by CYP3A4 and CYP1A2. Plasma protein binding is 98.9% for parent drug and 97% for the *N*-oxide. After a single 500 μg dose, the C_{max} for roflumilast was 5.3 μg/mL ($T_{max} = 1.3$ h) and for the *N*-oxide, 9.4 μg/mL ($T_{max} = 11$ h). AUCs were 35 and 350 μg·h/mL for the parent and *N*-oxide, respectively. Steady state concentrations were reached after 7 days of daily 500 μg doses of roflumilast, with the *N*-oxide C_{max} being twofold higher than with acute dosing. The absolute oral bioavailability was 80%, and terminal plasma half-lives were 18 h for the parent and 21 h for the *N*-oxide. There was no food effect on pharmacokinetics. Roflumilast and its metabolites are primarily excreted in the urine (70%). Roflumilast and its *N*-oxide metabolite are not inhibitors of CYP enzymes. Roflumilast is a weak inducer of CYP2B6, but not of other CYP isozymes. Roflumilast has been evaluated in several clinical trials involving over 9000 patients with COPD, including 6-month and 12-month Phase III trials [121]. Patients were allowed to continue with most other respiratory medications, including short-acting β2 agonists. Primary outcome measures were changes in FEV_1 and numbers of exacerbations. In the 6-month trials with patients with moderately severe disease, roflumilast-treated patients (250 or 500 μg daily oral dose) showed greater improvements in FEV_1 and quality of life than placebo-treated patients, but these changes did not reach clinical significance. Exacerbations were decreased with roflumilast treatment, but more patients discontinued treatment in the roflumilast arms than in the placebo group. In 12-month trials in patients with more severe COPD, daily treatment with 500 μg roflumilast statistically increased FEV_1 but did not change the rate of exacerbations overall. However, a subset of patients with more severe airway obstruction did have a significant reduction in exacerbation frequency. In 12-month trials in patients with chronic bronchitis, a 500-μg daily oral dose roflumilast also gave a significant improvement in FEV_1 and a reduction in exacerbation rate. Roflumilast was well tolerated, with slightly more discontinuations than placebo. Diarrhea, nausea, and headache were the most common reasons for discontinuation. Weight loss was also observed, particularly in patients with GI effects and severe COPD. The incidence of insomnia, anxiety, and depression was two to three times higher in patients who received 500 μg of roflumilast than those receiving 250 μg or placebo. Another potential adverse event was cancer, in that of the cancers observed during the trial, 60% were in the roflumilast-treated group compared to 40% in placebo. The significance of this finding is still unclear. In addition to the German approval in July 2010, the U.S. FDA approved roflumilast in March 2011. Roflumilast is an oral drug taken daily to decrease the frequency of flare-ups (exacerbations) or worsening of symptoms from severe COPD.

18. ROMIDEPSIN (ANTICANCER) [127–133]

Class	Histone deacetylase inhibitor
Country of origin	Japan
Originator	Fujisawa (Astellas Pharma)
First introduction	United States
Introduced by	Celgene
Trade name	Istodax®
CAS registry no.	128517-07-7
Molecular weight	540.7

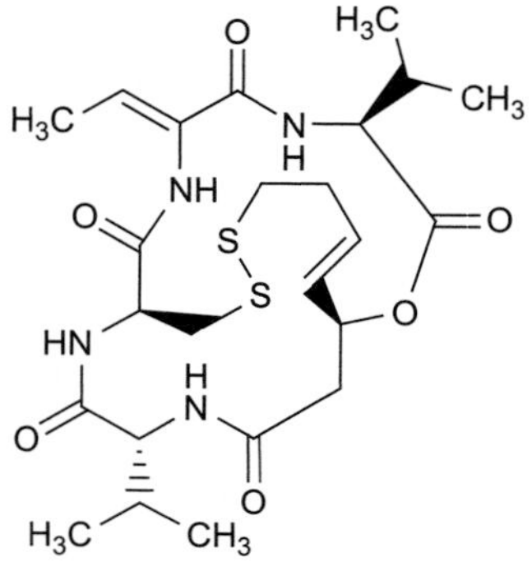

The U.S. FDA approved romidepsin (also referred to as FK228) in 2009 for the treatment of cutaneous T-cell lymphoma (CTCL) for patients who received at least one systemic therapy. The annual incidence of CTCL in the United States is 0.96 per 100,000 persons, and the overall prevalence is 16,000–20,000 cases. Early-stage CTCL is primarily managed by dermatologists using skin-directed therapies like phototherapy, topical agents, and local radiation. Besides romidepsin, there are three approved systemic therapies for advanced disease. Vorinostat and bexarotene are indicated for cutaneous manifestations of CTCL and not for cases with blood, lymph node, or visceral involvement. Denileukin diftitox is approved for CTCL patients expressing the CD25 epitope (30–50% of CTCL cases). However, the relapse rate for patients treated with most therapeutics for CTCL, regardless of the disease stage, is high, and newer therapeutic options for patients with relapsed or refractory CTCL are still needed. Romidepsin is a natural product that was first isolated from the fermentation broth of *C. violaceum*. Romidepsin is the second histone deacetylase (HDAC) inhibitor approved for CTCL, the other being vorinostat, which was approved by the FDA in 2006. Unlike vorinostat which is a pan-HDAC inhibitor, romidepsin shows modest selectivity for class I HDACs in *in vitro* assays. It has been shown that after romidepsin enters the cytoplasm, the disulfide bond is cleaved by glutathione to release the sulfhydryl group which chelates with the active site zinc of class I HDACs and inhibits the enzymatic activity at nanomolar concentrations [127]. Although romidepsin inhibits class I HDACs, it is 17–23 times less potent as the parent than the corresponding reduced form at each isozyme. For example, the IC_{50} of romidepsin at HDAC1 is 36 ± 16 nM while that of the reduced form is $IC_{50} = 1.6 \pm 0.9$ nM [127]. Romidepsin has also been shown to induce cell cycle arrest, differentiation, and apoptosis in tumor cells by mechanisms that cannot be completely explained by HDAC inhibition alone. The synthesis of romidepsin has been reported in the patent and open literature [128,129]. In the first reported total synthesis, Simon *et al.* [129a] employ the Carreira catalytic asymmetric aldol reaction to build the thiol containing β-hydroxy acid followed by a modified Mitsunobu macrolactonization and iodine-mediated intramolecular oxidative coupling to form the disulfide ring. The synthesis by Williams *et al.* [129b] utilizes Noyori's asymmetric transfer hydrogenation of a propargylic ketone as a key step in the synthesis of the β-hydroxy acid. Katoh *et al.* [129c] employ a Julia-Kocienski olefination of a 1,3-propanediol-derived sulfone as the critical step to access the β-hydroxy acid.

In vitro studies show that romidepsin induces apoptosis in a variety of tumor cells such as primary CLL cells, MM cell lines, primary myeloma cells, and small cell lung cancer cells. Preclinical *in vivo* studies demonstrated that romidepsin improved the survival rates of mice bearing murine ascitic tumors (P388 and L1210 leukemia's and B16 melanoma) and inhibited the growth of human solid tumors (Lu-65 and LC-6 lung carcinomas and SC-6 stomach adenocarcinoma) implanted in normal or nude mice. The pharmacokinetic parameters of romidepsin were evaluated at a dose of 17.8 mg/m^2 in a Phase I trial in cancer patients [130]. Following a 4-h IV infusion, romidepsin had a C_{max} of ~1.02 μM and an AUC of ~4.2 μM h, which increased in a dose-proportional fashion, and it showed a mean elimination half-life of ~41 h. The mean clearance was 10.5 L/h/m^2 and the terminal half-life was ~8.1 h. Romidepsin is extensively metabolized *in vitro* by CYP3A4 with minor contributions by CYP3A5, CYP1A1, CYP2B6, and CYP2C19. The human plasma protein binding of romidepsin ranges from 92% to 94% at concentrations of 50–1000 ng/mL, respectively, primarily because of binding to α1-acid-glycoprotein. Romidepsin approval for CTCL was based on two multicenter, single-arm clinical studies in patients with CTCL [131,132]. The first study had 96 patients with CTCL after failure of at least one systemic therapy. The second study comprising 71 patients was also in patients with CTCL who received at least two prior skin-directed therapies or one prior systemic therapy. Patients in these studies were given romidepsin at a starting dose of 14 mg/m^2 infused over 4 h on days 1, 8, and 15 every 28 days. The overall objective disease response (ODR) and complete response (CR) rates in both studies were similar (~34% and 6%, respectively). The median response duration was ~15 months in study I and ~11 months in study II. The median times for CR for studies I and II were 6 and 4 months, respectively. Most common serious adverse events associated with romidepsin treatment were neutropenia, lymphopenia, thrombocytopenia, anemia, fatigue, and nausea. The recommended dose of romidepsin is 14 mg/m^2 administered intravenously over 4 h on days 1 and 8 and 15 of a 28-day cycle [133].

19. SIPULEUCEL-T (ANTICANCER) [134–138]

Class	Therapeutic cancer vaccine for hormone-refractory prostrate cancer
Country of origin	United States
Originator	Dendreon
First introduction	United States
Introduced by	Dendreon
Trade name	Provenge®
CAS registry no.	917381-47-6

Sipuleucel-T is the first FDA-approved therapeutic cancer vaccine. It is approved for the treatment of asymptomatic or minimally symptomatic metastatic castration-resistant (hormone-refractory) prostate cancer.

Sipuleucel-T is described as a therapy rather than a precisely defined therapeutic agent because it is an autologous vaccine that is generated from each individual patient's own blood cells. Patient peripheral blood mononuclear cells (PBMCs) are isolated by leukophoresis and then cultured together with a recombinant fusion protein, comprising human prostatic acid phosphatase (a protein found in prostate and prostate cancer cells) fused at its COOH terminus to full length human GM-CSF *via* a gly-ser linker (PA2024, PAP-GM-CSF). The recombinant PA2024 molecule is largely removed by washing, or is catabolized by PBMC during culture, and it is not a significant component of the dosed vaccine. Preparations are produced fresh for each treatment by a manufacturing process that takes 3–4 days and includes *in vitro* culture for 36–44 h at 37 °C followed by washing, resuspension, and packaging. The resulting vaccine is delivered in an infusion bag comprising a 250-mL cell suspension in Ringer's lactate solution. The actual cellular composition of the infusion bag is incompletely defined. The release assay is a count of the number of activated antigen presenting cells, defined as positive for expression of the cell surface marker CD54 [134]. Each dose is required to contain at least 50 million CD54 positive cells. However, other lymphocytic cells (such as T cells, B cells, and NK cells) that could contribute to the activity of sipuleucel-T are found in the infusion bag, and the actual number and activation state of these cell types is not controlled and will vary for each individual manufactured dose. T cells may be a particularly important component of the activity, and T cell activation has been reported to be specifically induced by PA2024 during sipuleucel-T manufacture [135]. The therapeutic course for sipuleucel-T comprises three infusions, given at approximately 2-week intervals. Patients are premedicated approximately 30 min prior to each dose with oral acetaminophen and an antihistamine such as diphenhydramine to minimize infusion reactions. The safety and efficacy of this treatment schedule were examined in three randomized (2:1) Phase III trials enrolling 737 asymptomatic, or minimally symptomatic, patients with metastatic, castration-resistant prostate cancer. In each of these trials, the control group received placebo infusions comprising the patient's own leukophoresed PBMCs that had been cultured in the absence of recombinant PA2024. The placebo infusion bag was released with the same requirements, >50 million CD54 positive cells, as sipuleucel-T. Two of the Phase III trials were completed: the pivotal trial enrolling 512 patients [136] and a smaller trial enrolling 127 patients [137]. A third trial was not completed; however, 98 patients were enrolled and followed for time to progression and survival [138]. The predefined endpoint for the smaller non-pivotal trials was time to progression, and statistical significance was not reached; however, in the 127-patient trial, sipuleucel-T was observed to have a beneficial effect on overall survival, with a median survival difference of 4.5 months ($p = 0.01$, log-rank; HR, 1.70; 95% CI, 1.13–2.56). The data from the uncompleted trial were not inconsistent with a positive survival benefit for vaccine therapy [137]. The primary endpoint for the pivotal trial was overall survival, and sipuleucel-T was again found to have a beneficial effect, with a median survival difference of 4.1 months (25.8 months in the sipuleucel-T group *vs.* 21.7 months in the placebo group). There was an observed relative reduction of 22% in the risk of death as compared with the placebo group (HR, 0.78; 95% CI, 0.61–0.98; $p = 0.03$). The 36-month survival probability was 31.7% in the sipuleucel-T group versus 23.0% in the placebo group. The Kaplan–Meier estimated survival probability did not show a difference beyond 48 months; however, the number of patients still at risk at these time points is too small to draw conclusions. Consistent with the earlier trials, in the pivotal trial, the difference in time to progression for the sipuleucel-T and the placebo groups was not significant. Because of the difference in the mechanism of action compared to cytotoxic therapy, time to progression may not be an appropriate endpoint for a therapeutic cancer vaccine where activity is not expected until the host immune system becomes engaged. In rapidly progressing cancers, such as metastatic castration-resistant prostate cancer, this engagement may be too slow to see an effect on progression. Acute infusion reactions (reported within 1 day of infusion) are a frequent occurrence for sipuleucel-T therapy. The most common adverse events observed in the pivotal study for the sipuleucel-T group within 1 day after infusion were chills (in 51.2%), fever (22.5%), fatigue (16.0%), nausea (14.2%), and headache (10.7%). Grade 3 or higher adverse events within 1 day after infusion were reported in 6.8% of the sipuleucel-T group. These included chills (in four patients), fatigue (three patients), and back pain, hypertension, hypokalemia, and muscular weakness (in two patients each). Overall, only 3 of 338 patients (0.9%) in the sipuleucel-T group were unable to receive all three infusions because of acute infusion-related adverse events. In the pivotal trial, there was no statistically significant difference in cerebrovascular events, including hemorrhagic and ischemic strokes, between the sipuleucel-T group (2.4%) and placebo group (1.8%) ($p = 1.00$). Overall experience in controlled studies found cerebrovascular events observed in 3.5% of sipuleucel-T treated patients compared with 2.6% of patients in the control groups.

20. TESAMORELIN ACETATE (HIV LIPODYSTROPHY) [139–142]

Class	Growth hormone-releasing factor
Country of origin	Canada
Originator	Theratechnologies

First introduction	United States
Introduced by	Theratechnologies
Trade name	Egrifta™
CAS registry no.	804475-66-9
Molecular weight	5135.9 Da

44-Amino acid polypeptide-*trans*-3-hexenoyl-YADAIFTNSYRKV LGQLSARKLLQDIMSRQQGESNQERGARARL-NH_2

Tesamorelin acetate is an analog of growth hormone-releasing hormone (GHRH) that was approved in the United States in 2010 for treatment of lipodystrophy in HIV patients. With the advent of potent antiretroviral treatment (ART) in the 1990s as a treatment regime for HIV infection, some patients were observed to have either a loss or an accumulation of fat, termed lipodystrophy. Individuals with lipodystrophy can develop excess fat notably in the abdominal visceral adipose tissue (VAT), liver, trunk, and breasts. Up to 30% of patients undergoing ART experience increases in abdominal fat. Patients with HIV-associated lipodystrophy have an increased need for health services when compared to ART patients without lipodystrophy. The market size for the treatment of HIV-associated lipodystrophy in the United States is currently projected to be $800 million to $1.2 billion [139]. Lipodystrophy has both medical and social implications. Increased VAT has been associated with dyslipodemia, thereby increasing the risk of the progression of metabolic diseases. Excess VAT is associated with increased coronary artery calcification, increasing the risk of cardiovascular events. The pathogenesis of lipodystrophy is not clearly understood, but several biological processes have been implicated. Impaired fatty acid metabolism in the adipocyte, deficiencies in adiponectin, increases in leptin, and alteration in growth hormone secretion have all been studied. Decreased growth hormone (GH) levels are found in HIV patients with increased VAT. The impairment of the GH secretion through suppression of GH by elevated fatty acids and decreases in ghrelin levels may contribute to lipodystrophy. The observation of lower GH levels in patients with lipodystrophy and the observation that GH treatment to prevent HIV-related wasting results in improved lipodystrophy have led to efforts to target the GH axis to treat lipodystrophy [140,141].

Tesomorelin is an analog of GHRH. GHRH stimulates the synthesis and release of GH. In the pituitary, GH is secreted in a pulsatile manner. GH control is regulated by somatostatin and the negative feedback regulator, IGF-1. Direct administration of GH to patients with lipodystrophy decreases VAT but has associated side effects such as fluid retention and joint swelling. Because of the side effects associated with direct GH administration, GHRH represents a attractive mechanism for increasing GH levels. Tesomorelin is an analog of GHRH in which the N-terminal amino acid, Tyr, is amidated with a *trans*-3-hexenoyl group. Capping of the N-terminus protects GHRH from cleavage by DPP-4. Tesomorelin demonstrates enhanced stability compared with GHRH in animal models. After subcutaneous administration, GH levels were increased out to 8 h when tested in rats, dogs, and pigs. In healthy male volunteers, tesamorelin administered subcutaneously showed a linear increase in pharmacokinetic parameters when dosed at 0.5, 1, or 2 mg per day. In Phase II studies comparing 1 and 2 mg doses of tesamorelin, the 2 mg dose showed better efficacy at reducing trunk and visceral fat. IGF-1 levels were found to be significantly increased relative to placebo. In two Phase III clinical trials, tesamorelin was shown to significantly reduce VAT and to improve body image. Significant reductions in trunk fat were observed in the tesamorelin-treated group. Lean body mass increased and waist circumference decreased. Tesamorelin was found to affect metabolic biomarkers by reducing triglycerides and total cholesterol levels [141,142]. In November 2010, tesamorelin, a GRF analog, was approved by the U.S. FDA for the reduction of excess abdominal fat in HIV-infected patients with lipodystrophy. Tesamorelin is marketed under the trade name Egrifta™ by Theratechnologies and EMD Serono. Egrifta™ is available as a subcutaneously administered 2 mg dose. Egrifta™ is the first FDA-approved treatment specifically approved for lipodystrophy.

21. TICAGRELOR (ANTITHROMBOTIC) [143–150]

Class	$P2Y_{12}$ antagonist
Country of origin	United Kingdom
Originator	Astra-Zeneca
First introduction	European Union
Introduced by	Astra-Zeneca
Trade names	Brilique™ and Possia® in the European Union
CAS registry no.	274693-27-5
Molecular weight	522.57

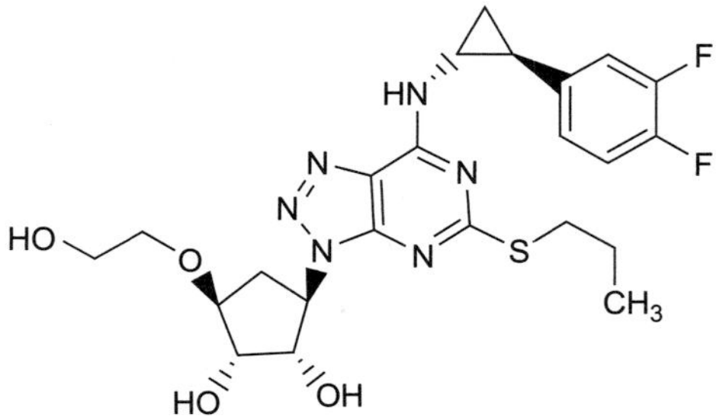

HN F F N N N N N S HO O CH3 HO OH

In December 2010, the $P2Y_{12}$ receptor antagonist ticagrelor (also known as AZD6140) was approved in Europe for the treatment of acute coronary syndrome (ACS), a condition that covers several clinical symptoms with the potential to cause acute myocardial ischemia (MI). The symptoms range from unstable angina, to non-Q wave infarctions and Q-wave myocardial infarctions, which can lead to life-threatening events. ACS is the leading cause of death in the United States and represents 50% of the deaths related to cardiovascular disease. In Europe, ACS affects an estimated 1.4 million people every year [143]. Despite the availability of current treatment options for ACS, data suggest that up to 15% of patients die within 1 year of their cardiovascular event. The economic impact of ACS is substantial and estimated to be $150 billion per year. The rupture of an atherosclerotic plaque in coronary arteries triggers events that lead to the formation of a platelet-rich thrombus plug or clot. This results in an ischemic event, myocardial infarction, or stroke and often is fatal. When a plaque ruptures, a cascade of biochemical events, termed platelet activation, is set in place. At initiation, several agonists of platelet activation are released; among these are thrombin and adenosine diphosphate (ADP). ADP binds to two purinergic receptors, the $P2Y_1$ and $P2Y_{12}$ receptors. The action of ADP binding to the $P2Y_{12}$ receptor results in activation of the GP IIb/IIIa (integrin) receptor. GP IIb/IIIa initiates and prolongs platelet aggregation, which in turn results in the cross-linking of platelets through fibrin and finally thrombus formation. Inhibition of ADP stimulation of the $P2Y_{12}$ receptor has been found to be an effective strategy for managing the atherothrombotic events associated with ACS and potentially resulting from percutaneous coronary intervention (PCI, stent implantation) [144,145].

The first $P2Y_{12}$ inhibitor drugs to reach the market were of the thienopyridine class represented by clopidogrel and prasugrel. Both are converted by CYP-mediated processes to an active metabolite that binds covalently and irreversibly to the $P2Y_{12}$ receptor. A second class of $P2Y_{12}$ receptor inhibitors has been developed that are reversible and bind directly to the receptor. Drug design efforts began with the structure of the endogenous antagonist adenosine triphosphate (ATP). ATP contains several functionalities which would in themselves make the discovery of an orally active drug difficult. These are the acidic and unstable triphosphate functionality, the ribose core and the purine-sugar bond. Astra-Zeneca researchers postulated that many of these structural features could be modified and replaced, resulting in the discovery of orally bioavailable antagonists of the $P2Y_{12}$ receptors. The phosphate moiety was addressed by replacing the phosphate with dicarboxylic acids, then monocarboxylic acids and ultimately with nonacidic, neutral groups, such as the hydroxyl-ethyl ether group found in ticagrelor. The purine group was replaced with a triazolopyrimidine, a known bioisostere, and through substitution on the amino group, molecules with potent binding were identified. In addition, the ribose sugar was replaced by a cyclopentane to prevent the possibility of enzymatic cleavage. The optimal molecule discovered from this medicinal chemistry effort is AZD6140, ticagrelor. The synthesis of ticagrelor is convergent. The cyclopentyl core is built up from mono-protected cyclopentene diol. After elaboration to add the adjacent diols and the ethyl alcohol *via* ether bond formation, one of the ring alcohols is converted to an amine group. This amine is then used to build up the triazolopyrimidine ring. Addition of the fully functionalized phenyl-cyclopropyl amine to chloro ethylthio-triazolopyrimidine completes the synthesis of ticagrelor. Ticagrelor binds reversibly to the $P2Y_{12}$ receptor and has been found to be a noncompetitive (with ADP) allosteric antagonist with rapid on-/off-rate kinetics [146].

The metabolism, distribution, and elimination profiles of ticagrelor have been published. In a study in six healthy male subjects, radiolabeled ticagrelor was found to reach T_{max} in 1.5 h. The major active metabolite formed, AR-C124910XX, results from O-deethylation of ticagrelor. AR-C124910XX circulates at 29% peak and 40% total exposure relative to ticagrelor. Very little parent drug and AR-C124910XX were found in the urine with the primary direct elimination pathway being in the feces. Secondary *O*-glucuronide metabolites and ticagrelor metabolites where the cyclopropyl-phenyl group is cleaved represent the metabolites excreted in the urine. A total of 10 metabolites of ticagrelor were reported [147]. Ticagrelor has been studied in several Phase II and Phase III studies. The DISPERSE II trial, a Phase III trial, compared ticagrelor to clopidogrel in non-ST-segment elevation myocardial infarction in 990 patients with atherosclerotic disease where efficacy and bleeding events were evaluated. In this study, ticagrelor lowered the incidence of MI and showed a low bleeding risk and effective inhibition of platelet aggregation. Ticagrelor was studied in patients with stable coronary artery disease in the ONSET/OFFSET (Phase III) trial. The purposes of this trial were to show how rapidly platelets were inhibited after dose administration and what the duration was for restoration of platelet function after drug discontinuation. The results of the study showed that ticagrelor had a faster and greater inhibition of platelet activation when compared with clopidogrel. After termination of drug administration, platelet function was restored within 5 days in the ticagrelor-treated groups compared to 7 days in the clopidogrel-treated group. The RESPOND trial (Phase III) was designed to look at the effectiveness of ticagrelor in patients who are nonresponders to clopidogrel. The outcome was that ticagrelor was effective in both patients that respond to clopidogrel and those that are nonresponsive to clopidogrel. The PLATO trial (Phase III) was a large $>18{,}000$ patient trial. Patients hospitalized for ST-elevation ACS with scheduled primary PCI or for non-ST-elevation ACS were evaluated for reduction in the event rate of MI, vascular-related death, stroke- and drug-related bleeding risk. Ticagrelor was found to reduce all primary outcome events, including MI, stroke, and death from vascular complication, ischemia, and stent thrombosis [147–150]. In December 2010, the EC granted marketing approval for Brilique™ (ticagrelor) for the prevention of atherothrombotic events in patients with ACS.

22. VERNAKALANT (ANTIARRHYTHMIC) [151–154]

Class	Atrial potassium channel blocker
Country of origin	Canada
Originator	Cardiome Pharma Corp.
First introduction	European Union
Introduced by	Merck and Cardiome Pharma Corp.
Trade name	Brinavess™
CAS registry no.	794466-70-9
	748810-28-8 (HCl)
Molecular weight	349.46

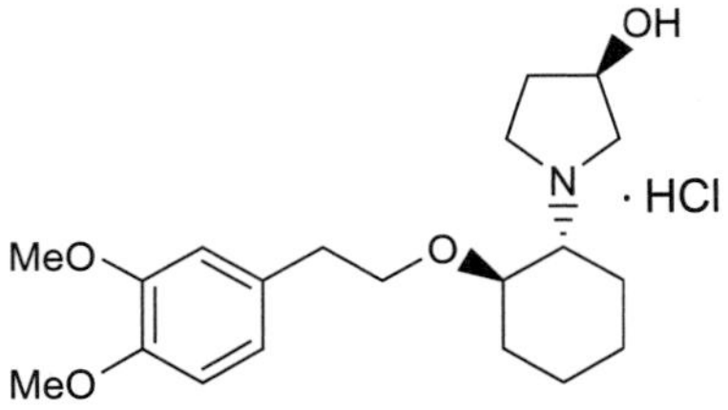

Vernakalant is a potassium channel blocker that was approved in Europe in 2010 for treatment of atrial fibrillation (AF), a condition of cardiac arrhythmia in which the atria of the heart beat irregularly due to changes in cardiac ion channel function and distribution. Patients with AF can have events that last from minutes to weeks to years depending on the severity of the condition and other complicating factors. AF affects quality of life by causing shortness of breath, chest pains, palpitations, and general weakness. AF is associated with congestive heart failure (CHF) and increases the risk of stroke through the increased chance of blood clot formation. The risk of AF rises with age. There are an estimated 2.2 million patients with AF in the United States and 4.5 million in the European Union. The annual U.S. sales for antiarrhythmics are $1.6 billion [151]. The worldwide market for treating AF is growing as the population ages and as the incidence of diabetes and hypertension increases. For the patient with AF, therapeutic intervention involves heart rate management (β-blockers or calcium-channel blockers), reducing the risk of stroke (warfarin or aspirin) and restoring sinus rhythm (SR) to normal (Class Ia, Ic, or III antiarrhythmic agent) [152]. The pathogenesis of AF involves a multitude of events which result in irregular impulses disrupting a normal heartbeat. These irregular impulses originate from the atria which are out of coordination with the ventricles of the heart. The changes in the atria are both electrical and structural. Currently, marketed antiarrhythmic agents vary in their safety and efficacy profile [152,153]. Antiarrhythmic agents target sodium and potassium ion channels of the heart. Side effects result from the lack of selectivity of agents for ion channels in atrial and ventricular tissues. This results in ventricular proarrhythmias and an increased risk of Torsades de Pointes (TdP). Class I antiarrhythmic drugs primarily affect the inward sodium current (I_{Na}), whereas Class III antiarrhythmic drugs target the rapidly activating I_{kr} and ultra rapidly activating (I_{kur}) delayed rectifier potassium currents. Both I_{Na} and I_{kr} are found in atrial and ventricular tissues, whereas I_{kur} is found only in the atria. Drugs with greater atrial selectivity have been sought by targeting I_{kur} (the Kv1.5 channel). Vernakalant has activity for cardiac Na+ and K+ channels and also for the atrial-selective Kv1.5 channel. Vernakalant (RSD1235) was characterized in animal models to assess efficacy, atrial selectivity, and reduction in side effects. In canine models of AF, vernakalant given intravenously at 1, 2, 4, and 8 mg/kg reduced in a dose-dependent manner the time of conversion of AF to normal SR with no increase in QT (hERG) intervals. Primates given IV vernakalant at 2.5, 5, and 10 mg/kg had dose-dependent increases in the atrial refractory period with fewer increases in the ventricular refractory period [151–153]. The absorption, metabolism, and distribution of vernakalant have been characterized by both oral and IV administration. IV administration of vernakalant showed linear pharmacokinetics over the dose range of 0.1–5 mg. The elimination half-life was found to be 2 h. Vernakalant is rapidly metabolized by CYP2D6 to the 4-*O*-demethyl metabolite (RDS-1385), which can form an inactive *O*-glucuronide. Vernakalant is cleared by both the liver and kidney. Vernakalant has been studied and characterized in several Phase III trials (ACT I, ACT II, and ACT III) where the drug was dosed as a 10-min IV infusion of 3 mg/kg followed by a second 10-min infusion at a dose of 2 mg/kg 15 min later if AF had not been terminated. In short duration AF patients (3 h to 7 days), 51.7% of the patients converted to normal SR versus 4% for the placebo-treated group. In patients with long duration AF (8–45 days), the effectiveness of vernakalant was lower with 7.9% achieving normal SR versus none in the placebo group converting to SR [152–154]. In September 2010, IV vernakalant (Brinavess™; Cardiome/Merck) was granted marketing authorization by the EC for rapid conversion of recent onset AF to SR in adults, for nonsurgery patients with AF of 7 days or less and for postcardiac surgery patients with AF of 3 days or less.

23. VINFLUNINE DITARTRATE (ANTICANCER) [155–162]

Class	Tubulin inhibitor
Country of origin	France
Originator	Pierre Fabre
First introduction	United Kingdom
Introduced by	Pierre Fabre
Trade name	Javlor®
CAS registry no.	162652-95-1
	194468-36-5 [tartrate (1:2)]
Molecular weight	816.9

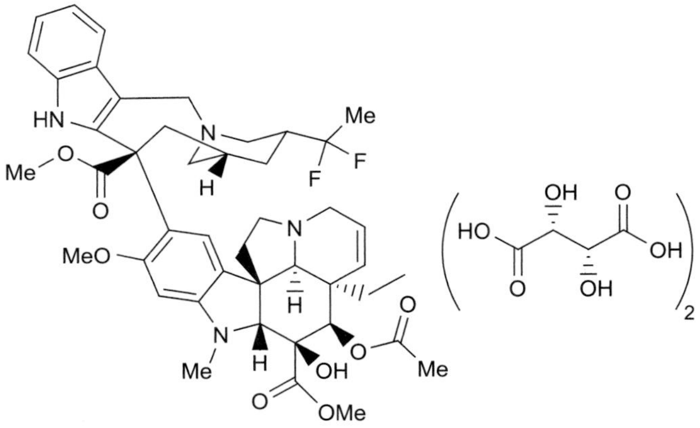

Vinflunine (also referred to as PM391) is a semisynthetic analog of the natural *vinca* alkaloids vinblastine and vincristine that was approved by the European Medicines Agency (EMEA) in 2009 for the treatment of adult patients with advanced or metastatic transitional cell carcinoma of the urothelial tract after failure of a prior platinum-containing regimen. Bladder cancer is a significant health problem in the European Union with 104,400 new cases (82,800 men and 21,600 women) identified in 2006. Cisplatin-based combination chemotherapy or single-agent-based chemotherapy in patients unable to receive cisplatin is the standard first-line treatment for metastatic bladder cancer. However, responses to cisplatin-based regimens are not durable and most patients usually experience disease progression or recurrence. Single-agent therapies such as gemcitabine, paclitaxel, docetaxel, ifosfamide, and ixabepilone have produced low to moderate response rates (0–20%) in patients with platinum-pretreated advanced bladder cancer. Combination therapies have produced higher response rates (30–60%) although this is accompanied with increasing toxicities. Therefore, there is need for newer chemotherapeutics with increased response rates and manageable toxicities. Vinflunine binds to the *vinca*-binding site on tubulin and inhibits microtubule dynamics and treadmilling and induces cell cycle arrest at the G2/M phase. This leads to the accumulation of cells during the mitotic phase and eventually cell death by apoptosis [155]. Compared to other *vinca* alkaloids, vinflunine exhibits the greatest tubulin-binding affinity leading to higher antitumor activity (*e.g.*, compared to vinorelbine) and reduced neurotoxicity. Vinflunine is synthesized from anhydrovinblastine in a superacid media (HF-SbF_5) and in the presence of a chlorinated solvent like $CHCl_3$ or CCl_4 [156,157]. The net result is a novel gem-difluorination at the allylic position and simultaneous reduction of the double bond, followed by ring contraction to yield vinflunine.

The significantly superior anticancer activity of vinflunine compared to its close analog vinorelbine was demonstrated upon intraperitoneal dosing in a range of murine and human tumor xenografts [158]. Preclinical *in vitro* studies demonstrated the cytotoxic ability of vinflunine to inhibit colon, prostrate, bladder, breast, and ovary cell lines with IC_{50} values ranging from 60 to 300 nM [159]. Like other *vinca* alkaloids, vinflunine is a substrate for P-gp-mediated drug resistance mechanisms, although the induction of drug resistance is significantly lower than that of vinorelbine or vincristine. In a Phase I clinical trial, vinflunine showed linear pharmacokinetics over a dose range of 30–400 mg/m^2 following IV administration on day 1 of a 3-week cycle to patients with solid tumors [160]. A dose-proportional increase in exposure was noted and no accumulation was seen between cycles. The terminal half-life of vinflunine is $\sim$40 h. The mean clearance and volume of distribution were $\sim$44 L/h and $\sim$2422 L/m^2, respectively. The human plasma protein binding ranges from 40% to 78%. Vinflunine is predominantly metabolized by CYP3A4; 4-*O*-deacetylvinflunine (DVFL) is the major circulating and active metabolite. The safety and efficacy of vinflunine were established in a randomized Phase III study comparing vinflunine + best supportive care (BSC) versus BSC alone in patients with advanced transitional cell carcinoma of the urothelial tract (TCCU) and who were previously treated with a first-line platinum-containing regimen [161]. Of the 370 patients, 253 were treated with vinflunine + BSC and 117 were treated with BSC. The study indicated that vinflunine treated patients showed improvement in the median overall survival of 6.9 months compared to 4.3 months for patients who were treated with BSC ($p = 0.040$). Other parameters such as overall response rate, disease control, and progression-free survival were all statistically significant favoring the vinflunine + BSC group ($p = 0.006$, 0.002, and 0.001, respectively). The most common serious adverse events associated with the vinflunine + BSC group treatment were neutropenia (50%), febrile neutropenia (6%) anemia, fatigue, and constipation. The recommended dose of vinflunine ditartrate is 320 mg/m^2 administered intravenously over 20 min once every 3 weeks [162].

24. ZUCAPSAICIN (ANALGESIC) [163–170]

Class	Transient receptor potential vanilloid type I channel activator
Country of origin	Germany
Originator	E Merck AG
First introduction	Canada
Introduced by	Winston
Trade name	Civanex®
CAS registry no.	25775-90-0
Molecular weight	305.4

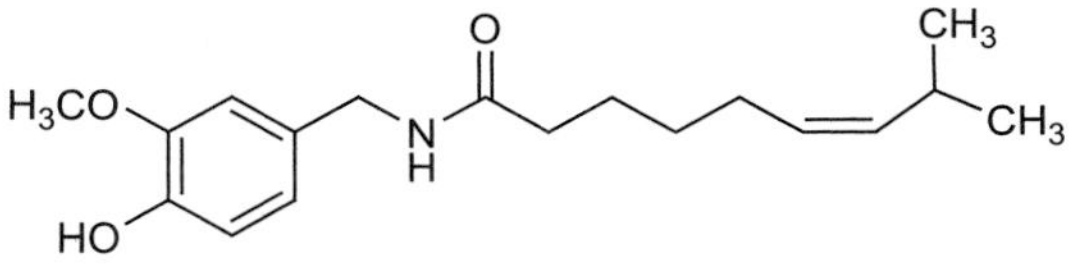

H_3CO HO N H O CH_3 CH_3

Zucapsaicin is a topical analgesic that was approved in Canada in July 2010 for use in conjunction with oral COX-2 inhibitors or NSAIDs to relieve severe pain in adults with osteoarthritis of the knee [163]. Zucapsaicin is the *cis*-isomer of the natural product capsaicin. Capsaicin is available without a prescription in creams, lotions, and patches for the treatment of neuropathic and musculoskeletal pain. Zucapsaicin is available as a 0.075% by weight cream. The advantages of zucapsaicin compared with capsaicin are reported to be a lesser degree of local irritation (stinging, burning, erythema) in patients and a greater degree of efficacy in preclinical animal models of pain [163,164]. The analgesic action of zucapsaicin and capsaicin is mediated through the transient receptor potential vanilloid type 1 (TRPV1) channel [165–168]. TRPV1 is a ligand-gated ion channel that is expressed in the spinal cord and brain and is localized on neurons in sensory ganglia, with peripheral projections to the skin, muscles, joints, and gut and central terminals to the spinal dorsal horn. Activation of TRPV1 triggers an influx of calcium and sodium ions, which initiates a cascade of events associated with pain transmission, including membrane depolarization, neuronal firing, and release of pain transmitters. Topical application of an agonist such as zucapsaicin initially activates TRPV1. The constant activation of the channel results in high intracellular levels of calcium, ultimately leading to desensitization to a variety of noxious stimuli through functional and morphological alterations to the peripheral ends of nerve fibers. The antinociceptive effects of zucapsaicin have been demonstrated with oral treatment in pain models in rats, including the formalin test, the thermal paw withdrawal test, and in the Chung model of neuropathic pain [169]. Zucapsaicin has been synthesized by coupling of vanillin amine with (Z)-8-methylnon-6-enoyl chloride [170]. In preclinical species, there was minimal absorption of zucapsaicin into systemic circulation following topical administration. The only evidence of a systemic effect was a decrease in mean body weight in male rats treated with 3.0% zucapsaicin cream. At the clinically relevant concentration of 0.075%, zucapsaicin cream produced slight erythema in minipigs and slight to moderate erythema in rats. In clinical studies, zucapsaicin serum concentration levels were below the level of quantitation for subjects receiving 0.075% cream applied three times daily to the knees for seven days. Zucapsaicin cream (0.075%) produced local reactions of burning and stinging that were tolerable and reversible. No serious adverse events were reported.

Evidence of efficacy for zucapsaicin was based on a single 12-week Phase III randomized, double-blinded, controlled study in subjects with osteoarthritis of the knee who were taking oral COX-2 inhibitors (42% of patients) or NSAIDs (58% of patients) [163]. Patients received applications of zucapsaicin as either a 0.075% by weight cream (the active arm) or a 0.01% by weight cream (the control arm) three times a day. Primary endpoints were knee pain assessment, physical function, and subject global evaluation. Patients with mild-to-moderate pain showed similar efficacy with both strengths of zucapsaicin cream. However, in patients experiencing severe pain, 0.075% zucapsaicin was significantly more efficacious in reducing pain than 0.01% zucapsaicin cream. Thus, patients who exhibited the best response to zucapsaicin were those in the worst condition as determined by baseline scores. Application site burning sensations were the most frequently reported adverse events and were predominantly mild to moderate. The number of withdrawals in the 0.075% zucapsaicin group was 7%, while fewer patient withdrawals occurred in the 0.01% zucapsaicin group (2%). On the basis of the safety and efficacy data, zucapsaicin was approved in Canada in 2010.

REFERENCES

[1] The collection of new therapeutic entities first launched in 2010 originated from the following sources: Prous Integrity Database; Thomson-Reuters Pipeline Database; The Pink Sheet; Drugs@FDA Website; FDA News Releases; IMS R&D Focus; Adis Business Intelligence R&D Insight; Pharmaprojects; (b) A. Mullard, *Nat. Rev. Drug Discov.*, 2011, **10**, 82.

[2] C. Origlieri and L. Bielory, *Expert Opin. Emerg.Drugs*, 2009, **14**, 523.
[3] S. J. Ono and K. Lane, *Drug Des. Dev. Ther.*, 2011, **5**, 77.
[4] H. Bohets, C. McGowan, G. Mannens, N. Schroeder, K. Edwards-Swanson and A. Shapiro, *J. Ocul. Pharm. Ther.*, 2011, **27**, 187.
[5] J. V. Greiner, K. Edwards-Swanson and A. Ingerman, *Clin. Ophthalmol.*, 2011, **5**, 87.
[6] G. Torkildsen and A. Shedden, *Curr. Med. Res. Opin.*, 2011, **27**, 623.
[7] B. Gallwitz and H. U. Haering, *Diabetes Obes. Metab.*, 2010, **12**, 1.
[8] J. J. Neumiller, L. Wood and R. K. Campbell, *Pharmacotherapy*, 2010, **30**, 463.
[9] J. Gerich, *Diabetes Res. Clin. Pract.*, 2010, **90**, 131.
[10] H. Zettl, M. Schubert-Zsilavecz and D. Steinhilber, *ChemMedChem*, 2010, **5**, 179.
[11] J. Feng, Z. Zhang, M. B. Wallace, J. A. Stafford, S. W. Kaldor, D. B. Kassel, M. Navre, L. Shi, R. J. Skene, T. Asakawa, K. Takeuchi, R. Xu, D. R. Webb and S. L. Gwaltney, *J. Med. Chem.*, 2007, **50**, 2297.
[12] B. Lee, L. Shi, D. B. Kassel, T. Asakawa, K. Takeuchi and R. J. Christopher, *Eur. J. Pharmacol.*, 2008, **589**, 306.
[13] R. Christopher and A. Karim, *Expert Rev. Clin. Pharmacol.*, 2009, **2**, 589.
[14] L. Scott and J. Lesley, *Drugs*, 2010, **70**, 2051.
[15] R. E. Pratley, *Expert Opin. Pharmacother.*, 2009, **10**, 503.
[16] A. del Cuvillo, J. Sastre, J. Montoro, I. Jauregui, I. Davila, M. Ferrer, J. Bartra, J. Mullol, A. Valero and J. Invest, *Allergol. Clin. Immunol.*, 2009, **19**(Suppl. 1), 11.
[17] R. Corcostegui, L. Labeaga, A. Innerarity, A. Berisa and A. Orjales, *Drugs Res. Dev.*, 2005, **6**, 371.
[18] R. Corcostegui, L. Labeaga, A. Innerarity, A. Berisa and A. Orjales, *Drugs Res. Dev.*, 2006, **7**, 219.
[19] N. Jauregizar, L. de la Fuente, M. L. Lucero, A. Sologuren, N. Leal and M. Rodriguez, *Clin. Pharmacokinet.*, 2009, **48**, 543.
[20] C. Garcia-Gea, J. Martinez-Colomer, R. M. Antonijoan, R. Valiente and M.-J. Barbanoj, *J. Clin. Psychopharmacol.*, 2008, **28**, 675.
[21] A. V. Orjales and V. Rubio-Royo, ES2048109, 1994.
[22] S. J. Collier, X. Wu, Z. Poh, G. A. Rajkumar and L. Yet, *Syn. Commun.*, 2011, **41**, 1394.
[23] E. Ferrer, R. Pandian, J. Bolos and R. Castaner, *Drugs Future*, 2010, **35**, 98.
[24] E. Didier, G. Oddon, D. Pauze, P. Leon and D. Riguet, *Patent Application WO9925704*, 1999.
[25] P. Vrignaud, P. Lejeune, F. Lavelle, D. Chaplin and M. C. Bissery, *Proc. Am. Assoc. Cancer Res.*, 2000, **41**, Abstract 1365.
[26] M. C. Bissery, H. Bouchard and H. Riou, *Proc. Am. Assoc. Cancer Res.*, 2000, **41**, Abstract 1364.
[27] A. C. Mita, L. J. Denis, E. K. Rowinsky, J. S. DeBono, A. D. Goetz, L. Ochoa, B. Forouzesh, M. Beeram, A. Patnaik, K. Molpus, D. Semiond, M. Besenval and A. W. Tolcher, *Clin. Cancer Res.*, 2009, **15**, 723.
[28] J. S. DeBono, S. Oudard, M. Ozguroglu, S. Hansen, J. H. Machiels, L. Shen, P. Matthews and A. O. Sartor, *J. Clin. Oncol.*, 2010, **28**(Suppl. 15), Abstract 4508.
[29] M. D. Galsky, A. Dritselis, P. Kirkpatrick and W. K. Oh, *Nat. Rev. Drug Dis.*, 2010, **9**, 677.
[30] B. P. Bouchet and C. M. Galmarini, *Drugs Today*, 2010, **46**, 735.
[31] D. Biek, I. A. Critchley, T. A. Riccobene and D. A. Thyne, *Antimicrob. Chemother.*, 2010, **65**(Suppl. 4), 9.
[32] G. G. Zhanel, G. Sniezek, F. Schwiezer, S. Zelenitsky, P. R. S. Lagace-Wiens, E. Rubenstein, A. S. Gin, D. J. Hoban and J. A. Karlowsky, *Drugs*, 2009, **69**, 809.
[33] M. E. Steed and M. J. Rybak, *Pharmacotherapy*, 2010, **30**, 375.

[34] R. M. Klevens, M. A. Morrison, J. Nadle, S. Petit, K. Gershman, S. Ray, L. H. Harrison, R. Lynfield, G. Dumyati, J. M. Townes, A. S. Craig, E. R. Zell, G. E. Fosheim, L. K. McDougal, R. B. Carey and S. K. Fridkin, *J. Am. Med. Assoc.*, 2007, **298**, 1763.
[35] R. Köck, K. Becker, B. Cookson, J. E. van Gemert-Pijnen, S. Harbarth, J. Kluytmans, M. Mielke, G. Peters, R. L. Skov, M. J. Struelens, E. Tacconelli, A. Navarro Torné, W. Witte and A. W. Friedrich, *Eur. Surveill.*, 2010, **15**, 19688.
[36] A. Villegas-Estrada, M. Lee, D. Hesek, S. B. Vakulenko and S. Mobashery, *J. Am. Chem. Soc.*, 2008, **130**, 9212.
[37] T. Ishikawa, N. Matsunaga, H. Tawada, N. Kuroda, Y. Nakayama, Y. Ishibashi, M. Tomimoto, Y. Ikeda, Y. Tagawa, Y. Iizawa, K. Okonogi, S. Hashiguchi and A. Miyake, *Bioorg. Med. Chem.*, 2003, **11**, 2427.
[38] Y. Wang, N. Mealy, N. Serradell, E. Rosa and J. Bolos, *Drugs Future*, 2008, **33**, 302.
[39] J. Boivin, L. Bunting, J. A. Collins and K. G. Nygren, *Hum. Reprod.*, 2007, **22**, 1506.
[40] P. Verbost, W. N. Sloot, U. M. Rose, R. de Leeuw, R. G. J. M. Hanssen and G. F. M. Verheijden, *Eur. J. Pharmacol.*, 2011, **651**, 227.
[41] D. Loutradis, P. Drakakis, A. Vlismas and A. Antsaklis, *Curr. Opin. Invest. Drugs (BioMed Cent.)*, 2009, **10**, 372.
[42] A. van Schanke, S. F. M. van de Wetering-Krebbers, E. Bos and W. N. Sloot, *Pharmacology*, 2010, **85**, 77.
[43] B. C. J. M. Fauser, B. M. J. L. Mannaerts, P. Devroey, A. Leader, I. Boime and D. T. Baird, *Hum. Reprod. Update*, 2009, **15**, 309.
[44] http://www.nationalmssociety.org/about-multiple-sclerosis.
[45] K. C. Hayes, *CNS Drug Rev.*, 2004, **10**, 295.
[46] J. M. Burton, C. M. Bell, S. E. Walker and P. W. O'Connor, *Neurology*, 2008, **71**, 1833.
[47] W. Smith, S. Swan, T. Marbury and H. Henney, *J. Clin. Pharmacol.*, 2010, **50**, 151.
[48] A. D. Goodman, T. R. Brown, L. B. Krupp, R. T. Schapiro, S. R. Schwid, R. Cohen, L. N. Marinucci and A. R. Blight, *Lancet*, 2009, **373**, 732.
[49] A. D. Goodman, S. R. Schwid and T. R. Brown, *Multiple Sclerosis*, 2008, **14**(Suppl. 1), S298.
[50] A. D. Goodman and M. Hyland, *Drugs Today*, 2010, **46**, 635.
[51] B. F. Boyce and L. Xing, *Arthritis Res. Ther.*, 2007, **9**(Suppl. 1), S1.
[52] L. Green, M. C. Hardy, C. E. Maynard-Currie, H. Tsuda, D. M. Louie, M. J. Mendez, H. Abderrahim, M. Noguchi, D. H. Smith, Y. Zeng, N. E. David, H. Sasai, D. Garza, D. G. Brenner, J. F. Hales, R. P. McGuinness, D. J. Capon, S. Klapholz and A. Jakobovits, *Nat. Genet.*, 1994, **7**, 13.
[53] P. J. Kostenuik, H. Q. Nguyen, J. McCabe, K. S. Warmington, C. Kurahara, N. Sun, C. Chen, L. Li, R. C. Cattley, G. Van, S. Scully, R. Elliott, M. Grisanti, S. Morony, H. L. Tan, F. Asuncion, X. Li, M. S. Ominsky, M. Stolina, D. Dwyer, W. C. Dougall, N. Hawkins, W. J. Boyle, W. S. Simonet and J. K. Sullivan, *J. Bone Miner. Res.*, 2009, **24**, 182.
[54] S. R. Cummings, J. San Martin, M. R. McClung, E. S. Siris, R. Eastell, I. R. Reid, P. Delmas, H. B. Zoog, M. Austin, A. Wang, S. Kutilek, S. Adami, J. Zanchetta, C. Libanati, S. Siddhanti and C. Christiansen, *N. Engl. J. Med.*, 2009, **361**, 756. Erratum in: *N. Engl. J. Med.*, 2009, **361**, 1914.
[55] D. L. Kendler, C. Roux, C. L. Benhamou, J. P. Brown, M. Lillestol, S. Siddhanti, H. S. Man, J. San Martin and H. G. Bone, *J. Bone Miner. Res.*, 2010, **25**, 72.
[56] J. P. Brown, R. L. Prince, C. Deal, R. R. Recker, D. P. Kiel, L. H. de Gregorio, P. Hadji, L. C. Hofbauer, J. M. Alvaro-Gracia, H. Wang, M. Austin, R. B. Wagman, R. Newmark, C. Libanati, J. San Martin and H. G. Bone, *J. Bone Miner. Res.*, 2009, **24**, 153.
[57] A. T. Stopeck, A. Lipton, J. J. Body, G. G. Steger, K. Tonkin, R. H. de Boer, M. Lichinitser, Y. Fujiwara, D. A. Yardley, M. Viniegra, M. Fan, Q. Jiang, R. Dansey, S. Jun and A. Braun, *J. Clin. Oncol.*, 2010, **28**, 5132.

[58] D. H. Henry, L. Costa, F. Goldwasser, V. Hirsh, V. Hungria, J. Prausova, G. V. Scagliotti, H. Sleeboom, A. Spencer, S. Vadhan-Raj, R. von Moos, W. Willenbacher, P. J. Woll, J. Wang, Q. Jiang, S. Jun, R. Dansey and H. Yeh, *J. Clin. Oncol.*, 2011, **29**, 1125.
[59] K. Fizazi, M. Carducci, M. Smith, R. Damião, J. Brown, L. Karsh, P. Milecki, N. Shore, M. Rader, H. Wang, Q. Jiang, S. Tadros, R. Dansey and C. Goessl, *Lancet*, 2011, **377**, 813.
[60] Santen Pharmaceuticals and Inspire Pharmaceuticals Press Release, April 2010.
[61] A. Peral, C. O. Dominguez-Godinez, G. Carracedo and J. Pintor, *Drug News Perspect.*, 2008, **21**, 166.
[62] K. K. Nichols, B. Yerxa and D. J. Kellerman, *Expert Opin. Invest. Drugs*, 2004, **13**, 47.
[63] J. Fischbarg, *Curr. Opin. Invest. Drugs*, 2003, **4**, 1377.
[64] W. Pendergast, B. R. Yerxa, J. G. Douglass and S. R. Shaver, *Bioorg. Med. Chem. Lett.*, 2001, **11**, 157.
[65] B. R. Yerxa and E. G. Brown, *US Patent 7528119*, 2009.
[66] FDA Dermatologic and Ophthalmic Drugs Advisory Committee Meeting Briefing Document, June 2009.
[67] Thomson Reuters Report for Diquafosol, updated March 14, 2011.
[68] A. Banerji, *Allergy Asthma Proc.*, 2010, **31**, 398.
[69] J. A. Bernstein and M. Qazi, *Expert Rev. Clin. Immunol.*, 2010, **6**, 29.
[70] K. P. Garnock-Jones, *Drugs*, 2010, **70**, 1423.
[71] US FDA. Center for Drug Evaluation and Research application number 125277: ecallantide summary review [online]. http://www.accessdata.fda.gov/drugsatfda_docs/nda/2009/125277s000SumR.pdf.
[72] Ecallantide (Kalbitor) Advisory Committee Briefing Document. http://www.fda.gov/downloads/advisorycommittees/committeesmeetingmaterials/drugs/pulmonary-allergydrugsadvisorycommittee/ucm170334.pdf.
[73] M. J. Towle, K. A. Salvato, J. Budrow, B. F. Wels, G. Kuznetsov, K. K. Aalfs, S. Welsh, W. Zheng, B. M. Seletsky, M. H. Palme, G. J. Habgood, L. A. Singer, L. V. Dipietro, Y. Wang, J. J. Chen, D. A. Quincy, A. Davis, K. Yoshimatsu, Y. Kishi, M. J. Yu and B. A. Littlefield, *Cancer Res.*, 2001, **61**, 1013.
[74] J. A. Smith, L. Wilson, O. Azarenko, X. Zhu, B. M. Lewis, B. A. Littlefield and M. A. Jordan, *Biochemistry*, 2010, **49**, 1331.
[75] C.-G. Dong, J. A. Henderson, Y. Kaburagi, T. Sasaki, D.-S. Kim, J. T. Kim, D. Urabe, H. Guo and Y. Kishi, *J. Am. Chem. Soc.*, 2009, **131**, 15642.
[76] Y.-R. Yang, D.-S. Kim and Y. Kishi, *Org. Lett.*, 2009, **11**, 4516.
[77] Y. Wang, N. Serradell, J. Bolos and E. Rosa, *Drugs Future*, 2007, **32**, 681.
[78] C. Twelves, D. Loesch, J. L. Blum, L. T. Vahdat, K. Petrakova, P. J. Chollet, C. E. Akerele, S. Seegobin, J. Wanders and J. Cortes, *Proc. Am. Soc. Clin. Oncol.*, 2010, **28**, 18s.
[79] P. G. Morris, *Anti-Cancer Drugs*, 2010, **21**, 885.
[80] K. Adachi, T. Kohara, N. Nakao, M. Arita, K. Chiba, T. Mishina, S. Sasaki and T. Fujita, *Bioorg. Med. Chem. Lett.*, 1995, **5**, 853.
[81] V. Brinkmann, M. D. Davis, C. E. Heise, R. Albert, S. Cottens, R. Hof, C. Bruns, E. Prieschl, T. Baumruker, P. Hiestand, C. A. Foster, M. Zollinger and K. R. Lynch, *J. Biol. Chem.*, 2002, **277**, 21453.
[82] S. Mandala, R. Hajdu, J. Bergstrom, El Quackenbush, J. Xie, J. Milligan, R. Thornton, G. J. Shei, D. Card, C. Keohane, M. Rosenbach, J. Hale, C. L. Lynch, K. Rupprecht, W. Parsons and H. Rosen, *Science*, 2002, **296**, 346.
[83] (a) M. Kiuchi, K. Adachi, T. Kohara, M. Minoguchi, T. Hanano, Y. Aoki, T. Mishina, M. Arita, N. Nakao, M. Ohtsuki, Y. Hoshino, K. Teshima, K. Chiba, S. Sasaki and T. Fujita, *J. Med. Chem.*, 2000, **43**, 2946 (b). M. Tadashi, O. Toshiyuki and A. Kunitomo,

JP 1999310556; (c) S. Sugiyama, S. Araj, M. Kiriyama and K. Ishi, *Chem. Pharm. Bull*, 2005, **53**, 100.
[84] L. Kappos, E. W. Radue, P. O'Connor, C. Polman, R. Hohlfeld, P. Calabresi, K. Selmaj, C. Agoropoulou, M. Leyk, L. Zheng-Auberson and P. Burtin, *N. Engl. J. Med.*, 2010, **362**, 387.
[85] J. A. Cohen, F. Barkhof, G. Comi, H. P. Hartung, B. O. Khatri, X. Montalban, J. Pelletier, R. Capra, P. Gallo, G. Izquierdo, K. Tiel-Wilck, A. de Vera, J. Jin, T. Stites, S. Wu, S. Aradhye and L. Kappos, *N. Engl. J. Med.*, 2010, **362**, 402.
[86] V. Brinkmann, A. Billich, T. Baumruker, P. Heining, R. Schmouder, G. Francis, S. Aradhye and P. Burtin, *Nat. Rev. Drug Discov.*, 2010, **9**, 883.
[87] D. M. Sobieraj, *Formulary*, 2010, **45**, 245.
[88] C. Reviriego, *Drugs Future*, 2010, **35**, 537.
[89] M. Yamashita, *Antiviral Chem. Chemother.*, 2010, **21**, 71.
[90] D. A. Boltz, J. R. Aldridge, R. G. Webster and E. A. Govorkova, *Drugs*, 2010, **70**, 1349.
[91] L. V. Gubareva, A. A. Trujillo, M. Okomo-Adhiambo, V. P. Mishin, V. M. Deyde, K. Sleeman, H. T. Nguyen, T. G. Sheu, R. J. Garten, M. W. Shaw, A. M. Fry and A. I. Klimov, *Antiviral Ther.*, 2010, **15**, 1151.
[92] T. Honda, S. Kubo, T. Masuda, M. Asai, Y. Kobayashi and M. Yamashita, *Bioorg. Med. Chem. Lett.*, 2009, **19**, 2938.
[93] M. Yamashita, T. Tomozawa, M. Kakuta, A. Tokumitsu, H. Nasu and S. Kubo, *Antimicrob. Agents Chemother.*, 2009, **53**, 186.
[94] M. Kiso, S. Kubo, M. Ozawa, Q. M. Le, C. A. Nidom, M. Yamashita and Y. Kawaoka, *PloS Pathog.*, 2010, **6**, e10000786.
[95] H. Ishizuka, S. Yoshiba, H. Okabe and K. Yoshihara, *J. Clin. Pharmacol.*, 2010, **50**, 1319.
[96] A. Watanabe, S. C. Chang, M. J. Kim, D. W. Chu and Y. Osashi, *Clin Infect. Dis.*, 2010, **51**, 1167.
[97] N. Sugaya and Y. Ohashi, *Antimicrob. Agents Chemother.*, 2010, **54**, 2575.
[98] C. R. Hopkins, *ACS Chem. Neurosci.*, 2011, **2**, 58.
[99] L. Citrome, *Int. J. Clin. Pract.*, 2011, **65**, 189.
[100] T. Ishibashi, T. Horisawa, K. Tokuda, T. Ishiyama, M. Ogasa, R. Tagashira, K. Matsumoto, H. Nishikawa, Y. Ueda, S. Toma, H. Oki, N. Tanno, I. Saji, A. Ito, Y. Ohno and M. Nakamura, *J. Pharmacol. Exp. Ther.*, 2010, **334**, 171.
[101] T. Ishiyama, K. Tokuda, K. Ishibashi, A. Ito, S. Toma and Y. Ohno, *Eur. J. Pharmacol.*, 2007, **572**, 160.
[102] T. Enomoto, T. Ishibashi, K. Tokuda, T. Ishiyama, S. Toma and A. Ito, *Behav. Brain Res.*, 2008, **186**, 197.
[103] P. Cole, N. Serradell, E. Rosa and J. Bolos, *Drugs Future*, 2008, **33**, 316.
[104] J. M. Meyer, A. D. Loebel and E. Schweizer, *Expert Opin. Invest. Drugs*, 2009, **18**, 1715.
[105] W. E. Fogler and I. J. Fidler, *Int. J. Immunopharmacol.*, 1987, **9**, 141.
[106] (a) T. Asano, A. McWatters, T. An, K. Matsushima and E. S. Kleinerman, *J. Pharmacol. Exp. Ther.*, 1994, **268**, 1032; (b) P. Dieter, P. Ambs, E. Fitzke, H. Hidaka, R. Hoffman and H. Schwende, *J. Immunol.*, 1995, **155**, 2595.
[107] W. Strober, P. J. Murray, A. Kitani and T. Watanabe, *Nat. Rev. Immunol.*, 2006, **6**, 9.
[108] J. R. Prous and J. Castaner, *Drugs Future*, 1989, **14**, 220.
[109] D. E. Brundish and R. Wade, *J. Label. Compd. Radiopharm.*, 1985, **22**, 29.
[110] P. A. Meyers, C. A. Scwartz, M. D. Krailo, J. H. Healey, M. L. Bernstein, D. Betcher, W. S. Ferguson, M. C. Gebhardt, A. M. Goorin, M. Harris, E. Kleinerman, M. P. Link, H. Nadel, M. Nieder, G. P. Siegal, M. A. Weiner, R. J. Wells, R. B. Womer and H. E. Grier, *J. Clin. Oncol.*, 2008, **26**, 633.
[111] K. Ando, K. Mori, N. Corradini, F. Redini and D. Heymann, *Expert Opin. Pharmacother.*, 2011, **12**, 285.
[112] P. M. Anderson, M. Tomaras and K. McConnell, *Drugs Today*, 2010, **46**, 327.

[113] D. Birnkrant and E. Cox, *N. Engl. J. Med.*, 2009, **361**, 2204.
[114] C. E. Mancuso, M. P. Gabay, L. M. Steinke and S. J. VanOsdol, *Ann. Pharmacother.*, 2010, **44**, 1240.
[115] S. Kohno, H. Kida, M. Mizuguchi and J. Shimada, *Antimicrob. Agents Chemother.*, 2010, **54**, 4568.
[116] F. Hayden, *Clin. Infect. Dis.*, 2009, **48**(Suppl. 1), S3.
[117] Y. S. Babu, P. Chand, S. Bantia, P. Kotian, A. Dehghani, Y. El-Kattan, T. H. Lin, T. L. Hutchison, A. J. Elliott, C. D. Parker, S. L. Ananth, L. L. Horn, G. W. Laver and J. A. Montgomery, *J. Med. Chem.*, 2000, **43**, 3482.
[118] P. Chand, P. L. Kotian, A. Dehghani, Y. El-Kattan, T. H. Lin, T. L. Hutchison, Y. S. Babu, S. Bantia, A. J. Elliott and J. A. Montgomery, *J. Med. Chem.*, 2001, **44**, 4379.
[119] P. Chand, *Expert Opin. Ther. Patents*, 2005, **15**, 1009.
[120] J. E. Hernandez, R. Adiga, R. Armstrong, J. Bazan, H. Bonilla, J. Bradley, R. Dretler, M. G. Ison, J. E. Mangino, S. Maroushek, A. K. Shetty, A. Wald, C. Ziebold, J. Elder, A. S. Hollister and W. Sheridan, on behalf of the eIND Peramivir Investigators, *Clin. Infect. Dis.*, 2011, **52**, 695.
[121] M. A. Giembycz and S. K. Field, *Drug Des. Dev. Ther.*, 2010, **4**, 147.
[122] M. Sanford, *Drugs*, 2010, **70**, 1615.
[123] G. Higgs, *Curr. Opin. Invest. Drugs*, 2010, **11**, 495.
[124] A. Hatselmann and C. Schudt, *J. Pharmacol. Exp. Ther.*, 2001, **297**, 267.
[125] S. K. Field, *Expert Opin. Invest. Drugs*, 2008, **17**, 811.
[126] L. A. Sorbera, P. A. Leeson and J. Castañer, *Drugs Future*, 2000, **25**, 1261.
[127] R. Furumai, A. Matsuyama, N. Kobashi, K. H. Lee, M. Nishiyama, H. Nakajima, A. Tanaka, Y. Komatsu, N. Nishino, M. Yoshida and S. Horinouchi, *Cancer Res.*, 2002, **62**, 4916.
[128] For example, see N. H. Vrolijk and G. L. Verdine, *Patent Application US20100093610*, 2010.
[129] (a) W. L. Khan, J. Wu, W. Xing and J. A. Simon, *J. Am. Chem. Soc.*, 1996, **118**, 7237; (b) T. J. Greshock, D. M. Johns, Y. Noguchi and R. M. Williams, *Org. Lett.*, 2008, **10**, 613; (c) K. Narita, T. Kikuchi, K. Watanabe, T. Takizawa, T. Oguchi, K. Kudo, K. Matsuhara, H. Abe, T. Yamori, M. Yoshida and T. Katoh, *Chem. Eur. J.*, 2009, **15**, 11174.
[130] V. Sandor, S. Bakke, R. W. Robey, M. H. Kang, M. V. Blagosklonny, J. Bender, R. Brooks, R. L. Piekarz, E. Tucker, W. D. Figg, K. K. Chan, B. Goldspiel, A. T. Fojo, S. P. Balcerzak and S. E. Bates, *Clin. Cancer Res.*, 2002, **8**, 718.
[131] M. Demierre, S. Whittaker, Y. Kim, E. Kim, R. Piekarz, M. Prince, J. Nichols, J. Balser, A. Prentice and S. Bates, *J. Clin. Oncol.*, 2009, **27**(Suppl. 15), Abstract 8546.
[132] R. L. Piekarz, R. Frye, M. Turner, J. J. Wright, S. L. Allen, M. H. Kirschbaum, J. Zain, H. M. Prince, J. P. Leonard, L. J. Geskin, C. Reeder, D. Joske, W. D. Figg, E. R. Gardner, S. M. Steinberg, E. S. Jaffe, M. S. Syevenson, S. Lade, T. Fojo and S. E. Bates, *J. Clin. Oncol.*, 2009, **27**, 5410.
[133] C. Campas-Moya, *Drugs Today*, 2009, **45**, 787.
[134] N. A. Sheikh and L. A. Jones, *Cancer Immunol. Immunother.*, 2008, **57**, 1381.
[135] N. A. Sheikh, J. D. Wesley, E. Chadwick, N. Perdue, C. P. dela Rosa, M. W. Frohlich, F. P. Stewart and D. L. Urdal, *J. Clin. Oncol.*, 2011, **29**(Suppl. 7), Abstract 155.
[136] P. W. Kantoff, C. S. Higano, N. D. Shore, E. R. Berger, E. J. Small, D. F. Penson, C. H. Redfern, A. C. Ferrari, R. Dreicer, R. B. Sims, Y. Xu, M. W. Frohlich and P. F. Schellhammer, the IMPACT Study Investigators, *N. Engl. J. Med.*, 2010, **363**, 411.
[137] E. J. Small, P. F. Schellhammer, C. S. Higano, C. H. Redfern, J. J. Nemunaitis, F. H. Valone, S. S. Verjee, L. A. Jones and R. M. Hershberg, *J. Clin. Oncol.*, 2006, **24**, 3089.
[138] C. S. Higano, P. F. Schellhammer, E. J. Small, P. A. Burch, J. Nemunaitis, L. Yuh, N. Provost and M. W. Frohlich, *Cancer*, 2009, **115**, 3670.
[139] V. L. Leung and M. J. Glesby, *Curr. Opin. Infect. Dis.*, 2011, **24**, 43.

[140] J. Falutz, J. C. Mamputu, D. Potvin, G. Moyle, G. Soulban, H. Loughrey, C. Marsolais, R. Turner and S. Grinspoon, *J. Clin. Endocrinol. Metab.*, 2010, **95**, 4291.
[141] C. Grunfeld, A. Dritselis and P. Kirkpatrick, *Nat. Rev. Drug Discov.*, 2011, **10**, 95.
[142] Y. Wang and B. Tomlinson, *Expert Opin. Invest. Drugs*, 2009, **18**, 303.
[143] Z. Zhao and M. Winget, *BMC Health Serv. Res.*, 2011, **11**, 35.
[144] E. Abergel and E. Nikolsky, *Vasc. Health Risk Manage.*, 2010, **6**, 963.
[145] L. Wallentin, *Eur. Heart J.*, 2009, **30**, 1964.
[146] B. Springthorpe, A. Bailey, P. Barton, T. N. Birkinshaw, R. V. Bonnert, R. C. Brown, D. Chapman, J. Dixon, S. D. Guile, R. G. Humphries, S. F. Hunt, F. Ince, A. H. Ingall, I. P. Kirk, P. D. Leeson, P. Leff, R. J. Lewis, B. P. Martin, D. F. McGinnity, M. P. Mortimore, S. W. Paine, G. Pairaudeau, A. Patel, A. J. Rigby, R. J. Riley, B. J. Teobald, W. Tomlinson, P. J. H. Webborn and P. A. Willis, *Bioorg. Med. Chem. Lett.*, 2007, **17**, 6013.
[147] R. Teng, S. Oliver, M. A. Hayes and K. Butler, *Drug Metab. Dispos.*, 2010, **38**, 1514.
[148] J. M. Siller-Matula and B. Jilma, *Future Cardiol.*, 2010, **6**, 753.
[149] D. L. Jennings, *Formulary*, 2010, **45**, 148.
[150] S. D. Anderson, N. K. Shah, J. Yim and B. J. Epstein, *Ann. Pharmacother.*, 2010, **44**, 524.
[151] J. R. Ehrlich and S. Nattel, *Drugs*, 2009, **69**, 757.
[152] J. W. M. Cheng and I. Rybak, *Clin. Med. Ther.*, 2009, **1**, 215.
[153] D. Dobrev, B. Hamad and P. Kirkpatrick, *Nat. Rev. Drug Discov.*, 2010, **9**, 915.
[154] G. E. Billman, *Curr. Opin. Invest. Drugs*, 2010, **11**, 1048.
[155] V. K. Ngan, K. Bellman, D. Panda, B. T. Hill, M. A. Jordan and L. Wilson, *Cancer Res.*, 2000, **60**, 5045.
[156] J. C. Jacquesy, J. Fahy, C. Berrier, D. Bigg, M. P. Jouannetaud, F. Zunino, A. Kruczynski and R. Kiss, Patent Application WO 95/03312, 1995.
[157] J. Fahy, A. Duflos, J. P. Ribet, J. C. Jacquesy, C. Berrier, M. P. Jouannetaud and F. Zunino, *J. Am. Chem. Soc.*, 1997, **119**, 8576.
[158] A. Kruczynski, F. Colpaert, J. P. Tarayre, P. Mouillard, J. Fahy and B. T. Hill, *Cancer Chemother. Pharmacol.*, 1998, **41**, 437.
[159] A. Kruczynski and B. T. Hill, *Crit. Rev. Oncol. Hematol.*, 2001, **40**, 159.
[160] J. Bennouna, P. Fumoleau, J. P. Armand, E. Raymond, M. Campone, F. M. Delgado, C. Puozzo and M. Marty, *Ann. Oncol.*, 2003, **14**, 630.
[161] J. Bellmunt, C. Theodore, T. Demkov, B. Komyakov, L. Sengelov, G. Daugaard, A. Caty, J. Carles, A. Jagiello-Gruszfeld, O. Karyakin, F. M. Delgado, P. Hurteloup, E. Winquist, N. Morsli, Y. Salhi, S. Culine and H. von der Maase, *J. Clin. Oncol.*, 2009, **27**, 4454.
[162] J. E. Frampton and M. D. Moen, *Drugs*, 2010, **70**, 1283.
[163] Summary Basis of Decision CIVANEX, *Health Canada*, 2010.
[164] (a) J. E. Bernstein, *US Patent 5063060*, 1991; (b) J. E. Bernstein, *US Patent 7244446*, 2007.
[165] M. J. Caterina and D. Julius, *Annu. Rev. Neurosci.*, 2001, **24**, 487.
[166] K. Bley, *Expert Opin. Invest. Drugs*, 2004, **13**, 1445.
[167] S. M. Westaway, *J. Med. Chem.*, 2007, **50**, 2589.
[168] H. Knotkova, M. Pappagallo and A. Szallasi, *Clin. J. Pain*, 2008, **24**, 142.
[169] X. Y. Hua, P. Chen, J. Hwang and T. L. Yaksh, *Pain*, 1997, **71**, 313.
[170] P. M. Gannett, D. L. Nagel, P. J. Reilly, T. Lawson, J. Sharpe and B. Toth, *J. Org. Chem.*, 1988, **53**, 1064.

KEYWORD INDEX, VOLUME 46

CUMULATIVE CHAPTER TITLES KEYWORD INDEX, VOLUME 1–46

CUMULATIVE NCE INTRODUCTION INDEX, 1983–2010

GENERIC NAME	INDICATION	YEAR INTRO.	ARMC VOL., (PAGE)
abacavir sulfate	antiviral	1999	35, 333
abarelix	anticancer	2004	40, 446
abatacept	antiarthritic	2006	42, 509
acarbose	antidiabetic	1990	26, 297
aceclofenac	antiinflammatory	1992	28, 325
acemannan	wound healing agent	2001	37, 259
acetohydroxamic acid	urinary tract/bladder disorders	1983	19, 313
acetorphan	antidiarrheal	1993	29, 332
acipimox	antihypercholesterolemic	1985	21, 323
acitretin	antipsoriasis	1989	25, 309
acrivastine	antiallergy	1988	24, 295
actarit	antiinflammatory	1994	30, 296
adalimumab	antiarthritic	2003	39, 267
adamantanium bromide	antibacterial	1984	20, 315
adefovir dipivoxil	antiviral	2002	38, 348
adrafinil	sleep disorders	1986	22, 315
AF-2259	antiinflammatory	1987	23, 325
afloqualone	muscle relaxant	1983	19, 313
agalsidase alfa	Fabry's disease	2001	37, 259
alacepril	antihypertensive	1988	24, 296
alcaftadine	ophthalmologic, allergic conjunctivitis	2010	46, 444
alclometasone dipropionate	antiinflammatory	1985	21, 323
alefacept	antipsoriasis	2003	39, 267
alemtuzumab	anticancer	2001	37, 260
alendronate sodium	osteoporosis	1993	29, 332
alfentanil hydrochloride	analgesic	1983	19, 314
alfuzosin hydrochloride	antihypertensive	1988	24, 296
alglucerase	Gaucher's disease	1991	27, 321
alglucosidase alfa	Pompe disease	2006	42, 511
aliskiren	antihypertensive	2007	43, 461
alitretinoin	anticancer	1999	35, 333
alminoprofen	analgesic	1983	19, 314
almotriptan	antimigraine	2000	36, 295
alogliptin	antidiabetic	2010	46, 446
alosetron hydrochloride	irritable bowel syndrome	2000	36, 295
alpha-1 antitrypsin	emphysema	1988	24, 297
alpidem	anxiolytic	1991	27, 322
alpiropride	antimigraine	1988	24, 296
alteplase	antithrombotic	1987	23, 326
alvimopan	post-operative ileus	2008	44, 584
ambrisentan	pulmonary hypertension	2007	43, 463
amfenac sodium	antiinflammatory	1986	22, 315
amifostine	cytoprotective	1995	31, 338
aminoprofen	antiinflammatory	1990	26, 298
amisulpride	antipsychotic	1986	22, 316

GENERIC NAME	INDICATION	YEAR INTRO.	ARMC VOL., (PAGE)
amlexanox	antiasthma	1987	23, 327
amlodipine besylate	antihypertensive	1990	26, 298
amorolfine hydrochloride	antifungal	1991	27, 322
amosulalol	antihypertensive	1988	24, 297
ampiroxicam	antiinflammatory	1994	30, 296
amprenavir	antiviral	1999	35, 334
amrinone	congestive heart failure	1983	19, 314
amrubicin hydrochloride	anticancer	2002	38, 349
amsacrine	anticancer	1987	23, 327
amtolmetin guacil	antiinflammatory	1993	29, 332
anagrelide hydrochloride	antithrombotic	1997	33, 328
anakinra	antiarthritic	2001	37, 261
anastrozole	anticancer	1995	31, 338
angiotensin II	anticancer adjuvant	1994	30, 296
anidulafungin	antifungal	2006	42, 512
aniracetam	cognition enhancer	1993	29, 333
anti-digoxin polyclonal antibody	antidote, digoxin poisoning	2002	38, 350
APD	osteoporosis	1987	23, 326
apraclonidine hydrochloride	antiglaucoma	1988	24, 297
aprepitant	antiemetic	2003	39, 268
APSAC	antithrombotic	1987	23, 326
aranidipine	antihypertensive	1996	32, 306
arbekacin	antibacterial	1990	26, 298
arformoterol	antiasthma	2007	43, 465
argatroban	antithrombotic	1990	26, 299
arglabin	anticancer	1999	35, 335
aripiprazole	antipsychotic	2002	38, 350
armodafinil	sleep disorders	2009	45, 478
arotinolol hydrochloride	antihypertensive	1986	22, 316
arteether	antimalarial	2000	36, 296
artemisinin	antimalarial	1987	23, 327
asenapine	antipsychotic	2009	45, 479
aspoxicillin	antibacterial	1987	23, 328
astemizole	antiallergy	1983	19, 314
astromycin sulfate	antibacterial	1985	21, 324
atazanavir	antiviral	2003	39, 269
atomoxetine	attention deficit hyperactivity disorder	2003	39, 270
atorvastatin calcium	antihypercholesterolemic	1997	33, 328
atosiban	premature labor	2000	36, 297
atovaquone	antiparasitic	1992	28, 326
auranofin	antiarthritic	1983	19, 314
azacitidine	anticancer	2004	40, 447
azelaic acid	acne	1989	25, 310
azelastine hydrochloride	antiallergy	1986	22, 316
azelnidipine	antihypertensive	2003	39, 270
azithromycin	antibacterial	1988	24, 298
azosemide	diuretic	1986	22, 316

GENERIC NAME	INDICATION	YEAR INTRO.	ARMC VOL., (PAGE)
aztreonam	antibacterial	1984	20, 315
balofloxacin	antibacterial	2002	38, 351
balsalazide disodium	ulcerative colitis	1997	33, 329
bambuterol	antiasthma	1990	26, 299
barnidipine hydrochloride	antihypertensive	1992	28, 326
beclobrate	antihypercholesterolemic	1986	22, 317
befunolol hydrochloride	antiglaucoma	1983	19, 315
belotecan	anticancer	2004	40, 449
benazepril hydrochloride	antihypertensive	1990	26, 299
benexate hydrochloride	antiulcer	1987	23, 328
benidipine hydrochloride	antihypertensive	1991	27, 322
beraprost sodium	antiplatelet	1992	28, 326
besifloxacin	antibacterial	2009	45, 482
betamethasone butyrate propionate	antiinflammatory	1994	30, 297
betaxolol hydrochloride	antihypertensive	1983	19, 315
betotastine besilate	antiallergy	2000	36, 297
bevacizumab	anticancer	2004	40, 450
bevantolol hydrochloride	antihypertensive	1987	23, 328
bexarotene	anticancer	2000	36, 298
biapenem	antibacterial	2002	38, 351
bicalutamide	anticancer	1995	31, 338
bifemelane hydrochloride	nootropic	1987	23, 329
bilastine	antiallergy	2010	46, 449
bimatoprost	antiglaucoma	2001	37, 261
binfonazole	sleep disorders	1983	19, 315
binifibrate	antihypercholesterolemic	1986	22, 317
biolimus drug-eluting stent	coronary artery disease, antirestenotic	2008	44, 586
bisantrene hydrochloride	anticancer	1990	26, 300
bisoprolol fumarate	antihypertensive	1986	22, 317
bivalirudin	antithrombotic	2000	36, 298
blonanserin	antipsychotic	2008	44, 587
bopindolol	antihypertensive	1985	21, 324
bortezomib	anticancer	2003	39, 271
bosentan	antihypertensive	2001	37, 262
brimonidine	antiglaucoma	1996	32, 306
brinzolamide	antiglaucoma	1998	34, 318
brodimoprin	antibacterial	1993	29, 333
bromfenac sodium	antiinflammatory	1997	33, 329
brotizolam	sleep disorders	1983	19, 315
brovincamine fumarate	cerebral vasodilator	1986	22, 317
bucillamine	immunomodulator	1987	23, 329
bucladesine sodium	congestive heart failure	1984	20, 316
budipine	Parkinson's disease	1997	33, 330
budralazine	antihypertensive	1983	19, 315
bulaquine	antimalarial	2000	36, 299
bunazosin hydrochloride	antihypertensive	1985	21, 324

GENERIC NAME	INDICATION	YEAR INTRO.	ARMC VOL., (PAGE)
bupropion hydrochloride	antidepressant	1989	25, 310
buserelin acetate	hormone therapy	1984	20, 316
buspirone hydrochloride	anxiolytic	1985	21, 324
butenafine hydrochloride	antifungal	1992	28, 327
butibufen	antiinflammatory	1992	28, 327
butoconazole	antifungal	1986	22, 318
butoctamide	sleep disorders	1984	20, 316
butyl flufenamate	antiinflammatory	1983	19, 316
cabazitaxel	anticancer	2010	46, 451
cabergoline	antiprolactin	1993	29, 334
cadexomer iodine	wound healing agent	1983	19, 316
cadralazine	antihypertensive	1988	24, 298
calcipotriol	antipsoriasis	1991	27, 323
camostat mesylate	anticancer	1985	21, 325
canakinumab	antiinflammatory	2009	45, 484
candesartan cilexetil	antihypertensive	1997	33, 330
capecitabine	anticancer	1998	34, 319
captopril	antihypertensive	1982	13, 086
carboplatin	antibacterial	1986	22, 318
carperitide	congestive heart failure	1995	31, 339
carumonam	antibacterial	1988	24, 298
carvedilol	antihypertensive	1991	27, 323
caspofungin acetate	antifungal	2001	37, 263
catumaxomab	anticancer	2009	45, 486
cefbuperazone sodium	antibacterial	1985	21, 325
cefcapene pivoxil	antibacterial	1997	33, 330
cefdinir	antibacterial	1991	27, 323
cefditoren pivoxil	antibacterial	1994	30, 297
cefepime	antibacterial	1993	29, 334
cefetamet pivoxil hydrochloride	antibacterial	1992	28, 327
cefixime	antibacterial	1987	23, 329
cefmenoxime hydrochloride	antibacterial	1983	19, 316
cefminox sodium	antibacterial	1987	23, 330
cefodizime sodium	antibacterial	1990	26, 300
cefonicid sodium	antibacterial	1984	20, 316
ceforanide	antibacterial	1984	20, 317
cefoselis	antibacterial	1998	34, 319
cefotetan disodium	antibacterial	1984	20, 317
cefotiam hexetil hydrochloride	antibacterial	1991	27, 324
cefozopran hydrochloride	antibacterial	1995	31, 339
cefpimizole	antibacterial	1987	23, 330
cefpiramide sodium	antibacterial	1985	21, 325
cefpirome sulfate	antibacterial	1992	28, 328
cefpodoxime proxetil	antibacterial	1989	25, 310
cefprozil	antibacterial	1992	28, 328
ceftaroline fosamil	antibacterial	2010	46, 453

GENERIC NAME	INDICATION	YEAR INTRO.	ARMC VOL., (PAGE)
ceftazidime	antibacterial	1983	19, 316
cefteram pivoxil	antibacterial	1987	23, 330
ceftibuten	antibacterial	1992	28, 329
ceftobiprole medocaril	antibacterial	2008	44, 589
cefuroxime axetil	antibacterial	1987	23, 331
cefuzonam sodium	antibacterial	1987	23, 331
celecoxib	antiarthritic	1999	35, 335
celiprolol hydrochloride	antihypertensive	1983	19, 317
centchroman	contraception	1991	27, 324
centoxin	immunomodulator	1991	27, 325
cerivastatin	antihypercholesterolemic	1997	33, 331
certolizumab pegol	irritable bowel syndrome	2008	44, 592
cetirizine hydrochloride	antiallergy	1987	23, 331
cetrorelix	infertility	1999	35, 336
cetuximab	anticancer	2003	39, 272
cevimeline hydrochloride	antixerostomia	2000	36, 299
chenodiol	gallstones	1983	19, 317
CHF-1301	Parkinson's disease	1999	35, 336
choline alfoscerate	cognition enhancer	1990	26, 300
choline fenofibrate	antihypercholesterolemic	2008	44, 594
cibenzoline	antiarrhythmic	1985	21, 325
ciclesonide	antiasthma	2005	41, 443
cicletanine	antihypertensive	1988	24, 299
cidofovir	antiviral	1996	32, 306
cilazapril	antihypertensive	1990	26, 301
cilostazol	antithrombotic	1988	24, 299
cimetropium bromide	antispasmodic	1985	21, 326
cinacalcet	hyperparathyroidism	2004	40, 451
cinildipine	antihypertensive	1995	31, 339
cinitapride	gastroprokinetic	1990	26, 301
cinolazepam	anxiolytic	1993	29, 334
ciprofibrate	antihypercholesterolemic	1985	21, 326
ciprofloxacin	antibacterial	1986	22, 318
cisapride	gastroprokinetic	1988	24, 299
cisatracurium besilate	muscle relaxant	1995	31, 340
citalopram	antidepressant	1989	25, 311
cladribine	anticancer	1993	29, 335
clarithromycin	antibacterial	1990	26, 302
clevidipine	antihypertensive	2008	44, 596
clevudine	antiviral	2007	43, 466
clobenoside	antiinflammatory	1988	24, 300
cloconazole hydrochloride	antifungal	1986	22, 318
clodronate disodium	calcium regulation	1986	22, 319
clofarabine	anticancer	2005	41, 444
clopidogrel hydrogensulfate	antithrombotic	1998	34, 320
cloricromen	antithrombotic	1991	27, 325
clospipramine hydrochloride	antipsychotic	1991	27, 325
colesevelam hydrochloride	antihypercholesterolemic	2000	36, 300

GENERIC NAME	INDICATION	YEAR INTRO.	ARMC VOL., (PAGE)
colestimide	antihypercholesterolemic	1999	35, 337
colforsin daropate hydrochloride	congestive heart failure	1999	35, 337
conivaptan	hyponatremia	2006	42, 514
corifollitropin alfa	infertility	2010	46, 455
crotelidae polyvalent immune fab	antidote, snake venom poisoning	2001	37, 263
cyclosporine	immunosuppressant	1983	19, 317
cytarabine ocfosfate	anticancer	1993	29, 335
dabigatran etexilate	anticoagulant	2008	44, 598
dalfampridine	multiple sclerosis	2010	46, 458
dalfopristin	antibacterial	1999	35, 338
dapiprazole hydrochloride	antiglaucoma	1987	23, 332
dapoxetine	premature ejaculation	2009	45, 488
daptomycin	antibacterial	2003	39, 272
darifenacin	urinary tract/bladder disorders	2005	41, 445
darunavir	antiviral	2006	42, 515
dasatinib	anticancer	2006	42, 517
decitabine	myelodysplastic syndromes	2006	42, 519
defeiprone	iron chelation therapy	1995	31, 340
deferasirox	iron chelation therapy	2005	41, 446
defibrotide	antithrombotic	1986	22, 319
deflazacort	antiinflammatory	1986	22, 319
degarelix acetate	anticancer	2009	45, 490
delapril	antihypertensive	1989	25, 311
delavirdine mesylate	antiviral	1997	33, 331
denileukin diftitox	anticancer	1999	35, 338
denopamine	congestive heart failure	1988	24, 300
denosumab	osteoporosis	2010	46, 459
deprodone propionate	antiinflammatory	1992	28, 329
desflurane	anesthetic	1992	28, 329
desloratadine	antiallergy	2001	37, 264
desvenlafaxine	antidepressant	2008	44, 600
dexfenfluramine	antiobesity	1997	33, 332
dexibuprofen	antiinflammatory	1994	30, 298
dexlansoprazole	antiulcer	2009	45, 492
dexmedetomidine hydrochloride	sleep disorders	2000	36, 301
dexmethylphenidate hydrochloride	attention deficit hyperactivity disorder	2002	38, 352
dexrazoxane	cardioprotective	1992	28, 330
dezocine	analgesic	1991	27, 326
diacerein	antiinflammatory	1985	21, 326
didanosine	antiviral	1991	27, 326
dilevalol	antihypertensive	1989	25, 311
diquafosol tetrasodium	ophthalmologic, dry eye	2010	46, 462
dirithromycin	antibacterial	1993	29, 336
disodium pamidronate	osteoporosis	1989	25, 312
divistyramine	antihypercholesterolemic	1984	20, 317
docarpamine	congestive heart failure	1994	30, 298

GENERIC NAME	INDICATION	YEAR INTRO.	ARMC VOL., (PAGE)
docetaxel	anticancer	1995	31, 341
dofetilide	antiarrhythmic	2000	36, 301
dolasetron mesylate	antiemetic	1998	34, 321
donepezil hydrochloride	Alzheimer's disease	1997	33, 332
dopexamine	congestive heart failure	1989	25, 312
doripenem	antibacterial	2005	41, 448
dornase alfa	cystic fibrosis	1994	30, 298
dorzolamide hydrochloride	antiglaucoma	1995	31, 341
dosmalfate	antiulcer	2000	36, 302
doxacurium chloride	muscle relaxant	1991	27, 326
doxazosin mesylate	antihypertensive	1988	24, 300
doxefazepam	anxiolytic	1985	21, 326
doxercalciferol	hyperparathyroidism	1999	35, 339
doxifluridine	anticancer	1987	23, 332
doxofylline	antiasthma	1985	21, 327
dronabinol	antiemetic	1986	22, 319
dronedarone	antiarrhythmic	2009	45, 495
drospirenone	contraception	2000	36, 302
drotrecogin alfa	antisepsis	2001	37, 265
droxicam	antiinflammatory	1990	26, 302
droxidopa	Parkinson's disease	1989	25, 312
duloxetine	antidepressant	2004	40, 452
dutasteride	benign prostatic hyperplasia	2002	38, 353
duteplase	anticoagulant	1995	31, 342
ebastine	antiallergy	1990	26, 302
eberconazole	antifungal	2005	41, 449
ebrotidine	antiulcer	1997	33, 333
ecabet sodium	antiulcer	1993	29, 336
ecallantide	angioedema, hereditary	2009	46, 464
eculizumab	hemoglobinuria	2007	43, 468
edaravone	neuroprotective	2001	37, 265
efalizumab	antipsoriasis	2003	39, 274
efavirenz	antiviral	1998	34, 321
efonidipine	antihypertensive	1994	30, 299
egualen sodium	antiulcer	2000	36, 303
eletriptan	antimigraine	2001	37, 266
eltrombopag	antithrombocytopenic	2009	45, 497
emedastine difumarate	antiallergy	1993	29, 336
emorfazone	analgesic	1984	20, 317
emtricitabine	antiviral	2003	39, 274
enalapril maleate	antihypertensive	1984	20, 317
enalaprilat	antihypertensive	1987	23, 332
encainide hydrochloride	antiarrhythmic	1987	23, 333
enfuvirtide	antiviral	2003	39, 275
enocitabine	anticancer	1983	19, 318
enoxacin	antibacterial	1986	22, 320
enoxaparin	anticoagulant	1987	23, 333
enoximone	congestive heart failure	1988	24, 301

GENERIC NAME	INDICATION	YEAR INTRO.	ARMC VOL., (PAGE)
enprostil	antiulcer	1985	21, 327
entacapone	Parkinson's disease	1998	34, 322
entecavir	antiviral	2005	41, 450
epalrestat	antidiabetic	1992	28, 330
eperisone hydrochloride	muscle relaxant	1983	19, 318
epidermal growth factor	wound healing agent	1987	23, 333
epinastine	antiallergy	1994	30, 299
epirubicin hydrochloride	anticancer	1984	20, 318
eplerenone	antihypertensive	2003	39, 276
epoprostenol sodium	antiplatelet	1983	19, 318
eprosartan	antihypertensive	1997	33, 333
eptazocine hydrobromide	analgesic	1987	23, 334
eptilfibatide	antithrombotic	1999	35, 340
erdosteine	expectorant	1995	31, 342
eribulin mesylate	anticancer	2010	46, 465
erlotinib	anticancer	2004	40, 454
ertapenem sodium	antibacterial	2002	38, 353
erythromycin acistrate	antibacterial	1988	24, 301
erythropoietin	hematopoietic	1988	24, 301
escitalopram oxolate	antidepressant	2002	38, 354
eslicarbazepine acetate	anticonvulsant	2009	45, 498
esmolol hydrochloride	antiarrhythmic	1987	23, 334
esomeprazole magnesium	antiulcer	2000	36, 303
eszopiclone	sleep disorders	2005	41, 451
ethyl icosapentate	antithrombotic	1990	26, 303
etizolam	anxiolytic	1984	20, 318
etodolac	antiinflammatory	1985	21, 327
etoricoxibe	antiarthritic	2002	38, 355
etravirine	antiviral	2008	44, 602
everolimus	immunosuppressant	2004	40, 455
exemestane	anticancer	2000	36, 304
exenatide	antidiabetic	2005	41, 452
exifone	cognition enhancer	1988	24, 302
ezetimibe	antihypercholesterolemic	2002	38, 355
factor VIIa	haemophilia	1996	32, 307
factor VIII	hemostatic	1992	28, 330
fadrozole hydrochloride	anticancer	1995	31, 342
falecalcitriol	hyperparathyroidism	2001	37, 266
famciclovir	antiviral	1994	30, 300
famotidine	antiulcer	1985	21, 327
fasudil hydrochloride	amyotrophic lateral sclerosis	1995	31, 343
febuxostat	gout	2009	45, 501
felbamate	anticonvulsant	1993	29, 337
felbinac	antiinflammatory	1986	22, 320
felodipine	antihypertensive	1988	24, 302
fenbuprol	biliary tract dysfunction	1983	19, 318
fenoldopam mesylate	antihypertensive	1998	34, 322
fenticonazole nitrate	antifungal	1987	23, 334
fesoterodine	urinary tract/bladder disorders	2008	44, 604

GENERIC NAME	INDICATION	YEAR INTRO.	ARMC VOL., (PAGE)
fexofenadine	antiallergy	1996	32, 307
filgrastim	immunostimulant	1991	27, 327
finasteride	benign prostatic hyperplasia	1992	28, 331
fingolimod	multiple sclerosis	2010	46, 468
fisalamine	antiinflammatory	1984	20, 318
fleroxacin	antibacterial	1992	28, 331
flomoxef sodium	antibacterial	1988	24, 302
flosequinan	congestive heart failure	1992	28, 331
fluconazole	antifungal	1988	24, 303
fludarabine phosphate	anticancer	1991	27, 327
flumazenil	antidote, benzodiazepine overdose	1987	23, 335
flunoxaprofen	antiinflammatory	1987	23, 335
fluoxetine hydrochloride	antidepressant	1986	22, 320
flupirtine maleate	analgesic	1985	21, 328
flurithromycin ethylsuccinate	antibacterial	1997	33, 333
flutamide	anticancer	1983	19, 318
flutazolam	anxiolytic	1984	20, 318
fluticasone furoate	antiallergy	2007	43, 469
fluticasone propionate	antiinflammatory	1990	26, 303
flutoprazepam	anxiolytic	1986	22, 320
flutrimazole	antifungal	1995	31, 343
flutropium bromide	antiasthma	1988	24, 303
fluvastatin	antihypercholesterolemic	1994	30, 300
fluvoxamine maleate	antidepressant	1983	19, 319
follitropin alfa	infertility	1996	32, 307
follitropin beta	infertility	1996	32, 308
fomepizole	antidote, ethylene glycol poisoning	1998	34, 323
fomivirsen sodium	antiviral	1998	34, 323
fondaparinux sodium	antithrombotic	2002	38, 356
formestane	anticancer	1993	29, 337
formoterol fumarate	chronic obstructive pulmonary disorder	1986	22, 321
fosamprenavir	antiviral	2003	39, 277
fosaprepitant dimeglumine	antiemetic	2008	44, 606
foscarnet sodium	antiviral	1989	25, 313
fosfluconazole	antifungal	2004	40, 457
fosfosal	analgesic	1984	20, 319
fosinopril sodium	antihypertensive	1991	27, 328
fosphenytoin sodium	anticonvulsant	1996	32, 308
fotemustine	anticancer	1989	25, 313
fropenam	antibacterial	1997	33, 334
frovatriptan	antimigraine	2002	38, 357
fudosteine	expectorant	2001	37, 267
fulveristrant	anticancer	2002	38, 357
gabapentin	anticonvulsant	1993	29, 338
gadoversetamide	diagnostic	2000	36, 304

GENERIC NAME	INDICATION	YEAR INTRO.	ARMC VOL., (PAGE)
gallium nitrate	calcium regulation	1991	27, 328
gallopamil hydrochloride	antianginal	1983	19, 3190
galsulfase	mucopolysaccharidosis VI	2005	41, 453
ganciclovir	antiviral	1988	24, 303
ganirelix acetate	infertility	2000	36, 305
garenoxacin	antibacterial	2007	43, 471
gatilfloxacin	antibacterial	1999	35, 340
gefitinib	anticancer	2002	38, 358
gemcitabine hydrochloride	anticancer	1995	31, 344
gemeprost	abortifacient	1983	19, 319
gemifloxacin	antibacterial	2004	40, 458
gemtuzumab ozogamicin	anticancer	2000	36, 306
gestodene	contraception	1987	23, 335
gestrinone	contraception	1986	22, 321
glatiramer acetate	multiple sclerosis	1997	33, 334
glimepiride	antidiabetic	1995	31, 344
glucagon, rDNA	antidiabetic	1993	29, 338
GMDP	immunostimulant	1996	32, 308
golimumab	antiinflammatory	2009	45, 503
goserelin	hormone therapy	1987	23, 336
granisetron hydrochloride	antiemetic	1991	27, 329
guanadrel sulfate	antihypertensive	1983	19, 319
gusperimus	immunosuppressant	1994	30, 300
halobetasol propionate	antiinflammatory	1991	27, 329
halofantrine	antimalarial	1988	24, 304
halometasone	antiinflammatory	1983	19, 320
histrelin	precocious puberty	1993	29, 338
hydrocortisone aceponate	antiinflammatory	1988	24, 304
hydrocortisone butyrate	antiinflammatory	1983	19, 320
ibandronic acid	osteoporosis	1996	32, 309
ibopamine hydrochloride	congestive heart failure	1984	20, 319
ibritunomab tiuxetan	anticancer	2002	38, 359
ibudilast	antiasthma	1989	25, 313
ibutilide fumarate	antiarrhythmic	1996	32, 309
icatibant	angioedema, hereditary	2008	44, 608
idarubicin hydrochloride	anticancer	1990	26, 303
idebenone	nootropic	1986	22, 321
idursulfase	mucopolysaccharidosis II, Hunter syndrome	2006	42, 520
iloprost	antiplatelet	1992	28, 332
imatinib mesylate	anticancer	2001	37, 267
imidafenacin	urinary tract/bladder disorders	2007	43, 472
imidapril hydrochloride	antihypertensive	1993	29, 339
imiglucerase	Gaucher's disease	1994	30, 301
imipenem/cilastatin	antibacterial	1985	21, 328
imiquimod	antiviral	1997	33, 335
incadronic acid	osteoporosis	1997	33, 335
indacaterol	chronic obstructive pulmonary disease	2009	45, 505
indalpine	antidepressant	1983	19, 320

GENERIC NAME	INDICATION	YEAR INTRO.	ARMC VOL., (PAGE)
indeloxazine hydrochloride	nootropic	1988	24, 304
indinavir sulfate	antiviral	1996	32, 310
indisetron	antiemetic	2004	40, 459
indobufen	antithrombotic	1984	20, 319
influenza virus, live	antiviral	2003	39, 277
insulin lispro	antidiabetic	1996	32, 310
interferon alfacon-1	antiviral	1997	33, 336
interferon gamma-1b	immunostimulant	1991	27, 329
interferon, b-1a	multiple sclerosis	1996	32, 311
interferon, b-1b	multiple sclerosis	1993	29, 339
interferon, gamma	antiinflammatory	1989	25, 314
interferon, gamma-1	anticancer	1992	28, 332
interleukin-2	anticancer	1989	25, 314
ioflupane	diagnostic	2000	36, 306
ipriflavone	osteoporosis	1989	25, 314
irbesartan	antihypertensive	1997	33, 336
irinotecan	anticancer	1994	30, 301
irsogladine	antiulcer	1989	25, 315
isepamicin	antibacterial	1988	24, 305
isofezolac	antiinflammatory	1984	20, 319
isoxicam	antiinflammatory	1983	19, 320
isradipine	antihypertensive	1989	25, 315
itopride hydrochloride	gastroprokinetic	1995	31, 344
itraconazole	antifungal	1988	24, 305
ivabradine	antianginal	2006	42, 522
ivermectin	antiparasitic	1987	23, 336
ixabepilone	anticancer	2007	43, 473
ketanserin	antihypertensive	1985	21, 328
ketorolac tromethamine	analgesic	1990	26, 304
kinetin	dermatologic, skin photodamage	1999	35, 341
lacidipine	antihypertensive	1991	27, 330
lacosamide	anticonvulsant	2008	44, 610
lafutidine	antiulcer	2000	36, 307
lamivudine	antiviral	1995	31, 345
lamotrigine	anticonvulsant	1990	26, 304
landiolol	antiarrhythmic	2002	38, 360
laninamivir octanoate	antiviral	2010	46, 470
lanoconazole	antifungal	1994	30, 302
lanreotide acetate	growth disorders	1995	31, 345
lansoprazole	antiulcer	1992	28, 332
lapatinib	anticancer	2007	43, 475
laronidase	mucopolysaccharidosis I	2003	39, 278
latanoprost	antiglaucoma	1996	32, 311
lefunomide	antiarthritic	1998	34, 324
lenalidomide	myelodysplastic syndromes, multiple myeloma	2006	42, 523
lenampicillin hydrochloride	antibacterial	1987	23, 336

GENERIC NAME	INDICATION	YEAR INTRO.	ARMC VOL., (PAGE)
lentinan	immunostimulant	1986	22, 322
lepirudin	anticoagulant	1997	33, 336
lercanidipine	antihypertensive	1997	33, 337
letrazole	anticancer	1996	32, 311
leuprolide acetate	hormone therapy	1984	20, 319
levacecarnine hydrochloride	cognition enhancer	1986	22, 322
levalbuterol hydrochloride	antiasthma	1999	35, 341
levetiracetam	anticonvulsant	2000	36, 307
levobunolol hydrochloride	antiglaucoma	1985	21, 328
levobupivacaine hydrochloride	anesthetic	2000	36, 308
levocabastine hydrochloride	antiallergy	1991	27, 330
levocetirizine	antiallergy	2001	37, 268
levodropropizine	antitussive	1988	24, 305
levofloxacin	antibacterial	1993	29, 340
levosimendan	congestive heart failure	2000	36, 308
lidamidine hydrochloride	antidiarrheal	1984	20, 320
limaprost	antithrombotic	1988	24, 306
linezolid	antibacterial	2000	36, 309
liraglutide	antidiabetic	2009	45, 507
liranaftate	antifungal	2000	36, 309
lisdexamfetamine	attention deficit hyperactivity disorder	2007	43, 477
lisinopril	antihypertensive	1987	23, 337
lobenzarit sodium	antiinflammatory	1986	22, 322
lodoxamide tromethamine	antiallergy	1992	28, 333
lomefloxacin	antibacterial	1989	25, 315
lomerizine hydrochloride	antimigraine	1999	35, 342
lonidamine	anticancer	1987	23, 337
lopinavir	antiviral	2000	36, 310
loprazolam mesylate	sleep disorders	1983	19, 321
loprinone hydrochloride	congestive heart failure	1996	32, 312
loracarbef	antibacterial	1992	28, 333
loratadine	antiallergy	1988	24, 306
lornoxicam	antiinflammatory	1997	33, 337
losartan	antihypertensive	1994	30, 302
loteprednol etabonate	antiallergy	1998	34, 324
lovastatin	antihypercholesterolemic	1987	23, 337
loxoprofen sodium	antiinflammatory	1986	22, 322
lulbiprostone	constipation	2006	42, 525
luliconazole	antifungal	2005	41, 454
lumiracoxib	antiinflammatory	2005	41, 455
lurasidone hydrochloride	antipsychotic	2010	46, 473
Lyme disease vaccine	Lyme disease	1999	35, 342
mabuterol hydrochloride	antiasthma	1986	22, 323
malotilate	hepatoprotective	1985	21, 329
manidipine hydrochloride	antihypertensive	1990	26, 304
maraviroc	antiviral	2007	43, 478

GENERIC NAME	INDICATION	YEAR INTRO.	ARMC VOL., (PAGE)
masoprocol	anticancer	1992	28, 333
maxacalcitol	hyperparathyroidism	2000	36, 310
mebefradil hydrochloride	antihypertensive	1997	33, 338
medifoxamine fumarate	antidepressant	1986	22, 323
mefloquine hydrochloride	antimalarial	1985	21, 329
meglutol	antihypercholesterolemic	1983	19, 321
melinamide	antihypercholesterolemic	1984	20, 320
meloxicam	antiarthritic	1996	32, 312
mepixanox	respiratory stimulant	1984	20, 320
meptazinol hydrochloride	analgesic	1983	19, 321
meropenem	antibacterial	1994	30, 303
metaclazepam	anxiolytic	1987	23, 338
metapramine	antidepressant	1984	20, 320
methylnaltrexone bromide	constipation	2008	44, 612
mexazolam	anxiolytic	1984	20, 321
micafungin	antifungal	2002	38, 360
mifamurtide	anticancer	2009	46, 476
mifepristone	abortifacient	1988	24, 306
miglitol	antidiabetic	1998	34, 325
miglustat	Gaucher's disease	2003	39, 279
milnacipran	antidepressant	1997	33, 338
milrinone	congestive heart failure	1989	25, 316
miltefosine	anticancer	1993	29, 340
minodronic acid	osteoporosis	2009	45, 509
miokamycin	antibacterial	1985	21, 329
mirtazapine	antidepressant	1994	30, 303
misoprostol	antiulcer	1985	21, 329
mitiglinide	antidiabetic	2004	40, 460
mitoxantrone hydrochloride	anticancer	1984	20, 321
mivacurium chloride	muscle relaxant	1992	28, 334
mivotilate	hepatoprotective	1999	35, 343
mizolastine	antiallergy	1998	34, 325
mizoribine	immunosuppressant	1984	20, 321
moclobemide	antidepressant	1990	26, 305
modafinil	sleep disorders	1994	30, 303
moexipril hydrochloride	antihypertensive	1995	31, 346
mofezolac	analgesic	1994	30, 304
mometasone furoate	antiinflammatory	1987	23, 338
montelukast sodium	antiasthma	1998	34, 326
moricizine hydrochloride	antiarrhythmic	1990	26, 305
mosapride citrate	gastroprokinetic	1998	34, 326
moxifloxacin hydrochloride	antibacterial	1999	35, 343
moxonidine	antihypertensive	1991	27, 330
mozavaptan	hyponatremia	2006	42, 527
mupirocin	antibacterial	1985	21, 330
muromonab-CD3	immunosuppressant	1986	22, 323
muzolimine	diuretic	1983	19, 321
mycophenolate mofetil	immunosuppressant	1995	31, 346

GENERIC NAME	INDICATION	YEAR INTRO.	ARMC VOL., (PAGE)
mycophenolate sodium	immunosuppressant	2003	39, 279
nabumetone	antiinflammatory	1985	21, 330
nadifloxacin	antibacterial	1993	29, 340
nafamostat mesylate	pancreatitis	1986	22, 323
nafarelin acetate	hormone therapy	1990	26, 306
naftifine hydrochloride	antifungal	1984	20, 321
naftopidil	urinary tract/bladder disorders	1999	35, 344
nalfurafine hydrochloride	pruritus	2009	45, 510
nalmefene hydrochloride	addiction, opioids	1995	31, 347
naltrexone hydrochloride	addiction, opioids	1984	20, 322
naratriptan hydrochloride	antimigraine	1997	33, 339
nartograstim	leukopenia	1994	30, 304
natalizumab	multiple sclerosis	2004	40, 462
nateglinide	antidiabetic	1999	35, 344
nazasetron	antiemetic	1994	30, 305
nebivolol	antihypertensive	1997	33, 339
nedaplatin	anticancer	1995	31, 347
nedocromil sodium	antiallergy	1986	22, 324
nefazodone	antidepressant	1994	30, 305
nelarabine	anticancer	2006	42, 528
nelfinavir mesylate	antiviral	1997	33, 340
neltenexine	cystic fibrosis	1993	29, 341
nemonapride	antipsychotic	1991	27, 331
nepafenac	antiinflammatory	2005	41, 456
neridronic acide	calcium regulation	2002	38, 361
nesiritide	congestive heart failure	2001	37, 269
neticonazole hydrochloride	antifungal	1993	29, 341
nevirapine	antiviral	1996	32, 313
nicorandil	antianginal	1984	20, 322
nif ekalant hydrochloride	antiarrhythmic	1999	35, 344
nilotinib	anticancer	2007	43, 480
nilutamide	anticancer	1987	23, 338
nilvadipine	antihypertensive	1989	25, 316
nimesulide	antiinflammatory	1985	21, 330
nimodipine	cerebral vasodilator	1985	21, 330
nimotuzumab	anticancer	2006	42, 529
nipradilol	antihypertensive	1988	24, 307
nisoldipine	antihypertensive	1990	26, 306
nitisinone	antityrosinaemia	2002	38, 361
nitrefazole	addiction, alcohol	1983	19, 322
nitrendipine	antihypertensive	1985	21, 331
nizatidine	antiulcer	1987	23, 339
nizofenzone	nootropic	1988	24, 307
nomegestrol acetate	contraception	1986	22, 324
norelgestromin	contraception	2002	38, 362
norfloxacin	antibacterial	1983	19, 322
norgestimate	contraception	1986	22, 324
OCT-43	anticancer	1999	35, 345
octreotide	growth disorders	1988	24, 307

GENERIC NAME	INDICATION	YEAR INTRO.	ARMC VOL., (PAGE)
ofatumumab	anticancer	2009	45, 512
ofloxacin	antibacterial	1985	21, 331
olanzapine	antipsychotic	1996	32, 313
olimesartan Medoxomil	antihypertensive	2002	38, 363
olopatadine hydrochloride	antiallergy	1997	33, 340
omalizumab allergic	antiasthma	2003	39, 280
omeprazole	antiulcer	1988	24, 308
ondansetron hydrochloride	antiemetic	1990	26, 306
OP-1	osteoinductor	2001	37, 269
orlistat	antiobesity	1998	34, 327
ornoprostil	antiulcer	1987	23, 339
osalazine sodium	antiinflammatory	1986	22, 324
oseltamivir phosphate	antiviral	1999	35, 346
oxaliplatin	anticancer	1996	32, 313
oxaprozin	antiinflammatory	1983	19, 322
oxcarbazepine	anticonvulsant	1990	26, 307
oxiconazole nitrate	antifungal	1983	19, 322
oxiracetam	cognition enhancer	1987	23, 339
oxitropium bromide	antiasthma	1983	19, 323
ozagrel sodium	antithrombotic	1988	24, 308
paclitaxal	anticancer	1993	29, 342
palifermin	mucositis	2005	41, 461
paliperidone	antipsychotic	2007	43, 482
palonosetron	antiemetic	2003	39, 281
panipenem/ betamipron carbapenem	antibacterial	1994	30, 305
panitumumab	anticancer	2006	42, 531
pantoprazole sodium	antiulcer	1995	30, 306
parecoxib sodium	analgesic	2002	38, 364
paricalcitol	hyperparathyroidism	1998	34, 327
parnaparin sodium	anticoagulant	1993	29, 342
paroxetine	antidepressant	1991	27, 331
pazopanib	anticancer	2009	45, 514
pazufloxacin	antibacterial	2002	38, 364
pefloxacin mesylate	antibacterial	1985	21, 331
pegademase bovine	immunostimulant	1990	26, 307
pegaptanib	ophthalmologic, macular degeneration	2005	41, 458
pegaspargase	anticancer	1994	30, 306
pegvisomant	growth disorders	2003	39, 281
pemetrexed	anticancer	2004	40, 463
pemirolast potassium	antiasthma	1991	27, 331
penciclovir	antiviral	1996	32, 314
pentostatin	anticancer	1992	28, 334
peramivir	antiviral	2010	46, 477
pergolide mesylate	Parkinson's disease	1988	24, 308
perindopril	antihypertensive	1988	24, 309
perospirone hydrochloride	antipsychotic	2001	37, 270
picotamide	antithrombotic	1987	23, 340

GENERIC NAME	INDICATION	YEAR INTRO.	ARMC VOL., (PAGE)
pidotimod	immunostimulant	1993	29, 343
piketoprofen	antiinflammatory	1984	20, 322
pilsicainide hydrochloride	antiarrhythmic	1991	27, 332
pimaprofen	antiinflammatory	1984	20, 322
pimecrolimus	immunosuppressant	2002	38, 365
pimobendan	congestive heart failure	1994	30, 307
pinacidil	antihypertensive	1987	23, 340
pioglitazone hydrochloride	antidiabetic	1999	35, 346
pirarubicin	anticancer	1988	24, 309
pirfenidone	pulmonary fibrosis, idiopathic	2008	44, 614
pirmenol	antiarrhythmic	1994	30, 307
piroxicam cinnamate	antiinflammatory	1988	24, 309
pitavastatin	antihypercholesterolemic	2003	39, 282
pivagabine	antidepressant	1997	33, 341
plaunotol	antiulcer	1987	23, 340
plerixafor hydrochloride	stem cell mobilizer	2009	45, 515
polaprezinc	antiulcer	1994	30, 307
porfimer sodium	anticancer	1993	29, 343
posaconazole	antifungal	2006	42, 532
pralatrexate	anticancer	2009	45, 517
pramipexole hydrochloride	Parkinson's disease	1997	33, 341
pramiracetam sulfate	cognition enhancer	1993	29, 343
pramlintide	antidiabetic	2005	41, 460
pranlukast	antiasthma	1995	31, 347
prasugrel	antiplatelet	2009	45, 519
pravastatin	antihypercholesterolemic	1989	25, 316
prednicarbate	antiinflammatory	1986	22, 325
pregabalin	anticonvulsant	2004	40, 464
prezatide copper acetate	wound healing agent	1996	32, 314
progabide	anticonvulsant	1985	21, 331
promegestrone	contraception	1983	19, 323
propacetamol hydrochloride	analgesic	1986	22, 325
propagermanium	antiviral	1994	30, 308
propentofylline propionate	cerebral vasodilator	1988	24, 310
propiverine hydrochloride	urinary tract/bladder disorders	1992	28, 335
propofol	anesthetic	1986	22, 325
prulifloxacin	antibacterial	2002	38, 366
pumactant	respiratory distress syndrome	1994	30, 308
quazepam	sleep disorders	1985	21, 332
quetiapine fumarate	antipsychotic	1997	33, 341
quinagolide	hyperprolactinemia	1994	30, 309
quinapril	antihypertensive	1989	25, 317
quinf amideamebicide	antiparasitic	1984	20, 322
quinupristin	antibacterial	1999	35, 338
rabeprazole sodium	antiulcer	1998	34, 328
raloxifene hydrochloride	osteoporosis	1998	34, 328
raltegravir	antiviral	2007	43, 484
raltitrexed	anticancer	1996	32, 315

GENERIC NAME	INDICATION	YEAR INTRO.	ARMC VOL., (PAGE)
ramatroban	antiallergy	2000	36, 311
ramelteon	sleep disorders	2005	41, 462
ramipril	antihypertensive	1989	25, 317
ramosetron	antiemetic	1996	32, 315
ranibizumab	ophthalmologic, macular degeneration	2006	42, 534
ranimustine	anticancer	1987	23, 341
ranitidine bismuth citrate	antiulcer	1995	31, 348
ranolazine	antianginal	2006	42, 535
rapacuronium bromide	muscle relaxant	1999	35, 347
rasagiline	Parkinson's disease	2005	41, 464
rebamipide	antiulcer	1990	26, 308
reboxetine	antidepressant	1997	33, 342
remifentanil hydrochloride	analgesic	1996	32, 316
remoxipride hydrochloride	antipsychotic	1990	26, 308
repaglinide	antidiabetic	1998	34, 329
repirinast	antiallergy	1987	23, 341
retapamulin	antibacterial	2007	43, 486
reteplase	antithrombotic	1996	32, 316
reviparin sodium	anticoagulant	1993	29, 344
rifabutin	antibacterial	1992	28, 335
rifapentine	antibacterial	1988	24, 310
rifaximin	antibacterial	1985	21, 332
rifaximin	antibacterial	1987	23, 341
rilmazafone	sleep disorders	1989	25, 317
rilmenidine	antihypertensive	1988	24, 310
rilonacept	genetic autoinflammatory syndromes	2008	44, 615
riluzole	amyotrophic lateral sclerosis	1996	32, 316
rimantadine hydrochloride	antiviral	1987	23, 342
rimexolone	antiinflammatory	1995	31, 348
rimonabant	antiobesity	2006	42, 537
risedronate sodium	osteoporosis	1998	34, 330
risperidone	antipsychotic	1993	29, 344
ritonavir	antiviral	1996	32, 317
rivaroxaban	anticoagulant	2008	44, 617
rivastigmin	Alzheimer's disease	1997	33, 342
rizatriptan benzoate	antimigraine	1998	34, 330
rocuronium bromide	muscle relaxant	1994	30, 309
rofecoxib	antiarthritic	1999	35, 347
roflumilast	chronic obstructive pulmonary disorder	2010	46, 480
rokitamycin	antibacterial	1986	22, 325
romidepsin	anticancer	2009	46, 482
romiplostim	antithrombocytopenic	2008	44, 619
romurtide	immunostimulant	1991	27, 332
ronafibrate	antihypercholesterolemic	1986	22, 326
ropinirole hydrochloride	Parkinson's disease	1996	32, 317
ropivacaine	anesthetic	1996	32, 318
rosaprostol	antiulcer	1985	21, 332

GENERIC NAME	INDICATION	YEAR INTRO.	ARMC VOL., (PAGE)
rosiglitazone maleate	antidiabetic	1999	35, 348
rosuvastatin	antihypercholesterolemic	2003	39, 283
rotigotine	Parkinson's disease	2006	42, 538
roxatidine acetate hydrochloride	antiulcer	1986	22, 326
roxithromycin	antiulcer	1987	23, 342
rufinamide	anticonvulsant	2007	43, 488
rufloxacin hydrochloride	antibacterial	1992	28, 335
rupatadine fumarate	antiallergy	2003	39, 284
RV-11	antibacterial	1989	25, 318
salmeterol hydroxynaphthoate	antiasthma	1990	26, 308
sapropterin hydrochloride	phenylketouria	1992	28, 336
saquinavir mesvlate	antiviral	1995	31, 349
sargramostim	immunostimulant	1991	27, 332
sarpogrelate hydrochloride	antithrombotic	1993	29, 344
saxagliptin	antidiabetic	2009	45, 521
schizophyllan	immunostimulant	1985	22, 326
seratrodast	antiasthma	1995	31, 349
sertaconazole nitrate	antifungal	1992	28, 336
sertindole	antipsychotic	1996	32, 318
setastine hydrochloride	antiallergy	1987	23, 342
setiptiline	antidepressant	1989	25, 318
setraline hydrochloride	antidepressant	1990	26, 309
sevoflurane	anesthetic	1990	26, 309
sibutramine	antiobesity	1998	34, 331
sildenafil citrate	male sexual dysfunction	1998	34, 331
silodosin	urinary tract/bladder disorders	2006	42, 540
simvastatin	antihypercholesterolemic	1988	24, 311
sipuleucel-t	anticancer	2010	46, 484
sitafloxacin hydrate	antibacterial	2008	44, 621
sitagliptin	antidiabetic	2006	42, 541
sitaxsentan	pulmonary hypertension	2006	42, 543
sivelestat	antiinflammatory	2002	38, 366
SKI-2053R	anticancer	1999	35, 348
sobuzoxane	anticancer	1994	30, 310
sodium cellulose phosphate	urinary tract/bladder disorders	1983	19, 323
sofalcone	antiulcer	1984	20, 323
solifenacin	urinary tract/bladder disorders	2004	40, 466
somatomedin-1	growth disorders	1994	30, 310
somatotropin	growth disorders	1994	30, 310
somatropin	growth disorders	1987	23, 343
sorafenib	anticancer	2005	41, 466
sorivudine	antiviral	1993	29, 345
sparfloxacin	antibacterial	1993	29, 345
spirapril hydrochloride	antihypertensive	1995	31, 349
spizofurone	antiulcer	1987	23, 343
stavudine	antiviral	1994	30, 311

GENERIC NAME	INDICATION	YEAR INTRO.	ARMC VOL., (PAGE)
strontium ranelate	osteoporosis	2004	40, 466
succimer	antidote, lead poisoning	1991	27, 333
sufentanil	analgesic	1983	19, 323
sugammadex	neuromuscular blockade, reversal	2008	44, 623
sulbactam sodium	antibacterial	1986	22, 326
sulconizole nitrate	antifungal	1985	21, 332
sultamycillin tosylate	antibacterial	1987	23, 343
sumatriptan succinate	antimigraine	1991	27, 333
sunitinib	anticancer	2006	42, 544
suplatast tosilate	antiallergy	1995	31, 350
suprofen	analgesic	1983	19, 324
surfactant TA	respiratory surfactant	1987	23, 344
tacalcitol	antipsoriasis	1993	29, 346
tacrine hydrochloride	Alzheimer's disease	1993	29, 346
tacrolimus	immunosuppressant	1993	29, 347
tadalafil	male sexual dysfunction	2003	39, 284
tafluprost	antiglaucoma	2008	44, 625
talaporfin sodium	anticancer	2004	40, 469
talipexole	Parkinson's disease	1996	32, 318
taltirelin	neurodegeneration	2000	36, 311
tamibarotene	anticancer	2005	41, 467
tamsulosin hydrochloride	benign prostatic hyperplasia	1993	29, 347
tandospirone	anxiolytic	1996	32, 319
tapentadol hydrochloride	analgesic	2009	45, 523
tasonermin	anticancer	1999	35, 349
tazanolast	antiallergy	1990	26, 309
tazarotene	antipsoriasis	1997	33, 343
tazobactam sodium	antibacterial	1992	28, 336
tegaserod maleate	irritable bowel syndrome	2001	37, 270
teicoplanin	antibacterial	1988	24, 311
telavancin	antibacterial	2009	45, 525
telbivudine	antiviral	2006	42, 546
telithromycin	antibacterial	2001	37, 271
telmesteine	expectorant	1992	28, 337
telmisartan	antihypertensive	1999	35, 349
temafloxacin hydrochloride	antibacterial	1991	27, 334
temocapril	antihypertensive	1994	30, 311
temocillin disodium	antibacterial	1984	20, 323
temoporphin	anticancer	2002	38, 367
temozolomide	anticancer	1999	35, 349
temsirolimus	anticancer	2007	43, 490
tenofovir disoproxil fumarate	antiviral	2001	37, 271
tenoxicam	antiinflammatory	1987	23, 344
teprenone	antiulcer	1984	20, 323
terazosin hydrochloride	antihypertensive	1984	20, 323
terbinafine hydrochloride	antifungal	1991	27, 334
terconazole	antifungal	1983	19, 324

GENERIC NAME	INDICATION	YEAR INTRO.	ARMC VOL., (PAGE)
tertatolol hydrochloride	antihypertensive	1987	23, 344
tesamorelin acetate	lipodystrophy	2010	46, 486
thrombin alfa	hemostatic	2008	44, 627
thrombomodulin, recombinant	anticoagulant	2008	44, 628
thymopentin	immunomodulator	1985	21, 333
tiagabine	anticonvulsant	1996	32, 319
tiamenidine hydrochloride	antihypertensive	1988	24, 311
tianeptine sodium	antidepressant	1983	19, 324
tibolone	hormone therapy	1988	24, 312
ticagrelor	antithrombotic	2010	46, 488
tigecycline	antibacterial	2005	41, 468
tilisolol hydrochloride	antihypertensive	1992	28, 337
tiludronate disodium	Paget's disease	1995	31, 350
timiperone	antipsychotic	1984	20, 323
tinazoline	nasal decongestant	1988	24, 312
tioconazole	antifungal	1983	19, 324
tiopronin	urolithiasis	1989	25, 318
tiotropium bromide	chronic obstructive pulmonary disorder	2002	38, 368
tipranavir	antiviral	2005	41, 470
tiquizium bromide	antispasmodic	1984	20, 324
tiracizine hydrochloride	antiarrhythmic	1990	26, 310
tirilazad mesylate	subarachnoid hemorrhage	1995	31, 351
tirofiban hydrochloride	antithrombotic	1998	34, 332
tiropramide hydrochloride	muscle relaxant	1983	19, 324
tizanidine	muscle relaxant	1984	20, 324
tolcapone	Parkinson's disease	1997	33, 343
toloxatone	antidepressant	1984	20, 324
tolrestat	antidiabetic	1989	25, 319
tolvaptan	hyponatremia	2009	45, 528
topiramate	anticonvulsant	1995	31, 351
topotecan hydrochloride	anticancer	1996	32, 320
torasemide	diuretic	1993	29, 348
toremifene	anticancer	1989	25, 319
tositumomab	anticancer	2003	39, 285
tosufloxacin tosylate	antibacterial	1990	26, 310
trabectedin	anticancer	2007	43, 492
trandolapril	antihypertensive	1993	29, 348
travoprost	antiglaucoma	2001	37, 272
treprostinil sodium	antihypertensive	2002	38, 368
tretinoin tocoferil	antiulcer	1993	29, 348
trientine hydrochloride	antidote, copper poisoning	1986	22, 327
trimazosin hydrochloride	antihypertensive	1985	21, 333
trimegestone	contraception	2001	37, 273
trimetrexate glucuronate	antifungal	1994	30, 312
troglitazone	antidiabetic	1997	33, 344
tropisetron	antiemetic	1992	28, 337
trovafloxacin mesylate	antibacterial	1998	34, 332

GENERIC NAME	INDICATION	YEAR INTRO.	ARMC VOL., (PAGE)
troxipide	antiulcer	1986	22, 327
ubenimex	immunostimulant	1987	23, 345
udenafil	male sexual dysfunction	2005	41, 472
ulipristal acetate	contraception	2009	45, 530
unoprostone isopropyl ester	antiglaucoma	1994	30, 312
ustekinumab	antipsoriasis	2009	45, 532
vadecoxib	antiarthritic	2002	38, 369
vaglancirclovir hydrochloride	antiviral	2001	37, 273
valaciclovir hydrochloride	antiviral	1995	31, 352
valrubicin	anticancer	1999	35, 350
valsartan	antihypertensive	1996	32, 320
vardenafil	male sexual dysfunction	2003	39, 286
varenicline	addiction, nicotine	2006	42, 547
venlafaxine	antidepressant	1994	30, 312
vernakalant	antiarrhythmic	2010	46, 491
verteporfin	ophthalmologic, macular degeneration	2000	36, 312
vesnarinone	congestive heart failure	1990	26, 310
vigabatrin	anticonvulsant	1989	25, 319
vildagliptin	antidiabetic	2007	43, 494
vinflunine	anticancer	2009	46, 493
vinorelbine	anticancer	1989	25, 320
voglibose	antidiabetic	1994	30, 313
voriconazole	antifungal	2002	38, 370
vorinostat	anticancer	2006	42, 549
xamoterol fumarate	congestive heart failure	1988	24, 312
ximelagatran	anticoagulant	2004	40, 470
zafirlukast	antiasthma	1996	32, 321
zalcitabine	antiviral	1992	28, 338
zaleplon	sleep disorders	1999	35, 351
zaltoprofen	antiinflammatory	1993	29, 349
zanamivir	antiviral	1999	35, 352
ziconotide	analgesic	2005	41, 473
zidovudine	antiviral	1987	23, 345
zileuton	antiasthma	1997	33, 344
zinostatin stimalamer	anticancer	1994	30, 313
ziprasidone hydrochloride	antipsychotic	2000	36, 312
zofenopril calcium	antihypertensive	2000	36, 313
zoledronate disodium	osteoporosis	2000	36, 314
zolpidem hemitartrate	sleep disorders	1988	24, 313
zomitriptan	antimigraine	1997	33, 345
zonisamide	anticonvulsant	1989	25, 320
zopiclone	sleep disorders	1986	22, 327
zucapsaicin	analgesic	2010	46, 495
zuclopenthixol acetate	antipsychotic	1987	23, 345

CUMULATIVE NCE INTRODUCTION INDEX, 1983–2010 (BY INDICATION)

GENERIC NAME	INDICATION	YEAR INTRO.	ARMC VOL., (PAGE)
gemeprost	abortifacient	1983	19 (319)
mifepristone	abortifacient	1988	24 (306)
azelaic acid	acne	1989	25 (310)
nitrefazole	addiction, alcohol	1983	19 (322)
varenicline	addiction, nicotine	2006	42 (547)
naltrexone hydrochloride	addiction, opioids	1984	20 (322)
nalmefene hydrochloride	addiction, opioids	1995	31 (347)
tacrine hydrochloride	Alzheimer's disease	1993	29 (346)
donepezil hydrochloride	Alzheimer's disease	1997	33 (332)
rivastigmin	Alzheimer's disease	1997	33 (342)
fasudil hydrochloride	amyotrophic lateral sclerosis	1995	31 (343)
riluzole	amyotrophic lateral sclerosis	1996	32 (316)
alfentanil hydrochloride	analgesic	1983	19 (314)
alminoprofen	analgesic	1983	19 (314)
meptazinol hydrochloride	analgesic	1983	19 (321)
sufentanil	analgesic	1983	19 (323)
suprofen	analgesic	1983	19 (324)
emorfazone	analgesic	1984	20 (317)
fosfosal	analgesic	1984	20 (319)
flupirtine maleate	analgesic	1985	21 (328)
propacetamol hydrochloride	analgesic	1986	22 (325)
eptazocine hydrobromide	analgesic	1987	23 (334)
ketorolac tromethamine	analgesic	1990	26 (304)
dezocine	analgesic	1991	27 (326)
mofezolac	analgesic	1994	30 (304)
remifentanil hydrochloride	analgesic	1996	32 (316)
parecoxib sodium	analgesic	2002	38 (364)
ziconotide	analgesic	2005	41 (473)
tapentadol hydrochloride	analgesic	2009	45 (523)
zucapsaicin	analgesic	2010	46 (495)
propofol	anesthetic	1986	22 (325)
sevoflurane	anesthetic	1990	26 (309)
desflurane	anesthetic	1992	28 (329)
ropivacaine	anesthetic	1996	32 (318)
levobupivacaine hydrochloride	anesthetic	2000	36 (308)
icatibant	angioedema, hereditary	2008	44 (608)
ecallantide	angioedema, hereditary	2009	46 (464)
astemizole	antiallergy	1983	19 (314)
azelastine hydrochloride	antiallergy	1986	22 (316)
nedocromil sodium	antiallergy	1986	22 (324)
cetirizine hydrochloride	antiallergy	1987	23 (331)
repirinast	antiallergy	1987	23 (341)
setastine hydrochloride	antiallergy	1987	23 (342)
acrivastine	antiallergy	1988	24 (295)
loratadine	antiallergy	1988	24 (306)

GENERIC NAME	INDICATION	YEAR INTRO.	ARMC VOL., (PAGE)
ebastine	antiallergy	1990	26 (302)
tazanolast	antiallergy	1990	26 (309)
levocabastine hydrochloride	antiallergy	1991	27 (330)
lodoxamide tromethamine	antiallergy	1992	28 (333)
emedastine difumarate	antiallergy	1993	29 (336)
epinastine	antiallergy	1994	30 (299)
suplatast tosilate	antiallergy	1995	31 (350)
fexofenadine	antiallergy	1996	32 (307)
olopatadine hydrochloride	antiallergy	1997	33 (340)
loteprednol etabonate	antiallergy	1998	34 (324)
mizolastine	antiallergy	1998	34 (325)
betotastine besilate	antiallergy	2000	36 (297)
ramatroban	antiallergy	2000	36 (311)
desloratadine	antiallergy	2001	37 (264)
levocetirizine	antiallergy	2001	37 (268)
rupatadine fumarate	antiallergy	2003	39 (284)
fluticasone furoate	antiallergy	2007	43 (469)
bilastine	antiallergy	2010	46 (449)
gallopamil hydrochloride	antianginal	1983	19 (3190)
nicorandil	antianginal	1984	20 (322)
ivabradine	antianginal	2006	42 (522)
ranolazine	antianginal	2006	42 (535)
cibenzoline	antiarrhythmic	1985	21 (325)
encainide hydrochloride	antiarrhythmic	1987	23 (333)
esmolol hydrochloride	antiarrhythmic	1987	23 (334)
moricizine hydrochloride	antiarrhythmic	1990	26 (305)
tiracizine hydrochloride	antiarrhythmic	1990	26 (310)
pilsicainide hydrochloride	antiarrhythmic	1991	27 (332)
pirmenol	antiarrhythmic	1994	30 (307)
ibutilide fumarate	antiarrhythmic	1996	32 (309)
nif ekalant hydrochloride	antiarrhythmic	1999	35 (344)
dofetilide	antiarrhythmic	2000	36 (301)
landiolol	antiarrhythmic	2002	38 (360)
dronedarone	antiarrhythmic	2009	45 (495)
vernakalant	antiarrhythmic	2010	46 (491)
auranofin	antiarthritic	1983	19 (314)
meloxicam	antiarthritic	1996	32 (312)
lefunomide	antiarthritic	1998	34 (324)
celecoxib	antiarthritic	1999	35 (335)
rofecoxib	antiarthritic	1999	35 (347)
anakinra	antiarthritic	2001	37 (261)
etoricoxibe	antiarthritic	2002	38 (355)
vadecoxib	antiarthritic	2002	38 (369)
adalimumab	antiarthritic	2003	39 (267)
abatacept	antiarthritic	2006	42 (509)
oxitropium bromide	antiasthma	1983	19 (323)
doxofylline	antiasthma	1985	21 (327)
mabuterol hydrochloride	antiasthma	1986	22 (323)
amlexanox	antiasthma	1987	23 (327)

GENERIC NAME	INDICATION	YEAR INTRO.	ARMC VOL., (PAGE)
flutropium bromide	antiasthma	1988	24 (303)
ibudilast	antiasthma	1989	25 (313)
bambuterol	antiasthma	1990	26 (299)
salmeterol hydroxynaphthoate	antiasthma	1990	26 (308)
pemirolast potassium	antiasthma	1991	27 (331)
pranlukast	antiasthma	1995	31 (347)
seratrodast	antiasthma	1995	31 (349)
zafirlukast	antiasthma	1996	32 (321)
zileuton	antiasthma	1997	33 (344)
montelukast sodium	antiasthma	1998	34 (326)
levalbuterol hydrochloride	antiasthma	1999	35 (341)
omalizumab allergic	antiasthma	2003	39 (280)
ciclesonide	antiasthma	2005	41 (443)
arformoterol	antiasthma	2007	43 (465)
cefmenoxime hydrochloride	antibacterial	1983	19 (316)
ceftazidime	antibacterial	1983	19 (316)
norfloxacin	antibacterial	1983	19 (322)
adamantanium bromide	antibacterial	1984	20 (315)
aztreonam	antibacterial	1984	20 (315)
cefonicid sodium	antibacterial	1984	20 (316)
ceforanide	antibacterial	1984	20 (317)
cefotetan disodium	antibacterial	1984	20 (317)
temocillin disodium	antibacterial	1984	20 (323)
astromycin sulfate	antibacterial	1985	21 (324)
cefbuperazone sodium	antibacterial	1985	21 (325)
cefpiramide sodium	antibacterial	1985	21 (325)
imipenem/cilastatin	antibacterial	1985	21 (328)
miokamycin	antibacterial	1985	21 (329)
mupirocin	antibacterial	1985	21 (330)
ofloxacin	antibacterial	1985	21 (331)
pefloxacin mesylate	antibacterial	1985	21 (331)
rifaximin	antibacterial	1985	21 (332)
carboplatin	antibacterial	1986	22 (318)
ciprofloxacin	antibacterial	1986	22 (318)
enoxacin	antibacterial	1986	22 (320)
rokitamycin	antibacterial	1986	22 (325)
sulbactam sodium	antibacterial	1986	22 (326)
aspoxicillin	antibacterial	1987	23 (328)
cefixime	antibacterial	1987	23 (329)
cefminox sodium	antibacterial	1987	23 (330)
cefpimizole	antibacterial	1987	23 (330)
cefteram pivoxil	antibacterial	1987	23 (330)
cefuroxime axetil	antibacterial	1987	23 (331)
cefuzonam sodium	antibacterial	1987	23 (331)
lenampicillin hydrochloride	antibacterial	1987	23 (336)
rifaximin	antibacterial	1987	23 (341)
sultamycillin tosylate	antibacterial	1987	23 (343)

GENERIC NAME	INDICATION	YEAR INTRO.	ARMC VOL., (PAGE)
azithromycin	antibacterial	1988	24 (298)
carumonam	antibacterial	1988	24 (298)
erythromycin acistrate	antibacterial	1988	24 (301)
flomoxef sodium	antibacterial	1988	24 (302)
isepamicin	antibacterial	1988	24 (305)
rifapentine	antibacterial	1988	24 (310)
teicoplanin	antibacterial	1988	24 (311)
cefpodoxime proxetil	antibacterial	1989	25 (310)
lomefloxacin	antibacterial	1989	25 (315)
RV-11	antibacterial	1989	25 (318)
arbekacin	antibacterial	1990	26 (298)
cefodizime sodium	antibacterial	1990	26 (300)
clarithromycin	antibacterial	1990	26 (302)
tosufloxacin tosylate	antibacterial	1990	26 (310)
cefdinir	antibacterial	1991	27 (323)
cefotiam hexetil hydrochloride	antibacterial	1991	27 (324)
temafloxacin hydrochloride	antibacterial	1991	27 (334)
cefetamet pivoxil hydrochloride	antibacterial	1992	28 (327)
cefpirome sulfate	antibacterial	1992	28 (328)
cefprozil	antibacterial	1992	28 (328)
ceftibuten	antibacterial	1992	28 (329)
fleroxacin	antibacterial	1992	28 (331)
loracarbef	antibacterial	1992	28 (333)
rifabutin	antibacterial	1992	28 (335)
rufloxacin hydrochloride	antibacterial	1992	28 (335)
tazobactam sodium	antibacterial	1992	28 (336)
brodimoprin	antibacterial	1993	29 (333)
cefepime	antibacterial	1993	29 (334)
dirithromycin	antibacterial	1993	29 (336)
levofloxacin	antibacterial	1993	29 (340)
nadifloxacin	antibacterial	1993	29 (340)
sparfloxacin	antibacterial	1993	29 (345)
cefditoren pivoxil	antibacterial	1994	30 (297)
meropenem	antibacterial	1994	30 (303)
panipenem/ betamipron carbapenem	antibacterial	1994	30 (305)
cefozopran hydrochloride	antibacterial	1995	31 (339)
cefcapene pivoxil	antibacterial	1997	33 (330)
flurithromycin ethylsuccinate	antibacterial	1997	33 (333)
fropenam	antibacterial	1997	33 (334)
cefoselis	antibacterial	1998	34 (319)
trovafloxacin mesylate	antibacterial	1998	34 (332)
dalfopristin	antibacterial	1999	35 (338)
gatilfloxacin	antibacterial	1999	35 (340)
moxifloxacin hydrochloride	antibacterial	1999	35 (343)

GENERIC NAME	INDICATION	YEAR INTRO.	ARMC VOL., (PAGE)
quinupristin	antibacterial	1999	35 (338)
linezolid	antibacterial	2000	36 (309)
telithromycin	antibacterial	2001	37 (271)
balofloxacin	antibacterial	2002	38 (351)
biapenem	antibacterial	2002	38 (351)
ertapenem sodium	antibacterial	2002	38 (353)
pazufloxacin	antibacterial	2002	38 (364)
prulifloxacin	antibacterial	2002	38 (366)
daptomycin	antibacterial	2003	39 (272)
gemifloxacin	antibacterial	2004	40 (458)
doripenem	antibacterial	2005	41 (448)
tigecycline	antibacterial	2005	41 (468)
garenoxacin	antibacterial	2007	43 (471)
retapamulin	antibacterial	2007	43 (486)
ceftobiprole medocaril	antibacterial	2008	44 (589)
sitafloxacin hydrate	antibacterial	2008	44 (621)
besifloxacin	antibacterial	2009	45 (482)
telavancin	antibacterial	2009	45 (525)
ceftaroline fosamil	antibacterial	2010	46 (453)
enocitabine	anticancer	1983	19 (318)
flutamide	anticancer	1983	19 (318)
epirubicin hydrochloride	anticancer	1984	20 (318)
mitoxantrone hydrochloride	anticancer	1984	20 (321)
camostat mesylate	anticancer	1985	21 (325)
amsacrine	anticancer	1987	23 (327)
doxifluridine	anticancer	1987	23 (332)
lonidamine	anticancer	1987	23 (337)
nilutamide	anticancer	1987	23 (338)
ranimustine	anticancer	1987	23 (341)
pirarubicin	anticancer	1988	24 (309)
fotemustine	anticancer	1989	25 (313)
interleukin-2	anticancer	1989	25 (314)
toremifene	anticancer	1989	25 (319)
vinorelbine	anticancer	1989	25 (320)
bisantrene hydrochloride	anticancer	1990	26 (300)
idarubicin hydrochloride	anticancer	1990	26 (303)
fludarabine phosphate	anticancer	1991	27 (327)
interferon, gamma-1	anticancer	1992	28 (332)
masoprocol	anticancer	1992	28 (333)
pentostatin	anticancer	1992	28 (334)
cladribine	anticancer	1993	29 (335)
cytarabine ocfosfate	anticancer	1993	29 (335)
formestane	anticancer	1993	29 (337)
miltefosine	anticancer	1993	29 (340)
paclitaxal	anticancer	1993	29 (342)
porfimer sodium	anticancer	1993	29 (343)
irinotecan	anticancer	1994	30 (301)
pegaspargase	anticancer	1994	30 (306)
sobuzoxane	anticancer	1994	30 (310)

GENERIC NAME	INDICATION	YEAR INTRO.	ARMC VOL., (PAGE)
zinostatin stimalamer	anticancer	1994	30 (313)
anastrozole	anticancer	1995	31 (338)
bicalutamide	anticancer	1995	31 (338)
docetaxel	anticancer	1995	31 (341)
fadrozole hydrochloride	anticancer	1995	31 (342)
gemcitabine hydrochloride	anticancer	1995	31 (344)
nedaplatin	anticancer	1995	31 (347)
letrazole	anticancer	1996	32 (311)
oxaliplatin	anticancer	1996	32 (313)
raltitrexed	anticancer	1996	32 (315)
topotecan hydrochloride	anticancer	1996	32 (320)
capecitabine	anticancer	1998	34 (319)
OCT-43	anticancer	1999	35 (345)
alitretinoin	anticancer	1999	35 (333)
arglabin	anticancer	1999	35 (335)
denileukin diftitox	anticancer	1999	35 (338)
SKI-2053R	anticancer	1999	35 (348)
tasonermin	anticancer	1999	35 (349)
temozolomide	anticancer	1999	35 (349)
valrubicin	anticancer	1999	35 (350)
bexarotene	anticancer	2000	36 (298)
exemestane	anticancer	2000	36 (304)
gemtuzumab ozogamicin	anticancer	2000	36 (306)
alemtuzumab	anticancer	2001	37 (260)
imatinib mesylate	anticancer	2001	37 (267)
amrubicin hydrochloride	anticancer	2002	38 (349)
fulveristrant	anticancer	2002	38 (357)
gefitinib	anticancer	2002	38 (358)
ibritunomab tiuxetan	anticancer	2002	38 (359)
temoporphin	anticancer	2002	38 (367)
bortezomib	anticancer	2003	39 (271)
cetuximab	anticancer	2003	39 (272)
tositumomab	anticancer	2003	39 (285)
abarelix	anticancer	2004	40 (446)
azacitidine	anticancer	2004	40 (447)
belotecan	anticancer	2004	40 (449)
bevacizumab	anticancer	2004	40 (450)
erlotinib	anticancer	2004	40 (454)
pemetrexed	anticancer	2004	40 (463)
talaporfin sodium	anticancer	2004	40 (469)
clofarabine	anticancer	2005	41 (444)
sorafenib	anticancer	2005	41 (466)
tamibarotene	anticancer	2005	41 (467)
dasatinib	anticancer	2006	42 (517)
nelarabine	anticancer	2006	42 (528)
nimotuzumab	anticancer	2006	42 (529)
panitumumab	anticancer	2006	42 (531)
sunitinib	anticancer	2006	42 (544)
vorinostat	anticancer	2006	42 (549)
ixabepilone	anticancer	2007	43 (473)

GENERIC NAME	INDICATION	YEAR INTRO.	ARMC VOL., (PAGE)
lapatinib	anticancer	2007	43 (475)
temsirolimus	anticancer	2007	43 (490)
trabectedin	anticancer	2007	43 (492)
catumaxomab	anticancer	2009	45 (486)
degarelix acetate	anticancer	2009	45 (490)
ofatumumab	anticancer	2009	45 (512)
pazopanib	anticancer	2009	45 (514)
pralatrexate	anticancer	2009	45 (517)
nilotinib	anticancer	2007	43 (480)
mifamurtide	anticancer	2009	46 (476)
romidepsin	anticancer	2009	46 (482)
vinflunine	anticancer	2009	46 (493)
cabazitaxel	anticancer	2010	46 (451)
eribulin mesylate	anticancer	2010	46 (465)
sipuleucel-t	anticancer	2010	46 (484)
angiotensin II	anticancer adjuvant	1994	30 (296)
enoxaparin	anticoagulant	1987	23 (333)
parnaparin sodium	anticoagulant	1993	29 (342)
reviparin sodium	anticoagulant	1993	29 (344)
duteplase	anticoagulant	1995	31 (342)
lepirudin	anticoagulant	1997	33 (336)
ximelagatran	anticoagulant	2004	40 (470)
dabigatran etexilate	anticoagulant	2008	44 (598)
rivaroxaban	anticoagulant	2008	44 (617)
thrombomodulin, recombinant	anticoagulant	2008	44 (628)
progabide	anticonvulsant	1985	21 (331)
vigabatrin	anticonvulsant	1989	25 (319)
zonisamide	anticonvulsant	1989	25 (320)
lamotrigine	anticonvulsant	1990	26 (304)
oxcarbazepine	anticonvulsant	1990	26 (307)
felbamate	anticonvulsant	1993	29 (337)
gabapentin	anticonvulsant	1993	29 (338)
topiramate	anticonvulsant	1995	31 (351)
fosphenytoin sodium	anticonvulsant	1996	32 (308)
tiagabine	anticonvulsant	1996	32 (319)
levetiracetam	anticonvulsant	2000	36 (307)
pregabalin	anticonvulsant	2004	40 (464)
rufinamide	anticonvulsant	2007	43 (488)
lacosamide	anticonvulsant	2008	44 (610)
eslicarbazepine acetate	anticonvulsant	2009	45 (498)
fluvoxamine maleate	antidepressant	1983	19 (319)
indalpine	antidepressant	1983	19 (320)
tianeptine sodium	antidepressant	1983	19 (324)
metapramine	antidepressant	1984	20 (320)
toloxatone	antidepressant	1984	20 (324)
fluoxetine hydrochloride	antidepressant	1986	22 (320)
medifoxamine fumarate	antidepressant	1986	22 (323)
bupropion hydrochloride	antidepressant	1989	25 (310)
citalopram	antidepressant	1989	25 (311)

GENERIC NAME	INDICATION	YEAR INTRO.	ARMC VOL., (PAGE)
setiptiline	antidepressant	1989	25 (318)
moclobemide	antidepressant	1990	26 (305)
setraline hydrochloride	antidepressant	1990	26 (309)
paroxetine	antidepressant	1991	27 (331)
mirtazapine	antidepressant	1994	30 (303)
nefazodone	antidepressant	1994	30 (305)
venlafaxine	antidepressant	1994	30 (312)
milnacipran	antidepressant	1997	33 (338)
pivagabine	antidepressant	1997	33 (341)
reboxetine	antidepressant	1997	33 (342)
escitalopram oxolate	antidepressant	2002	38 (354)
duloxetine	antidepressant	2004	40 (452)
desvenlafaxine	antidepressant	2008	44 (600)
tolrestat	antidiabetic	1989	25 (319)
acarbose	antidiabetic	1990	26 (297)
epalrestat	antidiabetic	1992	28 (330)
glucagon, rDNA	antidiabetic	1993	29 (338)
voglibose	antidiabetic	1994	30 (313)
glimepiride	antidiabetic	1995	31 (344)
insulin lispro	antidiabetic	1996	32 (310)
troglitazone	antidiabetic	1997	33 (344)
miglitol	antidiabetic	1998	34 (325)
repaglinide	antidiabetic	1998	34 (329)
nateglinide	antidiabetic	1999	35 (344)
pioglitazone hydrochloride	antidiabetic	1999	35 (346)
rosiglitazone maleate	antidiabetic	1999	35 (348)
mitiglinide	antidiabetic	2004	40 (460)
exenatide	antidiabetic	2005	41 (452)
pramlintide	antidiabetic	2005	41 (460)
sitagliptin	antidiabetic	2006	42 (541)
vildagliptin	antidiabetic	2007	43 (494)
liraglutide	antidiabetic	2009	45 (507)
saxagliptin	antidiabetic	2009	45 (521)
alogliptin	antidiabetic	2010	46 (446)
lidamidine hydrochloride	antidiarrheal	1984	20 (320)
acetorphan	antidiarrheal	1993	29 (332)
flumazenil	antidote, benzodiazepine overdose	1987	23 (335)
trientine hydrochloride	antidote, copper poisoning	1986	22 (327)
anti-digoxin polyclonal antibody	antidote, digoxin poisoning	2002	38 (350)
fomepizole	antidote, ethylene glycol poisoning	1998	34 (323)
succimer	antidote, lead poisoning	1991	27 (333)
crotelidae polyvalent immune fab	antidote, snake venom poisoning	2001	37 (263)
dronabinol	antiemetic	1986	22 (319)
ondansetron hydrochloride	antiemetic	1990	26 (306)
granisetron hydrochloride	antiemetic	1991	27 (329)
tropisetron	antiemetic	1992	28 (337)

GENERIC NAME	INDICATION	YEAR INTRO.	ARMC VOL., (PAGE)
nazasetron	antiemetic	1994	30 (305)
ramosetron	antiemetic	1996	32 (315)
dolasetron mesylate	antiemetic	1998	34 (321)
aprepitant	antiemetic	2003	39 (268)
palonosetron	antiemetic	2003	39 (281)
indisetron	antiemetic	2004	40 (459)
fosaprepitant dimeglumine	antiemetic	2008	44 (606)
oxiconazole nitrate	antifungal	1983	19 (322)
terconazole	antifungal	1983	19 (324)
tioconazole	antifungal	1983	19 (324)
naftifine hydrochloride	antifungal	1984	20 (321)
sulconizole nitrate	antifungal	1985	21 (332)
butoconazole	antifungal	1986	22 (318)
cloconazole hydrochloride	antifungal	1986	22 (318)
fenticonazole nitrate	antifungal	1987	23 (334)
fluconazole	antifungal	1988	24 (303)
itraconazole	antifungal	1988	24 (305)
amorolfine hydrochloride	antifungal	1991	27 (322)
terbinafine hydrochloride	antifungal	1991	27 (334)
butenafine hydrochloride	antifungal	1992	28 (327)
sertaconazole nitrate	antifungal	1992	28 (336)
neticonazole hydrochloride	antifungal	1993	29 (341)
lanoconazole	antifungal	1994	30 (302)
trimetrexate glucuronate	antifungal	1994	30 (312)
flutrimazole	antifungal	1995	31 (343)
liranaftate	antifungal	2000	36 (309)
caspofungin acetate	antifungal	2001	37 (263)
micafungin	antifungal	2002	38 (360)
voriconazole	antifungal	2002	38 (370)
fosfluconazole	antifungal	2004	40 (457)
eberconazole	antifungal	2005	41 (449)
luliconazole	antifungal	2005	41 (454)
anidulafungin	antifungal	2006	42 (512)
posaconazole	antifungal	2006	42 (532)
befunolol hydrochloride	antiglaucoma	1983	19 (315)
levobunolol hydrochloride	antiglaucoma	1985	21 (328)
dapiprazole hydrochloride	antiglaucoma	1987	23 (332)
apraclonidine hydrochloride	antiglaucoma	1988	24 (297)
unoprostone isopropyl ester	antiglaucoma	1994	30 (312)
dorzolamide hydrochloride	antiglaucoma	1995	31 (341)
brimonidine	antiglaucoma	1996	32 (306)
latanoprost	antiglaucoma	1996	32 (311)
brinzolamide	antiglaucoma	1998	34 (318)
bimatoprost	antiglaucoma	2001	37 (261)
travoprost	antiglaucoma	2001	37 (272)
tafluprost	antiglaucoma	2008	44 (625)
meglutol	antihypercholesterolemic	1983	19 (321)

GENERIC NAME	INDICATION	YEAR INTRO.	ARMC VOL., (PAGE)
divistyramine	antihypercholesterolemic	1984	20 (317)
melinamide	antihypercholesterolemic	1984	20 (320)
acipimox	antihypercholesterolemic	1985	21 (323)
ciprofibrate	antihypercholesterolemic	1985	21 (326)
beclobrate	antihypercholesterolemic	1986	22 (317)
binifibrate	antihypercholesterolemic	1986	22 (317)
ronafibrate	antihypercholesterolemic	1986	22 (326)
lovastatin	antihypercholesterolemic	1987	23 (337)
simvastatin	antihypercholesterolemic	1988	24 (311)
pravastatin	antihypercholesterolemic	1989	25 (316)
fluvastatin	antihypercholesterolemic	1994	30 (300)
atorvastatin calcium	antihypercholesterolemic	1997	33 (328)
cerivastatin	antihypercholesterolemic	1997	33 (331)
colestimide	antihypercholesterolemic	1999	35 (337)
colesevelam hydrochloride	antihypercholesterolemic	2000	36 (300)
ezetimibe	antihypercholesterolemic	2002	38 (355)
pitavastatin	antihypercholesterolemic	2003	39 (282)
rosuvastatin	antihypercholesterolemic	2003	39 (283)
choline fenofibrate	antihypercholesterolemic	2008	44 (594)
captopril	antihypertensive	1982	13 (086)
betaxolol hydrochloride	antihypertensive	1983	19 (315)
budralazine	antihypertensive	1983	19 (315)
celiprolol hydrochloride	antihypertensive	1983	19 (317)
guanadrel sulfate	antihypertensive	1983	19 (319)
enalapril maleate	antihypertensive	1984	20 (317)
terazosin hydrochloride	antihypertensive	1984	20 (323)
bopindolol	antihypertensive	1985	21 (324)
bunazosin hydrochloride	antihypertensive	1985	21 (324)
ketanserin	antihypertensive	1985	21 (328)
nitrendipine	antihypertensive	1985	21 (331)
trimazosin hydrochloride	antihypertensive	1985	21 (333)
arotinolol hydrochloride	antihypertensive	1986	22 (316)
bisoprolol fumarate	antihypertensive	1986	22 (317)
bevantolol hydrochloride	antihypertensive	1987	23 (328)
enalaprilat	antihypertensive	1987	23 (332)
lisinopril	antihypertensive	1987	23 (337)
pinacidil	antihypertensive	1987	23 (340)
tertatolol hydrochloride	antihypertensive	1987	23 (344)
alacepril	antihypertensive	1988	24 (296)
alfuzosin hydrochloride	antihypertensive	1988	24 (296)
amosulalol	antihypertensive	1988	24 (297)
cadralazine	antihypertensive	1988	24 (298)
cicletanine	antihypertensive	1988	24 (299)
doxazosin mesylate	antihypertensive	1988	24 (300)
felodipine	antihypertensive	1988	24 (302)
nipradilol	antihypertensive	1988	24 (307)
perindopril	antihypertensive	1988	24 (309)
rilmenidine	antihypertensive	1988	24 (310)
tiamenidine hydrochloride	antihypertensive	1988	24 (311)
delapril	antihypertensive	1989	25 (311)

GENERIC NAME	INDICATION	YEAR INTRO.	ARMC VOL., (PAGE)
dilevalol	antihypertensive	1989	25 (311)
isradipine	antihypertensive	1989	25 (315)
nilvadipine	antihypertensive	1989	25 (316)
quinapril	antihypertensive	1989	25 (317)
ramipril	antihypertensive	1989	25 (317)
amlodipine besylate	antihypertensive	1990	26 (298)
benazepril hydrochloride	antihypertensive	1990	26 (299)
cilazapril	antihypertensive	1990	26 (301)
manidipine hydrochloride	antihypertensive	1990	26 (304)
nisoldipine	antihypertensive	1990	26 (306)
benidipine hydrochloride	antihypertensive	1991	27 (322)
carvedilol	antihypertensive	1991	27 (323)
fosinopril sodium	antihypertensive	1991	27 (328)
lacidipine	antihypertensive	1991	27 (330)
moxonidine	antihypertensive	1991	27 (330)
barnidipine hydrochloride	antihypertensive	1992	28 (326)
tilisolol hydrochloride	antihypertensive	1992	28 (337)
imidapril hydrochloride	antihypertensive	1993	29 (339)
trandolapril	antihypertensive	1993	29 (348)
efonidipine	antihypertensive	1994	30 (299)
losartan	antihypertensive	1994	30 (302)
temocapril	antihypertensive	1994	30 (311)
cinildipine	antihypertensive	1995	31 (339)
moexipril hydrochloride	antihypertensive	1995	31 (346)
spirapril hydrochloride	antihypertensive	1995	31 (349)
aranidipine	antihypertensive	1996	32 (306)
valsartan	antihypertensive	1996	32 (320)
candesartan cilexetil	antihypertensive	1997	33 (330)
eprosartan	antihypertensive	1997	33 (333)
irbesartan	antihypertensive	1997	33 (336)
lercanidipine	antihypertensive	1997	33 (337)
mebefradil hydrochloride	antihypertensive	1997	33 (338)
nebivolol	antihypertensive	1997	33 (339)
fenoldopam mesylate	antihypertensive	1998	34 (322)
telmisartan	antihypertensive	1999	35 (349)
zofenopril calcium	antihypertensive	2000	36 (313)
bosentan	antihypertensive	2001	37 (262)
olimesartan Medoxomil	antihypertensive	2002	38 (363)
treprostinil sodium	antihypertensive	2002	38 (368)
azelnidipine	antihypertensive	2003	39 (270)
eplerenone	antihypertensive	2003	39 (276)
aliskiren	antihypertensive	2007	43 (461)
clevidipine	antihypertensive	2008	44 (596)
butyl flufenamate	antiinflammatory	1983	19 (316)
halometasone	antiinflammatory	1983	19 (320)
hydrocortisone butyrate	antiinflammatory	1983	19 (320)
isoxicam	antiinflammatory	1983	19 (320)
oxaprozin	antiinflammatory	1983	19 (322)
fisalamine	antiinflammatory	1984	20 (318)
isofezolac	antiinflammatory	1984	20 (319)

GENERIC NAME	INDICATION	YEAR INTRO.	ARMC VOL., (PAGE)
piketoprofen	antiinflammatory	1984	20 (322)
pimaprofen	antiinflammatory	1984	20 (322)
alclometasone dipropionate	antiinflammatory	1985	21 (323)
diacerein	antiinflammatory	1985	21 (326)
etodolac	antiinflammatory	1985	21 (327)
nabumetone	antiinflammatory	1985	21 (330)
nimesulide	antiinflammatory	1985	21 (330)
amfenac sodium	antiinflammatory	1986	22 (315)
deflazacort	antiinflammatory	1986	22 (319)
felbinac	antiinflammatory	1986	22 (320)
lobenzarit sodium	antiinflammatory	1986	22 (322)
loxoprofen sodium	antiinflammatory	1986	22 (322)
osalazine sodium	antiinflammatory	1986	22 (324)
prednicarbate	antiinflammatory	1986	22 (325)
AF-2259	antiinflammatory	1987	23 (325)
flunoxaprofen	antiinflammatory	1987	23 (335)
mometasone furoate	antiinflammatory	1987	23 (338)
tenoxicam	antiinflammatory	1987	23 (344)
clobenoside	antiinflammatory	1988	24 (300)
hydrocortisone aceponate	antiinflammatory	1988	24 (304)
piroxicam cinnamate	antiinflammatory	1988	24 (309)
interferon, gamma	antiinflammatory	1989	25 (314)
aminoprofen	antiinflammatory	1990	26 (298)
droxicam	antiinflammatory	1990	26 (302)
fluticasone propionate	antiinflammatory	1990	26 (303)
halobetasol propionate	antiinflammatory	1991	27 (329)
aceclofenac	antiinflammatory	1992	28 (325)
butibufen	antiinflammatory	1992	28 (327)
deprodone propionate	antiinflammatory	1992	28 (329)
amtolmetin guacil	antiinflammatory	1993	29 (332)
zaltoprofen	antiinflammatory	1993	29 (349)
actarit	antiinflammatory	1994	30 (296)
ampiroxicam	antiinflammatory	1994	30 (296)
betamethasone butyrate propionate	antiinflammatory	1994	30 (297)
dexibuprofen	antiinflammatory	1994	30 (298)
rimexolone	antiinflammatory	1995	31 (348)
bromfenac sodium	antiinflammatory	1997	33 (329)
lornoxicam	antiinflammatory	1997	33 (337)
sivelestat	antiinflammatory	2002	38 (366)
lumiracoxib	antiinflammatory	2005	41 (455)
nepafenac	antiinflammatory	2005	41 (456)
canakinumab	antiinflammatory	2009	45 (484)
golimumab	antiinflammatory	2009	45 (503)
mefloquine hydrochloride	antimalarial	1985	21 (329)
artemisinin	antimalarial	1987	23 (327)
halofantrine	antimalarial	1988	24 (304)
arteether	antimalarial	2000	36 (296)
bulaquine	antimalarial	2000	36 (299)

GENERIC NAME	INDICATION	YEAR INTRO.	ARMC VOL., (PAGE)
alpiropride	antimigraine	1988	24 (296)
sumatriptan succinate	antimigraine	1991	27 (333)
naratriptan hydrochloride	antimigraine	1997	33 (339)
zomitriptan	antimigraine	1997	33 (345)
rizatriptan benzoate	antimigraine	1998	34 (330)
lomerizine hydrochloride	antimigraine	1999	35 (342)
almotriptan	antimigraine	2000	36 (295)
eletriptan	antimigraine	2001	37 (266)
frovatriptan	antimigraine	2002	38 (357)
dexfenfluramine	antiobesity	1997	33 (332)
orlistat	antiobesity	1998	34 (327)
sibutramine	antiobesity	1998	34 (331)
rimonabant	antiobesity	2006	42 (537)
quinf amideamebicide	antiparasitic	1984	20 (322)
ivermectin	antiparasitic	1987	23 (336)
atovaquone	antiparasitic	1992	28 (326)
epoprostenol sodium	antiplatelet	1983	19 (318)
beraprost sodium	antiplatelet	1992	28 (326)
iloprost	antiplatelet	1992	28 (332)
prasugrel	antiplatelet	2009	45 (519)
cabergoline	antiprolactin	1993	29 (334)
acitretin	antipsoriasis	1989	25 (309)
calcipotriol	antipsoriasis	1991	27 (323)
tacalcitol	antipsoriasis	1993	29 (346)
tazarotene	antipsoriasis	1997	33 (343)
alefacept	antipsoriasis	2003	39 (267)
efalizumab	antipsoriasis	2003	39 (274)
ustekinumab	antipsoriasis	2009	45 (532)
timiperone	antipsychotic	1984	20 (323)
amisulpride	antipsychotic	1986	22 (316)
zuclopenthixol acetate	antipsychotic	1987	23 (345)
remoxipride hydrochloride	antipsychotic	1990	26 (308)
clospipramine hydrochloride	antipsychotic	1991	27 (325)
nemonapride	antipsychotic	1991	27 (331)
risperidone	antipsychotic	1993	29 (344)
olanzapine	antipsychotic	1996	32 (313)
sertindole	antipsychotic	1996	32 (318)
quetiapine fumarate	antipsychotic	1997	33 (341)
ziprasidone hydrochloride	antipsychotic	2000	36 (312)
perospirone hydrochloride	antipsychotic	2001	37 (270)
aripiprazole	antipsychotic	2002	38 (350)
paliperidone	antipsychotic	2007	43 (482)
blonanserin	antipsychotic	2008	44 (587)
asenapine	antipsychotic	2009	45 (479)
lurasidone hydrochloride	antipsychotic	2010	46 (473)
drotrecogin alfa	antisepsis	2001	37 (265)
tiquizium bromide	antispasmodic	1984	20 (324)
cimetropium bromide	antispasmodic	1985	21 (326)
romiplostim	antithrombocytopenic	2008	44 (619)

GENERIC NAME	INDICATION	YEAR INTRO.	ARMC VOL., (PAGE)
eltrombopag	antithrombocytopenic	2009	45 (497)
indobufen	antithrombotic	1984	20 (319)
defibrotide	antithrombotic	1986	22 (319)
alteplase	antithrombotic	1987	23 (326)
APSAC	antithrombotic	1987	23 (326)
picotamide	antithrombotic	1987	23 (340)
cilostazol	antithrombotic	1988	24 (299)
limaprost	antithrombotic	1988	24 (306)
ozagrel sodium	antithrombotic	1988	24 (308)
argatroban	antithrombotic	1990	26 (299)
ethyl icosapentate	antithrombotic	1990	26 (303)
cloricromen	antithrombotic	1991	27 (325)
sarpogrelate hydrochloride	antithrombotic	1993	29 (344)
reteplase	antithrombotic	1996	32 (316)
anagrelide hydrochloride	antithrombotic	1997	33 (328)
clopidogrel hydrogensulfate	antithrombotic	1998	34 (320)
tirofiban hydrochloride	antithrombotic	1998	34 (332)
eptilfibatide	antithrombotic	1999	35 (340)
bivalirudin	antithrombotic	2000	36 (298)
fondaparinux sodium	antithrombotic	2002	38 (356)
ticagrelor	antithrombotic	2010	46 (488)
levodropropizine	antitussive	1988	24 (305)
nitisinone	antityrosinaemia	2002	38 (361)
sofalcone	antiulcer	1984	20 (323)
teprenone	antiulcer	1984	20 (323)
enprostil	antiulcer	1985	21 (327)
famotidine	antiulcer	1985	21 (327)
misoprostol	antiulcer	1985	21 (329)
rosaprostol	antiulcer	1985	21 (332)
roxatidine acetate hydrochloride	antiulcer	1986	22 (326)
troxipide	antiulcer	1986	22 (327)
benexate hydrochloride	antiulcer	1987	23 (328)
nizatidine	antiulcer	1987	23 (339)
ornoprostil	antiulcer	1987	23 (339)
plaunotol	antiulcer	1987	23 (340)
roxithromycin	antiulcer	1987	23 (342)
spizofurone	antiulcer	1987	23 (343)
omeprazole	antiulcer	1988	24 (308)
irsogladine	antiulcer	1989	25 (315)
rebamipide	antiulcer	1990	26 (308)
lansoprazole	antiulcer	1992	28 (332)
ecabet sodium	antiulcer	1993	29 (336)
tretinoin tocoferil	antiulcer	1993	29 (348)
polaprezinc	antiulcer	1994	30 (307)
pantoprazole sodium	antiulcer	1995	30 (306)
ranitidine bismuth citrate	antiulcer	1995	31 (348)
ebrotidine	antiulcer	1997	33 (333)
rabeprazole sodium	antiulcer	1998	34 (328)

GENERIC NAME	INDICATION	YEAR INTRO.	ARMC VOL., (PAGE)
dosmalfate	antiulcer	2000	36 (302)
egualen sodium	antiulcer	2000	36 (303)
esomeprazole magnesium	antiulcer	2000	36 (303)
lafutidine	antiulcer	2000	36 (307)
dexlansoprazole	antiulcer	2009	45 (492)
rimantadine hydrochloride	antiviral	1987	23 (342)
zidovudine	antiviral	1987	23 (345)
ganciclovir	antiviral	1988	24 (303)
foscarnet sodium	antiviral	1989	25 (313)
didanosine	antiviral	1991	27 (326)
zalcitabine	antiviral	1992	28 (338)
sorivudine	antiviral	1993	29 (345)
famciclovir	antiviral	1994	30 (300)
propagermanium	antiviral	1994	30 (308)
stavudine	antiviral	1994	30 (311)
lamivudine	antiviral	1995	31 (345)
saquinavir mesvlate	antiviral	1995	31 (349)
valaciclovir hydrochloride	antiviral	1995	31 (352)
cidofovir	antiviral	1996	32 (306)
indinavir sulfate	antiviral	1996	32 (310)
nevirapine	antiviral	1996	32 (313)
penciclovir	antiviral	1996	32 (314)
ritonavir	antiviral	1996	32 (317)
delavirdine mesylate	antiviral	1997	33 (331)
imiquimod	antiviral	1997	33 (335)
interferon alfacon-1	antiviral	1997	33 (336)
nelfinavir mesylate	antiviral	1997	33 (340)
efavirenz	antiviral	1998	34 (321)
fomivirsen sodium	antiviral	1998	34 (323)
abacavir sulfate	antiviral	1999	35 (333)
amprenavir	antiviral	1999	35 (334)
oseltamivir phosphate	antiviral	1999	35 (346)
zanamivir	antiviral	1999	35 (352)
lopinavir	antiviral	2000	36 (310)
tenofovir disoproxil fumarate	antiviral	2001	37 (271)
vaglancirclovir hydrochloride	antiviral	2001	37 (273)
adefovir dipivoxil	antiviral	2002	38 (348)
atazanavir	antiviral	2003	39 (269)
emtricitabine	antiviral	2003	39 (274)
enfuvirtide	antiviral	2003	39 (275)
fosamprenavir	antiviral	2003	39 (277)
influenza virus, live	antiviral	2003	39 (277)
entecavir	antiviral	2005	41 (450)
tipranavir	antiviral	2005	41 (470)
darunavir	antiviral	2006	42 (515)
telbivudine	antiviral	2006	42 (546)
clevudine	antiviral	2007	43 (466)
maraviroc	antiviral	2007	43 (478)

GENERIC NAME	INDICATION	YEAR INTRO.	ARMC VOL., (PAGE)
raltegravir	antiviral	2007	43 (484)
etravirine	antiviral	2008	44 (602)
laninamivir octanoate	antiviral	2010	46 (470)
peramivir	antiviral	2010	46 (477)
cevimeline hydrochloride	antixerostomia	2000	36 (299)
etizolam	anxiolytic	1984	20 (318)
flutazolam	anxiolytic	1984	20 (318)
mexazolam	anxiolytic	1984	20 (321)
buspirone hydrochloride	anxiolytic	1985	21 (324)
doxefazepam	anxiolytic	1985	21 (326)
flutoprazepam	anxiolytic	1986	22 (320)
metaclazepam	anxiolytic	1987	23 (338)
alpidem	anxiolytic	1991	27 (322)
cinolazepam	anxiolytic	1993	29 (334)
tandospirone	anxiolytic	1996	32 (319)
dexmethylphenidate hydrochloride	attention deficit hyperactivity disorder	2002	38 (352)
atomoxetine	attention deficit hyperactivity disorder	2003	39 (270)
lisdexamfetamine	attention deficit hyperactivity disorder	2007	43 (477)
finasteride	benign prostatic hyperplasia	1992	28 (331)
tamsulosin hydrochloride	benign prostatic hyperplasia	1993	29 (347)
dutasteride	benign prostatic hyperplasia	2002	38 (353)
fenbuprol	biliary tract dysfunction	1983	19 (318)
clodronate disodium	calcium regulation	1986	22 (319)
gallium nitrate	calcium regulation	1991	27 (328)
neridronic acide	calcium regulation	2002	38 (361)
dexrazoxane	cardioprotective	1992	28 (330)
nimodipine	cerebral vasodilator	1985	21 (330)
brovincamine fumarate	cerebral vasodilator	1986	22 (317)
propentofylline propionate	cerebral vasodilator	1988	24 (310)
indacaterol	chronic obstructive pulmonary disease	2009	45 (505)
formoterol fumarate	chronic obstructive pulmonary disorder	1986	22 (321)
tiotropium bromide	chronic obstructive pulmonary disorder	2002	38 (368)
roflumilast	chronic obstructive pulmonary disorder	2010	46 (480)
levacecarnine hydrochloride	cognition enhancer	1986	22 (322)
oxiracetam	cognition enhancer	1987	23 (339)
exifone	cognition enhancer	1988	24 (302)
choline alfoscerate	cognition enhancer	1990	26 (300)
aniracetam	cognition enhancer	1993	29 (333)
pramiracetam sulfate	cognition enhancer	1993	29 (343)
amrinone	congestive heart failure	1983	19 (314)
bucladesine sodium	congestive heart failure	1984	20 (316)
ibopamine hydrochloride	congestive heart failure	1984	20 (319)

GENERIC NAME	INDICATION	YEAR INTRO.	ARMC VOL., (PAGE)
denopamine	congestive heart failure	1988	24 (300)
enoximone	congestive heart failure	1988	24 (301)
xamoterol fumarate	congestive heart failure	1988	24 (312)
dopexamine	congestive heart failure	1989	25 (312)
milrinone	congestive heart failure	1989	25 (316)
vesnarinone	congestive heart failure	1990	26 (310)
flosequinan	congestive heart failure	1992	28 (331)
docarpamine	congestive heart failure	1994	30 (298)
pimobendan	congestive heart failure	1994	30 (307)
carperitide	congestive heart failure	1995	31 (339)
loprinone hydrochloride	congestive heart failure	1996	32 (312)
colforsin daropate hydrochloride	congestive heart failure	1999	35 (337)
levosimendan	congestive heart failure	2000	36 (308)
nesiritide	congestive heart failure	2001	37 (269)
lulbiprostone	constipation	2006	42 (525)
methylnaltrexone bromide	constipation	2008	44 (612)
promegestrone	contraception	1983	19 (323)
gestrinone	contraception	1986	22 (321)
nomegestrol acetate	contraception	1986	22 (324)
norgestimate	contraception	1986	22 (324)
gestodene	contraception	1987	23 (335)
centchroman	contraception	1991	27 (324)
drospirenone	contraception	2000	36 (302)
trimegestone	contraception	2001	37 (273)
norelgestromin	contraception	2002	38 (362)
ulipristal acetate	contraception	2009	45 (530)
biolimus drug-eluting stent	coronary artery disease, antirestenotic	2008	44 (586)
neltenexine	cystic fibrosis	1993	29 (341)
dornase alfa	cystic fibrosis	1994	30 (298)
amifostine	cytoprotective	1995	31 (338)
kinetin	dermatologic, skin photodamage	1999	35 (341)
gadoversetamide	diagnostic	2000	36 (304)
ioflupane	diagnostic	2000	36 (306)
muzolimine	diuretic	1983	19 (321)
azosemide	diuretic	1986	22 (316)
torasemide	diuretic	1993	29 (348)
alpha-1 antitrypsin	emphysema	1988	24 (297)
telmesteine	expectorant	1992	28 (337)
erdosteine	expectorant	1995	31 (342)
fudosteine	expectorant	2001	37 (267)
agalsidase alfa	Fabry's disease	2001	37 (259)
chenodiol	gallstones	1983	19 (317)
cisapride	gastroprokinetic	1988	24 (299)
cinitapride	gastroprokinetic	1990	26 (301)
itopride hydrochloride	gastroprokinetic	1995	31 (344)
mosapride citrate	gastroprokinetic	1998	34 (326)
alglucerase	Gaucher's disease	1991	27 (321)
imiglucerase	Gaucher's disease	1994	30 (301)

GENERIC NAME	INDICATION	YEAR INTRO.	ARMC VOL., (PAGE)
miglustat	Gaucher's disease	2003	39 (279)
rilonacept	genetic autoinflammatory syndromes	2008	44 (615)
febuxostat	gout	2009	45 (501)
somatropin	growth disorders	1987	23 (343)
octreotide	growth disorders	1988	24 (307)
somatomedin-1	growth disorders	1994	30 (310)
somatotropin	growth disorders	1994	30 (310)
lanreotide acetate	growth disorders	1995	31 (345)
pegvisomant	growth disorders	2003	39 (281)
factor VIIa	haemophilia	1996	32 (307)
erythropoietin	hematopoietic	1988	24 (301)
eculizumab	hemoglobinuria	2007	43 (468)
factor VIII	hemostatic	1992	28 (330)
thrombin alfa	hemostatic	2008	44 (627)
malotilate	hepatoprotective	1985	21 (329)
mivotilate	hepatoprotective	1999	35 (343)
buserelin acetate	hormone therapy	1984	20 (316)
leuprolide acetate	hormone therapy	1984	20 (319)
goserelin	hormone therapy	1987	23 (336)
tibolone	hormone therapy	1988	24 (312)
nafarelin acetate	hormone therapy	1990	26 (306)
paricalcitol	hyperparathyroidism	1998	34 (327)
doxercalciferol	hyperparathyroidism	1999	35 (339)
maxacalcitol	hyperparathyroidism	2000	36 (310)
falecalcitriol	hyperparathyroidism	2001	37 (266)
cinacalcet	hyperparathyroidism	2004	40 (451)
quinagolide	hyperprolactinemia	1994	30 (309)
conivaptan	hyponatremia	2006	42 (514)
mozavaptan	hyponatremia	2006	42 (527)
tolvaptan	hyponatremia	2009	45 (528)
thymopentin	immunomodulator	1985	21 (333)
bucillamine	immunomodulator	1987	23 (329)
centoxin	immunomodulator	1991	27 (325)
schizophyllan	immunostimulant	1985	22 (326)
lentinan	immunostimulant	1986	22 (322)
ubenimex	immunostimulant	1987	23 (345)
pegademase bovine	immunostimulant	1990	26 (307)
filgrastim	immunostimulant	1991	27 (327)
interferon gamma-1b	immunostimulant	1991	27 (329)
romurtide	immunostimulant	1991	27 (332)
sargramostim	immunostimulant	1991	27 (332)
pidotimod	immunostimulant	1993	29 (343)
GMDP	immunostimulant	1996	32 (308)
cyclosporine	immunosuppressant	1983	19 (317)
mizoribine	immunosuppressant	1984	20 (321)
muromonab-CD3	immunosuppressant	1986	22 (323)
tacrolimus	immunosuppressant	1993	29 (347)
gusperimus	immunosuppressant	1994	30 (300)
mycophenolate mofetil	immunosuppressant	1995	31 (346)

GENERIC NAME	INDICATION	YEAR INTRO.	ARMC VOL., (PAGE)
pimecrolimus	immunosuppressant	2002	38 (365)
mycophenolate sodium	immunosuppressant	2003	39 (279)
everolimus	immunosuppressant	2004	40 (455)
follitropin alfa	infertility	1996	32 (307)
follitropin beta	infertility	1996	32 (308)
cetrorelix	infertility	1999	35 (336)
ganirelix acetate	infertility	2000	36 (305)
corifollitropin alfa	infertility	2010	46 (455)
defeiprone	iron chelation therapy	1995	31 (340)
deferasirox	iron chelation therapy	2005	41 (446)
alosetron hydrochloride	irritable bowel syndrome	2000	36 (295)
tegaserod maleate	irritable bowel syndrome	2001	37 (270)
certolizumab pegol	irritable bowel syndrome	2008	44 (592)
nartograstim	leukopenia	1994	30 (304)
tesamorelin acetate	lipodystrophy	2010	46 (486)
Lyme disease vaccine	Lyme disease	1999	35 (342)
sildenafil citrate	male sexual dysfunction	1998	34 (331)
tadalafil	male sexual dysfunction	2003	39 (284)
vardenafil	male sexual dysfunction	2003	39 (286)
udenafil	male sexual dysfunction	2005	41 (472)
laronidase	mucopolysaccharidosis I	2003	39 (278)
idursulfase	mucopolysaccharidosis II, Hunter syndrome	2006	42 (520)
galsulfase	mucopolysaccharidosis VI	2005	41 (453)
palifermin	mucositis	2005	41 (461)
interferon, b-1b	multiple sclerosis	1993	29 (339)
interferon, b-1a	multiple sclerosis	1996	32 (311)
glatiramer acetate	multiple sclerosis	1997	33 (334)
natalizumab	multiple sclerosis	2004	40 (462)
dalfampridine	multiple sclerosis	2010	46 (458)
fingolimod	multiple sclerosis	2010	46 (468)
afloqualone	muscle relaxant	1983	19 (313)
eperisone hydrochloride	muscle relaxant	1983	19 (318)
tiropramide hydrochloride	muscle relaxant	1983	19 (324)
tizanidine	muscle relaxant	1984	20 (324)
doxacurium chloride	muscle relaxant	1991	27 (326)
mivacurium chloride	muscle relaxant	1992	28 (334)
rocuronium bromide	muscle relaxant	1994	30 (309)
cisatracurium besilate	muscle relaxant	1995	31 (340)
rapacuronium bromide	muscle relaxant	1999	35 (347)
decitabine	myelodysplastic syndromes	2006	42 (519)
lenalidomide	myelodysplastic syndromes, multiple myeloma	2006	42 (523)
tinazoline	nasal decongestant	1988	24 (312)
taltirelin	neurodegeneration	2000	36 (311)
sugammadex	neuromuscular blockade, reversal	2008	44 (623)
edaravone	neuroprotective	2001	37 (265)
idebenone	nootropic	1986	22 (321)
bifemelane hydrochloride	nootropic	1987	23 (329)

GENERIC NAME	INDICATION	YEAR INTRO.	ARMC VOL., (PAGE)
indeloxazine hydrochloride	nootropic	1988	24 (304)
nizofenzone	nootropic	1988	24 (307)
alcaftadine	ophthalmologic, allergic conjunctivitis	2010	46 (444)
diquafosol tetrasodium	ophthalmologic, dry eye	2010	46 (462)
verteporfin	ophthalmologic, macular degeneration	2000	36 (312)
pegaptanib	ophthalmologic (macular degeneration)	2005	41 (458)
ranibizumab	ophthalmologic (macular degeneration)	2006	42 (534)
OP-1	osteoinductor	2001	37 (269)
APD	osteoporosis	1987	23 (326)
disodium pamidronate	osteoporosis	1989	25 (312)
ipriflavone	osteoporosis	1989	25 (314)
alendronate sodium	osteoporosis	1993	29 (332)
ibandronic acid	osteoporosis	1996	32 (309)
incadronic acid	osteoporosis	1997	33 (335)
raloxifene hydrochloride	osteoporosis	1998	34 (328)
risedronate sodium	osteoporosis	1998	34 (330)
zoledronate disodium	osteoporosis	2000	36 (314)
strontium ranelate	osteoporosis	2004	40 (466)
minodronic acid	osteoporosis	2009	45 (509)
denosumab	osteoporosis	2010	46 (459)
tiludronate disodium	Paget's disease	1995	31 (350)
nafamostat mesylate	pancreatitis	1986	22 (323)
pergolide mesylate	Parkinson's disease	1988	24 (308)
droxidopa	Parkinson's disease	1989	25 (312)
ropinirole hydrochloride	Parkinson's disease	1996	32 (317)
talipexole	Parkinson's disease	1996	32 (318)
budipine	Parkinson's disease	1997	33 (330)
pramipexole hydrochloride	Parkinson's disease	1997	33 (341)
tolcapone	Parkinson's disease	1997	33 (343)
entacapone	Parkinson's disease	1998	34 (322)
CHF-1301	Parkinson's disease	1999	35 (336)
rasagiline	Parkinson's disease	2005	41 (464)
rotigotine	Parkinson's disease	2006	42 (538)
sapropterin hydrochloride	phenylketouria	1992	28 (336)
alglucosidase alfa	Pompe disease	2006	42 (511)
alvimopan	post-operative ileus	2008	44 (584)
histrelin	precocious puberty	1993	29 (338)
dapoxetine	premature ejaculation	2009	45 (488)
atosiban	premature labor	2000	36 (297)
nalfurafine hydrochloride	pruritus	2009	45 (510)
pirfenidone	pulmonary fibrosis, idiopathic	2008	44 (614)
sitaxsentan	pulmonary hypertension	2006	42 (543)
ambrisentan	pulmonary hypertension	2007	43 (463)
pumactant	respiratory distress syndrome	1994	30 (308)
mepixanox	respiratory stimulant	1984	20 (320)

GENERIC NAME	INDICATION	YEAR INTRO.	ARMC VOL., (PAGE)
surfactant TA	respiratory surfactant	1987	23 (344)
binfonazole	sleep disorders	1983	19 (315)
brotizolam	sleep disorders	1983	19 (315)
loprazolam mesylate	sleep disorders	1983	19 (321)
butoctamide	sleep disorders	1984	20 (316)
quazepam	sleep disorders	1985	21 (332)
adrafinil	sleep disorders	1986	22 (315)
zopiclone	sleep disorders	1986	22 (327)
zolpidem hemitartrate	sleep disorders	1988	24 (313)
rilmazafone	sleep disorders	1989	25 (317)
modafinil	sleep disorders	1994	30 (303)
zaleplon	sleep disorders	1999	35 (351)
dexmedetomidine hydrochloride	sleep disorders	2000	36 (301)
eszopiclone	sleep disorders	2005	41 (451)
ramelteon	sleep disorders	2005	41 (462)
armodafinil	sleep disorders	2009	45 (478)
plerixafor hydrochloride	stem cell mobilizer	2009	45 (515)
tirilazad mesylate	subarachnoid hemorrhage	1995	31 (351)
balsalazide disodium	ulcerative colitis	1997	33 (329)
acetohydroxamic acid	urinary tract/bladder disorders	1983	19 (313)
sodium cellulose phosphate	urinary tract/bladder disorders	1983	19 (323)
propiverine hydrochloride	urinary tract/bladder disorders	1992	28 (335)
naftopidil	urinary tract/bladder disorders	1999	35 (344)
solifenacin	urinary tract/bladder disorders	2004	40 (466)
darifenacin	urinary tract/bladder disorders	2005	41 (445)
silodosin	urinary tract/bladder disorders	2006	42 (540)
imidafenacin	urinary tract/bladder disorders	2007	43 (472)
fesoterodine	urinary tract/bladder disorders	2008	44 (604)
tiopronin	urolithiasis	1989	25 (318)
cadexomer iodine	wound healing agent	1983	19 (316)
epidermal growth factor	wound healing agent	1987	23 (333)
prezatide copper acetate	wound healing agent	1996	32 (314)
acemannan	wound healing agent	2001	37 (259)

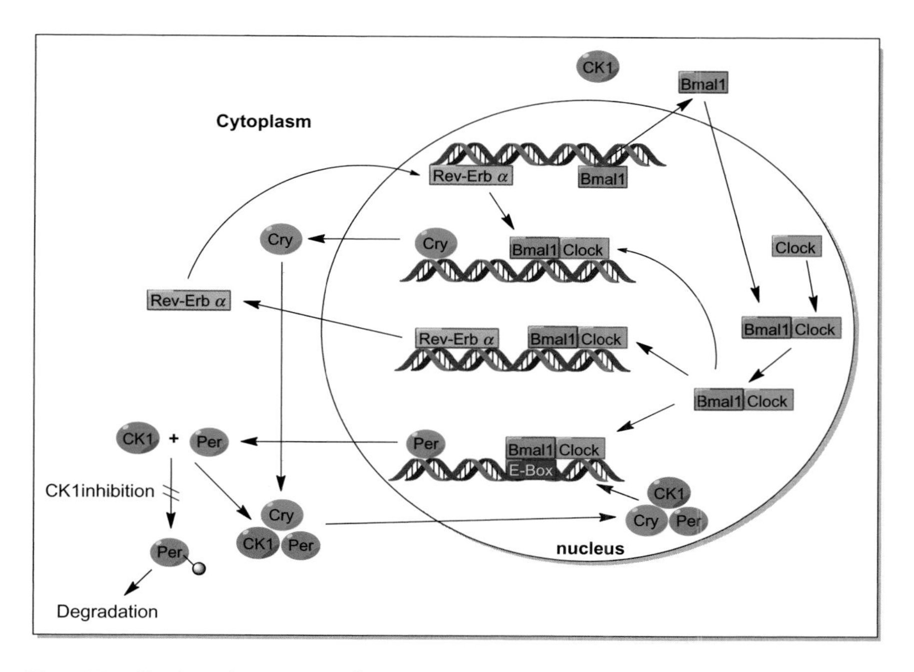

Plate 3.1 Clock cycle in mammalian SCN.

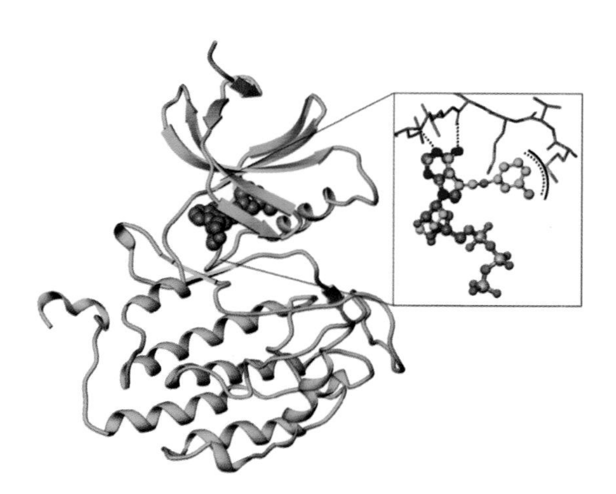

Plate 3.2 Model of PF-4800567 (**35**) bound in the ATP-binding site of CK1δ.

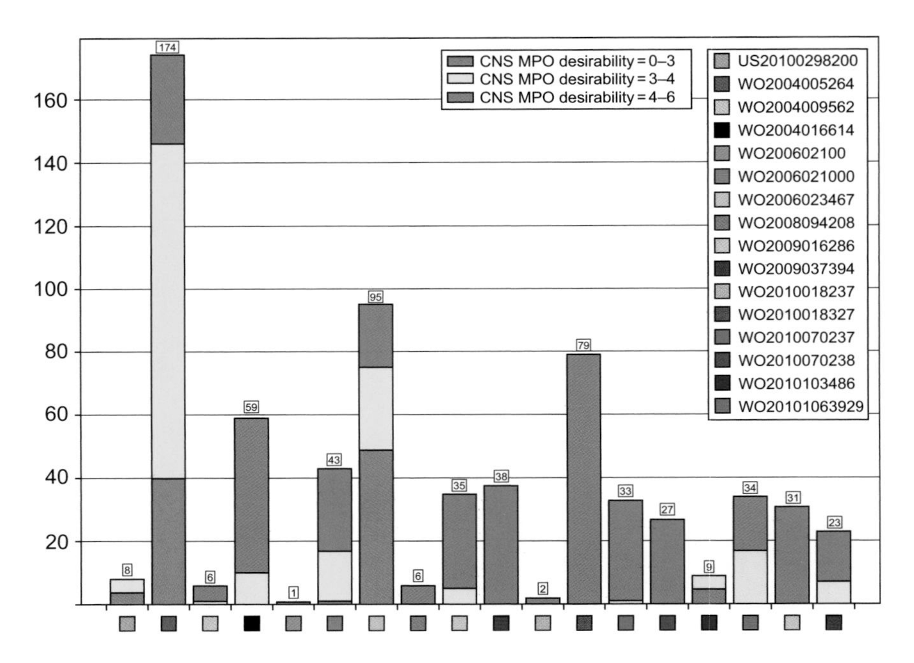

Plate 3.3 CNS MPO desirability analysis of 703 compounds representing CK1 patents (2004–2010).

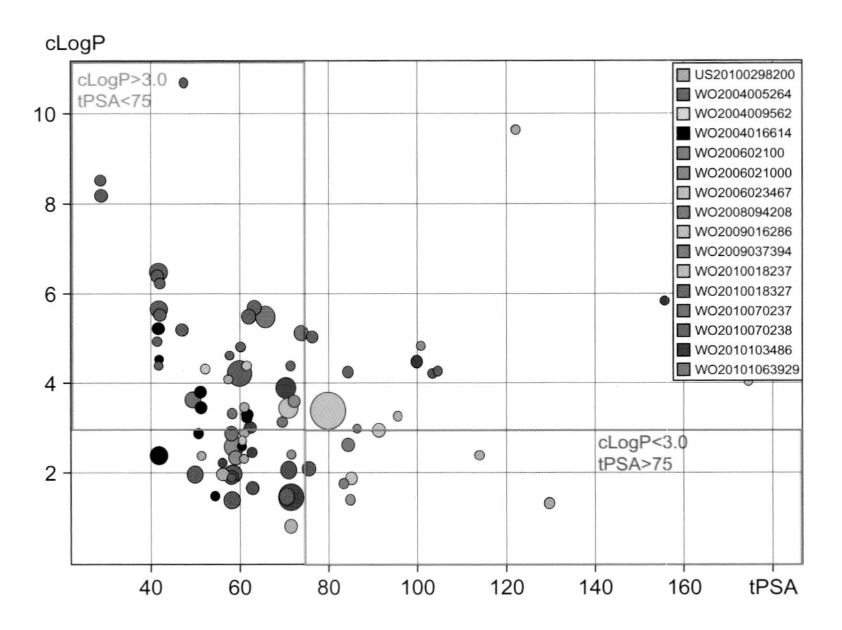

Plate 3.4 Toxicity plot (probability of *in vivo* toxicological findings) for 82 cluster centroids representing CK1 patent chemical space (2004–2010) sized by number of compounds in a cluster.

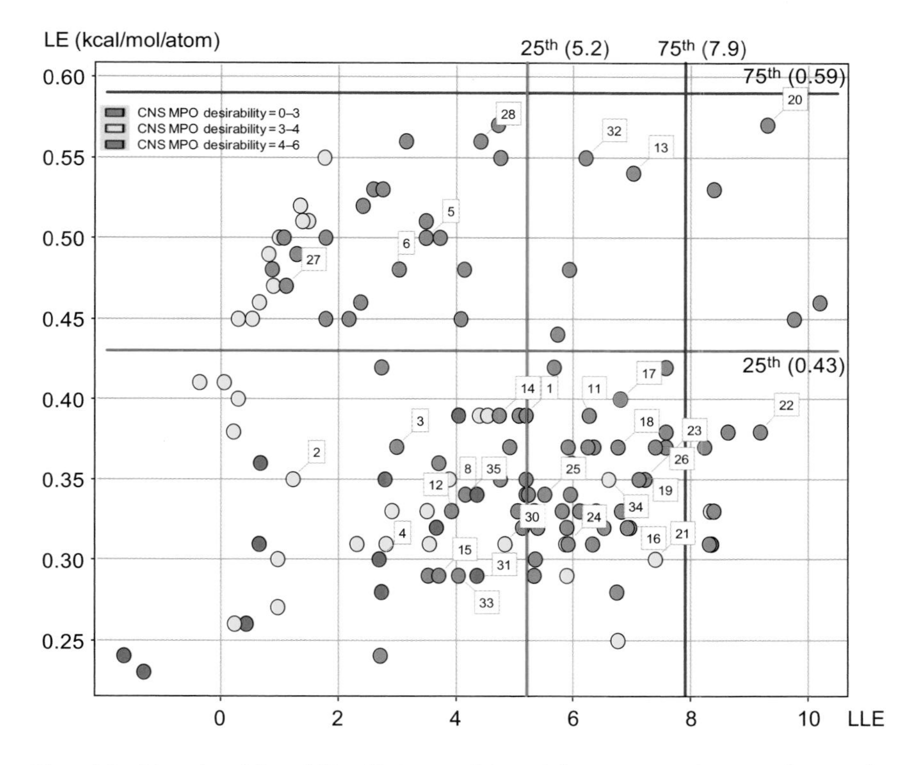

Plate 3.5 Ligand and lipophilic efficiency of CK1 inhibitors, using the LE and LLE 25th and 75th percentiles defined by marketed drug analysis [31]. Markers flagged by compound numbers in the body of the text.

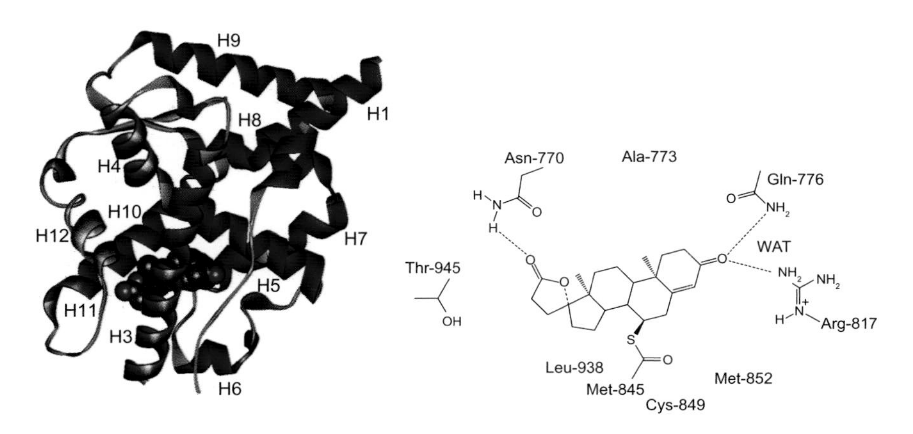

Plate 6.1 The 3D structure of the LBD of the MR with bound spironolactone (left) and schematic drawing of the binding mode (right).

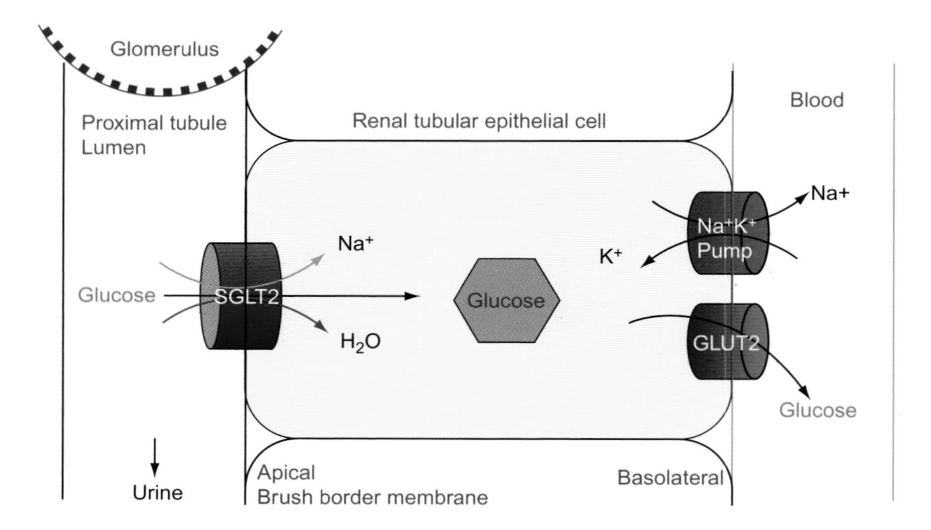

Plate 7.1 Tubular glucose reabsorption by SGLT2 in the kidney.

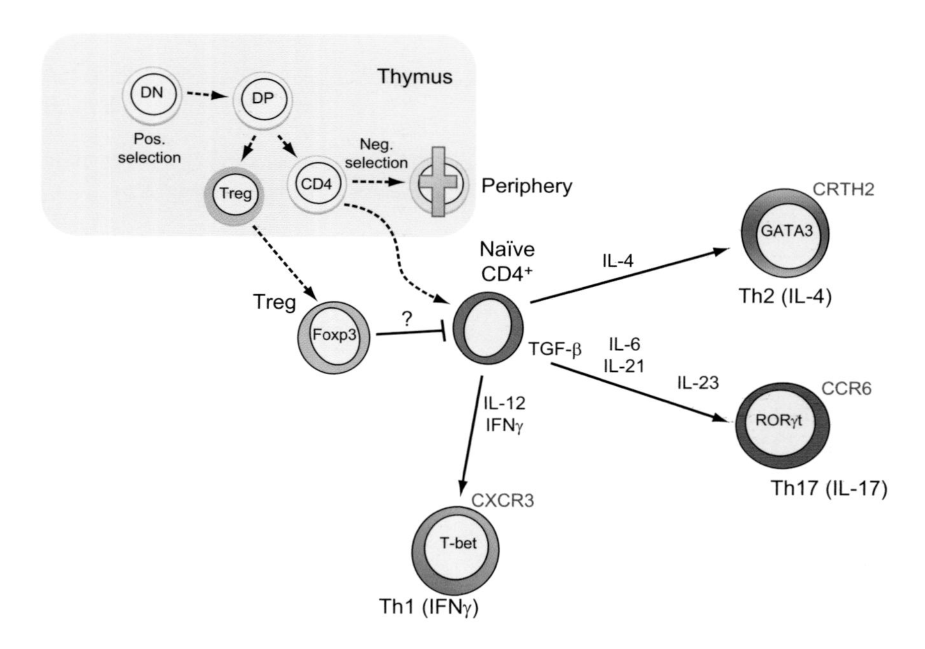

Plate 10.1 Model of T cell development and differentiation. Naïve T cells and Tregs develop in the thymus. Naïve T cells become activated by antigens in the periphery and can differentiate into one of three effector linages (*e.g.*, Th1, Th2, Th17). Regulatory T cells block the activation of bystander naïve T cells.

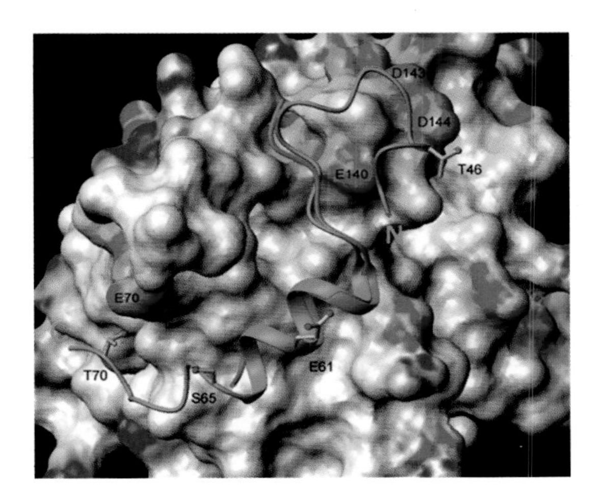

Plate 12.4 eIF4G- and 4E-BP-derived peptides (cyano and aquamarine, respectively) containing the helical conserved consensus motif, $Y(X)_4L\phi$, bind to the same hydrophobic patch on the convex dorsal surface on eIF4E.

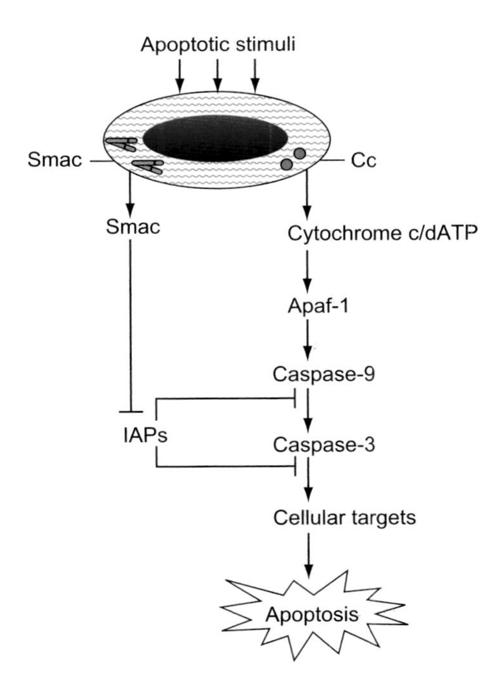

Plate 13.1 The intrinsic apoptotic cascade (adapted from Ref. [2] with permission from publisher).

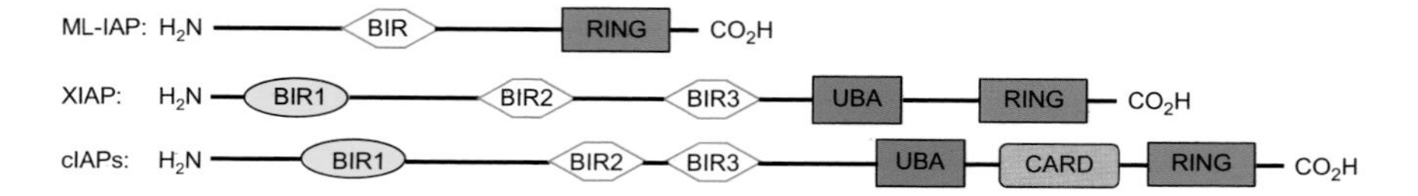

Plate 13.2 Domain architecture of ML-IAP, XIAP, and the cIAPs.

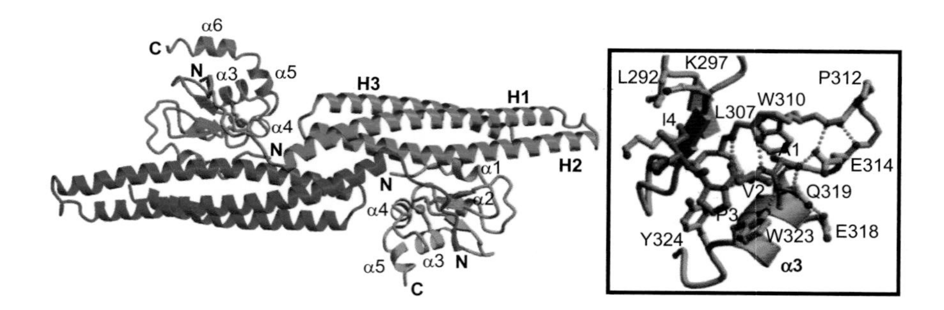

Plate 13.3 The structures of the Smac homodimer and AVPI bound to the XIAP BIR3 domain (adapted from Ref. [12] with permission from publisher).

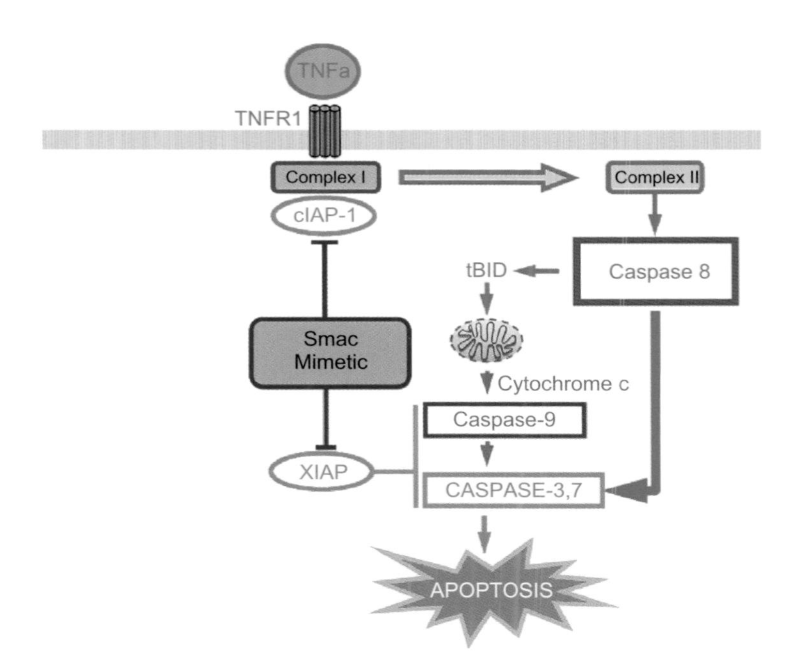

Plate 13.4 Schematic summary of Smac mimetic mechanism(s) of action.

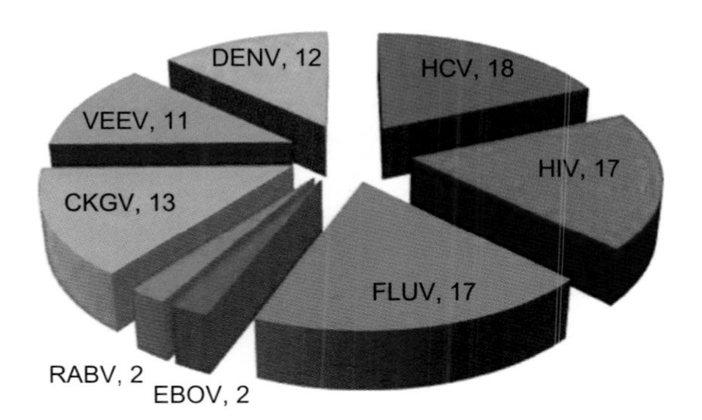

Plate 17.1 The number of distinct pre-lead chemical series identified from the cell-free screening system. All pre-leads display antiviral activity in the cell culture live virus assays.

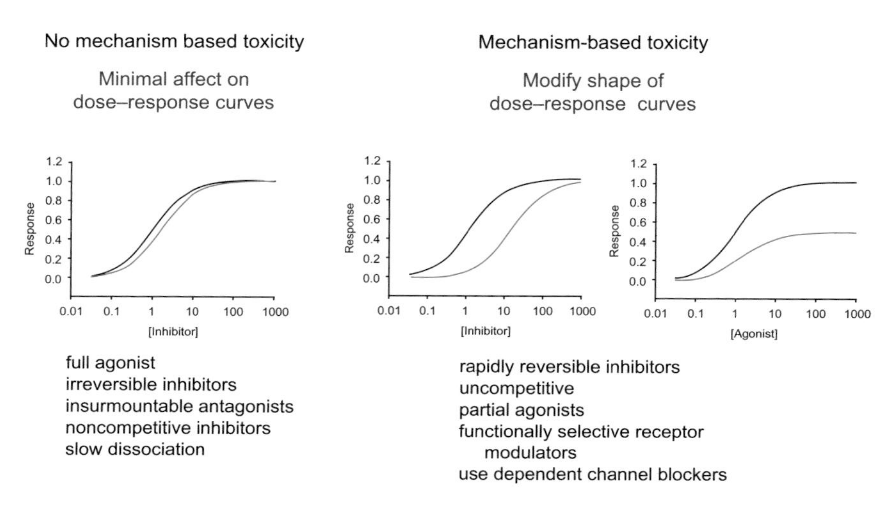

Plate 18.1 A primary driver of the impact of binding mechanism on the therapeutic index is the potential for mechanism-based toxicity. The curves show the relationship of concentration to binding (black) versus function (red). When there is no mechanism-based toxicity the binding should be efficiently coupled to function and the concentration–response curves will optimally be overlapping (left). When there is potential for mechanism-based toxicity, the functional curves may be shifted to the higher concentrations to limit mechanism-based toxicity (center); this is what would be expected with rapidly reversible competitive inhibitors at equilibrium. A decrease in maximal response as seen with partial agonists is another mechanism to minimize mechanism-based toxicity (right). Source: Swinney [2] reproduced with permission from Wolters Kluwer© 2008.